# ATZ/MTZ-Fachbuch

In der Reihe ATZ/MTZ-Fachbuch vermitteln Fachleute, Forscher und Entwickler aus Hochschule und Industrie Grundlagen, Theorien und Anwendungen der Fahrzeug- und Verkehrstechnik. Die komplexe Technik, die moderner Mobilität zugrunde liegt, bedarf eines immer größer werdenden Fundus an Informationen, um die Funktion und Arbeitsweise von Komponenten sowie Systemen zu verstehen. Fahrzeuge aller Verkehrsträger sind ebenso Teil der Reihe, wie Fragen zu Energieversorgung und Infrastruktur.
Das ATZ/MTZ-Fachbuch wendet sich an Ingenieure aller Mobilitätsfelder, an Studierende, Dozenten und Professoren. Die Reihe wendet sich auch an Praktiker aus der Fahrzeug- und Zulieferindustrie, an Gutachter und Sachverständige, aber auch an interessierte Laien, die anhand fundierter Informationen einen tiefen Einblick in die Fachgebiete der Mobilität bekommen wollen.

More information about this series at http://www.springer.com/series/12236

Rolf Isermann

# Automotive Control

## Modeling and Control of Vehicles

Springer

Rolf Isermann
Technische Universität Darmstadt
Darmstadt, Germany

ATZ/MTZ-Fachbuch
ISBN 978-3-642-39439-3 ISBN 978-3-642-39440-9 (eBook)
https://doi.org/10.1007/978-3-642-39440-9



This Springer imprint is published by the registered company Springer-Verlag GmbH, DE part of Springer Nature.
The registered company address is: Heidelberger Platz 3, 14197 Berlin, Germany

# Preface

The development of automobiles shows a continuous improvement of the chassis, body, powertrain, exterior design, and many others. However, the technical progress during the last decades is especially characterized by the electrification and electronification. The introduction of many sensors for physical quantities in the vehicle and for the detection of the surroundings in combination with electrical actuated brakes, steering systems, powertrains, and electronic control units enabled the realization of driver-assistance systems and increasingly automatic driving functions for the improvement of safety and driving comfort.

The design of the stationary and dynamic behavior of vehicles is in addition to classical engineering principles supported by mathematical models and simulation. Therefore, besides the design of the vehicle properties as for the design of advanced driver-assistance systems, detailed models for the longitudinal, lateral, and vertical behavior are required. This holds, for example, for the design of electronic stability control (ESC), or adaptive cruise control (ACC) or lane keeping control (LKC). The further steps with increasing degrees of automatic driving, from partial automation to high automation require precise and safe automatic longitudinal and lateral vehicle control systems which have to be based on actual mathematical models of the driving car.

This book treats after considering the E/E architecture first vehicle control structures and a workflow for control system design. Then mathematical models for the tire traction, the longitudinal, lateral, vertical, roll, and pitch dynamic behavior are established and compared with measurements. In order to obtain well-adapted models during driving, different parameter and state variable estimation methods are introduced and experimental results are shown. The vehicle models are based on physical properties, and it is one of the goals to obtain a good compromise between a clear interpretation and accuracy of the models.

The design and application of braking control systems is shown for hydraulic and electromechanical brakes, followed by hydraulic and electrical steering control systems and semi-active and active suspension systems.

Based on this background, the structures and signal flow graphs of advanced driver-assistance systems like TCS, ESC, and LKA are first represented. Then the different degrees of automatic driving are considered and the design of longitudinal velocity and distance control (ACC) is treated as well as the design of lateral path control, lane change control, and lane merging control.

For improving active safety anticollision systems with automatic braking and/or steering and for overtaking maneuvers are described including experimental test maneuvers. Then the development of automatic (autonomous) driving is considered and the required functional tasks from environmental detection, through data processing, perception, and situation awareness to motion planning and control are outlined.

The book is an introduction into modeling and automatic control of vehicles and its components with many practical examples and experimental test drive results. It is oriented to advanced students of control, electrical, mechanical, and automotive engineering and will also be useful for practicing engineers in the field of automotive development.

The author is grateful to his research associates, who have performed many theoretical and experimental research projects on the subject of this book since about 1990; among them are C. Ackermann, M. Bauer, J. B. Bechtloff, M. Beck, M. Börner, J. Busshardt, St. Drogies, D. Fischer, St. Germann, I. Halbe, Chr. Halfmann, H. Holzmann, Ph. Keßler, A. Khanafer, M. Kochem, R. Mannale, K. Schmitt, M. Schorn, R. Schwarz, S. Semmler, U. Stählin, St. Stölzl, H. Straky, T. Weispfenning, D. Wesemeier, and M. Würtenberger. Especially with regard to their continuous and detailed work on new methods and building up test benches, measurement, and computer equipment for research cars, the results of this book would not have been possible.

We also would like to thank the research organization Deutsche Forschungsgemeinschaft (DFG), the colleagues from the (DFG) funded special research program (Sonderforschungsbereich) IMES on Mechatronic Systems (1988–2001). Several results were obtained in cooperation with Adam Opel AG, Bosch Engineering GmbH, Continental-Teves AG, Daimler AG and during the Industry/University cooperation project PRORETA, funded by Continental AG, Frankfurt. We appreciate these cooperations strongly as they contributed greatly to our own research.

Finally, I would like to thank Brigitte Hoppe, S. Babilon, S. Rajurkar, T. Pham, and especially D. Timothy and A. Weber for the laborious and precise text setting, Sandra Schütz for drawing figures, Ilse Brauer for organizational and team work, and Springer Verlag for the excellent cooperation.

Darmstadt, Germany
January 2021

Rolf Isermann

# Contents

## Part IV Driver-Assistance Systems



## Part V Automatic Driving

# Symbols

## Coordinate System (Axis System)

| | |
|---|---|
| $X_E, Y_E, Z_E$ | Earth-fixed axis system (m) |
| $X_V, Y_V, Z_V$ | Vehicle axis system (m) |
| $X, Y, Z$ | Intermediate axis system (m) |
| $X_W, Y_W, Z_W$ | Wheel axis system (m) |
| $X_T, Y_T, Z_T$ | Tire axis system (m) |

## Vehicle kinematics

| | |
|---|---|
| $v$ | Vehicle velocity of the center of gravity, velocity over ground (m/s) |
| $v_X$ | Longitudinal vehicle velocity (m/s) |
| $v_Y$ | Lateral vehicle velocity (m/s) |
| $v_Z$ | Vertical vehicle velocity (m/s) |
| $v_h$ | Horizontal velocity (m/s) |
| $a_X$ | Longitudinal acceleration ($m/s^2$) |
| $a_Y$ | Lateral acceleration ($m/s^2$) |
| $a_Z$ | Vertical acceleration ($m/s^2$) |
| $a_c$ | Centripetal acceleration ($m/s^2$) |
| $a_t$ | Tangential acceleration ($m/s^2$) |
| $a_h$ | Horizontal acceleration ($m/s^2$) |
| $\psi$ | Yaw angle (rad) |
| $\theta$ | Pitch angle (rad) |
| $\varphi$ | Roll angle, relative to horizontal axis system (rad) |
| $\varphi_K$ | Roll angle, relative to road plane (rad) |
| $\beta$ | Vehicle side slip angle (rad) |
| $\psi$ | Yaw angle (rad) |
| $v$ | Course angle (rad) |
| $\dot{\psi}$ | Yaw angular rate (rad/s) |
| $\dot{\theta}$ | Pitch angular rate (rad/s) |
| $m$ | Total vehicle mass (kg) |

| | |
|---|---|
| $m_S$ | Sprung mass (kg) |
| $m_{Bo}$ | Body mass (kg) |
| $l$ | Wheel base (m) |
| $h_{CG}$ | Height of the center of gravity (m) |
| $h_{RC}$ | Height of roll axis (m) |
| $h_{PA}$ | Height of pitch axis (m) |
| $b$ | Wheel track (m) |
| $J_X$ | Moment of inertia around longitudinal (roll) axis ($kgm^2$) |
| $J_Y$ | Moment of inertia around lateral (pitch) axis ($kgm^2$) |
| $J_Z$ | Moment of inertia around vertical (yaw) axis ($kgm^2$) |
| $R$ | Vehicle motion radius for slow velocity (m) |
| $R_P$ | Vehicle motion radius for high velocity (momentary path radius) (m) |
| x | Longitudinal coordinate |
| y | Lateral coordinate |
| z | Vertical coordinate |
| $\kappa$ | Curvature of trajectory (1/m) |
| $\nu$ | Course angle (rad) |

## Forces and Moments

| | |
|---|---|
| $F_X$ | Longitudinal force (N) |
| $F_Y$ | Lateral force (N) |
| $F_Z$ | Vertical force (N) |
| $M_X$ | Roll moment (Nm) |
| $M_Y$ | Pitch moment (Nm) |
| $M_Z$ | Yaw moment (Nm) |

## Suspension

| | |
|---|---|
| $l$ | Wheel base (m) |
| $b$ | Track (m) |
| $c_B$ | Suspension spring stiffness (kN/m) |
| $d_B$ | Suspension damper coefficient (Ns/m) |
| $F_D$ | Damper force (N) |
| $F_S$ | Spring force (N) |
| $F_C$ | Damper friction force (N) |
| $F_{St}$ | Stabilizer force (N) |
| $\tau$ | Castor angle (rad) |
| $\varepsilon_V$ | Camber angle (rad) |
| $\delta$ | Wheel steer angle (rad) |
| $\Delta$ | Total static toe angle (rad) |
| $\delta_D$ | Dynamic reference steer angle (rad) |

| | |
|---|---|
| $\delta_H$ | Steering wheel angle (rad) |
| $M_H$ | Steering-wheel torque (Nm) |
| $i_S = \delta_H/\delta_f$ | Overall steering ratio |
| $z_B$ | Body deflection (m) |
| $z_W$ | Wheel deflection (m) |
| $z_T$ | Road side elevation ($m$) |

## Wheels and Tires

| | |
|---|---|
| $\alpha$ | Tire slip angle (rad) |
| $\beta_B$ | Brake pedal position (%) |
| $\varepsilon_V$ | Road plan camber angle (rad) |
| $r_{stat}$ | Static loaded tire radius (m) |
| $r_{dyn}$ | Dynamic rolling tire radius (m) |
| $F_R$ | Rolling resistance (N) |
| $F_{XW}$ | Longitudinal force at wheel (N) |
| $F_{YW}$ | Lateral force at wheel (N) |
| $F_{ZW}$ | Vertical force at wheel (N) |
| $F_{XT}$ | Longitudinal force at tire (N) |
| $F_{YT}$ | Lateral force at tire (N) |
| $F_{ZT}$ | Vertical force at tire (N) |
| $\delta_f$ | Front wheel steering angle (rad) |
| $\delta_r$ | Rear wheel steering angle (rad) |
| $c_\alpha$ | Cornering stiffness (N/rad) |
| $\delta_0$ | Toe angle (rad) |
| $c_W$ | Tire stiffness (kN/m) |
| $c_Y$ | Lateral tire stiffness (kN/m) |
| $m_W$ | Wheel mass (kg) |
| $M_{ds}$ | Drive shaft torque (Nm) |
| $M_B$ | Wheel braking torque (Nm) |
| $M_D$ | Wheel driving torque (Nm) |
| $M_{XW}$ | Overturning moment (Nm) |
| $M_{YW}$ | Rolling torque (Nm) |
| $M_{ZW}$ | Aligning torque (Nm) |
| $n_T$ | Caster offset (m) |
| $\mu_r es$ | Resulting combined friction coefficient |
| $\mu_X$ | Longitudinal friction coefficient |
| $\mu_{X,max}$ | Maximum longitudinal friction coefficient |
| $\mu_{X,lock}$ | Sliding braking friction coefficient |
| $\mu_{Y,lock}$ | Sliding lateral force coefficient |
| $\mu_Y$ | Lateral force coefficient |
| $\omega_W$ | Wheel angular velocity (rad/s) |
| $\omega_{W0}$ | Wheel angular velocity of straight free rolling wheel (rad/s) |

| | |
|---|---|
| $\omega_W$ | Wheel slip angular velocity (rad/s) |
| $\omega_W - \omega_{W0}$ | Longitudinal slip angular velocity (rad/s) |
| $S_X$ | Longitudinal tire slip |
| $S_Y$ | Tire side slip |
| $v_m$ | Circumferential tire velocity (m/s) |
| $v_{XT}$ | Longitudinal tire trajectory velocity (m/s) |
| $v_{YT}$ | Lateral tire trajectory velocity (m/s) |
| $v_T$ | Tire trajectory velocity (control center) (m/s) |
| $\partial F_{XW}/\partial S_X$ | Longitudinal force/longitudinal slip gradient (N) |

## Steering

| | |
|---|---|
| $\delta_{BS}$ | Ball screw nut angle (rad) |
| $\delta_H$ | Steering wheel angle (rad) |
| $\delta_f$ | Front wheel steering angle (rad) |
| $\delta_r$ | Rear wheel steering angle (rad) |
| $\delta_v$ | Hydraulic steering valve angle (rad) |
| $\delta_s$ | Steering shaft angle (rad) |
| $\Delta\delta_M$ | Steering shaft torque sensor twist angle (rad) |
| $r_R$ | Pinion radius (m) |
| $r_L$ | Turning radius of the wheel (m) |
| $z_R$ | Rack travel way (m) |
| $l_L$ | Steering lever (arm) length (m) |
| $J_{SW}$ | Inertia of steering wheel ($kgm^2$) |
| $m'_R$ | Representative mass for the steering rack, the tie rods, steering arms, and inertia of the turning wheels around the wheel carrier (kg) |
| $c_S$ | Overall stiffness of the steering (N/rad) |
| $c_R$ | Stiffness from the rack to wheel carrier (N/m) |
| $d_S$ | Viscous friction coefficient of the steering shaft (Ns/rad) |
| $d_R$ | Viscous friction coefficient of the rack to wheel carrier (Ns/m) |
| $f_{RC}$ | Dry friction coefficient of the rack to wheel carrier (Ns/m) |
| $i_S = \Delta\delta_H/\Delta\delta_f$ | Overall steering ratio |
| $M_H$ | Steering wheel torque (Nm) |
| $F_R$ | Rack force (N) |
| $F_{PA}$ | Power assistant force (N) |
| $M_M$ | Electric motor torque (Nm) |
| $n_M$ | Electric motor angular speed (rpm) |
| $\varphi_M$ | Electric motor rotor angle (rad) |
| $\Psi$ | Rotor flux of electric motor (Wb) |

## Chassis

| | |
|---|---|
| $A_V$ | Frontal vehicle area for air resistance ($m^2$) |
| $c_{WX}$ | Air drag coefficient |
| $c_{ZA}$ | Air lift coefficient frontal |
| $F_A$ | Air drag (N) |
| $F_R$ | Rolling resistance (N) |
| $F_{ZB}$ | Vertical force on the body (N) |
| $h_{CG}$ | Height of center of gravity (m) |
| $m_B$ | Body mass (kg) |
| $\rho_A$ | Air density ($kg/m^3$) |

## Powertrain and Drive Train

| | |
|---|---|
| $\alpha_{th}$ | Pedal position (%) |
| $M_{eng}$ | Engine torque (Nm) |
| $\omega_{eng}$ | Engine angular speed (rad/s) |
| $\alpha_{th}$ | Position accelerator pedal (%) |
| $\beta_B$ | Position brake pedal (%) |
| $M_d$ | Drive train torque (Nm) |
| $M_B$ | Braking, torque (Nm) |

## Road and Environment

| | |
|---|---|
| $\lambda$ | Road gradient angle, longitudinal (rad) |
| $\eta$ | Road gradient angle, lateral (rad) |
| $p$ | Longitudinal slope of a road (%) |
| $\kappa$ | Path curvature ($1/m$) |

## Subscripts

| | |
|---|---|
| air | Air |
| b | Brake |
| Bo | Body |
| Bp | Brake pedal |
| cl | Clutch |
| d | Driving |
| di | Differential gear |
| ds | Drive shaft |
| eng | Engine |
| f | Friction, front |

| | |
|---|---|
| h | Hand |
| $i\varepsilon[f, r]$ | Front or rear |
| $j\varepsilon[l, r]$ | Left or rear |
| l | Left |
| S | Steering |
| ped | Pedal |
| ps | Propeller shaft |
| pt | Powertrain |
| PT | Powertrain |
| r | Right, rear |
| R | Road |
| T | Tire |
| th | Throttle |
| tr | Transmission |
| v | Vehicle |
| V | Vehicle |
| W | Wheel |
| X | Direction of x-axis |
| Y | Direction of y-axis |
| Z | Direction of z-axis |

## Abbreviations, General

| | |
|---|---|
| AFS | Active front steering |
| CDC | Continuous damping control |
| CG | Center of gravity |
| ECU | Electronic control unit |
| EKF | Extended Kalman filter |
| EMB | Electro mechanical brake |
| EPS | Electrical power steering |
| GNSS | Global navigation satellite system |
| GPS | Global positioning system |
| HiL | Hardware-in-the-loop |
| HMI | Human machine interface |
| IC | Inertial center |
| IMU | Inertial measurement unit |
| KF | Kalman filter |
| KKF | Kinematic Kalman filter |
| LS | Least squares estimation |
| max | Maximal |
| min | Minimal |
| RC | Roll center |
| RLS | Recursive least squares estimation |
| RMS | Root mean square |

| | |
|---|---|
| SbW | Steer by wire |
| SG | Self-steering gradient |
| SiL | Software-in-the-loop |
| SVF | State variable filter |
| TMC | Tandem master cylinder |
| UKF | Unscented Kalman filter |

## Abbreviations for ADAS and Automatic Driving

| | |
|---|---|
| ABS | Anti-lock braking system |
| ACC | Adaptive cruise control |
| ACC-FR | Adaptive cruise control - full range |
| ADAS | Advanced driver-assistance systems |
| ADS | Automatic driving systems |
| AEB | Automatic emergency braking |
| AFL | Adaptive front lighting |
| APS | Automatic parking system |
| BA | Brake assist |
| BSW | Blind spot warning |
| CA | Conditional automation system |
| CAS | Collision avoidance system |
| CAV | Connected and autonomous vehicle |
| CC | Cruise control |
| DAS | Driver-assistance system |
| EBA | Emergency brake assist |
| ESC | Electronic stability control |
| FA | Full automation system |
| FCW | Forward collision warning |
| HA | High automation system |
| HDC | Hill descent control |
| LDC | Lane departure warning |
| LKA | Lane keeping assist |
| LKC | Lane keeping control |
| PA | Partial automation system |
| PAS | Park assistance system |
| PSA | Park steering assist |
| TCS | Traction control system |
| TSA | Traction stability assist |

# Introduction 1

The increasing *electrification* and *electronification* is a dominant feature of modern automotive developments. This is demonstrated by an increasing part of electrics/electronics (E/E) of the manufacturing costs from about 20% in 1995 to more than 35% in 2020. The electrics comprise primarily the electrical energy flows to the consumers through the energy board net. Frequently, former mechanical, pneumatic, or hydraulic actuated components of the chassis and the powertrain are replaced by electrical ones. The electronics are primarily used for the operation, and control and diagnosis functions. This enables the introduction of many new functions for the chassis and the powertrain, for driver-assistance systems and automatic driving.

## 1.1 Mechatronic Components and First Driver-Assistance Systems

Many automotive developments in the last three decades have been possible through an increasing number of *mechatronic components* in the powertrain and the chassis. Figure 1.1 gives some examples for engines, drive trains, suspensions, brakes, and steering systems. *Mechatronic systems* are characterized by an integration of mechanics and electronics, where the integration is between the components (hardware) and the information-driven functions (software). This development has a considerable influence on the design and operation of the *powertrain* consisting of the combustion engine and the drive train and the *chassis* with suspension, steering, and braking systems. In the case of hybrid drives, this includes also the electrical motor and the battery.

The mechatronic components replace formally pure mechanical, hydraulic, or pneumatic parts and use sensors with electrical outputs, actuators with electrical inputs, and digital electronics for control. The available electrical sensor measurements open access to internal functions and thus enable new possibilities not only for control but also for fault detection and diagnosis.



R. Isermann, *Automotive Control*, ATZ/MTZ-Fachbuch,
https://doi.org/10.1007/978-3-642-39440-9_1

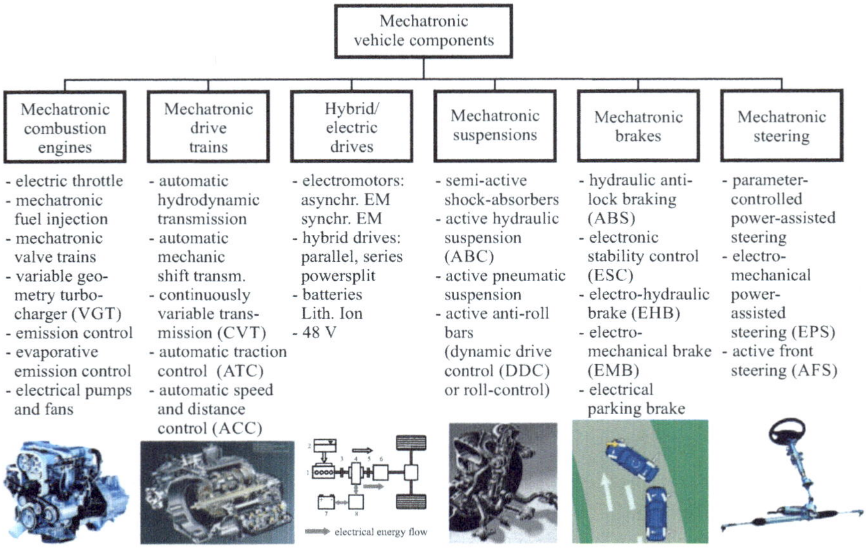


**Fig. 1.1** Mechatronic components and systems for automobiles and engines

The development of sensors, actuators, and electronic control for automobiles is depicted in Fig. 1.2. The first mechatronic components for the control of vehicles have been wheel speed sensors and electro-hydraulic switching valves for the brake system (1979), electrical throttle (1986), and semi-active shock absorbers (1988). These basic mechatronic components then allowed to develop control systems for anti-lock braking (ABS, 1979), automatic traction control (TCS, 1986), electronic stability control (ESC, 1995), and active body control (ABC, 1999). The next development steps were electric power steering (EPS, 1996) and active front steering (AFS, 2003). These mechatronic systems served mainly to increase safety and comfort and required many new sensors with electrical outputs, actuators with electrical inputs, and microcontroller-based electronic control units. Because they support the driver in performing driving maneuvers, they are *driver-assistance systems*. The addition of sensors for the surroundings enables parking-assistance systems which measure the parking space and adaptive cruise control (ACC, 1999) which measures the distance and relative velocity to vehicles in front by radar sensors improving mainly the comfort and convenience of driving. Driver-assistance systems for lane departure warning or control operate with video cameras and improve primarily safety. The introduction of ABS, ESC, TCS, and ACC has proven to be a significant success with regard to avoiding accidents.

A common feature of these developments is the increase of electrical sensors, actuators, and electronic control units, their coupling through cables and bus systems, and interconnection of the decentralized control units. Some of the vehicle control systems give commands to the engine control system, as, for example, TCS, ESC, and ACC.

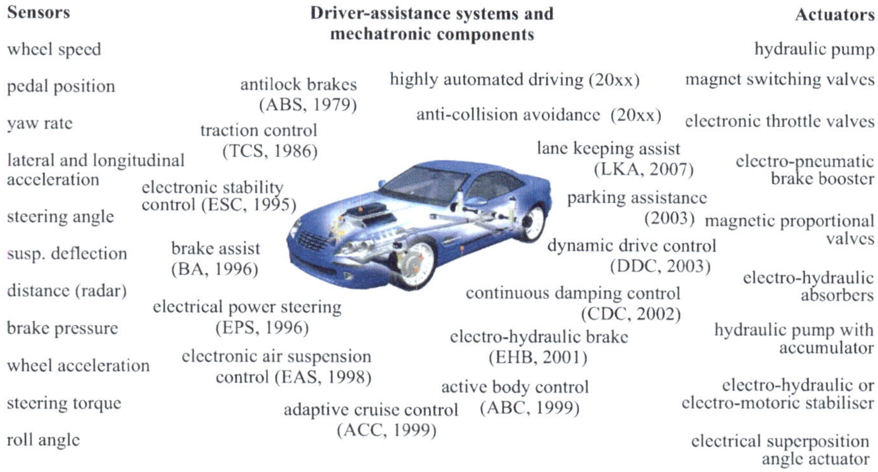


**Fig. 1.2** Examples for the introduction of sensors, actuators, and electronic control systems for automobiles

Parallel to the increase of electronic control functions for the chassis, the *engines and drive trains* have shown a similar development. This has to be seen together with the improvements of the combustion, fuel consumption and emission reductions, hybrid and electrical drives; see, e.g. Robert Bosch GmbH (2018), Guzzella and Onder (2010), and Isermann (2014).

## 1.2 Automatic Vehicle Control Developments

According to Donges (1982, 2016) and inspired by Rasmussen (1993), the guidance of motored vehicles can be divided into three levels; see Fig. 1.3 which is oriented according to the driver's task. The control inputs to the vehicle by the driver are the steering wheel angle, the accelerator pedal position, and brake pedal position. They are used to control the vehicle motion. The lowest level is a *stability level*, where the driver keeps the vehicle on a certain track by steering and at a certain speed by the two driving pedals. This includes a feedback control of a yaw angle relative to a driving space or road and a speed indicated by the speedometer such that the vehicle does not become either monotonically or oscillatory unstable. The next level is the *lane driving control level*. Here, the goal is to keep the vehicle on a certain road to control the lateral distance and the distance to other vehicles, to circumvent obstacles and to follow traffic lights and speed signs. Based on the optical information of the driver from the environment, this level gives reference variables for the control at the first level. The *navigation level* is a third level, where the driver selects a certain route within the road network and traffic situation. The selected roads are then given as references to the lane driving control level. The resulting three-level control system, depicted in Fig. 1.3, comprises several feedback loops. The controlled variables for the first level are the orientation of the vehicle on the road determined by the yaw

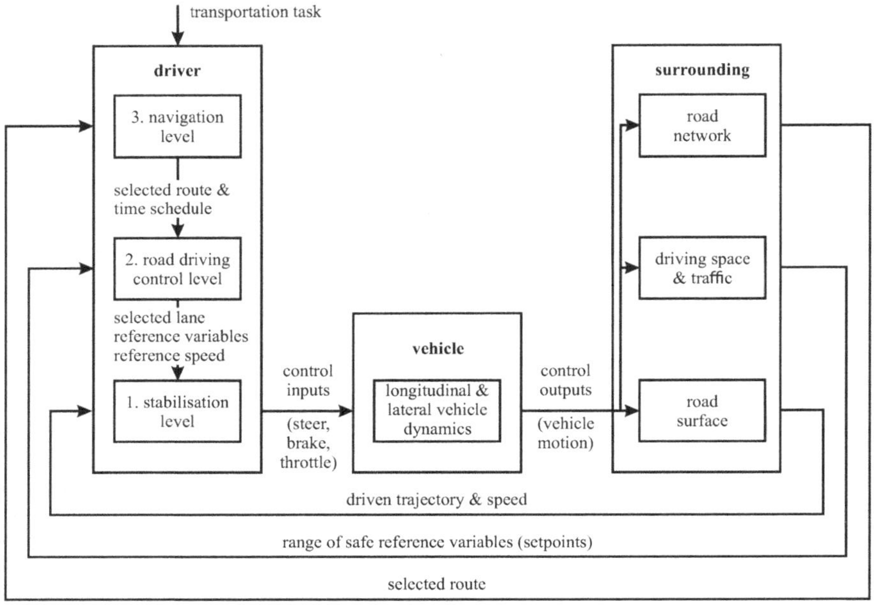


**Fig. 1.3** Three-level vehicle guidance (Donges 1982)

angle and speed within a driving space and the surrounding traffic. The second level consists of a range of traffic safe reference variables for controlling the course, the lateral distance, and the speed. For the third level, the reference variable is the selected road with the final goal and a time schedule.

According to Rasmussen (1993), the human driver tasks can be interpreted from a work psychological perspective and classified as human skill and practice, rule-based, and knowledge-based; see Donges (2016).

Until the 1970s, the vehicle control system consisted of the classical driver commands to the steering wheel, the engine throttle, the brake, and gear shifting. During the last three decades, modern automobiles obtained completely electronic controlled power trains, electronic controlled hydraulic power steering or electrical power steering systems, and electronic controlled hydraulic brake systems. This *electrification* and *electronification* was a prerequisite for the development and introduction of driver- assistance systems (DAS) and automatic driving systems (ADS); compare Fig. 1.4. With regard to their temporal introduction, one may subdivide them in sequential generations:

- First generation: lateral and longitudinal control:
  - ABS: anti-lock braking;
  - TCS: automatic traction control;
  - ESC: electronic stability control;
  - ACC: adaptive cruise control;
  - PAS: parking assist system.

- Second generation: Warning and careful actions for collision avoidance:
  - ACC-FSR: adaptive cruise control, full speed range, emergency stop;
  - LDW: lane departure warning;
  - LDC: lane departure control (one side braking or steering action);
  - BA: brake assist.
- Third generation:
  - ADAS: advanced driver-assistance system;
  - APS: automatic parking system;
  - CAS: collision avoidance system.
- Fourth generation: Automatic driving with limited degrees of automation:
  - APS: automatic parking system;
  - PA: partial automation system;
  - CA: conditional automation system.
- Fifth generation: Automatic driving:
  - HA: high automation system;
  - FA: full automation system.

A timeline of the developments, including some mechatronic components, the introduced sensors, and electromechanic and electro-fluidic actuators, is shown in Fig. 1.4.

The driver-assistance systems were developed sequentially to improve firstly safety and secondly comfort and are characterized by automatic control and warning functions (Winner et al. 2016; Isermann 2006; Robert Bosch GmbH 2018). They can be seen as steps toward automatic driving.

Table 1.1 classifies these systems according to different *degrees of automation*, normal and conflicting driving maneuvers, and role of driver and control systems (Gasser et al. 2012; Ruchatz 2013; Verband der Automobilindustrie 2015; SAE 2016).

Level 0 is the normal driving by the driver only, where the driver takes over all tasks of driving. Level 1 comprises the *driver-assistance systems* (DAS) which came into series application between 1979 and 2008, as depicted in Fig. 1.4. The driver has the full task of driving and is assisted in some (critical) driving situations. *Advanced Driver-Assistance Systems* (ADAS) are characterized by some automatic control functions for parking and longitudinal driving and slow driving in traffic jams. Therefore, the system is able to control the longitudinal and lateral motion automatically in very special cases. The driver has to be attentive all the time, supervise the system, and ready to take over if required

In the case of *partial automation* (PA), level 2, an autopilot controls the vehicle in several selected cases. Examples are automatic parking assist, jam traffic assist, and longitudinal and lateral control assist on highways and rural roads. The driver has to supervise the system all the time and has to take over driving immediately upon request by the system.

*Conditional automation* (CA) in level 3 has extended automatic control functions. The system performs lateral and longitudinal automatic driving on highways and rural roads, in traffic jams, and for parking. The driver does not need to monitor

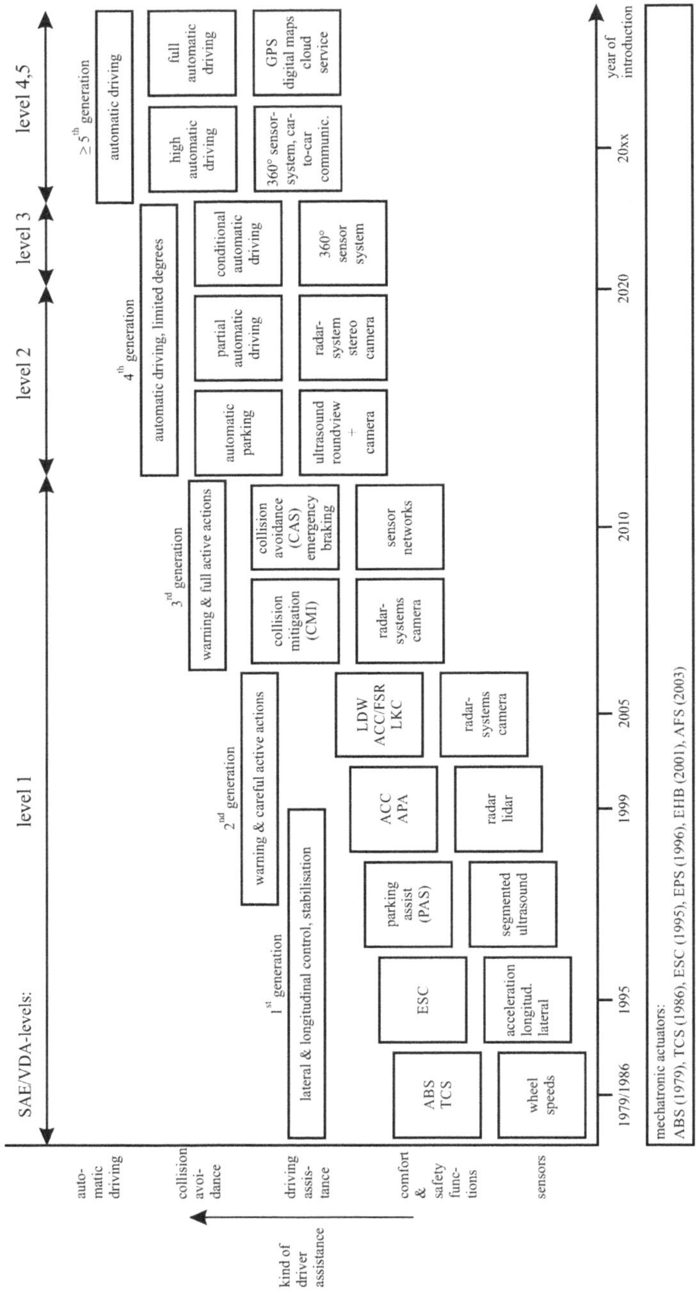


**Fig. 1.4** Time-dependent road map for driver-assistance systems and automatic driving

**Table 1.1** Degrees of automatic driving (Extension of Gasser et al. 2012; Ruchatz 2013; Verband der Automobilindustrie 2015; SAE 2016)

| | | LEVEL | DEGREES OF AUTOMATIC DRIVING 1 | | 2 | | 3 | 4 | 5 |
|---|---|---|---|---|---|---|---|---|---|
| | | GENERAL DESCRIPTION | Driver Assistance (DA) | | Partial Automation (PA) | | Conditional Automation (CA) | High Automation (HA) | Full Automation (FA) |
| MANEUVERS | NORMAL | KIND OF DRIVING | Eyes-ON | | Eyes-ON | | Eyes-OFF | Eyes-OFF | Eyes-OFF |
| | | | Hands-ON | | Hands temp. OFF | | Hands temp. OFF | Hands-OFF | Hands-OFF |
| | | EXECUTION OF DRIVING | Driver | | System | | System | System | System |
| | | MONITORING OF DRIVING | Driver | | Driver | | System | System | System |
| | | FALLBACK | Driver | | Driver | | Driver | System | System |
| | | DRIVING MODES | some | | some | | some | some | all |
| | | DIRECTION OF CONTROL | longitudinal | lateral | longitudinal | lateral | some | many | all |
| | | SLOW MANEUVERS | jam assist | parking assist | parking assist jam assist | | parking pilot | parking pilot | parking pilot |
| | | MANEUVERS WITH LIMITED DURATION | braking assist (ABS) | electronic stability control (ESC) | | lane change assist (LCA) | lane change pilot overtaking pilot | lane change pilot overtaking pilot crossing pilot | overall driving pilot (highway, rural road, city) |
| | | CONTINUOUS DRIVING | speed & distance control assist (ACC) | lane keeping assist (LKA) | highway distance & speed control (ACC-FR) | lane keeping control (LKC) | highway pilot rural raod pilot | highway pilot rural road pilot | |
| | CONFLICT | ACCIDENT AVOIDANCE MANEUVERS | warning | | warning | | warning | warning | warning |
| | | | emergency braking (AEB) | blindspot warning | emergency braking | evasive maneuver control | emergency maneuvers | emergency maneuvers | emergency maneuvers |
| DRIVER | | ROLE OF SYSTEM & DRIVER | • Driver performs continuously driving task.<br>• Assistance in special cases. | | • System performs longitudinal and lateral driving<br>• Defined use cases<br>• Driver monitors all time<br>• Driver performs remaining tasks | | • System performs longitudinal and lateral driving<br>• Defined use cases<br>• Driver does not monitor all time<br>• Driver resumes after request | • System performs longitudinal and lateral driving in all cases<br>• Driver is not required in defined use cases | • System performs driving in all situations during a trip<br>• Driver is not required |

the automatic driving all the time. The automatic system recognizes its limits and requires the driver to take over within a certain time period. Lane changing and overtaking may also be possible automatically.

*High automation* (HA) is dedicated to level 4. The degree of automation is further increased such that lateral and longitudinal driving are possible in many situations, but limited to use cases with regard to road type, speed ranges, and environmental conditions. The driver is not required in these use cases but still has to be in the vehicle. Valet parking (driverless parking) and urban automatic driving may be possible.

*Full automation* (FA), level 5, includes automatic control for all situations. The driver is then not required. Shuttle buses with low speed in protected areas belong to this level. However, for vehicles in normal dense traffic very high requirements have to be satisfied. Therefore it is present as an ideal case.

As Table 1.1 illustrates, the number of automatic driving modes increases from level 2 to level 5, and the monitoring tasks by the driver decrease from level 3 to level 5. Until level 3, the human driver is the fallback system in the case of malfunctions of the control systems. Especially for levels 4 and 5, *fault-tolerant systems* have to be implemented, as the driver is at least not required partially or completely.

The classified degrees of automatic driving are mainly dedicated to normal driving. However, the automation of driving may especially help in conflicts to initiate *accident avoiding maneuvers*; see Table 1.1 at the bottom. A lower degree of a collision avoidance maneuver is automatic emergency braking (AEB). The next degree is emergency braking and steering to stop and/or for the evasion of obstacles. In the case of fully automatic collision maneuvers, an automatic driving into a conflict-free space may be included. One goal of these collision avoidance system is to react faster than a normal human driver and to perform the best possible maneuver.

The steps toward high and full automatic driving need a multitude of investigations, testing, and changes of legislation.

## 1.3 Contents of the Book

A systematic design of vehicle control functions requires a control engineering view of the stationary and dynamic behavior of the vehicles and the *electronic architecture and control structure* of advanced vehicles, as considered in Part I. In order to describe the stationary and dynamic behavior, Part II is devoted to *modeling*. First, the development of mathematical models by theoretical modeling in using the physical laws or by experimental modeling with measurements and identification methods is considered in general. Then, the laws for the *tire traction and force transfer* are described. The modeling of the *longitudinal dynamic behavior* includes the powertrain dynamics with a combustion engine, clutch, transmission, and drive train. Then, different longitudinal vehicle models are given for multi-mass and two-mass systems. Based on these models, the acceleration and braking behavior of the vehicle is modeled in a simplified version without slip and with variable slip and vertical wheel forces.

The *lateral dynamic behavior* is treated by using the kinematic relations, dynamic one-track and two-track models in different representations. Based on the one-track model, the fundamental vehicle behavior for cornering is analyzed for stationary behavior with understeering and oversteering, and stability criteria are developed. The dynamic two-track models are used to describe the longitudinal, lateral, and vertical forces at all wheels and the dynamic behavior for the yaw behavior, also for road gradients. Then, simplified *roll and pitch models* are derived.

Because many model parameters are not precisely known or vary during driving, *identification and parameter-estimation methods* are described and are applied for the longitudinal, lateral, roll, and pitch behavior. This applies, for example, to vehicle mass, vehicle moments of inertia, tire friction parameters, and cornering stiffness.

In order to determine dynamic vehicle state variables, different *state variable estimation methods* are considered, such as state observers, Kalman Filters, and Extended Kalman Filters. These methods allow estimating the ground velocity, the vehicle slip angle, and roll and pitch angle during driving.

Part III is devoted to the *drive dynamic control of chassis components. Braking control* is treated for hydraulic brakes, pneumatic and electrical boosters, anti-lock braking (ABS) with switching and proportional valves, and electro-hydraulic (EHB) and electromechanical brakes (EMB).

The chapter on *steering control systems* includes modeling of mechanical, power-assisted hydraulic (HPS) and electrical power systems (EPS) and a consideration on fault-tolerant, redundant EPS.

The chapter on *suspension control* begins with passive suspensions and continues with theoretical and experimental modeling of semi-active and active suspensions and their control. This is followed by a section on *tire pressure monitoring* with wheel and suspension sensors.

Based on the developed drive dynamic models, Part IV considers various *driver-assistance systems*. An overview of passive and active driver-assistance systems is given and required sensor systems for the environment representation are treated.

The chapter on *advanced driver-assistance systems* (ADAS) for longitudinal and lateral guidance shows the functions and signal flow schemes for traction control (TCS), electronic stability control (ESC), and lane keeping assistance (LKA).

Part V treats *automatic driving*. A classification of *automatic driving functions* considers the degrees of automatic driving from level 1 (driver-assistance systems) to level 5 (full automation systems), typical driving maneuvers, including path and trajectory control. *Automatic longitudinal vehicle control* is subdivided into acceleration control with the drive train or by braking, velocity control, distance control, and adaptive cruise control (ACC). The treatment of *lateral vehicle control* begins with path control for straight lanes and lateral distance measurement beside and ahead of the vehicle, and the design of different feedforward and feedback controllers. This is followed by path control for curves and lane change control and merging control.

Then developments for *anticollision systems* are described. First approaches for *emergency braking* and *emergency evasion* are shown, including the automatic control systems and experimental test drives without driver intervention. Then an *anti-*

*collision overtaking assistance* with the prediction of the overtaking maneuver and warnings and emergency braking is presented to avoid heavy accidents.

The last chapter gives an introduction into *automatic driving* and looks briefly at some historic developments, considers the *main function modules*, like data and image processing and vehicle guidance, vehicle-borne sensor systems, vehicle state identification, and external support, like navigation systems and vehicle-to-x communication. A signal flow scheme illustrates the required function blocks and shows the many tasks, for example, for the perception of the environment, situation awareness, motion planning, and vehicle control.

Then the *progress by evolution* is considered, taking into account a stepwise increase of automatic functions based on proven previous functions. As the automation levels from conditional automation (level 3) upwards require *fault-tolerant systems*, a short introduction into redundant structures, degradation steps, and functional safety is given and some automotive examples are shown.

The field of vehicle modeling and control is treated in the following books: Johansson and Rantzer (2003), Kiencke and Nielsen (2005), Isermann (2006), Rajamani (2012), and Ulsoy et al. (2012); see also the books on vehicle dynamics, cited in Chap. 4. These books give different views into vehicle dynamics and control and have given hints for the treatments in this book.

This book is to a large extent based on several own research projects for the vehicle dynamic behavior and control and includes many experimental results with different passenger cars and test benches. Though extensive and complex commercial vehicle simulation systems exist, model-based control, supervision and diagnosis, and the design of components require the basic physical or semi-physical models with their constructive parameters. These models of components and vehicles are used to determine unknown parameters with parameter-estimation methods and state variables with state-estimation methods experimentally, either off-line or online during driving. Therefore, it is tried to keep the models and controllers as simple as possible, clearly arranged, and transparent, but describing the stationary and dynamic behavior precisely enough for the intended applications.

## References

Donges E (1982) Aspekte der aktiven Sicherheit bei der Führung von Personenkraftwagen. Automobil-Industrie 27(2):183–190

Donges E (2016) Driver behavior models. In: Winner H, Hakuli S, Lotz F, Singer C (eds) Handbook of driver assistance systems. Springer International Publishing AG, Cham, Switzerland, pp 19–33

Gasser TM, Arzt C, Ayoubi M, Bartels A, Bürkle L, Eier J, Flemisch F, Häcker D, Hesse T, Huber W (2012) Rechtsfolgen zunehmender Fahrzeugautomatisierung. Berichte der Bundesanstalt für Straßenwesen. BASt. Heft F 83, Bundesanstalt für Straßenwesen, Bergisch Gladbach

GmbH Robert Bosch (ed) (2018) Automotive handbook, 10th edn. Wiley, Chichester

Guzzella L, Onder C (2010) Introduction to modeling and control of internal combustion engine systems, 2nd edn. Springer, Berlin

Isermann R (2014) Engine modeling and control. Springer, Berlin

Isermann R (ed) (2006) Fahrdynamik-Regelung. Vieweg-Verlag, Wiesbaden

Johansson R, Rantzer A (eds) (2003) Nonlinear and hybrid systems in automotive control. Springer, London

Kiencke U, Nielsen L (2005) Automotive control systems: for engine, driveline, and vehicle, 2nd edn. Springer, Berlin

Rajamani R (2012) Vehicle dynamics and control, 2nd edn. Mechanical engineering Series, Springer, US, New York

Rasmussen J (1993) Diagnostic reasoning in action. IEEE Trans Syst Man Cybern 23(4):981–991

Ruchatz (2013) Vision und Möglichkeiten des automatischen Fahrens. In: 6. Fachtagung AUTOREG, (2013) VDI-Berichte 2196. Wiesloch, Germany

SAE (2016) Taxonomy and Definitions for Terms Related to Driving Automation Systems for On-Road Motor Vehicles. Standard J3016_201609. SAE International, Warrendale, Pennsylvania

Ulsoy A, Peng H, Çakmakci M (2012) Automotive control systems. Cambridge University Press, New York

Verband der Automobilindustrie (2015) Automatisierung - Von Fahrerassistenzsystemen zum automatisierten Fahren. Die Zukunft von gestern - heute Realität. VDA-Broschüre, Berlin

Winner H, Hakuli S, Lotz F, Singer C (eds) (2016) Handbook of driver assistance systems. Springer International Publishing AG, Cham

# Part I
# Electronic Architectures and Control Structures

# 2 Electrical and Electronic Architectures of Automobiles

In the frame of vehicle electrification and electronic management of the chassis and the powertrain, the structures of the electrical and electronic architecture have a significant influence on the realization of functions. Therefore, this chapter briefly considers the types of networks, the communication networks and bus systems, and the software structure of the electronic control units (ECU).

## 2.1 Types of Network Architectures

Modern automobiles comprise a large number of electrical and mechatronic components which are connected by cables and form different kinds of electrical and electronic networks. One distinguishes between the *electrical board net* which provides the electrical and electronic components with electrical energy from the engine-driven generator and the battery and an *electronic board net* which connects the sensors, electronic control units, switches, and actuators. These networks are also described by the abbreviation "E/E-system". Because of the increasing complexity in the frame of electrification and electronification of advanced automobiles, the description and also standardization of the emerging E/E-structures play an important role.

A representation of these structures is guided by special views and shows up in different architectures which describe the overall system in the form of composition plans with modules and their connections. One may distinguish following automotive architectures:

- electrical network architecture and
- electronic network architecture.


R. Isermann, *Automotive Control*, ATZ/MTZ-Fachbuch,
https://doi.org/10.1007/978-3-642-39440-9_2

These two basic architectures can be described with a focus on

- hardware architecture,
- software architecture,
- vehicle control architecture, and
- powertrain control architecture.

Generally, the architecture structures can be represented as (Robert Bosch GmbH 2011)

- function network,
- component network (hardware),
- circuit diagram,
- wiring harness, and
- installation space.

These structures are used with regard to different views of the overall system and show the various kinds of assembling and integration. In the following a focus is on the electronic hardware and software components and structures which are the basis for automotive control systems; see also Robert Bosch GmbH (2007), Robert Bosch GmbH (2011), Borgeest (2014), and Reif (2014).

## 2.2 Electronic Communication Networks

### 2.2.1 Network Nodes

The communication network in vehicles connects the sensors and actuators with the ECU's via so-called nodes. *Data buses* allow connecting these nodes via a single cable channel. This has the advantage that the data can be distributed to different receivers and space, and plugs and costs are saved. Also, computing power can be distributed to different ECU's, and computations by one ECU can be shared with other ECU's. However, the signal values have to be in digital form, such that analog sensor signals have to be digitized. In order to use the bus systems by all users, like suppliers and OEM's and service stations, they must be standardized. Because the transfer requirements and safety aspects in vehicles are different, different buses are applied. For example, for switching the windows, the data transfer rate is a few bit/s, for engine management some 100 kbit/s, and for video applications some Mbit/s are required. For safety-relevant components, the time duration of data transfer must be guaranteed.

In order to provide the access of microcomputers to the bus system and vice versa, an interface electronics is required which is called a *network node*; see Fig. 2.1. The incoming data from the bus are processed by a bus transceiver or bus driver and sent to a communication controller which is connected to the microcontroller executing the application program. Outgoing messages are sent from the microcontroller to

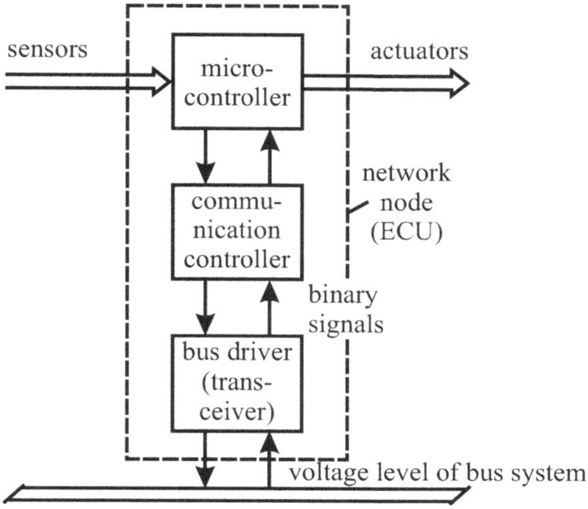


**Fig. 2.1** Network node in connection with a digital bus system; example for an ECU

the communication controller and the bus transceiver (driver) amplifies the signals, generates the required voltage, and transmits the data in the form of a serial bit stream on the bus line. Hence, the communication controller is responsible for sending and receiving the signals between the microcontroller and the bus.

*Sensors* are either directly connected to the ECU, and their signals are transferred through an analog-digital converter (ADC) to a conventional centralized data processing, or they already have an integrated digital circuit which performs signal conditioning and AD-conversion. This is then a decentralized solution and relieves the ECU. If this integrated sensor component is provided with a bus interface, it can be directly connected to the bus system and used by several ECU's.

*Actuators* need an electrical, pneumatic, or hydraulic auxiliary energy supply. They can be centrally commanded directly by the ECU, with electronic power circuits, pulse-width modulation (PWM) or switching drivers, or decentrally via the bus system. Some actuators are integrated mechatronic units with local integration of sensors, a microcontroller, and amplifiers, such as electrical power steering units, brakes with ABS, and active suspensions. These central or decentral structures of sensors and actuators play an important role for the kind of integration of active chassis and powertrain components into the vehicle networks.

### 2.2.2 Network Topologies

The network topology describes the arrangement of the network nodes and their connections. The *bus topology* connects all nodes by a single (linear) cable. Hence, it is a parallel structure, and a failing node does not directly affect the others. The system can be easily expanded, but if the bus fails, then the communication fails completely. A *star topology* has a central node to which other nodes are connected by individual lines. The advantages are a fast real-time operation capability and an easy expansion. If a node or a line fails, this does not affect the others. However, if the central node fails, the entire network fails. In the case of a *ring structure*, the nodes are serially connected by a bus ring. The messages are forwarded from one node to the other in one direction until they reach their destination. Hence, this network is relatively slow and if a node or the bus line fails, the communication fails. In a *mesh topology*, each node is connected to other nodes directly at the cost of more lines. This has the advantage to allow some redundancy in the transmission of messages in case of failing nodes or lines. *Hybrid topologies* combine different structures as a star-bus or star-ring topology.

### 2.2.3 Bus Systems

The application of different bus systems provides the framework for automotive networking. Therefore, their design and some properties are summarized.

The data communication of digital bus systems is based on the OSI reference model standardized by the International Standardization Organization (ISO). Seven hierarchical layers are defined, but they are not all used for vehicles until now. The following layers are distinguished, see Fig. 2.2,

- *Physical layer* defines the physical property of the transmission line, e.g. voltage difference on a twisted two-wire cable or on a single-variable cable with voltage to ground, or brightness of light for optical fiber cables.
- *Data link layer* cares about the right data transfer to the bus. The data are organized in frames which are defined by the bus protocol. Also, error correcting measures as checksums can be added.
- *Network layer* provides the data transfer between different users if several networks are connected. This layer ensures that the data is routed correctly to its destination.
- *Transport layer* cares about the transfer of the data error-free and in the right sequence between different users. Two users must first establish a connection via the network and can then interact with each other. The transport layer manages the sessions between users.
- *Application layer* provides functionalities as an interface to the application programs such that direct access to the application is possible. The data of this layer are forwarded to the transport layer.

The medium to higher levels like the session layer and the presentation layer of the OSI reference model are usually not used.

The data in bus systems are transmitted in packets, carrying all important information. The access of several users in a network at the same time is organized by the bus assignment or *arbitration*. This can be organized either centrally by the bus control or decentralized by the users. In the case of a *central bus arbitration*, the bus is assigned to the user, for instance, by status inquiry (polling) or after a determined,

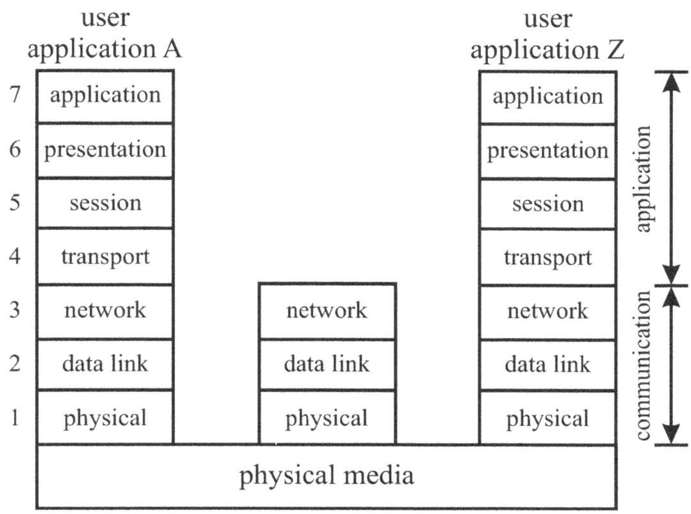


**Fig. 2.2** OSI reference model for digital buses, general software structure (ISO 7498-1 1994)

**Table 2.1** Classification of automotive bus systems (SAE 1994)

| Class | Data transfer rate | Applications | Representative |
|---|---|---|---|
| A | Low ≤ 10 kbit/s | Sensors, actuators | LIN bus |
| B | Medium ≤ 125 kbit/s | Control units comfort relevant | Low-speed CAN |
| C | High ≤ 1 Mbit/s | Real-time requirements powertrain, chassis | High-speed CAN |
| C+ | Very high ≤ 10 Mbit/s | Real-time requirements | FlexRay |
| D | Very high ≥ 10 Mbit/s | Multimedia, infotainment | MOST |

fixed time, called time control or time division on multiple access (TDMA). A *decentralized bus arbitration* shifts the decision to a user. One example is the multi-master concept. It can be controlled by using priorities, such that the master with the highest priority transmits a message first. Another example is token-passing, where the users are given cyclic arbitration to send their messages. In the case of the master-slave arbitration, a master node gives limited access to other nodes, the slaves.

According to SAE, the *automotive bus systems* can be divided in different classes; see Table 2.1. See also Navet and Simonot-Lion (2013), Zimmermann and Schmidgall (2014), and Wallentowitz and Reif (2011).

#### 2.2.3.1 CAN Bus

The Control Area Network (CAN) bus was introduced into series production in 1991 and has become a worldwide standard. It uses four levels of the OSI reference model, see Fig. 2.3, and is standardized in ISO 11898-1 (1999). The CAN bus requires only a twisted two-wire cable and has a relatively simple protocol with low computing demand. The CAN bus uses two binary states "dominant" = 0 and "recessive" = 1. The receiver of the node converts these logical states into voltage levels on the two lines CAN_H and CAN_L of the bus. The difference between these two voltages carries the information to be transferred. The high-speed and low-speed CAN buses have different *voltage levels*. In the recessive state, the high-speed CAN uses 2.5 V on both lines and in the dominant state 3.5 V on H and 1.5 V on L. The low-speed CAN has in the recessive state 0 V on H and 5 V on L and in the dominant state 3.6 V on H and 1.4 V on L; see Table 2.2. To avoid reflections, the bus lines are at each end terminated by a 120 Ohm resistance; see Fig. 2.4. The maximum bit rate depends on the bus length. For 40 m, a bit rate up to 1 Mbit/s is possible and for 100 m 500 kbit/s.

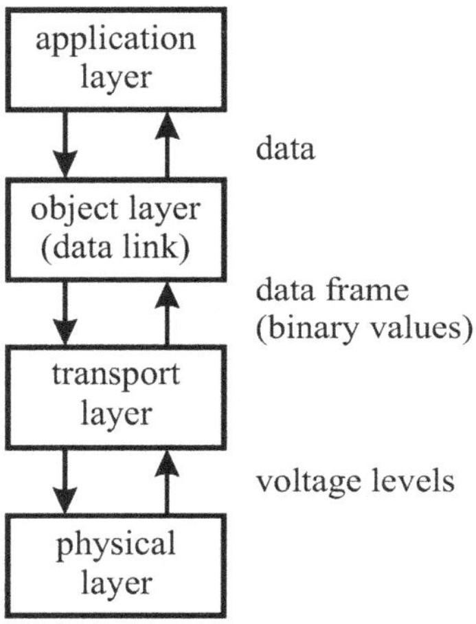


**Fig. 2.3** Protocol layers of the CAN bus (ISO 11898-1 1999)

**Table 2.2** Voltage levels of bus systems

| | CAN low-speed | CAN high-speed | LIN | FlexRay |
|---|---|---|---|---|
| Nominal | | 2.5 V | | 2.5 V |
| Dominant | H: 3.6 V<br>L: 1.4 V<br>$U_{bus} = 2.2\,V$ | H: 3.5 V<br>L: 1.5 V<br>$U_{bus} = 2.0\,V$ | $U_{bus} = 12\,V - U_{tol}$ | $U_{bus} = 1.6\,V$ |
| Recessive | H: 0 V<br>L: 5 V<br>$U_{bus} = -5\,V$ | H: 2.5 V<br>L: 2.5 V<br>$U_{bus} = 0\,V$ | $U_{bus} = 0\,V + U_{tol}$ | $U_{bus} = -1.6\,V$ |

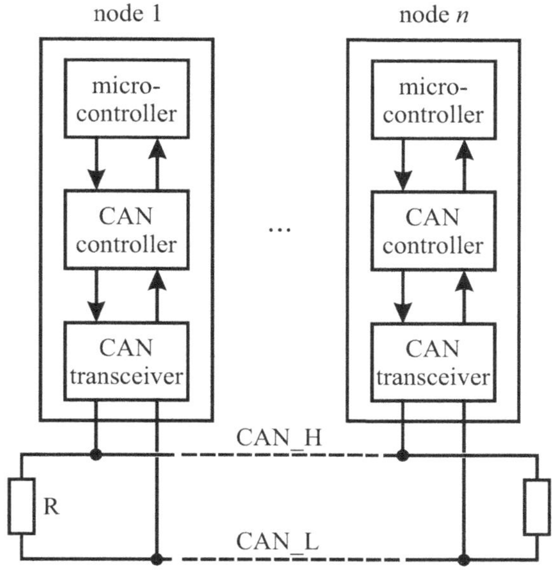


**Fig. 2.4** Scheme of a CAN bus system with hardware components for nodes as message exchanging units (ECU's): CAN controller: it generates bus data frame, includes transmit and receive buffer, acceptance checks, and provides binary signals. CAN transceiver: interface to CAN bus, generates bus voltage levels from the binary bit stream and vice versa

### CAN Software Structure

Because of the used *multi-master arbitration* Carrier Sense Multiple Access/Collision Avoidance (CSMA/CA), the protocol supports communication between network nodes without using a central management unit. Each node can send a message at any time. A message has an identifier marking its *content* and is broadcasted to all other nodes. But the message is only then taken over if the identifier is stored in the acceptance list of the node (acceptance checking). Therefore, each node checks the relevance of a bus message individually. The identifier has 11 bit for CAN 2.0 A, thus enabling 2048 different messages (or 29 bit in the extended version CAN 2.0 B). The identifier also determines the *priority*. All nodes compare their priority with the message presently on the bus, and the message with the highest priority (lowest binary value of the identifier) obtains the first access to the bus. If the other nodes have to send a message, they repeat their attempt as soon as the bus is free. Hence, the CAN bus is *event-triggered* and a certain message transfer time period is not guaranteed.

The *topology* is the linear bus structure, where the nodes operate in a parallel configuration. The number of connected nodes can be high; expansion to 32, 64, or until up to 110 nodes and cable lengths up to 500 m are possible.

The message format is a *data frame* of 130 bit (standard) or 150 bit (extended) and consists of seven fields; see Fig. 2.5: start of frame: 1 bit; arbitration field: 12 bit (message identifier); control field: 6 bit (number of data bytes); data field: 0–64 bit (0–8 bytes); CRC field: 16 bit (cyclic redundancy check with checksum); ACK field: 2 bit (acknowledgment of the transceivers from another node); end of frame field: 7 bit (end of message); interframe space: 3 bit (separation of frames).

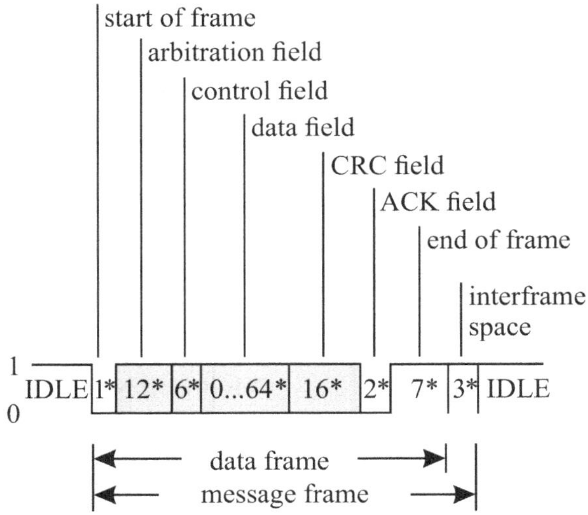


**Fig. 2.5** CAN message frame format, Robert Bosch GmbH (2011) ISO 11898. 0 dominant level, 1 recessive level, * number of bits

### CAN Hardware Structure

The CAN bus uses for software as well as for hardware the levels according to Fig. 2.3. The principle of the hardware realization is shown in Fig. 2.4. The data to and from the *application microcontroller* are received from or sent to an object layer, which corresponds to the OSI data link layer, and is implemented in the *CAN controller* hardware. It can be a stand-alone component coupled to the microcontroller by the inherent address and data bus, but can also be integrated. The CAN controller consists of a control unit and transmits and receives buffers. It selects the messages obtained from the bus according to the stored identifiers in the acceptance list and manages the messages due to their priorities. Therefore, several messages have to be stored and treated as communication objects. The CAN controller sends binary signals in the form of the CAN message frames as a bit stream to the *CAN transceiver* which as an interface generates the voltage levels CAN_H and CAN_L for the bus. CAN nodes can also be realized without a local microcontroller, for example, to connect sensors and actuators directly.

### 2.2.3.2 LIN Bus

The Local Interconnect Network (LIN) bus is a low-cost bus system to connect the sensors, actuators, and switches in the vehicle body as an alternative to the low-speed CAN. It is a local subsystem for installed mechatronic units used in doors, air conditioning, seat adjustment, etc. The LIN bus is limited to 20 kbit/s and maximal 16 nodes. It has a linear bus topology and uses an unshielded single wire. The communication operates with a master-slave concept in a time-synchronous manner defined by the master. The master is realized as a central control unit, which is also a gateway to other bus systems, such as Body CAN, Chassis CAN, and Diagnostic CAN.

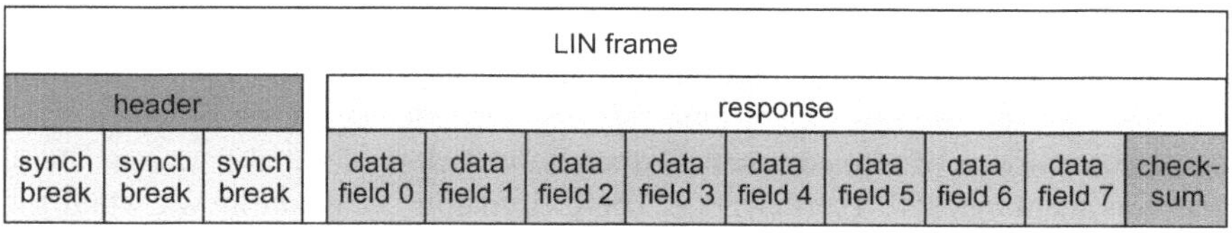


**Fig. 2.6** LIN message frame format (Reif 2014)

The LIN bus has two binary states: dominant level, 0 V, corresponds to logical 0, and recessive level, battery voltage, corresponds to logical 1. It allows larger voltage tolerances of 20% for sending and 40% for receiving. The bus access is controlled by the master only. It initiates messages to and from the slaves and can ask for data or gives commands from or to the slaves, or commands communication between two slaves. The LIN protocol is a defined frame and consists first of a header, containing synchronization break bits, synchronization field bits, and an identifier field, describing the contents of the message; see Fig. 2.6. The protocol is then followed by the response area, which contains the message. It consists of several data fields, which are dedicated to the slaves. The configurations of the LIN bus are performed by a LIN description file (Idf), which is used as a language tool.

#### 2.2.3.3 FlexRay Bus

The increasing real-time requirements for automatic open-loop and closed-loop control functions in modern vehicles need a bus system which is *time-triggered*, and which offers *high data rates* and *reliability*. Therefore, in 1999 a consortium of OEM's started to develop a corresponding bus system, called FlexRay (FlexRay 2004). This bus system has the following advantages: maximum data rate 10 Mbits/s, time-triggered with guaranteed transfer periods, also event-triggered with priorities, and application for different topologies.

The FlexRay bus uses a twisted two-wire cable, shielded or not shielded. One channel has then two wires (strands) determined as Bus-Plus (BP) and Bus-Minus (BM).The bus state follows from the difference voltage $U_{\text{bus}} = U_{\text{BP}} - U_{\text{BM}}$ of both wires, in order to lower the influence of electromagnetic disturbances. By applying different voltages to the bus, four states can be generated:

| | |
|---|---|
| Idle_LP: | BP = ± 0.2 V; BM = ± 0.2 V; (BP - BM) = 0 V) |
| | (low power consumption) |
| Idle: | BP = 2.5 V; BM = 2.5 V; (BP - BM) = 0 V) |
| 0 state: | (BP - BM) = 1.9 V - 3.1 V = - 1.6 V) |
| 1 state: | (BP - BM) = 3.1 V - 1.9 V = + 1.6 V) |

FlexRay networks can have flexible structures, thus as a linear bus, or star structure and their combinations. It can also operate with two channels where each channel has a separate cable. Thus, *redundant data transmission* is possible to build fault-tolerant systems for safety-critical functions.

**FlexRay Software Structure**

The deterministic time-controlled communication is reached by communication cycles of constant duration (about 1 ms); see Fig. 2.7. Each cycle begins with a static segment consisting of a fixed number of static time slots of equal duration. These time slots are assigned to a certain node. Then, a dynamic segment follows, where the bus access is controlled by priorities. Subsequently, a symbol window for sending a collision avoidance or test symbol is provided. To correct time deviations, the cycle is closed by a network idle time, for example, to shift the zero point of the cycle.

The nodes of the bus system have their own internal clocks. Because the access of the nodes is controlled by time slots, a synchronization of a standardized, global time of the bus network is required. This is reached by a clock control hierarchy with micro- and macroticks, derived from the oscillator of the nodes. Some nodes take the role of time generators and the other nodes synchronize their clocks by adapting the zero point and the rate. The synchronization procedure then adjusts the length of a macrotick (about 1 μs) to be the same for all nodes.

The message protocol of FlexRay is identical for the static and dynamic segments. It is divided into three sections: header, payload, and trailer; see Fig. 2.8:

*Header*: Reserved bit, payload preamble indicator (network information), null frame indicator (no update), sync frame indicator (use for synchronization), start-up indicator, frame ID (number of the slot to transfer this message), payload length (size of data), header cyclic redundancy check (CRC), cycle count (cycle number of sending network node).

*Payload*: Contains the user data as data blocks for the provided nodes with a maximum of 254 bytes.

*Trailer*: A 24-bit checksum acts as cyclic redundancy check (CRC) for the entire frame.

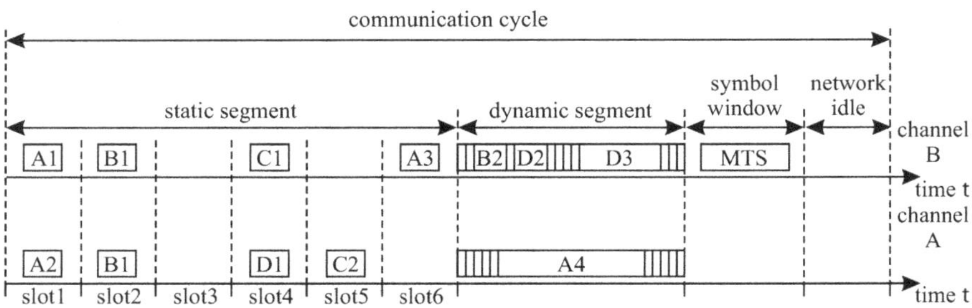


**Fig. 2.7** Communication cycle format of the FlexRey bus (Robert Bosch GmbH 2011). A1 node A transmits message 1; A2 node A transmits message 2; MTS Media Test Symbol

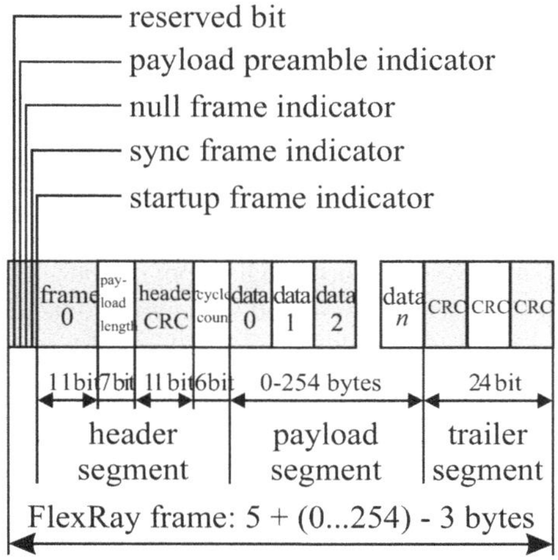


**Fig. 2.8** FlexRay message frame format (Robert Bosch GmbH 2011)

## FlexRay Hardware Structure

A FlexRay node (user, subscriber) consists of the microcontroller (host processor), a communication controller (CC), one bus driver (BD) for each channel, and optionally a bus guardian (BG) to observe the bus drivers as shown in Fig. 2.9.

The *microcontroller* is linked to the communication controller and obtains signals from the sensors connected to the bus and, after processing with control functions, sends commands to the bus-connected actuators. The *communication controller* processes all functions within the bus protocol. This includes synchronization with other nodes and generation of the macrotick signal (1 μs) and a bit stream. The conversion of the binary signals from the communication controller into the defined bus voltage levels is carried out by the bus driver in the form of a transceiver. It also connects network nodes to the channel with corresponding receivers and transmitters. A bus guardian is an optional part between the CC and the BD. It monitors the protocol

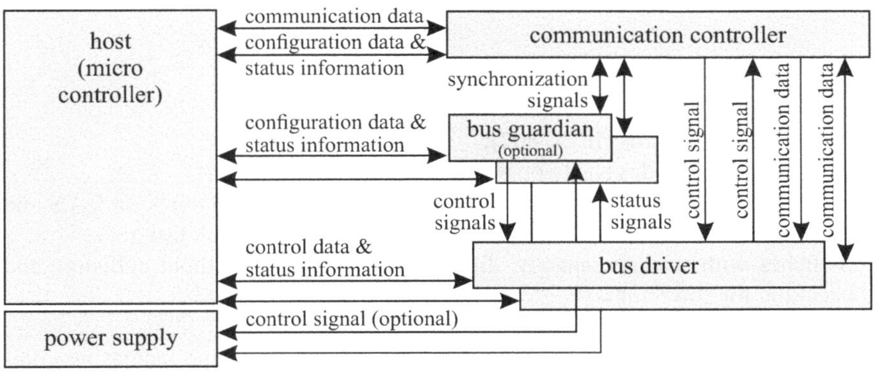


**Fig. 2.9** Hardware structure of a FlexRay node (Reif 2014)

messages of the CC and allows transmissions only if they are not faulty. By this way also, faulty nodes can be detected and countermeasures can be taken. The bus guardian can be part of a network node or can be implemented centrally to monitor all nodes.

A FlexRay bus system distinguishes several operating modes. In addition to normal operation, it supports sleep and standby modes. In the sleep mode, all node functions are deactivated to save energy. However, the bus drivers are always ready to receive wake-up signals from the bus. The required network nodes are initialized in the start-up phase and then synchronized.

The FlexRay bus was developed by the FlexRay group, founded in 2000 by the companies BMW, Daimler, Motorola, and Philips. One distinguishes core partners, premium associates, and associates. In 2004, General Motors, Bosch, and Volkswagen joined as core partners and Freescale took over the membership of Motorola. In 2006, NXP Semiconductors replaced Philips. In 2009, there were 28 premium members and 60 associate members. The FlexRay consortium ended end of 2009. Presently, the FlexRay standard specification becomes the ISO standard 17458-5.

The first series production vehicle with FlexRay was the BMW X5 in 2006, with an application for an adaptive damping system. The BMW 7 then introduced FlexRay in full use in 2008. Other cars followed, such as Audi A6, BMW 3, 7, Mercedes S-class, and Rolls-Royce Ghost.

Summing up, FlexRay is a deterministic, high-speed, and highly reliable bus system. It is time-controlled as well as event-controlled and can have two channels to allow a redundant message transfer. It includes a bus guardian with monitoring network access and with fault protection, automatic start-up, and initialization. Hence, it is applicable for safety-relevant driver-assistance systems with closed-loop control and for chassis components as part of safety-relevant systems.

#### 2.2.3.4 The Ethernet Bus

The Ethernet bus is together with the IP protocol a bus communication system between computers developed over years in the IT world. It has the following advantages:

- high transfer rate of 10 Mbit/s–10 Gbit/s;
- data transfer via two-wire cable possible;
- widely used standards (IEEE 802);
- addition of new nodes easily possible;
- comprehensive message length up to 1500 bytes instead of 8 bytes for CAN and 254 bytes for FlexRay. Therefore, the communication is much faster;
- Switches with storage capacity, direct message packets without collision, and checking for correctness.

The main reason to introduce Ethernet/IP is to cope with the increasing communication requests and to use the standards of the IT domain (Schaal 2012), like IEEE 802. It is not the goal to replace CAN, LIN, FlexRay, and MOST, because they have their special advantages in the ratio of cost to usage (Meier and Krieger 2013). However, this is on the cost of incompatibilities between the bus systems. Hence, for the increasing communication over the domains, a special bus-independent solution would be better. This is one of the goals of IP communication. The BroadR-Reach Ethernet standard is a physical layer standard worked out for automotive applications within the OPEN (One-pair Ethernet) Alliance Special Interest Group, founded in 2011 with currently 160 members. Ethernet uses several layers of the ISO/OSI reference model.

The applications are mostly in the area of ADAS, infotainment, and the fast flashing of software for the ECU's. Especially, the communication with several video cameras and the picture fusion (round view), networking with outside the vehicle, and software loading during manufacturing are of primary interest. It also mentioned using Ethernet/IP as a backbone for the communication between different domains. Ethernet is not a time-controlled real-time system. However, the real-time requirements in cars can be fulfilled (Gaus et al 2013; Trenkel and Wunner 2015).

### 2.2.4 Gateways

Modern vehicles have several networks like CAN, LIN, and FlexRay according to special applications and costs. Because many control functions of the chassis, body, and powertrain need vehicle-wide information, the different bus systems have to be coupled. However, as the bus systems have different protocols, one needs an *interpreter* which takes the data from one protocol and converts it to another one. This is performed by gateways, special electronic units. They can be localized centrally, connecting all domain-specific bus systems, or they can be distributed decentralized to connect two or more buses.

These gateways thus connect through buses like CAN, FlexRay, etc. different functional designed ECU's which have specific tasks like engine control, chassis control, driver-assistance control, infotainment, etc.

### 2.2.5 Electronic Network Architectures

The increasing number of electrical and electronic components and functions and the growing complexity require suitable electronic networks which connect the different ECU's, sensors, actuators, switches, etc. A conventional approach is a *functionally distributed E/E-architecture* with a modular structure; see Fig. 2.10. It is characterized by ECU's which have specific functions for the drive train, chassis, body, and infotainment with corresponding CAN and LIN buses. Each of the ECU's is coupled via the electrical board network with the affiliated sensors, actuators, and switches. The gateway connects the different specific bus systems. Figure 2.11 shows a more detailed example.

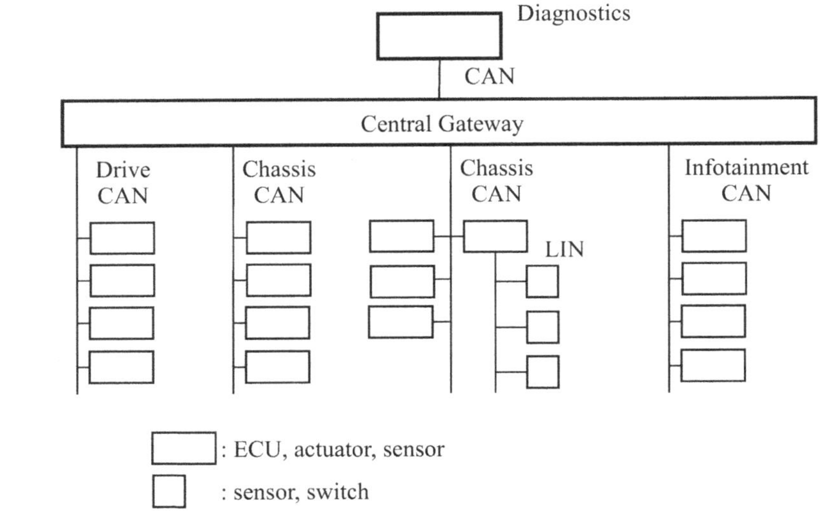


**Fig. 2.10** Conventional functionally distributed E/E-architecture

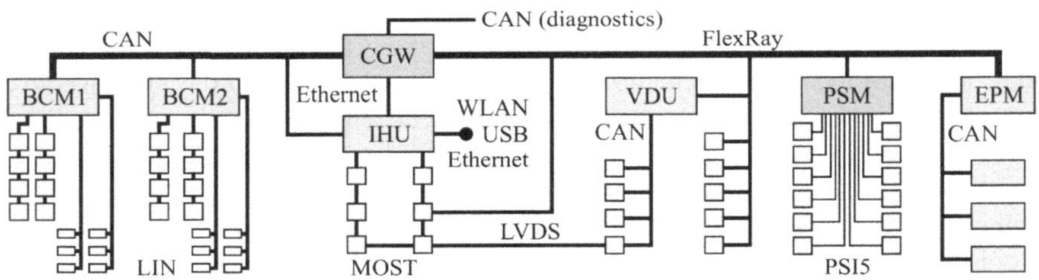


**Fig. 2.11** Typical electronic functional distributed architecture of a passenger car with a central gateway (Robert Bosch GmbH 2011). Possible scenario for a premium class vehicle. CAN: Controller Area Network; CGW: Central Gateway; BCM: Body Computer Module; IHU: Integrated Head Unit; VDU: Vehicle Dynamics Unit; PSM: Passive Safety Manager; EPM: Engine and Powertrain Manager; WLAN: Wireless Local Area Network; LIN: Local Interconnected Network; MOST: Media-Oriented System Transport; PSI: Peripheral Sensor Interface; LVDS: Low Voltage Differential Signaling

Modern vehicles have about 70−100 ECU's, cable lengths up to 8 m, and a total extent of about 1 km with weight up to 70 kg resulting in a complex board network with many branches.

As the number of electrical and electronic components further increases and also the interdependencies between the ECU-controlled functions, like powertrain and chassis (steering, braking), grouping of functions in domains may be favorized. This means that more functions are concentrated in domain ECU's which are more powerful components, e.g. for the powertrain, chassis, body, and infotainment; see Fig. 2.12. This is then called a *domain centralized E/E-architecture*; see Robert Bosch GmbH (2018) and Schulze and Liebetrau (2019). It allows a better vehicle-wide access to

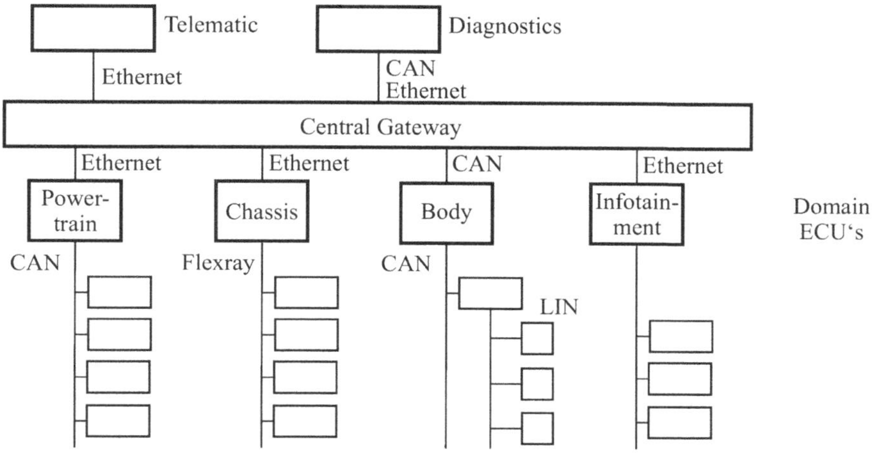


**Fig. 2.12** Domain centralized E/E-architecture

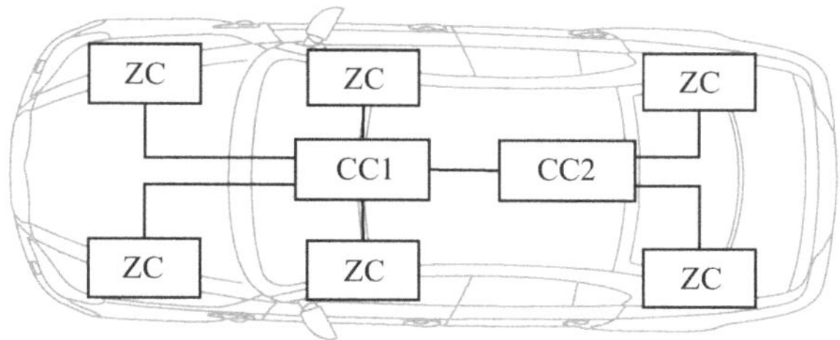


**Fig. 2.13** Geometrical zone-oriented E/E-architecture with zone controllers (ZC) and central computers (CC) (Maul et al 2018)

sensors, actuators, and information from the ECU's, as, for example, required for advanced driver-assistance systems (ADAS) like ACC, LDC, and APA. A centralized gateway connects the different bus systems and provides not only access to OBD-diagnostic tools but also connections to telematic functions (WLAN, LTE, cloud functions, over-the-air updates, and GPS).

Another development is to establish a *zone-oriented E/E-architecture* with locally oriented zone controllers (zone-ECU's) which are distributed over the vehicle, instead of the domains (Maul et al 2018). These zone controllers are standardized and repeatedly used in the vehicle. They are connected with two or more *central computers* via automotive Ethernet; see Fig. 2.13. This allows to come to a scalable architecture and to reduce cable lengths and weight, as sensors, switches, and actuators are not connected to a domain-specific ECU but to a next accommodated zone-ECU. The zone controllers may have a redundant voltage supply, and the central computers have safe operational functions (ISO 26262) for safety-critical tasks. This is also called a service-oriented EE-architecture (SOA).

## 2.3 Software Structure of the Electronic Control Units

### 2.3.1 Conventional Software Structure

The software architecture for automotive microcontrollers can be divided into a platform software and an application software. The platform software architecture was at least in parts standardized over years as shown in Fig. 2.14. One layer groups the software components of the input/output hardware in the form of a hardware abstraction layer (HAL) of a microcontroller. Other standardizations are the specification for a real-time operation system according to OSEK (**O**ffene **S**ysteme und Schnittstellen für die **E**lektronik in **K**raftfahrzeugen) and diagnostic protocols according to ISO. The OSEK/VDX organization was started in 1993 and has published several standards, e.g. for the operation system (OSEK-OS), the communication (OSEK-COM), network management (OSEK-NM), and runtime interface. In 2003, the efforts of the OSEK group were then continued by the AUTOSAR initiative.

### 2.3.2 Multilevel-Software Structure

Because many functions in advanced automobiles are realized by software in the various ECU's, the organization and development of automotive software have become a major challenge. As the electronic technology changes relatively fast and the software development is performed parallel with many teams, a clear and open structure

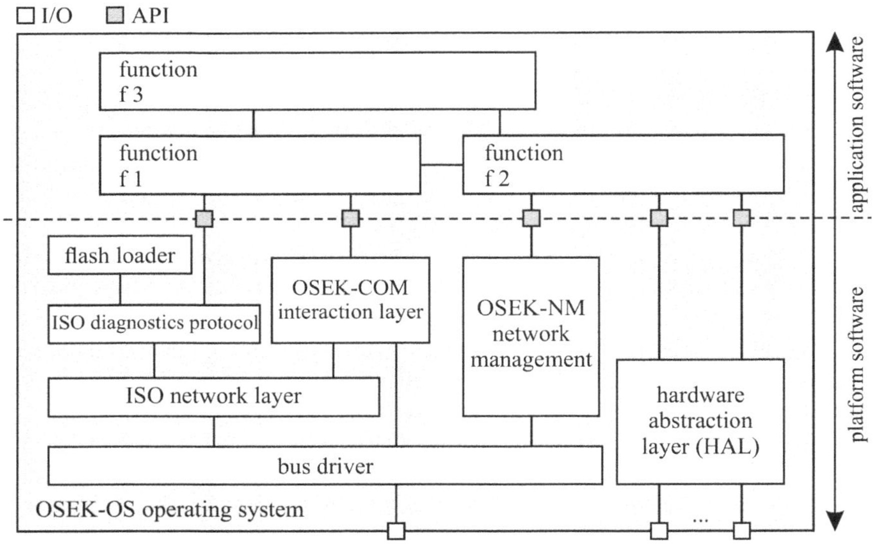


**Fig. 2.14** OSEK software architecture for microcontrollers and standardized components (Robert Bosch GmbH 2011)

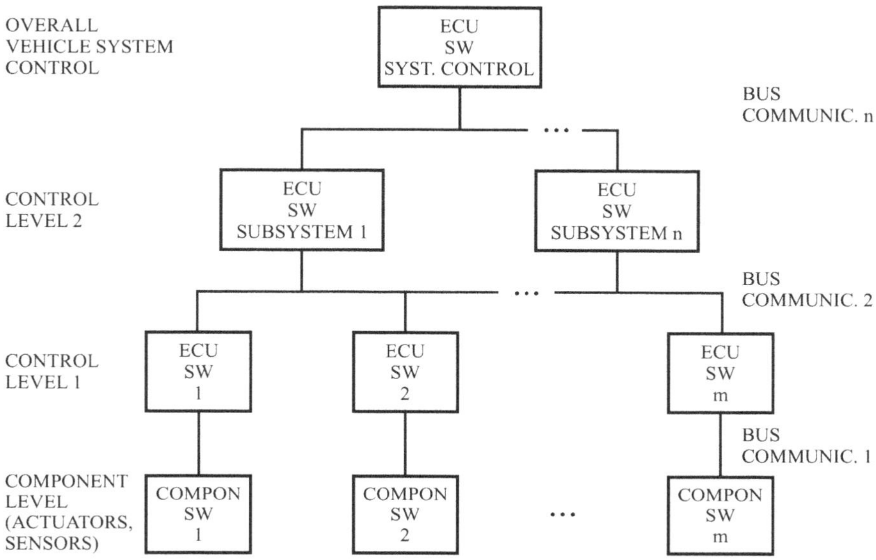


**Fig. 2.15** Functional multilevel software architecture (idealized) for automotive control

for the development and realization of software is required. The development may follow a V-model; see, e.g. Isermann (2005).

The software structure has to satisfy several requirements, like flexibility, expandability, re-usability, testability, and easy integration of modules. This has led to a hierarchically organized software structure as indicated in Fig. 2.15. At the (lowest) component level, the hardware-dependent software is implemented, like for special sensor and actuator control and communication. These components send their signals directly (centralized) or via a bus to the function control ECU with software level 1, like the subsystems braking control (ESC) or steering control (EPS). The function software for commanding several vehicle subsystems may then be centralized in a higher function control ECU with software level 2, like the expanded subsystem body control, vehicle dynamics control, powertrain control. The highest vehicle SW level then integrates several subsystem software functions in the sense of an overall vehicle control, like the different kinds of automatic driving for parking, cruising, and collision avoidance.

Generally, the software structure depends on the ECU and communication hardware structure, and is divided into a basic (hardware-dependent) and an application-specific (function-oriented) part.

### 2.3.3 AUTOSAR

The increasing electronic functions and complexity of advanced vehicles require more standardization and organization forms for the involved developing partners.

Therefore, an open and standardized automotive software architecture is jointly developed by automobile manufacturers, suppliers, and tool developers, since 2003, called **AUTOSAR** (**AUT**omotive **O**pen **S**ystem **A**rchitecture).

The goals are

- standardization and implementation of basic software functions;
- scalability to different vehicle and platform variants;
- transferability of functions within networks;
- integration of functional modules from multiple suppliers;
- high reliability, safety, and redundancy;
- maintainability, software updates, and software upgrades over vehicle lifetime;
- use of standard commercial hardware.

AUTOSAR supports an automated production of software for the ECU's from design to implementation. Instead of an ECU-based approach, a function-oriented approach is envisaged.

The AUTOSAR standard uses an architecture with different layers to decouple the functionality from the hardware and software services as depicted in Fig. 2.16:

*Basic Software Layer (BSW)*: This basic software layer provides hardware-dependent services, like communication (bus systems), ECU abstraction, and micro-controller abstraction. The communication to the next higher level is realized by standardized AUTOSAR interfaces.

*Runtime environment (RTE)*: The information exchange between the different software components is managed by the runtime environment. It serves as a middle-ware to decouple the application software from the hardware. The RTE also connects software modules from different ECU's. RTE together with the basis software thus realizes a *virtual function bus*, which is one major approach of AUTOSAR.

*Application layer*: This layer comprises all the application software components (SWC), which are mainly hardware-independent. These software parts represent the actual functionalities, are encapsulated and not standardized, but have standardized AUTOSAR interfaces to the RTE. Some software components offer connections to the sensors and actuators such that all other members of the RTE have access to these input and output devices by standardized interfaces.

This layered AUTOSAR structure is used in all ECU's, as illustrated in Fig. 2.17. It relaxes the design of the application software because the functions can be developed only by using the interfaces to other software modules. Then, one relies on the fact that the virtual function bus will provide all the required connections.

The AUTOSAR software development is supported by tool-chains, using XML for defining exchange formats and formal descriptions with UML.

AUTOSAR development is based on nine worldwide core members and a total of 146 members in 2012. The work is performed in different working groups. The first release 1.0 was in 2005, and 4.1 in 2013.

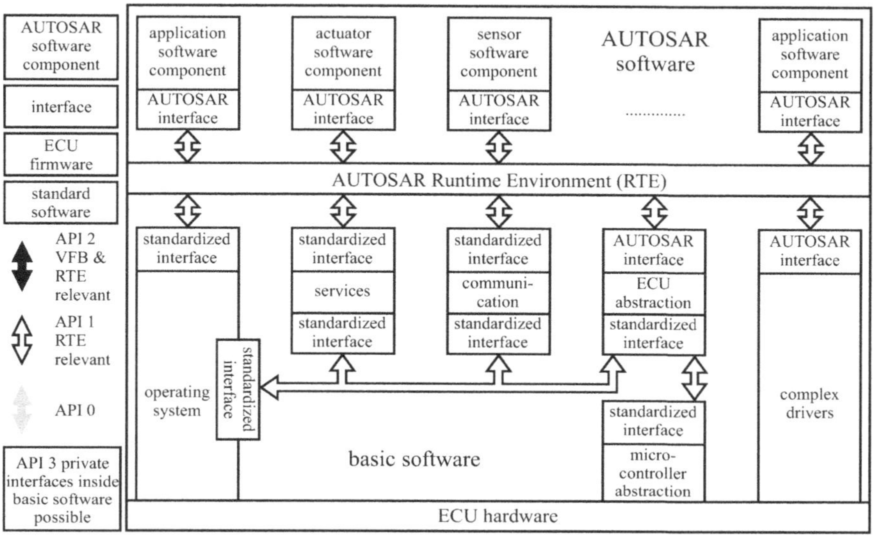


**Fig. 2.16** AUTOSAR software architecture of one ECU node (Kirschke-Biller 2011). VFB: virtual function bus; API: applications programming interface

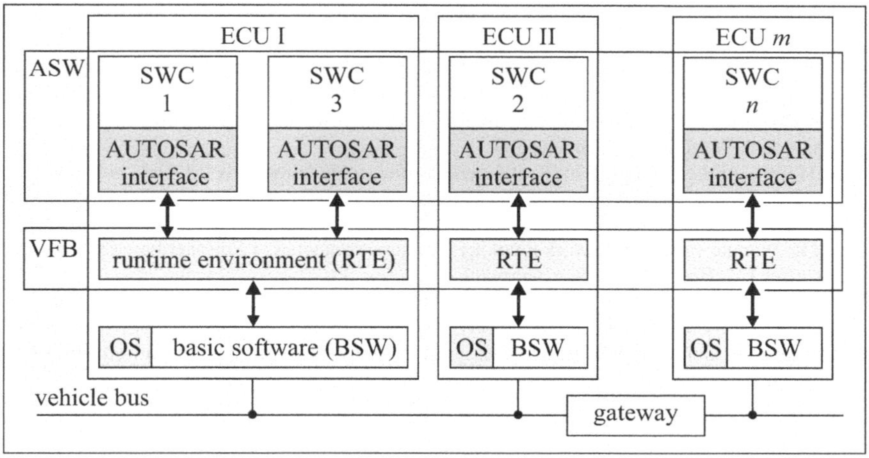


**Fig. 2.17** AUTOSAR software structure with several ECU's (Robert Bosch GmbH 2011). ASW: application software; SWM: software component; VFB: virtual function bus; RTE: runtime environment; BSW: basic software; OS: operating system

## References

Borgeest K (2014) Elektronik in der Fahrzeugtechnik, 4th edn. Springer Vieweg, Wiesbaden
FlexRay (2004) FlexRay communication system. Protocol, bus generation, physical layer specification. www.flexray.com
Gaus J, Krieger O, Stein S (2013) Automotive ethernet: service quality, security and testability. In: 16. Int. Kongress Elektronik im Kraftfahrzeug, VDI, Düsseldorf, vol VDI Bericht 2188, pp 173–183
Isermann R (2005) Mechatronic systems - fundamentals, 2nd edn. Springer, London
ISO 11898–1 (1999) Road vehicles - controller area network (CAN) - Part 1: data link layer and physical signaling. International Organization for Standardization, Geneva
ISO 7498–1 (1994) Information technology, basic reference model. International Organization for Standardization, Geneva
Kirschke-Biller F (2011) Autosar – a worldwide standard current developments, roll-out and outlook. In: 15th international VDI congress electronic systems for vehicles, Baden-Baden, Germany
Maul M, Becker G, Bernhard U (2018) Service oriented EE-zone architecture. ATZ electronics 1:36–41
Meier A, Krieger O (2013) IP-Kommunikation - Verwendung etablierter Standards statt Eigenentwicklung. In: 16. Internationaler Kongress Elektronik im Kraftfahrzeug. VDI-Bericht, vol 2188, pp 93–103
Navet N, Simonot-Lion F (2013) In-vehicle communication networks - a historical perspective and review. Internal Report and Chapter 4 in the automotive communication systems. In: Proceedings of the IEEE, special issue on industrial communication systems, vol 96, no 6, CRC Press, Taylor and Francis (2008). Technical report, Luxembourg
Reif K (2014) Automobilelektronik, 5th edn. Springer Vieweg, Wiesbaden
GmbH Robert Bosch (ed) (2007) Automotive electrics, automotive electronics. J. Wiley, Chichester, UK
GmbH Robert Bosch (ed) (2011) Automotive handbook, 8th edn. Bentley publishers, Cambridge
GmbH Robert Bosch (ed) (2018) Automotive handbook, 10th edn. J. Wiley, Chichester, England
SAE (1994) Vehicle architecture for data communication standards. Handbook. SAE J2507. SAE International, Warrendale, Pennsylvania
Schaal H (2012) Ethernet und IP im Kraftfahrzeug. Elektronik automotive 12(4):38–41
Schulze R, Liebetrau T (2019) Development platform for new network technologies in vehicle architectures. ATZ Electron 11:38–41
Trenkel K, Wunner P (2015) Network topology for chassis potential of Ethernet-based systems. In: ChassisTec Conference, München, Germany
Wallentowitz H, Reif K (2011) Handbuch Kraftfahrzeugelektronik - Grundlagen, Komponenten, Systeme, Anwendungen, 2nd edn. ATZ/MTZ-Fachbuch, Vieweg+Teubner Verlag, Wiesbaden
Zimmermann W, Schmidgall R (2014) Bussysteme in der Fahrzeugtechnik, 5th edn. Springer Fachmedien, Wiesbaden

# 3 Vehicle Control Structures

Because of the increasing complexity of vehicle control functions, a unified structure eases the understanding and development. Therefore, this chapter summarizes vehicle control structures and the model-based design workflow with different simulation tools.

## 3.1 Overall Vehicle Control Structures

Figure 3.1 shows the main signal flow from the driver's inputs acceleration pedal $\alpha_{\text{ped}}$, wheel steering angle $\delta_{\text{H}}$, braking pedal position $\beta_{\text{b}}$, transmission selector $u_{\text{tr}}$, and manual suspension selector $u_{\text{sp,H}}$ to the vehicle outputs like velocity $v$, yaw rate $\dot{\Psi}$, etc. of a conventional vehicle (with automatic transmission) controlled by the driver. The driver uses mainly his own sensors like eyes, ears, and sense of balance and the estimation of the velocity $v_{\text{e}}$ and receives as a feedback the longitudinal and lateral acceleration $a_{\text{X}}$ and $a_{\text{Y}}$ and a steering torque ("steering feel"). The main inputs act in a parallel way on the vehicle. However, vehicles have several crosscouplings to the outputs (as will be described by models later).

In order to obtain a systematic view of the conventional and advanced driver-assistance systems, a hierarchical arrangement in several control levels results with regard to the signal flow leading to a *multilevel control architecture* as shown in Fig. 3.2. This also includes the implementation of more drive dynamic and environmental sensors.

The *component control level* includes the engine, transmission, power steering, brake control (ABS), and suspension control. These control systems obtain usually command inputs from the driver through the pedals and the steering wheel and a switch for the suspension if no higher level driver-assistance systems are active. In



R. Isermann, *Automotive Control*, ATZ/MTZ-Fachbuch,
https://doi.org/10.1007/978-3-642-39440-9_3

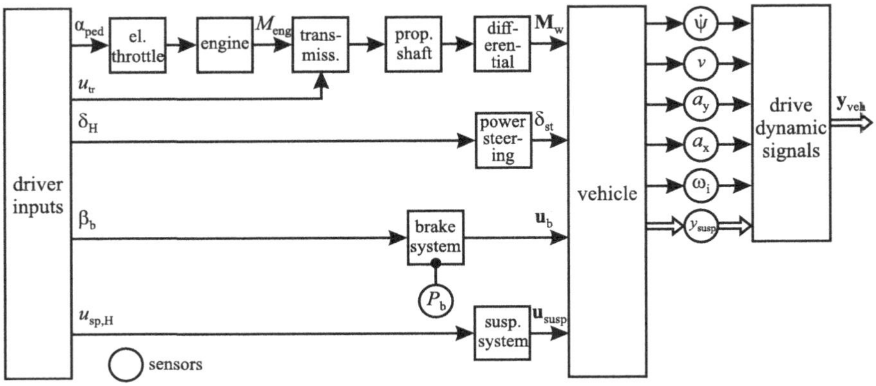


**Fig. 3.1** Main signal flow of a conventional vehicle

general, the component control systems for the chassis operate independently from each other.

The *driver-assistance systems* (DAS) ESC, ACC, LKC, and TCS are arranged in a vehicle control level 1. These control systems receive information from the drive dynamic and surrounding/environment sensors, like ultrasonic sensors, radar or lidar, and video cameras. They give commands to the component control systems and support the driver in guiding the vehicle. In general, they operate independently from each other.

According to the automotive vehicle control development and the different degrees of automation described in Chap. 1 and Table 1.1, corresponding vehicle control levels 2–5 can be distinguished. Thus, *partial automatic driving* (PA) is dedicated to level 2, with, for example, parking control, braking control, and lane keeping control.

*Conditional automatic driving* (CA) in level 3 includes longitudinal and lateral automatic driving on special roads like highways and rural roads, in traffic jams, and for parking.

For *high automatic driving* (HA) in level 4, the automatic control is further expanded to more use cases.

*Full automatic driving* (FA) then takes over automatic control for all situations.

In general, the higher levels use the functions of the lower levels and therefore act on the lower levels. However, the control functions of a lower level can also be improved and extended for a higher level automatic control. For example, the limiting conditions as road types, speed ranges, and environmental conditions may be extended.

Additional development of automated driving is the implementation of a communication level which includes navigation functions, digital cards, Internet connection, and cloud services. Their size depends on the corresponding automatic control functions and grows with higher levels.

The control functions of the component control level in Fig. 3.2 are mostly *independent* from each other and could therefore be developed and implemented indi-

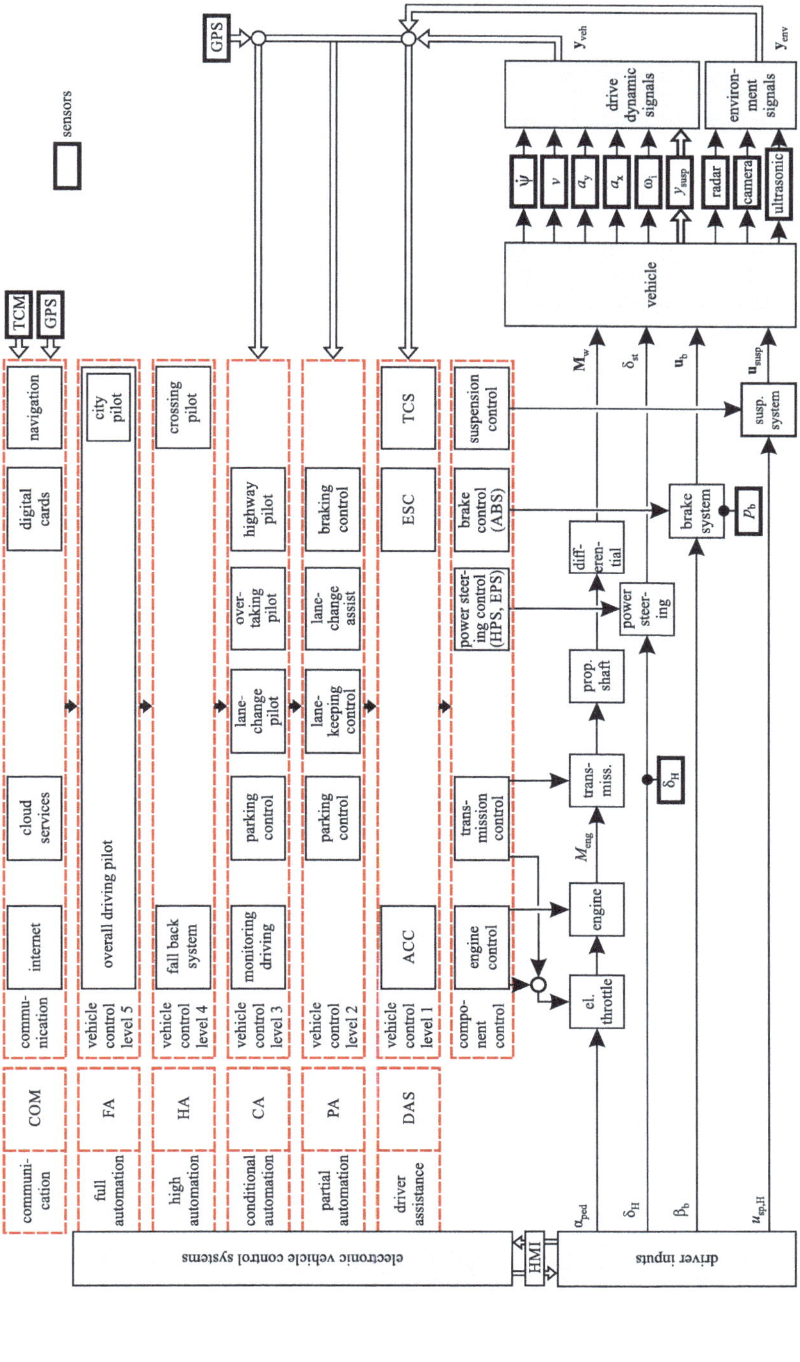


**Fig. 3.2** Vehicle control system with human driver inputs and arrangement in hierarchical levels

vidually and without disadvantages if delivered from different suppliers. This holds also for the control level 1 because these systems act mainly on one or two components. Hence, this *decoupled structure* or *parallel structure* is a way of "peaceful coexistence". However, adding more control systems in higher vehicle control levels simultaneously involves more and more control systems of the lower levels, such that at least a "cooperative coexistence" is advantageous (Reichart and Bielefeld 2009). Application examples are full braking and adjustment of the suspension dampers to hard damping and collision avoidance systems which combine emergency braking and emergency steering. Finally, higher automatic driving requires *centrally controlled overall systems* with coordinated control of engine, transmission, steering, and brakes, by using a fusion of all environment sensors. This can also be understood as a *general vehicle motion control* Therefore, the classical automotive structures will change. However, these developments will be based on conventional and well-proven concepts.

## 3.2 Control Structures of the Powertrain

### 3.2.1 Control Structure of Internal Combustion Engines

The control of the powertrain, consisting of the internal combustion engine and/or electric drive, the transmission and the propeller shaft, and differential and wheel shafts are integrated parts of the vehicle overall control. Therefore, the control structures of internal combustion engines and hybrid drives are briefly considered.

The control architecture of a *gasoline engine* is depicted in Fig. 3.3. Altogether, about seven main actuators manipulate the airflow and if provided the recirculated exhaust gas, and exhaust gas treatment through different control modules. The control modules are torque control, injection and air/fuel control, ignition control, knock control, air charge, and EGR control for normal operation of the warm engine. In addition, there are special control modules for the engine states cold start, warming-up, and idling.

A corresponding control architecture for *diesel engines* is shown in Fig. 3.4. For more details, refer to, for example, Robert Bosch GmbH (2018), Guzzella and Onder (2004), and Isermann (2014). These different structures are limited to control functions. However, diagnosis functions are also realized in the control modules.

With regard to vehicle control, the internal combustion engines are subsystems which provide a certain torque to the drive shaft with the accelerator pedal by the driver or electronically by, for example, the vehicle speed and distance control (ACC) or by a traction control system (TCS). This will be considered in detail in Chaps. 6 and 19, where the longitudinal vehicle behavior is considered.

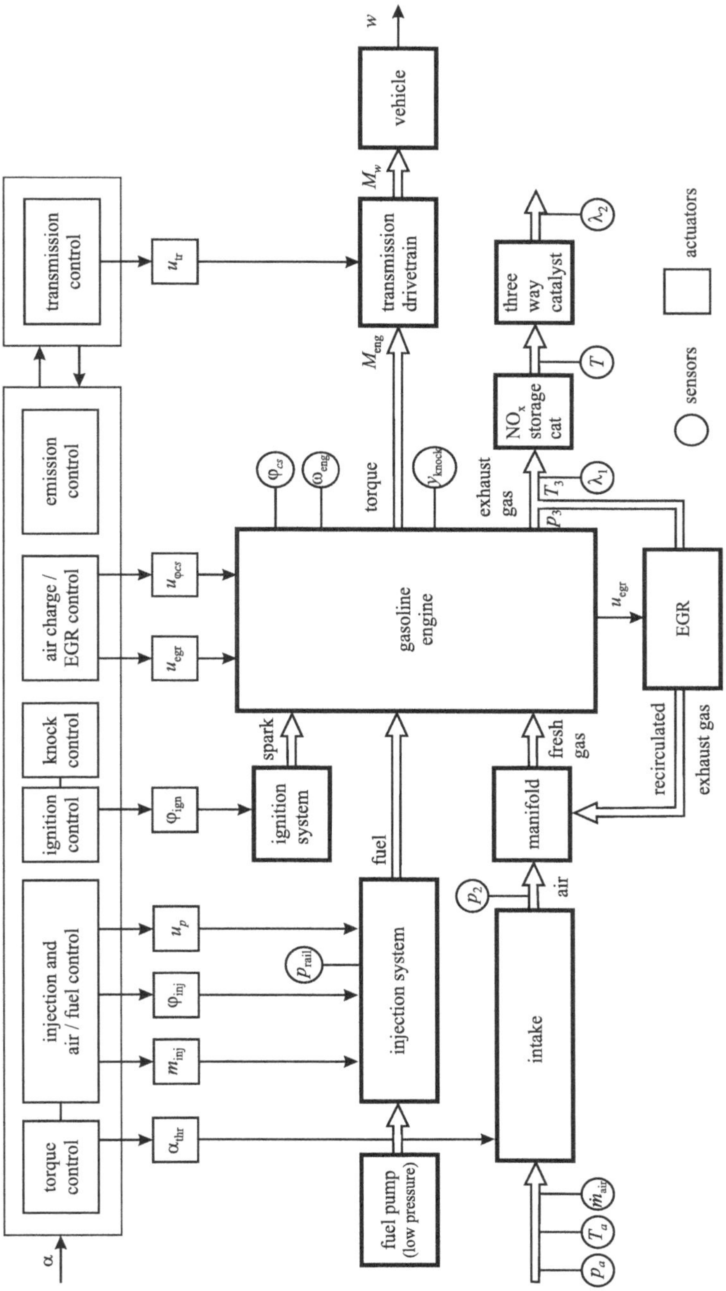


**Fig. 3.3** Control structure of a gasoline engine with direct injection, actuators, control modules, and some sensors (Isermann 2017)

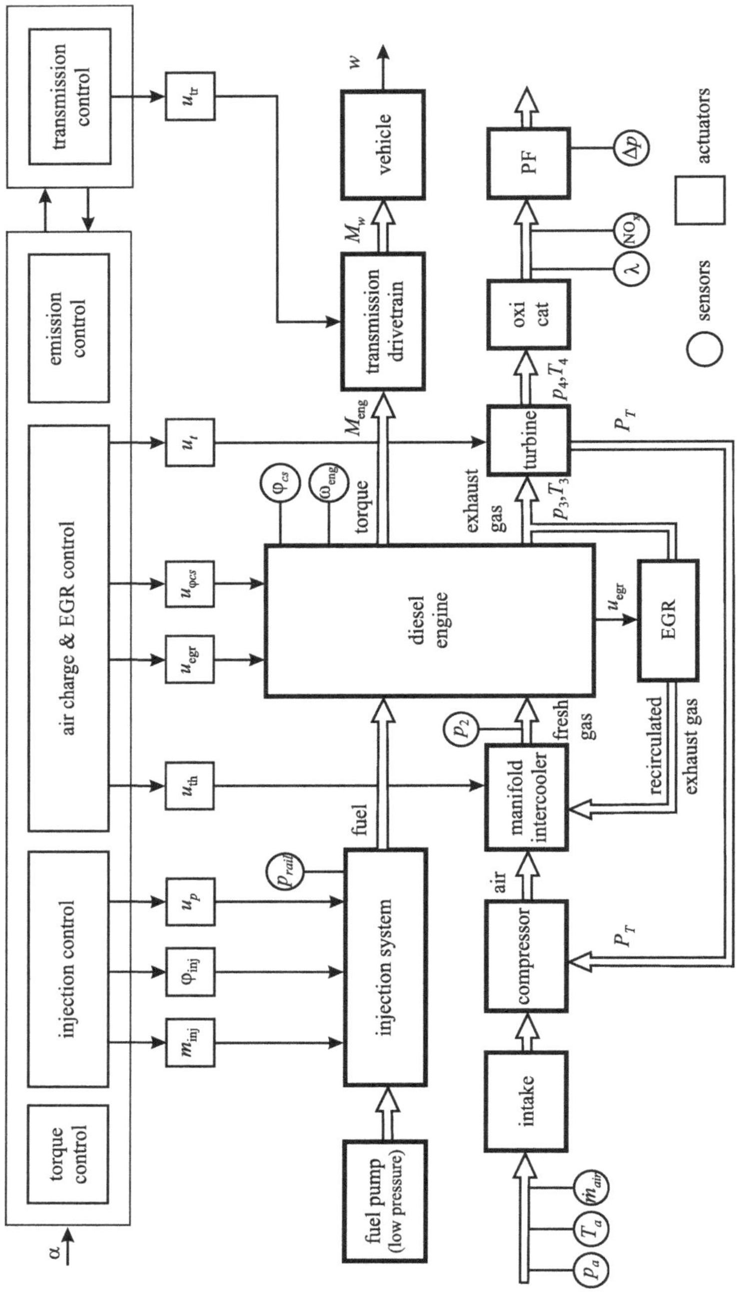


**Fig. 3.4** Control architecture of a diesel engine with common rail injection and turbocharger, actuators, control modules, and some sensors (Isermann 2017)

## 3.2.2 Control Structure of Hybrid Drives

A further step of the electrification of vehicles is the development of *hybrid and electrical drives*. The combination of internal combustion engines (ICE) and electrical motors (EM) allows a further saving of fuel consumption and emissions through the operation of the combustion engines in ranges of better specific fuel consumption in part load, regenerative braking, and electrical driving and boosting. Figure 3.5 shows three basic structures, according to the kind of energy flow.

A *series hybrid drive* is characterized by an ICE driving a generator (GEN) which charges a battery and which supplies the driving EM, acting on the wheels. The ICE is intended to operate stationarily with the best efficiency. Because of several energy conversions, this structure is usually not used for cars (except range extender configurations) but is, however, applied for locomotives and city buses.

The *parallel hybrid drive* allows the simultaneous (parallel) operation of the ICE and the EM. The EM is attached between the ICE and a gear or differential gear and either operates as a motor or charges the battery as a generator. If the EM is directly coupled with the flywheel, it is called starter/generator. However, if the EM

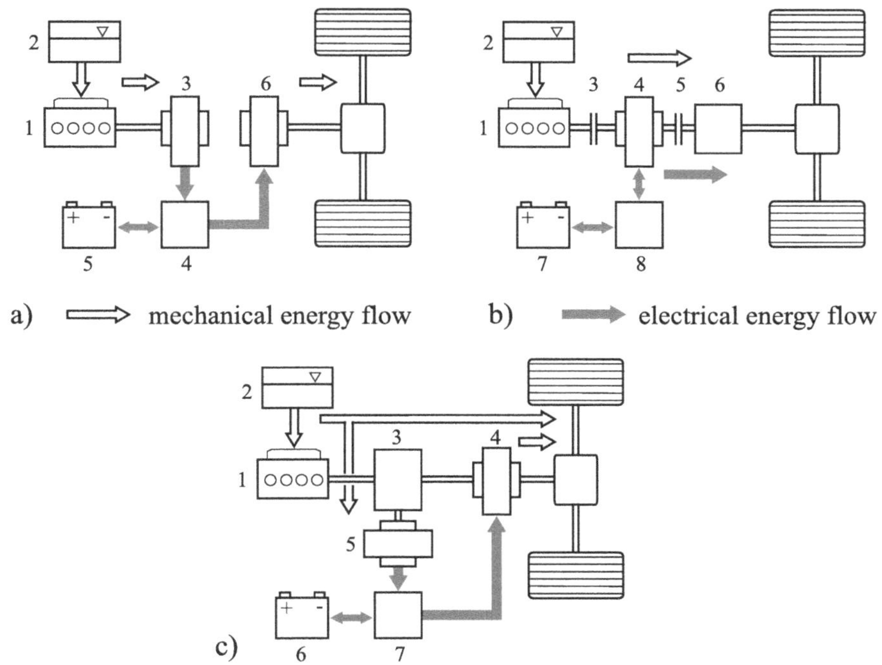


**Fig. 3.5** Basic structure of hybrid drives. **a** series hybrid: 1 ICE, 2 tank, 3 generator, 4 inverter, 5 battery, and 6 electromotor; **b** parallel hybrid: 1 ICE, 2 tank, 3 clutch 1, 4 electromotor, generator, 5 clutch 2, 6 gear, 7 battery, and 8 inverter; **c** power-split hybrid: 1 ICE, 2 tank, 3 planetary gear, 4 electromotor, 5 generator, 6 battery, and 7 inverter

is separated from the ICE by a second clutch, electrical driving and regenerative braking are possible without towing losses of the ICE.

A *power-split hybrid drive* is a combination of a series and parallel structure. A power-splitting planetary gear after the ICE allows supplying a part of the ICE power mechanically to the drive train and the other part to a generator. A direct transfer of the torque to the drive train is, because of the planetary gear, only possible if the generator consumes power. The generator power is then directly supplied to the EM and the drive train or it is used for charging the battery. This power-split hybrid drive allows a continuously controllable speed ratio between the ICE and the drive train, similar to a CVT-transmission.

Based on these energy flow structures, different hybridization degrees can be distinguished. The hybridization degree is understood as the ratio of the electrical power to the total power $H = P_{el}/P_{tot}$. *Micro hybrids* ($P_{el} < 5\,\mathrm{kW}$) typically have a parallel structure and a starter/generator at the crankshaft or at the belt drive. Start/stop operation allows to reduce fuel consumption at standstill. *Mild or medium hybrids* ($P_{el} < 15\,\mathrm{kW}$) usually have a parallel structure as well, with one or two clutches. Besides start/stop, start of driving, boosting, and regenerative braking become possible. *Strong hybrids* ($P_{el} \approx 30$ to $100\,\mathrm{kW}$) can either be realized as parallel or power-split structures and allow, with larger battery capacity, longer electrical driving.

The design of the components and the optimization of hybrid drives requires a mechatronic overall consideration from the beginning. Figure 3.6 shows a possible control architecture for a parallel hybrid drive. The engine control and transmission control have to be supplemented by an electrical drive control for the power electronics, electromotor/generator and clutches, a battery control, and a regenerative/friction

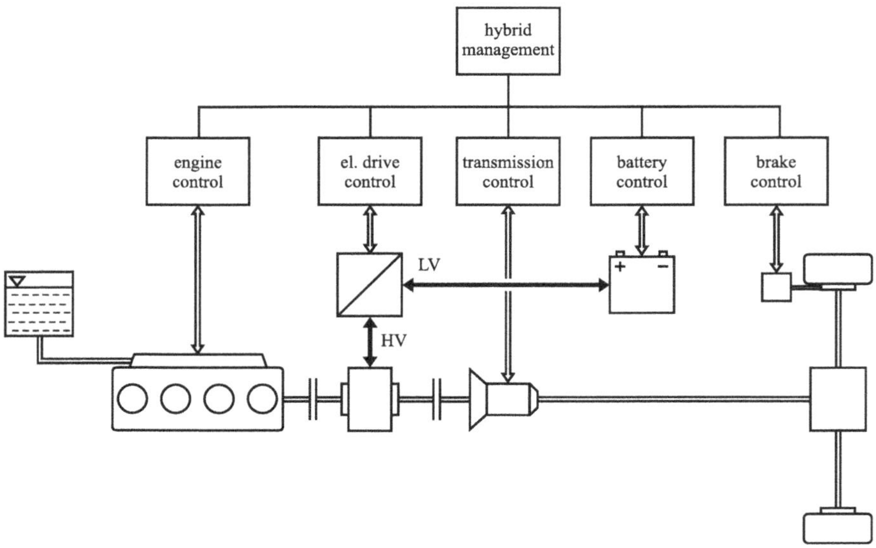


**Fig. 3.6** Overall control architecture of a parallel hybrid drive

brake control. These subsystem-oriented control systems are linked together with a data bus to a hybrid drive overall management system, where the operation of the engine and electromotor power, battery charging and discharging, and regenerative braking with the generator and friction brakes is optimized. Figure 3.7 depicts a scheme for the optimization of the torque contribution by the combustion engine and the electromotor, which may be a basis for off-line and online use with regard to a driving cycle (Kunkel 2015). Hence, the electronic control and management functions increase considerably. This holds also for the diagnosis functions with the many added low-voltage and high-voltage components and the battery, where the state-of-charge is part of the operation and has to be monitored continuously.

## 3.3 Design of Vehicle Control Systems

The design and implementation of vehicle control functions have developed into a sophisticated and labor-intensive procedure. This is for many reasons, among them the multi-variable complexity of modern automobiles, the increase of driver-assistance systems, the high performance requirements of suppliers, manufacturers and customers, and legislative certification limits for fuel consumption and emissions, safety, and competition. The development of partial, high, and finally full automatic driving requires the solution of a multitude of automatic control functions.

The following sections summarize some general procedures for control-function design, control-software generation, required computers and software tools, and test benches.

### 3.3.1 Vehicle-Oriented Electronic Control Design

The design and implementation of electronic control and diagnosis systems are highly interrelated with the design of the mechanics, mechanical, electrical, hydraulic, and pneumatic components. It belongs to the design of mechatronic systems and requires a systematic development across the classical boundaries. With regard to the timeline of the workflow, a simultaneous or concurrent engineering in different domains has to be performed; see, e.g. VDI 2206 (2003) and Isermann (2005).

#### 3.3.1.1 V-Development Model

The development of the electronic vehicle control system can, according to the design of the control-software functions for mechatronic systems, be divided into

1. control-system design:
   - control-function development;
   - control-software development;

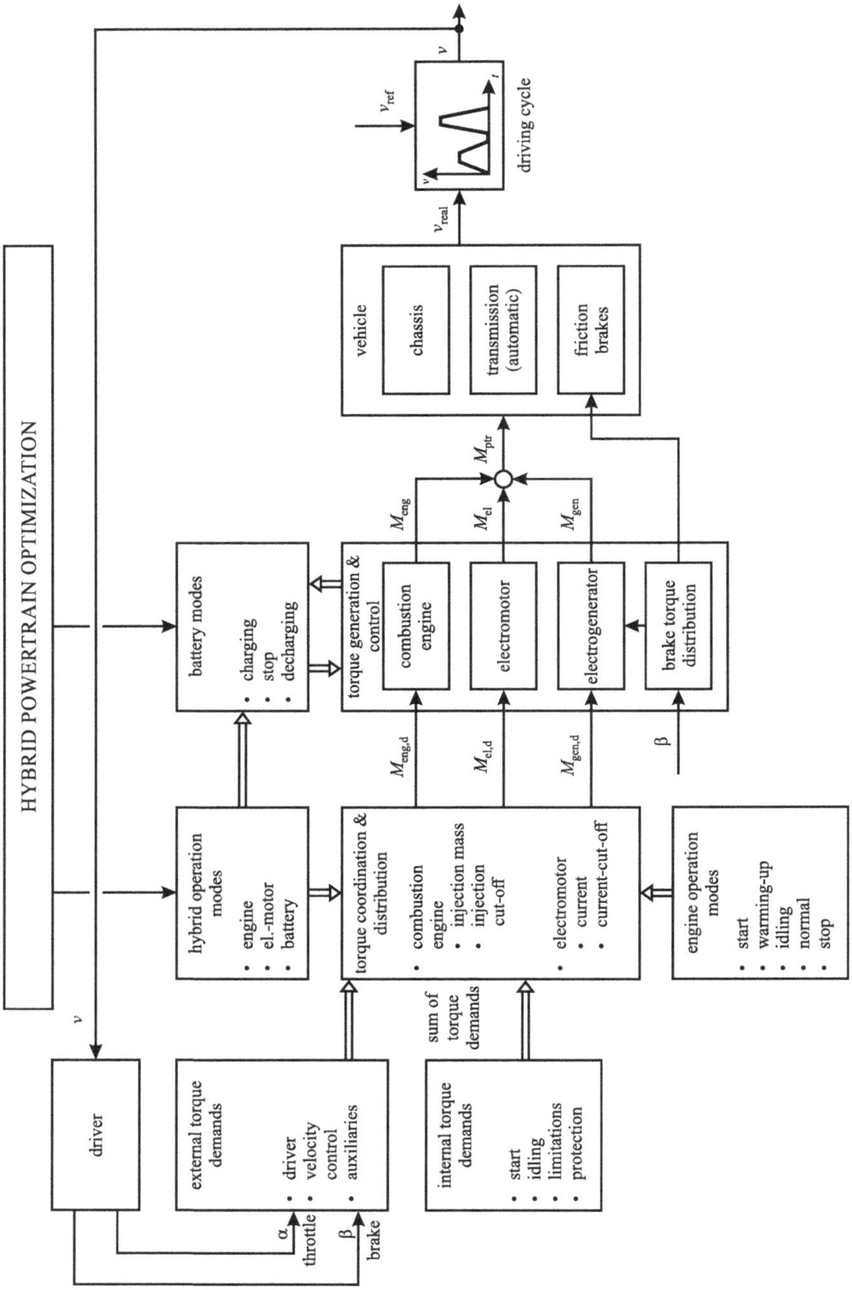


**Fig. 3.7** Signal flow modules for the control of hybrid powertrains

2. control-system integration:

   - component integration;
   - calibration;
   - performance testing.

This design procedure is highly interrelated and requires many iterative steps and special development tools. It can favorably be represented in a so-called V-model, see Fig. 3.8, which covers all aspects from the analysis of the user requirements to acceptance tests; see, e.g. VDI 2206 (2003), Schäuffele and Zurawka (2005), Isermann (2005), BRD (1997) with origins in Böhm (1979), STARTS Guide (1989), Bröhl (1995), and Droeschel and Wiemers (1999).

A corresponding V-model can be given for the hardware development of the electronic control unit (ECU) which is assumed here to exist already.

An alternative to the V-model is, e.g. the waterfall model, e.g. Royce (1970) which is organized sequentially in one direction with recursions. The V-model has the intention that the results, documents, and tests of the right branch correspond to the development procedures of the left branch. A further developed version with more flexibility is the V-model XT (BRD 2004) as discussed in Borgeest (2008).

Some important steps of the V-model for the development of control and software functions can be described as follows:

1. Requirements:

   - user requirements;
   - definition of general (overall) functions and data (rated values) of the final product (ECU);
   - general solution outline;
   - development and manufacturing costs;
   - timely development and milestones;
   - result: requirements document (does not include technical implementation).

2. Specifications:

   - definition of the product (ECU) that fulfills the requirements;
   - partitioning in manageable modules for control and diagnosis;
   - specification of features and data of the modules;
   - consideration of the sources, tools, and limitations for the development and final manufacturing and maintenance;
   - specification of hardware data;
   - specification of used software, compilers, and development systems;
   - result: specification document.

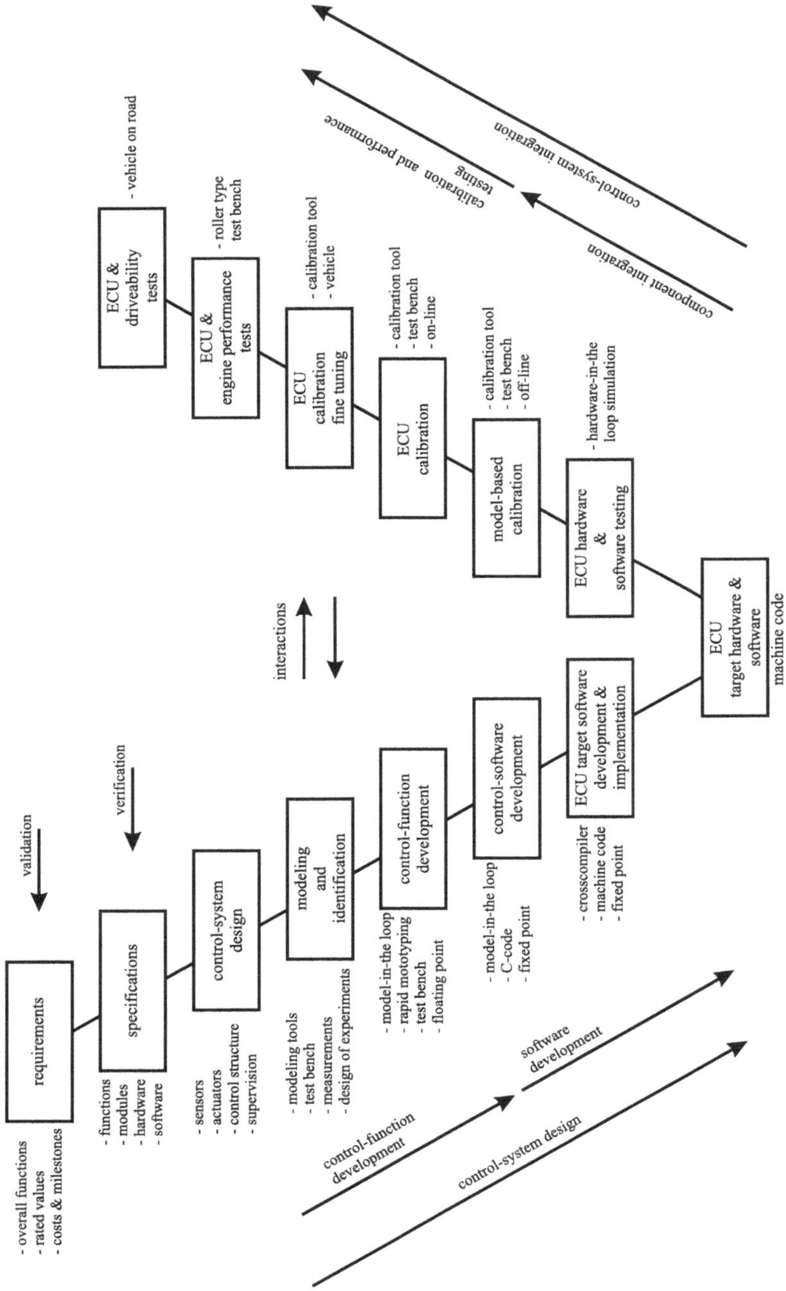


**Fig. 3.8** V-model for the development of control functions, control software, calibration, and performance tests

3. Control-system design:

   - detailed partitioning into electronic, mechanic, hydraulic, pneumatic, and thermal components with their auxiliary power supplies;
   - detailed fixing of type of sensors and actuators and their data;
   - detailed data of interfaces among ECU, sensors, and actuators;
   - task distribution between sensors and actuators with integrated electronics and ECU;
   - specification of power-related data;
   - hardware design: data of microprocessors, data storages, interfaces, bus systems, cabling, and plug systems;
   - control engineering design:
     - definition of sensor inputs and outputs to the actuators;
     - control-system structure: feedforward control (open loop) and feedback control (closed loop);
     - required engine and component models;
     - model-based design;
     - calibration (parametrization) methods;
     - supervision and diagnosis functions;
   - reliability and safety issues; FMEA (fault mode and effect analysis) studies for sensor, actuator, and ECU faults and failures;
   - result: control-system design document.

4. Modeling and identification:

   - required mathematical models of vehicles, sensors, actuators, powertrain, transmission;
   - theoretical/physical modeling;
   - experimental modeling;
   - use of modeling/identification tools;
   - measurement procedures for test benches, design of experiments;
   - kind of models: stationary (lookup tables, polynomials, neural networks); dynamic (differential equations, neural networks);
   - result: vehicle, powertrain, and component models.

5. Control-function development:

   - hierarchical control structure;
   - computer-supported design, manual design;
   - ECU states: from start-up to shut-off;
   - vehicle states (discrete): from start to shutoff;
   - control functions: vehicle-state-dependent, time-dependent;
   - sampling times and word length;
   - supervision and diagnosis functions;
   - model-in-the-loop simulation: vehicle models and ECU models;

- rapid control prototyping with development ECU, bypass computer and test bench;
- result: control structure and control algorithms.

6. Control-software development:

- software architecture layers, modules;
- software-component interfaces;
- high-level language: selection, floating point (e.g. C-code, MATLAB/ Simulink);
- availability of compilers for target software;
- implementation of control functions and modules into software structure;
- standardization and reuse of software modules;
- code optimization;
- testing of software modules with, for example, model-in-the-loop simulation;
- rapid control prototyping with bypass computer and test-bench experiments;
- result: control software (modules) in high-level language.

7. ECU target software development and implementation:

- transfer of high-level language control software into machine code with fix point arithmetic. Use of crosscompiler;
- test of control functions with simulated engine;
- software-in-the-loop simulation;
- hardware-in-the-loop simulation if real-time functions with components are of interest;
- result: implemented control software in target ECU.

8. ECU hardware and software testing:

- the ECU with target hardware and software undergoes intensive function tests;
- integration tests with simulated sensor signals and actuators in real time;
- hardware-in-the-loop simulation with simulated sensor output signals, simulated or real actuators, and real-time simulated comprehensive engine model;
- automated test-runs;
- testing of reaction to extreme speeds and loads (outside of normal operation);
- reliability and safety tests;
- tests for electromagnetic compatibility (EMC);
- result: verification that the ECU meets its specifications.

9. Calibration of the control functions:

- free parameters of control algorithms, characteristic curves, or lookup tables (maps) are adapted to the real vehicle and drive train;

- supported by calibration tool with editors at implementation level or physically defined level;
- off-line calibration;
- online calibration (test bench, real vehicle);
- basic calibration of the stationary behavior;
- dynamic calibration of the dynamic behavior;
- calibration by manual optimization;
- calibration by optimization with vehicle models;
- result: calibrated ECU control functions.

10. Calibration with fine-tuning:

- final fine-tuning of free parameters with the engine and transmission;
- driving experiments with the target vehicle;
- use of calibration tools, off-line or online;
- result: adapted ECU functions to transmission and vehicle. Verification of specifications.

11. Final ECU, driveability tests, and field tests:

- final vehicle control functions are tested with the target vehicle;
- performance tests for different loads and environmental conditions (summer, winter, weather);
- driveability tests;
- result: validation of requirements.

#### 3.3.1.2 Workflow for Control Development and Calibration

The *workflow for the control-function development* is depicted in Fig. 3.9 in more detail, also showing the use of software tools, computers, and test-bench experiments. It begins with physical and/or experimental modeling. The control development then comprises engine simulation, control-function development, and optimization, supported by an ECU prototype or prototyping computer for probing the control functions with the engine on the test bench. The next steps are then the software development and testing for the target ECU by using personal computers. Frequently, the ECU hardware and software testing is performed with hardware-in-the loop simulation, connecting the ECU with real actuators and sensor interfaces.

A corresponding *workflow for the model-based calibration of the control functions* is illustrated in Fig. 3.10. Based on the already gained physical and/or experimental engine models, the model-based calibration of the control functions is carried out. First, the basic stationary control functions, as, for example, reference variables dependent on the operation point, and then the calibration of the dynamic control functions, by using the dynamic engine, drive train models, and vehicle models, is executed.

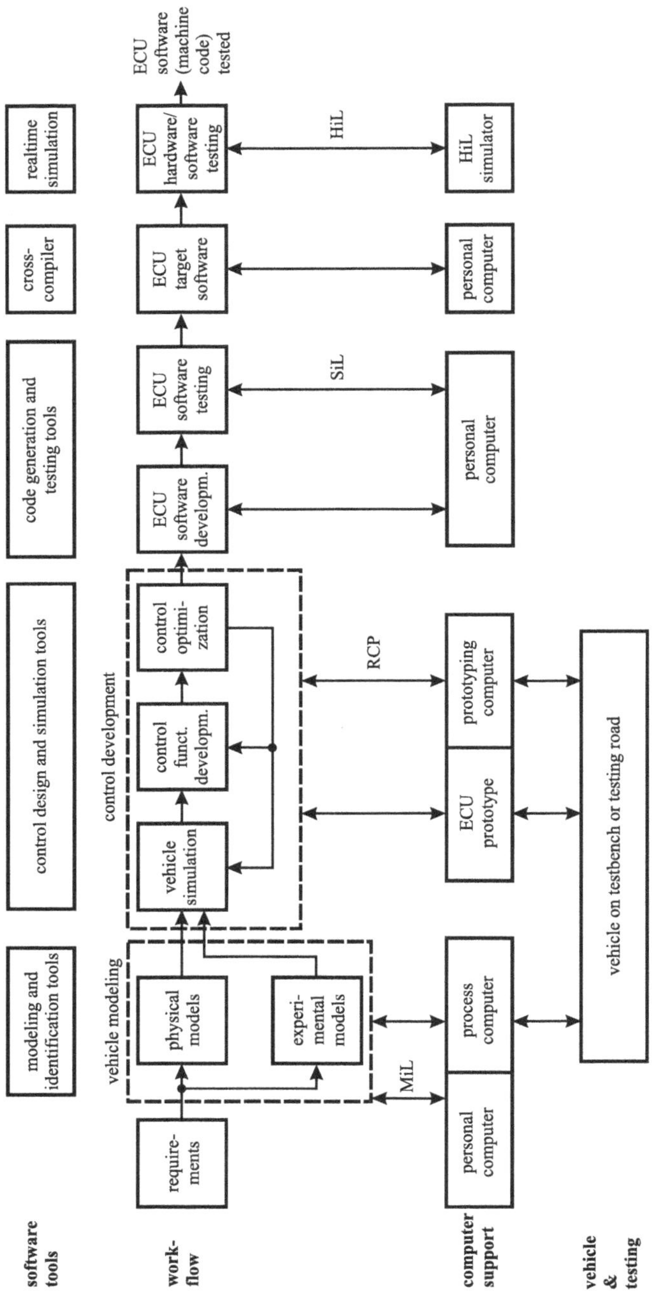


**Fig. 3.9** Workflow for control-function development

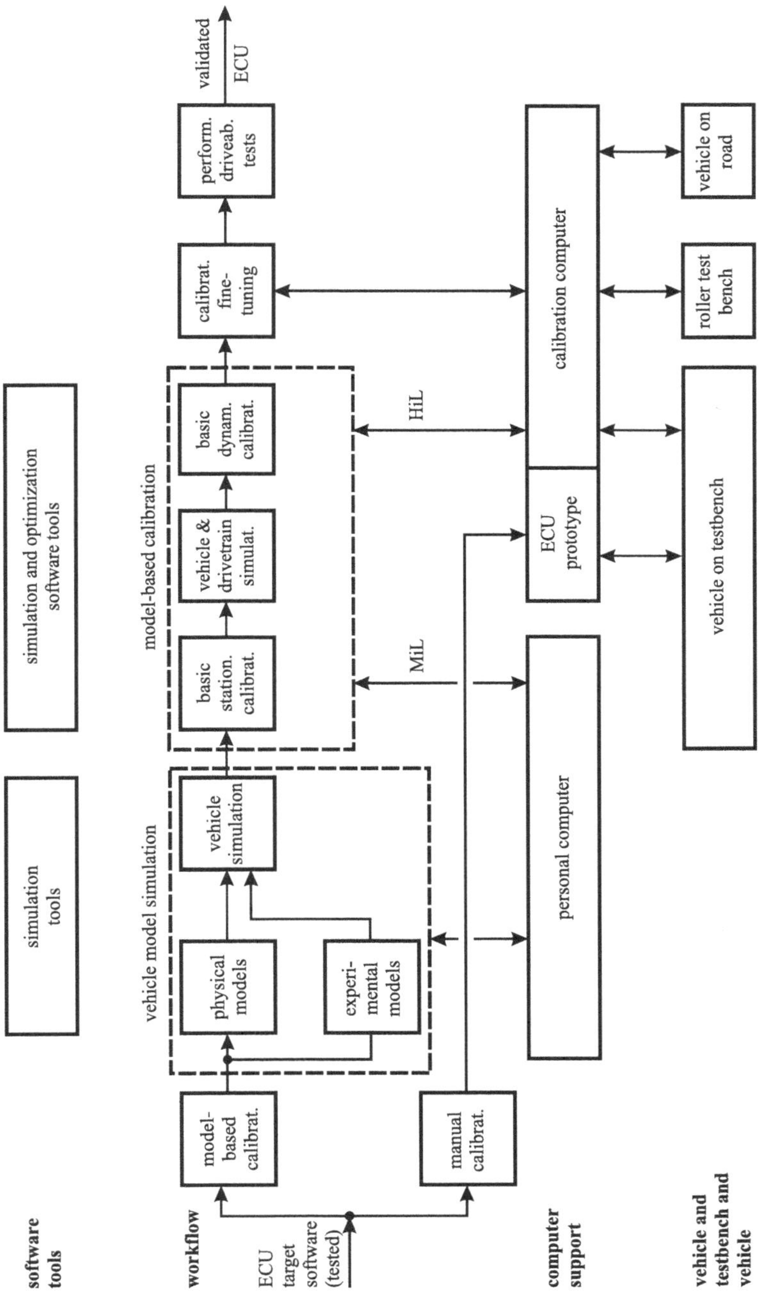


**Fig. 3.10** Workflow for model-based control-function calibration

These procedures are carried out either with personal computers or special calibration computers. The last step is a fine-tuning with the vehicle on a roller test bench and on the road.

The following sections consider the model-based control-function development and calibration in more detail. Some basic control structures and controllers are summarized in Appendix A.1.

## 3.3.2 Model-Based Control-Function Development with Special Design and Simulation Tools

A systematic and efficient development of control functions and their optimization and calibration requires special simulation methods and computers, which support the control-function development as well as calibration and software testing. This is part of the control-system integration, represented in the right branch of the V-development model in Fig. 3.8, and will be considered briefly in this section.

### 3.3.2.1 Model-in-the-Loop Simulation and Control Prototyping

The description of the overall procedure for vehicle control development in Sect. 3.3.1 has shown that different types of simulations are used for the development of control functions and control software; see Figs. 3.9 and 3.10.

A design of new control functions in an early development phase may be based on simulations with vehicle and engine models and an ECU model, i.e. control algorithms in a high-level language, as, for example, MATLAB-Simulink. This is called *model-in-the-loop simulation* (MiL). Both, the ECU and the vehicle are then represented as a model, i.e. a virtual picture of the real parts; see Fig. 3.11a.

If some control functions of a real development ECU can already be applied to the real vehicle on a test bench or on the road, because the real-time functions from a former, similar vehicle can be used, some new control functions may be tested as prototypes with a special real-time computer in parallel to the ECU. This is called *rapid control prototyping* (RCP). Frequently, the new control functions operate in a *bypass mode* and use the interfaces of the ECU to the sensors and the actuators; see Fig. 3.11b. The computing power for the experimental RCP computer exceeds that of the ECU and operates with a high-level language. Thus, the new control functions do not have to be implemented in machine code within the limited computer power and fix point restrictions of an ECU. This may save considerable development time by trying and testing new functions directly on a higher software level with the real engine. If a development ECU is not available, a powerful real-time computer can be used if the required sensor and actuator interfaces are implemented. It is then called *fullpass mode* (Schäuffele and Zurawka 2005).

### 3.3.2.2 Software-in-the-Loop and Hardware-in-the-Loop Simulation

Existing vehicle models in high-level language can be used for the validation of software functions as test candidates in an early development phase by *software-*

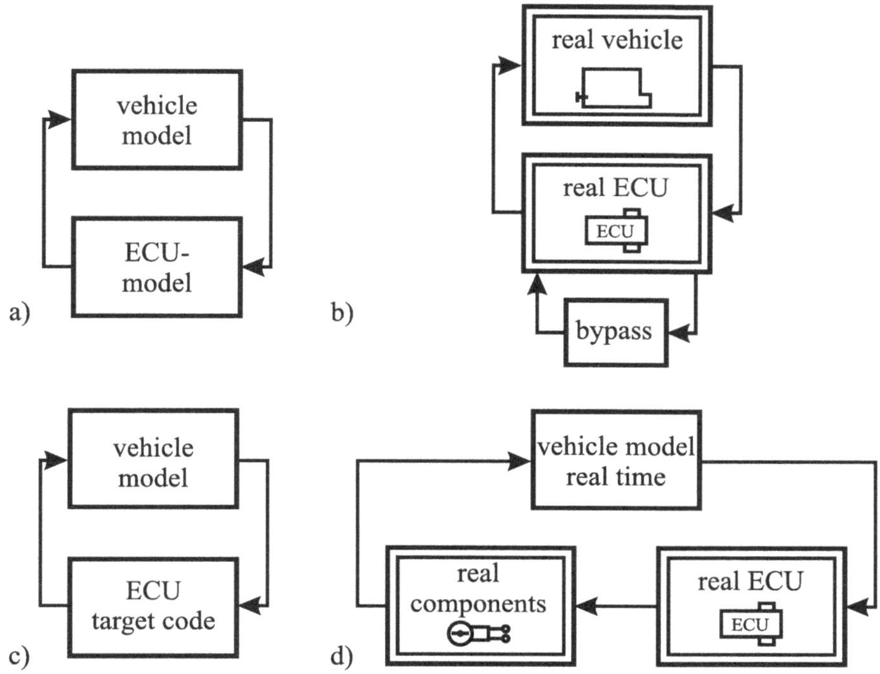


**Fig. 3.11** Different simulation and prototyping methods for control functions and software development. **a** Model-in-the-loop (MiL) simulation. **b** Rapid control prototyping (RCP). **c** Software-in-the-loop (SiL) simulation. **d** Hardware-in-the-loop (HiL) simulation

*in-the-loop simulation* (SiL); see Fig. 3.11c. The software functions may already be implemented with fix point or floating-point arithmetic and required interfaces, before they are implemented on the target ECU. Real-time behavior is not required.

For a final validation of control functions, the target ECU with its interfaces has to cooperate with real signals. In order not to use real vehicles on expensive test benches, real-time vehicle models are implemented in a powerful development computer. The sensor signals may be generated by special electronic modules and the output signals are frequently transferred to real actuators, like an electrical throttle, injection system, or steering actuator. Thus, the real ECU with implemented software operates with some real components, but with simulated real-time high- performance vehicle models and is known as *hardware-in-the loop simulation* (HiL); see Fig. 3.11d. The advantages are that, e.g. software functions can be tested under real-time constraints, validation tests are reproducible and can be automated, critical boundary conditions (high speed and high load) can be realized without being dangerous, the reaction to faults and failures can be investigated, onboard diagnosis functions can be tested, etc.; compare Sinsel (2000), Schaffnit (2002), and Zahn (2012). An alternative representation for MiL, RCP, and HiL is illustrated in Fig. 3.12.

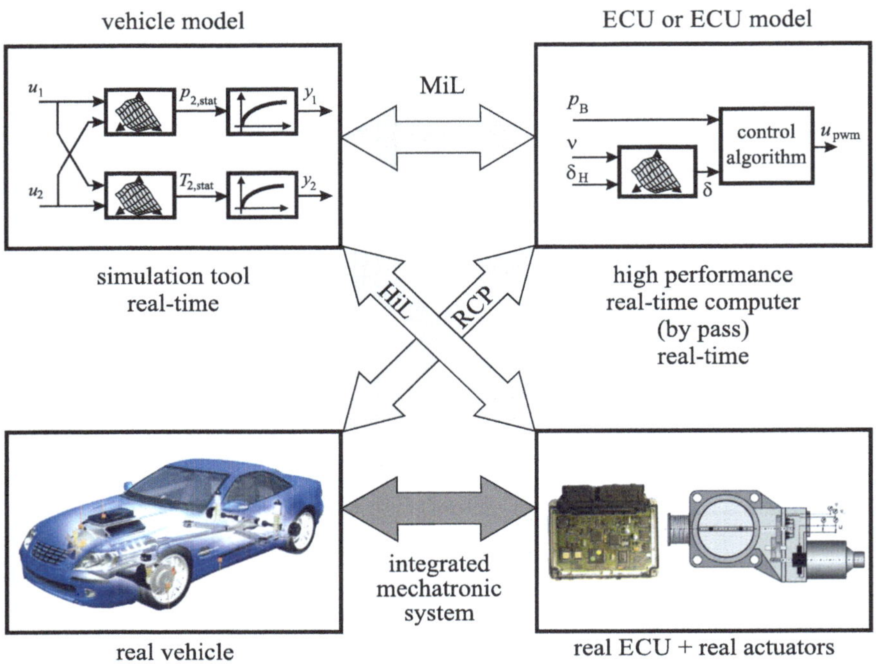


**Fig. 3.12** Different couplings between models and real parts for the development of control functions

This short summary of simulation and prototyping methods shows how the development of ECU control functions can be supported by using programmed dynamic engine and vehicle models of different granularity and different stages of the development of the control software, as depicted in Fig. 3.9. This is a basis for *virtual vehicle control development*. The computer-based and model-based development can also be used for a part of control calibration, as indicated in Fig. 3.10.

### 3.3.3 Control-Software Development

According to the V-development model in Fig. 3.8, the first steps are the control-system development and the control- function development in high-end software (e.g. MATLAB Simulink™, and Stateflow™). The next steps are then the *control-software development* and the *software implementation on the ECU* for series production. This is also depicted as part of the overall workflow in Fig. 3.9, using special software tools and computers for code generation and testing with compilers and real-time simulation. In the following, some remarks are given briefly for the software architecture code generation and software testing.

#### 3.3.3.1 Software Architecture

The design of the software architecture has to consider many aspects from software development to the requirements of the target microprocessor and include connected modules which have to be flexible with regard to continuous changes and variants. Several software layers have to be defined. The minimum is two layers, a platform software and an application software, as shown in Fig. 3.13 (Schäuffele and Zurawka 2005). The *platform software* is oriented to the ECU and comprises the operating system, communications, and network management according to OSEK/VDX (2005) standards and diagnostic protocols. OSEK stands for "Open Systems and Interfaces for Automotive Electronics" and is the result of a committee of major automotive manufacturers and component suppliers to support portability and re-usability of application software under real-time constraints, started in 1993. It also contains standardized flash memory programming procedures. The standardization of the platform software is additionally advantageous during software development with regard to software changes and parametrization. Interfaces for measurement and calibration via CAN protocols support the development phase as well (Borgeest 2008; Zimmerschied et al 2005). A hardware abstraction layer (HAL) gives access to the peripheral components of the ECU and is specified for the used microprocessors.

The *application software* can be designed by the vehicle manufacturer and contains vehicle- specific functions. Standardization takes place for control functions, ranging from lookup tables and their interpolation to dynamic control algorithms. The standardization is, e.g. treated in the MSR-MEGMA working group and ASAM (2012).

The *configuration* of standardized software components allows a specific application by using configuration tools. An automated configuration comprises, e.g. the handling of signals, messages, buses, nodes, and functions. It may contain export and import interfaces with data exchange formats and a documentation interface. More details like data models for engine and vehicle variants, storage in volatile (RAM) or nonvolatile memories (ROM, PROM, EPROM, or Flash memory), and description files for data structure can be found, e.g. in Schäuffele and Zurawka (2005).

Activities for an open industry standard of the automotive software architecture between suppliers and manufacturers are going on in the AUTOSAR consortium (AUTomotive Open System ARchitecture) since 2003 (AUTOSAR 2012; Heinecke et al 2004; ATZ extra 2013). One of the aims is an open and standardized automotive software architecture. The standard includes specifications describing software architecture components and defining their interfaces.

The AUTOSAR architecture separates the basis software from the application software and connects them by standardized interfaces; see Fig. 3.14. To master the complexity, several layers are defined (Wernicke and Rein 2007). The connection to the microcomputer is provided by the lowest level, the *microcontroller abstraction layer*. Here, the interfaces are defined to the memories, the I/O-drivers, their communication, and additional features which are not part of the microcontroller. The second layer is the *ECU abstraction layer*, comprising the hardware design of the ECU including the driver to external components. The *service layer* at the third level provides basic software modules like the operating system, memory administra-

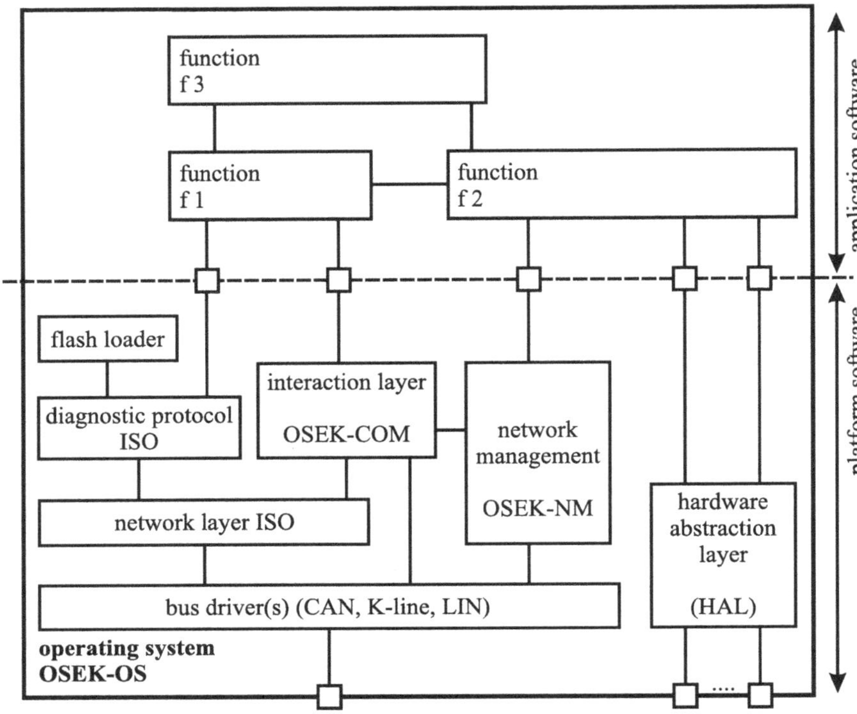


**Fig. 3.13** Software architecture composed of standardized software components (Schäuffele and Zurawka 2005)

tion, and bus communication. This layer is relatively independent of the ECU hardware. The fourth level is the *runtime environment* (RTE), which separates the basis software and application software and carries out the data exchange in both directions. Therefore, the application software components have standardized interfaces to the RTE. The RTE also integrates the application software components (SWCs); see Fig. 3.15. This separation and integration with standardized interfaces enable a hardware-independent software development. The application software components can therefore be transferred to other ECU's and reused.

A *virtual function bus* (VFB) connects the various software components during the *design* and allows a configuration independent of specific hardware. Thus, the SWCs are runnable entities and can be linked together for development and testing. An exchange of information becomes possible through standardized software input and output ports. Thus validation of the interaction of the SWCs and interfaces is possible before software implementation.

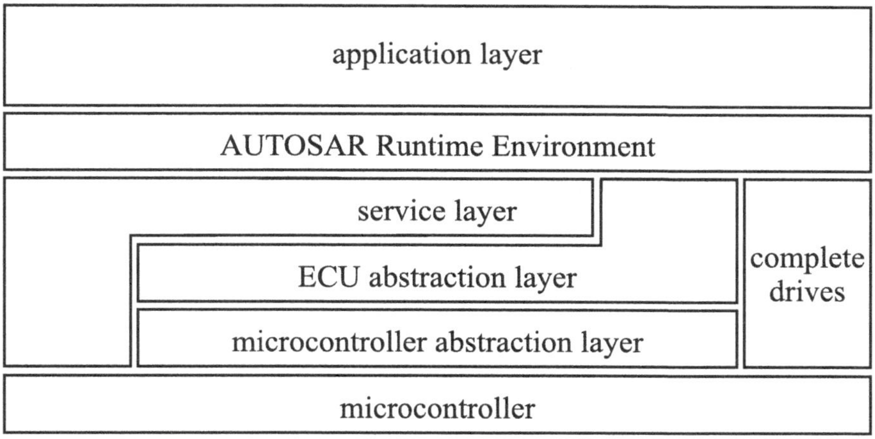


**Fig. 3.14** AUTOSAR layer structure of automotive software (Wernicke and Rein 2007)

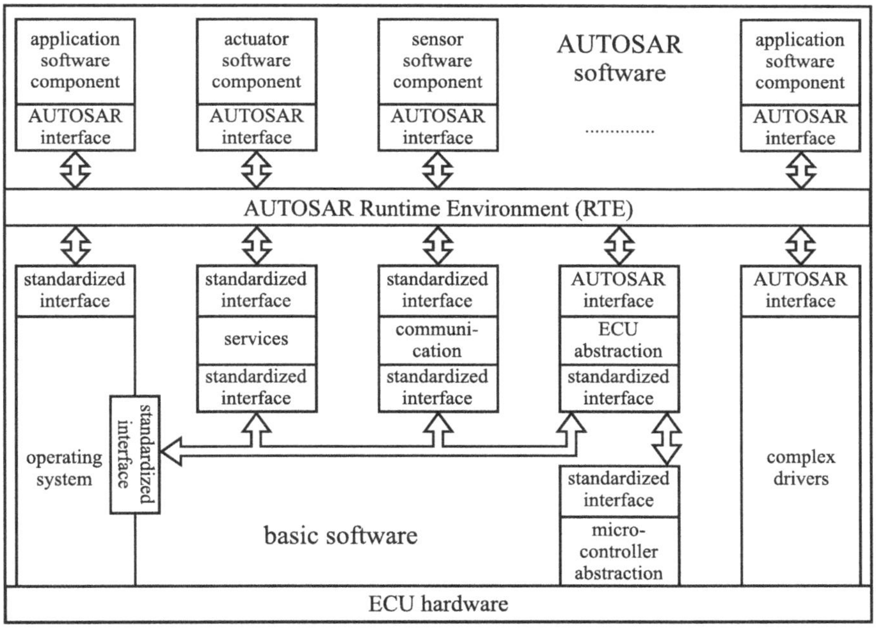


**Fig. 3.15** AUTOSAR software architecture components and interfaces (RTE: runtime environment) (Kirschke-Biller 2011)

#### 3.3.3.2 Code Generation

The control-software development can start if the control functions are ready and available as function blocks in a high- level software platform like MATLAB™ (2011), Simulink™ (2011), and Stateflow™ (2011) (The MathWorks (2011)). MATLAB™ is broadly used as an integrated function development environment for numerical calculations with a large library of mathematical design and analysis programs and special tool boxes. Simulink™ is an interactive development environment for modeling, analysis, and simulation with a graphical interface and is integrated into MATLAB™. It allows the handling of models and functions with block diagrams. Stateflow™ is an expansion of Simulink™ to operate with discrete event state charts and flow diagrams.

Based on the block-oriented control functions of this control-analysis-and simulation-oriented development environment, the series production code for the ECU is developed by using special tools. The control functions are specified in graphical form and they are converted in production C-code which runs on the target processor. Thereby, it is intended to reach a minimum of execution time, RAM and ROM resources, and stack size, compared to human programmer's abilities.

The process of code generation is performed in the following steps (Kiffmeier et al 1999), using a Software Development Tool (SDT) like Targetlink™ from dSpace or ASCET™ from ETAS or Real-Time Workshop™ from MathWorks.

1. The *control-function block* from Simulink has to be replaced by a corresponding SDT block, taken from a block library. These blocks manage data which is used for the production code, as scaling parameters and data types.
2. In the case of fixed-point arithmetic, *scaling parameters* are required. Therefore, maximal and minimal values of variables have to be known. They can be obtained from model-in-the-loop simulations to determine the range of variables, in order to prevent overflows. The scaling can be done manually or automatically by simulation. For floating-point arithmetic, no scaling is needed.
3. *Additional information* for production code generation has to be specified. This belongs to the partitioning of models or algorithms for, e.g. lookup table handling, like interpolation routines.
4. *Off-line simulations* on the host PC are performed with Simulink to detect problems with fixed-point arithmetic. The simulations are run with floating-point arithmetic.
5. The production C code is translated for the *target microprocessor* and loaded on an evaluation board. A communication link between the evaluation board and the host PC allows testing the microcontroller code together with a Simulink simulated process model. The code is readable by humans and the development is paralleled by documentation, which is automatically generated. Information on code size, execution times, and used RAM and ROM resources are also given. The result is a general portable ANSI C-code which runs on many microcomputers. However, some specific microcomputer adaptation always has to be taken into account. If manual coding has to be applied additionally, rules given by MSRA-C have to be followed. Special toolsets allow analyzing the timing behavior of the

real-time operating system application OSEK, providing runtime performance and possible modifications, before coding.

As the ECU-code must be prepared for the task of calibration, the variables are presented in standardized ASAM file format via a data dictionary. Further standards to be considered are OSEK/VDX and AUTOSAR. AUTOSAR structure elements are, for example, runnables, ports, and communication interfaces.

Special computers for simulation and control prototyping and software tools for control-software development are compiled in the catalogs dSpace (2017) and ETAS (2017).

As modern automotive control systems are interconnected by onboard data buses like CAN, FlexRay, and LIN, also the real-time cooperation of the vehicle ECU with other vehicle ECU's like for engine and transmission control, traction control (TCS), electronic stability control (ESC), and adaptive cruise control (ACC) has to be designed and monitored.

#### 3.3.3.3 Software Testing

The *software testing* is part of the verification comprised in the V-model for the development; see Fig. 3.8. *Verification* means that the developed functions meet the specifications, which are stated at the project start by the customer, by standards or legal regulations. More general is the *validation* where the final product is checked with regard to the overall requirements (Balzert 1998; Tran 2007); see the V-model in Sect. 3.3.1. Validation ensures that the final product meets the user's needs and includes that the specifications are correct; see, e.g. IEEE-STD 610.

Software testing can be understood as part of *analytical quality assurance* and contains the testing of the software. This quality assurance can be divided into analyzing methods (static) and testing methods (dynamic) (Liggesmeyer 2002; Thaller 2002). The *static, analyzing methods* can be applied early because no running software is required. Methods are manual inspection, visualization, and data flow analysis, usually by different persons. *Dynamic testing methods* operate with the running software and it is checked if the functions meet the specifications. One distinguishes functional tests which check the input/output behavior and consider the software program as a black-box and structural tests where the program is considered as a white-box. These structural tests can be divided into symbolic tests, diversification tests, and mutation tests (Schäfer 2012).

An application of software testing during the early phases of the development avoids too many iteration cycles. In order to integrate hardware components, hardware-in-the-loop simulation (HiL) is used. Then the real-time software on the target ECU is tested with simulated real-time models of the engine or the vehicle and real components like actuators or injection systems in the laboratory, such as avoiding tests with the real vehicle; see Sect. 3.3.2.

Table 3.1 gives a summary of the various development test stands and tools for the control development. Vehicle modeling is required in the beginning and performed online and off-line with the vehicle test bench. Control-function development is

**Table 3.1** Survey of test stands and tools for vehicle control development

| Control development steps | | Development test stands + tools | | | | | | |
|---|---|---|---|---|---|---|---|---|
| | | MiL | HiL | RCP | ECU | Vehicle test bench | | Vehicle on road |
| | | SiL (PC) | With real components | | Evaluation board | Roller dynamometer, (simulated vehicle driving forces) | Drive train test bench (engine, transmission driving axle) | Vehicle |
| Vehicle modeling | • physical | x | x | | | x | x | |
| | • experimentally | | | | | | | |
| | – stationary | x | | | | | | |
| | – dynamic | | x | x | | | | |
| Control-function development | • simulation | x | x | | | | | |
| | • development | x | | x | | x | x | x |
| | • optimization | | | x | | x | x | x |
| ECU software | • development | x | | | | | | |
| | • testing | x | x | | | x | x | x |
| | • target processor | | | | x | | | |
| Driveability | | | | | | x | x | x |
| Emission and fuel consumption certification | | | | | | x | | x |

mainly elaborated at test benches and with MiL, HiL, and RCP computers. Software development is usually done off-line with PCs and an ECU evaluation board. Fine-tuning for driveability and emission certification are worked out with the real vehicle.

## References


ATZ extra (2013) 10 years AUTOSAR. The worldwide automotive standard for E/E-systems. Special issue. Automobiltechnische Zeitschrift, Springer Vieweg/Springer Fachmedien Wiesbaden GmbH, Wiesbaden

AUTOSAR (2012) Automotive Open System Architecture. www.autosar.org

Balzert H (ed) (1998) Lehrbuch der Software-Technik, vol 2. Spektrum Akademischer Verlag, Heidelberg

Böhm B (1979) Guidelines for verifying and validating software requirements and design specifications. North-Holland, Euro-IFIP

Borgeest K (2008) Elektronik in der Fahrzeugtechnik. Vieweg, Wiesbaden

BRD (ed) (1997) V-Modell-Entwicklungsstandard für IT-Systeme des Bundes. http://www.v-modell.iabg.de/um97.htm

BRD (ed) (2004) V-Modell-XT. http://v-modell.iabg.de

Bröhl AP (ed) (1995) Das V-Modell - Der Standard für Softwareentwicklung, 2nd edn. Oldenbourg, München

Droeschel W, Wiemers M (eds) (1999) Das V-Modell 97 - Der Standard für die Entwicklung von IT-Systemen mit Anleitung für den Praxiseinsatz. Oldenburg Verlag, München

dSpace (2017) dSpace catalogue 2017. dSpace GmbH, Paderborn

ETAS (2017) ETAS catalogue 2017. ETAS GmbH, Stuttgart

Guzzella L, Onder C (2004) Introduction to modeling and control of internal combustion engine systems. Springer, Berlin

Heinecke H, Schnelle KP, Fennel H, Bortolazzi J, Lundh L, Leflour J, Maté JL, Nishikawa K, Scharnhorst T (2004) AUTomotive Open System ARchitecture-an industry-wide initiative to manage the complexity of emerging automotive E/E-architectures. In: SAE 2004 Convergence, pp 325–332

Isermann R (2005) Mechatronic Systems - Fundamentals, 2nd edn. Springer, London

Isermann R (2014) Engine Modeling and Control. Springer, Berlin

Isermann R (2017) Combustion engine diagnosis. Springer Vieweg, Berlin

Kiffmeier U, Köster L, Meyer M, Witke C (1999) Automatic prodective code generation for electronic control units. Automatisierungstechnik - at 47:295–304

Kirschke-Biller F (2011) Autosar – a worldwide standard current developments, roll-out and outlook. In: 15th International VDI congress electronic systems for vehicles, Baden-Baden, Germany

Kunkel F (2015) Optimaler Betrieb von Dieselhybridantrieben. Dissertation TU Darmstadt. epubli GmbH

Liggesmeyer P (ed) (2002) Software-Qualität: Testen. Spektrum Akademischer Verlag, Heidelberg, Analysieren und Verifizieren von Software

Reichart G, Bielefeld J (2009) Einflüsse von Fahrerassistenzsystemen auf die Systemarchitektur im Kraftfahrzeug. In: Winner H, Hakuli S, Wolf G (eds) Handbuch Fahrerassistenzsysteme, 1st edn, Vieweg+Teubner|GWV Fachverlage GmbH, Wiesbaden, pp 84–92

GmbH Robert Bosch (ed) (2018) Automotive handbook, 10th edn. J. Wiley, Chichester, England

Royce W (1970) Managing the development of large software projects. In: Proceed. IEEE, Wescon

Schäfer S (2012) Modellbasierte Steuerung des Kühlkreislaufes einer Brennstoffzelle mit automatisiertem Test der Software. Dissertation Technische Universität Darmstadt. Fortschr.-Ber. VDI Reihe 8, Nr. 1219 VDI Verlag, Düsseldorf

Schaffnit J (2002) Simulation und Control Prototyping zur Entwicklung von Steuergerätefunktionen für aufgeladene Nutzfahrzeug-Dieselmotoren. Dissertation Technische Universität Darmstadt. Fortschr.-Ber. VDI Reihe 12, 492. VDI Verlag, Düsseldorf

Schäuffele J, Zurawka T (2005) Automotive software engineering. SAE, Warrendale, PA
Sinsel S (2000) Echtzeitsimulation von Nutzfahrzeug-Dieselmotoren mit Turbolader zur Entwicklung von Motormanagementsystemen. Logos, Doctoral thesis. University of Technology, Darmstadt. Berlin
Guide STARTS (1989) The STARTS purchases Handbook: software tools for application to large real-time systems, 2nd edn. National Computing Centre Publications, Manchester
Thaller G (ed) (2002) Software-Test-Verifikation und Validation. Heinz-Heise-Verlag, Hannover
Tran E (2007) Verification/validation/certification. In: Koopman P (ed) Topics in dependable embedded software, Carnegie Mellon University. Pittsburgh, PA
VDI 2206, (2003) Design methodology for mechatronic systems. Beuth Verlag, Berlin
Wernicke M, Rein J (2007) Integration of existing ECU software in the autosar architecture. ATZ-Elektronik 1:20–25
Zahn S (2012) Arbeitsspielaufgelöste Modellbildung und Hardware-in-the-Loop-Simulation von Pkw-Dieselmotoren mit Abgasturbolader. Dissertation Technische Universität Darmstadt. Fortschr.-Ber. VDI Reihe 12, 760. VDI Verlag, Düsseldorf
Zimmerschied R, Weber M, Isermann R (2005) Statische und dynamische Motorvermessung zur Auslegung von Steuerkennfeldern - eine kurze Übersicht. Automatisierungstechnik - at 53(2):87–94

# Part II
# Modeling of Drive Dynamics

# 4 Vehicle Dynamics Modeling

Vehicle dynamics is concerned with the movements of the vehicle on a road surface. The stationary and dynamic behavior is based on the forces on the vehicle via the tires, gravitation, and aerodynamics. These forces act on the body, the tires, and the wheels.

A vehicle is composed of many components which move relative to each other. However, to simplify modeling stiff behavior is assumed for the body and the wheels. Therefore, the body is assumed to be rigid and is replaced as one lumped mass located in its center of gravity (CG) with body mass and moments of inertia of the vehicle. The wheels are considered as separate lumped masses, representing the unsprung mass, different from the body as sprung mass.

The treated vehicle dynamic models are derived with several simplifying assumptions and aim to be used for simulation, control design, and fault diagnosis, taking into account the limitations of real-time computing and computational expense of automotive electronic control units. Another important subject is the acquisition of the model parameters, either by construction data or on an experimental basis with identification and parameter estimation methods.

The symbols for the chassis and the body used follow in principle the conventions of ISO 8855/DIN 70000 wherever possible if they are not overlapping with symbols of other areas; see the list of symbols after the preface.

This chapter considers the used coordinate systems and a discussion on the various ways for modeling.



R. Isermann, *Automotive Control*, ATZ/MTZ-Fachbuch,
https://doi.org/10.1007/978-3-642-39440-9_4

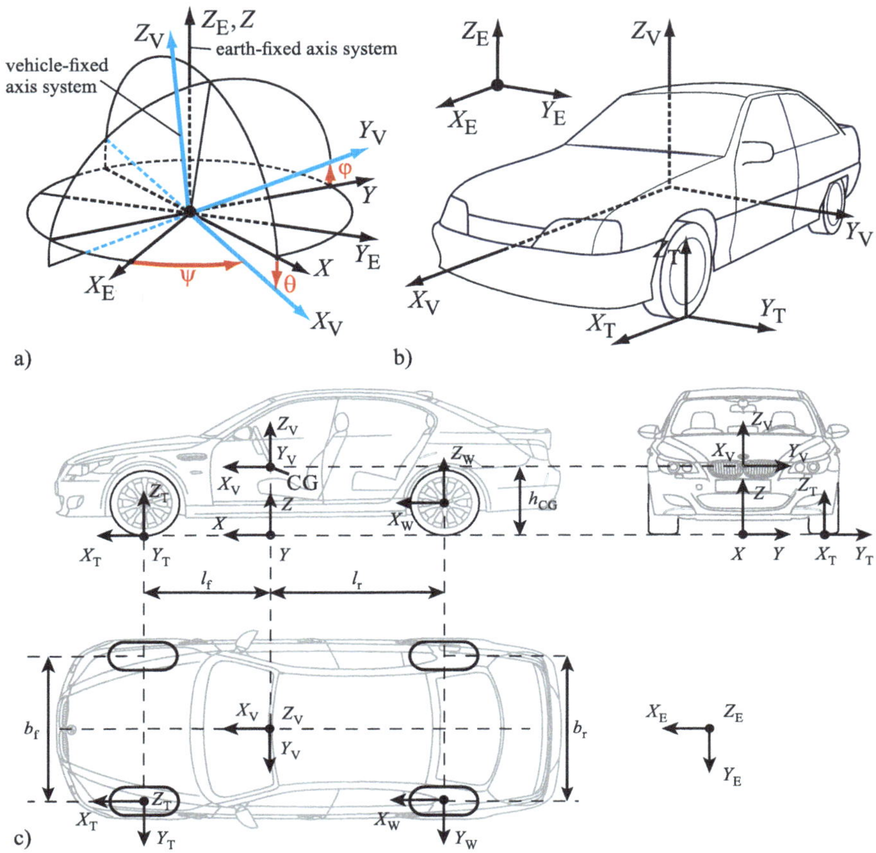


**Fig. 4.1** Coordinate systems according to ISO 8855 (2013). **a** Earth-fixed axis system ($X_E$, $Y_E$, $Z_E$) and horizontal axis system ($X$, $Y$, $Z$) (the vehicle axis system ($X_V$, $Y_V$, $Z_V$) is illustrated with the center of gravity in the same plane as the earth-fixed axis system, showing the Euler angles); **b** earth-fixed, vehicle-fixed, and tire-fixed axis systems; **c** three-side view of a vehicle with all three axis systems

## 4.1 Coordinate Systems

### 4.1.1 Definition of Coordinate Systems

Modeling vehicle dynamics requires different coordinate systems. According to ISO 8855 (2013) the following standardized coordinate systems are distinguished, see Fig. 4.1. The axis has capital letters like $X$, $Y$, $Z$ and the corresponding coordinates lower case letters, like $x$, $y$, and $z$. Right-hand orthogonal axis systems are used according to the cross-product (vector-product) $\vec{Z} = \vec{X} \times \vec{Y}$.

*Earth-fixed axis system*

The earth-fixed axis system is a right-hand orthogonal axis system with the coordinates $X_E$, $Y_E$, and $Z_E$. The $X_E$ and $Y_E$ coordinates are usually fixed on the road plane, e.g. at the beginning of a driving maneuver. It allows describing the vehicle movement relative to a point on the road.

*Vehicle axis system*
The vehicle-fixed right-hand orthogonal coordinate system has its origin in the center of gravity and has the coordinates $X_V$, $Y_V$, and $Z_V$. This coordinate system travels with the vehicle. Its movements relative to the earth's fixed axis system describe the behavior of the body in the space.

*Intermediate (horizontal) axis system*
The intermediate or horizontal axis system has the coordinates $X$, $Y$, and $Z$ where the $X$-axis is the projection of the fixed vehicle axis $X_V$ on the plane $X_E$, $Y_E$ of the earth-fixed axis. Hence, this coordinate system travels with the vehicle and its movements describe the behavior of the body relative to the plane $X_E$, $Y_E$. The origin can be placed under the center of gravity.

*Tire axis system*
Each tire obtains an own axis system $X_T$, $Y_T$, and $Z_T$. The origin is in the tire contact center. Its $X_T$, $Y_T$ plane lies in the plane of $X_E$, $Y_E$ with $X_T$ in the forward direction.

*Wheel axis system*
The wheel axis system $X_W$, $Y_W$, and $Z_W$ has its origin in the center of the wheel and with $Y_W$ parallel to the wheel turning axis.

*Angular motion variables*
The orientation of a vehicle and therefore the turning motion are described by the angles between the vehicle-fixed and the earth-fixed coordinate systems, resulting in

$\psi$: yaw angle, from axis $X_E$ to $X_V$ around the $Z_E$ axis;
$\theta$: pitch angle, from axis $X$ to $X_V$ around the $Y_V$ axis;
$\varphi$: roll angle, from axis $Y$ to $Y_V$ around the $X_V$ axis.

These three angles are called *Euler angles*. The sign of the angle is according to the right-hand rule; compare Fig. 4.1.

### 4.1.2 Transformations

The transformation for the angles from the earth fixed to the vehicle-fixed axis system is obtained by turning around the $Z_E$-axis (yawing), the new $Y$-axis (pitching), and the new $X_V$-axis (rolling) and follows Popp and Schiehlen (1993) and Schorn (2007), Fig. 4.2a,

$$\begin{bmatrix} X_V \\ Y_V \\ Z_V \end{bmatrix} = \mathbf{T}^{E \to V} \begin{bmatrix} X_E \\ Y_E \\ Z_E \end{bmatrix} \tag{4.1.1}$$

with

$$\mathbf{T}^{E\to V} = \underbrace{\begin{bmatrix} 1 & 0 & 0 \\ 0 & \cos\varphi & \sin\varphi \\ 0 & -\sin\varphi & \cos\varphi \end{bmatrix}}_{\text{rolling}} \underbrace{\begin{bmatrix} \cos\theta & 0 & -\sin\theta \\ 0 & 1 & 0 \\ \sin\theta & 0 & \cos\theta \end{bmatrix}}_{\text{pitching}} \underbrace{\begin{bmatrix} \cos\psi & \sin\psi & 0 \\ -\sin\psi & \cos\psi & 0 \\ 0 & 0 & 1 \end{bmatrix}}_{\text{yawing}}. \tag{4.1.2}$$

The wheel axis system ($X_W$, $Y_W$, $Z_W$) and the vehicle axis system ($X_V$, $Y_V$, $Z_V$) differ only by a parallel shift in the vertical axis, if it can be assumed that the camber angle is $\epsilon_V = 0$. For the transformation then only a turn around the $Z_W$ axis is required with the wheel steer angle $\delta$; see Fig. 4.2b.

$$\begin{bmatrix} X_T \\ Y_T \\ Z_T \end{bmatrix} = \mathbf{T}^{T\to W} \begin{bmatrix} X_V \\ Y_V \\ Z_V \end{bmatrix} \tag{4.1.3}$$

with

$$\mathbf{T}^{V\to T} = \begin{bmatrix} \cos\delta & \sin\delta & 0 \\ -\sin\delta & \cos\delta & 0 \\ 0 & 0 & 1 \end{bmatrix} \tag{4.1.4}$$

respectively

$$\mathbf{T}^{T\to V} = \begin{bmatrix} \cos\delta & -\sin\delta & 0 \\ \sin\delta & \cos\delta & 0 \\ 0 & 0 & 1 \end{bmatrix}. \tag{4.1.5}$$

Therefore, it holds for the effect of the tire forces on the vehicle, for example,

$$F_X = F_{XT}\cos\delta - F_{YT}\sin\delta \tag{4.1.6}$$

$$F_Y = F_{XT}\sin\delta + F_{YT}\cos\delta. \tag{4.1.7}$$

Figure 4.3 illustrates the different coordinate systems also for the case of a road plane elevation angle (slope), a road plane camber angle (banking angle), and yaw angle. Additionally, a vehicle-fixed sensor coordination system ($X_S$, $Y_S$, $Z_S$) is provided.

## 4.2 Model Building Approaches

The temporal behavior of technical systems, such as vehicles, power- and drive trains, and their components can be described with the help of system theory according to uniform methods. For this, however, mathematical models must exist for the static and dynamic behavior of the system components or the processes.

The derivation of mathematical models can take place in a theoretical (physical) or experimental way. Therefore, it is called theoretical or experimental analysis, respectively, modeling. For vehicles and drives, mathematical models for different physical areas have to be set up and combined, e.g. for mechanics, combustion,

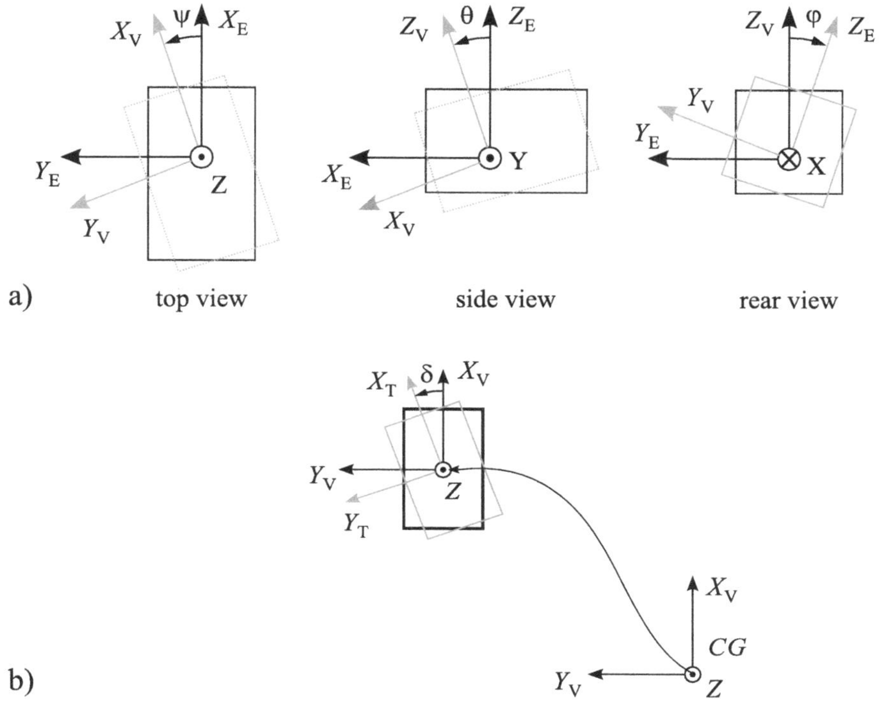


**Fig. 4.2** **a** Turnings of the earth-fixed and vehicle-fixed coordinate systems (Bauer 2015); **b** Turnings of the wheel axis system relative to the vehicle-fixed coordinate system (Bauer 2015)

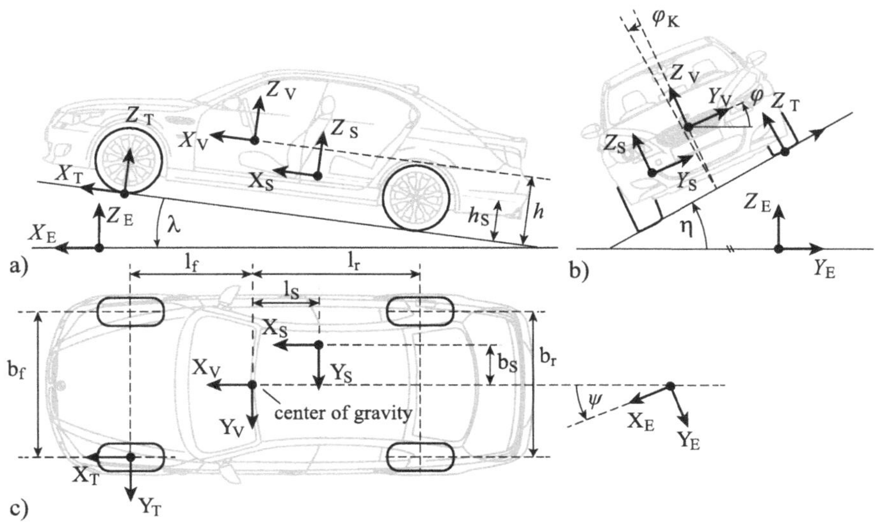


**Fig. 4.3** Coordinate systems **a** with elevation; **b** with banking; **c** with yaw angle (Bechtloff 2018)

thermodynamics, and electricity. The procedure during theoretical modeling is in principle known for the individual areas, and there also exist analogies for models between different areas. However, a basic, generally applicable methodology for theoretical modeling with an interdisciplinary view has several advantages, especially for applying computer-aided modeling. Therefore, a unified representation for modeling in different physical domains is briefly discussed, which is especially suited for mechatronic systems.

### 4.2.1 Theoretical and Experimental Modeling

The principles of theoretical modeling can follow a basic methodology; see MacFarlane (1970), Karnopp et al. (1990), Gawthrop and Smith (1996), and Isermann (2005). Fundamental equations are

1. Balance equations for stored masses, energies, and momentum;
2. Constitutive equations of special elements;
3. Phenomenological equations if irreversible processes take place;
4. Entropy balance equations if several irreversible processes are involved;
5. Connection equations.

In stating these equations, one has to distinguish between processes with distributed and lumped parameters. For *distributed parameters*, the dependency on space and time has to be considered. This usually leads to partial differential equations. If the space dependency is negligible, the process can be considered with *lumped parameters*, which leads to ordinary differential equations as a function of time. For vehicles, combustion engines and drive trains both types appear. However, one can frequently operate with lumped parameters.

By summarizing the basic equations of all process elements, one receives a *theoretical* or *physical process model* with a certain structure and certain parameters, if it can be solved explicitly. Frequently, this model is extensive and complicated, such that it must be simplified for further applications. The simplifications are made by linearization, reduction of the model order, or approximation of systems with distributed parameters by lumped parameters when limiting on fixed locations. But also if the set of equations cannot be solved explicitly, the individual equations supply important hints for the model structure. So, e.g. balance equations are always linear and some phenomenological equations are linear in wide areas. The constitutive equations often introduce nonlinear relations.

During experimental modeling, which is called *identification*, one obtains the mathematical model of a process from measurements. Here, one always proceeds from a priori knowledge, which was gained, e.g. from the theoretical analysis or from preceding measurements. Then, input and output signals are measured and evaluated by means of identification methods in such a way that the relation between the input and output signal is expressed in a mathematical model. The result of the identification then is an *experimental model*; see Åström and Eykhoff (1971), Eykhoff (1974), Ljung (1999), and Isermann and Münchhof (2011).

Theoretical and experimental modeling mutually complete themselves. The theoretical model contains the functional description between the physical data of the process and its parameters. Therefore, one will use this model, e.g. if the process is to be favorably designed with regard to dynamical behavior or if the process behavior has to be simulated before construction. The experimental model, on the other hand, contains parameters as numerical values whose functional relation with the physical basic data of the process remains unknown. In many cases, the real dynamic behavior can be described more exactly or it can be determined at smaller expenditure by experimentally obtained models which, e.g. is better suited for control design, the prediction of signals, or for fault detection.

A methodology for theoretical (physical) modeling is described in Isermann (2005), which can be applied generally for technical systems and thus holds also for vehicles and drive trains. As the basic equations for vehicle dynamics follow the Newton, Euler, and Lagrange laws, the fundamentals of mechanical systems are applied as treated, e.g. in Hagedorn (1990), Hauger et al. (1989), Meriam and Kraige (1982), Pfeiffer (1989), Schiehlen (1986), Sneck (1991), and Wells (1967). The mechanics of vehicles is especially treated in Popp and Schiehlen (1993) and the books on vehicle dynamics as Gillespie (1992), Kiencke and Nielsen (2000), Mitschke and Wallentowitz (2004), Rajamani (2006), Schramm et al. (2010), Heissing and Ersoy (2011), and those cited in later chapters.

### 4.2.2 Semi-physical Models

In general, theoretical end experimental modeling complement each other. The theoretical model contains the functional description between the physical/chemical variables and includes their parameters. The experimental model on the other hand delivers parameters as numerical values based on the real stationary and dynamic behavior.

Theoretical models are also called *"white-box models"* and experimental models *"black-box models"*. In many practical applications, one has to use a suitable combination of both ways; compare Fig. 4.4.

If the physical laws are known, but the parameters not at all or not precisely enough, the parameters have to be determined experimentally, e.g. by parameter estimation methods. The resulting models can be called *"light-gray models"*. If only physical-oriented if-then-rules are known, the model structure and the parameters have to be determined by experiments, leading to *"dark-gray models"*, for example, by fuzzy if-then-rules and parameter adjustment. Both gray models can also be called *semi-physical models*.

These types of combined theoretical and experimental models are frequently the result in modeling vehicles and their drives. For example, the tire forces are

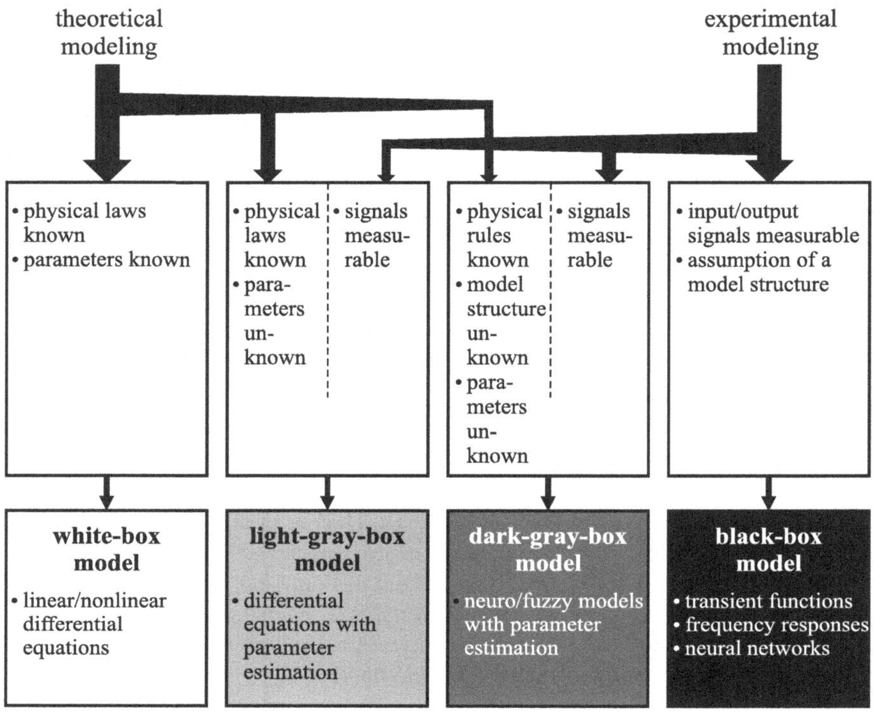


**Fig. 4.4** Different kinds of mathematical process models (Isermann and Münchhof 2011)

approximated by a simplified function and the parameters are estimated based on drive measurements. In the case of driver behavior or traffic situations, only some rules are known and experimental gained approximations are mainly used.

## References


Åström K, Eykhoff P (1971) System identification - a survey. Automatica 7(2):123–162

Bauer M (2015) Methoden zur modellbasierten Fahrdynamikanalyse und Bewertung von Fahrdynamikregelsystemen. Dissertation Technische Universität Darmstadt. Fortschr.-Ber. VDI Reihe 12, 792. VDI Verlag, Düsseldorf

Bechtloff J (2018) Schätzung des Schwimmwinkels und fahrdynamischer Parameter zur Verbesserung modellbasierter Fahrdynamikregelungen. Dissertation Technische Universität Darmstadt. Fortschr.-Ber. VDI Reihe 12, 809. VDI Verlag, Düsseldorf

Eykhoff P (1974) System identification. Wiley, London

Gawthrop P, Smith L (1996) Metamodelling: bond graphs and dynamic systems. Prentice Hall, Hemel Hempstead

Gillespie T (1992) Fundamentals of Vehicles Dynamics. SAE, Warrendale

Hagedorn P (1990) Technische Mechanik, vol 1–3. Harri Deutsch, Frankfurt

Hauger W, Schnell W, Gross D (1989) Technische Mechanik, 3rd edn. Springer, Berlin

Heissing B, Ersoy M (eds) (2011) Fahrwerkhandbuch: Grundlagen, Fahrdynamik, Komponenten, Systeme, Mechatronik. Perspektiven, ATZ/MTZ Fachbuch, Vieweg

Isermann R (2005) Mechatronic systems - fundamentals, 2nd edn. Springer, London
Isermann R, Münchhof M (2011) Identification of Dynamic Systems. Springer, Berlin
ISO 8855 (2013) Road vehicles - Vehicle dynamics and road-holding ability - Vocabulary. International Organization for Standardization, Geneva
Karnopp D, Margolis D, Rosenberg R (1990) System dynamics: a unified approach. Wiley, New York
Kiencke U, Nielsen L (2000) Automotive control systems. For engine, driveline and vehicle. Springer, Berlin
Ljung L (1999) System identification - theory for the user, 2nd edn. Prentice Hall, Upper Saddle River
MacFarlane A (ed) (1970) Dynamical system models. London, G.G, Harrop
Meriam J, Kraige L (1982) Engineering mechanics, vol 1 Statics, 2. Dynamics, 4th edn. Wiley, New York
Mitschke M, Wallentowitz H (2004) Dynamik der Kraftfahrzeuge, 4th edn. Springer, Berlin
Pfeiffer F (1989) Einführung in die Dynamik. Teubner Studienbücher Mechanik, Teubner, Stuttgart
Popp K, Schiehlen W (1993) Fahrzeugdynamik: Eine Einführung in die Dynamik des Systems Fahrzeug-Fahrweg. Teubner, Stuttgart
Rajamani R (2006) Vehicle dynamics and control. Springer, New York
Schiehlen W (1986) Technische Mechanik. Teubner, Stuttgart
Schorn M (2007) Quer- und Längsregelung eines Personenkraftwagens für ein Fahrerassistenzsystem zur Unfallvermeidung. Diss. Universität Darmstadt, Fortschr.-Ber. VDI Reihe 12, 651. VDI Verlag, Düsseldorf
Schramm D, Hiller M, Bardini R (2010) Modellbildung und Simulation der Dynamik von Kraftfahrzeugen. Springer, Berlin
Sneck H (1991) Machine dynamics of planar machinery. Prentice Hall, Englewood Ciffs
Wells D (1967) Lagrangian dynamics. Schaum's outline series. McGraw-Hill, New York

# Tire Traction and Force Transfer

# 5

The transfer of the forces and torques between the tire and the road surface is fundamental for all longitudinal and lateral motions of a vehicle. The physical effects of the force transfer depend on local properties at the contact patch. The amount of forces is a function of the frictional properties between the tire and the road surface.

Two main types of friction can be distinguished: an *adhesive friction* which depends on intermolecular adhesion and a *hysteretic friction* which depends on interlocking forces. Adhesive friction is effective on a molecular level in the contact area between the tire's tread surface and the road. It determines the main part of friction for dry roads. The hysteretic friction forces depend on the contact areas between the tread blocks of the tire and the roughness of the road surface and are determined by the viscoelastic properties of the tire rubber compound material; see, e.g. Mitschke and Wallentowitz (2014), Ersoy and Gies (2017), and Breuer and Bill (2017).

## 5.1 Longitudinal Tire Forces

If a longitudinal force is applied to a tire, e.g. by braking, the first part of the incoming contact patch sticks and the outgoing part slips relative to the road. The resulting measurable slip $\Delta v$ of the tire rotation velocity $r_{\mathrm{dyn}}\omega_{\mathrm{W}}$ compared to the vehicle velocity $v$ is due to a deformation part because of the elasticity of the tire treads and a relative movement part because of the partial gliding between the tire and the road (Kummer and Meyer 1967); see Fig. 5.1. The slip is usually expressed as a relative value with reference to the larger rotational wheel velocity $v_{\mathrm{XW}} = r_{\mathrm{dyn}}\omega_{\mathrm{W}}$ or longitudinal vehicle velocity $v_{\mathrm{XT}}$. Thus the slip in the longitudinal direction is defined for braking

$$S_{\mathrm{X,b}} = \frac{\Delta v_{\mathrm{XT}}}{v_{\mathrm{XT}}} = \frac{v_{\mathrm{XT}} - v_{\mathrm{XW}}}{v_{\mathrm{XT}}} = \frac{v_{\mathrm{XT}} - r_{\mathrm{dyn}}\omega_{\mathrm{W}}}{v_{\mathrm{XT}}} \tag{5.1.1}$$



R. Isermann, *Automotive Control*, ATZ/MTZ-Fachbuch,
https://doi.org/10.1007/978-3-642-39440-9_5

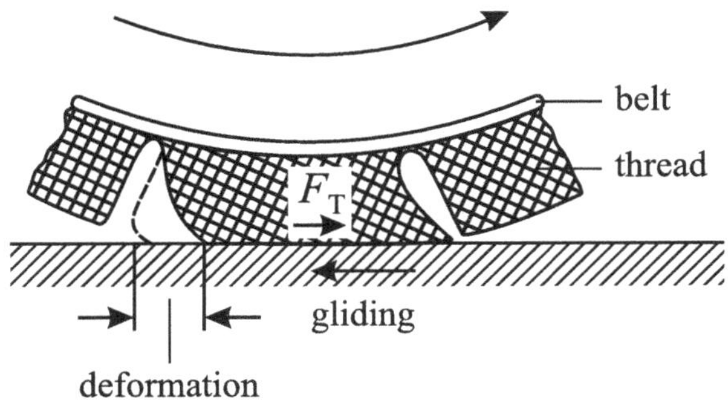


**Fig. 5.1** Elastic deformation and partial gliding of a tire (Mitschke and Wallentowitz 2014)

and for driving (traction)

$$S_{X,d} = \frac{\Delta v_{XT}}{v_{XW}} = \frac{v_{XW} - v_{XT}}{v_{XW}} = \frac{r_{dyn}\omega_W - v_{XT}}{r_{dyn}\omega_W} \tag{5.1.2}$$

or

$$S_{X,b,d} = \frac{\left|v_{XT} - r_{dyn}\omega_W\right|}{\max\left(v_{XT}, r_{dyn}\omega_W\right)} \tag{5.1.2a}$$

where $v_{XT}$ is the longitudinal tire traction velocity or the velocity of the wheel center respective vehicle velocity and $v_{XW}$ the rotating velocity of the wheel; see, e.g. Schramm et al. (2010). With this definition, the slip is positive for braking and driving, with $S_{x,b} = 1$ for a locked wheel during braking and $S_{X,d} = 1$ for a spinning wheel during driving.

Using this global definition of tire slip, a *longitudinal friction coefficient* is defined as

$$\mu_X(S_X) = \frac{F_{XT}}{F_{ZT}} \tag{5.1.3}$$

where $F_{XT}$ is the tangential tire force in the contact patch and $F_{ZT}$ the vertical force (load) of the tire.

The friction coefficient $\mu(S_X)$ depends on the slip as shown in Fig. 5.2. For small slips, it initially shows a linear behavior which is caused by the translatory deformation of the tire treads, increases with increasing slip to a maximal value $\mu_{max}$ at 10–30% slip where partial sticking and gliding in the contact patch appear, and then decreases to $\mu_{slip}$ until total slip $S_X = 1$ (100 %). This $\mu(S_X)$ characteristic is only valid for stationary conditions and depends on the vertical wheel load, tire pressure, vertical velocity, tire thread profile, and viscoelastic conditions, and on the road surface (roughness, wetness, and temperature). The $\mu(S_X)$ curves in graphical form or as lookup tables are nonparametric tire models and are determined for a specific tire with certain test conditions.

The friction characteristic $\mu(S_X)$ has a stable range for $0 \leq S_X \leq S_{Xcrit}$ and an unstable range for $S_{Xcrit} < S_X \leq 1$ with regard to the wheel dynamics during braking; see Sect. 5.5.

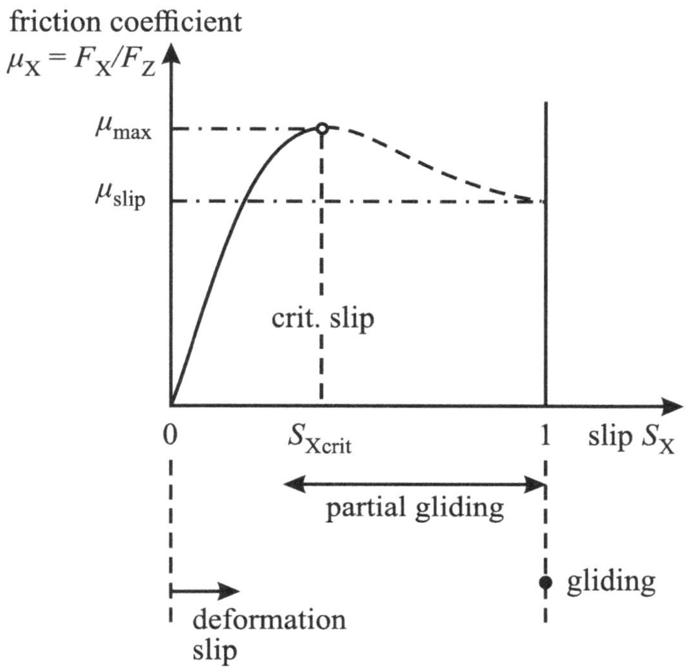


**Fig. 5.2** Friction coefficient $\mu_X$ in dependence on the tire slip (deformation slip, partial gliding slip, and total gliding slip), according to (Mitschke and Wallentowitz 2014)

Figure 5.3 depicts $\mu$-slip curves for different road surface conditions. It shows maximal values for dry asphalt, lower values for wet roads, and lowest values for snow and ice.

Different *mathematical models* for the longitudinal tire force have been obtained by approximation of the measured $\mu$-slip curves.

The *HSRI model* (Highway Safety Research Institute, Ann Arbor, USA) (Dugoff et al. 1969) approximates $\mu_X$ by two straight lines

$$\begin{aligned} \mu_X &= c_{\mu 0} S_X && S_X < S_{X,\max} \\ \mu_X &= \mu_{X0} - c_{\mu 1} S_X && S_X > S_{X,\max} \end{aligned} \tag{5.1.4}$$

where $S_{X,\max} = S_{Xcrit}$ is the slip for the maximum of $\mu_X(S_X)$. The initial gradient

$$c_{\mu 0} = \left. \frac{d\mu_X}{dS_X} \right|_{S_X \to 0} \tag{5.1.5}$$

is for dry and wet roads about the same (Mitschke and Wallentowitz 2014). The model according to Burckhardt (1993) uses the approximation

$$\mu_X = c_1 \left(1 - e^{c_2 S_X}\right) - c_3 S_X. \tag{5.1.6}$$

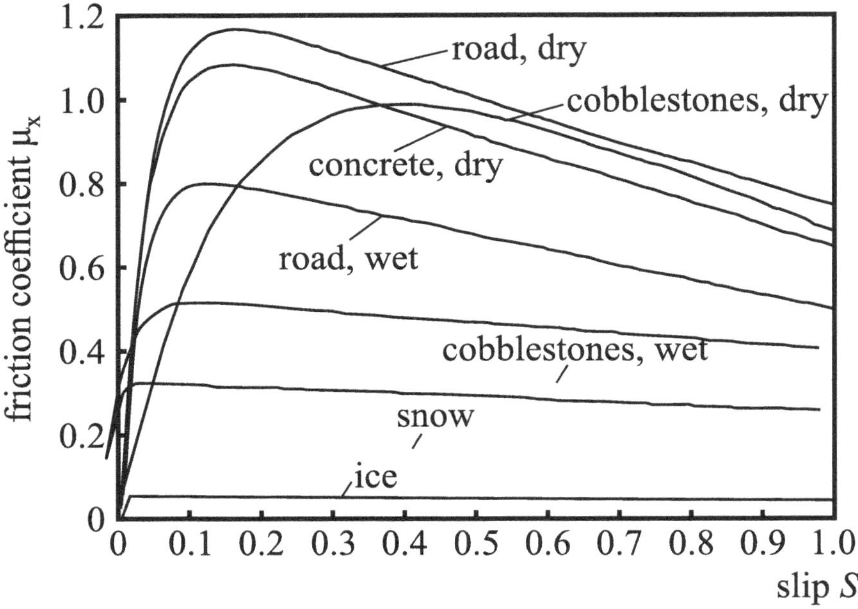


**Fig. 5.3** Typical friction coefficients $\mu$ in dependence on slip $S$ for different road surface conditions (Kiencke and Nielsen 2005)

The coefficients $c_1$ determine $\mu_{max}$, $c_2$ the position of the maximum, and $c_3$ the value for total slip $S_X = 1$; see also Daiss (1996). Burckhardt (1993), and Daiss (1996) present numbers of these parameters for different road conditions. A modified equation with five parameters is shown in Halfmann and Holzmann (2003).

More detailed tire models have been compiled by Pacejka (2012), called "magic formula". A basic equation is

$$F_{XT} = D_X \sin\left[C_X \arctan(B_X S_X - E_X(B_X S_X - \arctan(B_X S_X)))\right]. \quad (5.1.7)$$

Herewith

$$D_X = \mu_{X,\max} F_Z \quad (5.1.8)$$

determines the maximum longitudinal tire force.

$B_X$ determines the slope at zero slip (slip stiffness $C_{SX}$)

$$B_X = C_{SX}/C_X D_X. \quad (5.1.9)$$

The form coefficient

$$E_X = \frac{B_X S_{X,\text{crit}} - \tan\left(\frac{\pi}{2C_X}\right)}{B_X S_{X,\text{crit}} - \arctan(B_X S_{X,\text{crit}})} \quad (5.1.10)$$

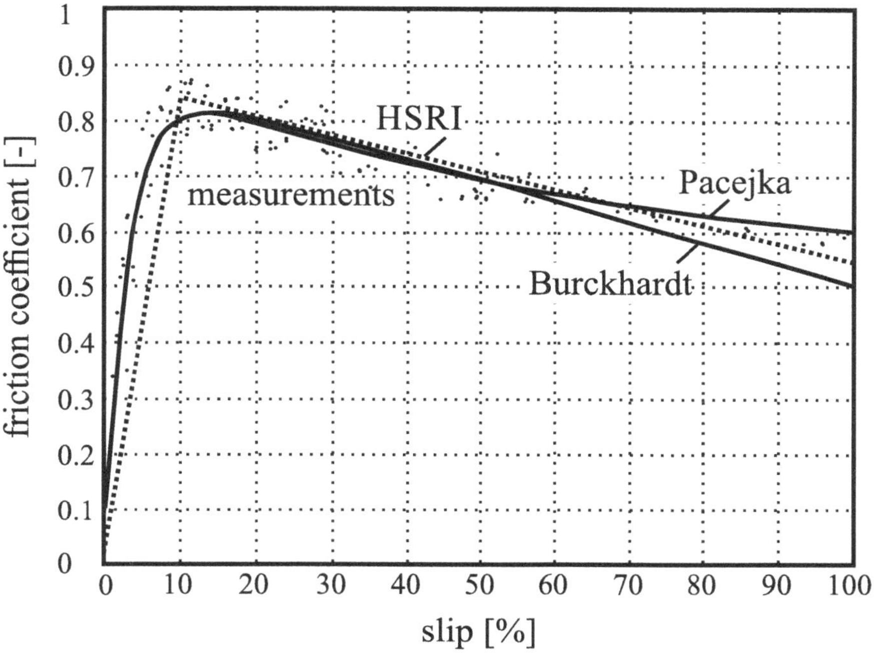


**Fig. 5.4** Comparison of different friction coefficient models in longitudinal direction with measurements (Germann 1997)

is determined such that the maximal tire force $F_{XT,max}$ is reached at the critical slip $S_{X,crit}$. $C_X$ is a free parameter and determines the force behavior after the maximum. Identified parameters of this tire model are, e.g. given in Würtenberger (1997).

Figure 5.4 shows a comparison of different friction coefficient-slip models with measured values illustrating a better agreement for the Burckhardt and the Pacejka model.

In order to obtain friction models which are more suitable for parameter estimation during driving maneuvers, algebraic approximation equations have been developed. They can be brought into a form which is linear in the parameters. Daiss (1996) has proposed

$$\mu_X(S_X) = c_{\mu 0} \frac{S_X}{1 + c_1 S_X + c_2 S_X^2}. \tag{5.1.11}$$

A comparison with the model from Burckhardt (1993), (5.1.6) shows a good agreement in a slip range $S_X < 0.6$. A similar approximation is given by Kiencke and Nielsen (2005)

$$\mu_X(S_X) = c_{\mu 0} \frac{1}{1 + c_1 S_X + c_2 S_X^2}. \tag{5.1.12}$$

These equations will be used for the experimental identification of friction coefficients in Sect. 11.6.

## 5.2 Lateral Tire Forces

Lateral tire forces appear during cornering and cause the tread blocks to be deflected relatively to the circumference of the tire, resulting in a shear stress of the rubber material. When the tread moves through the contact patch, the deflection decreases. In the case of higher lateral forces, the shear stress becomes larger than the maximal shear stress at the adhesion limit, which is described by the lateral friction coefficient $\mu_Y$, and the treads start gliding. The lateral tire slip is expressed by a tire side slip angle $\alpha$; see Fig. 5.5.

The tire moves with trajectory velocity $v_T$ and the velocity components of the wheel become

$$v_{XT} = v_T \cos\alpha; \qquad v_{YT} = v_T \sin\alpha. \tag{5.2.1}$$

The side slip or lateral slip is defined according to

$$S_Y = \frac{v_{YT}}{v_{XT}} = \frac{v_T \sin\alpha}{v_T \cos\alpha} = \tan\alpha \tag{5.2.2}$$

and becomes $S_Y = 1$ for $\alpha = 45°$. Under normal driving conditions, it holds $\alpha < 12°$. (Side slip for driving; see (5.3.4).)

Figure 5.6 depicts the typical course of the lateral force $F_{YT}$ in dependence on the side slip angle and the vertical wheel load. It is approximately linear for small

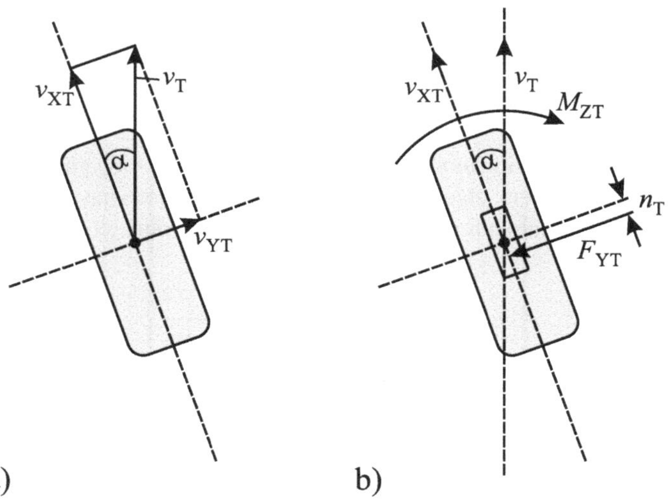


**Fig. 5.5** Rolling tire with slip angle $\alpha$: **a** velocities; **b** caster offset and aligning torque

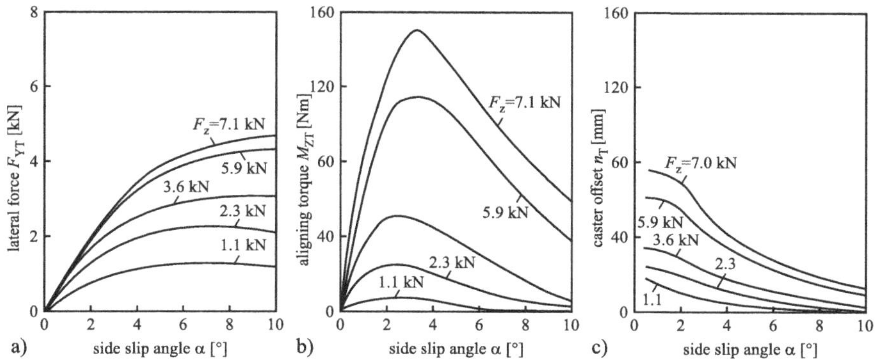


**Fig. 5.6** Lateral force, aligning torque, and caster offset in dependence on side slip angle and vertical wheel load (Mitschke and Wallentowitz 2014). Tire: 175/70 R; tire pressure 2.1 bar; camber 0°; rated wheel load 4.15 kN

values $\alpha \leq 3°$, then reaches a maximum at $\alpha = 6 \ldots 10°$ and stays then constant or decreases for larger side slip angles. The lateral forces increase with the vertical wheel load $F_Z$. For small side slip angles $\alpha = 0 \ldots 3°$, the lateral tire force can be described linearly by

$$F_{YT}(\alpha) = c_\alpha(F_{ZT})\alpha \tag{5.2.3}$$

where $c_\alpha$ is the *cornering stiffness*

$$c_\alpha = \left.\frac{dF_{YT}(\alpha)}{d\alpha}\right|_{\alpha=0°}. \tag{5.2.4}$$

The cornering stiffness $c_\alpha$ depends on the vertical wheel load $F_{ZT}$ and shows for small $F_{ZT}$ a linear behavior

$$c_\alpha = c_{\alpha 1} F_{ZT} \tag{5.2.5}$$

with superimposed quadratically decreasing values for larger $F_{ZT}$, which can be approximated by

$$c_\alpha = \left(c_{\alpha 1} - c_{\alpha 2}\frac{F_{ZT}}{F_{ZT,\text{rate}}}\right) F_{ZT} \tag{5.2.6}$$

where $F_{Z,\text{rate}}$ is a rated wheel load given from the manufacturer (Mitschke and Wallentowitz 2014).

According to the definition of the friction coefficient $\mu_X(S_X)$ in longitudinal direction, a *friction coefficient* can be defined in *lateral direction*

$$\mu_{YT}(S_Y) = \frac{F_{YT}}{F_{ZT}} \tag{5.2.7}$$

and with (5.2.4), one obtains for the cornering stiffness

$$c_\alpha = \left.\frac{d\mu_{\mathrm{YT}}(\alpha)}{d\alpha}\right|_{\alpha=0^\circ} F_{\mathrm{ZT}} \tag{5.2.8}$$

for small side slip angles. Furthermore, (5.2.3), (5.2.5), and (5.2.7) lead for small side slip angles and small vertical forces to

$$\mu_{\mathrm{YT}} = c_{\alpha 1}\alpha. \tag{5.2.9}$$

As the distribution of the shear stress in the contact patch is not symmetric around the rotational axis of the tire, the lateral force generates a torque about the vertical axis which is called *aligning torque* $M_{\mathrm{ZT}}$. It depends on the distance of the center of all lateral force components of the contact patch and the tire's lateral axis, which is called *caster offset* or pneumatic trail $n_{\mathrm{T}}$; see Fig. 5.5b. This caster offset depends strongly on the vertical wheel load and the side slip angle, hence it is $n_{\mathrm{T}}(F_{\mathrm{ZT},})$; see Fig. 5.6c. Because of the caster offset, the lateral force acts after the lateral axis and causes the aligning torque in the direction of reducing the slip angle, i.e. returning the wheel to straight driving. The aligning torque shows initially a linear behavior for small side slip angles $\alpha = 0^\circ \ldots 2^\circ$ and can be approximated by

$$M_{\mathrm{ZT}} = c_\alpha n_{\mathrm{T}}\alpha. \tag{5.2.10}$$

At about $\alpha = 3^\circ \ldots 6^\circ$, it reaches a maximum and then decreases for larger side slip angles toward zero, because the caster offset becomes smaller; see Fig. 5.6b and c.

*Mathematical models* for the lateral tire forces $F_{\mathrm{Y}}(\alpha)$ are similar to those in longitudinal direction. Table 5.1 shows some examples.

A comparison of the computational expense is made in Halfmann and Holzmann (2003). The ratio of computing time from HRSI: Burckhardt: Pacejka models is 1:2.5:4.6. All these models are nonlinear in the parameters and therefore not suitable for direct (non-iterative) parameter estimation. A broken rational (algebraic) equation according to Ammon (1997) is linear in the parameters and therefore better suited; see also modifications in Daiss (1996), Zomotor and Reimpell (1991), Zomotor (2002), and Börner (2004). Bauer (2015) proposes this model in the form

$$\mu_{\mathrm{YT}}(\alpha) = \frac{c_{\alpha 1}\alpha}{\left(\frac{|\alpha|}{\alpha_{\mathrm{crit}}} - 1\right)^2 + \frac{c_{\alpha 1}|\alpha|}{\mu_{\mathrm{YT,max}}}} \tag{5.2.11}$$

with $c_{\alpha 1} = c_\alpha / F_{\mathrm{ZT}}$; see (5.2.5). $\alpha_{\mathrm{crit}}$ is the side slip angle for $\mu_{\mathrm{YT,max}}$. The initial gradient is

$$\left.\frac{d\mu_{\mathrm{YT}}}{d\alpha}\right|_{\alpha=0^\circ} = c_{\alpha 1}. \tag{5.2.12}$$

Figure 5.7 shows a corresponding graph.

**Table 5.1** Mathematical tire models for lateral tire friction coefficients (Bauer 2015)

| Model type | Equation | Characteristic |
|---|---|---|
| Linear | $\mu_{YT}(\alpha) = c_\alpha \alpha$ | 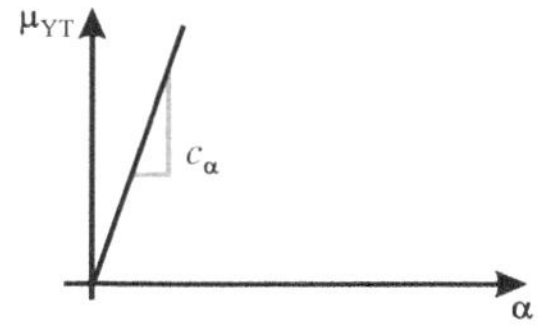 |
| HSRI Dugoff (1969) | $\mu_{YT}(\alpha) = \frac{c_\alpha \tan \alpha}{1 - \alpha} f(\theta_{HSRI})$<br>$f(\theta_{HSRI}) = \begin{cases} \theta_{HSRI}(2 - \theta_{HSRI}) & \theta_{HSRI} \le 1 \\ 1 & \theta_{HSRI} \ge 1 \end{cases}$<br>$\theta_{HSRI} = f(\alpha, \lambda, F_Z, v, c_\alpha, c_\lambda)$ | 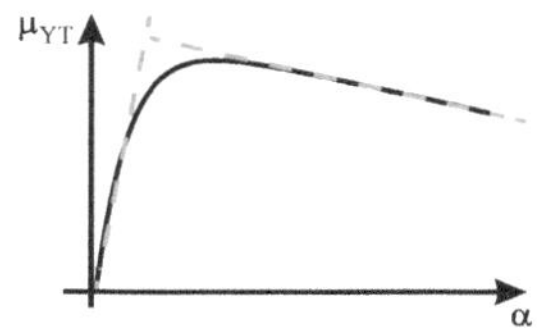 |
| Burckhardt (1993) | $\mu_{YT}(\alpha) = c_1\left(1 - \exp^{-c_2\alpha}\right) - c_3\alpha$<br>with influence of speed and load:<br>$\mu_{YT}(\alpha) = \left(c_1\left(1 - e^{-c_2\alpha}\right) - c_3\alpha\right)e^{-c_4\alpha v}\left(1 - c_5 F_Z^2\right)$ | 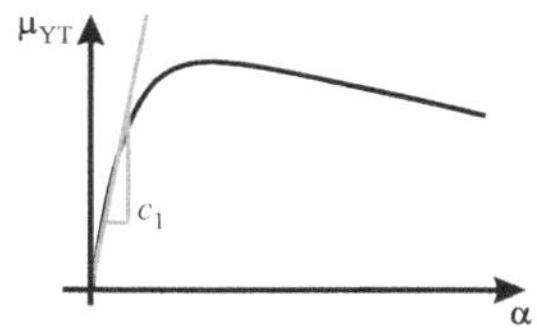 |
| Pacejka (2012) | $\mu_{YT}(\alpha) = D \sin(C \arctan(D\alpha - E(B\alpha - \arctan(B\alpha))))$<br>B: stiffness coefficient C: free parameter<br>D: maximum value E: form coefficient | 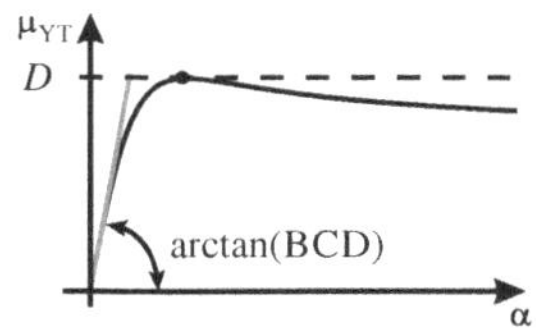 |
| Ammon (1997) | $\mu_{YT}(\alpha) = \frac{c_{\alpha 1}\alpha}{\left(\frac{\lvert\alpha\rvert}{\mu_{YT,max}\alpha_{crit,1}}\right)^2 + \frac{c_{\alpha 1}\lvert\alpha\rvert}{\mu_{YT,max}}}$ | 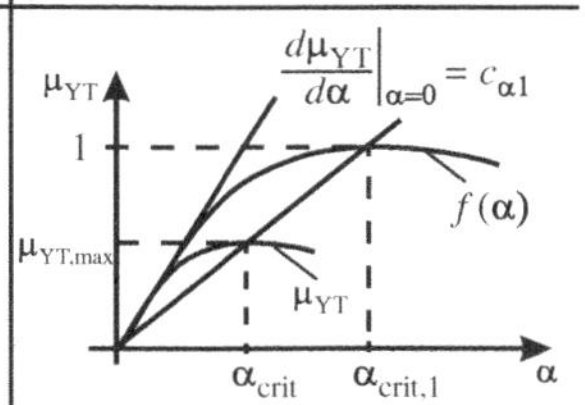 |
| Bauer (2015) | $\mu_{YT}(\alpha) = \frac{c_{\alpha 1}\alpha}{\left(\frac{\lvert\alpha\rvert}{\alpha_{crit}} - 1\right)^2 + \frac{c_{\alpha 1}\lvert\alpha\rvert}{\mu_{YT,max}}}$ | 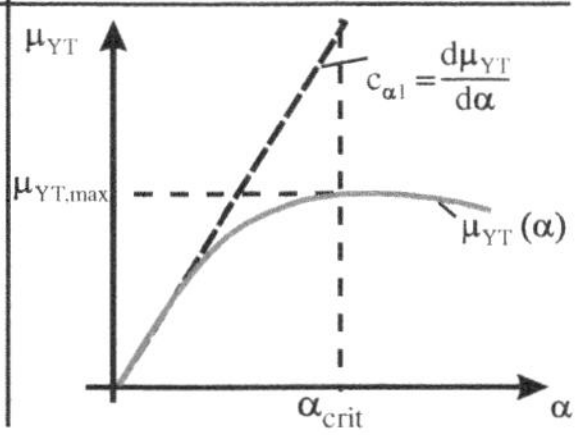 |

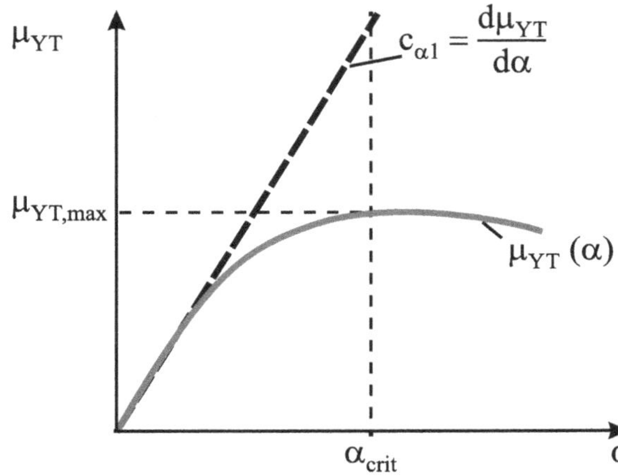


**Fig. 5.7** Broken rational tire model for the lateral behavior according to (5.2.11) (Bauer 2015)

$\alpha_{\mathrm{crit}}$ is proportional to $\mu_{\mathrm{YT,max}}$ (Ammon 1997; Zomotor et al. 1998). The parameters of the model (5.2.11) can be estimated online with a recursive least squares method; see Sects. 10.1.1 and 11.6.2.

Table 5.1 summarizes different approximation equations for the lateral tire friction coefficient. A version which is suitable for parameter estimation with the LS methods is (5.2.11), which after replacement of $\alpha$ by the slip value $S_{\mathrm{Y}}$ obtains the form, Bauer (2015),

$$\mu_{\mathrm{YT}} = \frac{c_{\alpha 1} S_{\mathrm{Y}}}{\left(\frac{S_{\mathrm{Y}}}{\mu_{\mathrm{YT,max}} S_{\mathrm{crit}}} - 1\right)^2} - \frac{c_{\alpha 1} S_{\mathrm{Y}}}{\mu_{\mathrm{YT,max}}} \tag{5.2.13}$$

with

$$c_{\alpha 1} = \frac{c_\alpha}{F_{\mathrm{YT}}}. \tag{5.2.14}$$

For small $S_{\mathrm{Y}}$, one obtains

$$\mu_{YT} = \mu_{\mathrm{YT}}|_{S_{\mathrm{Y}}=0} = c_{\alpha 1} S_{\mathrm{Y}} \tag{5.2.15}$$

which corresponds to (5.2.9).

The lateral tire forces depend in addition to $F_{\mathrm{ZT}}$ and $\alpha$ also on the *longitudinal tire slip* $S_{\mathrm{X}}$ and the road surface condition

$$F_{\mathrm{YT}} = f(\alpha, F_{\mathrm{ZT}}, S_{\mathrm{X}}, \mathrm{road}). \tag{5.2.16}$$

Figure 5.8 depicts some characteristic curves. Figure 5.8a shows that for $S_{\mathrm{X}} = 0 \ldots 10\,\%$ a maximal value $F_{\mathrm{YT,max}}(\alpha)$ exists and the lateral force $F_{\mathrm{YT}}(\alpha)$ decreases for larger longitudinal slip values $S_{\mathrm{x}} > 10\,\%$. Compared to a dry road, it becomes smaller for wet roads and increasing speed; see Fig. 5.8b. The maximal value $F_{\mathrm{YT,max}}$ depends nonlinearly on the vertical force $F_{\mathrm{ZT}}$ and is only proportional to $F_{\mathrm{ZT}}$ for smaller values; compare (5.2.5) and (5.2.6).

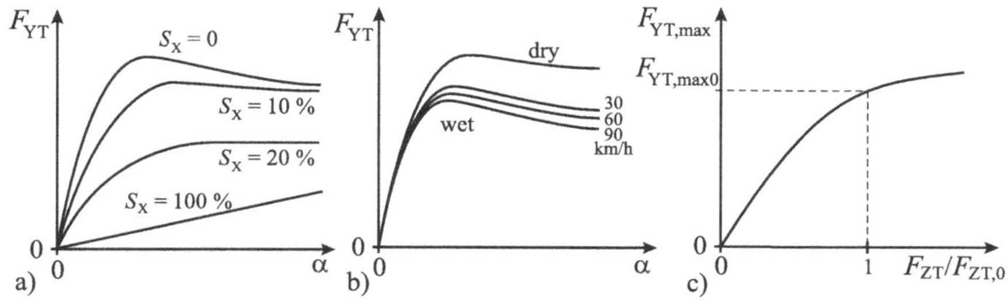


**Fig. 5.8** **a** Lateral tire force in dependence on the side slip angle $\alpha$ influenced by longitudinal slip $S_X$ and **b** road surface conditions. **c** Shows the maximal lateral force $F_{YT,max}$ in dependence on the vertical wheel force $F_{ZT}$. $F_{ZT,0}$ is the rated wheel load (Pacejka 2012; Bechtloff 2018)

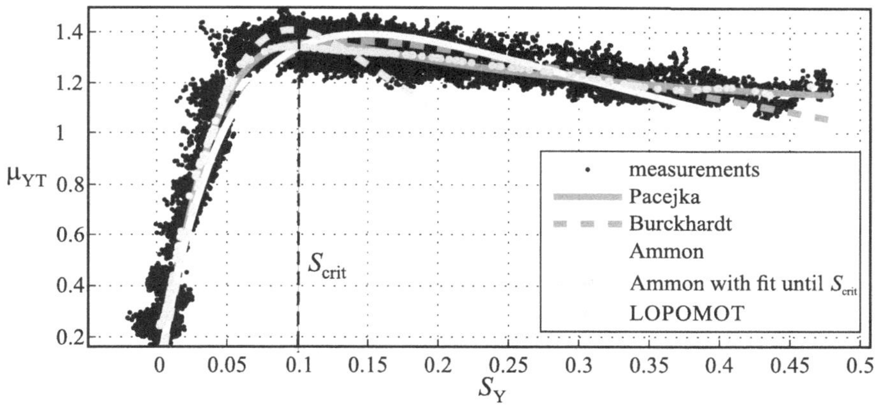


**Fig. 5.9** Comparison of the lateral friction coefficient $\mu_{YT}$ for different tire models for the tire 225-50-R17 (Bauer 2015)

A comparison of the lateral friction coefficient $\mu_{YT}(S_Y)$ of different tire models with measurements is depicted in Fig. 5.9. The parameters of the tire model equations were adapted by parameter estimation. The broken rational model according to Ammon (5.2.11) was fitted only until $S_Y = 0.2$, resulting in a good agreement with measurements until $S_{crit}$, but not beyond that. The models according to Pacejka and Burckhardt show a good agreement also for larger slip. The best approximation is obtained by a local polynomial tree model (LOPOMOT) according to Sequenz (2013). A summary and comparison of different tire models are also given in Einsle (2010).

The position of a tire relative to the road has a significant influence on the driving properties and handling of a vehicle. Therefore, the design of the suspension links and the rotation centers, the wheel travel curves, and the toe and camber angles are important parameters; see Fig. 5.10. Because of these kinematic properties, the toe angles $\delta_0$ change during wheel travels and influence the side slip angles $\alpha$.

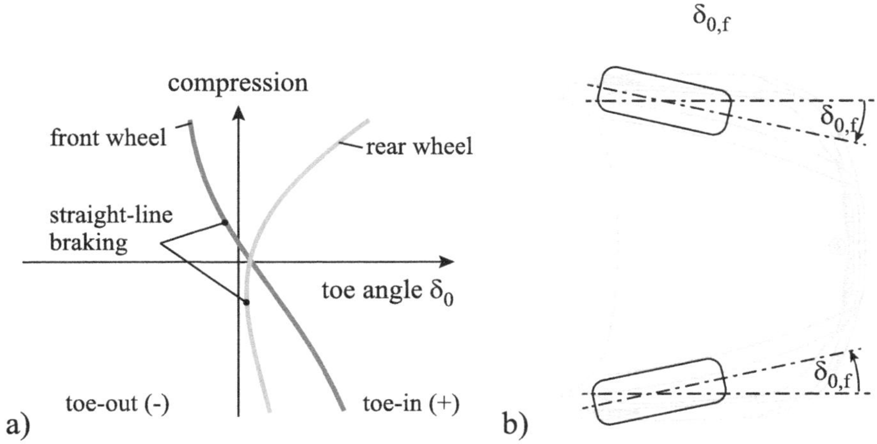


**Fig. 5.10** Wheel travel curves for front and rear wheels. **a** Toe angle in dependence of suspension compression. **b** Toe-in angles for a front axle (toe-in) (Heissing and Ersoy 2011)

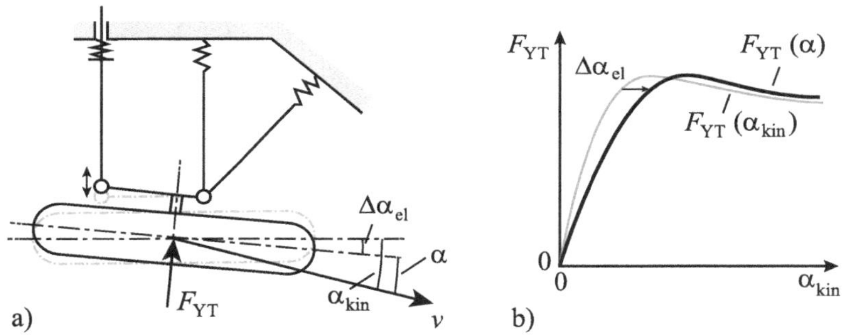


**Fig. 5.11** Elasto-kinematic changes of the side slip angles through lateral tire forces (Pacejka 2012; Bechtloff 2018)

The toe angle $\delta_0$ is not only influenced by the spring deflection but also by longitudinal and lateral tire forces because of deformations within the elasto-kinematic suspension, for example, by rubber bearings. As depicted in Fig. 5.11a, a lateral tire force $F_{YT}$ turns the wheel by the angle $\Delta\alpha_{el}$ and the effective side slip angle becomes smaller: $\alpha_{eff} = \alpha_{kin} - \Delta\alpha_{el}$. Therefore, the initial gradient of the lateral force characteristic decreases; see Fig. 5.11b (for the assumed suspension design with $\alpha_{kin}$).

## 5.3 Combined Longitudinal and Lateral Forces

For driving conditions with acceleration or braking during cornering as well longitudinal as lateral tire forces are generated. This leads to a combined slip and the transferable forces depend on the superposition of the shear stresses in the tire treads with regard to maximal transferable stress values.

Figure 5.12 depicts a schematic for a vehicle with longitudinal and lateral velocities $v_X$ and $v_Y$, resulting vehicle velocity $v$, and side slip angle $\beta$. A free rolling wheel then has a tire trajectory velocity $v_T$ with a tire longitudinal velocity $v_{XT}$, a tire lateral velocity $v_{YT}$, and a tire slip angle $\alpha$. The directions of $v$ and $v_T$ are different if a yaw turn with $\dot{\psi}$ arising, see Chap. 7, (7.2.20).

For the simultaneous slip in longitudinal and lateral directions, one finds different definitions in the literature. Burckhardt (1993), Kiencke and Nielsen (2005) calculate the longitudinal tire slip in the direction of the wheel motion $v_T$ (Reimpell and Sponagel 1995; Pacejka 2012; ISO 8855 2013) in the direction of the longitudinal wheel velocity $v_{XT}$. This last version is used in the following.

The longitudinal slip measure then uses as reference value, the larger value as for (5.1.1) and (5.1.2), i.e. $v_{XT}$ in the case of braking and $v_{XW}$ in the case of driving (Mitschke and Wallentowitz 2014).

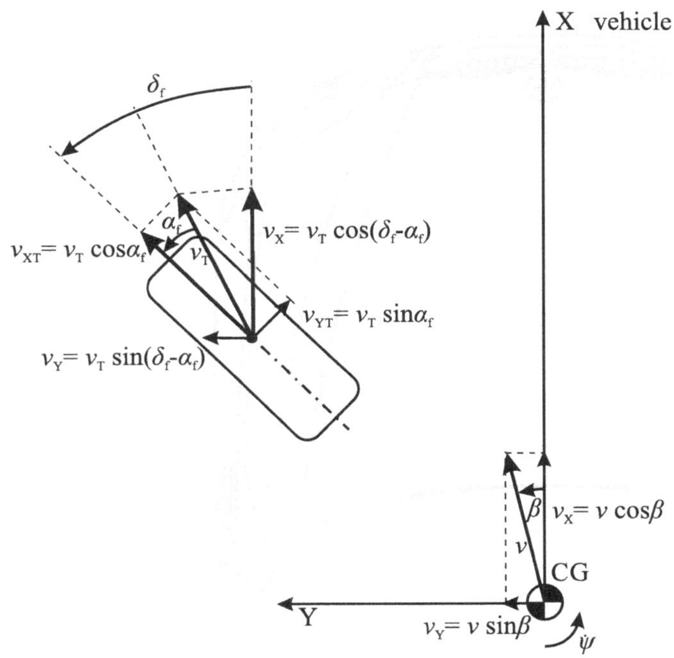


**Fig. 5.12** Velocity components of a free rolling front wheel with side slip of a cornering vehicle. $v_T$: tire trajectory velocity; $\alpha_f$: tire side slip angle; $v$: vehicle velocity; $v_X$: longitudinal vehicle velocity; $v_Y$: lateral vehicle velocity; $\beta$: vehicle slip angle; $\delta_f$: front steering angle

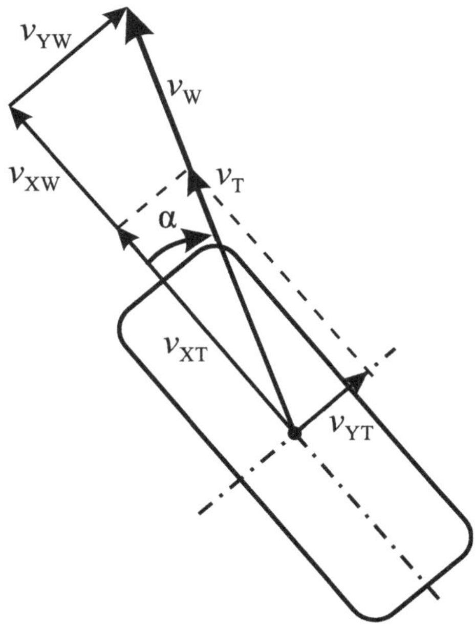


**Fig. 5.13** Velocity components for defining the tire slip in the case of combined longitudinal and lateral slips. Notations are for driving (wheel turns faster than vehicle $v_w > v_X$)

As shown in Fig. 5.13, the wheel and tire turns with rotational velocity $v_{XW} = r_{dyn}\omega_W$. The velocity $v_T$ of the tire in the tire/road contact patch is turned by the side slip angle $\alpha$ with regard to the longitudinal wheel axis $x_T$. Hence, the velocity in the longitudinal direction of the wheel is $v_{XT} = v_T \cos\alpha$ and in the lateral direction $v_{YT} = v_T \sin\alpha$.

Then, according to the definitions (5.1.1) and (5.1.2), one obtains for the longitudinal slip for braking

$$S_{X,b} = \frac{\Delta v_{XT}}{v_{XT}} = \frac{v_{XT} - v_{XW}}{v_{XT}} = \frac{v_{XT} - r_{dyn}\omega_W}{v_{XT}} \tag{5.3.1}$$

and for driving (traction)

$$S_{X,d} = \frac{\Delta v_{XT}}{v_{XW}} = \frac{v_{XW} - v_{XT}}{v_{XW}} = \frac{r_{dyn}\omega_W - v_T\cos\alpha}{r_{dyn}\omega_W} \tag{5.3.2}$$

such that $S_{X,b} = 0\ldots1$ and $S_{X,d} = 0\ldots1$. The lateral slip is according (5.2.2) for braking

$$S_{Y,b} = \frac{\Delta v_{YT}}{v_{XT}} = \frac{v_{YT}}{v_{XT}} = \frac{v_T\sin\alpha}{v_T\cos\alpha} = \tan\alpha, \tag{5.3.3}$$

and for driving

$$S_{Y,d} = \frac{\Delta v_{YT}}{v_{XW}} = \frac{v_T\sin\alpha}{r_{dyn}\omega_W}. \tag{5.3.4}$$

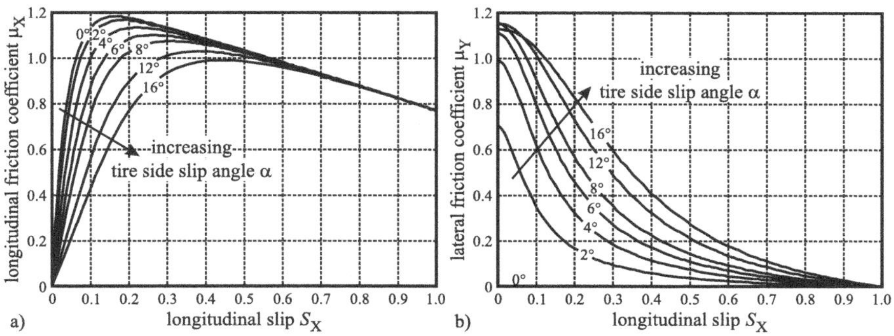


**Fig. 5.14** Friction coefficients in dependence on the longitudinal slip for different side slip angles (Kiencke and Nielsen 2005). **a** Longitudinal friction coefficient $\mu_X$. **b** Lateral friction coefficient $\mu_Y$

Using these definitions, a resultant slip becomes for braking or driving

$$S_{res} = \sqrt{S_X^2 + S_Y^2}. \tag{5.3.5}$$

Figure 5.14 depicts the longitudinal and lateral friction coefficients in dependence on the longitudinal side slip angles $\alpha$. $\mu_X(S_X, \alpha)$ shows only small dependence on $\alpha$ for $\alpha = 0° \ldots 2°$ but changes significantly for larger $\alpha = 8° \ldots 16°$. The cornering stiffnesses $c_\alpha$ and $\mu_{max}$ become smaller with increasing $\alpha$.

The lateral friction coefficient $\mu_Y(S_X, \alpha)$, Fig. 5.14b, increases with larger side slip angle from $\alpha = 0° \ldots 16°$ but decreases strongly with increasing longitudinal slip $S_X$. For larger slips $S_X > 0.4$, then $\mu_X$ has still significant values, but $\mu_Y$ tends to zero. Hence for $S_X = 0 \ldots 0.4$, the characteristics $\mu_X(S_X)$ and $\mu_Y(S_Y)$ show opposite behavior.

This can also be seen in Fig. 5.15 where $\mu_Y$ is plotted in dependence on $\mu_X$ for different slip angles. It shows that $\mu_X$ and $\mu_Y$ are related to each other. For small side slip angle, $\alpha = 2°$ $\mu_X$ increases for small $\mu_Y$, but then decreases strongly for larger $\mu_Y$, and for $\alpha = 16°$ $\mu_X$ increases moderately for small $\mu_Y$ and decreases strongly, now at higher values for $\mu_Y$.

Now, instead of the side slip angle $\alpha$ the lateral slip $S_Y$ is used, taking the relations (5.3.3) and (5.3.4) into account.

Due to (5.1.3) and (5.2.7), the generated tire forces are then

$$\begin{aligned} F_{XT} &= \mu_{XT}(S_X, S_Y) F_{ZT} \\ F_{YT} &= \mu_{YT}(S_X, S_Y) F_{ZT} \end{aligned} \tag{5.3.6}$$

and the transferable resulting force of a wheel with combined slips is

$$F_{res} = \sqrt{F_{XT}^2 + F_{YT}^2} = F_{ZT}\sqrt{\mu_{XT}^2(S_X, S_Y) + \mu_{YT}^2(S_X, S_Y)}. \tag{5.3.7}$$

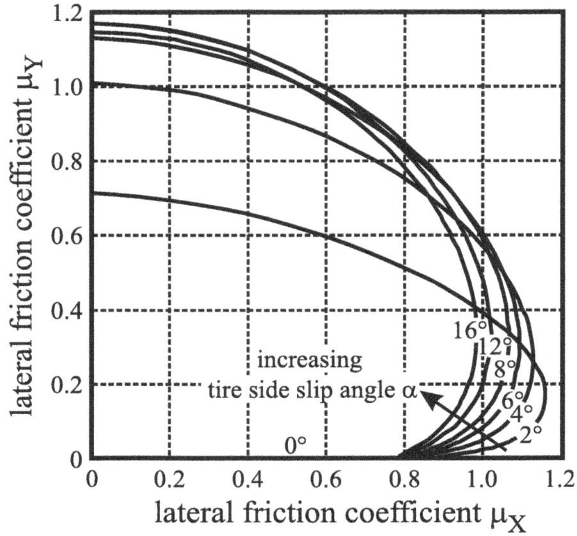


**Fig. 5.15** Lateral friction coefficient in dependance on the longitudinal friction coefficient for constant side slip angles (Kiencke and Nielsen 2005)

If a combined friction coefficient

$$\mu_{\mathrm{res}} = \sqrt{\mu_{\mathrm{XT}}^2 + \mu_{\mathrm{YT}}^2} \tag{5.3.8}$$

is defined, it follows

$$F_{\mathrm{res}} = \mu_{\mathrm{res}} F_{\mathrm{ZT}} \tag{5.3.9}$$

where the force components are

$$F_{\mathrm{XT}} = \mu_{\mathrm{res}} \frac{S_{\mathrm{X}}}{S_{\mathrm{res}}} F_{\mathrm{ZT}}; \quad F_{\mathrm{YT}} = \mu_{\mathrm{res}} \frac{S_{\mathrm{Y}}}{S_{\mathrm{res}}} F_{\mathrm{ZT}} \tag{5.3.10}$$

which can be shown after introducing (5.3.8) in (5.3.7) and use of (5.3.5).

Here it is assumed that $\mu_{\mathrm{res}}$ is not direction-dependent. Then it holds for the maximal transferable forces of a tire

$$F_{\mathrm{res,max}} = \mu_{\mathrm{res,max}} F_{\mathrm{ZT}} = \sqrt{F_{\mathrm{XT,max}}^2 + F_{\mathrm{YT,max}}^2}. \tag{5.3.11}$$

This means that the longitudinal and lateral forces of a tire have to stay within a circle, which is called the *Kamm circle*; see Fig. 5.16. Outside of the Kamm's circle, the tires start gliding.

For anisotropic tires with tread profiles, the maximal friction coefficients are different and they hold

$$\mu_{\mathrm{Y,max}} < \mu_{\mathrm{X,max}}. \tag{5.3.12}$$

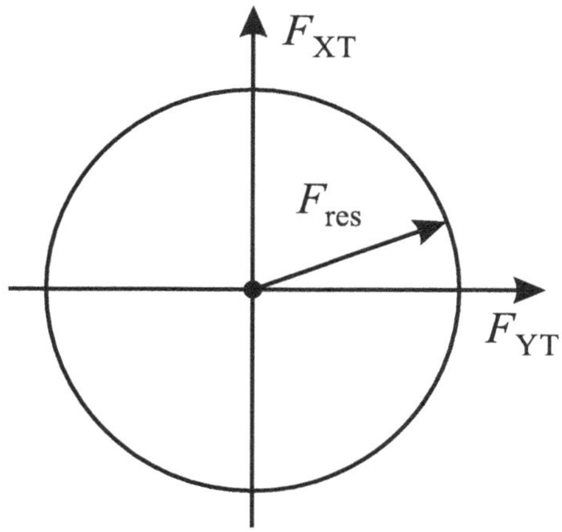


**Fig. 5.16** Kamm's circle for longitudinal and lateral tire forces

Then, one uses an attenuation factor

$$F_{YT} = k_Y \mu_{res} \frac{S_Y}{S_{res}} F_{ZT} \tag{5.3.13}$$

and the Kamm circle degenerates to an ellipse. Typical values are $k_Y = 0.8 \ldots 0.95$ (Kiencke and Nielsen 2005).

For *small longitudinal slip* $S_X$ and *lateral slip* $S_Y$, it holds, see (5.1.4) and (5.2.3), (5.2.6)

$$F_{XT} = \mu_X(S_X)|_{S_X \to 0} \cdot F_{ZT} = c_{\mu 0}\, S_X\, F_{ZT}, \tag{5.3.14}$$

$$F_{YT} = c_\alpha(F_{ZX})\alpha|_{\alpha \to 0} = c_{\alpha 1} \alpha F_{ZT}. \tag{5.3.15}$$

Assuming small $\alpha$ (5.2.2) leads to

$$S_Y = \tan \alpha|_{\alpha \to 0} = \alpha \tag{5.3.16}$$

and (5.3.15) becomes with

$$F_{YT} = c_{\alpha 1} S_Y F_{ZT}$$

and

$$\mu_{YT} = c_{\alpha 1} \alpha|_{\alpha \to 0} = c_{\alpha 1} S_Y|_{S_Y \to 0}. \tag{5.3.17}$$

Then (5.3.7) yields for small slips with (5.1.4)

$$\begin{aligned} F_{res} &= F_{ZT} \sqrt{(c_{\mu 0} S_X)^2 + (c_{\alpha 1} S_Y)^2} \\ &= F_{ZT} \sqrt{\mu_{XT}^2 + \mu_{YT}^2} = \mu_{res} F_{ZT}. \end{aligned} \tag{5.3.18}$$

This relation holds within the Kamm circle, using linear $\mu$-slip models.

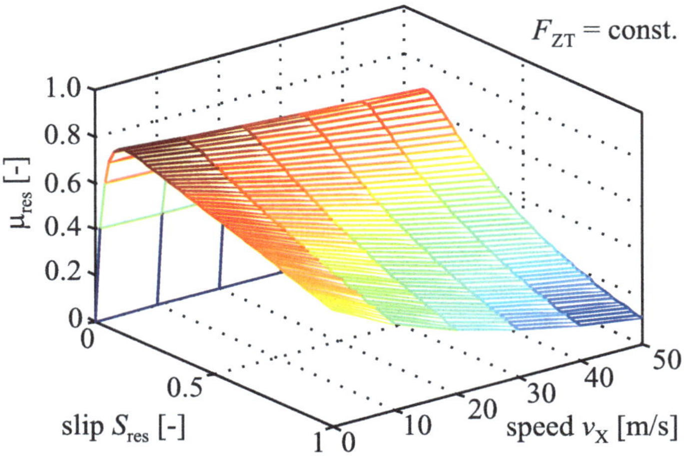


**Fig. 5.17** Lookup table for the resulting friction coefficient $\mu_{res}$ of a tire on wet asphalt, calculated with the tire model from Burckhardt (Halfmann and Holzmann 2003)

For the mathematical description of the friction coefficient $\mu$ in dependence on the influencing variables, different tire models were proposed for combined longitudinal and lateral forces similar to those in Sects. 5.1 and 5.2. The first way to express the $\mu_{res}(S_{res})$ characteristics as a lookup table is shown in Fig. 5.17.

The Burckhardt model is now expressed by

$$\mu_{res} = c_1 \left(1 - e^{c_2 S_{res}}\right) - c_3 S_{res}. \tag{5.3.19}$$

The coefficient $c_1$ determines $\mu_{max}$, $c_2$ the position of the maximum, and $c_3$ the value for $S_{res} = 1$; see also Daiss (1996) and Isermann (2006).

Another possibility is to use the broken rational algebraic model (5.2.11) by replacing $\alpha$ with $S_{res}$ and to use that $\alpha_{crit}$ is proportional to $\mu_{max}$, in the form, Bauer (2015), Breuer and Bill (2017)

$$\mu_{res}(S_{res}) = \frac{c_{res} S_{res}}{\left(\frac{S_{res}}{\mu_{max} S_{crit}} - 1\right)^2 - \frac{c_{res} S_{res}}{\mu_{max}}} \tag{5.3.20}$$

with

$$c_{res} = d\mu(S_{res})/\, dS_{res}|_{S_{res}=0} \,.$$

The resulting tire force then is, see (5.3.7),

$$F_{T,res} = \mu_{res}(S_{res}) F_{ZT}. \tag{5.3.21}$$

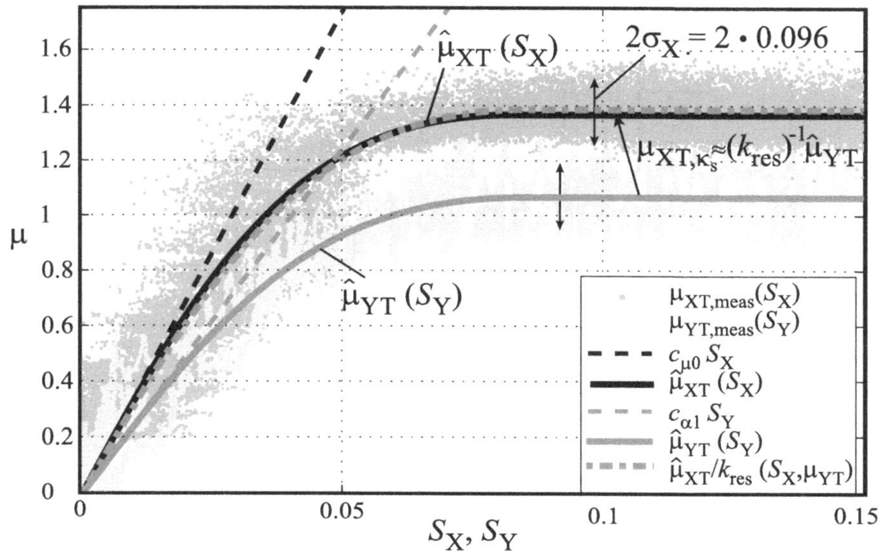


**Fig. 5.18** Measured friction coefficients $\mu_{XT}$ and $\mu_{YT}$ in dependence on (small) slip values $S_X$ and $S_Y$ with a tire measurement trailer and rim force measurement for the tire 225-50-R17 (Bauer 2015). Approximations with (5.3.20) and (5.3.22). $k_{res} = 0.77$ (Bauer 2015)

The longitudinal and lateral forces $F_{XT}$ and $F_{YT}$ then follow from (5.3.10). Their relation is for tread profiles expressed by

$$F_{YT} = k_{res} F_{XT} \tag{5.3.22}$$

with an attenuation factor $k_{res}$, compare (5.3.13), leading to a Kamm ellipse, with $k_{res} \approx 0.8$.

Figure 5.18 depicts measurements of friction coefficients $\mu_{XT}(S_X)$ and $\mu_{YT}(S_Y)$ separately and approximations with (5.3.20). The approximation of $\hat{\mu}_{XT}(S_X)$ is also determined from $\hat{\mu}_{YT}(S_Y)$ according to (5.3.22). The result shows that the longitudinal and lateral friction coefficients are for this tire and small slip values $S_X, S_Y \leq 0.15$ approximately related by a factor $k_{res}$.

The tire model according to Pacejka (2012) as described in (5.1.7) can be brought into the following general form:

$$y(x) = D \sin (C \arctan (Bx - E (Bx - \arctan Bx))) . \tag{5.3.23}$$

$y(x)$ stands for the longitudinal or lateral force or the aligning torque, $x$ for the longitudinal slip or slip angle, $D$ for the maximal force or torque, $C$ for the form of the characteristic, $E$ for the relaxation of the characteristic, and $B$ for the slope at zero slip (Lugaro et al. 2018). This tire model can also be used for combined longitudinal

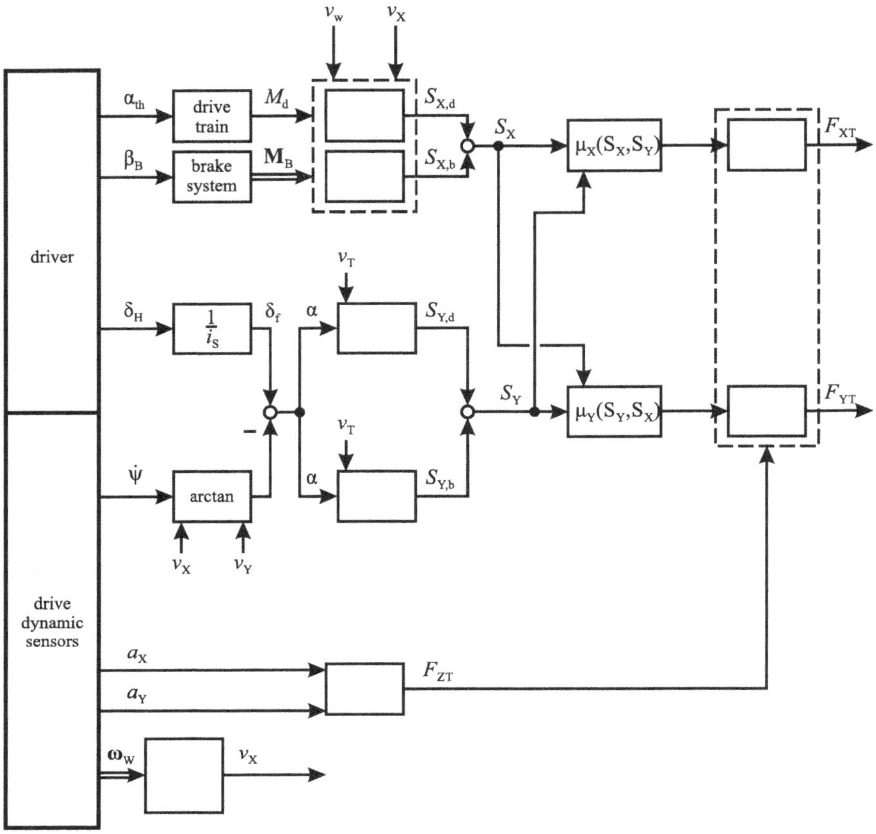


**Fig. 5.19** Signal flow for the determination of tire forces of one wheel with combined longitudinal and lateral slips for steering, acceleration, or braking. Use of longitudinal and lateral friction coefficients $\mu_X$ and $\mu_Y$

and lateral slips if $x$ is replaced by $S_{res}$; see Bechtloff (2018). The parameters have to be determined from measurements by approximation; see also Schramm et al. (2010).

A signal flow for the determination of the longitudinal and lateral forces according to (5.3.6) with combined slips $S_X$ and $S_Y$ is illustrated in Fig. 5.19. An alternative representation by using the resulting combined slip $S_{res}$ and combined friction coefficient $\mu_{res}$ is shown in Fig. 5.20.

More flexible tire models are compiled in computer simulation programs; see, e.g. Février et al. (2010). These models operate with shear stresses of the tire treads, torsion of the sidewalls, rubber-material dependent friction coefficients, and different temperatures within the tire.

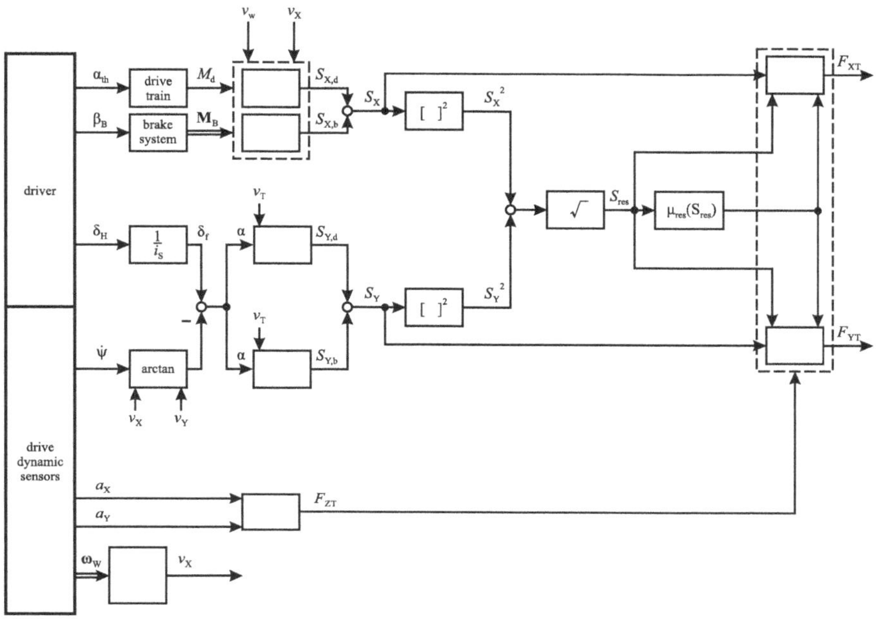


**Fig. 5.20** Signal flow for the determination of tire forces of one wheel with combined longitudinal and lateral slips for steering, acceleration, or braking. Use of a resulting combined friction coefficient $\mu_{res}$

## 5.4 Lateral Tire Dynamics

The lateral tire force $F_{YT} = c_\alpha \alpha$, see (5.2.3), is valid for a stationary behavior. After a change of the side slip angle $\alpha$, the tire builds up a deformation $\Delta y$ because of the lateral stiffness $c_Y$ of a tire, according to

$$\Delta F_{YT} = c_Y \Delta y_T; \tag{5.4.1}$$

see Fig. 5.21. However, the lateral deformation $\Delta y_T$ initially builds up with time, according to

$$\Delta y_T = v_{YT} \Delta t \tag{5.4.2}$$

where $v_{YT}$ is the lateral positive tire velocity over ground. For the lateral velocity holds assuming small side slip angles

$$v_{YT} = v_T \sin \alpha \approx v_T \alpha. \tag{5.4.3}$$

Hence, it follows

$$\Delta y_T = v_T \alpha \Delta t. \tag{5.4.4}$$

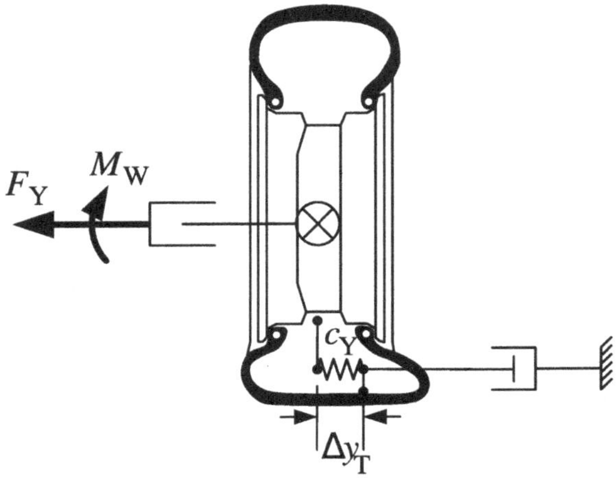


**Fig. 5.21** Lateral deformation of a tire due to lateral forces (Börner 2004)

Using the stiffness relation (5.4.1), one obtains

$$v_T \alpha \Delta t = \frac{\Delta F_{YT}}{c_Y}$$

and with $\Delta F_{YT} = c_\alpha \alpha$ the lateral force building up time becomes

$$\Delta t = T_{YT} = \frac{c_\alpha}{c_Y v_T}. \tag{5.4.5}$$

The time behavior for the lateral force $\Delta F_{YT}$ after a change of the side slip angle $\Delta\alpha$ is now approximated by a dynamic 1st order differential equation

$$T_{YT} \Delta \dot{F}_{YT}(t) + \Delta F_{YT}(t) = c_\alpha \Delta\alpha \tag{5.4.6}$$

with the transfer function

$$G_{TF}(s) = \frac{\Delta F_{YT}(s)}{\Delta\alpha(s)} = \frac{c_\alpha}{1 + T_{YT}s}. \tag{5.4.7}$$

For passenger cars with $c_\alpha = 80 \ldots 100$ kN/rad and $c_Y = 60 \ldots 200$ kN/rad, the time constant is for $v = 20$ m/s in the range of

$$T_{YT} = 0.02 \ldots 0.08\,\text{s}.$$

Considering (5.4.5) in the form

$$T_{YT} = \frac{l_{XT}}{v_T} \tag{5.4.8}$$

allows to interpret $l_{XT} = c_\alpha / c_Y$ as the initial settling run length of a tire, which is about 2/3 of the wheel circumference (Wang 1993; Mitschke and Wallentowitz

2014). With $l_{XT} = 4\pi r_{dyn}/3$, $r_{dyn} = 0.3\,m$, and $v = 20\,m/s$, it yields $T_{YT} = 0.063\,s$. For more details, see Einsle (2010).

In the context of the lateral vehicle behavior, this small time constant can usually be neglected.

## 5.5 Longitudinal Dynamic Wheel Models

The transfer of forces between the tire and a road surface is determined by the friction properties which depend on the slip as described in previous sections. For braking and straight driving, the longitudinal slip between the tire and the road is defined as

$$\begin{aligned} S_X &= \frac{\Delta v_X}{v_X} = \frac{\Delta v}{v} = \frac{v_X - \omega_W r_{dyn}}{v_X} \\ &= \left(1 - \frac{\omega_W r_{dyn}}{v_X}\right) \end{aligned} \tag{5.5.1}$$

with $\mu = \mu_X(S_X)$ such that $S_X > 0$ for braking; see (5.1.1). The braking force then becomes with (5.1.3) for one wheel

$$F_{XT} = \mu(S_X) F_{ZT} \tag{5.5.2}$$

with $\mu_X = \mu(S_X)$ as the friction coefficient depending on the longitudinal slip with typical characteristics according to Figs. 5.2 and 5.3. Different models can now be applied for approximating the $\mu(S_X)$ curves, as described in Sect. 5.1.

As the most simple approximation, the HRSI model (5.1.4) is now applied. The braking force in the contact patch is then for $S_X < S_{X,max}$

$$F_{XT} = c_{\mu 0} S_X F_{ZT}. \tag{5.5.3}$$

The braking force at the braking disk is according to (6.7.4)

$$F_B = F_F = 2 A_B \mu_B p_{Wcyl}, \tag{5.5.4}$$

the wheel/road traction torque

$$M_{XT} = F_{XT} r_{dyn}, \tag{5.5.5}$$

the braking torque

$$M_B = F_B r_B, \tag{5.5.6}$$

the driving torque, see (6.5.16),

$$M_{ds,out} = i_{tot} \eta_{tot} M_{eng} \tag{5.5.7}$$

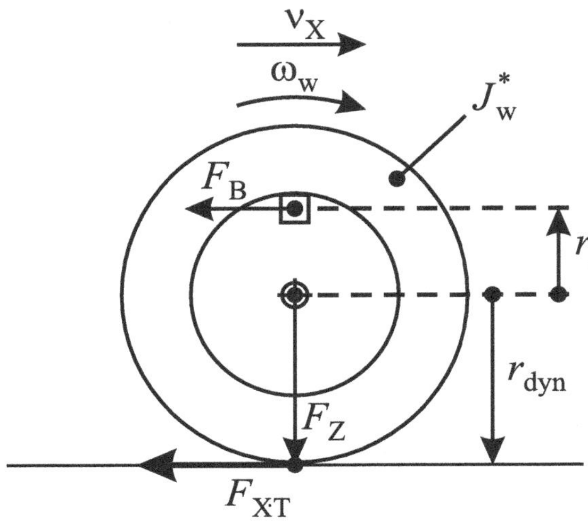


**Fig. 5.22** Wheel with braking forces

if the clutch is engaged, and the friction torque of the drive shaft and wheel bearings

$$M_{\mathrm{ds,F}} = d_{\mathrm{ds}} \omega_{\mathrm{W}} . \tag{5.5.8}$$

These terms lead to the torque balance equation of the braking wheel, compare Fig. 5.22,

$$J_{\mathrm{W}}^{*} \frac{d\omega_{\mathrm{W}}(t)}{dt} = M_{\mathrm{ds,out}} - M_{\mathrm{XT}}(t) - M_{\mathrm{B}}(t) - M_{\mathrm{ds,F}}(t) \tag{5.5.9}$$

where $J_{\mathrm{W}}^{*}$ is the moment of inertia. For a free rolling wheel, it is $J_{\mathrm{W}}^{*} = J_{\mathrm{W}}$ and in the case of a driving wheel according to (6.5.17) with opened clutch to the engine and assuming that for the other wheel of the driven axle it holds $\dot{\omega}_{\mathrm{W},2} = 0$, it is

$$J_{\mathrm{W}}^{*} = J_{\mathrm{tot}} = J_{\mathrm{W}} + i_{\mathrm{di}}^{2} \eta_{\mathrm{di}} J_{\mathrm{tr,di}} + i_{\mathrm{tr}}^{2} i_{\mathrm{di}}^{2} \eta_{\mathrm{tr}} \eta_{\mathrm{di}} J_{\mathrm{cl}} . \tag{5.5.10}$$

The dynamic behavior of the wheel now follows for braking with (5.5.5), (5.5.6), (5.5.2), and (5.5.8) with neglection of a driving torque

$$J_{\mathrm{W}}^{*} \frac{d\omega_{\mathrm{W}}(t)}{dt} - r_{\mathrm{dyn}} \mu_{\mathrm{X}}(S_{\mathrm{X}}(t)) F_{\mathrm{ZT}}(t) + d_{\mathrm{ds}} \omega_{\mathrm{W}}(t) = -M_{\mathrm{B}}(t) \tag{5.5.11}$$

and leads to the signal flow chart in Fig. 5.23 with $M_{\mathrm{B}}$ as input and $\omega_{\mathrm{W}}$ as output. It is characterized by the nonlinear feedback through the $\mu(S_{\mathrm{X}})$ curve. The feedback in Fig. 5.23 arises because of the slip-dependent tire/road braking force $F_{\mathrm{XT}}$; see (5.5.2). This feedback is negative if $\mathrm{d}\mu_{\mathrm{X}}/\mathrm{d}S_{\mathrm{X}} > 0$ and therefore the dynamic behavior of the braking wheel is stable. Hence, this part of the $\mu$-slip characteristic in Fig. 5.2 is called the stable range. If, however, $\mathrm{d}\mu_{\mathrm{X}}/\mathrm{d}S_{\mathrm{X}} < 0$ for $S_{\mathrm{X}} > S_{\mathrm{Xcrit}}$, the feedback

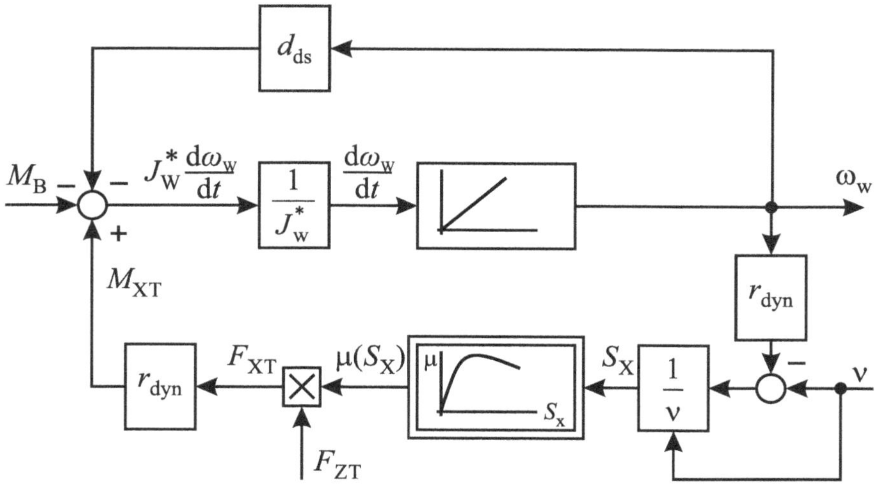


**Fig. 5.23** Signal flow chart for the angular speed $\omega_W$ of a braking wheel

becomes positive and the wheel dynamics are monotonic unstable. Applying the linear approximation of (5.1.4) it yields for small slips with (5.5.1)

$$J_W^* \frac{d\omega_W(t)}{dt} + \left( c_{\mu 0} F_{ZT}(t) \frac{r_{dyn}^2}{v(t)} + d_{ds} \right) \omega_W(t) - r_{dyn} c_{\mu 0} F_{ZT}(t) = -M_B(t) \tag{5.5.12}$$

which is a linear first-order differential equation with time-varying parameters.

Now, small deviations of the wheel velocity around an operation point for $v = \bar{v}$ are considered with

$$\omega_W = \bar{\omega}_W + \Delta\omega_W; \; M_B = \bar{M}_B + \Delta M_B; \; F_{ZT} = \bar{F}_{ZT} + \Delta F_{ZT}.$$

Then, (5.5.12) becomes by neglecting the small friction torque (5.5.8) and therefore $d_{ds} = 0$ and assuming a constant vertical force and therefore $\Delta F_{ZT} = 0$

$$J_W^* \frac{d\omega_W(t)}{dt} + c_{\mu 0} \bar{F}_{ZT} \frac{r_{dyn}^2}{\bar{v}} \Delta\omega_W(t) = -\Delta M_B(t), \tag{5.5.13}$$

and a simplified signal flow as depicted in Fig. 5.24 can be drawn. Again, a negative feedback and therefore stable behavior are obtained for $c_{\mu 0} > 0$ and monotonic unstable behavior for $c_{\mu 0} < 0$.

Compacting the parameters results in

$$T_{WB} \frac{d\omega_W(t)}{dt} + \Delta\omega_W(t) = K_{WB} \Delta M_B(t) \tag{5.5.14}$$

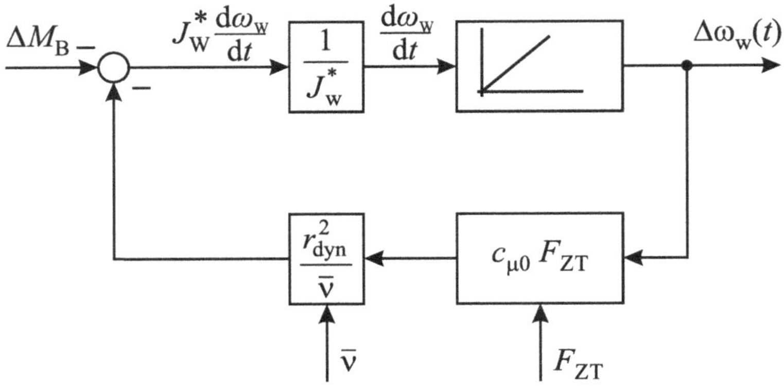


**Fig. 5.24** Simplified signal flow for the linearized braking wheel dynamics

or a first-order transfer function

$$G_{MB\omega}(s) = \frac{\Delta\omega_W(s)}{\Delta M_B(s)} = \frac{K_{WB}}{1 + T_{WB}s}. \tag{5.5.15}$$

Hence, a linear first-order behavior results in braking wheel gain and time constant

$$\begin{aligned} K_{WB} &= -\frac{\bar{v}}{c_{\mu 0}\bar{F}_{ZT} r_{dyn}^2} \\ T_{WB} &= \frac{J_W^* \bar{v}}{c_{\mu 0}\bar{F}_{ZT} r_{dyn}^2}. \end{aligned} \tag{5.5.16}$$

Both parameters are larger, the larger the vehicle velocity $\bar{v}$ and the smaller the product $c_{\mu 0}\bar{F}_{ZT}$. Therefore, the control behavior for the wheel speed depends not only on the gradient $c_{\mu 0}$ of the friction coefficient characteristic $\mu(S_X)$ but also on the product of $c_{\mu 0}$ and the vertical vehicle wheel force $F_{ZT}$. However, the initial acceleration of the wheel speed is

$$\left.\frac{d\omega_W(t)}{dt}\right|_{t\to 0} = -\frac{1}{J_W^*}\Delta M_B(t) \tag{5.5.17}$$

and depends therefore only on the moment of inertia $J_W^*$ of the wheel resp. wheel and drive train. If the vertical wheel force $F_{ZT}(t)$ in dependence on the braking acceleration $a_X$ is taken into account, as described in Sect. 6.5.4, the parameters (5.5.16) change with time. These dynamic wheel models are used for modeling the longitudinal behavior with the inclusion of the slip behavior in Sect. 6.7.2 and are a basis for ABS control, treated in Chap. 13.

## 5.6 Tire Forces for Aquaplaning

Aquaplaning (hydroplaning) of rolling tires occur if a layer of water builds up between the tire and the road surface. This phenomenon depends on the water thickness (WT), the speed $v$, tread depth (TD), and road texture. The grooves of a rubber tire are designed to disperse water from beneath the tire to result in high friction even in wet conditions. However, if a tire encounters more water as it can dissipate through the profile grooves, a water wedge arises between the tire and the road surface and the wheel loses friction forces in the longitudinal and lateral directions.

Detailed measurements with a special traction trailer are described in Klempau (2004). Figure 5.25 depicts the maximal friction coefficient in dependence on the water thickness for different speeds. For low speeds $v \leq 30$ km/h, the influence is small. However, for higher speeds $v \approx 60\ldots80$ km/h, the friction coefficient is $20\ldots40$ % smaller if WT = 1 mm and 50–80 % smaller if WT = 8 mm, for the investigated tire with a small tread depth TD = 2 mm.

The maximal lateral friction force in dependence on the speed for different water thickness is shown in Fig. 5.26. For a tire with small tread depth, it shows the strong loss of friction, about −25 % for WT = 1 mm and −55 % for WT = 7 mm at $v =$ 100 km/h.

For a large tread depth TD = 8 mm (new tire) and a water thickness for WT = 3 mm, the decreasing friction with increasing speed is far less significant (Klempau 2004). A medium tread depth of TD = 4 mm leads for $v = 90$ km/h to a loss of maximal friction in longitudinal direction of about −60 % for WT = 3 mm and in lateral direction −30 % for WT = 3 mm and −50 % for WT = 7 mm.

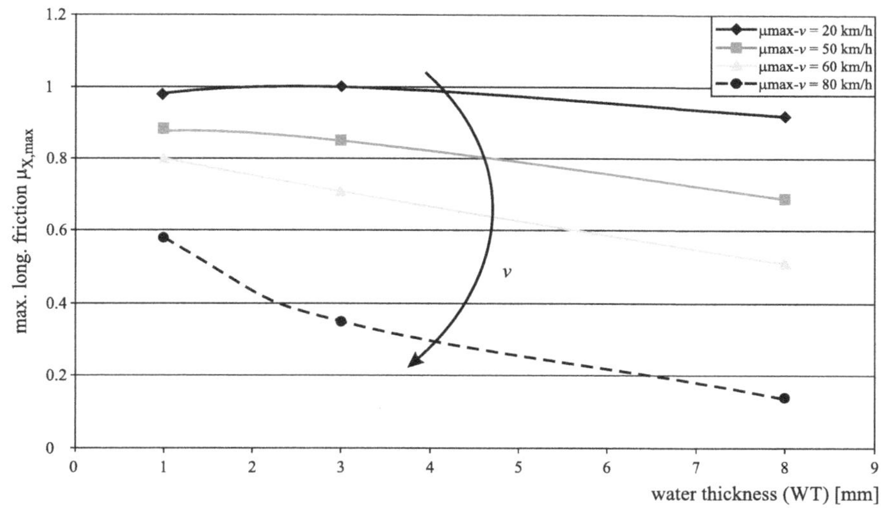


**Fig. 5.25** Maximal friction coefficient $\mu_{max}$ in longitudinal direction depending on water thickness (WT) and speed $v$ (Klempau 2004). (Tread depth TD = 2 mm). Tire 195/65 R15, summer compound. $F_Z = 5000$ N

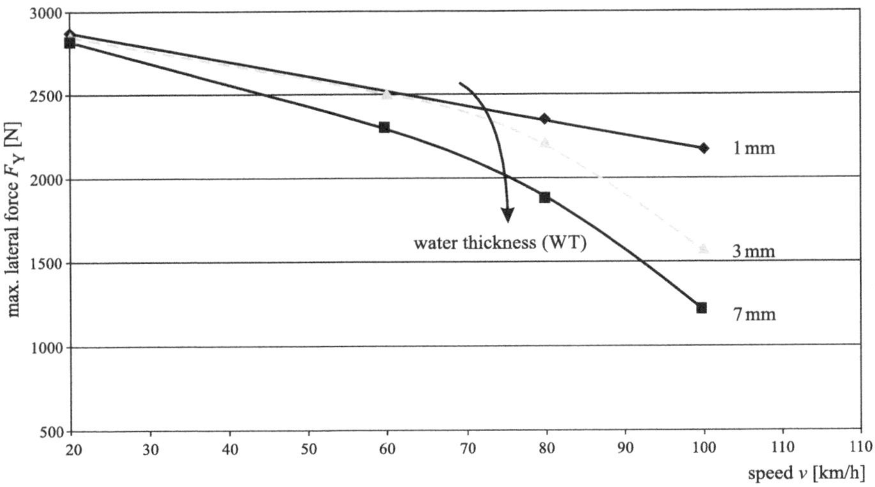


**Fig. 5.26** Maximal lateral force $F_{Y,max}$ depending on water thickness (WT) (Klempau 2004). (Tread depth TD = 2 mm). Tire: 185/65 R15, summer compound, $F_Z = 3500\,N$

A large influence on these values has also the roughness of the road itself, expressed by micro- and macro-texture. Mathematical models are treated in Gengenbach (1967) and Herrmann (2008).

## References


Ammon D (1997) Modellbildung und Systementwicklung in der Fahrzeugdynamik, 1st edn. Teubner, Stuttgart, B.G

Bauer M (2015) Methoden zur modellbasierten Fahrdynamikanalyse und Bewertung von Fahrdynamikregelsystemen. Dissertation Technische Universität Darmstadt. Fortschr.-Ber. VDI Reihe 12, 792. VDI Verlag, Düsseldorf

Bechtloff J (2018) Schätzung des Schwimmwinkels und fahrdynamischer Parameter zur Verbesserung modellbasierter Fahrdynamikregelungen. Dissertation Technische Universität Darmstadt. Fortschr.-Ber. VDI Reihe 12, 809. VDI Verlag, Düsseldorf

Börner M (2004) Adaptive Querdynamikmodelle für Personenkraftfahrzeuge - Fahrzustandserkennung und Sensorfehlertoleranz. Dissertation Technische Universität Darmstadt. Fortschr.-Ber. VDI Reihe 12, Nr. 563. VDI Verlag, Düsseldorf

Breuer B, Bill K (eds) (2017) Bremsenhandbuch: Grundlagen, Komponenten, Systeme, Fahrdynamik, 5th edn. Springer, Berlin

Burckhardt M (1993) Fahrwerktechnik: Radschlupf-Regelsysteme. Vogel-Verlag, Würzburg

Daiss A (1996) Beobachtung fahrdynamischer Zustände und Verbesserung einer ABS-und Fahrdynamikregelung. Diss. Universität Karlsruhe, Fortschr.-Ber. VDI Reihe 12, 283. VDI Verlag, Düsseldorf

Dugoff H, Fancher P, Segel L (1969) Tire performance characteristics affecting vehicle response to steering and braking control inputs, Highway Safety Research Institute, University of Michigan, Ann Arbor, National Bureau of Standards Control CST-460

Einsle S (2010) Analyse und Modellierung des Reifenübertragungsverhaltens bei transienten und extremen Fahrmanövern. Dissertation TU Dresden. Fakultät Verkehrswissenschaften, Dresden

Ersoy M, Gies S (eds) (2017) Fahrwerkhandbuch: Grundlagen, Fahrdynamik, Komponenten, Elektronische Systeme, Fahrerassistenz, Autonomes Fahren, Perspektiven, 5th edn. ATZ/MTZ Fachbuch, Springer Vieweg

Février P, Hague O, Schick B, Miquet C (2010) Advantages of a thermomechanical tire model for vehicle dynamics. ATZ Worldwide 112(7–8):33–37

Gengenbach W (1967) Das Verhalten von Kraft-Fahrzeugreifen auf trockener und nasser Fahrbahn. Dissertation. Institute for Transport Studies, University of Karlsruhe, Karlsruhe

Germann S (1997) Modellbildung und modellgestützte Regelung der Fahrzeuglängsdynamik. Dissertation Technische Universität Darmstadt. Fortschr.-Ber. VDI Reihe 12, 309. VDI Verlag, Düsseldorf

Halfmann C, Holzmann H (2003) Adaptive Modelle für die Kraftfahrzeugdynamik. VDI-Buch, Springer, Berlin

Heissing B, Ersoy M (eds) (2011) Fahrwerkhandbuch: Grundlagen, Fahrdynamik, Komponenten, Systeme, Mechatronik. Perspektiven, ATZ/MTZ Fachbuch, Vieweg

Herrmann S (2008) Simulation zum Wasserabfluss- und Aquaplaning-Verhalten auf Fahrbahnoberflächen. Dissertation. Institut für Straßen- und Verkehrswesen, Universität Stuttgart, Stuttgart

Isermann R (ed) (2006) Fahrdynamik-Regelung. Vieweg-Verlag, Wiesbaden, Wiesbaden

ISO 8855, (2013) Road vehicles - Vehicle dynamics and road-holding ability - Vocabulary. International Organization for Standardization, Geneva

Kiencke U, Nielsen L (2005) Automotive control systems: for engine, driveline, and vehicle, 2nd edn. Springer, Berlin

Klempau F (2004) Untersuchungen zum Aufbau eines Reibwertvorhersagesystems im fahrenden Fahrzeug. Dissertation Technische Universite Darmstadt. Fortschr.-Ber. VDI Reihe 12, 576. VDI Verlag, Düsseldorf

Kummer H, Meyer W (1967) Verbesserter Kraftschluß zwischen Reifen und Fahrbahn. Automobiltechnische Zeitschrift – ATZ pp 245–257 & 382–386

Lugaro C, Huismann S, van Putten S (2018) Optimized modeling of combined tire slip. ATZ Worldwide 120(6):18–25

Mitschke M, Wallentowitz H (2014) Dynamik der Kraftfahrzeuge, 5th edn. Springer, Berlin

Pacejka H (2012) Tyre and vehicle dynamics, 3rd edn. Butterworth Heinemann, Burlington

Reimpell J, Sponagel P (1995) Fahrwerktechnik: Reifen und Rder. Vogel, Würzburg

Schramm D, Hiller M, Bardini R (2010) Modellbildung und Simulation der Dynamik von Kraftfahrzeugen. Springer, Berlin

Sequenz H (2013) Emission modelling and model-based optimisation of the engine control. Dissertation Technische Universität Darmstadt. Fortschr.-Ber. VDI Reihe 12, 1222. VDI Verlag, Düsseldorf

Wang J (1993) A simulation model for the dynamic sideslip behavior of vehicles. Dissertation, Uni Karlsruhe

Würtenberger M (1997) Modellgestützte Verfahren zur Überwachung des Fahrzustandes eines Pkw. Diss. Universität Darmstadt, Fortschr.-Ber. VDI Reihe 12, 314, VDI Verlag, Düsseldorf

Zomotor A, Reimpell J (1991) Fahrwerktechnik: Fahrverhalten, 2nd edn. Vogel Buchverlag, Würzburg

Zomotor A, Braess HH, Rönitz R (1998) Verfahren und Kriterien zur Bewertung des Fahrverhaltens von Personenkraftwagen. ATZ-Automobiltechnische Zeitschrift 100(3):236–243

Zomotor Z (2002) Online-Identifikation der Fahrdynamik zur Bewertung des Fahrverhaltens von PKW. Dissertation Universität Stuttgart. Institut A für Mechanik, Stuttgart

# Longitudinal Vehicle Behavior 6

The longitudinal behavior of automobiles is a characteristic property which all drivers recognize immediately during acceleration, continuous driving, coasting, and braking. As will be shown in this chapter, many components from the combustion engine or electric motor through the torque converter and elastic drive train to the driven wheels act together. In order to understand their interactions and for the design of automatic control systems like anti-locking control (ABS), automatic velocity control (AVC), or adaptive cruise control (ACC) or automatic driving in general, static and dynamic process models are required. After summarizing the corresponding drive train components, the respective models are derived and compiled together with different kinds of granularity.

## 6.1 Vehicle Components for the Longitudinal Behavior of Vehicles

The drive trains of vehicles provide the forces and torques which are required for the movement. They consist, after the combustion engine, of a clutch, a manually or double-clutch shifted transmission or a converter-type automatic transmission, the drive shaft, the differential, the wheel shafts, and the wheels. Compare the schematic in Fig. 6.1 for a rear wheel-driven car. These drive train components are rotational multi-mass systems with different moments of inertia, stiffness, and damping coefficients; see Fig. 6.2. The design, simulation, and control of the driving behavior of a vehicle in a longitudinal direction require static and dynamic models of the drive train. They can be represented in the form of two-port systems from the internal combustion engine to the wheels and the chassis; see Fig. 6.3. These two-port systems operate with generalized power variables at their terminals, an "effort variable" $e$ like a force or torque, and a "flow variable" $f$ like a translatory velocity or



R. Isermann, *Automotive Control*, ATZ/MTZ-Fachbuch,
https://doi.org/10.1007/978-3-642-39440-9_6

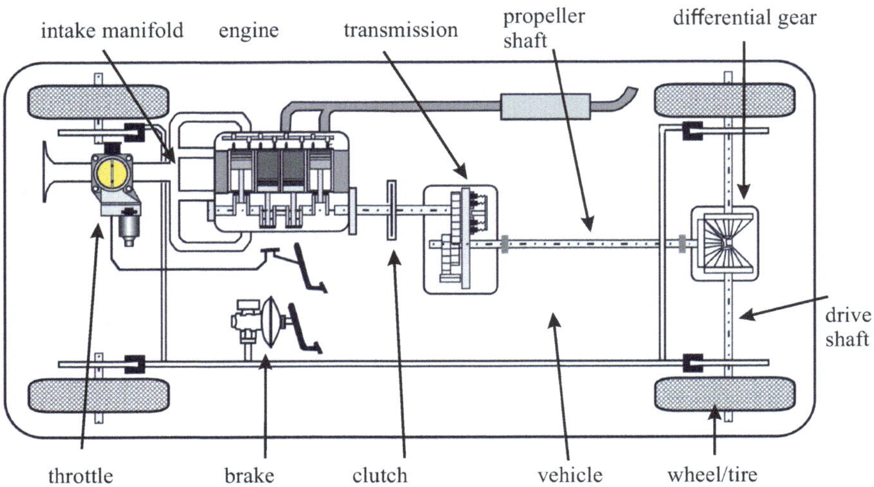


**Fig. 6.1** Schematic of a rear wheel-driven vehicle for driving and braking

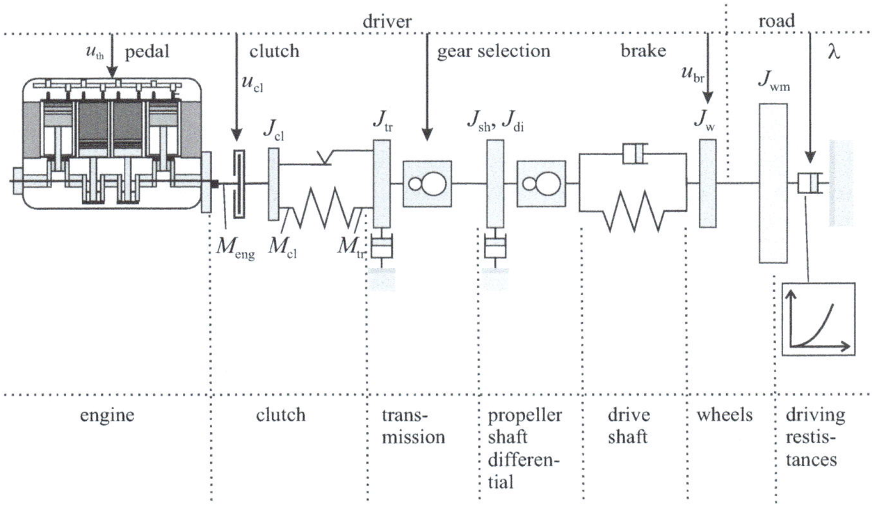


**Fig. 6.2** Schematic of a vehicle drive train (simplified)

angular velocity. Their product is a power $P = ef$; see, e.g. Karnopp and Rosenberg (1975), Karnopp et al. (1990), MacFarlane (1970), and Isermann (2005). One of these variables is an applied variable like the torque of the engine, the other is a dependent variable, like the angular engine velocity. Through chaining of these two-port elements, the overall behavior can be determined.

As the eigenfrequencies of these components differ by one or two orders of magnitude, some components can be lumped together, dependent on the application, thus leading to more simpler models. Also the brake system from the brake booster to the

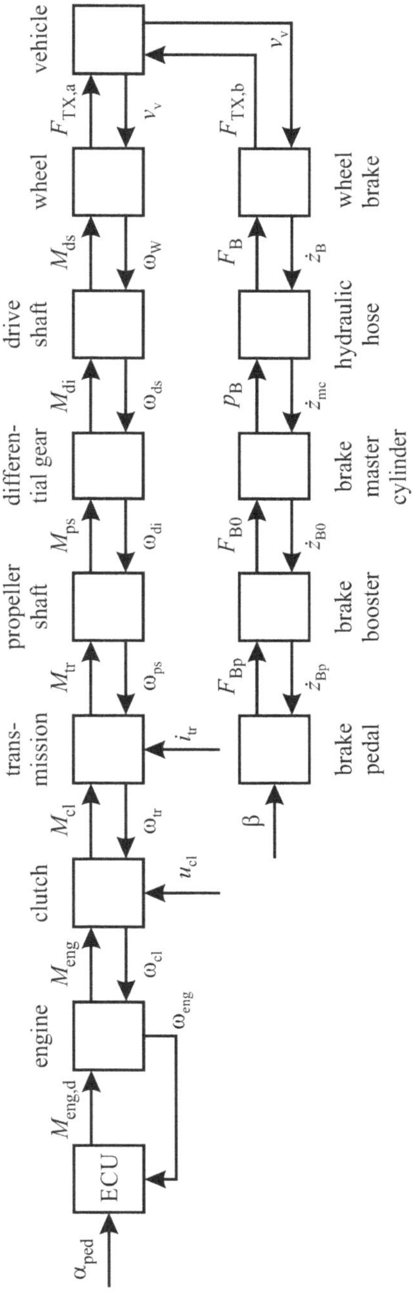


**Fig. 6.3** Drive train of a vehicle with two-port representation of the components

brakes and wheels can be presented as two-port systems, as illustrated in Fig. 6.3. The following sections treat the drive train components for modeling the longitudinal behavior, both for accelerating and driving and for braking. The variables and symbols in Sects. 6.2–6.4 are taken from Isermann (2014).

## 6.2 Internal Combustion Engine Models

### 6.2.1 Gasoline Engine

The generated torque of a gasoline engine at the clutch $M_{\mathrm{eng}}$ follows a signal flow as depicted in Fig. 6.4. The engine control system has a torque-oriented structure and distinguishes between an inner torque $M_{\mathrm{i}}$ in dependence on the cylinder charge (air mass) $m_{\mathrm{air}}$, and engine speed $n_{\mathrm{eng}}$ and a drag torque $M_{\mathrm{drg}}$ in dependence on cycle losses, friction, and engine-integrated auxiliary drives; see Robert Bosch GmbH (2018) and Isermann (2014). The mean value torque of the gasoline engine at the clutch is then

$$\overline{M}_{\mathrm{eng}}(m_{\mathrm{air}}, n_{\mathrm{eng}}, \lambda, \varphi_{\mathrm{ign}}) = \overline{M}_{\mathrm{i}}(m_{\mathrm{air}}, n_{\mathrm{eng}}, \lambda, \varphi_{\mathrm{ign}}) - \overline{M}_{\mathrm{drg}}(n_{\mathrm{eng}}). \quad (6.2.1)$$

The demanded inner torque from the driver, the transmission or the vehicle velocity control, or from demanded internal torques is the input to a torque coordinator, where the desired torque $M_{\mathrm{i,d}}$ is determined by priority selection. A torque conversion then calculates the required air mass $m_{\mathrm{air}}$ and throttle position $\alpha_{\mathrm{th}}$ by an inversion of the inner torque model. The injected fuel quantity $m_{\mathrm{f}}(m_{\mathrm{air}}, \lambda)$ is directly determined from the air mass for a given excess-air factor $\lambda$. Changes of the throttle position $\alpha_{\mathrm{th}}$ show a delay in the engine torque of some 100 ms because of the air mass storage in the manifold, and result therefore in a slower torque development. The fuel injection angle and corresponding ignition spark advance angle can be realized crank angle synchronously and influence the torque within some 10 ms. Hence, the engine torque can be manipulated by the throttle with slower dynamics, determining the torque in a larger time range, and it can be changed by ignition and/or injection timing in a short time range, e.g. reduced for gear shifting. The torque conversion block combines these actuator commands. For more details, see Isermann (2014), Chap. 7.

Figure 6.5 illustrates a simplified torque generation model. The main delay is caused by the intake manifold dynamics. Detailed models of the air path system are derived in Guzzella and Onder (2010) and Isermann (2014). The time constant of the intake manifold is described by

$$T_{\mathrm{int}} = \frac{2V_{\mathrm{int}}}{\eta_v V_{\mathrm{D}} n_{\mathrm{eng}}} \quad (6.2.2)$$

with $V_{\mathrm{int}}$ the intake manifold volume, $\eta_v$ the volumetric efficiency of the cylinder group, and $V_{\mathrm{D}}$ the displacement. Hence, $T_{\mathrm{int}}$ is the smaller $V_{\mathrm{int}}$ and the larger $n_{\mathrm{eng}}$. The position control of the electrical throttle valve introduces a further delay with a

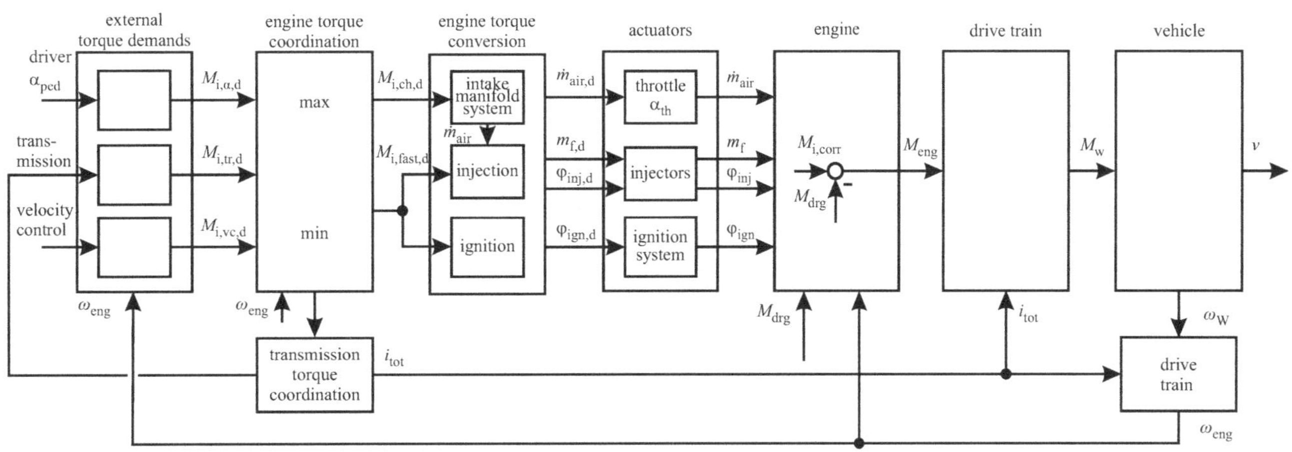


**Fig. 6.4** Signal flow for the torque generation of a naturally aspirated gasoline engine (powertrain model)

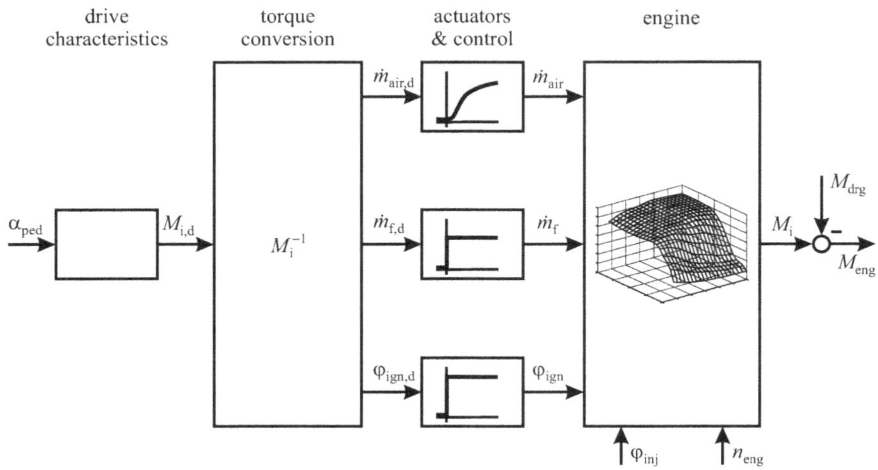


**Fig. 6.5** Simplified torque model for a naturally aspirated gasoline engine

time constant $T_{th}$. The fuel mass injection and ignition angle act only with negligible dynamics and are therefore assumed as proportional acting. A further delay may come from a dead time $T_d$ which results from ECU calculation times and a cycle time, depending on the change of the injection after a command signal, which is larger for port injection than for direct injection.

The inner torque generation of a gasoline engine without supercharging can therefore be approximated by a nonlinear map

$$\overline{M}_i(\dot{m}_{air}, n_{eng}) \tag{6.2.3}$$

dependent on optimal calibrated other manipulated variables like $\lambda$ and $\varphi_{ign}$. Linearization around an operating point $(\overline{M}_{eng}, \overline{n}_{eng})$ results for the airflow as output in the transfer function

$$G_{mm}(s) = \frac{\Delta\alpha_{th,d}(s)}{\Delta\dot{m}_{air,d}(s)} \cdot \frac{\Delta\dot{m}_{air}(s)}{\Delta\alpha_{th}(s)} \approx \frac{\Delta\dot{m}_{air}(s)}{\Delta\dot{m}_{air,d}(s)} = \frac{K_{th}K_{int}}{(1+T_{th}s)(1+T_{int}s)}e^{-T_d s} \tag{6.2.4}$$

and for the torque

$$G_{mM}(s) = \frac{\Delta M_i(s)}{\Delta\dot{m}_{air}(s)} \cdot \frac{\Delta\dot{m}_{air}(s)}{\Delta\dot{m}_{air,d}(s)} = \frac{\Delta M_i(s)}{\Delta\dot{m}_{air,d}(s)} = \frac{K_{mM}(\overline{M}_{eng}, n_{eng})}{(1+T_{th}s)(1+T_{int}s)}e^{-T_d s} \tag{6.2.5}$$

with $T_{th} \approx 50...100$ ms and $T_{int} \approx 50...500$ ms; compare Fig. 6.5.

Usually, the engine torque is given as a lookup table in the form of $M_{eng}(\alpha_{th}, n_{eng})$; see Fig. 6.6. The inner torque then follows due to (6.2.1) to

$$M_i(\alpha_{th}, n_{eng}) = M_{eng}(\alpha_{th}, n_{eng}) + M_{drg}(n_{eng}). \tag{6.2.6}$$

With $\alpha_{\mathrm{th}}$ as input, (6.2.4) becomes

$$G_{\mathrm{m}}(s) = \frac{\Delta \dot{m}_{\mathrm{air}}}{\Delta \alpha_{\mathrm{th}}(s)} = \frac{K_{\mathrm{m}}(\overline{M}_{\mathrm{i}}, n_{\mathrm{eng}})}{(1 + T_{\mathrm{th}}s)(1 + T_{\mathrm{int}}s)} e^{-T_{\mathrm{d}}s} \tag{6.2.7}$$

and instead of (6.2.5), it follows

$$G_{\mathrm{M}}(s) = \frac{\Delta M_{\mathrm{i}}(s)}{\Delta \alpha_{\mathrm{th}}(s)} = \frac{K_{\mathrm{M}}(\overline{M}_{\mathrm{i}}, n_{\mathrm{eng}})}{(1 + T_{\mathrm{th}}s)(1 + T_{\mathrm{int}}s)} e^{-T_{\mathrm{d}}s}. \tag{6.2.8}$$

Hence, the torque of a naturally aspirated gasoline engine can be described by second-order dynamics with a dead time.

### 6.2.2 Diesel Engine

The control system of diesel engines has also a torque-oriented structure. Different from naturally aspirated gasoline engines, a turbocharger is applied in a closed loop controlling the charging pressure $p_2$, and the air mass flowrate $\dot{m}_{\mathrm{air}}$ is controlled by manipulating the exhaust gas recirculation flow $\dot{m}_{egr}$ with the EGR-valve. These control loops lead to additional dynamics and make the torque response relatively fast. The mean value torque at the clutch is

$$\begin{aligned} &\overline{M}_{\mathrm{eng}}(m_{\mathrm{air}}, m_{\mathrm{egr}}, p_2, m_{\mathrm{f}}, \varphi_{\mathrm{inj}}, n_{\mathrm{eng}}) = \\ &\quad \overline{M}_{\mathrm{i}}(m_{\mathrm{f}}, \varphi_{\mathrm{inj}}, n_{\mathrm{eng}}) - M_{\mathrm{drg}}(n_{\mathrm{eng}}). \end{aligned} \tag{6.2.9}$$

As shown in Fig. 6.7, the external torque demands determine the demanded inner torques $M_{\mathrm{i,d}}$ from the accelerator pedal, the transmission, or the velocity control with a priority selection in the torque coordination block. An inversion of the inner torque lookup table in the torque conversion block then yields the desired injected fuel mass per stroke $m_{\mathrm{f,d}}$ and calibrated values for the desired values of the air mass $m_{\mathrm{air,d}}$, charging pressure $p_{2,\mathrm{d}}$, injection angle $\varphi_{\mathrm{inj,d}}$, and rail pressure $p_{\mathrm{rail,d}}$.

The air mass flow control loop with the manipulated exhaust gas recirculation flow and the charging pressure control loop with the manipulated turbocharger (either variable turbine or wastegate) then determine the dynamics of the cylinder air charge and the coordinated injected fuel mass. Both control loops are coupled, mutually. For more details, see Isermann (2014).

Figure 6.8 illustrates a simplified structure of the torque model. The generated inner torque can be represented by a nonlinear map

$$\overline{M}_{\mathrm{i}}(m_{\mathrm{f}}, n_{\mathrm{eng}}) \tag{6.2.10}$$

dependent on optimal calibrated manipulated variables, like $p_2$, $p_{\mathrm{rail}}$, $\varphi_{\mathrm{inj}}$, $\dot{m}_{\mathrm{egr}}$.

The delays of the torque generation are caused by the dynamics of the air path including the turbocharger expressed with a time constant $T_{\mathrm{int}}$ for the intake volume

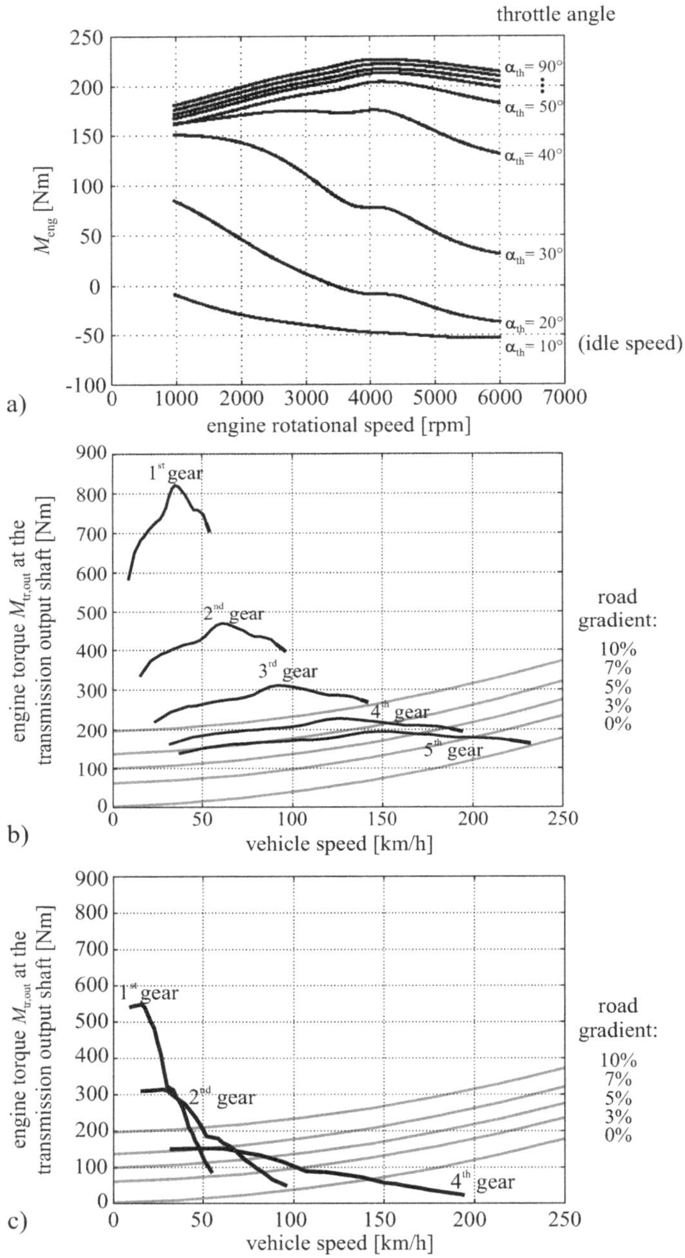


**Fig. 6.6** Torque characteristics of a six-cylinder gasoline engine: (a) torque at the clutch of a six-cylinder spark-ignition engine, as a function of the crankshaft rotational speed for various throttle angles $\alpha_{th}$; (b) torque characteristics at the transmission output shaft at full load ($\alpha = 90°$); (c) torque characteristics at the transmission output shaft at partial load ($\alpha = 30°$)

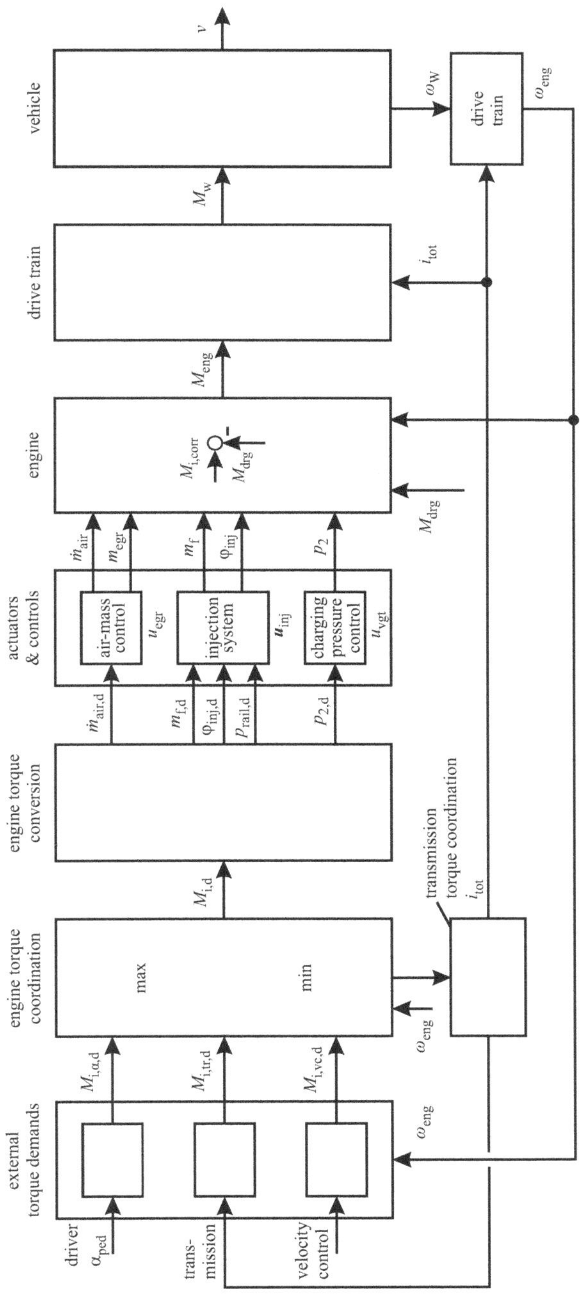


**Fig. 6.7** Signal flow for the torque generation of a diesel engine with turbocharger (powertrain model)

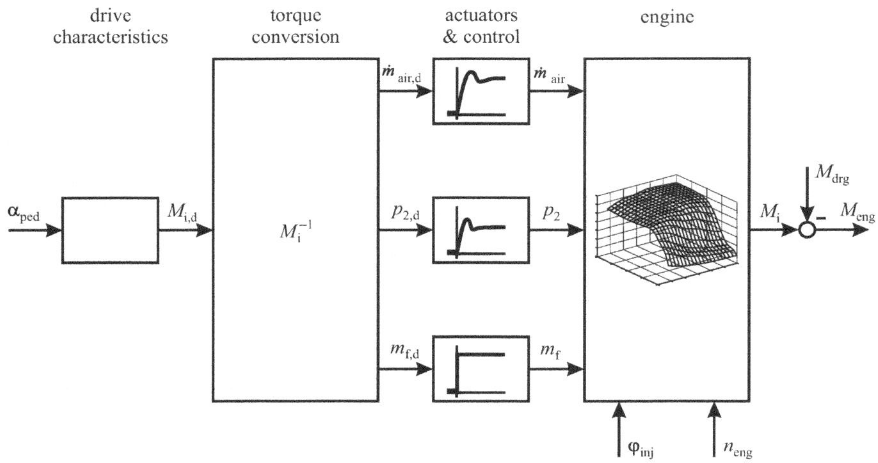


**Fig. 6.8** Simplified torque model for a turbo-charged diesel engine

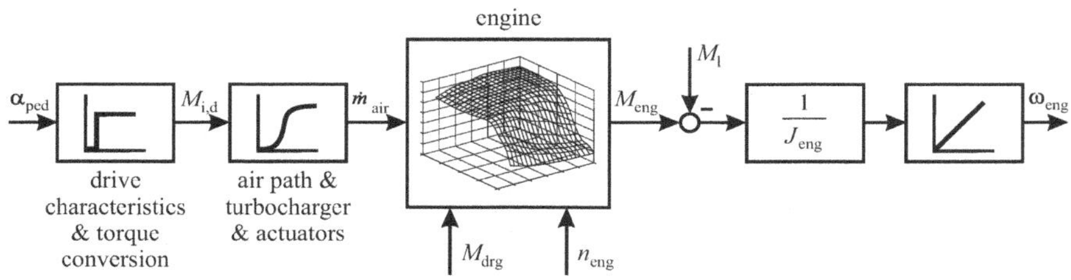


**Fig. 6.9** Simplified overall torque model for gasoline and diesel engine

and the turbocharger and $T_{act}$ for the turbocharger actuator. The transfer function can be approximated for a linearization around an operation point $(\overline{M}_{eng}, \overline{n}_{eng})$ by first-order elements

$$G_{Mm}(s) = \frac{\Delta m_{air}(s)}{\Delta M_{i,d}(s)} = \frac{K_{Mm}}{(1 + T_{act}s)(1 + T_{int}s)} e^{-T_d s}. \tag{6.2.11}$$

The dead time $T_d$ may result from the time of one cycle (e.g. 40 ms for 3000 rpm). ECU calculation time and can frequently be neglected. The inner torque at the output becomes

$$G_{MM}(s) = \frac{\Delta M_i(s)}{\Delta M_{i,d}(s)} = \frac{K_{MM}}{(1 + T_{act}s)(1 + T_{int}s)} e^{-T_d s}. \tag{6.2.12}$$

The time constants are for a four-cylinder diesel engine with $V_D = 1.91$ approximately $T_{act} \approx 50$ ms and $T_{int} \approx 0.2\ldots1$ s for $n_{eng} = 1800\ldots3000$ rpm; see Isermann (2014). The engine torque $M_{eng}$ follows with subtraction of the drag torque according to (6.2.9).

Comparing Figs. 6.5 and 6.8 shows that a general overall torque model for combustion engines can be obtained as depicted in Fig. 6.9. The nonlinear steady-state behavior of the torque is described by multi-dimensional maps

$$M_{\text{eng}}(\boldsymbol{u}_{\text{eng}}, n_{\text{eng}}, M_{\text{drg}}) \tag{6.2.13}$$

where the input vector is

$$\boldsymbol{u}_{\text{eng,gasoline}} = [\dot{m}_{\text{air}}, m_{\text{f}}, \varphi_{\text{ign}}, \varphi_{\text{inj}}] \tag{6.2.14}$$

$$\boldsymbol{u}_{\text{eng,diesel}} = \left[\dot{m}_{\text{air}}, m_{\text{f}}, p_2, \varphi_{\text{inj}}\right]. \tag{6.2.15}$$

The desired inner torque $M_{\text{i,d}}$ follows the pedal position $\alpha_{\text{ped}}$ according to the implemented drive characteristics and the torque conversion without delay. The main dynamic delays stem from the air path volume storage, inertia of the turbocharger, and actuator delays and can be approximated by the behavior of the delayed airflow $\dot{m}_{\text{air}}$ as a representative variable. Applying the dynamic models (6.2.4) and (6.2.11) leads to

$$G_{\text{m}}(s) = \frac{\Delta \dot{m}_{\text{air}}(s)}{\Delta \alpha_{\text{ped}}(s)} = \frac{K_{\text{m}}}{(1 + T_{\text{act}}s)(1 + T_{\text{int}}s)} e^{-T_{\text{d}}s}. \tag{6.2.16}$$

$\dot{m}_{\text{air}}$ is then a representative dynamic input to the steady-state map (6.2.13). $T_{\text{act}}$ is either the time constant of an electrical throttle for gasoline engines or of the turbocharger actuator for diesel engines. After linearization around an operating point $(\overline{M}_{\text{eng}}, \overline{n}_{\text{eng}})$, one obtains for the dynamic torque generation

$$G_{\text{M}}(s) = \frac{\Delta M_{\text{eng}}(s)}{\Delta \alpha_{\text{ped}}(s)} = G_{\text{m}}(s) K_{\text{mM}} \tag{6.2.17}$$

with

$$K_{\text{mM}} = \left. \frac{\Delta M_{\text{eng}}}{\Delta \dot{m}_{\text{air}}} \right|_{\overline{M}_{\text{eng}}, \overline{n}_{\text{eng}}}. \tag{6.2.18}$$

The time constants are for gasoline engines (4 cyl.) $T_{\text{act}} \approx 50...100$ ms, $T_{\text{int}} \approx 50...500$ ms, and $T_{\text{d}} \approx 30...100$ ms, and for diesel engines (4 cyl.) $T_{\text{act}} \approx 50$ ms, $T_{\text{int}} \approx 200...1000$ ms, and $T_{\text{d}} \approx 30...100$ ms. $T_{\text{int}}$ depends on the speed and $T_{\text{d}}$ on the speed and number of cylinders.

If the relatively small dead time $T_{\text{d}}$ can be omitted, the engine model (6.2.16) may be represented by second order

$$G_{\text{M}} = \frac{\Delta M_{\text{eng}}(s)}{\Delta \alpha_{\text{ped}}(s)} = \frac{K_{\text{M}}}{(1 + T'_{\text{act}}s)(1 + T_{\text{int}}s)} \tag{6.2.19}$$

with an increased time constant

$$T'_{\text{act}} = T_{\text{act}} + T_{\text{d}}$$

for the approximation of (6.2.16) with $T'_{act} \approx 80...150$ ms for diesel engines and the gain

$$K_M(\overline{M}_{eng}, \overline{n}_{eng}) = \frac{\Delta M_{eng}}{\Delta \alpha_{ped}} = K_M K_{mM}, \tag{6.2.20}$$

which follows from the engine lookup table.

Germann (1997) applied a parameter estimation for a six-cylinder gasoline engine with $V_D = 2.5$ l and 125 kW/265 Nm and obtained $T'_{act} = 50$ ms and $T_{int} = 250$ ms and $T'_{act} = 70$ ms and $T_{int} = 280$ ms for a six-cylinder engine with $V_D = 3$ l and 165 kW/265 Nm. A comparison with the model (6.2.16) including dead time shows a good agreement for $\omega \leq 30$ 1/s or $f \leq 5$ Hz of the frequency responses.

For applications where only the very low-frequency part is of interest, the model (6.2.19) may be further reduced to first order

$$G_M = \frac{\Delta M_{eng}(s)}{\Delta \alpha_{ped}(s)} = \frac{K_M}{(1 + T''_{act} s)} \tag{6.2.21}$$

with

$$T''_{act} = T_{int} + T_{act} + T_d.$$

The torque $M_{eng}$ acts on the crankshaft and flywheel of the engine. An angular momentum balance yields

$$J_{eng} \frac{d\omega_e(t)}{dt} = 2\pi J_{eng} \frac{dn_{eng}(t)}{dt} = M_{eng}(n_{eng}, t) - M_l(n_{eng}, t) \tag{6.2.22}$$

with $J_{eng}$ the moment of inertia of the crankshaft, pistons and flywheel, and $M_l$ a counteracting load torque; see Fig. 6.9. The load torque is either the clutch torque $M_l = M_{cl,cd}$ from a shifted transmission, see Sect. 6.3.1, or the impeller torque $M_l = M_p$ from a hydrodynamic torque converter; see Sect. 6.4. Thus, the acceleration $d\omega_{eng}/dt$ of the engine is proportional to the difference between the engine and the load torque. $M_{eng}$ acts as an applied variable on the drive train and the angular engine velocity $\omega_{eng} = 2\pi n_{eng}$ is a dependable variable, based on the feedback of the driving resistances from the vehicle and drive train, as discussed in the next section.

## 6.3 Drive Train with Friction Clutch and Shifted Transmission

For passenger cars, mainly two kinds of transmissions can be distinguished:

- manually shifted and automatic double-clutch transmissons with spur gears;
- automatic transmissions with torque converters and planetary gears.

The manually shifted transmissions require a friction clutch as the connecting component between the flywheel and the gear sets. One distinguishes dry-plate friction

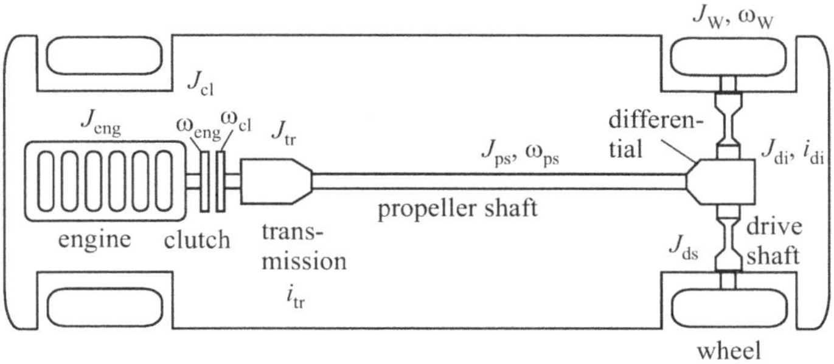


**Fig. 6.10** Scheme of a vehicle with rotational masses of the powertrain

clutches as a standard solution for manually shifted transmissions and wet-plate friction clutches, which are used for double-clutch transmissions.

Automatic transmissions have generally a hydraulic torque converter followed by hydraulically shifted planetary gear sets.

In the following, a manual or automatic gear shift transmission is considered. It is assumed that the components of the drive train can be represented as in Fig. 6.10. Two dominating elasticities are supposed, one at the clutch or converter and one at the drive shaft between the differential and the rear wheels. This leads to a mass-spring-damper system with the following turning and moving masses: engine and primary part of the clutch or converter, secondary part of the clutch or converter, transmission, propeller shaft and differential, drive shafts, wheels, and vehicle mass.

It is shown in Kiencke and Nielsen (2000) that the main contribution to drive train dynamics stems from the drive shaft which is subject to a large torsion, due to the high transmission rates of the shifted gear and the differential gear. The transmission, propeller shaft, and differential are in the following assumed to be stiff.

### 6.3.1 Dry-Plate Friction Clutch

The mean value torque balance at the flywheel yields according to (6.2.21)

$$J_{eng}\dot{\omega}_{eng}(t) = M_{eng}(t) - M_{cl,f}(t) \tag{6.3.1}$$

with

$J_{eng}$: moment of inertia of the engine
$M_{eng}$: mean engine torque at clutch, operating-point-dependent
$M_{cl,f}$: clutch torque during gliding

The transmitted torque of the clutch depends on its state: open, engaged, or gliding. During gliding, dry friction can be assumed:

$$M_{cl,f}(u_{cl}, \Delta\omega_{cl}, t) = M_{cl,0}(u_{cl})\,\mathrm{sign}(\omega_{eng}(t) - \omega_{cl}(t)) \tag{6.3.2}$$

with

$M_{cl,0}$: dry friction coefficient of the clutch
$u_{cl}$: clutch pedal position
$\Delta\omega_{cl} = \omega_{eng} - \omega_{cl}$: difference speed engine/clutch

As the clutch possesses spring-damper elements, the torque balance at the output of the clutch becomes

$$M_{cl,out}(u_{cl}, \Delta\omega_{cl}, t) = M_{cl,c}(\Delta\varphi_{cl}, t) + M_{cl,d}(\Delta\dot{\varphi}, t) \tag{6.3.3}$$

with a torque generated by the clutch springs and dampers

$$\begin{aligned} M_{cl,c}(\Delta\varphi_{cl}, t) &= c_{cl}(\varphi_{cl} - \varphi_{tr,in}) \\ M_{cl,d}(\Delta\dot{\varphi}_{cl}, t) &= d_{cl}(\dot{\varphi}_{cl} - \dot{\varphi}_{tr,in}) \end{aligned} \tag{6.3.4}$$

$\varphi_{cl}$: clutch output angle
$\varphi_{tr,in}$: transmission input angle
$c_{cl}$: clutch spring constant
$d_{cl}$: clutch damping coefficient

Note that the clutch usually contains different springs in series, which leads to a nonlinear characteristic $M_{cl,c}(\Delta\varphi_{cl})$; see, e.g. Kiencke and Nielsen (2005). Hence, in the gliding phase the torque balance at the clutch yields

$$J_{cl}\ddot{\varphi}_{cl}(t) = M_{cl,f}(u_{cl}, \Delta\omega_{cl}, t) - M_{cl,out}(t) \tag{6.3.5}$$

with $J_{cl}$: moment of clutch inertia on transmission side.

These equations lead to the parts "engine" and "clutch" in the signal flow scheme of Fig. 6.11.

### 6.3.2 Shifted Transmission, Propeller Shaft, and Differential

Assuming that the shifted transmission, propeller shaft, and differential can be treated as stiff, compared to the other drive train elements, it follows

$$J_{tr,di}(i_{tr})\ddot{\varphi}_{tr,in} = M_{tr,in}(t) - M_{tr,d}(t) - \frac{1}{i_{tr}i_{di}} M_{di,out}(t) \tag{6.3.6}$$

with

$J_{tr,di}$: moment of inertia of shifted transmission, shaft and differential
$M_{tr,in} = M_{cl,out}$: torque at transmission input
$M_{di,out} = M_{ds,in}$: torque at differential output
$M_{tr,d}$: friction torque of transmission

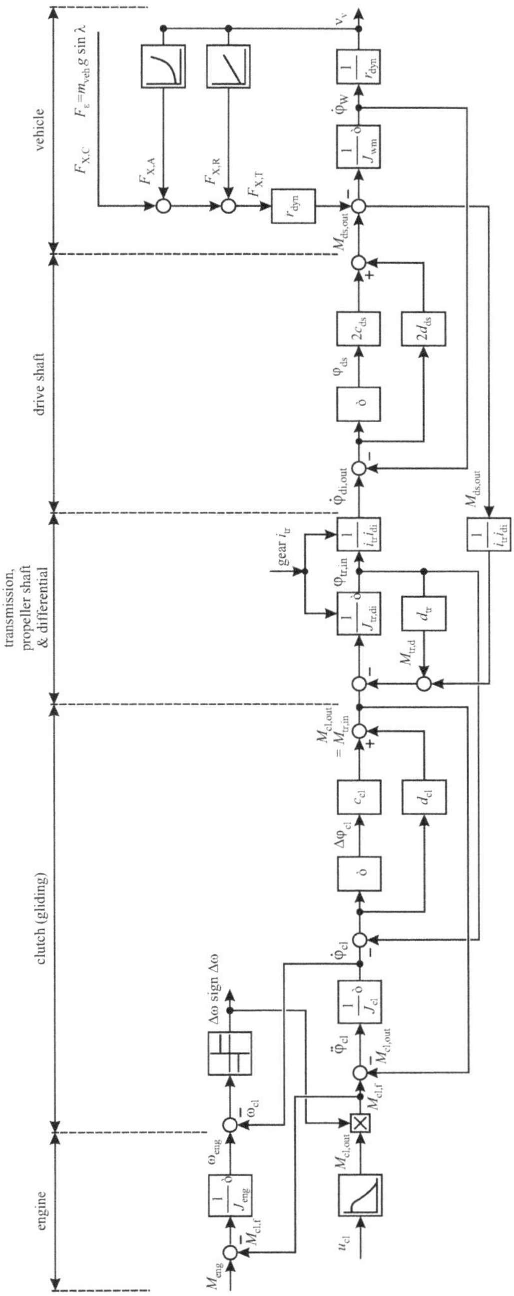


**Fig. 6.11** Signal flow scheme for the dynamic behavior of the drive train with shifted transmission and clutch in gliding operation

and transmission ratios and torque ratios

$$\begin{aligned} i_{\text{tr}} &= \frac{\dot{\varphi}_{\text{tr,in}}}{\dot{\varphi}_{\text{ps}}} = \frac{\text{input speed transmission}}{\text{propeller shaft speed}} \\ i_{\text{di}} &= \frac{\dot{\varphi}_{\text{ps}}}{\dot{\varphi}_{\text{di,out}}} = \frac{\text{propeller shaft speed}}{\text{differential speed at output}} \\ M_{\text{ps}} &= i_{\text{tr}} M_{\text{tr,in}}\ ; \quad M_{\text{di}} = i_{\text{di}} M_{\text{ps}}. \end{aligned}$$

If the shifted spur gear set has two axles with moments of inertia $J_{\text{tr},1}$ and $J_{\text{tr},2}$, the propeller shaft has $J_{\text{ps}}$, and the differential $J_{\text{di}}$, then the lumped overall moment of inertia is

$$J_{\text{tr,di}} = J_{\text{tr},1} + \frac{1}{i^2_{\text{tr}}}(J_{\text{tr},2} + J_{\text{ps}} + \frac{1}{i^2_{\text{di}}} J_{\text{di}}). \tag{6.3.7}$$

The friction torque of the transmission (bearings, oil) is assumed to be proportional to the angular speed

$$M_{\text{tr,d}} = d_{\text{tr}} \dot{\varphi}_{\text{tr}}. \tag{6.3.8}$$

A corresponding signal flow for this part of the drive train is shown in Fig. 6.11.

### 6.3.3 Drive Shaft and Wheels

The torque balance of the drive shaft, the two wheels with dynamic roll radius $r_{\text{dyn}}$, and the vehicle with mass $m_{\text{v}}$ yields

$$J_{\text{wm}} \ddot{\varphi}_{\text{W}}(t) = M_{\text{ds,in}}(t) - F_{\text{XT}}(t) r_{\text{dyn}} \tag{6.3.9}$$

with the output torque of the drive shaft

$$M_{\text{W}}(t) = M_{\text{ds,out}}(t) = M_{\text{ds,in}} + 2c_{\text{ds}}(\varphi_{\text{di,out}}(t) - \varphi_{\text{W}}(t)) + 2d_{\text{ds}}(\dot{\varphi}_{\text{di,out}}(t) - \dot{\varphi}_{\text{W}}(t)) \tag{6.3.10}$$

and the longitudinal vehicle drive force

$$F_{\text{XT}}(t) = m_{\text{v}} \frac{d v_{\text{v}}(t)}{dt} + F_{\text{res}}(t) \tag{6.3.11}$$

with the resistance forces for air drag, rolling, and climbing

$$F_{\text{X,res}} = F_{\text{X,A}} + F_{\text{X,R}} + F_{\text{X,C}}. \tag{6.3.12}$$

The equations for these vehicle driving forces are treated in Sect. 6.5. Insertion into (6.3.9), assuming no slip between tire and road, the torque balance becomes $v_{\text{v}} = r_{\text{dyn}} \omega_{\text{W}} = r_{\text{dyn}} \dot{\varphi}_{\text{W}}$

$$J_{\text{wm}} \ddot{\varphi}_{\text{W}}(t) = M_{\text{ds,out}}(t) - r_{\text{dyn}}(F_{\text{X,A}}(t) + F_{\text{X,R}}(t) + F_{\text{X,C}}(t)) \tag{6.3.13}$$

with replacement moment of inertia for two wheels (one driven axle) and mass of the vehicle

$$J_{\mathrm{wm}} = 2J_{\mathrm{W}} + m_{\mathrm{V}} r_{\mathrm{dyn}}^2. \tag{6.3.14}$$

The signal flow for the drive shaft, the wheels, and the vehicle forms the last part in Fig. 6.11. Figure 6.11 summarizes the relations of all components of the considered drive train for the case that the clutch is in gliding operation. If the clutch is fully engaged, the engine torque acts directly on the secondary side of the clutch, leading to the signal flow scheme of Fig. 6.12. Then a *three-mass system* with two elasticities results.

## 6.4 Drive Train with Automatic Hydrodynamic Transmission

Automatic transmissions for cars are frequently a combination of a hydrodynamic torque converter and several planetary gear sets which are shifted by wet friction clutches and brake bands.

The torque converter consists of an impeller or hydrodynamic pump on the engine side and a turbine on the drive side. In between is a stator which supports the torque conversion; see Fig. 6.13. The impeller converts mechanical energy by fluid dynamic forces to the encapsulated oil, which is converted back to mechanical energy in the turbine wheel. The stator diverts the oil back from the turbine to the impeller and by appropriate design of the vanes, the torque of the turbine becomes larger than that of the impeller, with a torque conversion rate

$$\frac{M_{\mathrm{t}}}{M_{\mathrm{p}}} = \mu \left( \frac{\omega_{\mathrm{t}}}{\omega_{\mathrm{p}}} \right). \tag{6.4.1}$$

See, e.g. Robert Bosch GmbH (2018), Dach et al. (2001), Naunheimer et al. (2007), and Germann (1997).

This conversion rate decays approximately linearly with the speed ratio $\nu = \omega_{\mathrm{t}}/\omega_{\mathrm{p}}$; see Fig. 6.14. For $\nu = 0$, the torque multiplication is in the range of $\mu = 1.7...2.5$.

The impeller input torque $M_{\mathrm{p}}$ follows

$$M_{\mathrm{p}} = \lambda \left( \frac{\omega_{\mathrm{t}}}{\omega_{\mathrm{p}}} \right) \rho D^5 \omega_{\mathrm{p}}^2 = k_{\mathrm{p}} \left( \frac{\omega_{\mathrm{t}}}{\omega_{\mathrm{p}}} \right) \omega_{\mathrm{p}}^2 \tag{6.4.2}$$

where $\lambda$ is a performance index, $\rho$ the density of the oil, D the circuit diameter, and $\omega_{\mathrm{p}}$ the impeller angular velocity. Then, the turbine torque at the output of the converter becomes

$$M_{\mathrm{t}} = \mu \left( \frac{\omega_{\mathrm{t}}}{\omega_{\mathrm{p}}} \right) k_{\mathrm{p}} \left( \frac{\omega_{\mathrm{t}}}{\omega_{\mathrm{p}}} \right) \omega_{\mathrm{p}}^2. \tag{6.4.3}$$

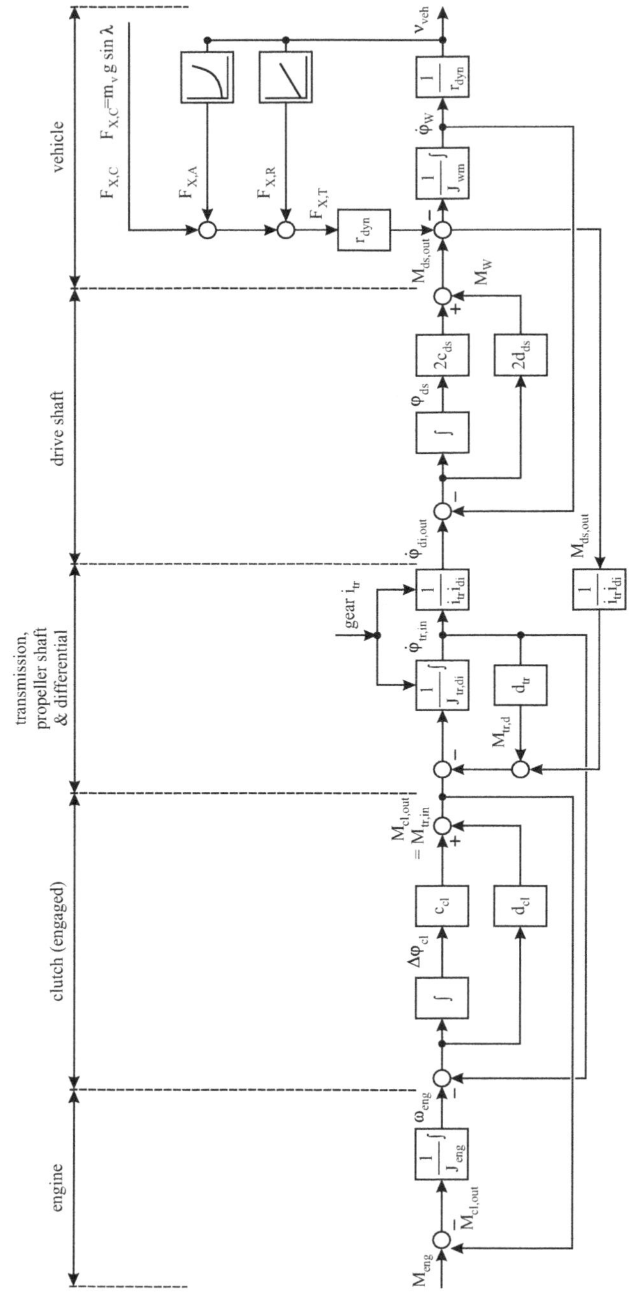


**Fig. 6.12** Signal flow scheme for the dynamic behavior of the drive train with shifted transmission and engaged clutch

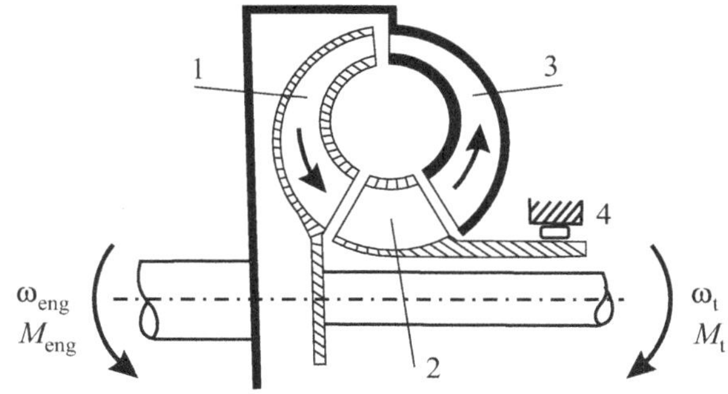


**Fig. 6.13** Scheme of a hydrodynamic converter. 1 turbine, 2 stator, 3 impeller, and 4 one-way clutch

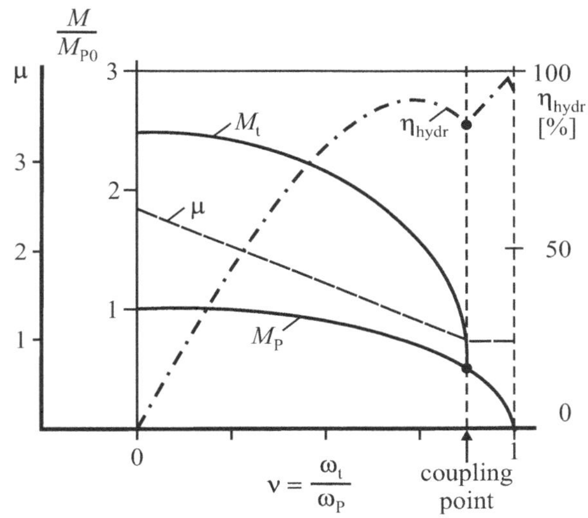


**Fig. 6.14** Characteristics of a trilock converter for constant engine speed; see Robert Bosch GmbH (2011) and Dach et al. (2001)

The torque can only be transferred for $\omega_p > \omega_t$, i.e. by a slip

$$s = \frac{\omega_p - \omega_t}{\omega_p} = 1 - \nu. \tag{6.4.4}$$

The hydraulic efficiency is

$$\begin{aligned} \eta_{hyd} &= \frac{P_{out}}{P_{in}} = \frac{M_t \omega_t}{M_p \omega_p} = \mu \left( \frac{\omega_t}{\omega_p} \right) \nu \\ &= \mu(1 - s). \end{aligned} \tag{6.4.5}$$

It is zero for $\nu = 0$ with full slip $s = 1$ and reaches a maximum of about 95 % at smaller slips of 2–10 %.

The main characteristics of a fluid dynamic torque converter are shown in Fig. 6.14. Because of the not turning and by the casing supported stator-wheel, it is $M_t > M_p$ until the coupling point, where the stator begins to turn and the automatic

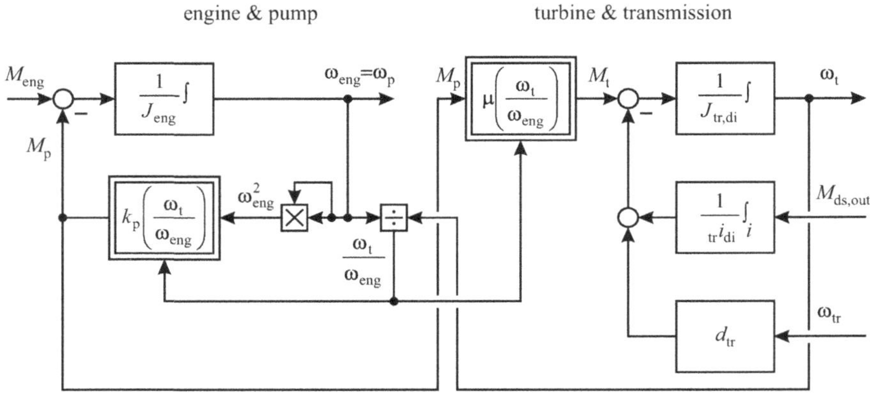


**Fig. 6.15** Signal flow of a hydrodynamic torque converter with slip

transmission acts as a hydraulic clutch. The transmitted torque tends to $M_t \to 0$ if $\nu \to 1$, and $s \to 0$.

An angular momentum balance of the torque converter yields with $\omega_{eng} = \omega_p$ and $J_{eng}$ the engine moment of inertia

$$\begin{aligned} J_{eng}\frac{d\omega_{eng}(t)}{dt} &= M_{eng}(\omega_{eng}, t) - M_p\left(\omega_{eng}, \omega_t, t\right) \\ &= M_{eng}(\omega_{eng}, t) - k_p\left(\frac{\omega_t}{\omega_{eng}}\right)\omega_{eng}^2 \end{aligned} \quad (6.4.6)$$

which is a nonlinear differential equation of first order.

The angular momentum balance of the turbine with the attached transmission becomes

$$\begin{aligned} J_{tr,di}\frac{d\omega_t(t)}{dt} &= M_t(t) - \frac{1}{i_{tr}i_{di}}M_{ds,out} - M_{tr,d}(t) \\ &= \mu\left(\frac{\omega_t}{\omega_{eng}}\right)M_p(t) - \frac{1}{i_{tr}i_{di}}M_{ds,out} - M_{tr,d}(t) \end{aligned} \quad (6.4.7)$$

where $J_{tr,di}$ is the moment of inertia of the transmission, propeller shaft, and differential, $M_{ds,out}$ is the counteracting torque with $i_{tr}i_{di}$ the acting gear ratio, and $M_{tr,d}$ is the friction torque of the transmission; compare (6.3.6) to (6.3.8).

Figure 6.15 depicts the signal flow of both balance equations. Due to the nonlinear relations between the impeller and turbine torque and the terms $(\omega_t/\omega_{eng})$ and $\omega_{eng}^2$, a strongly nonlinear behavior between $M_{eng}$ as input and $M_t$ or $\omega_t$ as output results. The signal flow chart of Fig. 6.11 holds also for the automatic hydrodynamic transmissions, if the models for "engine" and "clutch" are replaced by Fig. 6.15 and $M_{tr,in} = M_t$. Linearization of (6.4.2) and (6.4.3) around an operation point $(\overline{M}_{eng}, \overline{\omega}_{eng})$ leads for the impeller and turbine torque with regard to the angular

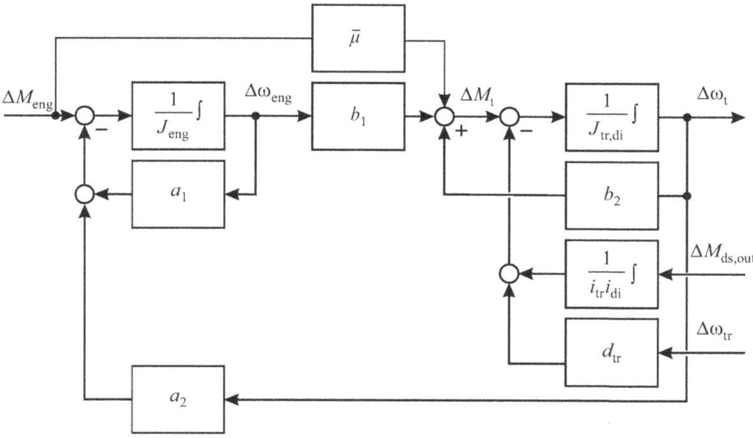


**Fig. 6.16** Linearized signal flow for a hydrodynamic torque converter with slip

speed $\omega_\mathrm{p} = \omega_\mathrm{eng}$ and $M_\mathrm{p} = M_\mathrm{eng}$

$$\begin{aligned}\Delta M_\mathrm{eng} &= \left(\frac{\partial k_\mathrm{p}}{\partial \omega_{eng}}\Delta\omega_\mathrm{eng} + \frac{\partial k_\mathrm{p}}{\partial \omega_\mathrm{t}}\Delta\omega_\mathrm{t}\right)\overline{\omega}_\mathrm{eng}^2 + 2\overline{k}_\mathrm{p}\overline{\omega}_\mathrm{eng}\Delta\omega_\mathrm{eng} \\ &= a_1\Delta\omega_\mathrm{eng} + a_2\Delta\omega_\mathrm{t},\end{aligned} \tag{6.4.8}$$

$$\begin{aligned}\Delta M_\mathrm{t} &= \left(\frac{\partial \mu}{\partial \omega_{eng}}\Delta\omega_\mathrm{eng} + \frac{\partial \mu}{\partial \omega_\mathrm{t}}\Delta\omega_\mathrm{t}\right)\overline{M}_\mathrm{eng} + \overline{\mu}\Delta M_\mathrm{eng} \\ &= b_1\Delta\omega_\mathrm{eng} + b_2\Delta\omega_\mathrm{t} + \bar{\mu}\Delta M_\mathrm{eng}.\end{aligned} \tag{6.4.9}$$

Introduction into (6.4.6) and (6.4.7) yields

$$J_\mathrm{eng}\frac{d\omega_\mathrm{eng}(t)}{dt} + a_1\Delta\omega_\mathrm{eng}(t) = \Delta M_\mathrm{eng}(t) - a_2\Delta\omega_\mathrm{t}(t), \tag{6.4.10}$$

$$\begin{aligned}&J_\mathrm{tr,di}\frac{d\omega_\mathrm{t}(t)}{dt} - b_2\Delta\omega_\mathrm{t} = -\frac{1}{i_\mathrm{tr}i_\mathrm{di}}\Delta M_\mathrm{ds,out} + b_1\Delta\omega_\mathrm{eng}(t) \\ &-d_\mathrm{tr}\Delta\omega_\mathrm{tr}(t) + \bar{\mu}\Delta M_\mathrm{eng}.\end{aligned} \tag{6.4.11}$$

Hence, one obtains two coupled differential equations of first order; see the signal flow in Fig. 6.16.

Laplace transformation of (6.4.9) and (6.4.10) yields

$$\begin{aligned}\Delta M_\mathrm{t}(s) &= \left(\frac{b_1}{(J_\mathrm{eng}s + a_1)} + \bar{\mu}\right)\Delta M_\mathrm{eng}(s) - \left(\frac{b_1 a_2}{J_\mathrm{eng}s + 1} + b_2\right)\Delta\omega_\mathrm{t} \\ &- \frac{1}{i_\mathrm{tr}i_\mathrm{di}}\Delta M_\mathrm{ds,out} - d_\mathrm{tr}\Delta\omega_\mathrm{tr}.\end{aligned} \tag{6.4.12}$$

Hence, the turbine torque follows a change of the engine torque according to a first-order delay, for $\omega_t = \text{const}$ with time constant

$$T_{Mt} = \frac{J_{eng}}{a_1}. \quad (6.4.13)$$

As Fig. 6.16 indicates, the turbine speed $\Delta\omega_t$ is delayed by a further first-order delay with time constant

$$T_{!t} = \frac{J_{tr,di}}{-b_2} \quad (6.4.14)$$

which depends on the moment of inertia $J_{tr,di}$ of the drive shaft and is therefore larger than $T_{Mt}$. (It holds $b_2 = \partial\mu/\partial\omega_t < 0$.)

As advanced automatic transmissions operate with a *lockup clutch*, the impeller and the turbine are locked together briefly after start-up. Then (6.4.7) can be used with $M_t = M_{eng}$ as the engine torque acts directly on the input of the transmission gear set. If the lockup clutch can be modeled with spring and damping coefficients as (6.3.4), then Fig. 6.12 is also valid for the automatic hydrodynamic transmission.

## 6.5 Longitudinal Vehicle Model

To model the longitudinal behavior of a vehicle, it is first assumed that the vehicle drives straight forward. The goal is to describe the behavior for acceleration and for braking. First, the general vehicle model is considered which is linked to the elastic multi-mass drive train. Then a simplified vehicle model is derived assuming a stiff drive train. This is followed by taking into account the vertical wheel forces for stationary and dynamic behavior which influence the wheel slip and the transferred longitudinal tire forces.

### 6.5.1 Basic Longitudinal Vehicle Model

For straight driving, the wheel steering angle is $\delta_f = 0$ and the side slip angle is $\beta = 0$. The momentum balance for the center of gravity in the x-direction then becomes

$$m_V \frac{dv(t)}{dt} = F_{XT}(t) - \sum F_{X,res}(t) \quad (6.5.1)$$

where $m_V$ is the mass of the vehicle (chassis, body, powertrain, and load) and $v = v_X$. Thus, the difference between the tire forces $F_{XT}$ and the resistance forces $F_{X,res}$ determines the acceleration of the vehicle. For the tangential tire contact forces with $i = 4$ wheels holds

$$\begin{aligned} F_{XT} &= \sum_i F_{XT,i} \\ &= \sum_{i=l}^{r} F_{XT,fi} + \sum_{i=l}^{r} F_{XT,ri}. \end{aligned} \quad (6.5.2)$$

These forces are generated either by the powertrain or the brakes. The resistance forces are

$$\sum F_{\mathrm{X,res}} = F_{\mathrm{X,R}} + F_{\mathrm{X,A}} + F_{\mathrm{X,C}}. \tag{6.5.3}$$

The rolling force for all wheels is described by

$$F_{\mathrm{X,R},i} = (f_{\mathrm{R0}} + f_{\mathrm{R1}}\upsilon + f_{\mathrm{R4}}\upsilon^4)_i F_{\mathrm{Z}i} \tag{6.5.4}$$

with $F_{\mathrm{Z}i}$ the vertical wheel force. For the aerodynamic drag holds

$$F_{\mathrm{X,A}} = \frac{\rho_{\mathrm{A}}}{2} c_{\mathrm{wx}} A_{\mathrm{X}} \upsilon_{\mathrm{res}}^2 = \frac{c_{\mathrm{air}}}{2} \upsilon_{\mathrm{res}}^2 \tag{6.5.5}$$

with $\rho_{\mathrm{A}}$ the air density, $c_{\mathrm{wx}}$ the air drag coefficient, $A_{\mathrm{X}}$ the frontal vehicle area, and the relative velocity to the air velocity

$$\upsilon_{\mathrm{res}} = \upsilon_{\mathrm{X,res}} = \upsilon_{\mathrm{X}} - \upsilon_{\mathrm{X,air}}. \tag{6.5.6}$$

The climbing resistance becomes

$$F_{\mathrm{X,C}} = m_{\mathrm{V}} g \sin(\lambda) \approx m_{\mathrm{V}} g \lambda \tag{6.5.7}$$

with $\lambda$ the longitudinal slope (road gradient) angle. If the slope $p$ of a road is defined by the ratio of vertical to horizontal distances, it holds

$$p = \tan(\lambda) \quad [\%] \tag{6.5.8}$$

and because of $\sin\lambda \approx \tan\lambda$, one can take $\lambda \approx p$. For more details and parameters, see Reimpell and Hoseus (1992), Gillespie (1992), Popp and Schiehlen (1993), Mitschke and Wallentowitz (2004), and Heissing and Ersoy (2011).

The basic longitudinal vehicle model (6.5.1)–(6.5.7) can in the case of accelerating with the throttle input $\alpha$ be connected to the three-mass drive train models of Sects. 6.3 and 6.4 by applying for a rear-driven vehicle

$$F_{\mathrm{XT}} = \sum_i F_{\mathrm{XT},i} = \sum_{i=l}^{r} F_{\mathrm{XT,r}i} = F_{\mathrm{XT,rl}} + F_{\mathrm{XT,rr}} \tag{6.5.9}$$

and for a front-driven vehicle

$$F_{\mathrm{XT}} = \sum_i F_{\mathrm{XT},i} = \sum_{i=l}^{r} F_{\mathrm{XT,f}i} = F_{\mathrm{XT,fl}} + F_{\mathrm{XT,fr}}. \tag{6.5.10}$$

These equations then lead to the part "vehicle" in the signal flow charts Figs. 6.4 and 6.7 with $F_{\mathrm{XT}} = M_{\mathrm{w}}/r_{\mathrm{dyn}}$. Together with the three-mass model of the drive train described in Sects. 6.2 and 6.3, the overall longitudinal vehicle model results, taking the elasticities of the drive train into account.

**Table 6.1** Influence of gear on the change of vehicle speed as a result of a change in the road gradient (at throttle angle $\alpha = 30°$)

| Road gradient | | 5% | 7% | Change of vehicle speed [km/h] |
|---|---|---|---|---|
| Vehicle speed [km/h] | Fourth gear | 86 | 53 | 33 |
| | Second gear | 73 | 65 | 8 |
| | First gear | 51 | 47 | 4 |

**Example 6.1** (Characteristics of a passenger car) Stationary driving follows from (6.5.1)

$$\overline{F}_{XT} = \sum \overline{F}_{X,res}. \tag{6.5.11}$$

Figure 6.6 depicts the torque characteristics of a gasoline engine (6 cylinders, 2.5l, 125 kW at 6000 rpm, max. torque 227 Nm, and vehicle mass 1250 kg) and the required torques at the transmission output $M_{tr,out}$ for the compensation of the driving resistances in dependence on the velocity for different road slopes and gears. The torque characteristics at the clutch for various throttle angles $\alpha$ indicate that the torque curve at full load ($\alpha = 90°$) is quite flat, whereas for smaller throttle angles, the engine torque decreases with increasing engine speed. In Fig. 6.6b), the engine torque at the transmission output shaft for full load and various road gradients is given. For a road gradient of 0%, the point of intersection of the characteristic curve of the fifth gear gives the maximum speed of 233 km/h. Maximum speed in the fifth gear reduces with increasing road gradients of 3% and 5%, respectively, to 200 km/h and 170 km/h.

In the fifth gear, a further increase of the road gradient from 5% to 7% leads to a speed drop from 170 km/h to values between 85 km/h and 50 km/h. Exact values cannot be given due to the parallel curves in this region. A downshift to the fourth gear results in a higher speed of 187 km/h at the road gradient of 5%, and reduces to 153 km/h at 7% road gradient, i.e. the speed loss is lesser than in the fifth gear. For the first, second, and third gears with full load, no points of intersection exist for the considered road gradients of up to 10% and the considered engine speed, which is limited by the overspeed governor. The respective torque difference is available for accelerating the vehicle.

Figure 6.6c gives the torque characteristics for partial load (throttle angle 30°). Due to the larger decline of the engine torque at partial load, stable operating conditions also exist for the lower gears; for better clarity, the characteristic curves of the first, second, and fourth gears only are depicted.

The points of intersection indicate the stationary speed values shown in Table 6.1. Due to the steeper curve shapes of the driving torque after the transmission, the lower the gear, the smaller is the influence of a change of the road gradient on a change of the vehicle speed.

### 6.5.2 Simplified Vehicle Model with Stiff Powertrain (One-Mass System)

If an approximate model for the longitudinal motion of a vehicle is sufficient, the elasticities of the powertrain can be neglected and the powertrain can be represented as one stiff rotating multi-mass system. This applies to the case of an engaged clutch or an engaged lockup clutch. The accelerated mass is then described by a total "representative" mass

$$m_{\text{tot}} = m_{\text{V}} + m_{\text{pt}} \tag{6.5.12}$$

where $m_{\text{V}}$ is the mass of the vehicle and $m_{\text{pt}}$ is a value for the rotational masses of the powertrain.

Instead of the friction models used in Sect. 6.3, now efficiencies

$$\eta = \frac{M_{\text{out}}}{M_{\text{in}}} \tag{6.5.13}$$

are used for the components, with $\eta_{\text{tr}}$ for the transmission and $\eta_{\text{dif}}$ for the differential. The torque balances for the engine, the transmission, and the differential then become for an accelerating powertrain

$$\begin{aligned} J_{\text{eng}} \frac{d\omega_{\text{eng}}}{dt} &= M_{\text{eng}} - M_{\text{tr,in}} \\ J_{\text{tr,di}} \frac{d\omega_{\text{tr}}}{dt} &= M_{\text{tr,in}} - M_{\text{di,out}} \\ 2J_{\text{W}} \frac{d\omega_{\text{W}}}{dt} &= M_{\text{di,out}} - \sum_{i=r}^{l} M_{\text{w},i} \end{aligned} \tag{6.5.14}$$

where $M_{\text{W},i}$ is the torque resulting from tire forces $F_{\text{XT},i}$ at one wheel. Inserting

$$\begin{aligned} M_{\text{tr,out}} &= i_{\text{tr}} \eta_{\text{tr}} M_{\text{tr,in}}; \quad M_{\text{di,out}} = i_{\text{di}} \eta_{\text{di}} M_{\text{tr,out}} \\ \omega_{\text{eng}} &= i_{\text{tr}} \omega_{\text{tr}}; \quad \omega_{\text{tr}} = i_{\text{di}} \omega_{\text{W}}, \end{aligned} \tag{6.5.15}$$

one obtains from the equation system (6.5.14) for an accelerating powertrain

$$\begin{aligned} J_{\text{tot}} \frac{d\omega_{\text{W}}}{dt} &= i_{\text{tr}} i_{\text{di}} \eta_{\text{tr}} \eta_{\text{di}} M_{\text{eng}} - \sum_{i=r}^{l} M_{\text{W},i} \\ &= i_{\text{tot}} \eta_{\text{tot}} M_{\text{eng}} - \sum_{i=r}^{l} M_{\text{W},i} \end{aligned} \tag{6.5.16}$$

with the representative moment of inertia for the rotational masses of the powertrain

$$J_{\text{tot}} = 2J_{\text{W}} + i_{\text{di}}^2 \eta_{\text{di}} J_{\text{tr,di}} + i_{\text{tr}}^2 i_{\text{di}}^2 \eta_{\text{tr}} \eta_{\text{di}} J_{\text{eng}}. \tag{6.5.17}$$

Thus $J_{\mathrm{tot}}$ determines the rotational masses for accelerations $d\omega_{\mathrm{W}}/dt$ either by the engine torque $M_{\mathrm{eng}}$ or by wheel torques $M_{\mathrm{W},i}$. The vehicle velocity is related to the wheel velocity, assuming no slip, by

$$v = \omega_{\mathrm{W}} r_{\mathrm{dyn}} \text{ or } \dot{v} = \dot{\omega}_{\mathrm{W}} r_{\mathrm{dyn}}. \tag{6.5.18}$$

The torque at one wheel holds

$$M_{\mathrm{W},i} = r_{\mathrm{dyn}} F_{\mathrm{XT},i}. \tag{6.5.19}$$

Insertion into (6.5.16) allows to determine the tire forces with engine torque as input

$$\sum_{i=r}^{l} F_{\mathrm{T},i} = \frac{1}{r_{\mathrm{dyn}}} \left( i_{\mathrm{tot}} \eta_{\mathrm{tot}} M_{\mathrm{eng}} - J_{\mathrm{tot}} \frac{d\omega_{\mathrm{W}}}{dt} \right). \tag{6.5.20}$$

Hence, the force balance of the vehicle according to (6.5.1) becomes now, including the rotational masses and assuming no tire slip with $v = r_{\mathrm{dyn}} \omega_{\mathrm{W}}$,

$$m_{\mathrm{tot}} \frac{dv(t)}{dt} = i_{\mathrm{tot}} \eta_{\mathrm{tot}} \frac{M_{\mathrm{eng}}(t)}{r_{\mathrm{dyn}}} - \sum F_{\mathrm{X,res}}(t) \tag{6.5.21}$$

where the rotational masses of the powertrain and the wheels are

$$m_{\mathrm{pt}} = \frac{J_{\mathrm{tot}}}{r_{\mathrm{dyn}}^2}. \tag{6.5.22}$$

The combined total mass then is with (6.5.12)

$$m_{\mathrm{tot}} = m_{\mathrm{V}} + \frac{J_{\mathrm{tot}}}{r_{\mathrm{dyn}}^2} = \lambda_{\mathrm{m}} m_{\mathrm{V}} \tag{6.5.23}$$

$$\lambda_{\mathrm{m}} = 1 + \frac{J_{\mathrm{tot}}}{r_{\mathrm{dyn}}^2 m_{\mathrm{V}}}, \tag{6.5.24}$$

where $m_{\mathrm{V}}$ includes the non-driven wheels and $\lambda_{\mathrm{m}}$ is a rotational mass factor. Depending on the drive train ratio $i_{\mathrm{tot}} \approx 3...18$ with the selected gear ratio $i_{\mathrm{tr}}$, the rotational mass factor shows values in the range of $\lambda_{\mathrm{m}} \approx 1.05...1.5$; see Heissing and Ersoy (2011). The resulting acceleration of a vehicle can now be determined by (6.5.21). The signal flow chart of Fig. 6.17 summarizes the different equations and terms for acceleration and braking.

Figure 6.18 shows the range of dominant angular frequencies of the powertrain subsystems. They range from about $\omega_{\mathrm{c}} = 500$ to 0.03 1/s or $f = 80$ to 0.005 Hz with dominating time constants of $T_{\mathrm{c}} = 2$ ms to 30 s, hence, the characteristic frequency span over four decades. Therefore, it depends on the application, which frequencies are most relevant. The higher frequencies determine more arising oscillations, whereas the lower frequencies are more relevant for the driving dynamics.

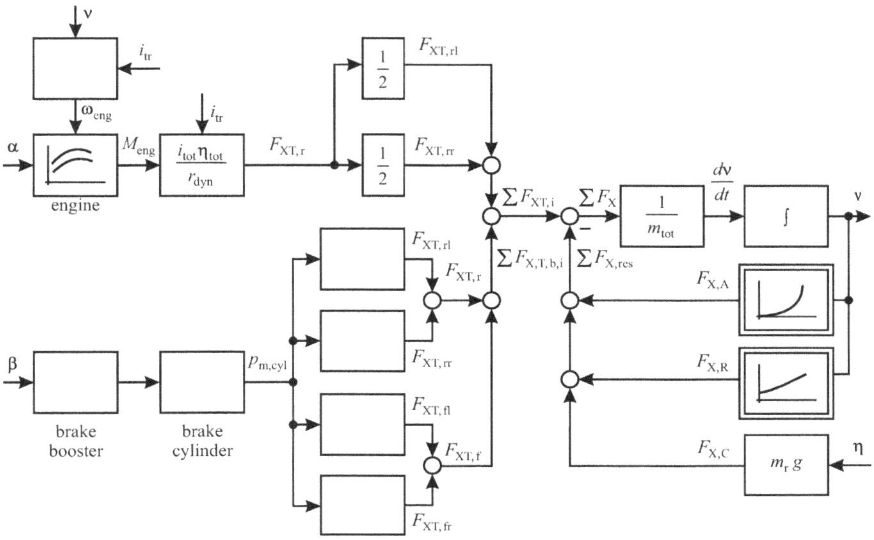


**Fig. 6.17** Signal flow for the longitudinal behavior of a vehicle with rear-driven axle and stiff drive train (one-mass model) for acceleration and braking

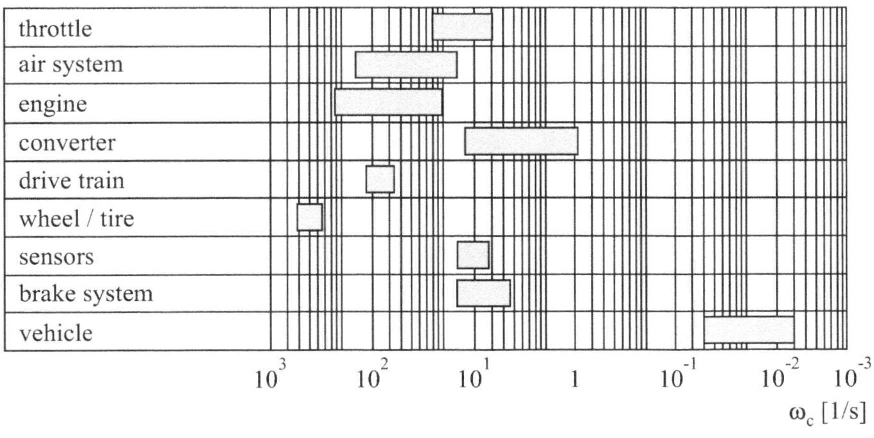


**Fig. 6.18** Angular frequency ranges of powertrain subsystems with a gasoline engine. $\omega_c = 1/T$ corner frequency (cutoff frequency); see Germann (1997)

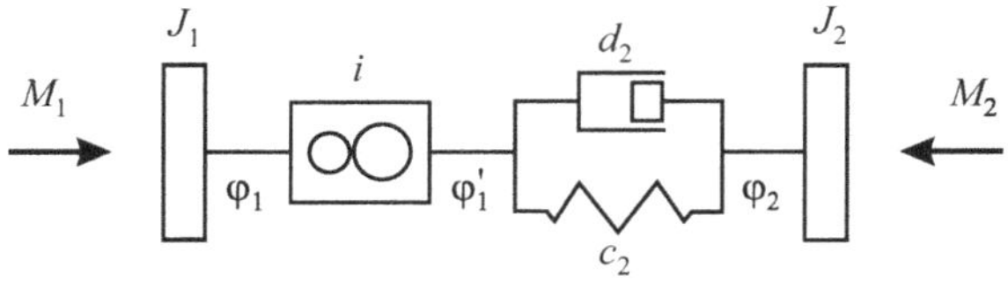


**Fig. 6.19** Simplified representation of the drive train as a two-mass system

### 6.5.3 The Drive Train As a Two-Mass-System

For some applications, it may be sufficient to simplify the drive train to a two-mass system with the main elasticity at the drive shaft; see Fig. 6.19. One can then, for example, directly obtain an approximation of the undamped natural frequency of the drive train. The ratio of inertia of the engine, clutch, gear gets, propeller shaft, and differential are then lumped together, where the dominating term comes from the engine

$$J_1 \approx J_{\text{eng}}. \tag{6.5.25}$$

The moment of inertia of the driven wheels and the vehicle follows according to (6.5.22)

$$J_2 = J_{\text{wm}} = 2J_{\text{W}} + m_{\text{V}} r_{\text{dyn}}^2. \tag{6.5.26}$$

Both masses are connected with the overall transmission ratio $i = i_{\text{tr}} i_{\text{di}}$, the stiffness and elasticity of the drive shaft $c_2 = 2c_{\text{ds}}$ and $d_2 = 2d_{\text{ds}}$.

Then, the following torque balances can be stated, compare Fig. 6.19 with $M_1 = M_{\text{eng}}$,

$$J_1 \ddot{\varphi}_1 = M_1 - \frac{1}{i}\left[c_2\left(\varphi_1' - \varphi_2\right) + d_2\left(\dot{\varphi}_1' - \dot{\varphi}_2\right)\right], \tag{6.5.27}$$

$$J_2 \ddot{\varphi}_2 = c_2\left(\varphi_1' - \varphi_2\right) + d_2\left(\dot{\varphi}_1' - \dot{\varphi}_2\right) - M_2. \tag{6.5.28}$$

Introducing the torsion angle

$$\Delta\varphi = \varphi_1' - \varphi_2 \tag{6.5.29}$$

and subtracting (6.5.27) and (6.5.29) leads to, compare Isermann (2005),

$$J_{\text{eff}} \Delta\ddot{\varphi}(t) + d_2 \Delta\dot{\varphi}(t) + c_2 \Delta\varphi(t) = k_1 M_1(t) + k_2 M_2(t) \tag{6.5.30}$$

with

$$\begin{aligned} J_{\text{eff}} &= \frac{J_1 J_2 i^2}{J_2 + i^2 J_1} \\ k_1 &= \frac{J_1}{J_2 + i^2 J_1} \\ k_2 &= \frac{J_2}{i\left(J_2 + i^2 J_1\right)}. \end{aligned}$$

**Table 6.2** Undamped natural buckling (surge) oscillations for a passenger car; see Pfeiffer (1997)

| Gear | 1 | 2 | 3 | 4 | 5 |
|---|---|---|---|---|---|
| $f_0$ [Hz] | 2.9 | 4.6 | 6.3 | 8.0 | 8.9 |

The undamped natural frequency of the drive train then becomes

$$\omega_0^2 = \frac{c_2}{J_{\text{eff}}} = c_2 \frac{J_2 + i^2 J_1}{J_1 J_2 i^2} = c_2 \left( \frac{1}{J_1 i^2} + \frac{1}{J_2} \right). \tag{6.5.31}$$

Applying the data of a compact car, the frequencies of Table 6.2 result; see Pfeiffer (1997).

The transmission ratio has a significant influence on the buckling frequency. The lower the gear, the lower the buckling frequency. Typical frequencies are in the range of 2 Hz for the 1st gear and 9 Hz for the 5th gear; see Reik et al. (1990).

Backlash in the transmission, shaft, or differential has a further influence on the buckling oscillations. This leads to larger amplitudes and smaller damping; see Pfeiffer (1997).

The buckling oscillations can be damped by appropriate feedback controllers in the ECU with the engine speed as the preferred controlled variable; see Schmidt (1995), Kiencke and Nielsen (2000), and Isermann (2014). Applications of this two-mass drive train model for the acceleration of trucks are shown in Sinsel (2000), Schaffnit (2002), and Isermann (2014).

At least upper class vehicles have *two-mass flywheels* which consist of a smaller primary and a larger secondary rotational mass connected by a spring-damper system. Their application results in a considerable reduction of torsion oscillations from the engine to the drive line; see Reik et al. (1990), Walter et al (2007). One reason is that the resonance frequency through the large secondary rotational mass is shifted to values smaller than the idling speed. In that case, the starter has to be relatively strong to move the engine fast through the low resonance frequency; see Reik (1998). Buckling drive line oscillations can still occur, because they depend on the drive line elasticities. Because of the good damping of the two-mass flywheel, the excitation from the engine side is smaller.

### 6.5.4 Vertical Wheel Forces for Stationary and Dynamic Behavior

The vertical wheel forces or ground contact forces $F_{Z,i}$ depend for the *holding vehicle* only on the position of the center of gravity (CG). In the case of a driving vehicle with constant speed, aerodynamic lift forces $F_{ZA}$ have to be added. If dynamic forces from the suspensions are neglected, a force balance in vertical direction yields on an even road

$$m_V g = \overline{F_{Zf}} + \overline{F_{Zr}} - F_{ZA}(v) \tag{6.5.32}$$

where $\overline{F_{Zf}}$ and $\overline{F_{Zr}}$ are the static vertical forces for the front and rear axle. In the following, $F_{ZA} = 0$ is assumed. A torque balance around the CG yields

$$F_{Zf} l_f - F_{Zr} l_r = 0; \tag{6.5.33}$$

see Fig. 4.1b. Insertion of (6.5.33) in (6.5.32) results in the vertical wheel forces per axle

$$\begin{aligned} \overline{F_{Zf}} &= m_V g \frac{l_r}{l} \\ \overline{F_{Zr}} &= m_V g \frac{l_f}{l} \end{aligned} \tag{6.5.34}$$

with the wheel base $l = l_f + l_r$.

In the case of a *road with gradient* and slope angle $\lambda$, the torque balance around the rear tire contact point yields

$$m_V g l_r \cos\lambda - m_V g h_{CG} \sin\lambda = F_{Zf} l \tag{6.5.35}$$

and the vertical wheel force for the front axle becomes

$$\overline{F_{Zf}} = m_V g \left( \frac{l_r}{l} \cos\lambda - \frac{h_{CG}}{l} \sin\lambda \right) \tag{6.5.36}$$

where $h_{CG}$ is the height of the CG over the ground. Correspondingly, a torque balance around the front tire contact point results in

$$\overline{F_{Zr}} = m_V g \left( \frac{l_f}{l} \cos\lambda + \frac{h_{CG}}{l} \sin\lambda \right). \tag{6.5.37}$$

In the case of an *acceleration* $a_X$ *in longitudinal direction*, a pitch torque arises which changes the vertical axle forces. A torque balance around the rear tire contact point to the road then becomes for an even road, neglecting the dynamics of the suspension,

$$-m_V a_X h_{CG} + m_V g l_r - F_{Zf} l = 0. \tag{6.5.38}$$

The front axle vertical force then follows to

$$F_{Zf} = \overline{F_{Zf}} + F_{Zf}(a_X) = m_V \left( \frac{l_r}{l} g - \frac{h_{CG}}{l} a_X \right) \tag{6.5.39}$$

correspondingly it holds for the rear axle vertical force

$$F_{Zr} = \overline{F_{Zr}} + F_{Zr}(a_X) = m_V \left( \frac{l_f}{l} g + \frac{h_{CG}}{l} a_X \right). \tag{6.5.40}$$

Hence, for a positive acceleration the vertical force at the front axle decreases and it increases at the rear axle.

For a road with a longitudinal slope, (6.5.38) results with (6.5.35) in

$$-m_V a_X h_{CG} - m_V g h_{CG} \sin\lambda + m_V g l_r \cos\lambda - F_{Zf} l = 0 \tag{6.5.41}$$

and therefore

$$F_{Zf} = m_V \left( \frac{l_r}{l} g \cos\lambda - \frac{h_{CG}}{l} (a_X + g \sin\lambda) \right). \tag{6.5.42}$$

Correspondingly the vertical force on the rear axle becomes

$$F_{Zr} = m_V \left( \frac{l_f}{l} g \cos\lambda + \frac{h_{CG}}{l} (a_X + g \sin\lambda) \right). \tag{6.5.43}$$

The vertical wheel forces are then for a symmetric car in lateral direction

$$\begin{aligned} F_{Zfl} &= F_{Zfr} = F_{Zf}/2 \\ F_{Zrl} &= F_{Zrr} = F_{Zr}/2. \end{aligned} \tag{6.5.44}$$

The pitch torque through the longitudinal acceleration results in a turn around the pitch axis, which is different from the position of the CG. This is considered in Chap. 9.

An *acceleration* $a_Y$ *in the lateral direction*, e.g. by cornering to the right, causes a roll torque which changes the vertical wheel forces in addition to the static ones due to (6.5.34). The lateral force on the CG is

$$F_Y = -m_V a_Y. \tag{6.5.45}$$

It is now assumed that this acceleration force is distributed to the front and rear axle as the static vertical wheel forces (6.5.34).

$$\begin{aligned} F_{Yf} &= -m_V a_Y \frac{l_r}{l} \\ F_{Yr} &= -m_V a_Y \frac{l_f}{l}. \end{aligned} \tag{6.5.46}$$

A torque balance around the right rear wheel contact point then yields

$$-m_V a_Y \frac{l_f}{l} h_{CG} - m_V g \frac{l_r}{l} \frac{b_r}{2} + F_{Zrl} b_r = 0 \tag{6.5.47}$$

with $b_r$ the wheel track of the rear axle. The vertical force on the left rear wheel then results in

$$F_{Zrl} = m_V \frac{l_f}{l} \left( \frac{g}{2} + \frac{h_{CG}}{b_r} a_Y \right). \tag{6.5.48}$$

Similarly, one obtains for the right rear wheel

$$F_{Zrr} = m_V \frac{l_f}{l} \left( \frac{g}{2} - \frac{h_{CG}}{b_r} a_Y \right) \tag{6.5.49}$$

and for the front wheels with wheel base $b_f$

$$F_{Zfl} = m_V \frac{l_r}{l} \left( \frac{g}{2} + \frac{h_{CG}}{b_f} a_Y \right), \tag{6.5.50}$$

$$F_{Zfr} = m_V \frac{l_r}{l} \left( \frac{g}{2} - \frac{h_{CG}}{b_f} a_Y \right). \tag{6.5.51}$$

The roll torque through the lateral acceleration causes a twist around the roll axis, which is usually different from the position of CG; see Chap. 9.

## 6.6 Acceleration Behavior

### 6.6.1 Simplified Acceleration Model

The resulting acceleration of a vehicle with the assumption of a stiff powertrain follows from (6.5.21), (6.5.3) to (6.5.7) and yields for the case of an *engine torque input* and no braking forces

$$m_{tot} \frac{dv(t)}{dt} = F_{XT}(t) - F_{X,R}(t) - F_{X,A}(t) - F_{X,C}(t) \tag{6.6.1}$$

with the driving tire forces

$$F_{XT}(t) = \sum_i F_{XT,i}(t) = \frac{i_{tot}\eta_{tot}}{r_{dyn}} M_{eng}(t). \tag{6.6.2}$$

Herewith, $m_{tot}$ includes the vehicle mass and the rotational masses according (6.5.23). A corresponding signal flow scheme is illustrated in Fig. 6.20. With the driving force $\sum F_{XT,i}$ as input, it consists of a linear feedforward branch with non-linear feedback paths by the air and rolling resistance forces. However, the generation of the driving force is nonlinear because of the engine characteristics.

The determination of the driving forces according to (6.6.2) supposes that there holds a fixed ratio between engine speed and wheel speed and that the acting engine torque $M_{eng}(t)$ determines the tire forces $F_{XT}(t)$ without tire/road slip, i.e. complete adhesion between the tire and the road. This can be assumed for small to medium accelerations. For higher accelerations, the slip and vertical force-dependent tire forces have to be taken into account; see Sect. 6.7.2 and Fig. 6.22.

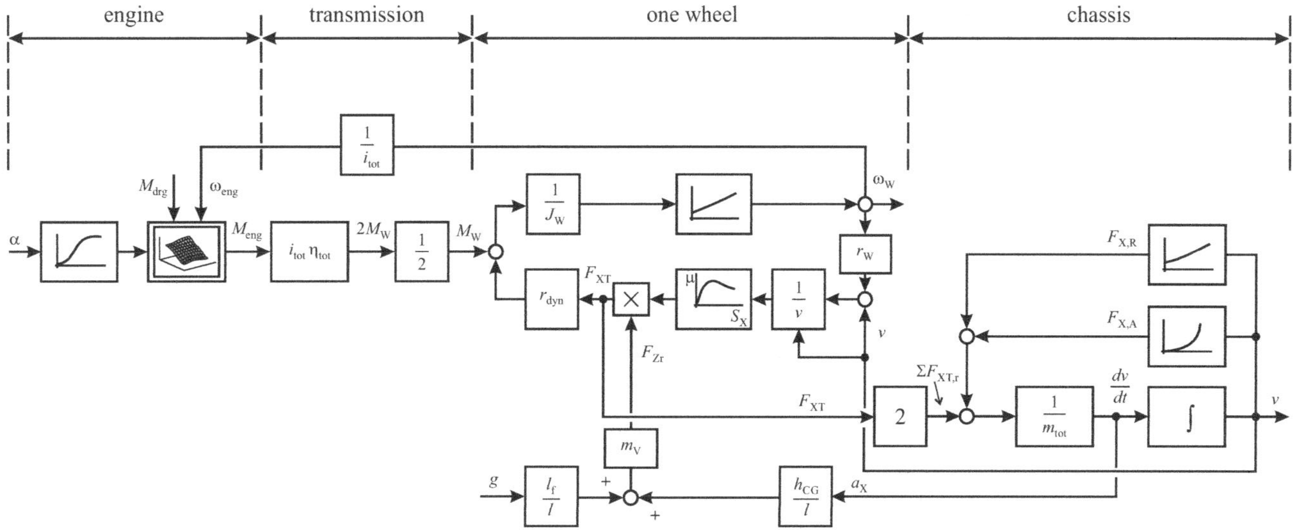


**Fig. 6.20** Signal flow for the longitudinal acceleration of a vehicle with rear-driven axle, stiff powertrain, and dynamic wheel model with slip and dynamic vertical force. Even road: $\lambda = 0$

In order to simplify the nonlinear vehicle equation (6.6.1), small changes $\Delta M_{\mathrm{eng}}$ around a steady-state $\left[\overline{M}_{\mathrm{eng}}, \overline{v}, \overline{i}_{\mathrm{tr}}\right]$ are considered. Linearization of the air drag and rolling resistance results with (6.5.4) and (6.5.5) in

$$\Delta F_{\mathrm{X,A}} = \rho_{\mathrm{A}} c_{\mathrm{wx}} A_{\mathrm{X}} \overline{v} \Delta v = c_{\mathrm{A}} \overline{v} \Delta v \tag{6.6.3}$$

$$\Delta F_{\mathrm{X,R}} = f_{\mathrm{R1}} F_{\mathrm{Z}} \Delta v = c_{\mathrm{R}} \Delta v, \tag{6.6.4}$$

where $f_{\mathrm{R1}}$ is the velocity-dependent rolling force coefficient for all wheels. For an even road with $F_{\mathrm{X,c}} = 0$, then one obtains for (6.6.1)

$$m_{\mathrm{tot}} \frac{dv(t)}{dt} + c_{\mathrm{R}v} \Delta v(t) = c_{\mathrm{MF}} \Delta M_{\mathrm{eng},}(t) \tag{6.6.5}$$

with

$$c_{\mathrm{R}v} = c_{\mathrm{A}} \overline{v} + c_{\mathrm{R}} \tag{6.6.6}$$

$$c_{\mathrm{MF}} = \frac{i_{\mathrm{tot}} \eta_{\mathrm{tot}}}{r_{\mathrm{dyn}}} \tag{6.6.7}$$

and the transfer function

$$G_{\mathrm{M}v}(s) = \frac{\Delta v(s)}{\Delta M_{\mathrm{eng}}(s)} = \frac{K_{\mathrm{M}v}}{(1 + T_{\mathrm{acc}} s)} \tag{6.6.8}$$

with

$$K_{\mathrm{M}v} = \frac{c_{\mathrm{MF}}}{c_{\mathrm{R}v}}, \tag{6.6.9}$$

$$T_{\mathrm{acc}} = \frac{m_{\mathrm{tot}}}{c_{\mathrm{R}v}}. \tag{6.6.10}$$

For small changes of the engine torque $\Delta M_{\mathrm{eng}}$ after a change $\Delta\alpha = \Delta\alpha_{\mathrm{ped}}$ of the accelerator pedal, the simplified second-order model (6.2.19) is used. Combining with (6.6.8), this leads to

$$G_{\alpha v}(s) = \frac{\Delta v(s)}{\Delta \alpha_{\mathrm{ped}}(s)} = \frac{K_{\alpha v}}{(1 + T'_{\mathrm{act}} s)(1 + T_{\mathrm{int}} s)(1 + T_{\mathrm{acc}} s)} \tag{6.6.11}$$

with the gain, see also (6.2.20),

$$K_{\alpha v} = K_{\mathrm{M}} K_{\mathrm{M}v}. \tag{6.6.12}$$

Hence, a third-order model results with a larger time constant $T_{\mathrm{acc}}$ and two smaller time constants $T'_{\mathrm{act}}$ and $T_{\mathrm{int}}$.

The parameters depend on the operating point $\left(\overline{M}_{\mathrm{eng}}, \overline{n}_{\mathrm{eng}}\right)$ of the powertrain, $\overline{c}_{\mathrm{MF}}$ of the direction, and $\overline{c}_{\mathrm{R}v}$ of the vehicle. Figure 6.21 depicts a corresponding

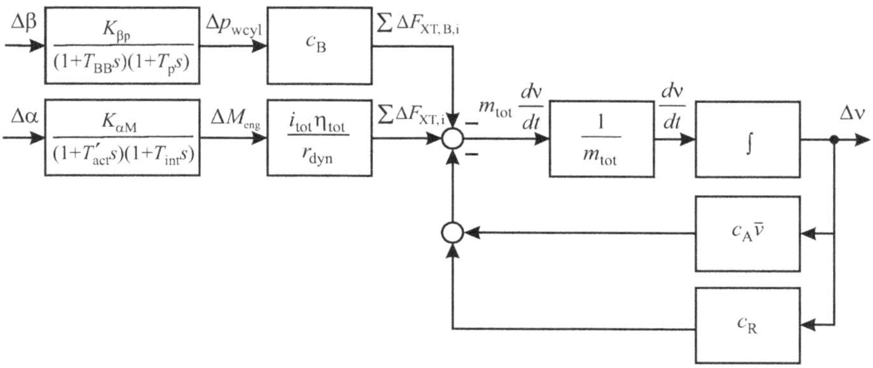


**Fig. 6.21** Signal flow chart for the longitudinal acceleration of a vehicle with stiff powertrain by changing the throttle position $\alpha$ or the brake pedal position $\beta$. Linearization around a steady-state operation point

signal flow chart. The resistance parameter $c_{Rv}$ forms a feedback to the integrator with the mass storage $m_{tot}$, thus creating a proportional behavior with a first-order delay. The gain is larger, the larger the gear ratio $i_{tr}$, the steeper the $c_{eng,}$ and the smaller the $\overline{v}$, and the time constant is smaller, the larger the $i_{tr}$ and the smaller the $\overline{v}$.

It follows from (6.6.5) that the initial acceleration $\dot{v}(t=0)$ is larger, the larger the $i_{tr}$ (the lower the gear), the larger the $c_{eng,}$, and the smaller the $m_{tot}$.

## 6.6.2 Acceleration Models with Variable Slip and Vertical Forces

If for the determination of larger vehicle accelerations the tire/road slip and the dynamic vertical forces have to be taken into account, the longitudinal tire force $F_{XT}(t)$ in (6.6.1) has to be calculated with a dynamic wheel model according to Sect. 5.5. Therefore, the tire force follows from (5.5.2)

$$F_{XT} = \mu(S_X)F_Z$$

with the slip $S_X$ from (5.5.1) and the dynamic wheel model (5.5.9). Figure 6.20 depicts a signal flow chart where the signal flow of Fig. 5.23 is added, assuming a stiff drive train and using (6.5.39) and (6.5.40) for the static and dynamic vertical wheel force through a pitching torque. It is assumed that both rear-driven tires have equal slip and traction forces. Dynamic oscillations of the suspension are neglected.

## 6.7 Braking Behavior

### 6.7.1 Simplified Braking Model

In the case of braking, (6.5.1) to (6.5.7) and (6.5.21) lead to

$$m_{\text{tot}}\frac{dv(t)}{dt} = -\sum_i F_{\text{XT,B},i} - F_{\text{X,A}}(t) - F_{\text{X,R}}(t) - F_{\text{X,C}}(t) + \frac{i_{\text{tot}}\eta_{\text{tot}}}{r_{\text{dyn}}} M_{\text{eng,b}}(t) \tag{6.7.1}$$

where $F_{\text{XT,B},i}$ are the wheel braking forces and $M_{\text{eng,b}} < 0$ is a drag torque of the engine if the clutch is engaged. If the clutch is not engaged, $M_{\text{eng,b}} = 0$ has to be set and for $m_{\text{tot}}$ the ratio of inertia $J_{\text{tot}}$ has to be taken with $J_{\text{eng}} = 0$; see (6.5.17) and (6.5.23). Models of the hydraulic braking system are treated in Chap. 13.

In the following, the dynamics between the position $\beta$ of the braking pedal, the resulting brake pressure $p_{\text{mcyl}}$ in the tandem master cylinder, and the pressure $p_{\text{wcyl}}$ in the wheel brake cylinders are neglected (the time constants are about 50–250 ms). These dynamics are considered in Chap. 13.

The friction force $F_{\text{f}}$ of a brake depends on the clamping force $F_{\text{c}}$ by

$$F_{\text{f}} = C^* F_{\text{c}} \tag{6.7.2}$$

where $C^*$ is an amplification factor. The clamping force holds

$$F_{\text{c}} = A_{\text{B}}\, p_{\text{wcyl}} \tag{6.7.3}$$

where $A_{\text{B}}$ is the *brake piston surface* area and $p_{\text{wcyl}}$ the hydraulic pressure in the wheel brake cylinder. The friction force of a brake disk with friction coefficient $\mu_{\text{B}}$ is

$$F_{\text{f}} = 2\mu_{\text{B}} F_{\text{c}} = 2A_{\text{B}}\mu_{\text{B}}\, p_{\text{wcyl}} \tag{6.7.4}$$

which leads to $C^* = 2\mu_{\text{B}}$. The braking torque of one wheel then follows

$$M_{\text{WB},i} = 2A_{\text{B}} r_{\text{B}} \mu_{\text{B}}\, p_{\text{wcyl},i} = c_{\text{WB},i}\, p_{\text{wcyl},i} \tag{6.7.5}$$

where $r_{\text{B}}$ is a representation radius of the brake disk. The braking pressure in the master cylinder is related to the pedal position $\beta$ with a characteristic $p_{\text{mcyl}}(\beta)$. Then one obtains for the braking force in the contact patch of one tire with $c_{\text{B},i} = c_{\text{WB},i}/r_{\text{dyn}}$

$$F_{\text{XT,B},i} = \frac{M_{\text{WB},i}}{r_{\text{dyn}}} = c_{\text{B},i}\, p_{\text{wcyl},i} = c_{\text{B},i}\, p_{\text{mcyl}}(\beta). \tag{6.7.6}$$

This assumes that the tire braking force only depends on the brake fluid pressure with a constant amplification factor $c_{\text{B},i}$ and that the brake fluid pressure is the same for all wheels. However, this holds only approximately for *small braking accelerations* $a_{\text{X}}$.

The resulting braking forces of the front wheel brakes are

$$F_{\mathrm{XT,B,f}} = \sum_{i=l}^{r} c_{\mathrm{B},i,\mathrm{f}}\, p_{\mathrm{mcyl}}(\beta) \tag{6.7.7}$$

and of the rear wheel brakes

$$F_{\mathrm{XT,B,r}} = \sum_{i=l}^{r} c_{\mathrm{a},i} c_{\mathrm{B},i,\mathrm{r}}\, p_{\mathrm{wcyl}}(\beta) = c_{\mathrm{B,r}}\, p_{\mathrm{wcyl}}(\beta) \tag{6.7.8}$$

where $c_{\mathrm{a},i}$ is an attenuation factor for lower brake pressure at the rear wheels in the case of a braking force limiter or because of smaller brake disks. The sum of the braking tire forces then becomes

$$F_{\mathrm{XTB}\,\Sigma} = \sum_{i} F_{\mathrm{XT,B}i} = c_{\mathrm{B}}\, p_{\mathrm{wcyl}}(\beta) \tag{6.7.9}$$

with $c_{\mathrm{B}} = 2(c_{\mathrm{B,f}} + c_{\mathrm{B,r}})$. Linearization of (6.7.1) around the operation point $\left[\overline{p}_{\mathrm{mcyl}}, \overline{v}, \overline{i}_{\mathrm{tr}}\right]$ and neglecting the engine drag torque $M_{\mathrm{eng,b}}$ results with (6.7.7) in

$$m_{\mathrm{tot}} \frac{dv(t)}{dt} + c_{\mathrm{R}v} \Delta v(t) = -\Delta F_{\mathrm{XTB}\,\Sigma}(t) \tag{6.7.10}$$

or

$$G_{\mathrm{p}v}(s) = \frac{\Delta v(s)}{\Delta F_{\mathrm{XTB}\,\Sigma}(s)} = -\frac{K_{\mathrm{F}v}}{1 + T_{\mathrm{acc}} s} \tag{6.7.11}$$

with

$$K_{\mathrm{F}v} = \frac{1}{c_{\mathrm{R}v}}$$
$$T_{\mathrm{acc}} = \frac{m_{\mathrm{tot}}}{c_{\mathrm{R}v}}.$$

The dynamic behavior between a small brake pedal change $\Delta\beta$ and a change of the sum of wheel braking forces $\Delta F_{\mathrm{XTB}\,\Sigma}$ can be modeled by (13.2.40) with a second-order system having two time constants. This leads together with (6.7.11) to

$$\begin{aligned} G_{\beta v}(s) &= \frac{\Delta v(s)}{\Delta \beta(s)} = -\frac{K_{\beta v}}{(1 + T_{\mathrm{BB}} s)(1 + T_{\mathrm{p}}' s)(1 + T_{\mathrm{acc}} s)} \\ K_{\beta v} &= K_{\beta \mathrm{F}} K_{\mathrm{F}v}\ ;\ K_{\beta \mathrm{F}} = c_{\mathrm{B}} K_{\beta \mathrm{p}}\ ;\ K_{\beta \mathrm{p}} = \frac{\Delta p_{\mathrm{wcyl}}}{\Delta \beta}. \end{aligned} \tag{6.7.12}$$

Also for braking, a third-order model results with one large time constant and two small time constants.

The linearized model should only be used for small changes of the brake pedal. For a complete braking maneuver, the nonlinear equation (6.7.1) has to be applied.

However, in the case of strong braking where the air drag, rolling resistance, and engine drag torque are small compared to the braking forces $F_{\mathrm{XTB}\sum}$ at the wheels, (6.7.1) can be reduced to

$$m_{\mathrm{tot}}\frac{dv(t)}{dt} = -F_{\mathrm{XTB}\sum}(t) \tag{6.7.13}$$

or as a transfer function after linearization

$$G_{\mathrm{F}v}(s) = \frac{\Delta v(s)}{\Delta F_{\mathrm{XTB}\sum}(s)} = -\frac{1}{m_{\mathrm{tot}}s}. \tag{6.7.14}$$

The deceleration is then approximately

$$a_{\mathrm{X}}(t) = \frac{dv(t)}{dt} = -\frac{F_{\mathrm{XTB}\sum}(t)}{m_{\mathrm{tot}}} \tag{6.7.15}$$

and therefore proportional to the braking force.

### 6.7.2 Braking Models with Variable Slip and Vertical Forces

Corresponding to the acceleration model shown in Fig. 6.20, a braking model with tire/road slip and dynamic vertical forces for the rear and front wheels is illustrated in Fig. 6.22. Again, the longitudinal tire forces $F_{\mathrm{XT}}(t)$ (5.5.2) are determined with the dynamic wheel models derived in Sect. 5.5 and depicted in Fig. 5.23.

It is assumed that the left and right wheels of the rear and front axle have the same braking torque and slip. Dynamic oscillations of the suspensions are neglected. The signal flow shows the negative and positive feedback of the deceleration on the vertical tire forces which leads to a decrease of braking forces of the rear wheels and an increase of braking forces of the front wheels through the pitching torque with $a_{\mathrm{X}} < 0$. These relations, which take the tire/road slip into account, have to be applied for *higher braking accelerations*.

If the braking forces are determined by the tire/road friction forces

$$\begin{aligned} F_{\mathrm{XTf}} &= \mu_{\mathrm{Xf}} F_{\mathrm{Zf}}, \\ F_{\mathrm{XTr}} &= \mu_{\mathrm{Xr}} F_{\mathrm{Zr}}, \end{aligned} \tag{6.7.16}$$

then (6.6.1) becomes assuming that the resistance and drag forces are negligible small compared to the brake forces, by using (6.5.39) and (6.5.40) for the vertical axle forces

$$m_{\mathrm{tot}} a_{\mathrm{X}} = -\left(\mu_{\mathrm{Xf}}\frac{l_{\mathrm{r}}}{l} + \mu_{\mathrm{Xr}}\frac{l_{\mathrm{f}}}{l}\right) m_{\mathrm{v}} g - (-\mu_{\mathrm{Xf}} + \mu_{\mathrm{Xr}}) m_{\mathrm{v}} \frac{h_{\mathrm{CG}}}{l} a_{\mathrm{X}}. \tag{6.7.17}$$

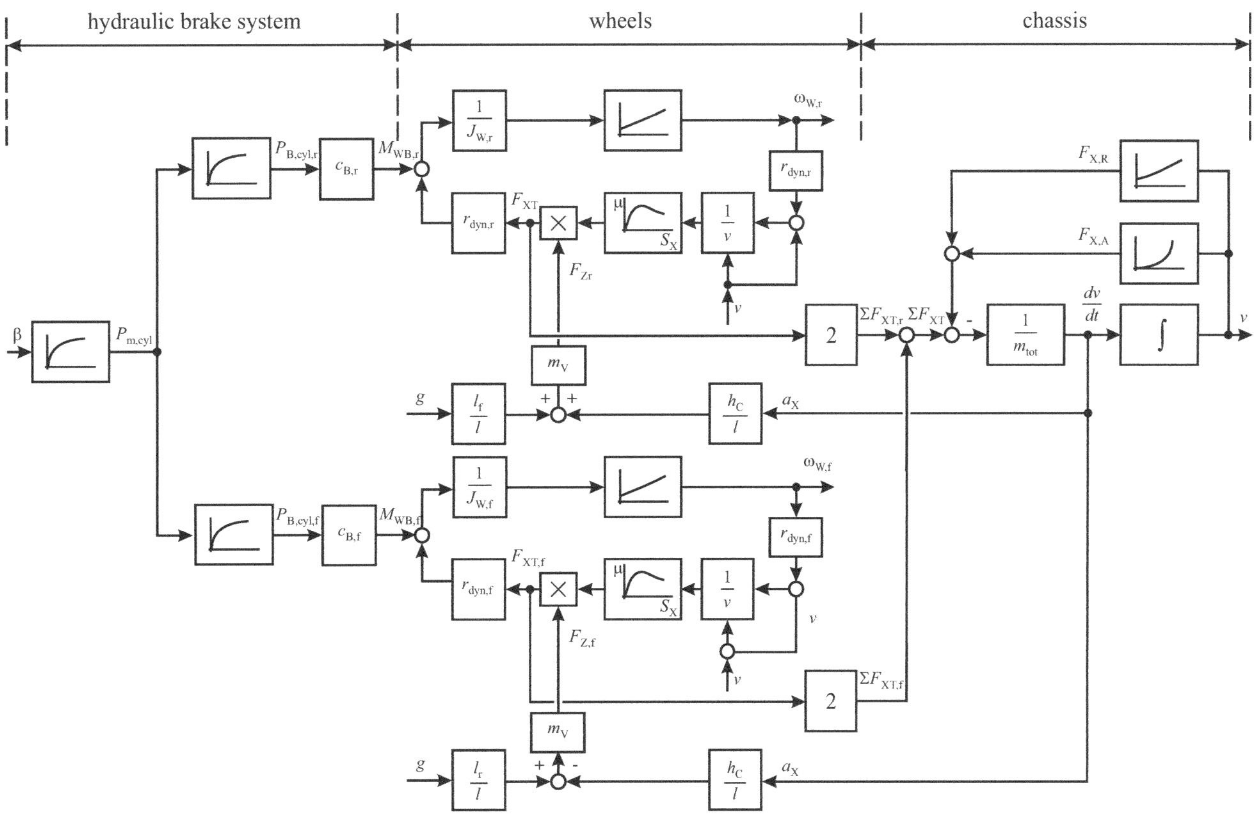


**Fig. 6.22** Signal flow for longitudinal deceleration of a vehicle by braking with rear and front wheels, dynamic wheel models with slip and dynamic vertical forces. Even road: $\lambda = 0$

If the friction coefficients for the front and rear wheels are equal $\mu_{\mathrm{Xf}} = \mu_{\mathrm{Xr}} = \mu_{\mathrm{X}}$, it holds with the rotational mass factor $\lambda_{\mathrm{m}}$, (6.5.23),

$$a_{\mathrm{X}} = \mu_{\mathrm{X}} \frac{g}{\lambda_{\mathrm{m}}} \tag{6.7.18}$$

and if braking with the maximal $\mu_{\mathrm{Xmax}}$, the maximal deceleration becomes

$$a_{\mathrm{X}} = -\mu_{\mathrm{Xmax}} \frac{g}{\lambda_{\mathrm{m}}}. \tag{6.7.19}$$

The vehicle velocity then decreases linearly from initial velocity $v_0$

$$v(t) = v_0 - a_{\mathrm{X}} t \tag{6.7.20}$$

and the braking time to $v = 0$ is

$$t_{\mathrm{stop}} = \frac{v_0}{a_{\mathrm{X}}}. \tag{6.7.21}$$

The brake distance becomes

$$\begin{aligned} x_{\mathrm{bd}} &= \int_0^{t_{\mathrm{stop}}} v(t) dt = \left( v_0 t - a_{\mathrm{X}} \frac{t^2}{2} \right) \Big|_0^{t_{\mathrm{stop}}} \\ &= \frac{v_0^2}{2 a_{\mathrm{X}}}. \end{aligned} \tag{6.7.22}$$

Inserting the maximal deceleration (6.7.19) yields the minimal braking distance under ideal conditions

$$x_{\mathrm{bd,min}} = \frac{\lambda_{\mathrm{m}}}{2 \mu_{\mathrm{Xmax}} g} v_0^2. \tag{6.7.23}$$

*Summarizing*, this chapter on the longitudinal vehicle behavior has shown that several torque generating, torque converting, and elastomechanical torque transmitting components act together and allow, with corresponding process models and signal flow representations, to indicate the operation-point-and frequency-dependent characteristics. The model descriptions give hints for the constructive design and allow a systematic approach for the design of automatic control systems like ABC, ESC, and ACC. A recent review of different longitudinal models is presented by Figel et al. (2019).

## References

Dach H, Gruhle W, Köpf P (2001) PKW-Automatikgetriebe, 2nd edn. Moderne Industrie, Landsberg/Lech

Figel KJ, Schultalbers M, Svaricek F (2019) Review and experimental evaluation of models for drivability simulation with focus on tire modeling. Forsch Ingenieurwes 83:105–118

Germann S (1997) Modellbildung und modellgestützte Regelung der Fahrzeuglängsdynamik. Dissertation Technische Universität Darmstadt. Fortschr.-Ber. VDI Reihe 12, 309. VDI Verlag, Düsseldorf

Gillespie T (1992) Fundamentals of vehicles dynamics. SAE, Warrendale

Guzzella L, Onder C (2010) Introduction to modeling and control of internal combustion engine systems, 2nd edn. Springer, Berlin

Heissing B, Ersoy M (eds) (2011) Fahrwerkhandbuch: Grundlagen, Fahrdynamik, Komponenten, Systeme, Mechatronik. Perspektiven, ATZ/MTZ Fachbuch, Vieweg

Isermann R (2005) Mechatronic systems - fundamentals, 2nd edn. Springer, London

Isermann R (2014) Engine modeling and control. Springer, Berlin

Karnopp D, Rosenberg R (1975) System dynamics: a unified approach. Wiley, New York

Karnopp D, Margolis D, Rosenberg R (1990) System dynamics: a unified approach. Wiley, New York

Kiencke U, Nielsen L (2000) Automotive control systems. For engine, driveline and vehicle. Springer, Berlin

Kiencke U, Nielsen L (2005) Automotive control systems: for engine, driveline, and vehicle, 2nd edn. Springer, Berlin

MacFarlane A (ed) (1970) Dynamical system models. G.G. Harrop, London

Mitschke M, Wallentowitz H (2004) Dynamik der Kraftfahrzeuge, 4th edn. Springer, Berlin

Naunheimer H, Bertsche B, Lechner G (2007) Fahrzeuggetriebe: Grundlagen, Auswahl, Auslegung und Konstruktion, 2nd edn. Springer, Berlin

Pfeiffer K (1997) Fahrsimulation eines Kraftfahrzeuges mit einem dynamischen Motorenprüfstand. Dissertation Technische Hochschule Darmstadt. Fortschr.-Ber. VDI Reihe 12, 336, VDI Verlag, Düsseldorf

Popp K, Schiehlen W (1993) Fahrzeugdynamik: Eine Einführung in die Dynamik des Systems Fahrzeug-Fahrweg. Teubner, Stuttgart

Reik W (1998) Höherer Komfort und weniger Geräusch durch das Zweimassenschwungrad. ATZ - Automobiltechnische Zeitschrift 100(12):896–900

Reik W, Albers A, Schnurr M (1990) Torsionsschwingungen im Antreibsstrang von Kraftfahrzeugen. In: 4. Internationales LuK-Kolloquium, Baden-Baden, Germany

Reimpell J, Hoseus K (1992) Fahrzeugtechnik: Fahrzeugmechanik. Vogel, Würzburg

GmbH RB (ed) (2011) Automotive handbook, 8th edn. Bentley Publishers, Cambridge

GmbH RB (ed) (2018) Automotive handbook, 10th edn. Wiley, Chichester

Schaffnit J (2002) Simulation und Control Prototyping zur Entwicklung von Steuergerätefunktionen für aufgeladene Nutzfahrzeug-Dieselmotoren. Dissertation Technische Universität Darmstadt. Fortschr.-Ber. VDI Reihe 12, 492. VDI Verlag, Düsseldorf

Schmidt C (1995) Digitale kurbelwinkelsynchrone Modellbildung und Drehschwingungsdämpfung eines Dieselmotors mit Last. Dissertation Technische Hochschule Darmstadt. Fortschr.-Ber. VDI Reihe 12, 253. VDI Verlag, Düsseldorf

Sinsel S (2000) Echtzeitsimulation von Nutzfahrzeug-Dieselmotoren mit Turbolader zur Entwicklung von Motormanagementsystemen. Logos, Doctoral thesis. University of Technology, Darmstadt. Berlin

Walter A, Kiencke U, Jones S, Winkler T (2007) The DMB as a virtual sensor. MTZ worldwide 68(6):24–27

# Lateral Vehicle Behavior

# 7

In order to describe and analyze the stationary and dynamic behavior of driving vehicles, the motions in horizontal and vertical directions and around the three axes have to be considered. Regarding the lateral behavior, first a one-track model is considered, where the two wheels of an axle are concentrated in one middle wheel to result in a most simple approach. This begins with kinematic models, considering geometric relations. Then, the dynamic behavior is treated with steering angles as inputs and yaw rate, slip angle and lateral velocity as outputs. These one-track models allow to analyze some main properties with regard to steering. The next step is the development of two-track models, where the forces on all four wheels in horizontal and vertical direction are taken into account. Therefore, the tire-road behavior of the individual wheels is considered with regard to lateral and longitudinal forces and passive suspensions as connections between the wheels and the body. As different outputs for the lateral behavior are possible and roads without or with slopes have to be considered, different one-track and two-track models are derived.

Figure 7.1 depicts a survey scheme for the treated lateral vehicle models. The models may be represented with yaw rate and lateral velocity as outputs or with yaw rate and slip angle. An early publication for the one-track model is Riekert and Schunck (1940). A more recent treatment of one- and two-track models can be found in Zomotor and Reimpell (1991), Popp and Schiehlen (1993), Ammon (1997), van Zanten et al. (1998), Kiencke and Nielsen (2005), van Zanten (2006), Schramm et al. (2010), and Mitschke and Wallentowitz (2014).

## 7.1 Kinematic Models for Lateral Behavior

The most simple models for the lateral behavior of vehicles are obtained for stationary cornering on a horizontal road plane using geometrical relationships. The vehicle is presented by a one-track model (bicycle model), where the two front and rear wheels are represented by one central wheel each; see Fig. 7.2. The front wheel steering



R. Isermann, *Automotive Control*, ATZ/MTZ-Fachbuch,
https://doi.org/10.1007/978-3-642-39440-9_7

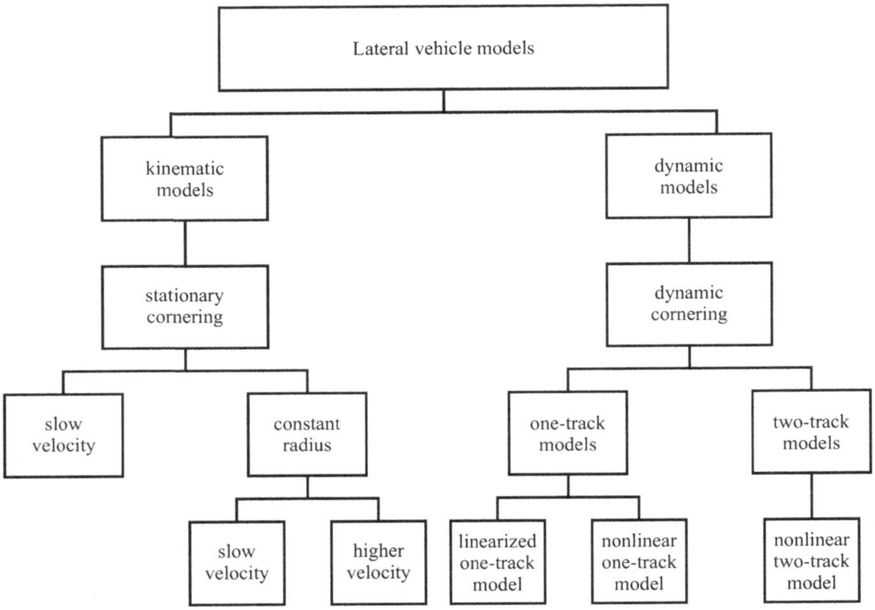


| | | | | | |
|---|---|---|---|---|---|
| **model representation:** | | | | | |
| yaw rate/ velocity | X | X | X | | X |
| yaw rate/ slip angle | | | X | X | X |

**Fig. 7.1** Survey scheme of lateral vehicle models

angle is $\delta_f$, relative to the longitudinal axis $X$. The center of gravity CG then moves around the instantaneous center of motion or *momentary pole* MP with velocity $v$. For *low velocities* $v$ a planar motion can be assumed where each wheel moves in the direction of the wheel. Hence, the MP is located at the intersection point of the lines perpendicular to the orientation of the two wheels. The velocity $v$ at the CG then shows an angle $\beta$ with the longitudinal axis which is called the *slip angle* of the vehicle. Partitioning the vector $v$ in its directions $v_X$ and $v_Y$ leads to the relation

$$\beta = \arctan\left(\frac{v_Y}{v_X}\right), \tag{7.1.1}$$

which can for small $\beta$ be approximated by

$$\beta \approx \frac{v_Y}{v_X}. \tag{7.1.1a}$$

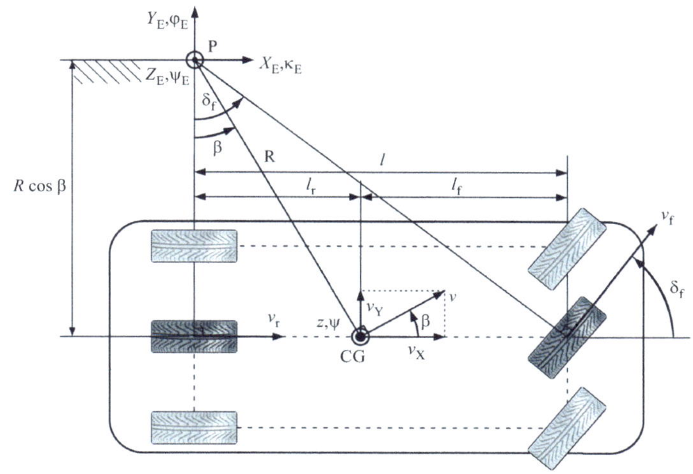


**Fig. 7.2** Scheme of a vehicle for stationary cornering with kinematic variables and low velocity (no tire slip angles)

Because of the planar motion it holds for the yaw rate in the CG

$$\dot{\psi} = \frac{v}{R}, \tag{7.1.2}$$

with $R$ the *turning radius*. If the vehicle corners with *higher velocity*, the lateral tire forces generate side slip angles $\alpha_f$ and $\alpha_r$ and the momentary pole changes to MP, see Fig. 7.3, as intersection point between the lines perpendicular to the wheel velocities. The momentum path radius (curvature radius) from the MP to the CG then becomes $R_P$.

The slip angle follows (7.1.1) and the yaw rate becomes

$$\dot{\psi} = \frac{v}{R_P}, \tag{7.1.3}$$

with $R_P$ as *path radius*. This relation is also valid for each wheel such that the wheel velocities become

$$v_{i,j} = R_{Pij}\,\dot{\psi} \quad i \in \{\text{fl, fr}\},\ j \in \{\text{rl, rr}\}. \tag{7.1.4}$$

The radii $R_{Pij}$ are not known in general. However, taking the difference of the wheel velocities it holds for the front wheels

$$v_{fr} - v_{fl} = \dot{\psi}\,(R_{P,fr} - R_{P,fl}) = \dot{\psi}\,\Delta R_P. \tag{7.1.5}$$

Replacing $\Delta R_P$ with the distance $b_f$ of the front wheels leads to

$$\dot{\psi} = \frac{v_{fr} - v_{fl}}{b_f} = \frac{r_{dyn}}{b_f}(\omega_{fr} - \omega_{fl}). \tag{7.1.6}$$

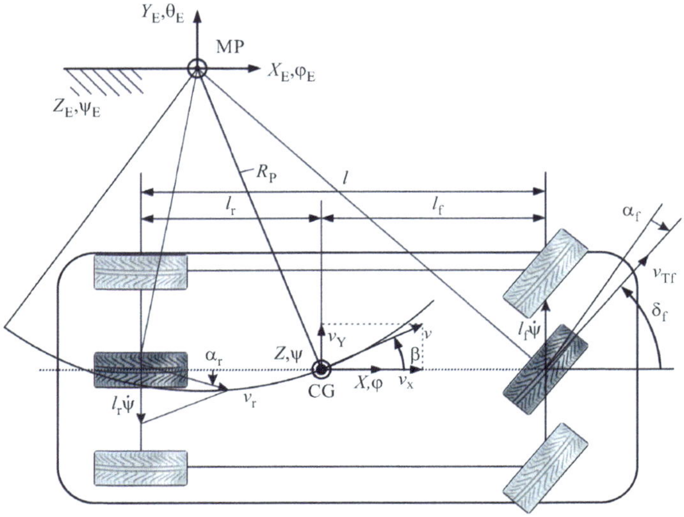


**Fig. 7.3** Scheme of a vehicle for stationary cornering with kinematic variables for higher velocity (with tire slip angles)

A corresponding relation holds for the rear wheels.

Another kinematic relation follows for the lateral (centripetal) acceleration in the CG

$$a_c = \frac{v^2}{R_P} = v\,\dot{\psi}. \tag{7.1.7}$$

The lateral acceleration in vehicle coordinates is

$$a_Y = a_c \cos\beta. \tag{7.1.8}$$

For small slip angles and $\cos\beta \approx 1$ follows

$$\dot{\psi} = \frac{a_Y}{v}. \tag{7.1.9}$$

The relationships (7.1.6) or (7.1.9) can be used to determine the yaw rate $\dot{\psi}$ from measurements of the wheel speeds or the lateral vehicle acceleration. Similar kinematic relations exist for $a_Y$ and $\delta_f$; see Table 7.1, Halbe and Isermann (2007).

## 7.2 Dynamic One-Track Models

### 7.2.1 Nonlinear One-Track Model

To derive the lateral dynamic models, a coordinate system is fixed to the vehicle's center of gravity (CG) and Newton's laws are applied. Roll, pitch, bounce, and

**Table 7.1** Kinematic relations for lateral vehicle dynamics, $\delta_f = \delta_H / i_s$

| | Yaw rate $\dot{\psi}$ | Lateral acceleration $a_y$ | Steering wheel angle $\delta_h$ |
|---|---|---|---|
| 1 | $\frac{r_{dyn}(\omega_{fr} - \omega_{fl})}{b_f}$ | $\frac{r_{dyn}(\omega_{fr} - \omega_{fl})}{b_f} v$ | $\frac{(l_f + l_r) i_S}{v} \cdot \frac{(r_{dyn}(\omega_{fr} - \omega_{fl})}{b_f}$ |
| 2 | $\frac{r_{dyn}(\omega_{rr} - \omega_{rl})}{b_r}$ | $\frac{r_{dyn}(\omega_{rr} - \omega_{rl})}{b_r} v$ | $\frac{(l_f + l_r) i_S}{v} \cdot \frac{r_{dyn}(\omega_{rr} - \omega_{rl})}{b_f}$ |
| 3 | $\frac{\ddot{y}}{v}$ | $\dot{\psi} v$ | $\frac{(l_f + l_r) i_S}{v} \cdot \frac{\ddot{y}}{v}$ |
| 4 | $\frac{v}{(l_f + l_r) i_S} \delta_H$ | $\frac{v^2}{(l_f + l_r) i_S} \delta_H$ | $\frac{(l_f + l_r) i_S}{v} \cdot \dot{\psi}$ |

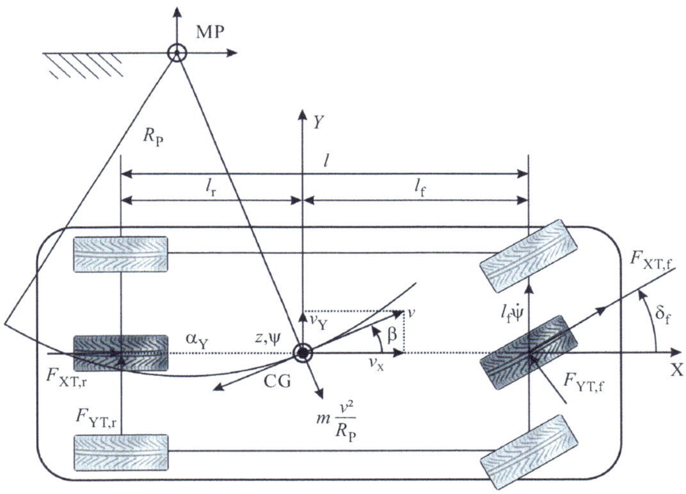


**Fig. 7.4** Scheme for modeling the lateral vehicle behavior with a one-track model indicating the resulting forces

acceleration dynamics are neglected to reduce the model to a *one-track model* with two degrees of freedom: the lateral position and the yaw angle; see Fig. 7.4. Further simplifications assume that, each axle shares the same steering angles for each wheel. The basic idea of a one-track model goes back to Riekert and Schunck (1940).

If the vehicle *corners slowly*, then it turns around the momentary pole MP for $v \to 0$ which is the intersection point of the rear axle prolongation and the line perpendicular to the front wheels, Fig. 7.2. For the steering angle holds $\tan \delta_f = l/R \cos \beta \approx l/R$ for large $R \gg l$ or small $\beta$ which can be approximated for small $\delta_f$ by the so called *Ackermann angle*

$$\delta_A = \frac{l}{R}. \tag{7.2.1}$$

In the case of *higher speed cornering*, the arising centrifugal force must be compensated by lateral tire forces. As this requires side slip angles $\alpha_r$ and $\alpha_f$ of the rear and front wheel the vehicle turns on a momentary pole MP for $v > 0$, Fig. 7.3, with the path radius $R_P$ and a corresponding path curvature

$$\kappa = \frac{1}{R_P}. \tag{7.2.2}$$

The velocity of the CG is the horizontal velocity $v_h$ over ground

$$\vec{v} = \vec{v}_h = \vec{v}_X + \vec{v}_Y, \tag{7.2.3}$$

and the yaw rate of the CG is

$$\dot{\psi} = \frac{v_h}{R_P}. \tag{7.2.4}$$

The centripetal acceleration then becomes

$$a_c = \frac{v_h^2}{R_P}. \tag{7.2.5}$$

The heading angle or course angle $\nu$ of the CG in the earth fixed coordinate system $(X_E, Y_E)$ is then the direction of the vehicle velocity vector $\vec{v}_h$

$$\nu = \psi + \beta, \tag{7.2.6}$$

where the yaw angle $\psi$ is the angle between the $X_E$-axis and the $X$-axis. This results from the superposition of the kinematic slip angle $\beta$ and an additional yaw angle $\psi$ because of the wheels side slip. If both, the yaw angle $\psi(t)$ and the slip angle $\beta(t)$ change with time, the angular velocity around the vertical axis is

$$\omega_Z = \dot{\nu} = \dot{\psi} + \dot{\beta}. \tag{7.2.7}$$

It is now assumed that the velocity of the center of gravity is $v = \text{const}$. The centrifugal acceleration of the CG is then with $v = R_P\omega_Z$ and (7.2.7)

$$a_c = a_Y = \frac{v^2}{R_P} = v\omega_Z = v\left(\dot{\psi} + \dot{\beta}\right); \tag{7.2.8}$$

see also Mitschke and Wallentowitz (2014) and Schramm et al. (2010).

The velocity components of the CG then become according to Fig 7.3

$$\begin{aligned} v_{XE} &= v_X = v\cos(\psi + \beta), \\ v_{YE} &= v_Y = v\sin(\psi + \beta). \end{aligned} \tag{7.2.9}$$

For *stationary cornering* with $\beta = \text{const.}$ and $\dot{\beta} = 0$ it follows

$$a_c = v\dot{\psi} = \frac{v^2}{R_P}. \tag{7.2.10}$$

**(a) Yaw Rate/Velocity Representation**

For a cornering vehicle, according to Fig. 7.4, the motion equations follow from two force balances and one torque balance. This leads directly to a *yaw rate/velocity representation* of the one-track model.

1. Longitudinal direction:

$$\begin{aligned}\dot{v}_X &= \frac{1}{m}\left[F_{X,f} + F_{X,r} - F_{XA} - F_{XR}\right] \\ &= \frac{1}{m}\left[F_{XT,f}\cos\delta_f - F_{YT,f}\sin\delta_f + F_{XT,r} - F_{XA} - F_{XR}\right].\end{aligned} \tag{7.2.11}$$

2. Lateral direction:

$$\begin{aligned}\dot{v}_Y &= \frac{1}{m}\left[F_{Y,f} + F_{Y,r} - F_{YA}\right] \\ &= \frac{1}{m}\left[F_{YT,f}\cos\delta_f + F_{YT,r} + F_{XT,f}\sin\delta_f - F_{YA}\right].\end{aligned} \tag{7.2.12}$$

3. Vertical axis: yaw angle $\psi$

$$\begin{aligned}\ddot{\psi} &= \frac{1}{J_Z}\left[F_{Y,f}l_f - F_{Y,r}l_r\right] \\ &= \frac{1}{J_Z}\left[\left(F_{YT,f}\cos\delta_f + F_{XT,f}\sin\delta_f\right)l_f - \left(F_{YT,r}l_r - F_{YA}d_A\right)\right].\end{aligned} \tag{7.2.13}$$

The distance between the center of gravity and the center of lateral air forces is described with $d_A$.

For the air drag forces holds, Reimpell and Hoseus (1992)

$$F_{XA} = \frac{\rho_A}{2}c_{WX}A_X v_{res,X}^2, \tag{7.2.14}$$

$$F_{YA} = \frac{\rho_A}{2}c_{WY}A_Y v_{res,Y}^2, \tag{7.2.15}$$

with

$$\begin{aligned}v_{res,X} &= v_X - v_{AX}, \\ v_{res,Y} &= v_Y - v_{AY},\end{aligned} \tag{7.2.16}$$

where $c_{WX}$ and $c_{WY}$ are drag coefficients and $v_A$ is the wind velocity. The rolling resistance can be determined by

$$F_{XR} = (f_{R0} + f_{R1}v + f_{R4}v^4)F_Z; \tag{7.2.17}$$

see Mitschke and Wallentowitz (2014).

The lateral tire forces depend on the side slip angle $\alpha$ and the vertical tire force $F_Z$; see Chap. 5. For small side slip angles $\alpha$, the lateral tire force is proportional to the side slip angle $\alpha$

$$F_Y = c_\alpha(F_Z)\,\alpha, \tag{7.2.18}$$

where $c_\alpha$ is the cornering stiffness. For the side slip angle holds; see (5.2.2)

$$\alpha = \arctan\left(\frac{v_{YT}}{v_{XT}}\right), \tag{7.2.19}$$

at each wheel.

The side slip angle $\alpha$ of one wheel depends on the steering angle $\delta_f$, the vehicle slip angle $\beta$, and the yaw rate $\dot{\psi}$. A relation between the tire side slip angles and the vehicle slip angles of all wheels can be established by considering the longitudinal and lateral velocity components of the center of gravity and the tires; see Figs. 5.12 and 7.3, Mitschke and Wallentowitz (2014)

$$\begin{aligned} v_X &= v\cos\beta = v_{Tr}\cos\alpha_r, \\ v_X &= v\cos\beta = v_{Tf}\cos(\delta_f - \alpha_f), \\ v_Y &= v_{Tr}\sin\alpha_r = l_r\dot{\psi} - v\sin\beta, \\ v_Y &= v_{Tf}\sin(\delta_f - \alpha_f) = l_f\dot{\psi} + v\sin\beta. \end{aligned} \tag{7.2.20}$$

This leads for the rear and front axis to

$$\tan\alpha_r = \frac{l_r\dot{\psi} - v\sin\beta}{v\cos\beta},$$

$$\tan(\delta_f - \alpha_f) = \frac{l_f\dot{\psi} + v\sin\beta}{v\cos\beta},$$

or to express the tire side slip angles by $\delta_f$, $\beta$ and $v$

$$\begin{aligned} \alpha_r &= \arctan\left(\frac{l_r\dot{\psi} - v\sin\beta}{v\cos\beta}\right), \\ \alpha_f &= \delta_f - \arctan\left(\frac{l_f\dot{\psi} - v\sin\beta}{v\cos\beta}\right). \end{aligned} \tag{7.2.21}$$

Thus, the nonlinear one-track model in the yaw rate/velocity formulation is based on (7.2.11)–(7.2.19) and (7.2.21).

### (b) Yaw Rate/Slip Angle Representation

If the longitudinal and lateral velocity $v_X$ and $v_Y$ in (7.2.11)–(7.2.13) are replaced by (7.2.9), the velocity $v$ of the CG and the slip angle $\beta$ are introduced, this leads to the balance equations in a *yaw rate/slip angle formulation*. For a cornering vehicle, according to Fig. 7.4, the motion equations follow from *two force balances* and *one torque balance*

1. Longitudinal direction $x$:

$$-m\dot{v}\cos\beta + m\frac{v^2}{R_P}\sin\beta + F_{Xr} + F_{Xf}\cos\delta - F_{Yf}\sin\delta_f - F_{XA} - F_{XR} = 0. \tag{7.2.22}$$

2. Lateral direction $y$:

$$-m\frac{v^2}{R_P}\cos\beta - m\dot{v}\sin\beta + F_{Yr} + F_{Xf}\sin\delta_f + F_{Yf}\cos\delta_f + F_{YA} = 0. \tag{7.2.23}$$

3. Vertical axis:

$$-J_Z\ddot{\psi} + (F_{Yf}\cos\delta_f + F_{Xf}\sin\delta_f)\,l_f - F_{Yr}l_r - F_{YA}d_A = 0. \tag{7.2.24}$$

### 7.2.2 Linearized One-Track Model

The basic motion equations (7.2.11)–(7.2.13) are now simplified by assuming small steering angles and side slip angles. Then, approximations $\sin\delta \approx 0$, $\cos\delta \approx 1$, $\sin\beta \approx \beta$, $\cos\beta \approx 1$ can be made. The side slip angle of the front and rear wheel then becomes

$$\alpha_f = \delta_f - \arctan\left(\frac{l_f\dot{\psi} + v\sin\beta}{v\cos\beta}\right) \approx \delta_f - \beta - \frac{l_f\dot{\psi}}{v} \tag{7.2.25}$$

$$\alpha_r = \arctan\left(\frac{l_r\dot{\psi} - v\sin\beta}{v\cos\beta}\right) \approx -\beta + \frac{l_r\dot{\psi}}{v}. \tag{7.2.26}$$

These approximations are valid for the slip angles $\beta \leq 5°$; see Kiencke and Nielsen (2005).

The cornering stiffness $c_\alpha$ depends on the lateral friction coefficient $\mu_Y$, wheel vertical force $F_Z$, camber, toe-in, tire pressure, etc.; see Sect. 5.2 for more details.

Using these approximations, the motion equations (7.2.22)–(7.2.24) then become

$$m\dot{v} + m\frac{v^2}{R_P}\beta + F_{Xr} + F_{Xf} - F_{Yf}\delta_f - F_{XA} - F_{XR} = 0, \tag{7.2.27}$$

$$-m\frac{v^2}{R_P} - m\dot{v}\beta + F_{Yr} + F_{Xf}\delta_f + F_{Yf} + F_{YA} = 0, \tag{7.2.28}$$

$$-J_Z\ddot{\psi} + (F_{Yf} + F_{Xf}\delta_f)\,l_f - F_{Yr}l_r - F_{YA}d_A = 0. \tag{7.2.29}$$

Usually, the tire forces $F_{YT,i}$ in the wheel coordinate system have to be transformed into a vehicle coordinate system via the relations (4.1.6), (4.1.7) as later shown for the two-track model; see Sect. 7.3. However, because of the assumption of small steering angle, it is $F_{YT,i} \approx F_{Y,i}$ and this transformation is not required.

**(a) Yaw Rate/Slip Angle Representation (Stationary Cornering)**
Introducing (7.2.8), (7.2.18), (7.2.25), and (7.2.26) into (7.2.27)–(7.2.29) yields

$$m\dot{v} = F_{\mathrm{Xr}} + F_{\mathrm{Xf}} - F_{\mathrm{XA}} - F_{\mathrm{XR}}, \tag{7.2.30}$$

$$mv(\dot{\psi} + \dot{\beta}) = -m\dot{v}\beta + c_{\alpha \mathrm{r}}\left(-\beta + \frac{l_{\mathrm{r}}\dot{\psi}}{v}\right) + c_{\alpha \mathrm{f}}\left(\delta_{\mathrm{f}} - \beta - \frac{l_{\mathrm{f}}\dot{\psi}}{v}\right) + F_{\mathrm{Xf}}\delta_{\mathrm{f}} + F_{\mathrm{YA}} \tag{7.2.31}$$

$$J_{\mathrm{Z}}\ddot{\psi} = c_{\alpha \mathrm{f}}\left(\delta_{\mathrm{f}} - \beta - \frac{l_{\mathrm{f}}\dot{\psi}}{v}\right)l_{\mathrm{f}} - c_{\alpha \mathrm{r}}\left(-\beta + \frac{l_{\mathrm{r}}\dot{\psi}}{v}\right)l_{\mathrm{r}} + F_{\mathrm{Xf}}\delta_{\mathrm{f}}l_{\mathrm{f}} - F_{\mathrm{YA}}\, d_{\mathrm{A}}. \tag{7.2.32}$$

It is now assumed that the vehicle drives with a constant operating point $v = \text{const.}$ Hence, (7.2.30) becomes

$$m\dot{v} = F_{\mathrm{Xr}} + F_{\mathrm{Xf}} - F_{\mathrm{XA}} - F_{\mathrm{XR}} = 0.$$

The longitudinal driving forces $F_{\mathrm{Xr}}$ and $F_{\mathrm{Xf}}$ are then equal to the resistance forces $F_{\mathrm{XA}}$ and $F_{\mathrm{XR}}$ and have no influence on the lateral behavior.

Equation (7.2.30) is required for the determination of the velocity. Neglecting the lateral air drag forces $F_{\mathrm{YA}}$, (7.2.31) and (7.2.32) can be represented in state-space form

$$\begin{aligned} \dot{\mathbf{x}}(t) &= \mathbf{A}(t)\,\mathbf{x}(t) + \mathbf{b}(t)\,u(t), \\ \mathbf{y}(t) &= \mathbf{C}(t)\,\mathbf{x}(t), \end{aligned} \tag{7.2.33}$$

where a *first state-space representation* with the *slip angle* and *yaw rate as state variables* is

$$\begin{aligned} &\dot{\mathbf{x}}(t) = \begin{bmatrix} \dot{\beta}(t) \\ \ddot{\psi}(t) \end{bmatrix} \quad \mathbf{y}(t) = \mathbf{x}(t) = \begin{bmatrix} \beta(t) \\ \dot{\psi}(t) \end{bmatrix} \quad u(t) = \delta_{\mathrm{f}}(t) = \delta_{\mathrm{H}}(t)/i_{\mathrm{S}}, \\ &\mathbf{A} = \begin{bmatrix} -\dfrac{c_{\alpha \mathrm{f}} + c_{\alpha \mathrm{r}}}{mv} & \dfrac{c_{\alpha \mathrm{r}}l_{\mathrm{r}} - c_{\alpha \mathrm{f}}l_{\mathrm{f}}}{mv^2} - 1 \\ \dfrac{c_{\alpha \mathrm{r}}l_{\mathrm{r}} - c_{\alpha \mathrm{f}}l_{\mathrm{f}}}{J_{\mathrm{Z}}} & -\dfrac{c_{\alpha \mathrm{r}}l_{\mathrm{r}}^2 + c_{\alpha \mathrm{f}}l_{\mathrm{f}}^2}{J_{\mathrm{Z}}v}, \end{bmatrix} \\ &\mathbf{b} = \begin{bmatrix} \dfrac{c_{\alpha \mathrm{f}}}{m\,v} \\ \dfrac{c_{\alpha \mathrm{f}}l_{\mathrm{f}}}{J_{\mathrm{Z}}} \end{bmatrix} \quad \mathbf{C} = \begin{bmatrix} 1 & 0 \\ 0 & 1 \end{bmatrix}. \end{aligned} \tag{7.2.34}$$

The lateral acceleration follows from (7.2.8) and (7.2.32)

$$a_{\mathrm{Y}} = v(\dot{\psi} + \dot{\beta}) = -\frac{c_{\alpha \mathrm{f}} + c_{\alpha \mathrm{r}}}{m}\beta + \frac{c_{\alpha \mathrm{r}}l_{\mathrm{r}} - c_{\alpha \mathrm{f}}l_{\mathrm{f}}}{mv}\dot{\psi} + \frac{c_{\alpha \mathrm{f}}}{m}\delta_{\mathrm{f}}, \tag{7.2.35}$$

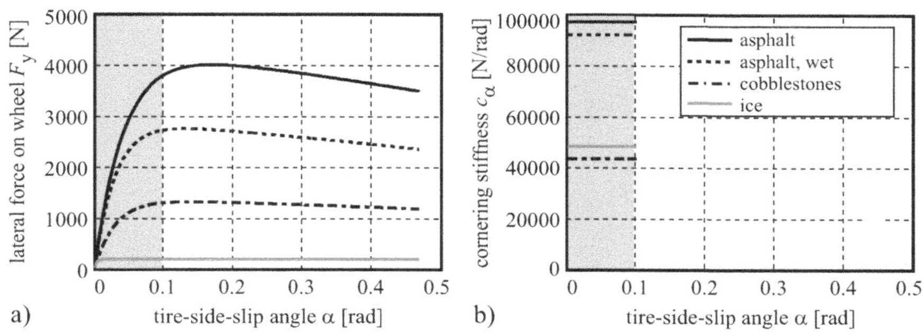


**Fig. 7.5** **a** Tire force in dependence on the tire side slip angle $\alpha$; **b** Cornering stiffness $c_\alpha$ in dependence on the tire side slip angle $\alpha$, Börner (2004)

which is in state-space form in addition to (7.2.34)

$$\underbrace{\begin{bmatrix}\dot{\beta}\\ \ddot{\psi}\\ a_{\mathrm{Y}}\end{bmatrix}}_{\mathbf{y}'} = \underbrace{\begin{bmatrix}1 & 0\\ 0 & 1\\ -\dfrac{c_{\alpha\mathrm{f}}+c_{\alpha\mathrm{r}}}{m} & \dfrac{c_{\alpha\mathrm{r}}l_{\mathrm{r}}-c_{\alpha\mathrm{f}}l_{\mathrm{f}}}{mv}\end{bmatrix}}_{\mathbf{C}'}\underbrace{\begin{bmatrix}\beta\\ \dot{\psi}\end{bmatrix}}_{\mathbf{x}} + \underbrace{\begin{bmatrix}0\\ 0\\ \dfrac{c_{\alpha\mathrm{f}}}{m}\end{bmatrix}}_{\mathbf{d}'}\delta_{\mathrm{f}}. \tag{7.2.36}$$

The one-track model (7.2.34) contains parameters which depend on the velocity $v$ and the side slip coefficients $c_{\alpha\mathrm{f}}$ and $c_{\alpha\mathrm{r}}$. If these coefficients change with time, the model has linear, time-variant behavior. Experiments have shown that it is a good approximation for lateral accelerations $a_{\mathrm{Y}} \leq 0.4\,\mathrm{g}$ and small side slip angles $\alpha < 5°$ on dry asphalt roads; see e.g. Milliken and Milliken (1995), Börner (2004), and Fig. 7.5.

Equation (7.2.30) is not included in this one-track model because of the assumption $\dot{v} = 0$. It is, however, for simulation required to determine $v$, the operation point.

Assuming stationary cornering with $\dot{v} = 0$, small angles and neglect of lateral air forces, the motion equations (7.2.31) and (7.2.32) can be simplified to

$$-mv(\dot{\beta}+\dot{\psi}) = F_{\mathrm{Y,r}} + F_{\mathrm{Y,f}} \tag{7.2.37}$$

$$J_{\mathrm{Z}}\ddot{\psi} = F_{\mathrm{Y,f}}l_{\mathrm{f}} - F_{\mathrm{Y,r}}l_{\mathrm{r}}, \tag{7.2.38}$$

or

$$\dot{\beta} = \frac{1}{mv}(F_{\mathrm{Y,f}} + F_{\mathrm{Y,r}}) - \dot{\psi} \tag{7.2.39}$$

$$\ddot{\psi} = \frac{1}{J_{\mathrm{Z}}}(F_{\mathrm{Y,f}}l_{\mathrm{f}} - F_{\mathrm{Y,r}}l_{\mathrm{r}}), \tag{7.2.40}$$

with the lateral tire forces from (7.2.18), (7.2.25), and (7.2.26)

$$F_{\mathrm{Y,f}} = c_{\alpha\mathrm{f}}\left(\delta_{\mathrm{f}} - \beta - \frac{l_{\mathrm{f}}\dot{\psi}}{v}\right) \tag{7.2.41}$$

$$F_{\mathrm{Y,r}} = c_{\alpha\mathrm{r}}\left(-\beta + \frac{l_{\mathrm{r}}\dot{\psi}}{v}\right). \tag{7.2.42}$$

A state variable model, according to (7.2.34), then contains the state vector

$$\mathbf{x} = \left[\beta\ \dot{\psi}\right]^{\mathrm{T}},$$

as (7.2.34).

These equations allow to develop a signal flow diagram with $\delta_{\mathrm{f}}$ as input and $\beta$ and $\dot{\psi}$ as outputs; see Fig. 7.6.

It shows that after an input excitation with the steering angle $\delta_{\mathrm{f}}$ the yaw rate $\dot{\psi}$ generates a negative feedback via $F_{\mathrm{Y,f}}$ and a positive feedback via $F_{\mathrm{Y,r}}$. Hence, a negative feedback from the lateral front wheel force, which damps and stabilizes, and a positive feedback from the lateral rear wheel force, which leads to less damping and destabilizes, act together. The slip angle $\beta$ shows two negative, stabilizing feedbacks. In addition, $\dot{\psi}$ acts negative on $\dot{\beta}$, thus reducing the vehicle slip angle. If during stationary cornering with $\delta_{\mathrm{f}} = \text{const.}$, $\dot{\psi} = \text{const.}$ and $\beta = \text{const.}$, suddenly the cornering stiffness of the rear tire $c_{\alpha\mathrm{r}}$ becomes smaller and $c_{\alpha\mathrm{f}}$ stays constant, it follows $\ddot{\psi} > 0$ and $\psi$ increases, thus resulting in an oversteering and a destabilizing reaction of the yaw rate $\dot{\psi}$. This underlines the importance of the rear wheel cornering stiffness $c_{\alpha\mathrm{r}}$.

**(b) Yaw Rate/Velocity Representation (Stationary Cornering)**
Assuming constant velocity $\dot{v} = 0$ and use of (7.2.8), neglecting $F_{\mathrm{YA}}$ and replacing $\beta$ by

$$\beta = \arctan\frac{v_{\mathrm{Y}}}{v_{\mathrm{X}}} \approx \frac{v_{\mathrm{Y}}}{v_{\mathrm{X}}} \approx \frac{v_{\mathrm{Y}}}{v},$$

which leads to the side slip angles

$$\begin{aligned}\alpha_{\mathrm{f}} &\approx \delta - \frac{v_{\mathrm{Y}}}{v} - l_{\mathrm{f}}\frac{\dot{\psi}}{v},\\ \alpha_{\mathrm{r}} &\approx -\frac{v_{\mathrm{Y}}}{v} + l_{\mathrm{r}}\frac{\dot{\psi}}{v},\end{aligned} \tag{7.2.43}$$

results with (7.2.18) and (7.2.30)–(7.2.32) in

$$m\dot{v} = F_{\mathrm{Xr}} + F_{\mathrm{Xf}} - F_{\mathrm{XA}} - F_{\mathrm{R}} = 0, \tag{7.2.44}$$

$$\dot{v}_{\mathrm{Y}} = \frac{1}{m}\left[c_{\alpha\mathrm{f}}\left(\delta - \frac{v_{\mathrm{Y}}}{v} - \frac{l_{\mathrm{f}}\dot{\psi}}{v}\right) + c_{\alpha\mathrm{r}}\left(-\frac{v_{\mathrm{Y}}}{v} + \frac{l_{\mathrm{r}}\dot{\psi}}{v}\right)\right], \tag{7.2.45}$$

$$\ddot{\psi} = \frac{1}{J_{\mathrm{z}}}\left[c_{\alpha\mathrm{f}}l_{\mathrm{f}}\left(\delta - \frac{v_{\mathrm{Y}}}{v} - \frac{l_{\mathrm{f}}\dot{\psi}}{v}\right) - c_{\alpha\mathrm{r}}l_{\mathrm{r}}\left(-\frac{v_{\mathrm{Y}}}{v} + l_{\mathrm{r}}\frac{\dot{\psi}}{v}\right)\right]. \tag{7.2.46}$$

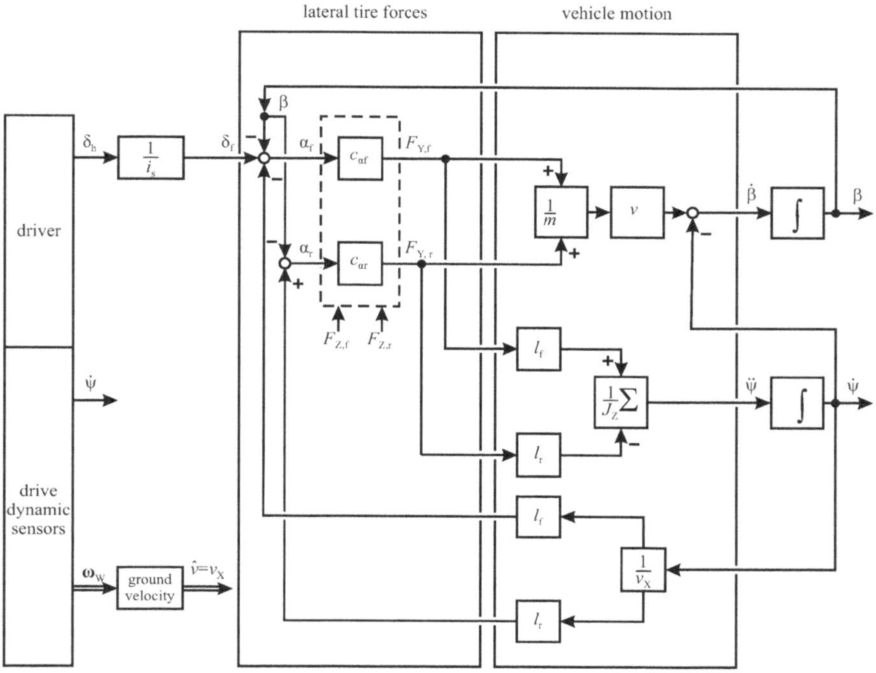


**Fig. 7.6** Signal flow of a linearized one-track model for steering and stationary cornering. Yaw rate/slip angle representation

The last two equations then give the state-space model

$$\begin{bmatrix} \ddot y \\ \ddot\psi \end{bmatrix} = \begin{bmatrix} -\frac{c_{\alpha f}+c_{\alpha r}}{mv} & -\frac{c_{\alpha r}+c_{\alpha f}l_f}{mv} \\ \frac{c_{\alpha r}l_r - c_{\alpha f}l_f}{J_Z v} & -\frac{c_{\alpha r}l_r^2 + c_{\alpha f}l_f^2}{J_Z v} \end{bmatrix} \begin{bmatrix} \dot y \\ \dot\psi \end{bmatrix} + \begin{bmatrix} \frac{c_{\alpha f}}{m} \\ \frac{c_{\alpha f}l_f}{J_Z} \end{bmatrix} \delta_f . \tag{7.2.47}$$

Thus, the state variable vector consists of the yaw rate and the lateral velocity

$$\mathbf{x} = [\dot y \ \dot\psi]^T, \tag{7.2.48}$$

and the state variable form becomes

$$\begin{aligned} \dot{\mathbf{x}}(t) &= \mathbf{A}'\mathbf{x}(t) + \mathbf{b}u(t) \\ \mathbf{y}(t) &= \mathbf{C}\mathbf{x}(t), \end{aligned} \tag{7.2.49}$$

as (7.2.33), but other $\mathbf{A}'$.

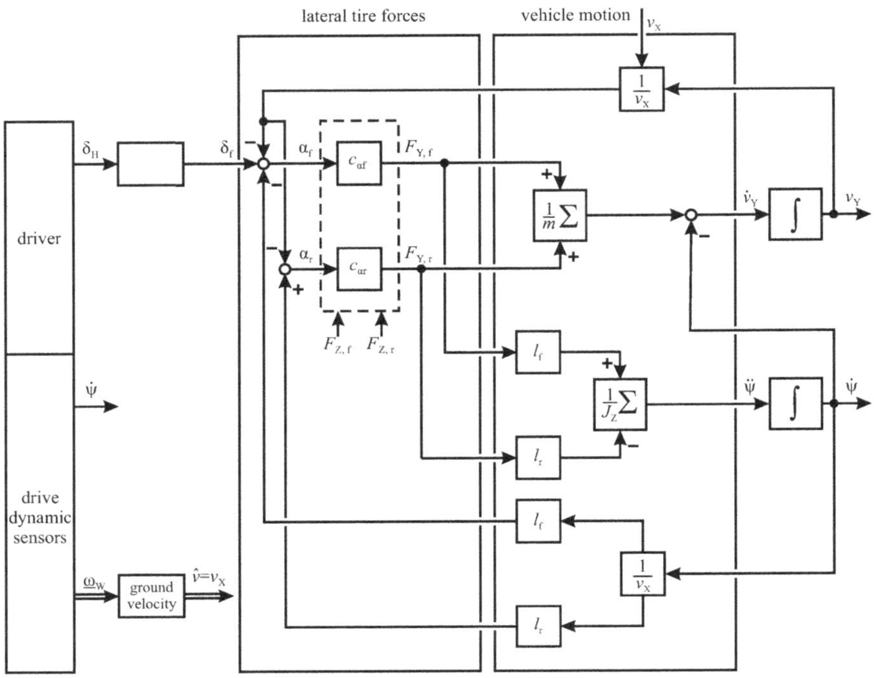


**Fig. 7.7** Signal flow of a linearized one-track model for steering and stationary cornering. Yaw rate/velocity representation

Based on (7.2.45) and (7.2.46), the signal flow of Fig. 7.7 can be drawn. Compared to Fig. 7.6 $\beta$ is replaced by $v_Y/v$. Again, $\dot{\psi}$ generates a positive and a negative feedback. The lateral velocity $v_Y$ shows two negative feedbacks, as the slip angle $\beta$ in Fig. 7.5.

### 7.2.3 Parameter Variations

To discuss the influence of parameter variations on the lateral vehicle behavior, the yaw rate transfer function $G_1(s)$ between steering wheel angle $\delta_H$ and yaw rate $\dot{\psi}$ with constant parameters, following from (7.2.31) and (7.2.33), is considered, Börner (2004)

$$G_1(s) = \frac{\dot{\psi}(s)}{\delta_H(s)} = \frac{\Delta\dot{\psi}(s)}{\Delta\delta_H(s)} = \frac{b_0 + b_1 s}{a_0 + a_1 s + a_2 s^2}, \tag{7.2.50}$$

where

$$
\begin{aligned}
b_0 &= \left.\frac{\dot{\psi}}{\delta_H}\right|_{\text{stat}} = \frac{1}{c_{\alpha f}\, c_{\alpha r}\, l^2 + m\, v^2 (l_r\, c_{\alpha r} - l_f\, c_{\alpha f})} \frac{c_{\alpha f}\, c_{\alpha r}\, v\, l}{i_S} \\
b_1 &= \frac{m\, v\, l_f}{c_{\alpha r}\, l} b_0 \\
a_0 &= 1 \\
a_1 &= \frac{2D}{\omega_0} = \frac{J_Z\, v (c_{\alpha f} + c_{\alpha r}) + m\, v (l_f^2\, c_{\alpha f} + l_r^2\, c_{\alpha r})}{c_{\alpha f}\, c_{\alpha r}\, l^2 + m\, v^2 (l_r\, c_{\alpha r} - l_f\, c_{\alpha f})} \\
a_2 &= \frac{1}{\omega_0^2} = \frac{J_Z\, m\, v^2}{c_{\alpha f}\, c_{\alpha r}\, l^2 + m\, v^2 (l_r\, c_{\alpha r} - l_f\, c_{\alpha f})},
\end{aligned}
$$

$\omega_0$ is the natural frequency and $D$ is a damping factor of the second order system. Hence, the lateral behavior and also its stability depends as follows on parameters:

$$
\text{stability} = f\left(\underbrace{m(t),\, J_Z(t),\, l_f(t),\, l_r(t)}_{\text{vehicle design and loading}},\, \underbrace{c_{\alpha f},\, c_{\alpha r},\, v(t)}_{\text{driving condition}}\right).
$$

The transfer function between the steering wheel angle and the vehicle slip angle $\beta$ is due to Börner (2004)

$$
G_2(s) = \frac{\Delta\beta(s)}{\Delta\delta_H(s)} = \frac{b_{0\beta} + b_{1\beta} s}{(a_0 + a_1 s + a_2 s^2)}, \tag{7.2.51}
$$

with

$$
\begin{aligned}
b_{0\beta} &= \frac{l_r}{v}\left(1 - \frac{m l_f v^2}{c_{\alpha r} l_r l}\right) b_0, \\
b_{1\beta} &= \frac{J_Z v}{c_{\alpha r} l_r l - l_f m v^2} b_0.
\end{aligned}
$$

In the following, the poles and zeros of the transfer function $G_1(s)$ are studied for varying velocity $v$ and then for varying front and rear wheel cornering stiffness $c_{\alpha f}$ and $c_{\alpha r}$. The results of these investigations give a first insight into the transient and stability behavior of the one-track model; Börner et al. (2002) and Börner (2006). Hereby, $c_{\alpha r} = 1.2 c_{\alpha f}$ was assumed. The remaining vehicle parameters were set to the values according to Table 7.2.

**(a) Velocity Variation**

The poles are real for $v < 20\,\text{km/h}$, become conjugate complex for $v > 20\,\text{km/h}$ and less damped with higher eigenfrequency until $v = 180\,\text{km/h}$, but stay stable; see Fig. 7.8a. The corresponding step responses of the transfer function $G_1(s)$ are depicted in Fig. 7.8b.

**Table 7.2** Vehicle parameters, medium size passenger car

| Symbol | Value | Unit | Description |
|---|---|---|---|
| $i_S$ | 13.5 | Ackermann (2016) | Steering gear ratio |
| $c_{\alpha f}$ | 80 000 | [N/rad] | Cornering stiffness at front wheel |
| $c_{\alpha r}$ | 100 000 | [N/rad] | Cornering stiffness at rear wheel |
| $l_f$ | 1.3 | [m] | Distance CG - front axle |
| $l_r$ | 1.45 | [m] | Distance CG - rear axle |
| $m$ | 1450 | [kg] | Vehicle mass |
| $J_Z$ | 1920 | [kg m$^2$] | Moment of inertia vertical axis |
| $v$ | 50 | [km/h] | Vehicle CG velocity |
| $v_{ch}$ | 31.8 | [m/s] | Characteristic velocity |
| SG | $2.7 \cdot 10^{-3}$ | [rad s$^2$/m] | Steering gradient |

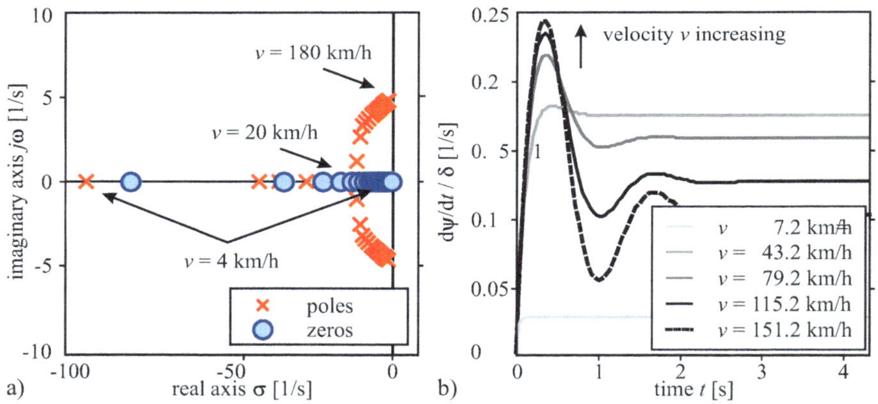


**Fig. 7.8** **a** Poles and zeros of the yaw rate frequency response $G_1(i\omega)$ for changing velocity $v \in [7.2\,\text{km/h};\ 180\,\text{km/h}]$ and constant cornering stiffness $c_{\alpha r} = 1.2 c_{\alpha f} = 60000$ [N/rad]; **b** Step responses for step input of the steering angle $\delta_H$ and different velocity $v$

### (b) Cornering Stiffness Variation

While the front and rear wheel cornering stiffness $c_\alpha$ are increased in the range between 0 to 100 000 N/rad, the vehicle behaves stable for positive $c_\alpha$, see Fig. 7.9a, and becomes unstable for $c_\alpha = 0$ N/rad with a double pole at the coordinate-origin.

The cornering stiffness becomes small for icy roads, but never negative. The zeros of the system are always on the left real axis. With decreasing front and rear stiffness $c_\alpha$ the gain also decreases and the dynamics become less damped. If $c_{\alpha r}$ is reduced, the gain of (7.2.50) becomes smaller and the yaw balance less damped; see Fig. 7.9b.

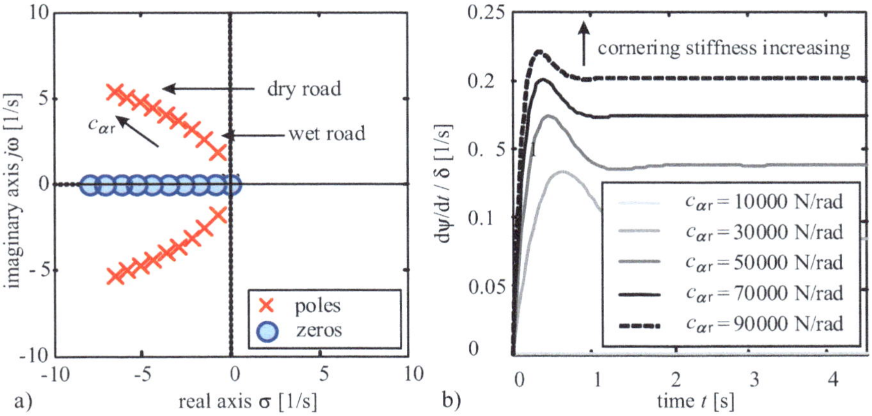


**Fig. 7.9 a** Poles and zeros of $G_1(i\omega)$ for different cornering stiffness $c_{\alpha r} = 1.2c_{\alpha f} \in$ [0 N/rad; 100 000 N/rad] and constant velocity $v = 14\,[\mathrm{m/s}] \approx 50\,[\mathrm{km/h}]$; **b** Step responses for step inputs of the steering angle $\delta_{\mathrm{H}}$ and different rear wheel cornering stiffness $c_{\alpha r}$

Hence, both the velocity and the cornering stiffness have a significant influence on the gain and damping of the steering behavior.

Figure 7.10 depicts the placement of poles and zeros and the frequency responses and Fig. 7.11 the corresponding step and impulse responses for changes of the steering angle at $v = 50\,\mathrm{km/h}$ and vehicle parameters according to Table 7.2 for different outputs. The *yaw rate transfer function* has a well-damped conjugate complex pole pair and a zero and the transient function looks approximately like a first order behavior with a settling time of about 0.3 s. In addition to the (same) pole pair, the *lateral acceleration transfer function* has conjugate complex zeros, and therefore, larger amplitudes and small phases for higher frequencies. This leads to an instantaneous step of the transient response. However, if a dynamic tire side force generation is added, which was neglected in the one-track model, one obtains the dashed course. The *transfer function for the slip angle* shows besides the well-damped pole pair a zero in the right half of the $s$-plane and therefore a non-minimum phase behavior. Hence, the transient function first attains positive and then negative values. Simulated step responses for different velocities, different loading and different tire pressures are shown in Halfmann and Holzmann (2003); see also Sect. 7.3.5.

### 7.2.4 Characteristic Velocity and Stability

Based on the linearized one-track model, some characteristic properties of the steering or cornering behavior of front-steered vehicles are discussed.

An important parameter for the cornering behavior of a vehicle is the *characteristic velocity*, which is defined as; see Mitschke and Wallentowitz (2014) and Schramm et al. (2010)

$$v_{\mathrm{ch}}^2 = \frac{c_{\alpha \mathrm{f}}\, c_{\alpha \mathrm{r}}\, l^2}{m(c_{\alpha \mathrm{r}}\, l_{\mathrm{r}} - c_{\alpha \mathrm{f}}\, l_{\mathrm{f}})}. \tag{7.2.52}$$

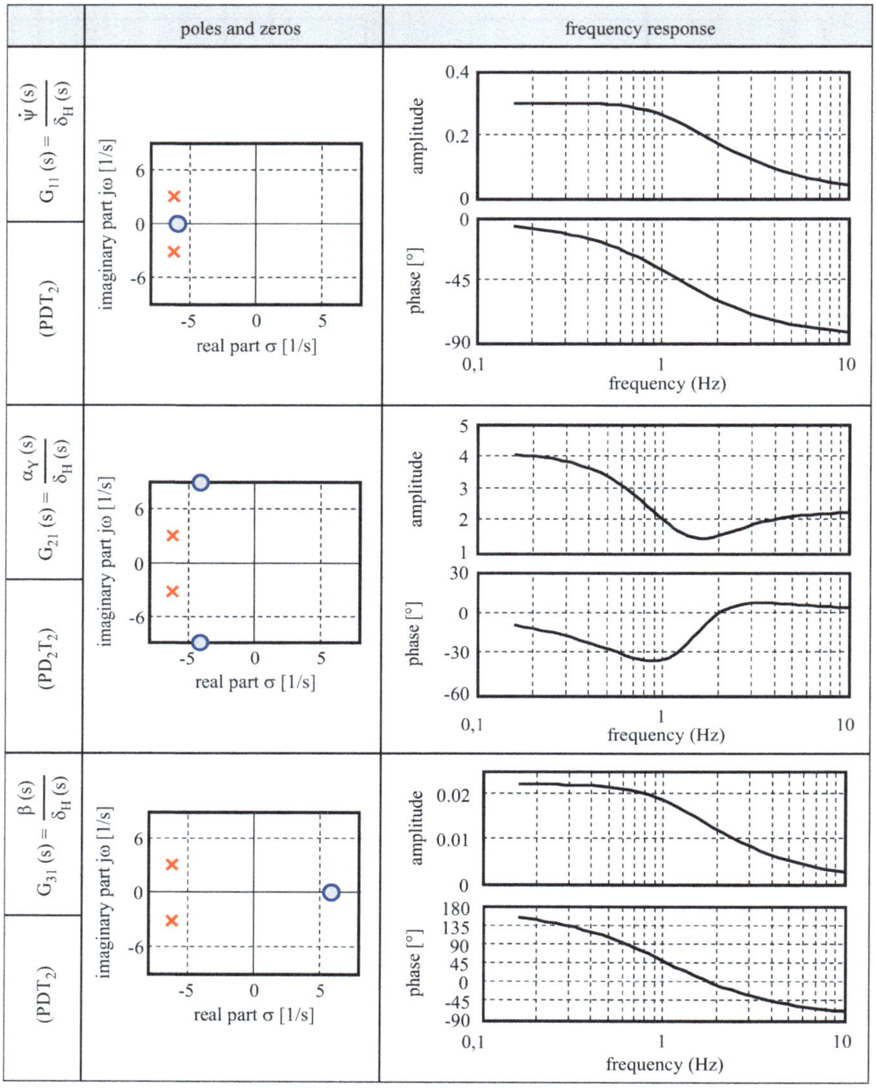


**Fig. 7.10** Poles, zeros and frequency responses $G(i\omega)$ for different outputs of the one-track model

For this speed, the yaw rate gain (7.2.54) for stationary corning obtains a maximal value

$$\frac{d}{dv}\left(\frac{\dot{\psi}}{\delta_{\mathrm{H}}}\right) = 0.$$

This characteristic velocity is usually considered as a constant parameter. Milliken and Milliken (1995) pointed out that "some North American cars have a characteristic

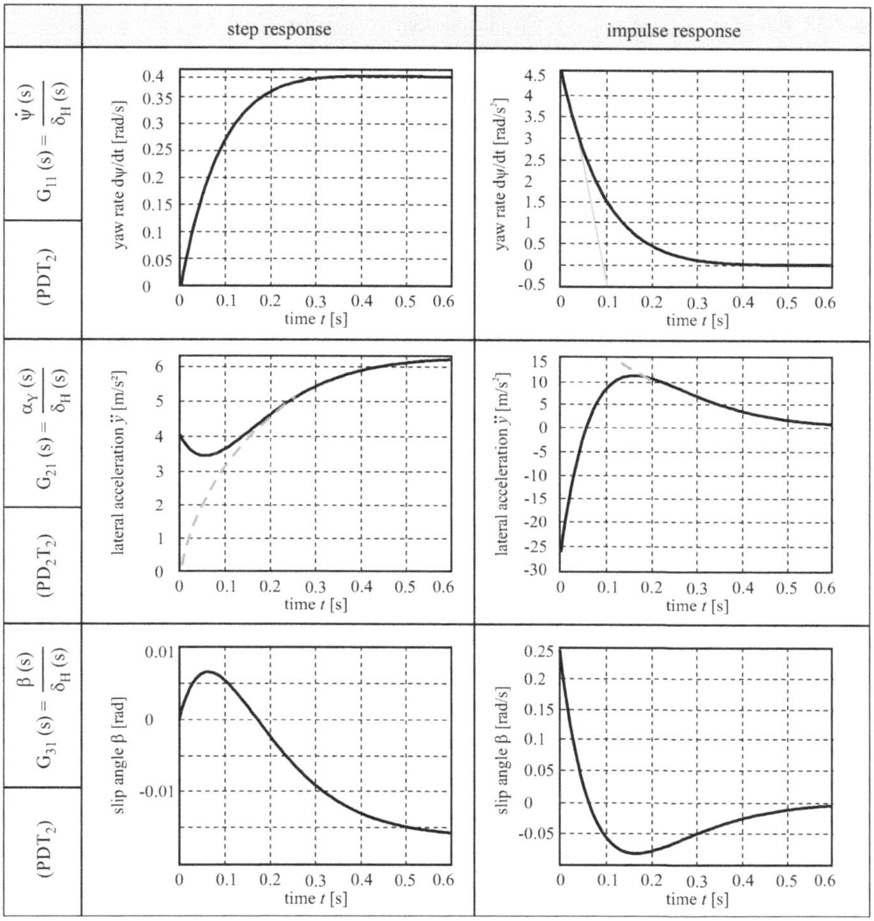


**Fig. 7.11** Step responses and impulse responses for different outputs of the one-track model, $v = 50\,\mathrm{km/h}$

velocity of 40 mph" ($\sim$ 64 km/h). Daiss (1996) describes that today's passenger cars are constructed with a characteristic velocity from 68 to 112 km/h. A value of 68 km/h corresponds to a limousine and 112 km/h to a sports car; see Fig. 7.12. The smaller the characteristic velocity the higher the tendency to understeer.

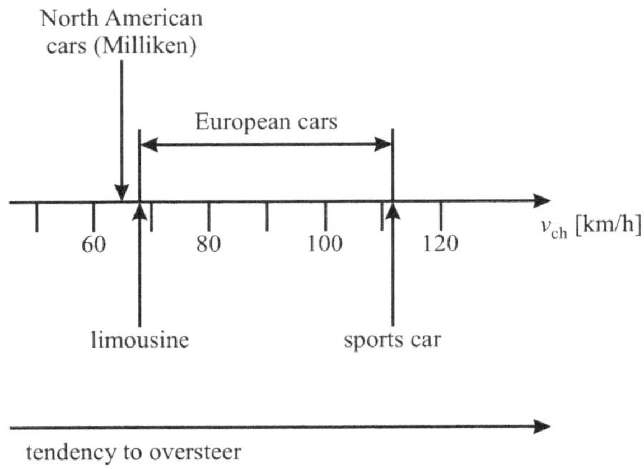


**Fig. 7.12** Range of characteristic velocities

Based on the linear state equation of the one-track model (7.2.34), the characteristic equation $\det(s\mathbf{I} - \mathbf{A}) = 0$ of the lateral vehicle dynamics becomes

$$s^2 + \underbrace{\frac{(J_Z + m\, l_f^2)\, c_{\alpha f} + (J_Z + m\, l_r^2)\, c_{\alpha r}}{J_Z\, m\, v}}_{a_1} s + \underbrace{\frac{c_{\alpha f}\, c_{\alpha r}(l_f + l_r)^2 + m\, v^2 (c_{\alpha r}\, l_r - c_{\alpha f}\, l_f)}{J_Z\, m\, v^2}}_{a_0} = 0. \tag{7.2.53}$$

According to the first condition of the Hurwitz stability criterion, stability requires that $a_1 > 0$ and $a_0 > 0$. As $a_1 > 0$ is always satisfied, because no negative values arise, only $a_0$ has to be considered. With the characteristic velocity (7.2.51), the following stability condition results:

$$\underbrace{c_{\alpha f}\, c_{\alpha r}(l_f + l_r)^2 + m\, v^2 (c_{\alpha r}\, l_r - c_{\alpha f}\, l_f)}_{a_0} > 0$$

$$\Rightarrow 1 + \frac{v^2}{v_{ch}^2} > 0. \tag{7.2.54}$$

Due to (7.2.51), the quadratic characteristic velocity $v_{ch}^2$ can be positive or negative in dependence on $c_{\alpha r}\, l_r$ and $c_{\alpha f}\, l_f$

$$v_{ch}^2 \begin{cases} < 0 & \forall\ c_{\alpha r}\, l_r < c_{\alpha f}\, l_f \\ = \infty & \forall\ c_{\alpha r}\, l_r = c_{\alpha f}\, l_f \\ > 0 & \forall\ c_{\alpha r}\, l_r > c_{\alpha f}\, l_f \end{cases} \quad \text{or} \quad \frac{1}{v_{ch}^2} = \begin{cases} < 0 \\ 0 \\ > 0 \end{cases} \tag{7.2.55}$$

This leads directly to the stability conditions pointed out in Table 7.3. Hence, the vehicle becomes monotonically unstable if

$$v^2 > -v_{ch}^2. \tag{7.2.56}$$

### 7.2.5 Stationary Cornering

To investigate the vehicle behavior for stationary cornering with constant radius $R_P$ it is now assumed that

$$\delta_H = \text{const.}, \quad \dot{\psi} = \text{const.} \quad v = \text{const.}, \quad \text{and} \quad \beta = \text{const.}$$

The dynamic one-track model (7.2.47) then leads to algebraic relationships. For this case of stationary cornering, one obtains from $b_0$ in (7.2.50) and (7.2.51) for the *yaw rate gain*

$$\frac{\dot{\psi}}{\delta_H} = \frac{1}{i_S\, l} \frac{v}{1 + \left(\frac{v}{v_{ch}}\right)^2}. \tag{7.2.57}$$

The lateral acceleration with $\dot{\beta} = 0$ from (7.2.8) is $a_Y = v\dot{\psi}$ and leads then to

$$\frac{a_y}{\delta_H} = \frac{1}{i_S\, l} \frac{v^2}{1 + \left(\frac{v}{v_{ch}}\right)^2}. \tag{7.2.58}$$

Further, (7.2.51) yields

$$\frac{\beta}{\delta_H} = \frac{l_r}{i_S l} \frac{1 - \frac{m l_f}{c_{\alpha r} l_r l} v^2}{1 + \left(\frac{v}{v_{ch}}\right)^2}. \tag{7.2.59}$$

Hence, the yaw rate gain is for small velocities proportional to $v$. From $d\dot{\psi}/d\delta_H = 0$ follows that it has a maximum for $v = v_{ch}$

$$\frac{\dot{\psi}(v)}{\delta_H} |_{max} = \frac{1}{i_s l} \frac{v_{ch}}{2}. \tag{7.2.60a}$$

This means that the yaw rate increases for $v = 0 \ldots v_{ch}$ and then decreases for $v > v_{ch}$; see van Zanten (2006). The lateral acceleration is for small $v$ proportional to $v^2$. The slip angle is for $v = 0$

$$\beta_0 = \frac{l_r}{i_s l} \delta_H = \frac{l_r}{l} \delta_f = \frac{l_r}{R_P} = \delta_A, \tag{7.2.60b}$$

positive and becomes negative for

$$v^2 > \frac{l l_r}{m l_f} c_{\alpha r}. \tag{7.2.61}$$

It only depends on the rear cornering stiffness $c_{\alpha r}$; see Mitschke and Wallentowitz (2014).

**Table 7.3** Stability conditions (one-track model)

| Stability condition | Driving situation |
|---|---|
| $1+\frac{v^2}{v_{ch}^2} > 0$ or $v^2 < -v_{ch}^2$ | Stable |
| $1+\frac{v^2}{v_{ch}^2} = 0$ or $v^2 = -v_{ch}^2$ | Indifferent |
| $1+\frac{v^2}{v_{ch}^2} < 0$ or $v^2 > -v_{ch}^2$ | Unstable |

With the measured steering wheel angle $\delta_H$ as input, the measured velocity $v$ and yaw rate $\dot{\psi}$ or lateral acceleration $a_Y$, the quadratic characteristic velocity $v_{ch}^2$ follows from (7.2.57)

$$v_{ch}^2 = -\frac{v^2}{1 - \frac{\delta_H v}{\dot{\psi} i_S l}}, \tag{7.2.62}$$

or from (7.2.58)

$$v_{ch}^2 = -\frac{v^2}{1 - \frac{\delta_H v^2}{a_y i_S l}}. \tag{7.2.63}$$

The yaw rate $\dot{\psi}$ can be expressed by the vehicles velocity $v$ and constant radius $R_P$ to the turn center

$$\dot{\psi} = \frac{v}{R_P}. \tag{7.2.64}$$

The steering wheel angle $\delta_H$ then follows from (7.2.62) as

$$\delta_H = \frac{i_S l}{R_P}\left(1 + \left(\frac{v}{v_{ch}}\right)^2\right). \tag{7.2.65}$$

Neutral steering is defined as

$$\delta_{H,0} = \frac{i_S\, l}{R_P} = \delta_A i_S, \tag{7.2.66}$$

where $\delta_A = l/R_P$ corresponds to the Ackermann angle; see (7.2.1). Then the right and left turning torque terms $c_{\alpha r}\, l_r = c_{\alpha f}\, l_f$ are equal, see (7.2.40)–(7.2.42) or Fig. 7.6, which means also $v_{ch}^2 = \infty$; see (7.2.52). Further with the curvature $\kappa = \frac{1}{R_P}$ as input, the *steering gain for neutral steering* with regard to the curvature follows as

$$K_{st,0} = \frac{\delta_{H,0}}{\kappa} = i_S\, l, \tag{7.2.67}$$

and introduction into (7.2.65) results in the *steering gain*

$$K_{st} = \frac{\delta_H}{\kappa} = K_{st,0} + \frac{i_S\, l}{v_{ch}^2} v^2,$$

or the *steering gain ratio*

$$\frac{K_{\mathrm{st}}}{K_{\mathrm{st},0}} = 1 + \frac{v^2}{v_{\mathrm{ch}}^2}, \tag{7.2.68}$$

and with (7.2.65)

$$\delta_{\mathrm{H}} = \delta_{\mathrm{H},0}(1 + \frac{v^2}{v_{\mathrm{ch}}^2}). \tag{7.2.69}$$

For a certain vehicle, the characteristic velocity $v_{\mathrm{ch}}$ is a fixed parameter with a given front and rear cornering stiffness $c_{\alpha\mathrm{f}}$ and $c_{\alpha\mathrm{r}}$ and can, therefore, be considered as a constant value for discussing the steering behavior. Introducing $v^2 = R_{\mathrm{P}} a_{\mathrm{y}}$ for small slip angles leads to the steering wheel angle

$$\delta_{\mathrm{H}} = \delta_{\mathrm{H},0}\left(1 + R_{\mathrm{P}}\frac{a_{\mathrm{Y}}}{v_{\mathrm{ch}}^2}\right) = \frac{i_{\mathrm{S}} l}{R_{\mathrm{P}}} + i_{\mathrm{S}}\frac{l}{v_{\mathrm{ch}}^2} a_{\mathrm{Y}}, \tag{7.2.70}$$

and the front wheel steering angle with $\delta_{\mathrm{f}} = \delta_{\mathrm{H}}/i_{\mathrm{s}}$

$$\delta_{\mathrm{f}} = \frac{l}{R_{\mathrm{P}}} + \frac{l}{v_{\mathrm{ch}}^2} a_{\mathrm{Y}} = \delta_{\mathrm{A}} + \frac{l}{v_{\mathrm{ch}}^2} a_{\mathrm{Y}}. \tag{7.2.71}$$

The required steering wheel angle $\delta_{\mathrm{H}}$ changes linearly with $a_{\mathrm{Y}}$, respectively, $v^2$. The gradient of this relation is

$$\frac{d\delta_{\mathrm{H}}}{da_{\mathrm{Y}}} = \delta_{\mathrm{H},0} R_{\mathrm{P}} \frac{1}{v_{\mathrm{ch}}^2} = \frac{i_{\mathrm{s}}\, l}{v_{\mathrm{ch}}^2}. \tag{7.2.72}$$

If it is positive, the vehicle is called to have *understeer behavior*, because the steering wheel angle has to be increased with the lateral acceleration or $v^2$ to drive with constant radius $R_{\mathrm{P}}$.

According to ISO 8855 an understeer/oversteer gradient is defined as

$$\frac{d\delta}{da_{\mathrm{Y}}} = \frac{d\delta}{da_{\mathrm{Y}}} - \frac{d\delta_{\mathrm{D}}}{da_{\mathrm{Y}}}, \tag{7.2.73}$$

where $d\delta_{\mathrm{D}}/da_{\mathrm{Y}}$ is the Ackermann steer angle gradient with $\delta_{\mathrm{D}} = l/R_{\mathrm{P}} = \delta_{\mathrm{A}}$. Hence, the *understeer/oversteer gradient* or *steer gradient* follows for $d\delta_{\mathrm{D}}/da_{\mathrm{Y}} = 0$ or for $R_{\mathrm{P}} = \mathrm{const.}$ to

$$\mathrm{SG} = \frac{d\delta}{da_{\mathrm{Y}}} = \frac{l}{v_{\mathrm{ch}}^2}, \tag{7.2.74}$$

and is, therefore, inverse proportional to the square of the characteristic velocity $v_{\mathrm{ch}}$. The steering gain expresses neutral-, under-, and oversteering as given in Table 7.4.

Figure 7.13 shows the driving conditions and the stable and unstable region for the case of stationary cornering. Hence, vehicles show understeering behavior for

**Table 7.4** Driving conditions

| Driving situation | Characteristic velocity | Steering gain |
|---|---|---|
| Neutralsteering | $v_{ch}^2 \to \pm\infty$ | $K_{st} = K_{st,0}$ |
| Understeering | $v_{ch}^2 > 0$ | $K_{st} > K_{st,0}$ |
| | $v_{ch}^2 > 0$ and $v_{ch}^2 = v^2$ | $K_{st} = 2K_{st,0}$ |
| Indifferent | $v_{ch}^2 = 0$ | $K_{st} \to \infty$ |
| Oversteering | $v_{ch}^2 < 0$ | $K_{st} < K_{st,0}$ |
| | $v_{ch}^2 < 0$ and $v^2 = -v_{ch}^2$ | $K_{st} = 0$ |
| Countersteering | $v_{ch}^2 < 0$ and $v^2 > -v_{ch}^2$ | $K_{st} < 0$ |

$v_{ch}^2 > 0$, oversteering for $v_{ch}^2 < 0$, and countersteering for $v^2 > -v_{ch}^2$. Comparing the condition for countersteering with the stability condition $v^2 > -v_{ch}^2$ (7.2.56) for the one-track model for any steering angle input, see Table 7.3, shows the same result. Therefore, countersteering means a dynamic unstable driving situation.

With (7.2.73) the *yaw rate gain* (7.2.57) for the velocity $v$ can also be expressed by the steer gradient

$$\frac{\dot{\psi}}{\delta_H} = \frac{1}{i_s} \frac{v}{l + \text{SG}\, v^2}. \tag{7.2.75}$$

It is small for positive and large SG (understeering), large for negative SG and infinite for $\text{SG} = -l/v^2$, i.e. indifferent behavior. For $\text{SG} < -l/v^2$ countersteering results. Introducing (7.2.74) in (7.2.71) gives another equation for the front wheel angle

$$\delta_f = \delta_A + \text{SG} a_Y. \tag{7.2.76}$$

The evaluation of the considered dynamic lateral models did not take into account the dynamic behavior between $\delta_H(t)$ and $\delta_f(t)$ of the steering system. These models are discussed in Chaps. 14 and 20.

### 7.2.6 Comparison with Measurements

A comparison of the simulated and measured behavior for a one-track model is depicted in Fig. 7.14 for a Volkswagen Golf IV and a slalom drive, Schorn (2007). The parameters of the vehicle are summarized in Table 7.5. The used input variables are the steering angle $\delta_H$ and the velocity $v$. Directly measured were $\dot{\psi}$ and $\beta$. $a_Y$ was calculated from $a_Y = v(\dot{\beta} + \dot{\psi})$.

The errors for the yaw rate and the lateral acceleration show small to medium values, and are larger for the slip angle. The output signals are comparable to the results with a two-track model in Fig. 7.18. In Chap. 12, it will be shown that, especially the slip angle can considerably better be determined by using state observers and state-estimation methods.

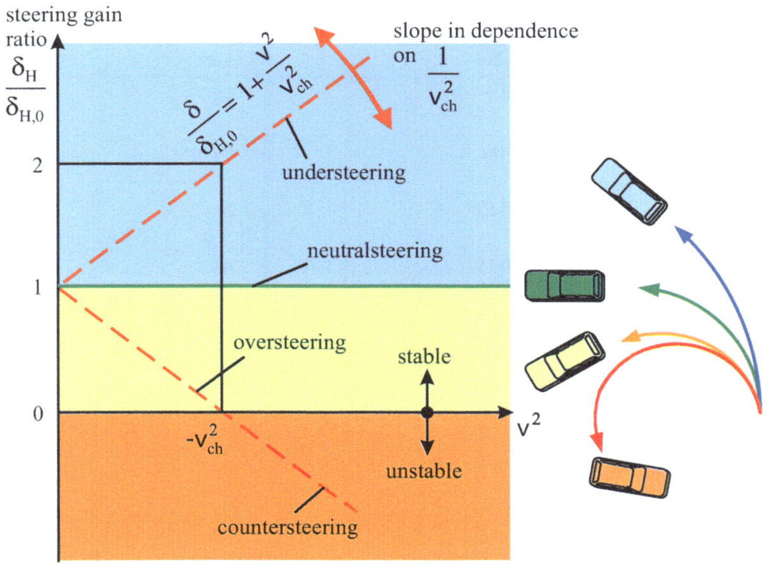


**Fig. 7.13** Steering gain ratio with areas of special steering behavior for stationary cornering with constant radius, linear range

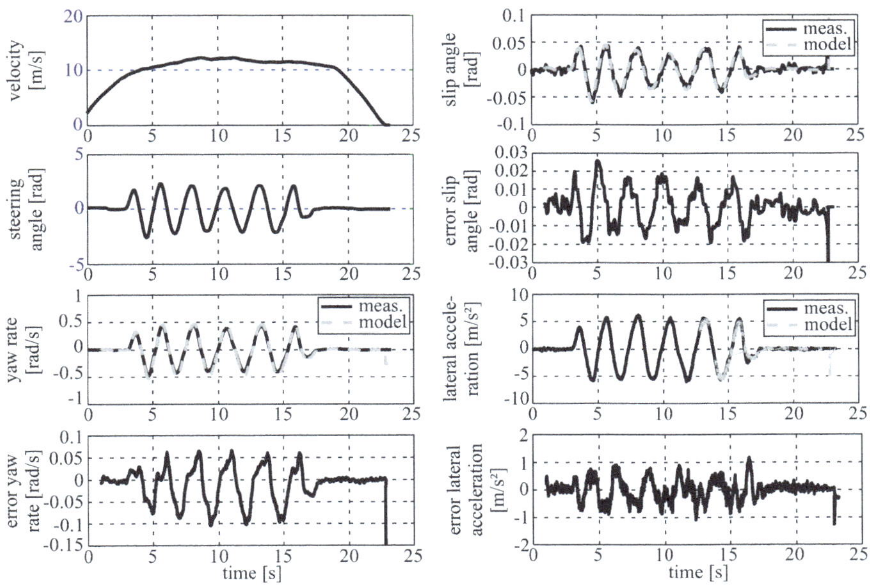


**Fig. 7.14** Measured and simulated variables for a linear one-track model of a Volkswagen Golf IV, Schorn (2007)

**Table 7.5** Vehicle parameters of Volkswagen Golf IV (2000), Schorn (2007)

| Symbol | Value | Unit |
|---|---|---|
| $m$ | 1505 | kg |
| $J_z$ | 3700 | kg m$^2$ |
| $l$ | 2.525 | m |
| $l_\mathrm{f}$ | 0.975 | m |
| $l_\mathrm{r}$ | 1.55 | m |
| $b_\mathrm{f}$ | 1.51 | m |
| $b_\mathrm{r}$ | 1.49 | m |
| $c_{\alpha\mathrm{r}}$ | 120 000 | N/rad |
| $c_{\alpha\mathrm{f}}$ | 100 000 | N/rad |
| $i_\mathrm{s}$ | 16.5 | – |
| $v_\mathrm{ch}$ | 23.9 | m/s |
| SG | $4.4\times 10^{-3}$ | rad s$^2$/m |

### 7.2.7 Wheel Slip Angle Difference Model

The force balance equation for the center of gravity in lateral direction (7.2.37) with (7.2.8) yields with assumption of small steering angles $\delta$ and slip angles $\beta$ for stationary cornering with path radius $R_\mathrm{P}$

$$m\frac{v^2}{R_\mathrm{P}} = F_{\mathrm{Y,r}} + F_{\mathrm{Y,f}}. \tag{7.2.77}$$

For the lateral forces holds

$$F_{\mathrm{Y,r}} = \frac{l_\mathrm{f}}{l}m\frac{v^2}{R_\mathrm{P}}\ ; \ \ F_{\mathrm{Y,f}} = \frac{l_\mathrm{r}}{l}m\frac{v^2}{R_\mathrm{P}}, \tag{7.2.78}$$

and

$$F_{\mathrm{Y,r}} = c_{\alpha\mathrm{r}}\,\alpha_\mathrm{r}\ ; \quad F_{\mathrm{Y,f}} = c_{\alpha\mathrm{f}}\,\alpha_\mathrm{f}. \tag{7.2.79}$$

Hence

$$\alpha_\mathrm{r} = m\frac{v^2}{R_\mathrm{P}l}\frac{l_\mathrm{f}}{c_{\alpha\mathrm{r}}}\ ; \quad \alpha_\mathrm{f} = m\frac{v^2}{R_\mathrm{P}l}\frac{l_\mathrm{r}}{c_{\alpha\mathrm{f}}}, \tag{7.2.80}$$

**Table 7.6** Driving conditions in dependence on the wheel slip angle difference

| Driving situation | Understeer gradient | Wheel slip difference angle |
|---|---|---|
| Understeering | $SG > 0$ | $\alpha_f - \alpha_r > 0$ |
| Neutral steering | $SG = 0$ | $\alpha_f - \alpha_r = 0$ |
| Oversteering | $SG < 0$ | $\alpha_f - \alpha_r < 0$ |
| Countersteering | $SG < 0$ | $\alpha_f - \alpha_r < -\frac{l}{R_P}$ or $\alpha_f - \alpha_r < -\delta_A$ |

and

$$\alpha_f - \alpha_r = m \frac{v^2 l}{R_P} \frac{(c_{\alpha r} l_r - c_{\alpha f} l_f)}{c_{\alpha f}\, c_{\alpha r}\, l^2} = \frac{l}{R_P} \frac{v^2}{v_{ch}^2}, \tag{7.2.81}$$

or

$$\Delta\alpha = \alpha_f - \alpha_r = \frac{l}{v_{ch}^2} a_Y = SG\, a_Y. \tag{7.2.82}$$

For stationary cornering and constant steer gradient SG, the slip angle difference $\Delta\alpha$ is proportional to the lateral acceleration $a_Y$. Hence, this holds under the condition that the characteristic velocity $v_{ch}^2$ is constant, and therefore, $c_{\alpha r}$ and $c_{\alpha f}$ are constant; see (7.2.52). The front wheel steering angle (7.2.76) can then also be described by

$$\delta_f = \frac{l}{R_P} + (\alpha_f - \alpha_r) = \delta_A + \Delta\alpha. \tag{7.2.83}$$

Hence, the steering angle at the front wheels is equal to the Ackermann angle $\delta_A$ plus the difference $\Delta\alpha$ of the front and rear wheel slip angle.

Using the above relations, the steering behavior can also be expressed by the steer gradient and the slip angle difference; see Table 7.6. Hence, for understeering holds $\alpha_r < \alpha_f$ for oversteering $\alpha_r > \alpha_f$ and for countersteering $\alpha_r > \alpha_f + \delta_A$, i.e. unstable behavior.

The lateral tire forces in the linear range can be expressed, see (5.2.5) and (7.2.18), by

$$F_{YT} = c_\alpha \alpha = c_{\alpha 1}\, F_Z\, \alpha, \tag{7.2.84}$$

or by

$$F_{YT} = \mu_Y F_Z, \tag{7.2.85}$$

where $\mu_y$ is the side force coefficient or side friction coefficient of a tire (5.2.7). This leads to

$$\alpha = \left(\frac{\mu_Y}{c_{\alpha 1}}\right). \tag{7.2.86}$$

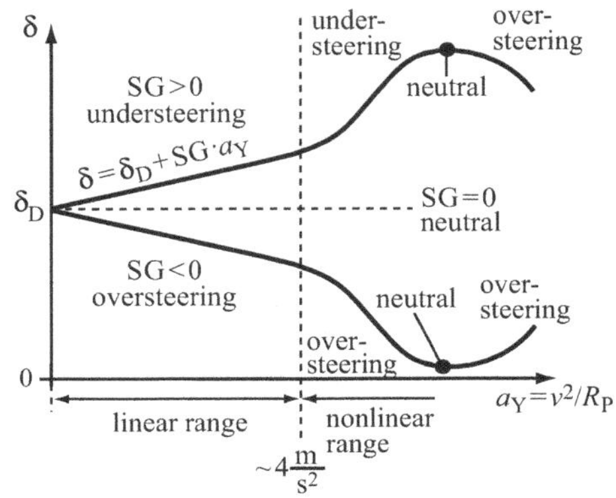


**Fig. 7.15** Steering angle in dependance on the lateral acceleration, for driving with constant curvature, with linear and nonlinear range, Schorn (2007)

Both the coefficients are a property of the tire/road conditions. Hence, the slip difference angle can also be described by

$$\Delta\alpha = \alpha_f - \alpha_r = \left(\frac{\mu_Y}{c_{\alpha 1}}\right)_f - \left(\frac{\mu_Y}{c_{\alpha 1}}\right)_r . \tag{7.2.87}$$

The dependence on the wheel load $F_z$ vanishes. If $c_{\alpha 1,r} > c_{\alpha 1,f}$ and $\mu_{Y,f} = \mu_{Y,r}$ then with $\Delta\alpha > 0$ the vehicle has understeering and for $c_{\alpha 1,r} < c_{\alpha 1,f}$ oversteering behavior.

The treated linear one-track model can be applied for lateral accelerations $a_Y < 4\,\mathrm{m/s^2}$ on dry asphalt, Milliken and Milliken (1995). If the lateral acceleration exceeds this value, the linear range is left. With increasing speed the vehicle first stays understeering, then may become neutral steering and oversteering, as illustrated in Fig. 7.15. Oversteering vehicles show a corresponding behavior.

## 7.3 Dynamic Two-Track Models

To model the dynamic behavior of a real vehicle in more detail, the acting forces on all four wheels have to be considered. Figure 7.16 illustrates the geometrical measures. This leads to a two-track model, which takes into account the main forces and moments acting on the wheels and the vehicle body. Therefore, the translatory behavior has to be described by Newton's second law of momentum for masses and the rotational behavior by the Euler's equation for the angular momentum of a gyroscope; see e.g. Zomotor and Reimpell (1991), Popp and Schiehlen (1993), Ammon (1997), and Kiencke and Nielsen (2000). The following derivations leans on Schorn (2007).

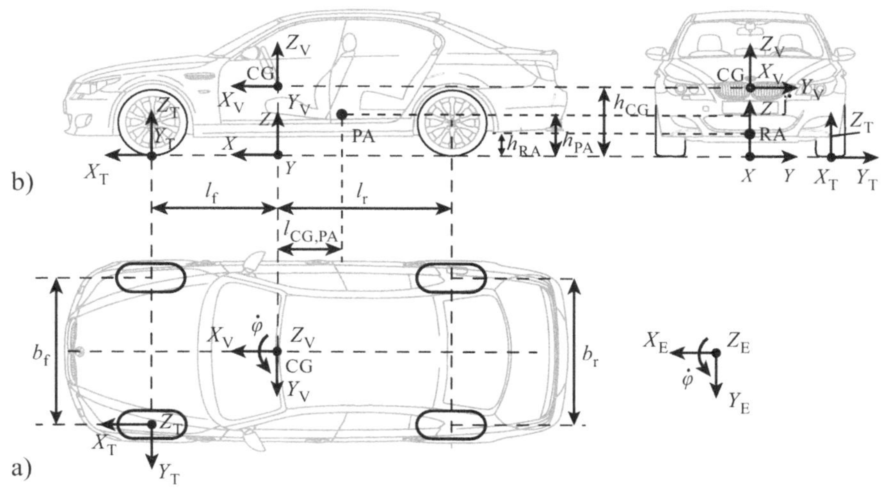


**Fig. 7.16** Scheme of a vehicle with four wheels, two tracks, and geometric measures. **a** horizontal motion. **b** vertical, pitch and roll motion, Schorn (2007), Bauer (2015)

### 7.3.1 General Two-Track Model

The resulting forces for the *translatory behavior* follows with Newton's second law of momentum from

$$\mathbf{F} = m\left(\frac{d\mathbf{v}}{dt} + \boldsymbol{\omega} \times \mathbf{v}\right), \tag{7.3.1}$$

where $m = m_V$ is the mass of the vehicle.

The term $d\mathbf{v}/dt$ describes the acceleration of the center of gravity (CG) with the vehicle-fixed axis system $X_V$, $Y_V$, and $Z_V$. The term $\boldsymbol{\omega} \times \mathbf{v}$ models the centripetal acceleration of a moved body in the space.

The motion variables are

$$\begin{aligned} \mathbf{v}^T &= [v_X \ v_Y \ v_Z], \\ \boldsymbol{\omega} &= [\dot{\varphi} \ \dot{\theta}, \ \dot{\psi}] \end{aligned} \tag{7.3.2}$$

and for the forces on the CG holds

$$\mathbf{F}^T = [F_X \ F_Y \ F_Z]. \tag{7.3.3}$$

Then, (7.3.1) yields for the forces on the CG in three directions for the *translatory behavior*

$$\mathbf{F} = \begin{bmatrix} F_X \\ F_Y \\ F_Z \end{bmatrix} = \begin{bmatrix} m & 0 & 0 \\ 0 & m & 0 \\ 0 & 0 & m \end{bmatrix} \left( \begin{bmatrix} \dot{v}_X \\ \dot{v}_Y \\ \dot{v}_Z \end{bmatrix} + \begin{bmatrix} \dot{\theta} v_Z - \dot{\psi} v_Y \\ \dot{\psi} v_X - \dot{\varphi} v_Z \\ \dot{\varphi} v_Y - \dot{\theta} v_X \end{bmatrix} \right). \tag{7.3.4}$$

The forces can be described by the body forces resulting from the tire forces, compare Figs. 7.16 and 7.19

$$\mathbf{F} = \begin{bmatrix} F_{X,fl} + F_{X,fr} + F_{X,rl} + F_{X,rr} - m_0 g \sin(\lambda) - F_{XA} - F_{xR} \\ F_{Y,fl} + F_{Y,fr} + F_{Y,rl} + F_{Y,rr} - m_0 g \sin(\eta) - F_{YA} \\ F_{Z,fl} + F_{Z,fr} + F_{Z,rl} + F_{Z,rr} - m_0 g + (c_{A,f} + c_{A,r}) v_{X,res}^2 \end{bmatrix}. \tag{7.3.5}$$

Herewith, the road gradient angle (slope) in longitudinal and lateral direction is described by $\lambda$ and $\eta$. The air drag forces, the air velocities and the rolling resistance force follow from (6.5.4) to (6.5.6). $m_0$ is the body mass, and $c_{A,f}$ and $c_{A,r}$ are the lift coefficients by the air flow with regard to the front and rear axle, Willumeit (2013).

The *rotational behavior* is covered by Euler's equations for the angular momentum

$$\mathbf{M} = \dot{\mathbf{L}} + (\boldsymbol{\omega}_V \times \mathbf{L},) \tag{7.3.6}$$

with the rotational impulse

$$\mathbf{L} = \mathbf{J}\boldsymbol{\omega}_V, \tag{7.3.7}$$

where $\mathbf{J}$ is the inertia tensor and $\boldsymbol{\omega}_V$ the rotational vector of the vehicle.

This leads to

$$\begin{bmatrix} J_X & 0 & 0 \\ 0 & J_Y & 0 \\ 0 & 0 & J_Z \end{bmatrix} \begin{bmatrix} \ddot{\varphi} \\ \ddot{\theta} \\ \ddot{\psi} \end{bmatrix} = \begin{bmatrix} M_X \\ M_Y \\ M_Z \end{bmatrix} + \begin{bmatrix} \dot{\theta}\dot{\psi}(J_Y - J_Z) \\ \dot{\psi}\dot{\varphi}(J_Z - J_X) \\ \dot{\varphi}\dot{\theta}(J_X - J_Y) \end{bmatrix}. \tag{7.3.8}$$

$J_X, J_Y, and J_Z$ are the moments of inertia and $M_X, M_Y, and M_Z$ the moments (torques) around the roll axis RA, the pitch axis PA, and the vertical axis through the CG; see Fig. 7.16.

For these moments acting on the body holds

$$\begin{aligned} M_X &= (F_{Z,fl} - F_{Z,fr})\frac{b_f}{2} + (F_{Z,rl} - F_{Z,rr})\frac{b_r}{2} - (h_{CG} - h_{RA})F_{Y,CG}, \\ M_Y &= (F_{Z,rl} + F_{Z,rr})l_r - (F_{Z,fl} + F_{Z,fr})l_f + (h_{CG} - h_{PA})F_{X,CG}, \\ &\quad + (h_{CG} - h_{XA})F_{XA} + (l_r c_{A,r} - l_f c_{A,f})v_{X,res}^2, \\ M_Z &= (F_{X,fr} - F_{X,fl})\frac{b_f}{2} + (F_{X,rr} - F_{X,rl})\frac{b_r}{2} + F_{Y,fl}(l_f - n_{tfl}) \\ &\quad + F_{Y,fr}(l_f - n_{tfr}) - F_{Y,rl}(l_r - n_{trl}) - F_{Y,rr}(l_r - n_{trr}), \end{aligned} \tag{7.3.9}$$

with the centrifugal forces on the CG

$$F_{X,CG} = -ma_X \; ; \quad F_{Y,CG} = -ma_Y, \tag{7.3.10}$$

causing pitch and roll motions. The pitch and roll rotations are around the pitch axis PA and roll axis RA and can be calculated based on geometric relations; Halfmann and Holzmann (2003). The roll axis is assumed to be in the vehicle center plane and

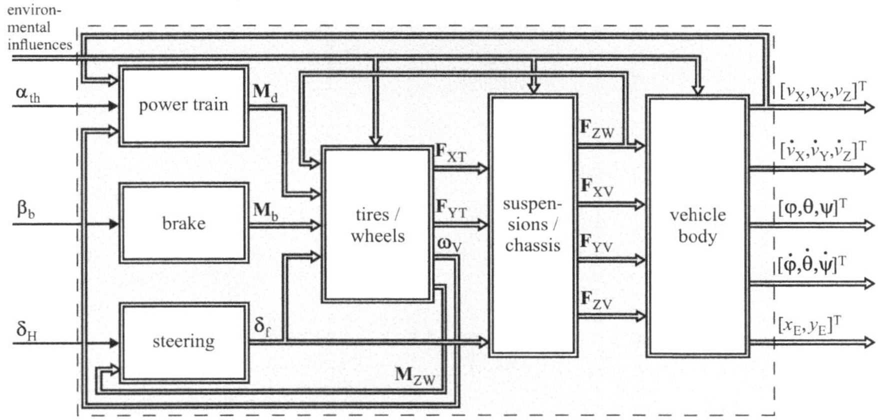


**Fig. 7.17** Overall signal flow block diagram for the two-track model of a vehicle, Schorn (2007)

the pitch axis parallel to the $Y_V$ axis. The heights of turning points are $h_{CG}$ for the CG, $h_{RA}$ and $h_{PA}$ for the roll and pitch axis; see Fig. 7.16. $n_{t,fl...rr}$ are the caster offsets of the tires at ground and $c_{A,f}$ and $c_{A,r}$ the lift coefficients for the front and rear axle.

Summarizing, the horizontal wheel forces on the body $F_{X,fl...rr}$ and $F_{Y,fl...rr}$ in longitudinal and lateral direction, the forces of the wheel and suspension $F_{Z,fl...rr}$ in vertical direction and the resistance and gravity forces determine the translatory motion of the vehicle CG according to (7.3.1)–(7.3.5). The rotational motion around the CG, RA, and PA by the wheel and suspension forces and the longitudinal and lateral acceleration follow (7.3.8) and (7.3.9). The forces acting on the "vehicle body" are illustrated in the overall signal flow block diagram Fig. 7.17. Because the wheel forces are the inputs for both equations (7.3.4) and (7.3.9) the translatory and the rotational behaviors of the vehicle are coupled. The forces generated by the tires and transferred to the vehicle body by the wheels and suspensions are considered in the next sections.

The required forces $F_{X,i} = F_{XT,i}$ and $F_{Y,i} = F_{YT,i}$ for $i = fl, ..., rr$ from the tires through the wheels to the body follow by applying the equations referenced in Table 7.7, see also next sections. To solve the equation systems, partially iterative solutions are required. Thus, the outputs of the block "tires/wheels" in Fig. 7.17 for the two-track model can be determined.

By properly combining the models of this section with the models of the power train, brake system and steering system, treated in Chaps. 6, 10, and 11 the overall two-track model according to the structure of Fig. 7.17 can be compiled together. The next sections treat nonlinear two-track models with yaw rate and lateral velocity and with yaw rate and slip angle as output variables.

The vehicle dynamic models contain many parameters. Some of them, as the geometrical quantities and masses can be obtained directly from construction data or measurements. The parameters for the tire-road behavior are more difficult to gain. Chapters 10, 11, and 12 illustrate in detail how many of the vehicle's parameters can

**Table 7.7** Calculation of tire forces with $i = fl, \ldots, rr$ for the two-track model

| Symbol | Variable | Input variables | Equations | (a) Driving condition (b) output variables |
|---|---|---|---|---|
| $\omega_{W,i}$ | Wheel velocity | $M_d$ or $M_b$ | (5.5.12) | (a) Longitudinal acceleration or braking |
| $v_X$ | Vehicle velocity | $M_{eng}$ or $M_b$ | (6.5.1), (6.6.1) | (b) Determination of wheel speeds and velocity |
| $S_{X,i}$ | Tire slip | $v_X$, $\omega_{W,i}$ (5.1.1), (5.1.2) | (a) Longitudinal acceleration or braking | |
| $\mu_{X,i}$ | Friction coefficient | $S_{X,i}$ | (5.1.3) | (b) Determination of tire forces for longitudinal direction |
| $F_{XT,i}$ | Tire force | $\mu_{X,i}$ , $F_{Zi}$ | (5.3.10) | |
| $c_{\alpha,i}$ | Cornering stiffness | $\mu_Y$ , $F_{Z,i}$ | (5.2.4) | (a) Cornering with acceleration or braking |
| $\alpha_i$ | Side slip angle | $S_{Y,i}$ | (5.2.2) | (b) Determination of tire forces in lateral and longitudinal direction |
| $S_{Y,i}$ | Tire slip | $v_X$, $\omega_{W,i}$ , $\alpha_i$ | (5.3.3) | |
| $S_{res,i}$ | Resulting tire slip | $S_{X,i}$, $S_{Y,i}$ | (5.3.5) | |
| $F_{XT,i}$ | Tire force | $\mu_{res,i}$ , $F_{Z,i}$ | (5.3.8), (5.3.10) | |
| $F_{YT,i}$ | Tire force | $\mu_{res,i}$, $F_{Z,i}$, $c_Y$ | (5.3.8), (5.3.10) | |

be obtained experimentally by applying parameter and state-estimation methods in combination with one-track and two-track models.

The mass moments for inertia have to be calculated with compartment mass models or determined by special experiments. Kiencke and Nielsen (2005) give for middle class passenger cars following estimations

$$J_X = m_b i_X^2; \quad J_Y = m_b i_Y^2; \quad J_Z = m i_Z^2, \tag{7.3.11}$$

where $m_b$ is the body mass (without unsprung suspension masses), $m$ the complete vehicle mass, and $i_j$ the radii of rotations, with

$$i_X = 0.56 \ldots 0.65; \quad i_Y = 1.13 \ldots 1.21,$$

$$i_Z = 1.18 \ldots 1.20 \text{ [m]},$$

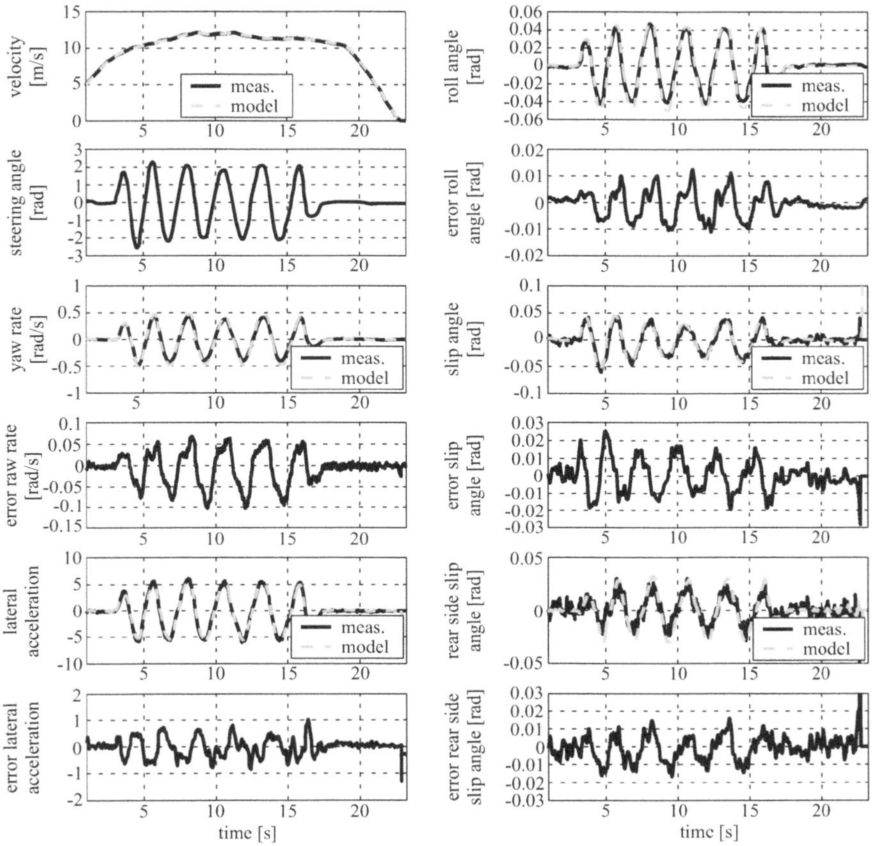


**Fig. 7.18** Comparison of a simulated two-track model with the measurements for a slalom maneuver with a VW Golf IV, Schorn (2007)

dependent on loading.

Figure 7.18 shows as a first example a comparison of the measured and with a two-track model simulated behavior for a slalom maneuver with some of the parameters given in Table 7.5, Schorn (2007). The available on-board measurements with an inertial measurement unit (IMU) mounted close to the CG were complemented by an optical Correvit sensor, such that the slip angle and the side slip angle of the rear wheels could be determined. The plots show that the results for $\dot{\psi}$, $a_{\mathrm{y}}$, and $\theta$ show a reasonable agreement and are comparable to those of the one-track model, Fig. 7.14, but that the results for $\beta$ and $\alpha_{\mathrm{r}}$ have to be improved with more precise parameters and state estimation, as will be shown in Chaps. 11 and 12.

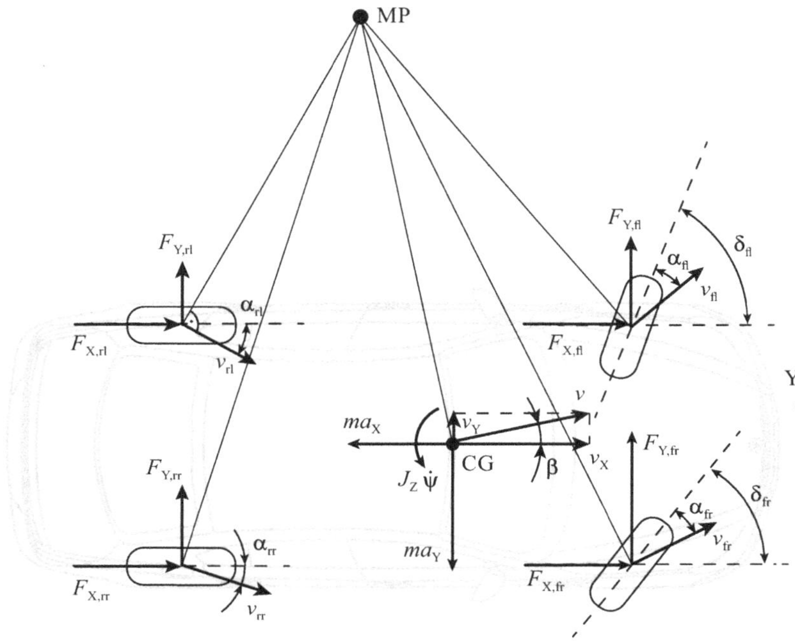


**Fig. 7.19** Scheme for the two-track model in the X-Y plane with acting forces, Bauer (2015)

### 7.3.2 Simplified Two-Track Model, Even Road Plane

**(a) Yaw Rate/Velocity Representation**

The general dynamic vehicle models (7.3.1)–(7.3.8) for the translatory and rotational behavior take into account the speed along all three axes $v_X$, $v_Y$, and $v_Z$ and the rotations around the pitch axis, roll axis, and vertical axis $\dot{\varphi}$, $\dot{\theta}$, and $\dot{\Psi}$. The corresponding equations can be simplified if the pitch and roll motions are neglected and only the motions in the horizontal X, Y level (ground level) are considered, Bauer (2015); see Fig. 7.19.

Then it follows from (7.3.4), (7.3.5) and (7.3.8) with $\lambda_X = 0$; $\eta_Y = 0$; $n_{ti} = 0$

$$\dot{v}_X = \frac{1}{m}[F_{X,fl} + F_{X,fr} + F_{X,rl} + F_{X,rr} - F_{XA} - F_{XR}] + \dot{\psi} v_Y, \quad (7.3.12)$$

$$\dot{v}_Y = \frac{1}{m}[F_{Y,fl} + F_{Y,fr} + F_{Y,rl} + F_{Y,rr} - F_{YA}] - \dot{\psi} v_X, \quad (7.3.13)$$

$$\ddot{\psi} = \frac{1}{J_Z}[\left(F_{X,fr} - F_{X,fl}\right)\frac{b_f}{2} + \left(F_{X,rr} - F_{X,rl}\right)\frac{b_r}{2} + \left(F_{Y,fl} + F_{Y,fr}\right) l_f$$
$$- \left(F_{Y,rl} + F_{Y,rr}\right) l_r]. \quad (7.3.14)$$

The longitudinal and lateral forces $F_{X,ij} = F_{XW,ij}$ and $F_{Y,ij} = F_{YW,ij}$ are the acting forces from the wheels on the vehicle body. They are generated through the longitudinal and lateral tire forces $F_{XT,ij}$ and $F_{YT,ij}$.

The calculation of the individual tire forces requires the individual tire slip values due to Chap. 5. The necessary wheel velocities at the ground contact patch are

$$\mathbf{v}_{T,ij} = \mathbf{v}_{ij} = \left[ v_{X,ij} \;\; v_{Y,ij} \right]. \tag{7.3.15}$$

According to (5.1.3), the longitudinal tire forces in the direction of the wheel rotation, i.e. in wheel coordinates, are

$$F_{XT,ij} = \mu_{X,ij} \left( S_{X,ij} \right) F_{Z,ij}, \tag{7.3.16}$$

and the lateral tire forces perpendicular to the wheel rotation; see (5.2.3)

$$F_{YT,ij} = c_{\alpha,ij} \left( F_{Z,ij} \right) \alpha_{ij}, \tag{7.3.17}$$

assuming that both wheel forces can be determined separately. The calculation of the tire slips $S_{X,ij}$ due to (5.1.1) and (5.1.2) and $S_{Y,ij}$ due to (5.2.2), require the tire velocities at the ground contact patch

$$\mathbf{v}_{T,ij} = \mathbf{v}_{ij} \begin{pmatrix} v_{XT,ij} \\ v_{YT,ij} \end{pmatrix}. \tag{7.3.18}$$

They can be determined from the yaw rate $\dot{\psi}$, the longitudinal and lateral velocities of the CG $v_X$ and $v_Y$ by; see Fig. 7.19

$$\begin{aligned} \mathbf{v}_{T,fl} &= \begin{bmatrix} v_X - \dot{\psi} b_f/2 \\ v_Y + \dot{\psi} l_f \end{bmatrix}, \quad & \mathbf{v}_{T,fr} &= \begin{bmatrix} v_X + \dot{\psi} b_f/2 \\ v_Y + \dot{\psi} l_f \end{bmatrix}, \\ \mathbf{v}_{T,rl} &= \begin{bmatrix} v_X - \dot{\psi} b_r/2 \\ v_Y - \dot{\psi} l_r \end{bmatrix}, \quad & \mathbf{v}_{T,rr} &= \begin{bmatrix} v_X + \dot{\psi} b_r/2 \\ v_Y - \dot{\psi} l_r \end{bmatrix}. \end{aligned} \tag{7.3.19}$$

The wheel side slip angles follow according to (7.3.19)

$$\alpha_{ij} = \delta_{ij} - \arctan \left( \frac{v_{YT,ij}}{v_{XT,ij}} \right) \quad \text{with i} \in [\text{f,r}]\,; \; \text{j} \in [\text{l,r,}\,] \tag{7.3.20}$$

and for the individual wheels with averaged steering angle at the front axle $\delta_f = (\delta_{fl} + \delta_{fr})/2$ by inserting the components of (7.3.19) one obtains

$$\begin{aligned} \alpha_{fl} &= \delta_f - \arctan \left( \frac{v_Y + l_f \dot{\psi}}{v_X - \frac{b_f}{2} \dot{\psi}} \right), \quad & \alpha_{fr} &= \delta_f - \arctan \left( \frac{v_Y + l_f \dot{\psi}}{v_X + \frac{b_f}{2} \dot{\psi}} \right), \\ \alpha_{rl} &= - \arctan \left( \frac{v_Y - l_r \dot{\psi}}{v_X - \frac{b_r}{2} \dot{\psi}} \right), \quad & \alpha_{rr} &= - \arctan \left( \frac{v_Y - l_r \dot{\psi}}{v_X + \frac{b_r}{2} \dot{\psi}} \right). \end{aligned} \tag{7.3.21}$$

The determination of the vertical forces $F_{Z,ij}$ of each wheel for the calculation of the tire forces $F_{XT,ij}$ and $F_{YT,ij}$ can be obtained from a quasi-stationary cornering model according to Kiencke and Nielsen (2005) by using the accelerations of the CG as inputs and neglecting roll and pitch dynamics, compare (6.5.48)–(6.5.51), but here cornering to the left side

$$\begin{aligned}
F_{Z,fl} &= m\left(\frac{l_r}{l}g - \frac{h_{CG}}{l}a_X\right)\left(\frac{1}{2} - \frac{h_{CG}}{b_f}\frac{a_Y}{g}\right),\\
F_{Z,fr} &= m\left(\frac{l_r}{l}g - \frac{h_{CG}}{l}a_X\right)\left(\frac{1}{2} + \frac{h_{CG}}{b_f}\frac{a_Y}{g}\right),\\
F_{Z,rl} &= m\left(\frac{l_f}{l}g + \frac{h_{CG}}{l}a_X\right)\left(\frac{1}{2} - \frac{h_{CG}}{b_r}\frac{a_Y}{g}\right),\\
F_{Z,rr} &= m\left(\frac{l_f}{l}g + \frac{h_{CG}}{l}a_X\right)\left(\frac{1}{2} + \frac{h_{CG}}{b_r}\frac{a_Y}{g}\right).
\end{aligned} \tag{7.3.22}$$

In this way, the individual longitudinal and lateral tire forces (7.3.16) and (7.3.17) related to the wheel rotation

$$\mathbf{F}_{T,ij}^{T} = \left[F_{XT,ij} \;\; F_{YT,ij},\right] \tag{7.3.23}$$

are obtained. These forces in the wheel coordinates have to be transformed into the longitudinal and lateral chassis (vehicle) coordinates

$$\mathbf{F}_{ij}^{T} = \left[F_{X,ij} \;\; F_{Y,ij};\right] \tag{7.3.24}$$

see (4.1.6), (4.1.7), and Fig. 4.2b.

For a vehicle with front steering and averaged steering angle

$$\delta_f = \frac{1}{2}\left(\delta_{fr} + \delta_{fl}\right), \tag{7.3.25}$$

it holds for the vehicle forces in dependence on the tire forces; see Figs. 7.19 and 7.4

$$F_X = \left(F_{XT,fl} + F_{XT,fr}\right)\cos\delta_f - \left(F_{YT,fl} + F_{YT,fr}\right)\sin\delta_f + \left(F_{XT,rl} + F_{XT,rr}\right). \tag{7.3.26}$$

$$F_Y = \left(F_{YT,fl} + F_{YT,fr}\right)\cos\delta_f + \left(F_{XT,fl} + F_{XT,fr}\right)\sin\delta_f + \left(F_{YT,rl} + F_{YT,rr}\right). \tag{7.3.27}$$

The torque on the vehicle around the vertical axis follows

$$\begin{aligned}
M_Z &= F_{Y,f}l_f - F_{Y,r}l_r + \Delta F_{X,f}\frac{b_f}{2} - \Delta F_{X,r}\frac{b_r}{2}\\
&= \left[\left(F_{YT,fl} + F_{YT,fr}\right)\cos\delta_f + \left(F_{XT,fl} + F_{XT,fr}\right)\sin\delta_f\right] l_f\\
&\quad - \left(F_{YT,rl} + F_{YT,rr}\right) l_r\\
&\quad + \left[\left(F_{XT,fr} - F_{XT,fl}\right)\cos\delta_f + \left(F_{YT,fr} - F_{YT,fl}\right)\sin\delta_f\right]\frac{b_f}{2}\\
&\quad + \left[\left(F_{XT,rr} - F_{XT,rl}\right)\right]\frac{b_r}{2}.
\end{aligned} \tag{7.3.28}$$

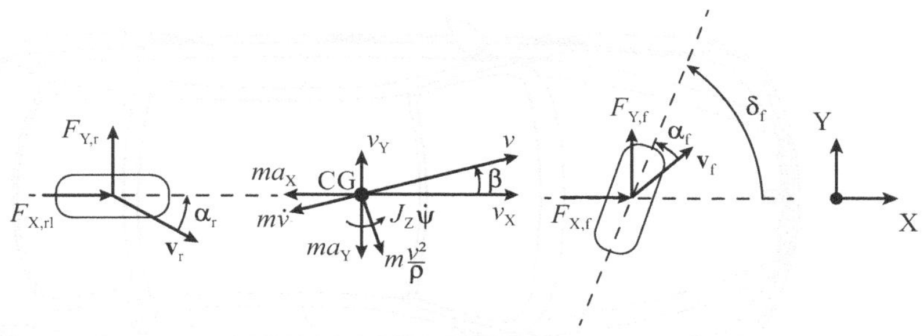


**Fig. 7.20** Scheme for the one-track model in the X-Y plane following from Fig. 7.19, Bauer (2015)

The corresponding equations for front and rear wheel steering are given in Sect. 7.3.3.

Based on the two-track model (7.3.12)–(7.3.19) the *one-track model* follows if the two wheels of one axle are replaced by one centered wheel; see Fig. 7.20. The balance equations for the forces and the torque then lead to (7.2.11)–(7.2.13). Assuming small steering angles, using (7.1.1a) to replace $\beta$, introducing (7.2.8) to replace $v^2/R_P$, and small angles $\delta$ and $\beta$, introducing (7.2.25)–(7.2.26) leads to the one-track model (7.2.34), with the steering angle $\delta_f$ as input and $\beta$ and $\dot{\psi}$ as state variables.

Figure 7.21 illustrates the main signal flow of the treated two-track model for the case that the longitudinal and lateral tire forces are determined separately from each other, summarizing the calculations from (7.3.12) to (7.3.28). The inputs are the measured values of $\alpha_{th}$, $\beta_b$, and $\delta_f$ from the driver, measurements of the drive dynamic sensors for $a_X$, $a_Y$, and $\dot{\psi}$ and the wheel angular velocities $\omega_w$. Outputs are then the calculated variables $\dot{v}_X$, $\dot{v}_Y$ and $\ddot{\psi}$. The forces from the tires and chassis and other variables are summarized as vectors. The signal flow shows the generation of the longitudinal tire forces $F_{\mathbf{XT}}$ using the longitudinal slips $\mathbf{S}_{XT}$ and longitudinal friction coefficients $\boldsymbol{\mu}_{XT}$. The lateral tire forces $\mathbf{F}_{YT}$ are determined with the tire side slip angles and cornering stiffnesses $\mathbf{c}_\alpha$. The vertical forces $\mathbf{F}_{ZT}$ are calculated with the measured acceleration $a_X$ and $a_Y$ and are required to determine the longitudinal and lateral tire forces $\mathbf{F}_{XT}$ and $\mathbf{F}_{YT}$.

The rear wheel tire forces $\mathbf{F}_{XT}$ and $F_{\mathbf{YT}}$ are oriented to the wheel rotation (wheel coordinates) and are then transformed into the longitudinal and lateral forces $\mathbf{F}_X$ and $\mathbf{F}_Y$ of the vehicle coordinate system which determine the vehicle motion, and thus the accelerations $\dot{v}_X$ and $\dot{v}_Y$ and the yaw angle accelerations $\ddot{\psi}$. The integration of these accelerations then delivers the vehicle state variables $v_X$, $v_Y$, and $\dot{\psi}$.

The signal flow shows that the outputs $v_X$ and $v_Y$, i.e. the determined velocities of the center of gravity are required to determine the longitudinal slip $\mathbf{S}_{XT}$ and side slip angles of the wheels, and thus generate several feedback paths from the outputs to the internal variables. This fact can be described as process-determined state variable feedback influencing the dynamic overall behavior of the vehicle.

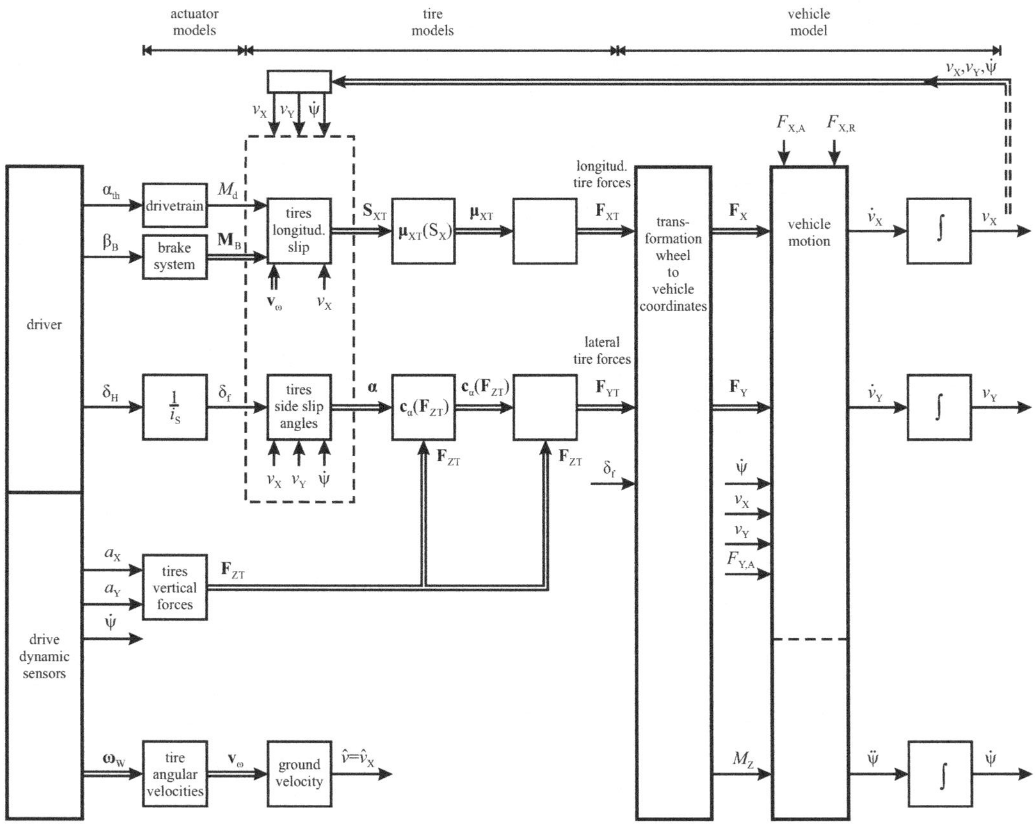


**Fig. 7.21** Signal flow of a nonlinear two-track model for steering and acceleration. Yaw rate/velocity representation with separate models for longitudinal and lateral tire forces

A more detailed signal flow of the two-track model is depicted in Fig. 7.22. The longitudinal slips and tire forces and the (small) side slip angles $\alpha_{ij}$ and lateral forces $F_{FT,ij}$ are now inserted for four wheels, as well as the vertical forces $F_{ZT,ij}$.

The lateral tire forces in Fig. 7.22 are determined with the cornering stiffnesses $c_{\alpha,ij}$ which are valid for small side slip angles $\alpha_{ij}$; see (5.2.3). For *larger side slip angles*, however, it is more appropriate to operate with the lateral friction coefficients $\mu_{YT,ij}$; see (5.2.7) and (5.2.11). For these cases, Fig. 7.22 is changed to Fig. 7.23.

For *combining steering and acceleration* or *braking* with larger values for the acceleration $a_X$ and $a_Y$ in longitudinal and lateral direction the friction coefficients $\mu_X$ and $\mu_Y$ behave differently and depend on each other as described in Sect. 5.3. It is then more realistic to calculate the longitudinal and lateral tire forces $\mathbf{F}_{XT}$ and $F_{YT}$ with friction coefficients, which take into account both slip values in form of $S_{res}$, see (5.3.5), and determine the tire forces with (5.3.8).

An alternative is to use $\mu_X(S_X, \alpha)$ and $\mu_Y(S_X, \alpha)$ if the tire characteristics as in Fig. 5.14 are known. Compared to the separate determination of the tire forces with $\mu_X$ and $c_\alpha(F_{ZT})$ the combined version uses $\mu_Y$ instead of $c_\alpha$. Figure 7.24 shows the corresponding signal flow with vector symbols for the different variables as Fig. 7.21.

**(b) Yaw Rate/Slip Angle Representation**

The yaw rate/velocity representation of the simplified two-track model with motions in the horizontal level was obtained with the force and torque balance equations (7.3.12)–(7.3.14) of the center of gravity with tire forces in the longitudinal and lateral direction of the vehicle. As the direction of the velocity vector $\mathbf{v}$ during cornering is turned by the side slip angle $\beta$ relative to the x-axis of the vehicle, the equations can be modified in replacing the velocities $v_X$ and $v_Y$ by

$$v_X = v\cos\beta; \quad v_Y = v\sin\beta; \tag{7.3.29}$$

see Fig. 7.3. For the acceleration of the CG follows from (7.3.4) with $\dot{\theta} = 0$, $\dot{\varphi} = 0$ and $a_X = F_X/m$, $a_Y = F_Y/m$

$$\begin{aligned} a_X &= \dot{v}_X - \dot{\psi} v_Y, \\ a_Y &= \dot{v}_Y + \dot{\psi} v_X. \end{aligned} \tag{7.3.30}$$

Using (7.3.29) to introduce $v$ and $\beta$ it follows

$$\begin{aligned} \dot{v}_X &= \dot{v}\cos\beta - v\dot{\beta}\sin\beta, \\ \dot{v}_Y &= \dot{v}\sin\beta + v\dot{\beta}\cos\beta. \end{aligned} \tag{7.3.31}$$

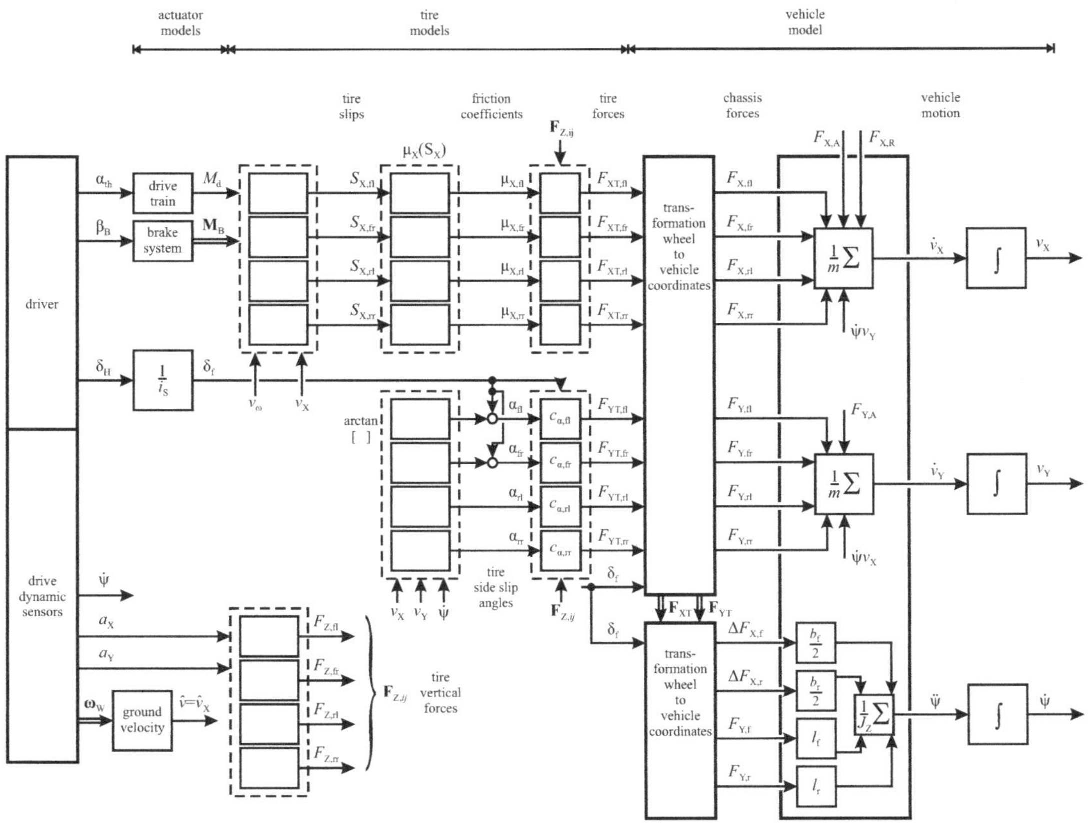


**Fig. 7.22** Signal flow of a two-track model for steering and acceleration. Yaw rate/velocity representation with four wheels and separate models for longitudinal and lateral tire forces with side slip angle $\alpha$

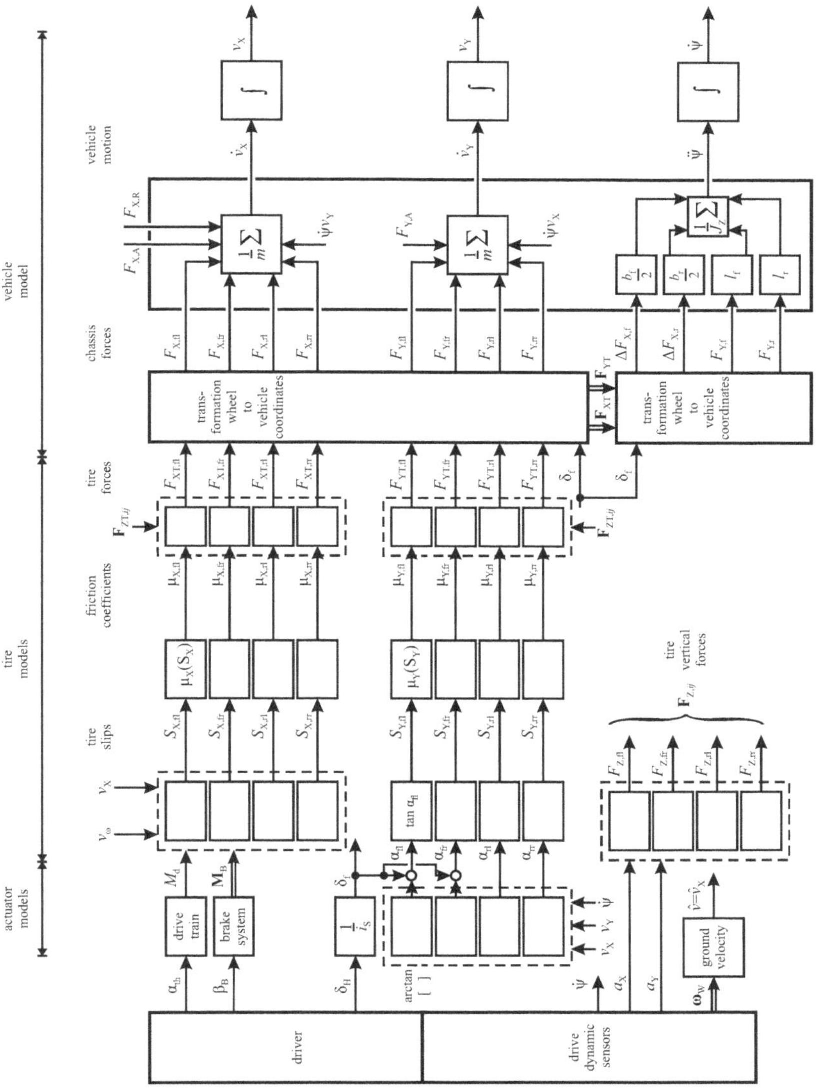


**Fig. 7.23** Signal flow of a two-track model for steering and acceleration. Yaw rate/velocity representation with four wheels and separate models for longitudinal and lateral tire forces with friction coefficients $\mu_Y$

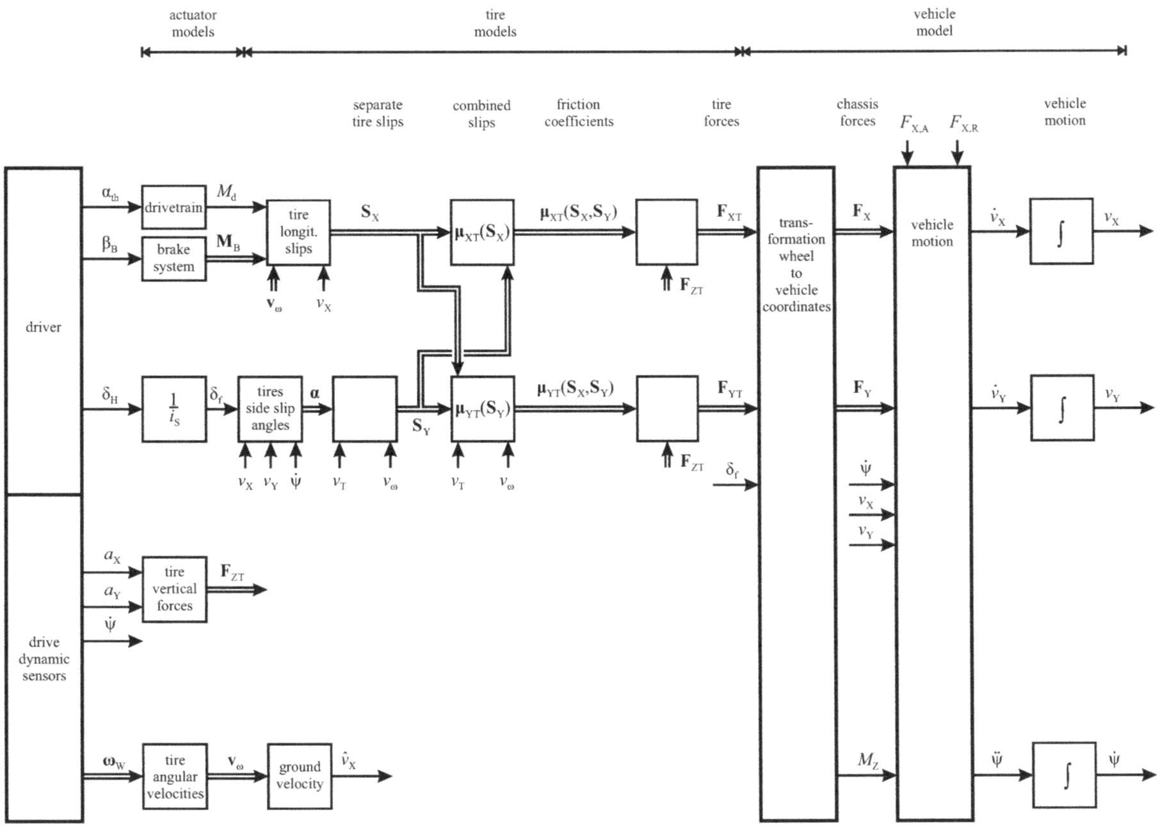


**Fig. 7.24** Signal flow of a nonlinear two-track model for steering and acceleration. Yaw rate/velocity representation with combined tire friction models

Introduction of (7.3.30) and (7.3.31) in (7.3.12)–(7.3.14) leads to

$$
\begin{aligned}
& m(\dot{v}\cos\beta - v\dot{\beta}\sin\beta - \dot{\psi}v\sin\beta) \\
&\qquad = F_{\mathrm{X,fl}} + F_{\mathrm{X,fr}} + F_{\mathrm{X,rl}} + F_{\mathrm{X,rr}} - F_{\mathrm{XA}} - F_{\mathrm{XR}}, \qquad (7.3.32)\\
& m(\dot{v}\sin\beta + v\dot{\beta}\cos\beta + \dot{\psi}v\cos\beta) \\
&\qquad = F_{\mathrm{Y,fl}} + F_{\mathrm{Y,fr}} + F_{\mathrm{Y,rl}} + F_{\mathrm{Y,rr}} - F_{\mathrm{YA}}, \qquad (7.3.33)\\
& J_{\mathrm{Z}}\ddot{\psi} = (F_{\mathrm{X,fr}} - F_{\mathrm{X,fl}})\frac{b_{\mathrm{f}}}{2} + (F_{\mathrm{X,rr}} - F_{\mathrm{X,rl}})\frac{b_{\mathrm{r}}}{2} \\
&\qquad + (F_{\mathrm{Y,fl}} + F_{\mathrm{Y,fr}})l_{\mathrm{f}} - (F_{\mathrm{Y,rl}} + F_{\mathrm{Y,rr}})l_{\mathrm{r}}. \qquad (7.3.34)
\end{aligned}
$$

These equations are now solved for $\dot{v}$, $\dot{\beta}$, and $\ddot{\psi}$ with several calculation steps, see Ackermann (2016) and Bechtloff (2018), resulting in

$$
\begin{aligned}
\dot{v} = \frac{1}{m}\big[&(F_{\mathrm{X,fl}} + F_{\mathrm{X,fr}} + F_{\mathrm{X,rl}} + F_{\mathrm{X,rr}} - F_{\mathrm{XA}} - F_{\mathrm{XR}})\cos\beta \\
&+(F_{\mathrm{Y,fl}} + F_{\mathrm{Y,fr}} + F_{\mathrm{Y,rl}} + F_{\mathrm{Y,rr}} - F_{\mathrm{YA}})\sin\beta\big],
\end{aligned}
\qquad (7.3.35)
$$

$$
\begin{aligned}
\dot{\beta} = &\frac{(F_{\mathrm{Y,fl}} + F_{\mathrm{Y,fr}} + F_{\mathrm{Y,rl}} - F_{\mathrm{Y,rr}} - F_{\mathrm{YA}})\cos\beta}{mv} \\
&- \frac{(F_{\mathrm{X,fl}} + F_{\mathrm{X,fr}} + F_{\mathrm{X,rl}} + F_{\mathrm{X,rr}} - F_{\mathrm{XA}} - F_{\mathrm{XR}})\sin\beta}{mv} - \dot{\psi},
\end{aligned}
\qquad (7.3.36)
$$

$$
\begin{aligned}
\ddot{\psi} = \frac{1}{J_{\mathrm{Z}}}\Big[&(F_{\mathrm{Y,fl}} + F_{\mathrm{Y,fr}})l_{\mathrm{f}} - (F_{\mathrm{Y,rl}} + F_{\mathrm{Y,rr}})l_{\mathrm{r}} + (F_{\mathrm{X,rr}} - F_{\mathrm{X,rl}})\frac{b_{\mathrm{r}}}{2} \\
&+(F_{\mathrm{X,fr}} - F_{\mathrm{X,fl}})\frac{b_{\mathrm{r}}}{2}\Big].
\end{aligned}
\qquad (7.3.37)
$$

The longitudinal and lateral tire forces are now replaced by the forces in the tire coordinates

$$
\begin{aligned}
F_{\mathrm{X,fl}} &= F_{\mathrm{XT,fl}}\cos\delta_{\mathrm{f}} + F_{\mathrm{YT,fl}}\sin\delta_{\mathrm{f}}, \quad & F_{\mathrm{Y,fl}} &= -F_{\mathrm{YT,fr}}\sin\delta_{\mathrm{f}} + F_{\mathrm{YT,fr}}\cos\delta_{\mathrm{f}},\\
F_{\mathrm{X,fr}} &= F_{\mathrm{XT,fr}}\cos\delta_{\mathrm{f}} + F_{\mathrm{YT,fr}}\sin\delta_{\mathrm{f}}, \quad & F_{\mathrm{Y,fr}} &= -F_{\mathrm{XT,fr}}\sin\delta_{\mathrm{f}} + F_{\mathrm{YT,fr}}\cos\delta_{\mathrm{f}},\\
F_{\mathrm{X,rl}} &= F_{\mathrm{XT,rl}}, \quad F_{\mathrm{Y,rl}} = F_{\mathrm{YT,fl}}, & F_{\mathrm{X,rr}} &= F_{\mathrm{XT,rr}}, \quad F_{\mathrm{Y,rr}} = F_{\mathrm{YT,rr}}.
\end{aligned}
\qquad (7.3.38)
$$

This leads to the two-track model

$$
\begin{aligned}
\dot{v} = \frac{1}{m}\big[&(F_{\mathrm{XT,fl}} + F_{\mathrm{XT,fr}} - F_{\mathrm{XA}} - F_{\mathrm{XR}})\cos(\beta - \delta_{\mathrm{f}}) \\
&+ (F_{\mathrm{YT,fl}} + F_{\mathrm{YT,fr}} - F_{\mathrm{YA}})\sin(\beta - \delta_{\mathrm{f}}) + (F_{\mathrm{XT,rl}} + F_{\mathrm{XT,rr}} - F_{\mathrm{XA}} - F_{\mathrm{XR}})\cos\beta \\
&+ (F_{\mathrm{YT,rl}} + F_{\mathrm{YT,rr}} - F_{\mathrm{YA}})\sin\beta\big],
\end{aligned}
\qquad (7.3.39)
$$

$$\begin{aligned}\dot{\beta} = \frac{1}{mv} \big[ &(F_{\text{XT,fl}} + F_{\text{XT,fr}} - F_{\text{XA}} - F_{\text{XR}}) \sin(\beta - \delta_f) \\ &+ (F_{\text{YT,fl}} + F_{\text{YT,fr}} - F_{\text{YA}}) \cos(\beta - \delta_{\text{f}}) + (F_{\text{XT,rl}} + F_{\text{XT,rr}} - F_{\text{XA}} - F_{\text{XR}}) \sin\beta + \\ &(F_{\text{YT,rl}} + F_{\text{YT,rr}} - F_{\text{YA}}) \cos\beta \big] - \dot{\psi},\end{aligned} \tag{7.3.40}$$

$$\begin{aligned}\ddot{\psi} = \frac{1}{J_{\text{Z}}} \big[ &l_{\text{f}} (F_{\text{YT,fl}} \cos\delta_{\text{f}} + F_{\text{YT,fr}} \cos\delta_{\text{f}} + F_{\text{XT,fl}} \sin\delta_f + F_{\text{XT,fr}} \sin\delta_{\text{f}}) \\ &- l_{\text{r}} (F_{\text{YT,rl}} + F_{\text{YT,rr}}) + \frac{b_{\text{f}}}{2} \big( F_{\text{XT,fr}} \cos\delta_{\text{f}} + F_{\text{YT,fr}} \sin\delta_{\text{f}} - F_{\text{XT,fl}} \cos\delta_f \\ &- F_{\text{YT,fl}} \sin\delta_{\text{f}} \big) + \frac{b_{\text{r}}}{2} (F_{\text{XT,rr}} - F_{\text{XT,rl}}) \big]\end{aligned} \tag{7.3.41}$$

Figure 7.25 depicts a signal flow diagram based on (7.3.35)–(7.3.37) and (7.3.38). The tire forces are determined by using appropriate tire models treated in Chap. 5. The longitudinal forces $F_{\text{XT,ij}}$ may additionally take into account forces from braking or acceleration; see also Table 7.7. The signal flow in Fig. 7.25 shows, that the part for the vehicle model is strongly interrelated with regard to the outputs $v$ and $\beta$. Hence, the signal flow of the yaw rate/velocity representation of Fig. 7.22 is more straight forward.

Summarizing, the simplified two-track model in the yaw rate/slip representation for the motions in the horizontal X, Y level according to Fig. 7.19 describes the vehicle motion with (7.3.39)–(7.3.41) with the output

$$\dot{\mathbf{x}} = \left[ \dot{v}\ \dot{\beta}\ \ddot{\psi} \right]^{\text{T}},$$

and the state variable vector

$$\mathbf{x} = \left[ v\ \beta\ \dot{\psi} \right]^{\text{T}}. \tag{7.3.42}$$

This state variable representation then corresponds to the one-track model (7.2.34), however, with variable velocity $v_{\text{X}}$

### 7.3.3 Two-Track Model with Road Gradients, Front and Rear Wheel Steering

**(a) Yaw Rate/Velocity Representation**

Based on the equations for the *translatory behavior* (7.3.4) and (7.3.5) for the acceleration one obtains, Bechtloff (2018), now with longitudinal road gradient $\lambda$ and lateral road gradient $\eta$; see Figs. 7.26 and 7.27

$$\dot{v}_{\text{X}} = \frac{F_{\text{X}}}{m} - \omega_{\text{Y}} v_{\text{Z}} + \omega_{\text{Z}} v_{\text{Y}} + g \sin\lambda, \tag{7.3.43}$$

$$\dot{v}_{\text{Y}} = \frac{F_{\text{Y}}}{m} + \omega_{\text{X}} v_{\text{Z}} - \omega_{\text{Z}} v_{\text{X}} + g \cos\lambda \sin\varphi, \tag{7.3.44}$$

$$\dot{v}_{\text{Z}} = \frac{F_{\text{Z}}}{m} - \omega_{\text{X}} v_{\text{Y}} + \omega_{\text{Y}} v_{\text{X}} - g \cos\eta \cos\lambda, \tag{7.3.45}$$

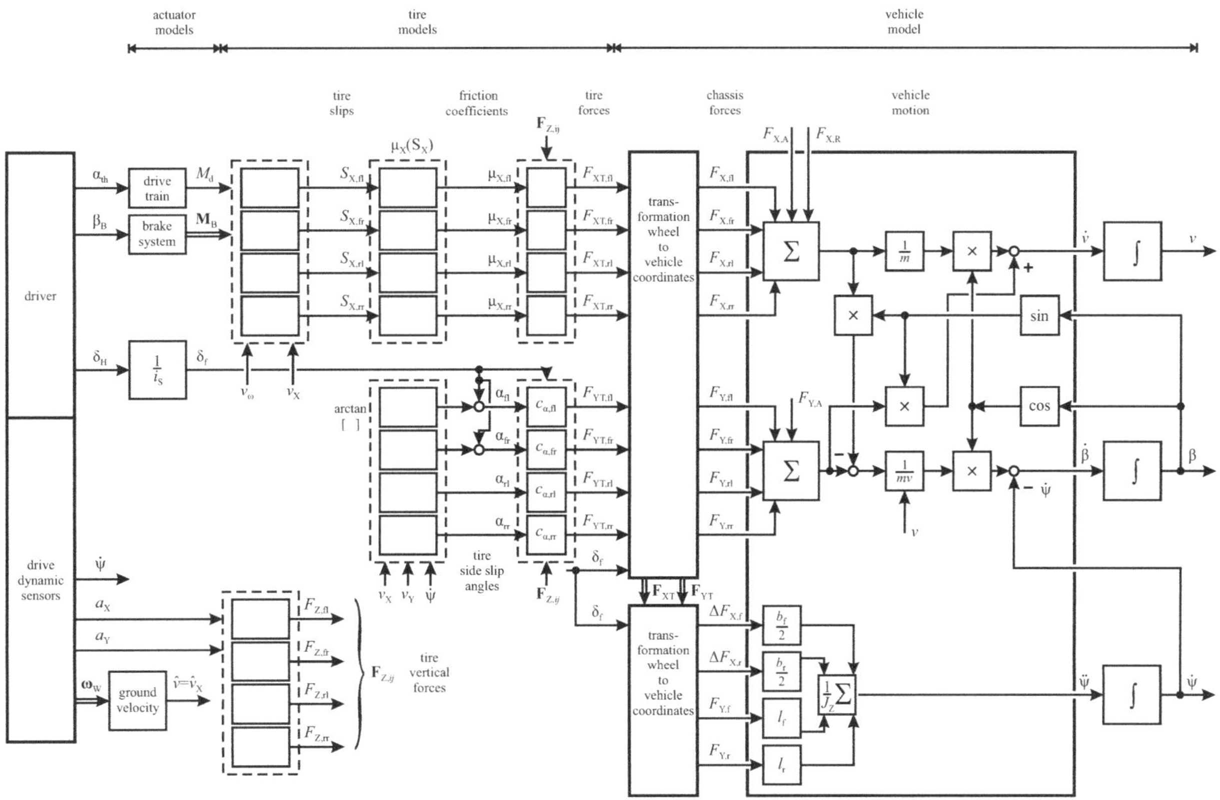


**Fig. 7.25** Signal flow of a two-track model for steering and acceleration. Yaw rate/velocity representation with four wheels and separate models for longitudinal and lateral tire forces with side slip angles $\alpha$

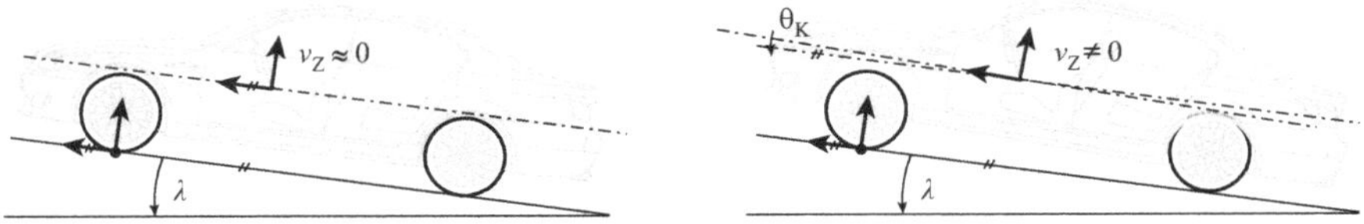


**Fig. 7.26** Vehicle-fixed coordinate system for longitudinal road gradient, Bechtloff (2018)

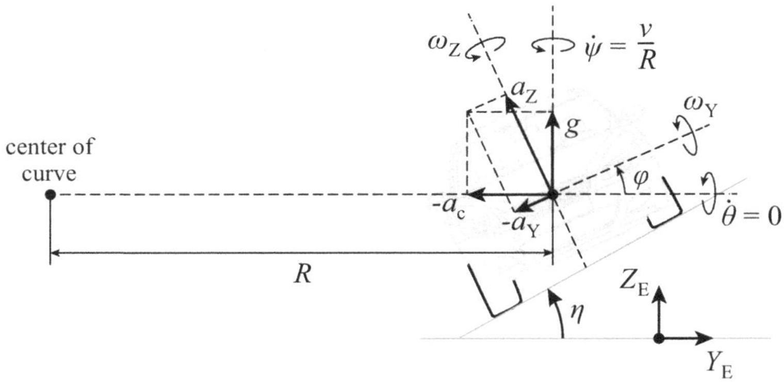


**Fig. 7.27** Vehicle-fixed coordinate system for lateral road gradient (banked corner), stationary driving, right curvature, negative lateral acceleration, Bechtloff (2018). Coordinates according to ISO 8855 (2013)

$\omega_X = \dot{\varphi}$, $\omega_Y = \dot{\theta}$, $\omega_Z = \dot{\psi}$ are the angular velocities around the vehicle-fixed axis according to ISO 8855 (2013). $F_X$, $F_Y$ and $F_Z$ are the sums of all forces acting on the body; see (7.3.5).

Assuming small vertical velocities $v_Z \approx 0$ and $\dot{v}_Z \approx 0$ these equations simplify to

$$\dot{v}_X = \frac{F_X}{m} + \dot{\psi}_Z v_Y + g \sin \lambda, \tag{7.3.46}$$

$$\dot{v}_Y = \frac{F_Y}{m} - \dot{\psi} - g \cos \eta \cos \lambda. \tag{7.3.47}$$

For the longitudinal force follows, taking into account a front steering angle $\delta_f$ and a rear steering angle $\delta_r$; see Fig. 7.28, and $\delta_{fl} = \delta_{fr} = \delta_f$

$$\begin{aligned} F_X = {} & (F_{XT,fl} + F_{XT,fr}) \cos \delta_f - (F_{YT,fl} + F_{YT,fr}) \sin \delta_f \\ & + (F_{XT,rl} + F_{XT,rr}) \cos \delta_r - (F_{YT,rl} + F_{YT,rr}) \sin \delta_r \\ & - F_{XA} - F_{XR}. \end{aligned} \tag{7.3.48}$$

The lateral force becomes with roll angle $\varphi$

$$\begin{aligned} F_Y = {} & \cos \varphi \big[ (F_{YT,fl} + F_{YT,fr}) \cos \delta_f + (F_{XT,fl} + F_{XT,fr}) \sin \delta_f \\ & + (F_{YT,rl} + F_{YT,rr}) \cos \delta_r + (F_{XT,rl} + F_{XT,rr}) \sin \delta_r \big] \\ & + mg \sin \varphi. \end{aligned} \tag{7.3.49}$$

The rotational behavior around the vertical axis results from the last row of (7.3.8)

$$\ddot{\psi} = \frac{1}{J_Z}(M_Z + \dot{\varphi}\dot{\theta}(J_X - J_Y)). \tag{7.3.50}$$

The torque around the vertical axis follows from (7.3.9) to

$$\begin{aligned} M_Z = &\left[(F_{YT,fl} + F_{YT,fr}) \cos \delta_f + (F_{XT,fl} + F_{XT,fr}) \sin \delta_f\right] l_f \\ &- \left[(F_{YT,rl} + F_{YT,rr}) \cos \delta_r + (F_{XT,rl} + F_{XT,rr}) \sin \delta_r\right] l_r \\ &+ \left[(F_{XT,fr} - F_{XT,fl}) \cos \delta_f - (F_{XT,fr} - F_{XT,fl}) \sin \delta_f\right] \frac{b_f}{2} \\ &+ \left[(F_{XT,rr} - F_{XT,rl}) \cos \delta_r - (F_{XT,rr} - F_{XT,rl}) \sin \delta_r\right] \frac{b_r}{2}. \end{aligned} \tag{7.3.51}$$

Bechtloff (2018) has shown with experimental drives on a banked corner of 30°, that the gyroscopic term in (7.3.50) can be neglected, such that is holds for the yaw angle velocity

$$\ddot{\psi} = \frac{1}{J_Z} M_Z. \tag{7.3.52}$$

The determination of the lateral tire forces results with the slip angles $\alpha_{ij}$ from (7.3.17).

The longitudinal and lateral tire forces are determined with the slip values (5.1.1), (5.1.2), (5.2.2) and are calculated from the angular wheel velocities and the ground speed, see also Bechtloff (2018), for the corrections with the steering angles.

Summarizing, the two-track model with road gradients, front and rear wheel steering in the yaw rate/velocity representation is described by (7.3.46)–(7.3.52) and determines the state variables longitudinal velocity, lateral velocity, and yaw rate

$$\mathbf{x} = \left[v_X \; v_Y \; \dot{\psi}\right]^T. \tag{7.3.53}$$

### 7.3.4 Nonlinear One-Track Model with Road Gradients

The equations of the two-track model can be used to derive a one-track model which takes into account the road gradient angles in longitudinal and lateral direction, $\lambda$ and $\eta$, front and rear wheel steering, and gyroscopic effects, Bechtloff (2018).

**(a) Yaw Rate/Velocity Representation**

With the assumption of small vertical velocities $v_Z \approx 0$ and $\dot{v}_Z \approx 0$ Eq. (7.3.4) leads to the force balances for the CG

$$\dot{v}_X = \frac{1}{m}(F_X - F_{XA} - F_{XR}) + \dot{\psi}_Z v_Y + g \sin \lambda, \tag{7.3.54}$$

$$\dot{v}_Y = \frac{1}{m}(F_Y - F_{YA}) - \dot{\psi}_Z v_X - g \cos \lambda \sin \eta, \tag{7.3.55}$$

where $\dot{\psi}$ is the rotational velocity around the vertical body axis, compare Figs. 7.26 and 7.27. The four wheel forces of Fig. 7.28 are replaced by *two axle forces* (one-track); see Fig. 7.29

$$F_{\mathrm{XT,f}} = F_{\mathrm{XT,fl}} + F_{\mathrm{XT,fr}}; \quad F_{\mathrm{XT,r}} = F_{\mathrm{XT,rl}} + F_{\mathrm{XT,rr}}, \tag{7.3.56}$$

$$F_{\mathrm{YT,f}} = F_{\mathrm{YT,fl}} + F_{\mathrm{YT,fr}}; \quad F_{\mathrm{YT,r}} = F_{\mathrm{YT,rl}} + F_{\mathrm{YT,rr}}, \tag{7.3.57}$$

and therefore

$$\begin{aligned} F_{\mathrm{XT}} &= F_{\mathrm{XT,f}} + F_{\mathrm{XT,r}}, \\ F_{\mathrm{YT}} &= F_{\mathrm{YT,f}} + F_{\mathrm{YT,r}}. \end{aligned} \tag{7.3.58}$$

Analogue to (7.3.48) it follows for the *longitudinal forces*

$$\begin{aligned} F_{\mathrm{X}} = {} & F_{\mathrm{XT,f}} \cos\delta_{\mathrm{f}} - F_{\mathrm{YT,f}} \sin\delta_{\mathrm{f}} + F_{\mathrm{XT,r}} \cos\delta_{\mathrm{r}} \\ & - F_{\mathrm{YT,r}} \sin\delta_{\mathrm{r}} - F_{\mathrm{XA}} - F_{\mathrm{XR}}, \end{aligned} \tag{7.3.59}$$

and corresponding to (7.3.49) for the *lateral forces*

$$\begin{aligned} F_{\mathrm{Y}} &= F_{\mathrm{Y,f}} + F_{\mathrm{Y,r}} + mg \sin\varphi \\ &= \left(F_{\mathrm{YT,f}} \cos\delta_{\mathrm{f}} + F_{\mathrm{XT,f}} \sin\delta_{\mathrm{f}}\right) \cos\varphi \\ &+ \left(F_{\mathrm{YT,r}} \cos\delta_{\mathrm{r}} + F_{\mathrm{XT,r}} \sin\delta_{\mathrm{r}}\right) \cos\varphi + mg \sin\varphi, \end{aligned} \tag{7.3.60}$$

for the *rotational behavior* around the vertical axis (7.3.52) is valid. This leads to

$$\ddot{\psi} = \frac{1}{J_{\mathrm{Z}}} M_{\mathrm{Z}} = \frac{1}{J_{\mathrm{Z}}} \left(F_{\mathrm{YT,f}}\, l_{\mathrm{f}} - F_{\mathrm{YT,r}}, l_{\mathrm{r}}\right) + M_{\mathrm{Z,ESC}}. \tag{7.3.61}$$

The term $M_{\mathrm{Z,ESC}}$ becomes active if for ESC wheel brake interventions are activated. Then it holds

$$M_{\mathrm{Z,ESC}} = \left(F_{\mathrm{XT,fr}} - F_{\mathrm{XT,fl}}\right) \frac{b_{\mathrm{f}}}{2} + \left(F_{\mathrm{XT,rr}} - F_{\mathrm{XT,rl}}\right) \frac{b_{\mathrm{f}}}{2}, \tag{7.3.62}$$

with $F_{\mathrm{XT,ij}}$ as longitudinal tire forces by individual braking; see Sects. 6.7 and 7.2.

The side slip angles have now be taken for the centered wheels of each axis; see Fig. 7.29

$$\alpha_{\mathrm{f}} = \delta_{\mathrm{f}} - \arctan\left(\frac{v_{\mathrm{Y}} + l_{\mathrm{f}}\dot{\psi}}{v_{\mathrm{X}}}\right), \tag{7.3.63}$$

$$\alpha_{\mathrm{r}} = \delta_{\mathrm{r}} - \arctan\left(\frac{v_{\mathrm{Y}} - l_{\mathrm{f}}\dot{\psi}}{v_{\mathrm{X}}}\right). \tag{7.3.64}$$

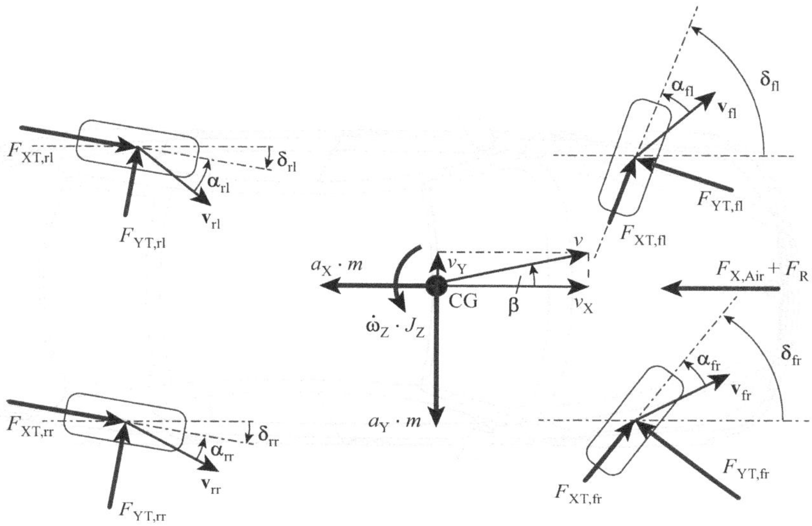


**Fig. 7.28** Scheme for the two-track model in the X-Y plane with front and rear wheel steering, Bechtloff (2018)

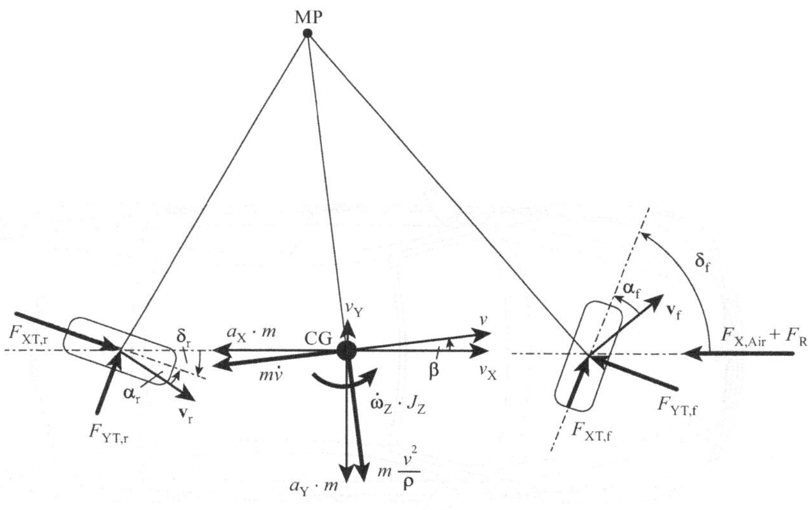


**Fig. 7.29** Scheme for a one-track model in the X-Y plane with front and rear wheel steering, Bechtloff (2018)

The tire slip values of the front and rear wheels are replaced by a weighted average

$$S_{X,f} = \frac{S_{X,fl} F_{Z,fl} + S_{X,fr} F_{Z,fr}}{F_{Z,fl} + F_{Z,fr}}, \tag{7.3.65}$$

$$S_{X,r} = \frac{S_{X,rl} F_{Z,rl} + S_{X,rr} F_{Z,rr}}{F_{Z,rl} + F_{Z,rr}}, \tag{7.3.66}$$

with the quasi-stationary vertical tire forces according to (7.3.22).

The vertical forces of the axles for a longitudinal acceleration $a_X$ can be simplified and follow from (7.3.22) with $a_Y = 0$

$$\begin{aligned} F_{Z,f} &= \frac{m}{l}(l_r\, g - h_{CG}\, a_X), \\ F_{Z,r} &= \frac{m}{l}(l_f\, g + h_{CG}\, a_X). \end{aligned} \tag{7.3.67}$$

Hence, the nonlinear one-track model with road gradients consists in the *yaw rate/velocity representation* by the determination of the state vector

$$\mathbf{x} = [v_X\ v_Y\ \dot{\psi}]^T, \tag{7.3.68}$$

and use of (7.3.54) and (7.3.55) with the tire forces (7.3.58), (7.3.59), and (7.3.60).

The vertical forces $F_{Z,ij}$ from the body to the wheels have been modeled for quasi-stationary behavior. In reality the dynamics of the suspensions have to be taken into account; see Chap. 8. Simplified models of passive suspensions are described in Sect 8.2.

Hence, this extended one-track model operates with all four tire forces $F_{XT,ij}$ and $F_{YT,ij}$ summarizing them to axle tire forces $F_{XT}$ and $F_{YT}$ (7.3.58)–(7.3.60) with simplified vertical forces (7.3.67) for the determination of the longitudinal and lateral acceleration $\dot{v}_X$ and $\dot{v}_Y$ (7.3.54) and (7.3.55).

Compared to the one-track model in Sect. 7.2.1, the determination of the yaw angle acceleration $\ddot{\psi}$ uses the tire forces $F_{XT,ij}$ and $F_{YT,ij}$ of all four wheels and not only those of two centered wheels. This allows then the addition of wheel individual torques $M_{Z,ESC,ij}$ from ESC interventions.

The signal flow is similar to Fig. 7.22, however, with summarized tire forces for one axis and neglecting of differences of left and right tire forces $\Delta F_{X,f}$ and $\Delta F_{X,r}$ for the determination of $\ddot{\psi}$.

**(b) Yaw Rate/Slip Angle Representation**

The longitudinal and lateral velocity of the CG can be expressed by the ground velocity $v$ and the slip angle

$$v_X = v \cos\beta;\ \ v_Y = v \sin\beta. \tag{7.3.69}$$

Their derivatives are

$$\begin{aligned} \dot{v}_{\mathrm{X}} &= -v \sin \dot{\beta} + \dot{v} \cos \beta, \\ \dot{v}_{\mathrm{Y}} &= -v \cos \dot{\beta} + \dot{v} \sin \beta. \end{aligned} \tag{7.3.70}$$

Introducing these terms in (7.3.39) and (7.3.40) leads after some calculation steps

$$\begin{aligned} \dot{v} &= \left(\frac{F_{\mathrm{X}}}{m} + g \sin \lambda\right) \cos \beta + \left(\frac{F_{\mathrm{Y}}}{m} - g \cos \lambda \sin \eta\right) \sin \beta \\ \dot{\beta} &= \left(\frac{F_{\mathrm{Y}}}{m} - g \cos \lambda \cdot \sin \eta\right) \frac{1}{v} \cos \beta - \left(\frac{F_{\mathrm{X}}}{m} + g \sin \lambda\right) \frac{1}{v} \sin \beta - \dot{\psi} \\ \dot{\omega}_{\mathrm{Z}} &= \frac{1}{J_{\mathrm{Z}}} M_{\mathrm{Z}}; \end{aligned} \tag{7.3.71}$$

see the derivation in Bechtloff (2018). $F_{\mathrm{X}}$ and $F_{\mathrm{Y}}$ follow from (7.3.26) and (7.3.27). The nonlinear one-track model with road gradients can then be described by the state vector

$$\mathbf{x} = \left[v \ \beta \ \dot{\psi}\right]^{\mathrm{T}}. \tag{7.3.72}$$

Experimental results with lateral dynamic models are illustrated in Chap. 12.

### 7.3.5 Comparison of Different Lateral Vehicle Models

In order to compare some lateral vehicle models by simulations following models have been used, Hoffmann (2018)

(A) Two-track model (7.3.12)–(7.3.28) (yaw rate / velocity representation),
(B) Linear one-track model (7.2.34),
(C) Nonlinear one-track model (7.3.54)–(7.3.68) (yaw rate / velocity representation for even road, $\lambda = 0$, $\mu = 0$).

The calculation of the tire forces $F_{\mathrm{XT,ij}}$ and $F_{\mathrm{YT,ij}}$ was performed with the nonlinear slip based models (5.2.13) and combined longitudinal and lateral slip (5.3.1)–(5.3.8) and (5.3.18). The applied vehicle parameters are given in Table 7.8. The simulation results are shown for a step function of $\delta_{\mathrm{H}} = 60°$ with $d\delta_{\mathrm{H}}/dt = 200\,°/\mathrm{s}$ and a velocity of $v = 75\,\mathrm{km/h}$, such that a required lateral acceleration of $a_{\mathrm{Y}} = 4\,\mathrm{m/s^2}$ is reached, according to ISO 7401 (2011)

To obtain reference values for the simulations the commercial multi-body simulation tool ASM (dSPACE automotive simulation) was used.

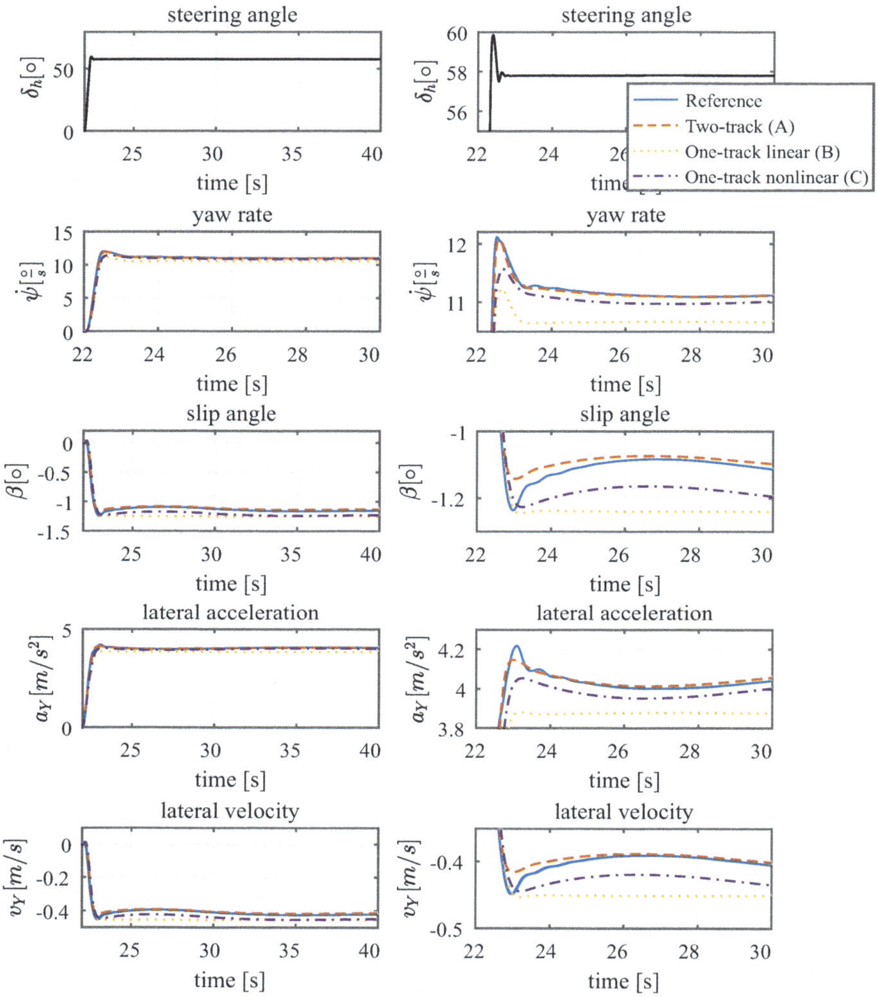


**Fig. 7.30** Simulated transient responses of different lateral vehicle models for a step input of the steering angle, $v = 75\,\mathrm{km/h}$. The right hand side shows a zoom

Figure 7.30 depicts the transient responses. All signal courses show overshoots. The best agreement with the reference is obtained for the two-track model and the nonlinear one-track model. Larger differences shows the linear one-track model. However, the overall behavior is relatively good for all models, with time averaged deviations for the slip angle $\beta$ of 2.4° for A, $-11, 5°$ for B, and $-10.5°$ for C.

Further simulations have been made for parameter variations. Then only one parameter was changed and the others are not changed, i.e. are the nominal values. Figure 7.31 shows the influence of the cornering stiffness $c_{\alpha,\mathrm{f}}$ of the front wheel(s). An increase of $c_{\alpha,\mathrm{f}}$ increases the gain of the yaw rate $\dot{\psi}$ and also increases the gain

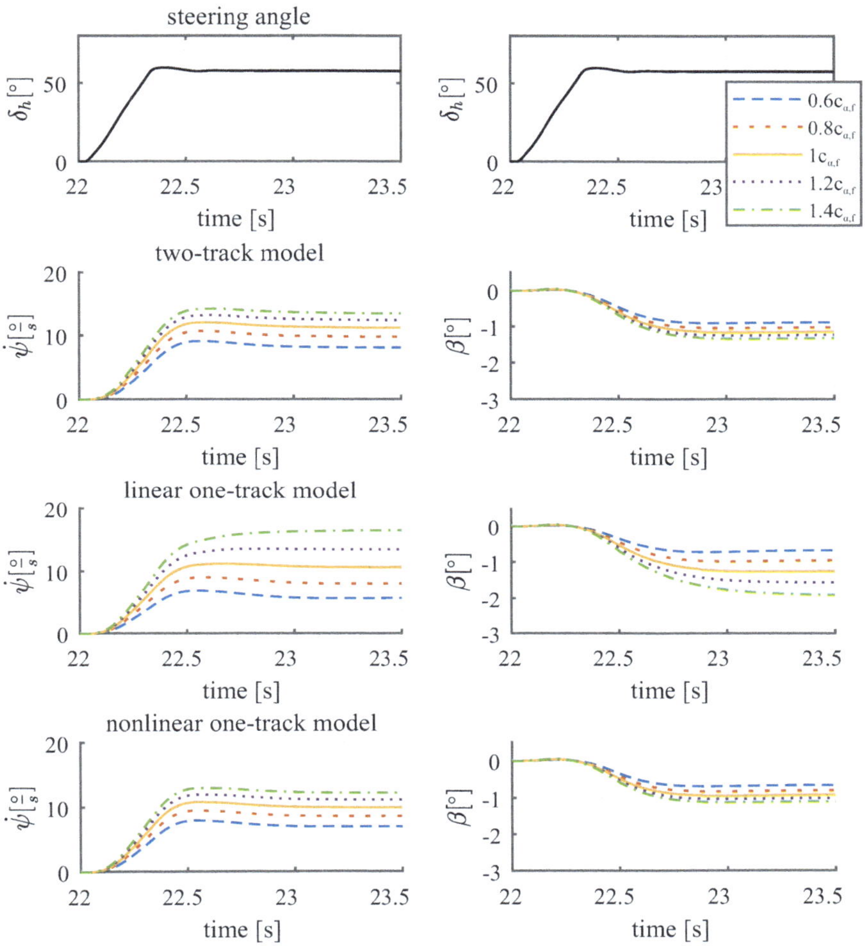


**Fig. 7.31** Simulated transient responses of the yaw rate $\dot{\psi}$ and the slip angle $\beta$ for different lateral vehicle models and changes of the lateral cornering stiffness $c_{\alpha,\mathrm{f}}$

of the slip angle $\beta$, i.e. the vehicle becomes more agile. The changes of the linear one-track model are larger than for the other two models.

The influence of the cornering stiffness $c_{\alpha,\mathrm{r}}$ of the rear wheel(s) is illustrated in Fig. 7.32. An increase of $c_{\alpha,\mathrm{r}}$ reduces slightly the yaw rate gain and the slip angle gain, i.e. shows opposite behavior as $c_{\alpha,\mathrm{f}}$. However, a reduction of $c_{\alpha,\mathrm{r}}$ results in a significantly larger effect in the other direction and leads for the linear one-track model even to monotonic unstable behavior. The effect of the changes of $c_{\alpha,\mathrm{r}}$ is larger for the linear one-track model compared to the other two models.

Figure 7.33 depicts the influence of the vehicle velocity $v$ on the transient functions for the linear one-track model. According to (7.2.57) and (7.2.59), the gain increases

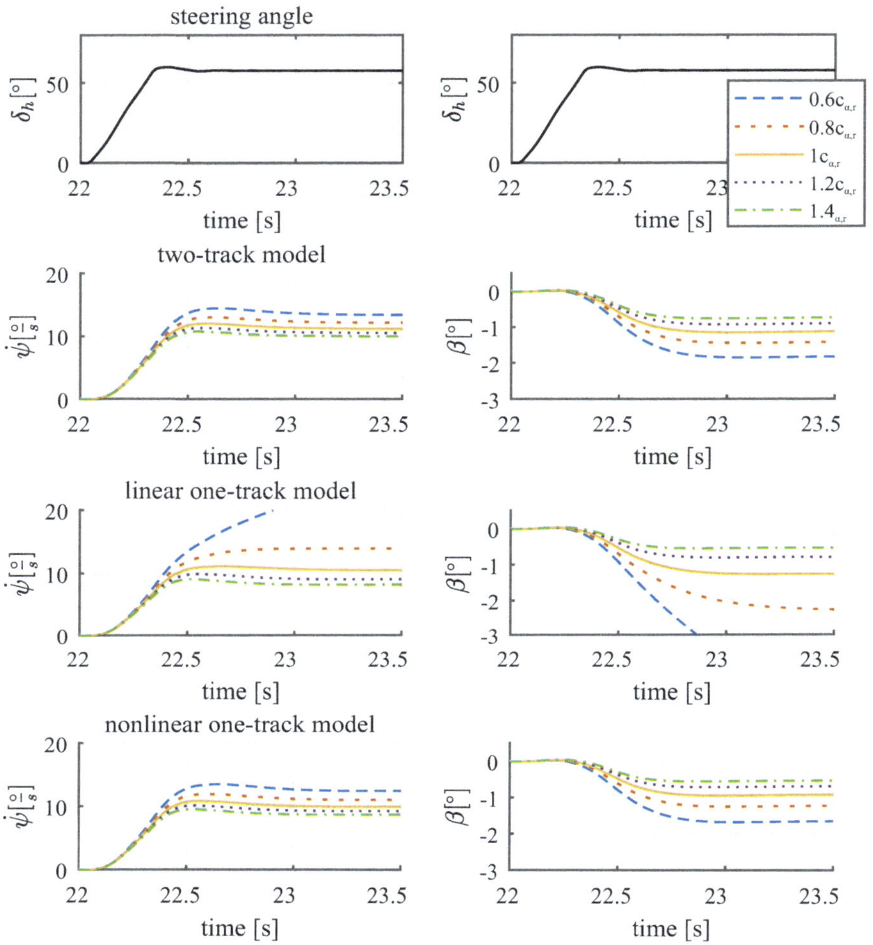


**Fig. 7.32** Simulated transient responses of the yaw rate $\dot{\psi}$ and the slip angle $\beta$ for different lateral vehicle models and changes of the lateral cornering stiffness $c_{\alpha,\mathrm{r}}$

with $v$ for the yaw rate, the lateral acceleration, and the slip angle. The overshoot of the yaw rate increases for larger velocities.

These simulations show that the influence of parameter changes is larger for the linear one-track model than for the nonlinear one-track and the two-track model. The sensitivity to not exactly known parameters is smaller for the two-track model and larger for the nonlinear one-track model, especially for the rear wheel cornering stiffness.

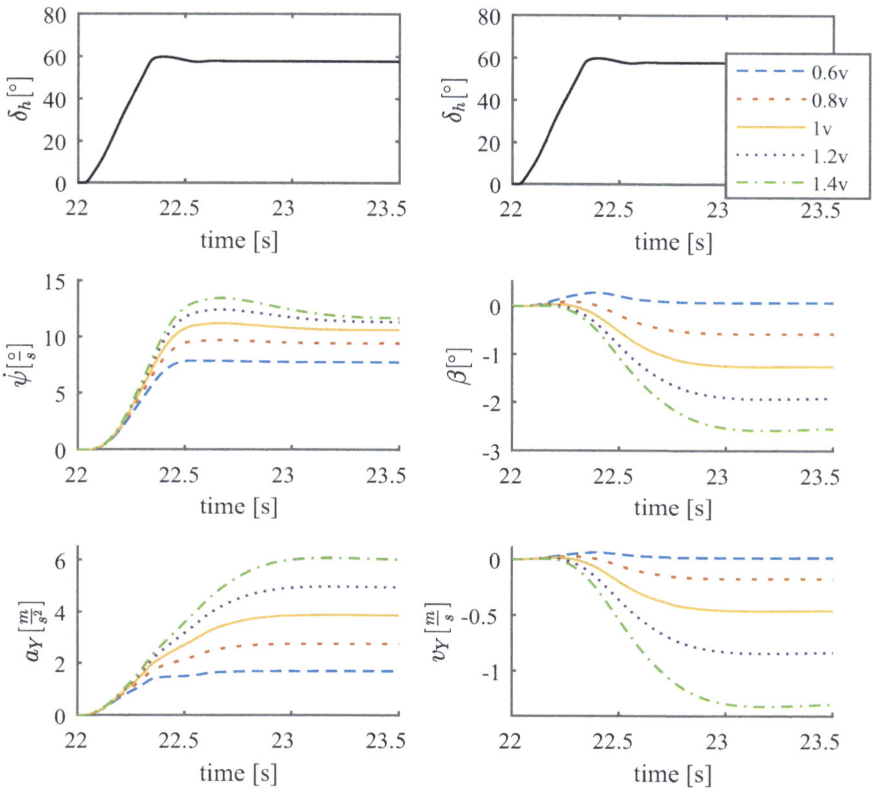


**Fig. 7.33** Simulated transient responses with the linear one-track model for different vehicle velocities, $v = 75\,\mathrm{km/h}$

### 7.3.6 Effect of Parameter Variations on the Lateral Behavior

To show the effects of parameter and state variable changes on the drive dynamic behavior, some simulations with two-track models are illustrated. The results are based on adaptive two-track models, which adapt the theoretical two-track model by a serially coupled local linear neuronal net model (LOLIMOT) to the measured behavior of the real vehicle, Holzmann (2001), Halfmann (2001), Halfmann and Holzmann (2003). Some data of the investigated front driven car, are: $m = 1265\,\mathrm{kg}$ (empty), $l = 2.6\,\mathrm{m}$.

The amplitude of the frequency response in the range of 1.3–1.6 Hz of the yaw rate behavior increases with the velocity, whereas the phase angle increases mainly for higher frequencies, Fig. 7.34. The transient functions in Fig. 7.35 for step inputs of the steering angle show for higher velocity $v_X > 40\,\mathrm{km/h}$ an increasing overshoot of the yaw rate, an increasing lateral acceleration and higher roll angles with larger oscillations.

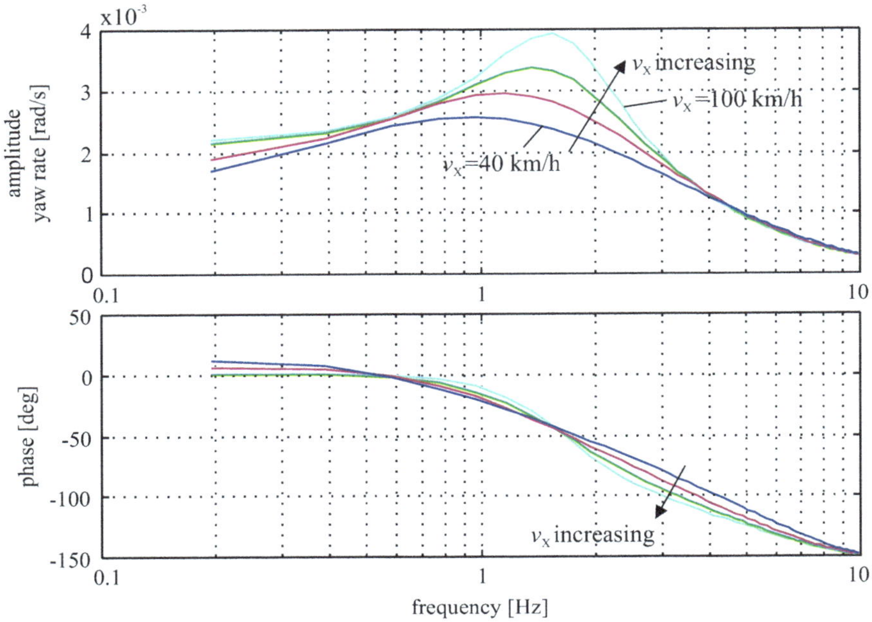


**Fig. 7.34** Frequency response for the yaw rate $\dot{\psi}$ for sinusoidal excitation of steering wheel angle with amplitude $\delta_{H0}$= 60° and velocities $v_x$ = 40, 60, 80, and 100 km/h (simulations)

**Table 7.8** Vehicle parameters used for simulation, higher class passenger car

| Symbol | Value | Unit | Description |
|---|---|---|---|
| $c_{\alpha f}$ | 98 063 | [N/rad] | Cornering stiffness at front wheel |
| $c_{\alpha r}$ | 106 624 | [N/rad] | Cornering stiffness at rear wheel |
| $l_f$ | 1.42 | [m] | Distance CG - front axle |
| $l_r$ | 1.55 | [m] | Distance CG - rear axle |
| $m$ | 2036 | [kg] | Vehicle mass |
| $J_Z$ | 3417 | [kg m$^2$] | Moment of inertia vertical axis |
| $J_X$ | 680 | [kg m$^2$] | Moment of inertia longitudinal axis |
| $J_Y$ | 2100 | [kg m$^2$] | Moment of inertia lateral axis |
| $v_{ch}$ | 61 | [km/h] | Characteristic velocity |

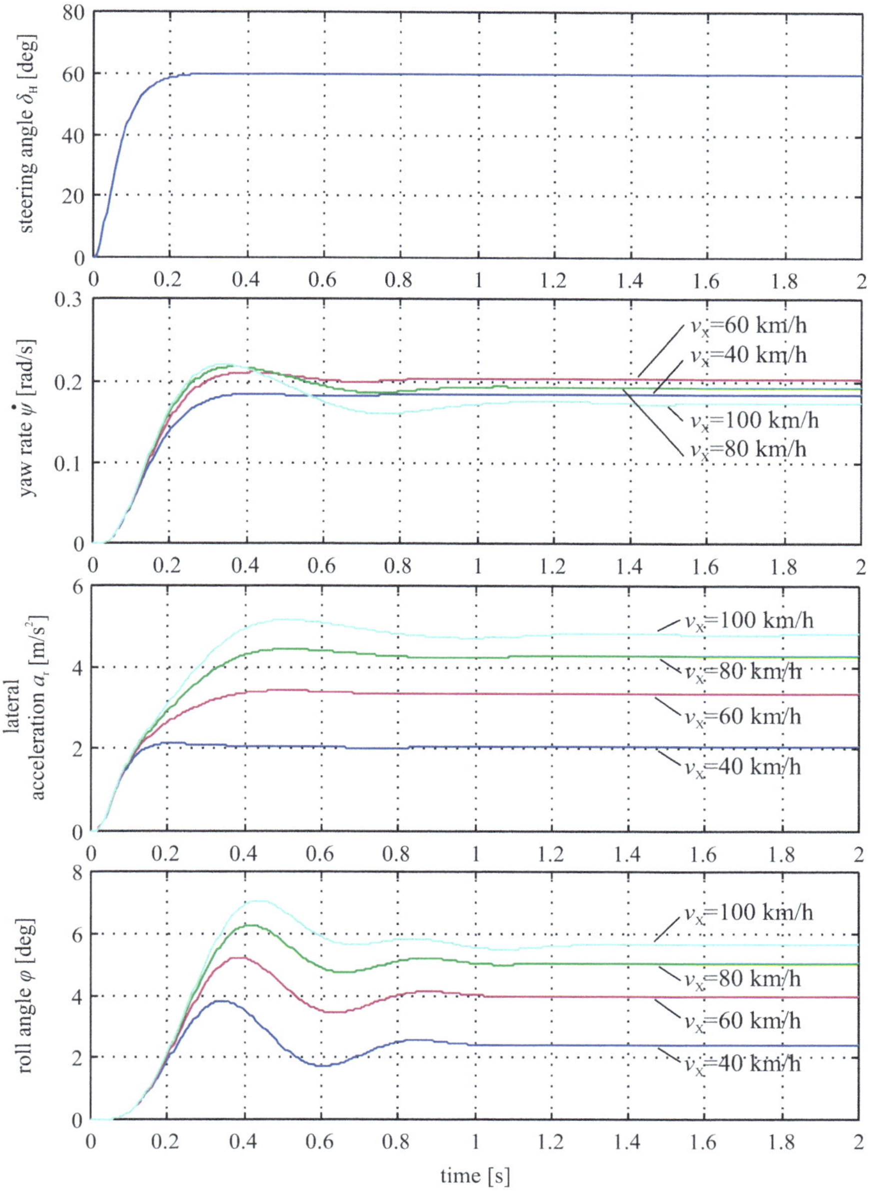


**Fig. 7.35** Transient functions for a step input of the steering wheel angle and different velocities $\upsilon_x = 40, 60, 80$, and 100 km/h (simulations)

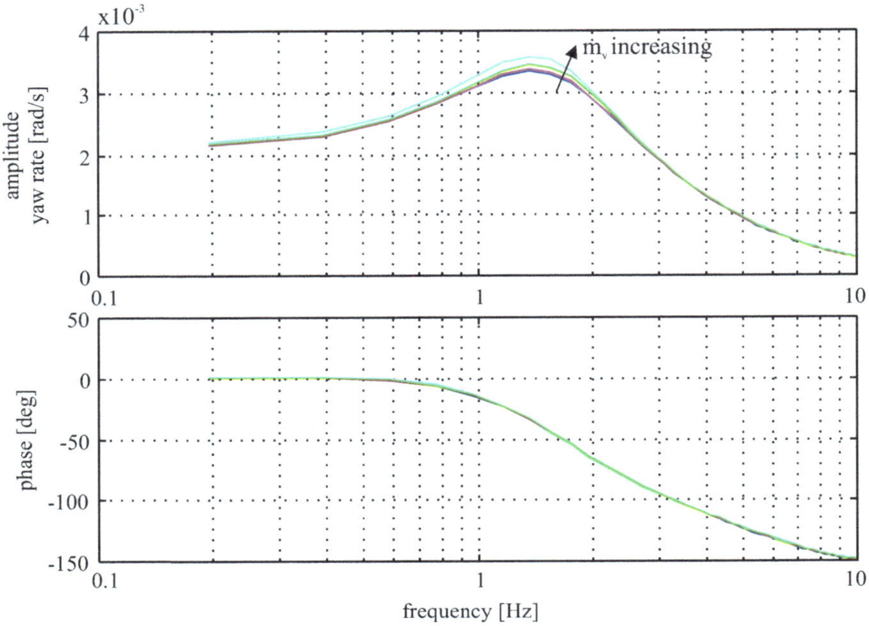


**Fig. 7.36** Frequency response of the yaw rate $\dot{\psi}$ for sinusodial changes of the steering wheel with amplitude $\delta_{H0} = 60°$ and changes of vehicle mass (simulations)

To illustrate the effect of an increase of the load mass (laden state), the simulations assumed following cases: (a) only driver, (b) driver and 1 person (front seat), (c) driver and 3 persons, (d) driver, 3 persons, and loaded trunk until permissible loaded weight. Figure 7.36 indicates that the resonance peak increases about 8%, but that the resonance frequency and phase hardly change. The effects are more visible in the transient functions, Fig. 7.37. With increasing laden weight the yaw rate, lateral acceleration and roll angle present stronger but damped oscillations until 4 persons. However, with additional 150 kg baggage in the trunk until maximal allowed weight the oscillation increases significantly and have a low damping for the yaw rate and lateral acceleration. The roll angle illustrates for this vehicle a very strong oscillation with low damping.

These simulations for a compact car indicate that the lateral behavior varies mainly with increasing velocity, increasing load mass, and mass distribution.

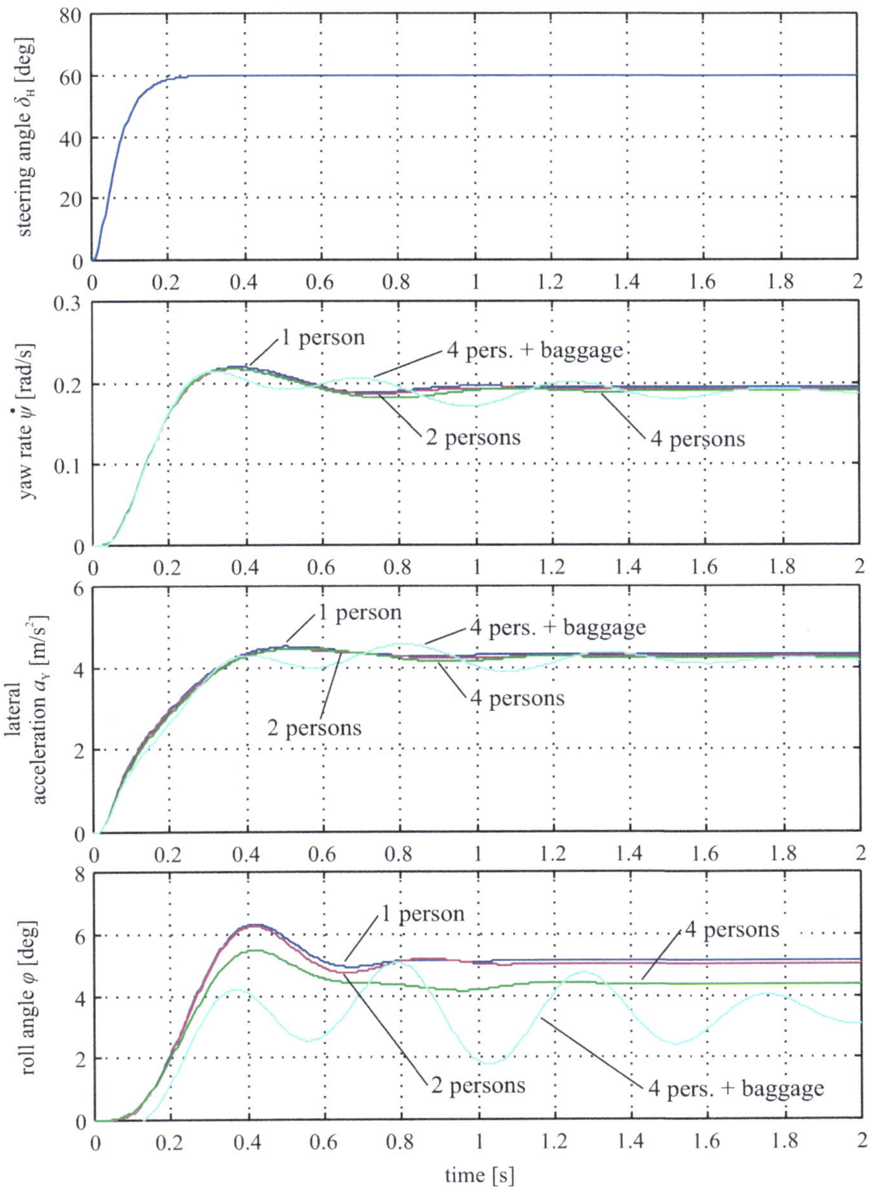


**Fig. 7.37** Transient functions for a step input of the steering wheel angle and different laden states (simulations)

## References


Ackermann C (2016) Neue Ansätze für Fahrerassistenzsysteme mit Regelung der Längs- und Querdynamik. Dissertation Technische Universität Darmstadt. Fortschr.-Ber. VDI Reihe 8, 1249. VDI Verlag, Düsseldorf

Ammon D (1997) Modellbildung und Systementwicklung in der Fahrzeugdynamik, 1st edn. Teubner, Stuttgart, B.G

Bauer M (2015) Methoden zur modellbasierten Fahrdynamikanalyse und Bewertung von Fahrdynamikregelsystemen. Dissertation Technische Universität Darmstadt. Fortschr.-Ber. VDI Reihe 12, 792. VDI Verlag, Düsseldorf

Bechtloff J (2018) Schätzung des Schwimmwinkels und fahrdynamischer Parameter zur Verbesserung modellbasierter Fahrdynamikregelungen. Dissertation Technische Universität Darmstadt. Fortschr.-Ber. VDI Reihe 12, 809. VDI Verlag, Düsseldorf

Börner M (2004) Adaptive Querdynamikmodelle für Personenkraftfahrzeuge - Fahrzustandserkennung und Sensorfehlertoleranz. Dissertation Technische Universität Darmstadt. Fortschr.-Ber. VDI Reihe 12, Nr. 563. VDI Verlag, Düsseldorf

Börner M (2006) Modellierung. Analyse und Simulation der Fahrzeugquerdynamik, Vieweg, Wiesbaden, pp 47–70

Börner M, Andréani L, Albertos P, Isermann R (2002) Detection of lateral vehicle driving conditions based on the characteristic velocity. In: IFAC World Congress 2002, Barcelona, Spain

Daiss A (1996) Beobachtung fahrdynamischer Zustände und Verbesserung einer ABS-und Fahrdynamikregelung. Diss. Universität Karlsruhe, Fortschr.-Ber. VDI Reihe 12, 283. VDI Verlag, Düsseldorf

Halbe I, Isermann R (2007) A model-based fault-tolerant sensor platform for vehicle dynamics control. In: Proceedings of the 5th symposium in advances in automotive control, seascape resort aptos, CA, USA, pp 509–516

Halfmann C (2001) Adaptive semiphysikalische Echtzeitsimulation der Kraftfahrzeugdynamik im bewegten Fahrzeug. Dissertation Technische Universität Darmstadt. Fortschr.-Ber. VDI Reihe 12, 467. VDI Verlag, Düsseldorf

Halfmann C, Holzmann H (2003) Adaptive Modelle für die Kraftfahrzeugdynamik. VDI-Buch, Springer, Berlin

Hoffmann M (2018) Modellgestützte Spurregelung eines PKW. Technische Universität Darmstadt, Institut für Automatisierungstechnik und Mechatronik

Holzmann H (2001) Adaptive Kraftfahrzeugdynamik-Echtzeitsimulation mit Hybriden Modellen. Dissertation Technische Universität Darmstadt. Fortschr.-Ber. VDI Reihe 12, 465. VDI Verlag, Düsseldorf

ISO 7401 (2011) Road vehicles - lateral transient response test methods: open loop test methods. In: International organization for standardization, Geneva

ISO 8855 (2013) Road vehicles - vehicle dynamics and road-holding ability - vocabulary. In: International organization for standardization, Geneva

Kiencke U, Nielsen L (2000) Automotive control systems. For engine, driveline and vehicle. Springer, Berlin

Kiencke U, Nielsen L (2005) Automotive control systems: for engine, driveline, and vehicle, 2nd edn. Springer, Berlin

Milliken W, Milliken D (1995) Race car vehicle dynamics. In: SAE international, Warrendale

Mitschke M, Wallentowitz H (2014) Dynamik der Kraftfahrzeuge, 5th edn. Springer, Berlin

Popp K, Schiehlen W (1993) Fahrzeugdynamik: Eine Einführung in die Dynamik des Systems Fahrzeug-Fahrweg. Teubner, Stuttgart

Reimpell J, Hoseus K (1992) Fahrzeugtechnik: Fahrzeugmechanik. Vogel, Würzburg

Riekert P, Schunck TE (1940) Zur Fahrmechanik des gummibereiften Kraftfahrzeugs. Ingenieur-Archiv 11(3):210–224

Schorn M (2007) Quer- und Längsregelung eines Personenkraftwagens für ein Fahrerassistenzsystem zur Unfallvermeidung. Diss. Universität Darmstadt, Fortschr.-Ber. VDI Reihe 12, 651. VDI Verlag, Düsseldorf

Schramm D, Hiller M, Bardini R (2010) Modellbildung und Simulation der Dynamik von Kraftfahrzeugen. Springer, Berlin

Willumeit HP (2013) Modelle und Modellierungsverfahren in der Fahrzeugdynamik, reprint 1st, edition edn. Teubner, Stuttgart, B.G

van Zanten A (2006) Elektronisches Stabilitätsprogramm (ESP). In: chapter 8 in Isermann R (ed) Fahrdynamik-Regelung, Vieweg, Wiesbaden, pp 169–211

van Zanten A, Erhardt R, Landesfeind K, Pfaff G (1998) VDC systems development and perspective. SAE Trans 107:424–444

Zomotor A, Reimpell J (1991) Fahrwerktechnik: Fahrverhalten, 2nd edn. Vogel Buchverlag, Würzburg

# Vertical Vehicle Behavior

# 8

The vertical behavior of driving cars is mainly determined by the suspension systems of the individual wheel's respective axles and the road roughness. The suspension systems characterize the driving comfort of the passengers and have a significant influence on driving safety. Therefore, firstly, a measure for driving comfort is considered, which is followed by a description of suspension components and mathematical models of passive suspensions which are also required for lateral two-track models, rolling and pitching behavior. Active suspensions are treated in Chap. 15.

## 8.1 Vehicle Suspensions

### 8.1.1 Driving Comfort and Safety

The vehicle suspension system carries the vehicle-body and transmits the forces between body and road. Therefore, the suspension system is responsible for the ride quality and influences considerably the drive stability; Reimpell et al. (2001), Heissing and Ersoy (2011). The task of the spring is to carry the body mass and to isolate the body from road disturbances. The damper contributes to both drive safety and drive quality. Its task is the damping of body and wheel oscillations. A non-bouncing wheel is the condition for transferring road-contact forces. Considering the vertical dynamics and taking into account the vehicle's symmetry, a suspension can in a first step be reduced to the so-called quarter-car model as shown in Fig. 8.1. In its simplest form, the suspension system consists of the spring and the damper. A single spring models the tire. Modeling the tire damping is usually negligible. This basic model satisfies the most common simulation and control applications; Bußhardt (1995).



R. Isermann, *Automotive Control*, ATZ/MTZ-Fachbuch,
https://doi.org/10.1007/978-3-642-39440-9_8

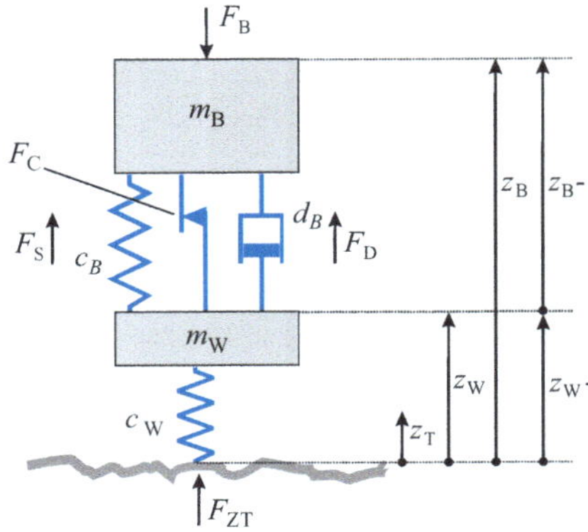


**Fig. 8.1** One-dimensional vertical suspension representation—the quarter-car model

*Driving safety* is the result of a harmonious suspension design in terms of wheel suspension, springing, steering, and braking, and is reflected in optimum vehicle dynamics. In order to describe driving comfort and safety mathematically, quantitative values are introduced. The acceleration of the body $\ddot{z}_B$ is an obvious quantity for the motion and vibration of the car body, and therefore, the passengers comfort. Thus, the effective value of the normalized body acceleration for a quarter-car model is chosen as a measure for comfort

$$\ddot{z}_{B,rel,eff} = \sqrt{\frac{1}{T}\int_{t=0}^{T}\left(\frac{\ddot{z}_B}{g}\right)^2 dt}. \tag{8.1.1}$$

As the dynamic vertical force at tire $F_{ZT}(t)$ determines the road-contact changes of the forces between tire and road, it is a quantity for the safety. Therefore, the effective value of the *normalized dynamic tire force* is applied as measure for *safety*

$$F_{ZTdyn,eff} = \sqrt{\frac{1}{T}\int_{t=0}^{T}\left(\frac{F_{ZTdyn}}{F_{ZTstat}}\right)^2 dt}. \tag{8.1.2}$$

The momentum balances for the body mass part and the wheel mass of the car suspension assuming a linear behavior are

$$m_B\ddot{z}_B = c_B\,(z_W - z_B) + d_B\,(\dot{z}_W - \dot{z}_B) - F_B, \tag{8.1.3}$$

$$m_W\ddot{z}_W = c_W\,(z_T - z_W) - c_B\,(z_W - z_B) - d_B\,(\dot{z}_W - \dot{z}_B)\,. \tag{8.1.4}$$

For more details; see Sect. 8.2.

This leads to an overall model of 4th order. Based on these simplified equations, Figs. 8.2 and 8.3 show the magnitudes of different frequency responses for a suspension with parameters as in Table 8.1 and changes of the parameters $d_B$ and $c_B$.

The effects on the body acceleration and dynamic vertical tire forces are summarized for wheel side excitation in Table 8.2 and for body side excitation in Table 8.3.

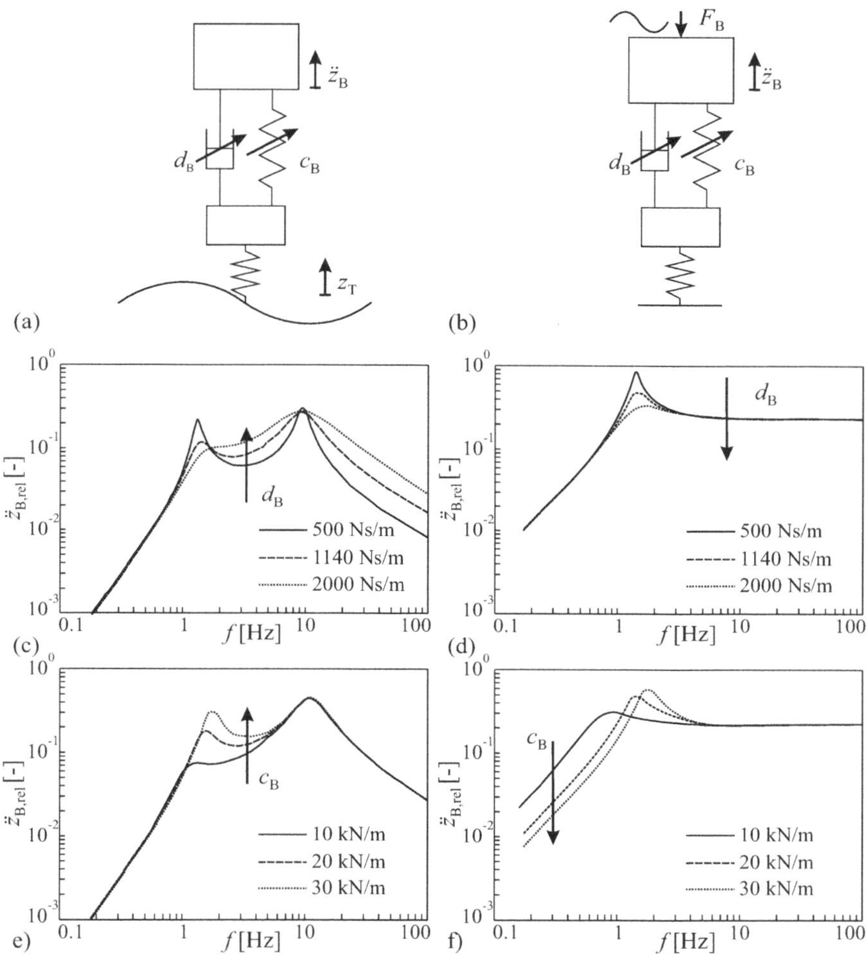


**Fig. 8.2** Linear suspension model and frequency response magnitudes of relative body acceleration (comfort) $\ddot{z}_{B,rel} = \ddot{z}_B/g$, Bußhardt (1995). **a** wheel side excitation $z_T(t) = 1\,\text{cm}\,\sin(2\pi f t)$; **b** body side excitation $F_B(t) = 500\,\text{N}\,\sin(2\pi f t)$; **c**, **d** variation of damping coefficient $d_B$; **e**, **f** variation of spring stiffness $c_B$

**Table 8.1** Parameters of a quarter-car suspension; see Mitschke (1984), Bußhardt (1995)

| | | |
|---|---|---|
| Body mass | $m_B$ | 256 kg |
| Wheel mass | $m_W$ | 31 kg |
| Body spring stiffness | $c_B$ | 20.2 kN/m |
| Body damper coefficient | $d_B$ | 1140 Ns/m |
| Tire stiffness | $c_W$ | 128 kN/m |

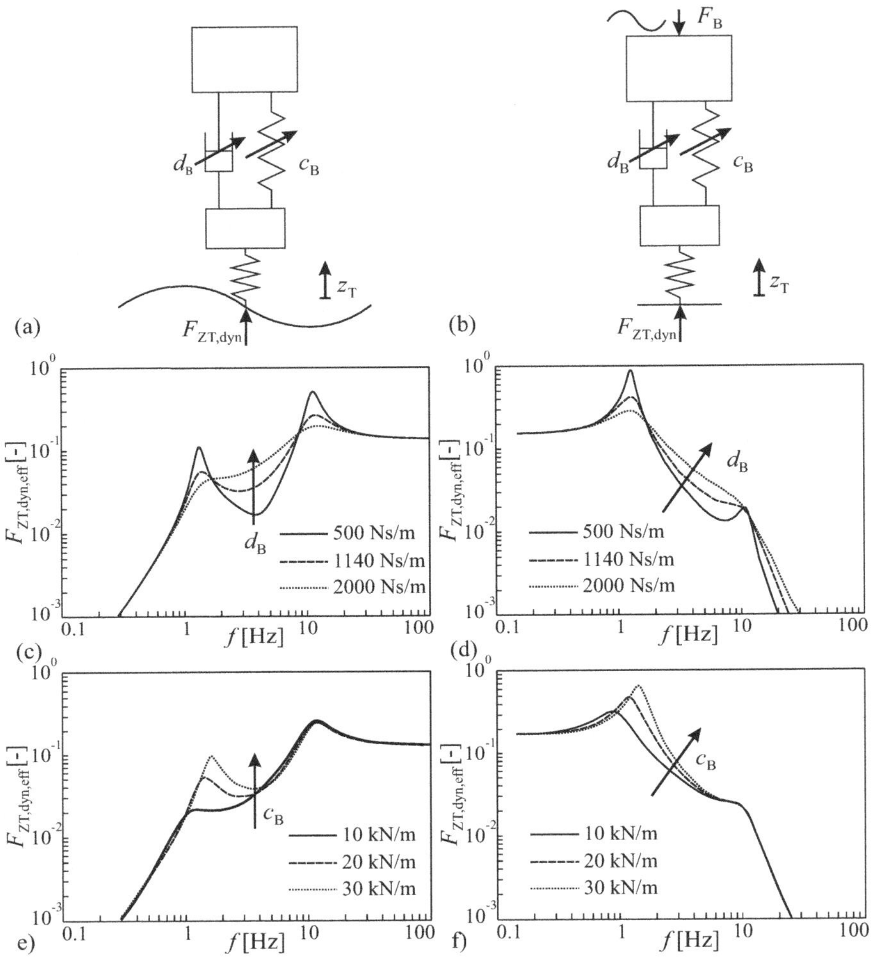


**Fig. 8.3** Linear suspension model and frequency response magnitudes of relative dynamic vertical tire forces $F_{ZT,dyn,rel} = F_{ZT,dyn}/F_{ZT,stat}$ (safety) Bußhardt (1995). **a** wheel side excitation $z_T(t) =$ 3 cm $\sin(2\pi f t)$; **b** body side excitation $F_B(t) = 500$ N $\sin(2\pi f t)$; **c**, **d** variation of damping coefficient $d_B$; **e**, **f** variation of spring stiffness $c_B$

It can be seen for changes of the damping coefficient $+\Delta d_B$ and wheel side excitation improvements for low and high frequencies lead to a deterioration at middle frequencies. These results show that the effect of changes of suspension parameters depends on the frequency range of excitation. Further, improvement in one frequency range is combined with a deterioration in another frequency range (like for $+\Delta d_B$). Hence, the adaption of suspension parameters should be made in dependance on the excitation frequencies, which is for a given road and wheel side excitation a function of the vehicle velocity.

**Table 8.2** Changes of frequency response magnitude for wheel side excitation. ++ strong increase; + increase; −− strong decrease; − decrease; 0 no change

| Frequency | Body acceleration | | Dynamic tire forces | |
|---|---|---|---|---|
| Range | $+\Delta d_B$ | $+\Delta c_B$ | $+\Delta d_B$ | $+\Delta c_B$ |
| low 1−3 Hz | − | + | − | ++ |
| medium 3−1 Hz | ++ | + | ++ | + |
| high 10−15 Hz | - | 0 | − | 0 |

**Table 8.3** Changes of frequency response magnitude for body side excitation. ++ strong increase; + increase; −− strong decrease; − decrease; 0 no change

| Frequency | Body acceleration | | Dynamic tire forces | |
|---|---|---|---|---|
| Range | $+\Delta d_B$ | $+\Delta c_B$ | $+\Delta d_B$ | $+\Delta c_B$ |
| low 1−3 Hz | − | ++ | − | ++ |
| medium 3−1 Hz | 0 | 0 | + | + |
| high 10−15 Hz | 0 | 0 | + | 0 |

However, normal uneven roads generate according to a velocity dependent power spectral density a distribution of frequencies within a range of about 1−30 Hz, such that many frequencies of the suspension frequency response are excited simultaneously. Humans are especially sensitive within 4–8 Hz, Heissing and Ersoy (2011). In general, one can state the following requirements for the design of suspensions.

In order to improve the ride quality, it is important to isolate the body, also called sprung mass, from the road disturbances and to decrease the resonance peak of the sprung mass near 1−2 Hz, which is known to be a sensitive frequency to the human body. A further goal is to keep the tire in contact with the road surface, and therefore, to decrease the resonance peak near 10 Hz, which is the resonance frequency of the wheel, i.e. the unsprung mass. For a given suspension spring, a better isolation of the sprung mass from road disturbances can be achieved with a soft damping by allowing a larger suspension deflection. However, a better road contact can be achieved with a strong damping preventing unnecessary suspension deflections. Therefore, the ride quality and the drive safety are two conflicting criteria. Figure 8.4 illustrates this conflict, showing the variation of drive safety and comfort with the changing vehicle parameters body mass, stiffness, and damping in the "conflict diagram". This diagram presents the vehicle's properties as a point in the $\ddot{z}_B - F_{Zdyn}$ diagram.

The fixed setting of a passive suspension system is a compromise between comfort and safety for any given input set of road conditions. An adaptation to changing road and load conditions can be reached with semi-active and active suspensions, as treated in Chap. 15.

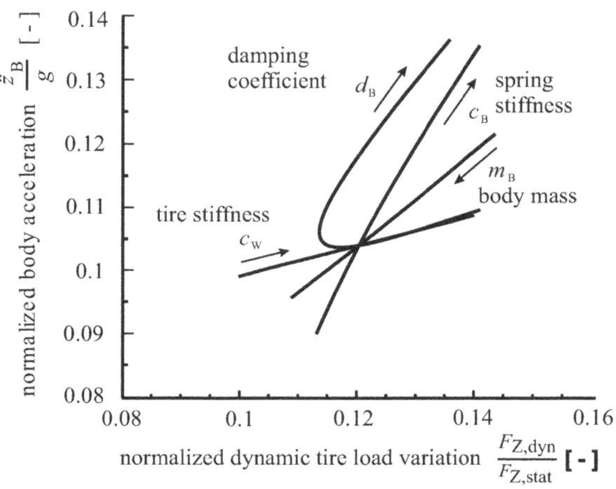


**Fig. 8.4** Influence of vehicle parameters, quarter-car simulations for a given road profile, Bußhardt (1995)

## 8.1.2 Suspension Components

The basic principles of suspension systems are presented briefly, in order to give insight in the function and the related restrictions for the application.

### (a) Automotive Spring Elements

In the field of passenger vehicles, four main types of spring elements are used; conventional steel springs in form of coil springs, torsion bars, air springs, and hydro pneumatic suspensions; see Fig. 8.5.

*Coil springs* are simple springs without self-damping. On account on their simplicity, they have low weight and are maintenance-free. Only limited space is required and dampers can be mounted within the spring. Their characteristic is linear, but can also be designed progressively. *Torsion bars* are made of round bar stock or flat steel. They are wear and maintenance-free. *Air springs* provide a high driving comfort and enable load-leveling. It is the only spring concept, where the natural frequency of the body remains constant under load, but the force characteristics are nonlinear due to the gas properties. The springing air volume is captured in roll or toroid bellows. Therefore, the wheel travel must be defined by separate mechanical guidance systems. In the case of hydro-pneumatic suspensions, a gas volume in an accumulator determines the response characteristics. The fluid compresses the gas according to the wheel load. The connection between the spring strut and the accumulator contains additional valves for the control of the damping function. The performance curves are nonlinear due to the gas properties. This system is favored by Citroën.

### (b) Automotive Dampers

Basically, one distinguishes *single-tube* and *twin-tube* shock absorbers; Fig. 8.6.

In the case of *single-tube shock absorbers*, a sliding separating-piston and gas cushion form the gas-pressure damper. Its advantage is the fact that it is easy to tailor to specific applications, as the large piston diameter allows low working pressures

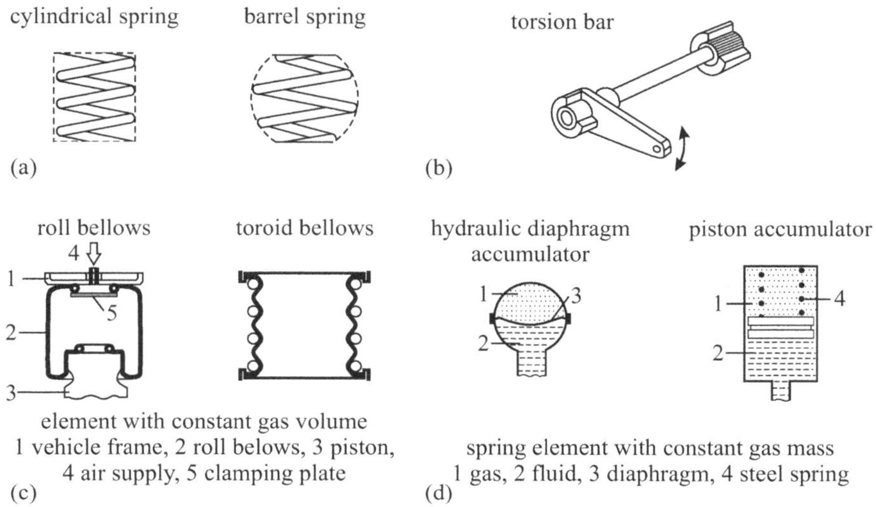


**Fig. 8.5** Spring systems: Steel spring, torsion bar, air spring, hydro pneumatic spring according to Robert Bosch GmbH (1996)

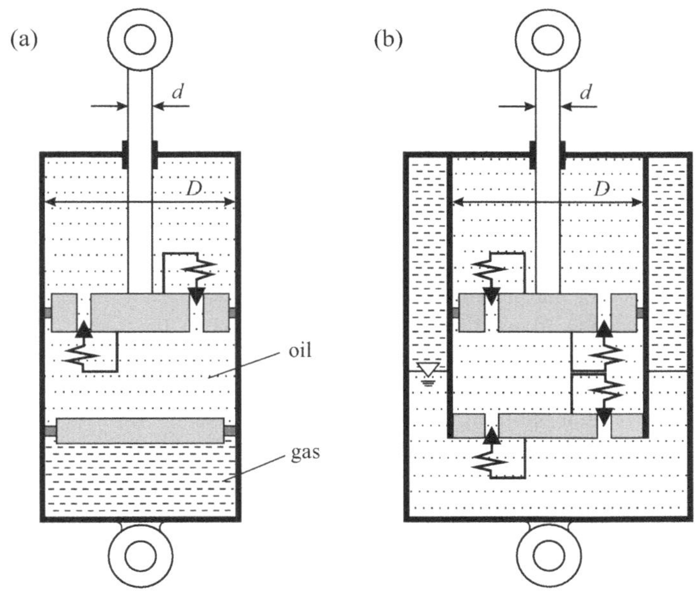


**Fig. 8.6** Schematic of a single-tube and twin-tube absorber according to Robert Bosch GmbH (2011). **a** Single-tube absorber; **b** twin-tube absorber

and sufficient space for valves and passages. Further advantages are that the heat is dissipated directly via the outer tube and that the shock absorber can be installed in any position. Whereas the disadvantages are the length and the sensitivity of the outer tube to damages from stone throw, etc., as it acts as guide cylinder for the piston. Also, the piston-rod seal is subjected to the damping pressure. *Twin-tube shock absorbers* are available as atmospheric or low-pressure twin-tube type. The advantages are firstly insensitivity to external damage, so supplementary mechanical measures can be taken at the outer tube in restricted bodywork areas. Secondly, this shock absorber type is short, as the balance chamber is next to the working cylinder. Disadvantages include the sensitivity to overloading which results in damping ceases and restricted installation positions. The damping characteristics are the result of the cumulative function of orifice damping and of the spring-loaded valves, which closes the passage. The spring responds to pressure by increasing the free aperture of the outlet orifice. The piston bore and spring can be specifically tailored to provide linear to mildly digressive damping curves. The compression forces are frequently only 30... 50% of those for the rebound mode, Robert Bosch GmbH (2011), i.e. if damper is moving out. The reason is that the absorber forces in the rebound stage are higher as in the compression stage.

### (c) Adjustable Dampers

The simplest adaptive dampers are *manually adjustable dampers*. The aim is the adaptation of the damping to various long distance driving situations with varying loads or road windings (e.g. holiday trip, mountain courses). The most common adjustment method is the twisting of the piston or the piston shell resulting in a change of the throttle's properties. The so-called *load dependent damping variation* is based on the alteration of the deflection in case of body mass changes. This effect is used to vary the damping coefficient in order to achieve a nearly constant body-damping ratio. In the case of load-leveling systems, the static deflection does not change. In this case, the increased pressure of the leveling air spring or hydraulic system can be used as manipulation variable for damper adjustment. This pressure directs a control valve through a barometric cell changing the damping. For the sake of completeness, the *deflection-dependent dampers* are mentioned. These variable dampers increase their damping ratio for deflections near the limit stops. This aims at a soft stop at the limits of the deflection and lead to a dissipation of the energy in contrast to steel spring stops. Thus, this system is not an adaptive systems in the sense of an adaptation to driving situations and excitations, as treated in Chap. 15.

## 8.2 Passive Suspension Models

Physical models for the investigation of the vertical dynamics of suspension systems are most commonly built on the quarter car. Greater accuracy is achieved by extensions to a half car, e.g. Krtolica and Hrovat (1990) or full car, e.g. Weispfenning (1996). This aims to describe correctly the roll and pitch motion of the car and the

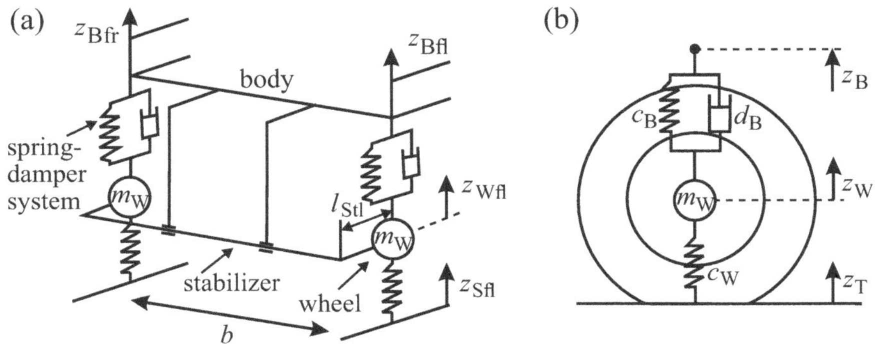


**Fig. 8.7** Scheme of suspensions. **a** front axle with torsion stabilizer; **b** suspension of one wheel

connected inertia forces; see Chap. 9. The stabilizers also can be involved in the modeling. In addition, the lever arm of the wishbones and the related forces can be considered, Streiter (1996).

In the following, suspensions models are considered. A schematic of the suspension is shown in Fig. 8.7. The sprung and unsprung mass displacements are $z_B$ and $z_W$, respectively. The damper generates the force $F_D$ and the spring generates the force $F_S$. The dynamic tire load is $F_{Zdyn}$. $F_C$ describes the friction of the damper and $F_{st}$ the stabilizer force. Body mass $m_B$, tire stiffness $c_W$, and damping coefficient $d_B$ are commonly not constant. Apart from semi-active changes of spring and damper, the wheel mass $m_W$ and the body spring coefficient $c_B$ can be considered to be constant.

### 8.2.1 Linear Suspension Model

The force balance of the corresponding body is described by

$$\begin{aligned} m_{B\nu}\ddot{z}_B(t) &= F_S(t) + F_D(t) + F_C(t) + F_{St,\circ}(t) - F_{B\nu}(t), \\ \nu &= \text{fl, fr, rl, rr} \end{aligned} \tag{8.2.1}$$

and the force balance of one wheel by

$$m_W\ddot{z}_W(t) = F_{Zdyn}(t) - F_S(t) - F_C(t) - F_D(t). \tag{8.2.2}$$

The vertical force for the suspension $\nu$ in the steady state

$$F_{B\nu} = \bar{F}_{ZB\nu} = m_{B\nu}g, \tag{8.2.3}$$

is the weight of the corresponding body mass, e.g.

$$m_{B\nu} = m_B\,(l_\circ/l)\,/2. \tag{8.2.4}$$

Assuming a *linear behavior* and neglecting dry friction in the damper and bearings, the spring, damper, and stabilizer contribute with the following forces; compare Fig. 8.1

$$\begin{aligned} F_S &= c_B\,(z_W - z_B)\,, \\ F_D &= d_B\,(\dot{z}_W - \dot{z}_B)\,, \\ F_{St,r} &= c_{St}\,((z_W - z_B)_l - (z_W - z_B)_r)\,, \\ F_{St,l} &= c_{St}\,((z_W - z_B)_r - (z_W - z_B)_l)\,, \end{aligned} \tag{8.2.5}$$

with the spring stiffness $c_B$, damper coefficient $d_B$, and stabilizer stiffness $c_{St}$; see Sect. 9.1.

The *dynamic behavior of the body* follows from (8.2.1) by inserting the forces (8.2.5) and neglecting stabilizer forces, $F_{St,\circ} = 0$, and the dry damper friction

$$m_{B\nu}\ddot{z}_B(t) = c_B((z_W(t) - z_B(t)) + d_B((\dot{z}_W(t) - \dot{z}_B(t)) - m_{B\nu}g. \tag{8.2.6}$$

For the steady-state follows

$$F_{B\nu} = c_B\,(\bar{z}_W - \bar{z}_B)_\nu = m_{B\nu}g, \tag{8.2.7}$$

and for small changes around the steady state

$$\begin{aligned} z_B(t) &= \bar{z}_B + \Delta z_B(t);\; z_W(t) = \bar{z}_W + \Delta z_W(t) \\ \dot{z}(t) &= d\Delta z(t)/dt, \end{aligned} \tag{8.2.8}$$

one obtains, after using (8.2.7)

$$m_{B\nu}\ddot{z}_B(t) + d_B\dot{z}_B(t) + c_B(t)\Delta z_B(t) = c_B\Delta z_W(t) + d_B\dot{z}_W(t). \tag{8.2.9}$$

This second order differential equation describes the body motion $\Delta z_B(t)$ after excitation from the wheel motion $\Delta z_W(t)$ of a suspension $\nu$.

The *dynamic behavior of one wheel* results from (8.2.2) after inserting the forces (8.2.5) and without stabilizer forces

$$\begin{aligned} m_W\ddot{z}_W(t) &= c_W\,(z_T(t) - z_W(t)) - c_B(t)\,(z_W(t) - z_B(t)) \\ &\quad - d_B\,(\dot{z}_W(t) - \dot{z}_B(t))\,, \end{aligned} \tag{8.2.10}$$

where $c_W$ is the tire stiffness, neglecting (small) tire damping. For the steady-state follows

$$c_W\,(\bar{z}_T - \bar{z}_W) = c_B\,(\bar{z}_W - \bar{z}_B) = \bar{F}_{ZT} = \bar{F}_{ZB} = m_{B\nu}g. \tag{8.2.11}$$

Hence, the vertical tire and spring forces in the steady state are identical if no dynamic excitations occur.

After insertion of (8.2.8) in (8.2.10) one obtains assuming small vertical motions

$$\begin{aligned} m_W \ddot{z}_W(t) + d_B \dot{z}_W(t) + (c_W + c_B)\, \Delta z_W(t) \\ = c_W \Delta z_T(t) + c_B \Delta z_B(t) + d_B \dot{z}_B(t). \end{aligned} \tag{8.2.12}$$

This second order differential equation models the wheel motion after excitation from the road surface $\Delta z_T(t)$ and the corresponding body $\Delta z_B(t)$.

The *dynamic vertical tire forces* follow from (8.2.10) and (8.2.11)

$$\begin{aligned} F_{ZT}(t) &= c_W \left(\Delta z_T(t) - \Delta z_W(t)\right) \\ &= m_W \ddot{z}_W(t) + c_B \left(\Delta z_W(t) - \Delta z_B(t)\right) \\ &+ d_B \left(\dot{z}_W(t) - \dot{z}_B(t)\right). \end{aligned} \tag{8.2.13}$$

If the effect of vehicle acceleration $a_Y$ and $a_Y$, and therefore, pitching and rolling have to be taken into account, the vertical tire forces follow according to (8.2.13).

These suspension models deliver the vertical forces $F_Z$ for the lateral vehicle models in Chap. 7.

### 8.2.2 Nonlinear Suspension Models

The characteristic operating curve of steel springs show a smooth progressive relationship. However, they can be approximated by a linear curve in their operating point.

The strongly degressive course of the damper's characteristic curve, see Fig. 8.9, is usually divided up into a constant, a linear, and a nonlinear square component. Kraus and Dantele (1970) explains this fact by a complex hydraulic analysis of a damper: the nonlinearity results from the opening of the piston valve and the stiffness of its spring. The damper characteristic can be approximated using the linear and nonlinear damping parameters $d_{Bl}$ and $d_{Bnl}$ and a constant friction force $F_{fc}$. These parameters reflect the constructive parameters of the damper

$$\begin{aligned} F_D = d_{Bl}\left(\dot{z}_W - \dot{z}_B\right) + F_{fc}\, \mathrm{sign}\left(\dot{z}_W - \dot{z}_B\right) \\ + d_{Bnl} \left|\left(\dot{z}_W - \dot{z}_B\right)\right|^{\frac{2}{1+2n}} \mathrm{sign}\left(\dot{z}_W - \dot{z}_B\right). \end{aligned} \tag{8.2.14}$$

The constant $n$ is usually set to $n = 1.5$; Majjad (1997). Another representation of $F_D$ is a piecewise linearization of the curve into the segments $\kappa$

$$F_{D\kappa} = d_{B\kappa}\left(\dot{z}_W - \dot{z}_B\right) + F_{gl\kappa}, \tag{8.2.15}$$

where $F_{gl\kappa}$ is a parameter needed for the bumpless segmentation of the function. Another method of handling the nonlinear characteristic of the damper is the application of neuronal network models. For more details; see e.g. Halfmann (2001). However, this approximation does not comprise physical parameters.

Equation (8.2.14) inserted in (8.2.6) leads to the *nonlinear model of a quarter-car suspension*

$$\begin{aligned} m_{\mathrm{Bi}}\ddot{z}_{\mathrm{B}} &= c_{\mathrm{B}}\left(z_{\mathrm{W}}-z_{\mathrm{B}}\right)+d_{\mathrm{Bl}}\left(\dot{z}_{\mathrm{W}}-\dot{z}_{\mathrm{B}}\right)+d_{\mathrm{Bnl}}\sqrt{|\dot{z}_{\mathrm{W}}-\dot{z}_{\mathrm{B}}|}\operatorname{sign}\left(\dot{z}_{\mathrm{W}}-\dot{z}_{\mathrm{B}}\right) \\ &\quad + F_{\mathrm{fc}}\operatorname{sign}\left(\dot{z}_{\mathrm{W}}-\dot{z}_{\mathrm{B}}\right)-F_{\mathrm{B}\nu}, \end{aligned} \tag{8.2.16}$$

$$\begin{aligned} m_W\ddot{z}_W &= c_W\left(z_{\mathrm{T}}-z_{\mathrm{W}}\right)-c_B\left(z_{\mathrm{W}}-z_{\mathrm{B}}\right)+d_{Bl}\left(\dot{z}_{\mathrm{W}}-\dot{z}_{\mathrm{B}}\right) \\ &\quad + d_{\mathrm{Bnl}}\sqrt{|\dot{z}_{\mathrm{W}}-\dot{z}_{\mathrm{B}}|}\operatorname{sign}\left(\dot{z}_{\mathrm{W}}-\dot{z}_{\mathrm{B}}\right)-F_{\mathrm{fc}}\operatorname{sign}\left(\dot{z}_{\mathrm{W}}-\dot{z}_{\mathrm{B}}\right). \end{aligned} \tag{8.2.17}$$

The application of the *piecewise linear curve* results in the equation system

$$m_{\mathrm{Bi}}\ddot{z}_{\mathrm{B}} = c_{\mathrm{B}}\left(z_{\mathrm{W}}-z_{\mathrm{B}}\right)+d_{\mathrm{B}\kappa}\left(\dot{z}_{\mathrm{W}}-\dot{z}_{\mathrm{B}}\right)+F_{\mathrm{gl}\kappa}-F_{\mathrm{B}\kappa} \quad \text{for } \kappa=1\ldots N, \tag{8.2.18}$$

$$m_{\mathrm{W}}\ddot{z}_{\mathrm{W}} = c_{\mathrm{W}}\left(z_{\mathrm{T}}-z_{\mathrm{W}}\right)+c_{\mathrm{B}}\left(z_{\mathrm{W}}-z_{\mathrm{B}}\right)+d_{\mathrm{B}\kappa}\left(\dot{z}_{\mathrm{W}}-\dot{z}_{\mathrm{B}}\right)-F_{\mathrm{gl}\kappa} \quad \text{for } \kappa=1\ldots N. \tag{8.2.19}$$

The validity of these models could be verified by comparing simulations with accompanying measurements on a test rig, Bußhardt (1995), see next section.

## 8.3 Parameter Identification of Semi-active Suspensions

### 8.3.1 Parameter Identification of a Quarter-Car Suspension

Because only very few measured variables will be available in suspensions, it is not possible to estimate all six parameters of a quarter-car model. Hence, some parameters must be set as known fixed values. Through simple conversion of (8.2.1) and (8.2.2), various sets of estimation equations and sensor setups can be found. The resulting arrangements for the different estimation equations are given below; see also Table 8.4

$$\left(z_{\mathrm{W}}-z_{\mathrm{B}}\right) = -\frac{d_{\mathrm{B}}}{c_{\mathrm{B}}}\left(\dot{z}_{\mathrm{W}}-\dot{z}_{\mathrm{B}}\right)+\frac{m_{\mathrm{B}}}{c_{\mathrm{B}}}\ddot{z}_{\mathrm{B}}-\frac{1}{c_{\mathrm{B}}}F_{\mathrm{gl}}, \tag{8.3.1}$$

$$\left(z_{\mathrm{W}}-z_{\mathrm{B}}\right) = -\frac{d_{\mathrm{B}}}{c_{\mathrm{B}}}\left(\dot{z}_{\mathrm{W}}-\dot{z}_{\mathrm{B}}\right)-\frac{m_{\mathrm{W}}}{c_{\mathrm{B}}}\ddot{z}_{\mathrm{W}}+\frac{c_{\mathrm{W}}}{c_{\mathrm{B}}}\left(z_{\mathrm{T}}-z_{\mathrm{W}}\right)-\frac{1}{c_{\mathrm{B}}}F_{\mathrm{gl}}, \tag{8.3.2}$$

$$\begin{aligned} \left(z_{\mathrm{W}}-z_{\mathrm{B}}\right) = &-\frac{d_{\mathrm{B}}}{c_{\mathrm{B}}}\left(\dot{z}_{\mathrm{W}}-\dot{z}_{\mathrm{B}}\right)-\frac{m_{\mathrm{W}}m_{\mathrm{B}}}{c_{\mathrm{B}}\left(m_{\mathrm{W}}+m_{\mathrm{B}}\right)}F_{\mathrm{gl}} \\ &-\frac{m_{\mathrm{W}}m_{\mathrm{B}}}{c_{\mathrm{B}}\left(m_{\mathrm{W}}+m_{\mathrm{B}}\right)}\left(\ddot{z}_{\mathrm{W}}-\ddot{z}_{\mathrm{B}}\right)+\frac{c_{\mathrm{W}}m_{\mathrm{B}}}{c_{\mathrm{B}}\left(m_{\mathrm{W}}+m_{\mathrm{B}}\right)}\left(z_T-z_{\mathrm{W}}\right), \end{aligned} \tag{8.3.3}$$

$$\left(z_{\mathrm{W}}-z_{\mathrm{B}}\right) = -\frac{d_{\mathrm{B}}}{c_{\mathrm{B}}}\left(\dot{z}_{\mathrm{W}}-\dot{z}_{\mathrm{B}}\right)-\frac{m_{\mathrm{W}}}{c_{\mathrm{B}}}\left(\ddot{z}_{\mathrm{W}}-\ddot{z}_{\mathrm{B}}\right)+\frac{m_{\mathrm{B}}}{c_{\mathrm{B}}}\ddot{z}_{\mathrm{W}}+\frac{1}{c_{\mathrm{B}}}F_{\mathrm{gl}}, \tag{8.3.4}$$

**Table 8.4** Various estimation equations for the quarter-car parameters

| Equation | Measurable sensor signals | Known parameters | Parameter estimation | Properties |
|---|---|---|---|---|
| (8.3.1) | $z_W - z_B$<br>$\ddot{z}_B$ | $c_B$ | $d_B, m_B, F_{gl}$ | + easily measurable quantities<br>+ $c_B$ is practically constant |
| (8.3.2) | $z_W - z_B$<br>$\ddot{z}_W$<br>$z_T - z_W$ | $m_W$ | $d_B, c_B$<br>$c_W, F_{gl}$ | + $c_W$ is estimated<br>− road deflection is hardly measurable |
| (8.3.3) | $z_W - z_B$<br>$z_T - z_W$ | $c_B, c_W$ | $d_B, m_B,$<br>$c_W, F_{gl}$ | + $c_B$ and $m_W$ are practically constant<br>+ $c_W$ is estimated<br>− road deflection is hardly measurable |
| (8.3.4) | $z_W - z_B$<br>$\ddot{z}_W$ | $c_B$ | $d_b, m_B, F_{gl}$ | + easily measurable quantities<br>+ $c_B$ is practically constant |
| (8.3.5)<br>(8.3.6) | $F_S/F_D$<br>$\ddot{z}_W$ | – | $d_B, m_B,$<br>$c_B, F_{gl}$ | + no pre-information is necessary<br>+ all parameters are measurable<br>− $F_S, F_D$ are generally not measurable, expect for air springs |

$$F_S = \frac{d_B}{c_B}\dot{F}_S - \frac{m_B}{c_B}(\ddot{z}_W - \ddot{z}_B) + m_B\ddot{z}_W - F_{gl}, \tag{8.3.5}$$

$$F_D = \frac{d_B}{c_B}\dot{F}_D - \frac{m_B}{c_B}\ddot{F}_D + \frac{m_B d_B}{c_B}\dddot{z}_W - F_{gl}. \tag{8.3.6}$$

Practically, mostly only the body acceleration and the deflection are measured. However, measuring instruments for the road/wheel distance are meanwhile discussed; e.g. Nomura and Takeshita (1993) and measurement of wheel acceleration is used for semi-active dampers; Kutsche and Rappelt (2006) and Causemann (2008). Therefore, (8.3.1) and eventually (8.3.4) are suitable for the identification. The values of the wheel mass and the body mass can be assumed to be constant. Supplementary, the variable body mass can be estimated out of the static spring deflection. In order to compensate for the nonlinearity of damper characteristics, a piecewise estimation of the damper force is assumed. This division is dependent on the actual values of $(\dot{z}_W - \dot{z}_B)$ and results in a structure-variable system. For the valid range of each single model, a set of parameters can be identified. The estimation equation for each model arises in analogy to (8.3.1) and (8.2.15) to

$$(z_W - z_B) = -\frac{d_{B\kappa}}{c_B}(\dot{z}_W - \dot{z}_B) + \frac{m_B}{c_B}\ddot{z}_B - \frac{1}{c_B}F_{gl\kappa} \quad \text{for } \kappa = 1...m, \tag{8.3.7}$$

or in a general form to

$$\begin{aligned} &y(t) = -a_{1\kappa}\dot{y}(t) + b_0 u(t) + c_{gl\kappa} \quad \text{for } \kappa = 1...m, \\ &\text{with } y(t) = z_W(t) - z_B(t) \text{ and } u(t) = \ddot{z}_B(t). \end{aligned} \tag{8.3.8}$$

Based on the measurements of $y(t)$ and $u(t)$, the unknown parameters can be estimated by recursive least squares algorithms (RLS or DSFI); see Isermann and Münchhof (2011). Alternatively, the identification can be based on the nonlinear equation (8.2.16).

The identification methods were applied to a test rig, shown in Fig. 8.8, which is equipped with a continuously adjustable damper. The damping is controlled by a magnetic valve, which opens or closes a bypass continuously. The test rig was constructed primarily for investigations on semi-active, parameter-adaptive suspension control; Bußhardt (1995).

Figure 8.9 shows the estimated damping curve for different damper magnetic valve currents. Because rising damper current opens the bypass the damping sinks. The damping curve was divided into four sections, two for each direction of motion. It can be seen that the damping curves at different damper currents can be clearly distinguished. Application of parameter estimation with the nonlinear equations led to approximately the same results.

### 8.3.2 Parameter Identification of a Driving Vehicle

Both, the suspension as well as the tire pressure have a large influence on the vehicle dynamics and are thus highly safety critical. In the following, identification techniques will be developed that allow to identify the characteristics of the vehicle suspension (and the tire in Chap. 15) and can serve for supervision of these components.

For application either in a service station, for example, for technical inspection, or in a driving state, it is important to use easily measurable variables. If the methods

**Fig. 8.8** Quarter-car test rig

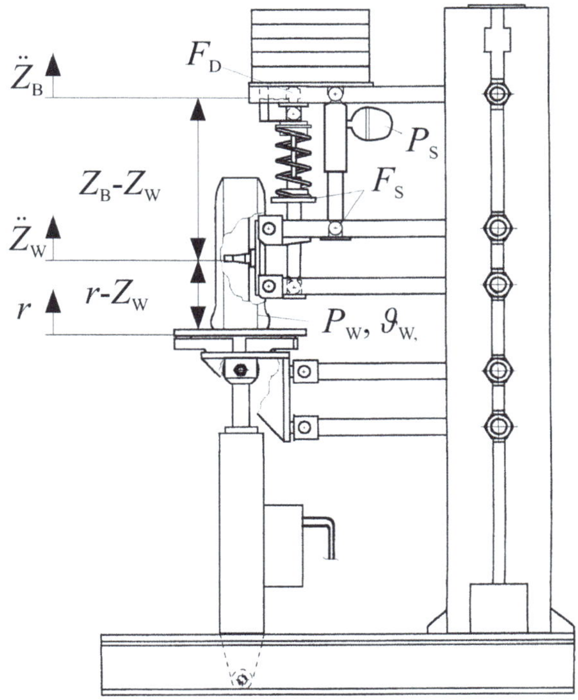


**Fig. 8.9** Estimated damping characteristics of the adjustable damper

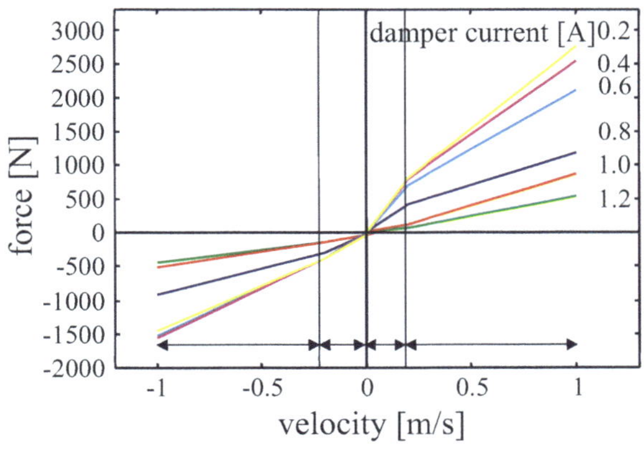


should be used for technical inspection, then the additional sensors must be easily mountable to the car. For on-board identification, the existing variables for suspension control should be used. Variables which meet these requirements are the vertical accelerations of body and wheel, $\ddot{z}_B$ and $\ddot{z}_W$, and the suspension deflection $z_W - z_B$. Another important point is that, the methods should require only little a priori knowledge about the type of car.

A scheme for a simplified model of a car suspension system, a quarter-car model, is shown in Fig. 8.1. The following equations follow from force balances, compare (8.2.6) and (8.2.10) with excitation from the road; see Fig. 8.7

$$m_B \ddot{z}_B(t) = c_B\big(z_W(t) - z_B(t)\big) + d_B\big(\dot{z}_W(t) - \dot{z}_B(t)\big), \tag{8.3.9}$$

$$\begin{aligned} m_W \ddot{z}_W(t) = &- c_B\big(z_W(t) - z_B(t)\big) - d_B\big(\dot{z}_W(t) - \dot{z}_B(t)\big) \\ &+ c_W\big(z_T(t) - z_W(t)\big). \end{aligned} \tag{8.3.10}$$

The small damping of the wheel is usually negligible.

In general, the relationship between force and velocity of a shock absorber is nonlinear. It is usually degressive and depends strongly on the direction of motion of the piston. In addition, the Coulomb friction of the damper should be taken into account.To approximate this behavior, the characteristic damper curve can be divided into $m$ sections as a function of the piston velocity; see (8.2.18). Considering $m$ sections, the following equation can be obtained

$$\ddot{z}_B = \frac{d_{B,\kappa}}{m_B(\dot{z}_W - \dot{z}_B) + \frac{c_B}{m_B}(z_W - z_B) + \frac{1}{m_B} F_{fc,\kappa}}, \quad \kappa = 1, \ldots, m\,, \tag{8.3.11}$$

$F_{fc,\kappa}$ denotes the force generated by Coulomb friction and $d_{B,\kappa}$ the damping coefficient for each section. Using (8.3.11), the damping curve can be estimated with a standard parameter estimation algorithm measuring the body acceleration $\ddot{z}_B$ and suspension deflection $z_W - z_B$. The velocity $\dot{z}_W - \dot{z}_B$ can be obtained by numerical differentiation. In addition, either the body mass $m_B$ or the spring stiffness $c_B$ can be estimated. One of the both parameters must be known a priori. Using (8.3.9) and (8.3.10), other equations for parameter estimation can be obtained, e.g. (8.3.12) which can be used to estimate the tire stiffness $c_W$ additionally

$$z_W - z_B = -\frac{d_{B,\kappa}}{m_B}(\dot{z}_W - \dot{z}_B) - \frac{c_B}{m_B}(z_W - z_B) + \frac{c_W}{c_B}(r - z_W) - \frac{1}{c_W} F_{fc,\kappa}. \tag{8.3.12}$$

The disadvantage of this equation is the necessity to measure the distance between road and wheel ($z_T - z_W$), see also Bußhardt (1995) and Weispfenning (1996) for modeling and identification of the automotive suspension.

To test the above method in a driving car, a medium class car, an Opel Omega A 2.0i was equipped with sensors to measure the vertical acceleration of body and wheel, as well as the suspension deflections. To realize different damping coefficients, the car is equipped with adjustable shock absorbers at the rear axle, which can be varied in three steps. In Fig. 8.10, the course of the estimated damping coefficients at different damper settings is given for driving over boards of height 2 cm.

After approximately 2.5 s, the estimated values converge to their final values. The estimated damping coefficients differ approximately 10% from the directly measured ones. In Fig. 8.11, the estimated characteristic curves at the different damper settings are shown. The different settings are separable and the different damping

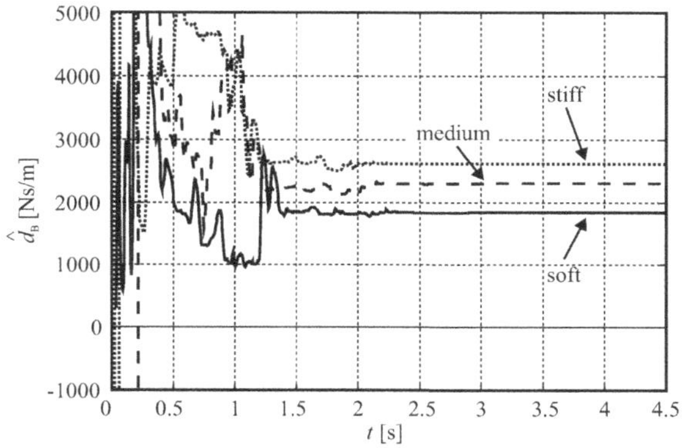


**Fig. 8.10** Estimated damping coefficients for different damper settings for the driving car with about 30 km/h

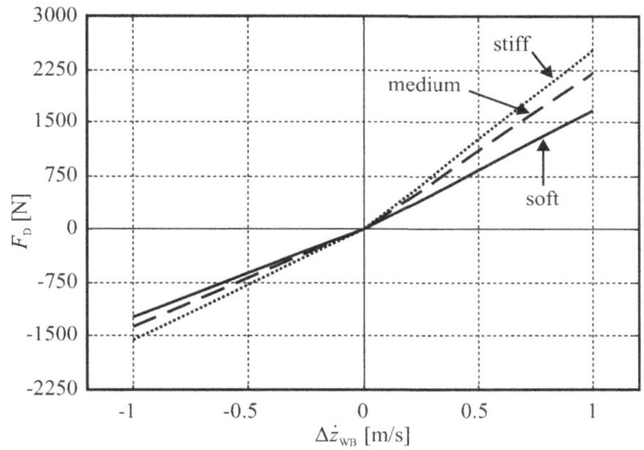


**Fig. 8.11** Estimated characteristic damping characteristics for different damper settings for the driving car

characteristics in compression and rebound is clearly visible, although the effect is not as strong as in the directly measured characteristic curve. More results are given in Börner et al (2001).

Next, the damping characteristics of the shock absorber were adjusted during a driving maneuver. Recursive parameter estimation then allow to adapt the damping coefficient accordingly. Figure 8.12 illustrates the suspension deflection $z_W - z_B$, the first derivative of the suspension deflection calculated with a state variable filter $\dot{z}_W - \dot{z}_B$, and the wheel acceleration $\ddot{z}_W$ for the right rear wheel during a highway test drive. After 30, 60, 90, and 120 s, a change of the shock absorber damping was made.

Several estimations have shown that the recursive least squares algorithm (RLS) with exponential forgetting factor received very good results. This recursive parameter estimation is able to adapt to the different damping settings in about 10 s; see Fig. 8.13.

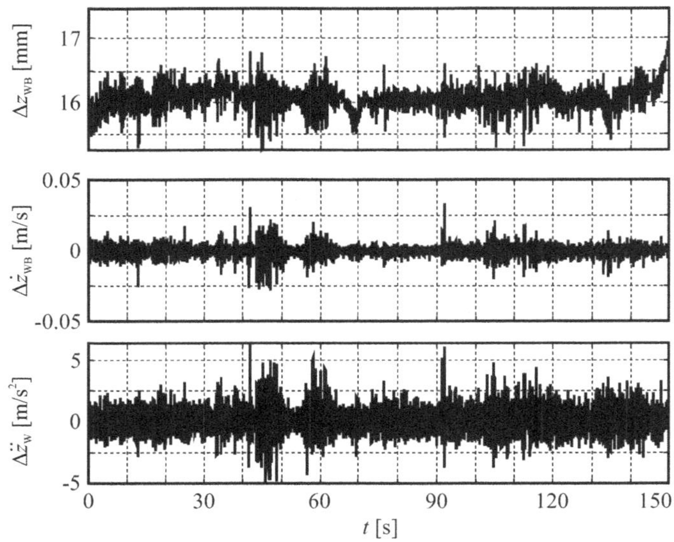


**Fig. 8.12** Measured signals on a highway with variation of the damper configuration

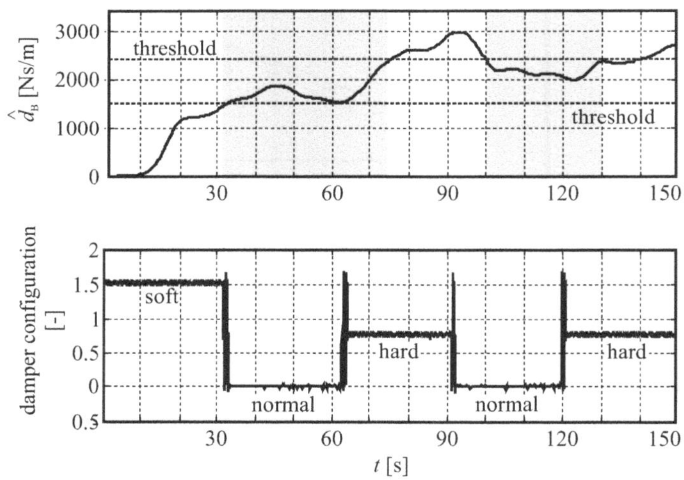


**Fig. 8.13** Parameter estimates of the damping coefficient with RLS and forgetting factor. $0 < t <$ 30 s: soft damping configuration, 30 s $< t <$ 60 s and 90 s $< t <$ 120 s: medium damping configuration (normal situation), 60 s $< t <$ 90 s and 120 s $< t <$ 150 s: hard damping coefficient

## References


Börner M, Zele M, Isermann R (2001) Comparison of different fault-detection algorithms for active body control components: automotive suspension system. In: American control conference, Arlington, VA, USA

Bußhardt J (1995) Selbsteinstellende Feder-Dämpfer-Last-Systeme für Kraftfahrzeuge. Dissertation Technische Hochschule Darmstadt. Fortschr.-Ber. VDI Reihe 12, 240. VDI Verlag, Düsseldorf

Causemann P (2008) Kraftfahrzeugstoäd$\beta$mpfer. Verlag moderne industrie AG, Landsberg/Lech

Halfmann C (2001) Adaptive semiphysikalische Echtzeitsimulation der Kraftfahrzeugdynamik im bewegten Fahrzeug. Dissertation Technische Universität Darmstadt. Fortschr.-Ber. VDI Reihe 12, 467. VDI Verlag, Düsseldorf

Heissing B, Ersoy M (eds) (2011) Fahrwerkhandbuch: Grundlagen, Fahrdynamik, Komponenten, Systeme, Mechatronik. Perspektiven, ATZ/MTZ Fachbuch, Vieweg

Isermann R, Münchhof M (2011) Identification of dynamic systems. Springer, Berlin, Heidelberg

Kraus H, Dantele H (1970) Berechnung und Messung von Stoßdämpfer-Ventilen. Automobiltechnische Zeitschrift 75(7):234–237

Krtolica R, Hrovat D (1990) Optimal active suspension control based on a half-car model. In: 29th IEEE conference decision and control, Honolulu, HI, USA, pp 2238–2243

Kutsche T, Rappelt S (2006) Semiaktive Stoßdämpfer und aktive Radaufhängungen. Vieweg, Wiesbaden, pp 252–264

Majjad R (1997) Estimation of suspension parameters. In: Proceedings of the 1997 IEEE international conference on control applications, Hartford, CT, USA, pp 522–527

Mitschke M (1984) Comparison of different mathematical models of oscillations of passenger cars. In: Proceedings of the 8th IAVSD symposium, the dynamics of vehicles and railway tracks, Cambridge, Massachusetts

Nomura S, Takeshita S (1993) New GALANT & ETERNA. JSAE-Paper no. 9300873. JSAE Rev 14(1):71–77

Reimpell J, Stoll H, Betzler J (2001) The automotive chassis, 2nd edn. Reed Educational and Professional Publishing Ltd, Oxford

Robert Bosch GmbH (1996) Automotive Handbook, 4th edn. Robert Bosch GmbH, Stuttgart

GmbH Robert Bosch (ed) (2011) Automotive handbook, 8th edn. Bentley publishers, Cambridge

Streiter R (1996) Entwicklung und Realisierung eines analytischen Regelkonzeptes für eine aktive Federung. Technische Universität, Berlin

Weispfenning T (1996) Fault detection and diagnosis of components of the vehicle vertical dynamics. In: 1st international conference on control and diagnostics in automotive applications, Genua, Italy

# 9 Roll and Pitch Dynamic Behavior

The roll and pitch behavior is already included in the basic dynamic equations (7.3.4)–(7.3.8) of the two-track models; see also Fig. 7.17. In the following, the roll and pitch dynamic behavior is considered separately by neglecting the couplings between the motions around the roll and pitch axis. The resulting models are required to obtain simplified models and as a basis for the estimation of the various parameters.

## 9.1 Roll Dynamic Model

Figure 9.1 shows a schematic to model the roll dynamic behavior. The lateral acceleration $a_Y$ acts on the body and it is assumed that the resulting lateral force $F_{Y,CG} = m_B a_Y$ can be lumped into the center of gravity CG, which has a distance $h_{CG}$ from the ground. The forces of the suspensions may be represented by (8.2.5) assuming linear behavior. With $b_f$ and $b_r$ as the effective track of the front and rear axle and $h_{RA}$ the distance of the roll axis from ground, a torque balance equation for the roll axis is for small roll angles $\varphi$ (positive for right turning),

$$J_X \ddot{\varphi}(t) = (F_{Zfl}(t) - F_{Zfr}(t)) \frac{b_f}{2} + (F_{Zrl}(t) - F_{Zrr}(t)) \frac{b_r}{2} - m_B (h_{CG} - h_{RA}) a_Y(t). \tag{9.1.1}$$

Herewith, it is assumed that the center of gravity and the roll axis are positioned on top of each other. For the body forces through the spring, dampers and stabilizer of one suspension holds

$$F_{ZBij} = F_{Sij} + F_{Dij} + F_{St,j} \quad (i = f, r;\; j = l, r). \tag{9.1.2}$$

Assuming linear behavior for the spring and damper one obtains

$$F_{ZBj} = c_{Bi}(z_W - z_B)_j + d_{Bi}(\dot{z}_W - \dot{z}_B)_j, \tag{9.1.3}$$



R. Isermann, *Automotive Control*, ATZ/MTZ-Fachbuch,
https://doi.org/10.1007/978-3-642-39440-9_9

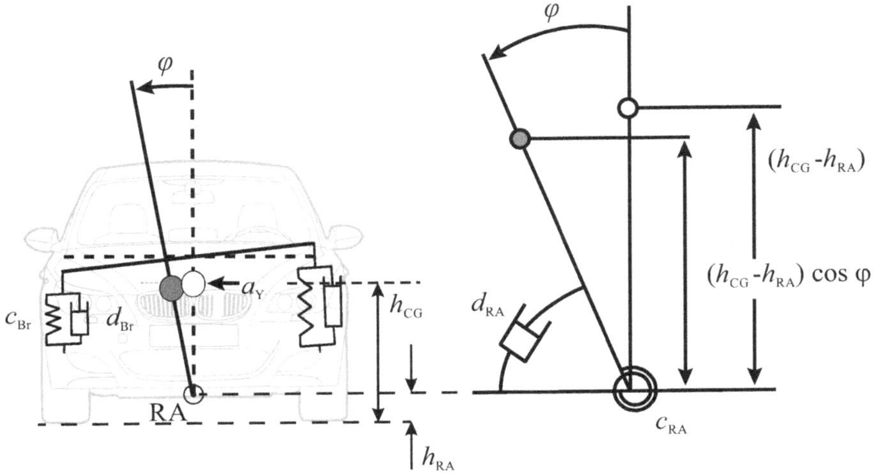


**Fig. 9.1** Scheme for the torques around the roll axis RA

and with $\Delta z_i = (z_W - z_B)_i$,

$$F_{ZBi} = c_{Bi}\Delta z_i + d_{Bi}\Delta \dot{z}_i. \tag{9.1.4}$$

A stabilizer contributes additional forces on the wheels depending on the torque of the torsion bar; see Fig. 8.7. A torsion angle $\Delta\gamma$ generates a torsion torque, assuming linear behavior

$$M_{St} = c_T \Delta\gamma, \tag{9.1.5}$$

where the torsional stiffness follows from

$$c_T = GI/l_{st}, \tag{9.1.6}$$

with $G$ the shear modulus, $I$ the polar moment of inertia and $l_{St} \approx b$ the length of the torsion bar; see e.g. Isermann (2003).

If $l_{St}$ is the length of the stabilizer lever to the suspension, see Fig. 8.7, the relation between the difference of suspension travel is $\Delta z = \Delta z_l - \Delta z_r$, the torsion angle $\Delta\gamma$ is

$$\Delta\gamma = \frac{\Delta z}{l_{St}}, \tag{9.1.7}$$

and the vertical force on the wheel becomes

$$F_{St} = \frac{M_{St}}{l_{St}} = c_T\frac{\Delta\gamma}{l_{St}} = c_T\frac{\Delta z}{l_{St}^2} = c_{St}\Delta z, \tag{9.1.8}$$

with the stabilizer stiffness coefficient

$$c_{St} = c_T/l_{St}^2. \tag{9.1.9}$$

A stabilizer force becomes active for the difference between the right and left suspension travel. Hence, for the right and left side holds

$$\begin{aligned} F_{\mathrm{St,r}} &= c_{\mathrm{St}}\,(z_{\mathrm{W}} - z_{\mathrm{B}})_{\mathrm{l}} - c_{\mathrm{St}}\,(z_{\mathrm{W}} - z_{\mathrm{B}})_{\mathrm{r}}\,, \\ F_{\mathrm{St,l}} &= c_{\mathrm{St}}\,(z_{\mathrm{W}} - z_{\mathrm{B}})_{\mathrm{r}} - c_{\mathrm{St}}\,(z_{\mathrm{W}} - z_{\mathrm{B}})_{\mathrm{l}}\,. \end{aligned} \tag{9.1.10}$$

The relation between the roll angle $\varphi$ and a suspension deflection is

$$\varphi = \frac{(z_{\mathrm{W}} - z_{\mathrm{B}})_{\mathrm{r}}}{b_{\mathrm{r}}/2} = \frac{(z_{\mathrm{W}} - z_{\mathrm{B}})_{\mathrm{f}}}{b_{\mathrm{f}}/2}. \tag{9.1.11}$$

The stabilizer forces for the front and rear axle then are

$$\begin{aligned} F_{\mathrm{St,fj}} &= c_{\mathrm{St,f}}\Delta z_{\mathrm{fj}} \qquad j = l, r, \\ F_{\mathrm{St,rj}} &= c_{\mathrm{St,r}}\Delta z_{\mathrm{rj}} \qquad j = l, r. \end{aligned} \tag{9.1.12}$$

Assuming that the suspension travels during rolling are $\Delta z_{\mathrm{fj}} = \Delta z_{\mathrm{rj}} = \Delta z$, the roll angle follows with $b = B_{\mathrm{r}} = b_{\mathrm{f}}$

$$\varphi = \frac{(\Delta z)}{b/2}.. \tag{9.1.13}$$

The vertical forces from the suspension on the body in the torque balance equation (9.1.1) are the sum of the forces from the spring, damper, and stabilizer

$$F_{\mathrm{ZB}} = F_{\mathrm{S}} + F_{\mathrm{D}} + F_{\mathrm{St,j}} \quad (j = l, r), \tag{9.1.14}$$

and follow from (8.2.5).

Introduction into (9.1.1) leads with $\Delta z_{\mathrm{i}} = \varphi b_{\mathrm{i}}/2$ to

$$\begin{aligned} J_{\mathrm{x}}\ddot{\varphi}(t) =& -2\left(c_{\mathrm{Br}}\frac{b_{\mathrm{r}}}{2} + c_{\mathrm{Bf}}\frac{b_{\mathrm{f}}}{2} + c_{\mathrm{St,r}}\frac{b_{\mathrm{r}}}{2} + c_{\mathrm{St,f}}\frac{b_{\mathrm{f}}}{2}\right)\varphi(t), \\ & -2\left(d_{\mathrm{Br}}\frac{b_{\mathrm{r}}}{2} + d_{\mathrm{Bf}}\frac{b_{\mathrm{f}}}{2}\right)\dot{\varphi}(t), \\ & - m_{\mathrm{B}}\,(h_{\mathrm{CG}} - h_{\mathrm{RA}})\,a_{\mathrm{Y}}(t)\cos\varphi(t), \\ =& -c_{\mathrm{rol}}\varphi(t) - d_{\mathrm{rol}}\dot{\varphi}(t) - k_{\mathrm{rol}}a_{\mathrm{Y}}(t), \end{aligned} \tag{9.1.15}$$

with the coefficients, assuming $\cos\varphi \approx 1$ for small $\varphi$

$$\begin{aligned} c_{\mathrm{rol}} &= (c_{\mathrm{Br}}b_{\mathrm{r}} + c_{\mathrm{Bf}}b_{\mathrm{f}} + c_{\mathrm{Str}}b_{\mathrm{r}} + c_{\mathrm{Stf}}b_{\mathrm{f}})\,, \\ d_{\mathrm{rol}} &= (d_{\mathrm{Br}}b_{\mathrm{r}} + d_{\mathrm{Bf}}b_{\mathrm{f}})\,, \\ k_{\mathrm{rol}} &= m_{\mathrm{B}}\,(h_{\mathrm{CG}} - h_{\mathrm{RA}})\,. \end{aligned} \tag{9.1.16}$$

The application of the Laplace transform results in the transfer function for the roll angle

$$G_{\mathrm{ay'}}(s) = \frac{\varphi(s)}{a_{\mathrm{Y}}(s)} = \frac{-k_{\mathrm{rol}}}{J_{\mathrm{x}}s^2 + d_{\mathrm{rol}}s + c_{\mathrm{rol}}}. \tag{9.1.17}$$

It is of second order and its eigenfrequency and damping are characterized by

$$\begin{aligned} \omega_{\mathrm{o,rol}} &= \sqrt{\frac{c_{\mathrm{rol}}}{J_{\mathrm{x}}}} && \text{(undamped natural frequency)}, \\ \omega_{\mathrm{e,rol}} &= \sqrt{\frac{c_{\mathrm{rol}}}{J_{\mathrm{x}}} - \frac{d_{\mathrm{rol}}^2}{4J_{\mathrm{x}}}} && \text{(damped natural frequency, } D < 1\text{)}, \\ \zeta &= \frac{d_{\mathrm{rol}}}{2} \frac{1}{\sqrt{c_{\mathrm{rol}} J_{\mathrm{x}}}} && \text{(damped ratio)}. \end{aligned} \tag{9.1.18}$$

Similar models are derived in Würtenberger (1997), Zomotor et al (1998), Halfmann and Holzmann (2003), Halbe (2008), Ding et al (2004), Hac et al (2004), Bauer (2015).

The ratio of inertia can be roughly estimated by $J_{\mathrm{x}} \approx m_{\mathrm{B}} i_{\mathrm{x}}^2$ according to (7.3.11). The experimental determination of the unknown parameters is illustrated in Sect. 11.8.

## 9.2 Pitch Dynamic Model

Corresponding to the roll dynamic model, it is assumed that a longitudinal acceleration $a_{\mathrm{X}}$ acts on the vehicle, respectively, the CG, leading to a pitch angle $\theta$ around the pitch axis, which is located parallel to the front and rear axle and has the height $h_{\mathrm{PA}}$ from ground and a backward distance $l_{\mathrm{CG,PA}}$ from the center of gravity; see Fig. 9.2. A torque balance around the PA then becomes

$$\begin{aligned} J_{\mathrm{Y}}\ddot{\theta} &= 2F_{\mathrm{Zr}}\left(l_{\mathrm{r}} - l_{\mathrm{CG,PA}}\right) - 2F_{\mathrm{Zf}}\left(l_{\mathrm{f}} + l_{\mathrm{CG,PA}}\right) \\ &\quad - m_{\mathrm{B}}\left(h_{\mathrm{CG}} - h_{\mathrm{PA}}\right)a_{\mathrm{X}} - m_{\mathrm{B}}g\, l_{\mathrm{CG,PA}}, \end{aligned} \tag{9.2.1}$$

with the vertical forces from the suspension according to (8.2.5). Simplifying the suspension forces to

$$F_{\mathrm{Zj}} = c_{\mathrm{Bj}}\left(z_{\mathrm{Wj}} - z_{\mathrm{Bj}}\right) + d_{\mathrm{Bj}}\left(\dot{z}_{\mathrm{Wj}} - \dot{z}_{\mathrm{Bj}}\right). \tag{9.2.2}$$

It is now assumed that $z_{\mathrm{W}} = 0$ and that the horizontal distance between the CG and the pitch axis PA is $l_{\mathrm{CG,PA}} = 0$. Then it follows for the small pitching angles

$$\Delta z_{\mathrm{Bf}} = l_{\mathrm{f}} \sin\theta \approx l_{\mathrm{f}}\theta; \;\; \Delta z_{\mathrm{Br}} \approx l_{\mathrm{r}}\theta. \tag{9.2.3}$$

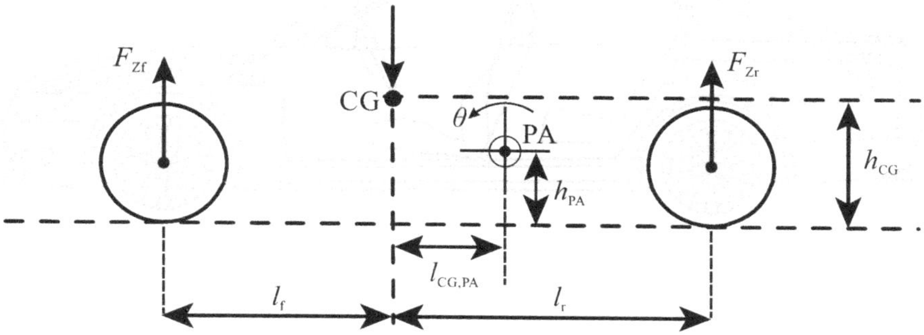


**Fig. 9.2** Scheme for the torques around the pitch axis PA

Equation (9.2.1) becomes then

$$\begin{aligned} J_Y\ddot{\theta} &= 2\left(c_{Br}l_r + c_{Bl}l_f\right)\theta - 2\left(d_{Br}l_r + d_{Bl}l_f\right)\dot{\theta} \\ &\quad - m_B\left(h_{CG} - h_{PA}\right)a_X \\ &= -c_{pit}\theta - d_{pit}\dot{\theta} - k_{pit}a_X, \end{aligned} \tag{9.2.4}$$

with

$$\begin{aligned} c_{pit} &= 2\left(c_{Br}l_r + c_{Bl}l_f\right), \\ d_{pit} &= 2\left(d_{Br}l_r + d_{Bl}l_f\right), \\ k_{pit} &= m_B\left(h_{CG} - h_{PA}\right). \end{aligned}$$

The transfer function for the pitch angle then becomes

$$G_{ax\theta}(s) = \frac{\theta(s)}{a_X(s)} = \frac{-k_{pit}}{J_y s^2 + d_{pit}s + c_{pit}}. \tag{9.2.5}$$

It is as the roll dynamic behavior of second order with the characteristic values

$$\begin{aligned} \omega_{o,rol} &= \sqrt{\frac{c_{pit}}{J_y}}, \\ \omega_{e,rol} &= \sqrt{\frac{c_{pit}}{J_y} - \frac{d_{pit}^2}{4J_y^2}}, \\ \zeta &= \frac{d_{pit}}{2}\frac{1}{\sqrt{c_{pit}J_y}}. \end{aligned} \tag{9.2.6}$$

The damping depends on the shock absorbers and the friction in the suspension bearings.

The deflection of the suspensions follow from (9.2.3) and the *changes of the vertical forces* on the front and rear axle through a positive acceleration $a_X$ and therefore $\Delta\theta < 0$ are

$$\Delta F_{Zf} = c_{Bf} l_f \Delta\theta + d_{Bf} l_f \dot{\theta}, \tag{9.2.7}$$

$$\Delta F_{Zr} = -c_{Br} l_r \Delta\theta - d_{Br} l_r \dot{\theta}. \tag{9.2.8}$$

Corresponding models are published by Zomotor et al (1998), Germann (1997), and Bauer (2015).

The experimental identification of unknown parameters is described in Sect. 11.8.

## References

Bauer M (2015) Methoden zur modellbasierten Fahrdynamikanalyse und Bewertung von Fahrdynamikregelsystemen. Dissertation Technische Universität Darmstadt. Fortschr.-Ber. VDI Reihe 12, 792. VDI Verlag, Düsseldorf

Ding E, Massel T, Arndt M (2004) Fault-tolerant roll rate sensor monitoring. In: Proceedings of AVEC, advanced vehicle control symposium, pp 585–590

Germann S (1997) Modellbildung und modellgestützte Regelung der Fahrzeuglängsdynamik. Dissertation Technische Universität Darmstadt. Fortschr.-Ber. VDI Reihe 12, 309. VDI Verlag, Düsseldorf

Hac A, Borwn T, Martens J (2004) Detection of vehicle rollover. Warrendale, PA, vol SAE 2004-01-1757

Halbe I (2008) Modellgestützte Sensoreninformationsplattform für die Quer- und Längsdynamik von Kraftfahrzeugen: Anwendungen zur Fehlerdiagnose und Fehlertoleranz. Diss. TU Darmstadt, Fortschr.-Ber. VDI Reihe 12, 680. VDI Verlag, Düsseldorf

Halfmann C, Holzmann H (2003) Adaptive Modelle für die Kraftfahrzeugdynamik. VDI-Buch, Springer, Berlin

Isermann R (2003) Mechatronic systems - fundamentals. Springer, London

Würtenberger M (1997) Modellgestützte Verfahren zur Überwachung des Fahrzustandes eines Pkw. Diss. Universität Darmstadt, Fortschr.-Ber. VDI Reihe 12, 314, VDI Verlag, Düsseldorf

Zomotor A, Braess HH, Rönitz R (1998) Verfahren und Kriterien zur Bewertung des Fahrverhaltens von Personenkraftwagen. ATZ-Automobiltechnische Zeitschrift 100(3):236–243

# 10 Parameter and State-Estimation Methods for Vehicle Dynamics

The derived mathematical models of vehicle dynamics are mostly based on theoretical physical models and contain *parameters* $\boldsymbol{\theta}$ which are partially known or not known at all. Therefore, they have to be determined experimentally by measurements of input and output signals and parameter identification or *parameter-estimation methods* have to be applied, see Fig. 10.1.

Because not all interesting time-depending *variables* $\mathbf{x}(t)$ are directly measurable, they have to be reconstructed by using mathematical models with known parameters and measurable input and output variables. If the measurements are only little influenced by disturbances, the state reconstruction can be made with *state observers*. However, if the disturbances have a considerable effect on the measurements, *state-estimation methods* have to be applied.

In the following two sections, a brief introduction to these methods is given. Figure 10.1 gives an overview of the modeling procedure. The input signals are usually

$$\mathbf{u}^{\mathrm{T}} = [\delta_{\mathrm{H}} \alpha \beta] \, ,$$

where $\delta_{\mathrm{H}}$ is the steering angle, $\alpha$ is the accelerator throttle position and $\beta$ the brake pedal position. The ESC provides in general the measured output variables

$$\mathbf{y}_{\mathrm{ESC}}^{\mathrm{T}} = \left[\delta_{\mathrm{H}}, \dot{\psi}, a_{\mathrm{X}}, a_{\mathrm{Y}}, \omega_{\mathrm{Tij}}\right] .$$

These signals are available onboard in series vehicles.

For research cars, an inertial measurement unit (IMU) may be installed, providing

$$\mathbf{y}_{\mathrm{IMU}}^{\mathrm{T}} = \left[a_{\mathrm{X}}, a_{\mathrm{Y}}, a_{\mathrm{Z}}, \dot{\varphi}, \dot{\psi}, \dot{\theta}\right] .$$



R. Isermann, *Automotive Control*, ATZ/MTZ-Fachbuch,
https://doi.org/10.1007/978-3-642-39440-9_10

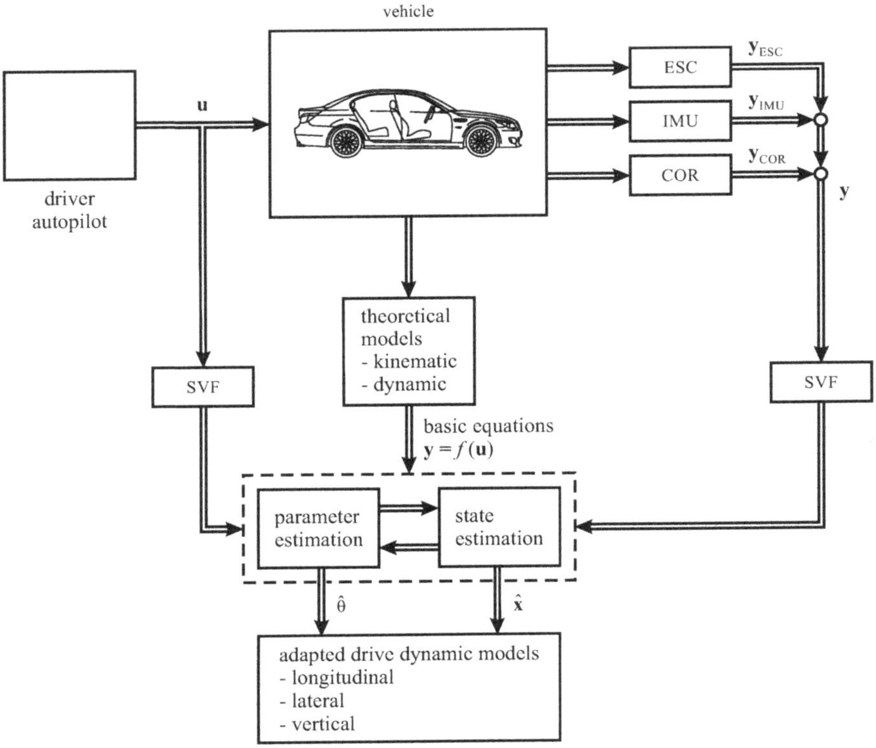


**Fig. 10.1** Schematic for the modeling procedure for drive dynamics of vehicles. ESC: Electronic Stability Control; IMU: Inertial Measurement Unit; COR: CORREVIT Sensor for $v$ and $\beta$

Further measurements for research may be obtained from 3D-GPS systems, Bauer (2015)

$$\mathbf{y}_{\mathrm{GPS}}^{\mathrm{T}} = [X_{\mathrm{E}}, Y_{\mathrm{E}}, v, \psi, \varphi, \theta] .$$

The measured input and output signals are usually sent to (identical) *state variable filters* in order to suppress noise effects and to generate derivatives like $\ddot{\psi}, \ddot{\varphi}, and \ddot{\theta}$.

Generally, first the unknown parameters $\boldsymbol{\theta}$ have to be estimated which are required for the state estimation of $\mathbf{x}(t)$. However, in some cases, the parameter estimation requires variables which are not measurable with onboard sensors, like the side slip angle $\beta$ or the roll angle $\varphi$. Then, parameter and state estimation have to operate in parallel. An alternative is to include the parameter estimation into the state estimation, e.g. with Kalman filters. The goal is finally to generate *adapted drive dynamic models* during driving and onboard which can be used online for adaptive, model-based drive dynamic control systems or for supervision and fault diagnosis. However, for research vehicles, additional measurements like with IMU or GPS may be used for developing additional models with online or offline data processing.

## 10.1 Parameter-Estimation Methods

The identification of dynamic models depends on the type of mathematical models and the kind of used input signals. A quite simple way is to evaluate transient functions after step inputs by using characteristic parameters like gains and time constants. However, for noisy measurements, other methods have to be applied. An impulse response can, for example, be obtained for stationary stochastic or pseudo stochastic input signals by the determination of *correlation functions*. A *frequency response* is the result of a Fourier analysis for periodic input signals. In both cases, *nonparametric models* are obtained. However, if theoretical models with a parametric structure already exist, as for driving dynamics of cars, then *parameter-estimation methods* should be applied.

The basic methods of least squares for the estimation of parameters $\boldsymbol{\theta}$ is firstly considered for the case of linear dynamic processes with sampled discrete-time measurements of one input signal $u(k)$ and one output signal $y(k)$. Then, the method is described for continuous-time signals $u(t)$ and $y(t)$. For a detailed treatment; see e.g. Isermann (2006) or Isermann and Münchhof (2011).

### 10.1.1 Method of Least Squares Parameter Estimation (LS), Discrete Time

It is assumed that the process can be described by the linear difference equation

$$\begin{aligned} &y_u(k) + a_1 y_u(k-1) + \ldots + a_m\, y_u(k-m) \\ &= b_1\, u(k-d-1) + \ldots + b_m u\, (k-d-m). \end{aligned} \tag{10.1.1}$$

Here

$$\begin{aligned} u(k) &= U(k) - U_{00} \\ y_u(k) &= Y_u(k) - Y_{00}, \end{aligned} \tag{10.1.2}$$

are the deviations of the absolute signals $U(k)$ and $Y_u(k)$ from the operating point described by $U_{00}$ and $Y_{00}$, $k$ is the discrete time $k = t/T_0 = 0, 1, 2, \ldots$, $T_0$ is the sampling time and $d = T_t/T_0 = 0, 1, 2, \ldots$ is the discrete dead-time of the process. The corresponding transfer function in the $z$-domain is

$$\begin{aligned} G_P(z) &= \frac{y_u(z)}{u(z)} = \frac{B(z^{-1})}{A(z^{-1})} z^{-d} \\ &= \frac{b_1\, z^{-1} + \ldots + b_m z^{-m}}{1 + a_1\, z^{-1} + \ldots + a_m z^{-m}} z^{-d}. \end{aligned} \tag{10.1.3}$$

The measured signal contains a stationary, stochastic disturbance

$$y(k) = y_u\,(k) + n\,(k) \quad \text{with} \quad E\,\{n(k)\} = 0. \tag{10.1.4}$$

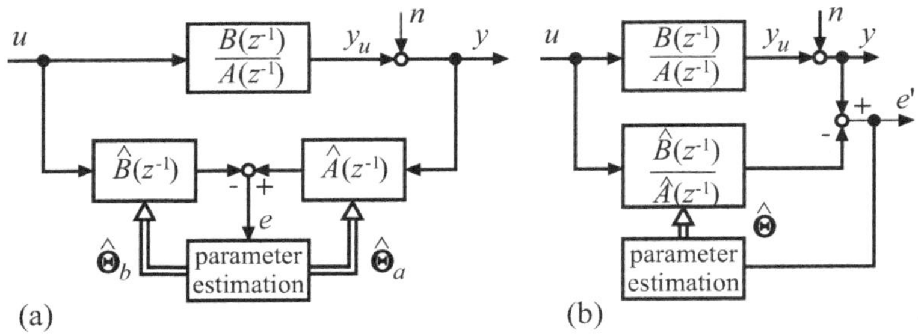


**Fig. 10.2** Model structures for parameter estimation. **a** Equation error; **b** output error

The task is to determine the unknown parameters $a_i$ and $b_i$ from $N$ measured input and output signal data points; see Fig. 10.2.

**(a) Equation Error Methods**
Let the model parameters obtained from the data up to the sample $(k-1)$ be denoted by $\hat{a}_i$ and $\hat{b}_i$. Then, (10.1.1) becomes in the presence of a disturbed output signal

$$\begin{gathered} y(k) + \hat{a}_1\, y(k-1) \ldots + \hat{a}_m\, y(k-m) \\ -\hat{b}_1\, u(k-d-1) - \ldots - \hat{b}_m u(k-d-m) = e(k), \end{gathered} \tag{10.1.5}$$

where the *equation error* (residual) $e(t)$ is introduced instead of "0". This error corresponds to a generalized error; see Fig. 10.2. This can be seen by rewriting (10.1.5), compare Fig. 10.2a.

$$\hat{A}\,(z^{-1})y(z) - \hat{B}(z^{-1})\, z^{-d}\, u(z) = e\,(z), \tag{10.1.6}$$

$e$ is linearly dependent on the parameters sought for (*linear in the parameters*).

From (10.1.5), $\hat{y}(k|k-1)$ can be interpreted as the one-step-ahead prediction, based on the measurements up to sample $(k-1)$

$$\hat{y}(k|k-1) = \boldsymbol{\psi}^T(k)\hat{\boldsymbol{\theta}}, \tag{10.1.7}$$

with the data vector

$$\boldsymbol{\psi}^T(k) = \begin{bmatrix} -y\,(k-1) \ldots - y\,(k-m) | \; u(k-d-1) \ldots \\ u(k-d-m) \end{bmatrix}, \tag{10.1.8}$$

and the parameter vector

$$\hat{\boldsymbol{\theta}} = [\hat{a}_1 \ldots \hat{a}_m \mid \hat{b}_1 \ldots \hat{b}_m]^T. \tag{10.1.9}$$

Consequently, (10.1.5) can be written as

$$y(k) = \boldsymbol{\psi}^T(k)\hat{\boldsymbol{\theta}} + e(k). \tag{10.1.10}$$

The measured signals for $k = m + d, \ldots, m + d + N$ are written in vectors, e.g.

$$\mathbf{y}^T(m + d + N) = [y(m + d) \ldots y(m + d + N)]. \tag{10.1.11}$$

Then

$$\mathbf{y}(m + d + n) = \boldsymbol{\Psi}(m + d + N)\hat{\boldsymbol{\theta}} + \mathbf{e}(m + d + N), \tag{10.1.12}$$

where $\boldsymbol{\Psi}$ is a $((N + 1) \times 2m)$-data matrix. Minimizing the sum of errors squared

$$V = \sum_{k=m+d}^{m+d+N} e^2(k) = \mathbf{e}^T(m + d + N)\,\mathbf{e}(m + d + N), \tag{10.1.13}$$

yields

$$\left.\frac{dV}{d\boldsymbol{\theta}}\right|_{\theta=\hat{\theta}} = -2\boldsymbol{\Psi}^T[\mathbf{y} - \boldsymbol{\Psi}\hat{\boldsymbol{\theta}}] = \mathbf{0}, \tag{10.1.14}$$

for the unknown parameters. From this, the (nonrecursive) estimation equation of the least squares (LS) method can be obtained

$$\hat{\boldsymbol{\theta}} = [\boldsymbol{\Psi}^T\boldsymbol{\Psi}]^{-1}\boldsymbol{\Psi}^T\mathbf{y}. \tag{10.1.15}$$

The covariance matrix

$$\mathbf{P} = [\boldsymbol{\Psi}^T\,\boldsymbol{\Psi}]^{-1}, \tag{10.1.16}$$

has the dimension $(2m, 2m)$. The inverse exists if and only if

$$\det\,[\boldsymbol{\Psi}^T\,\boldsymbol{\Psi}] = \det\,\mathbf{P}^{-1} \neq 0. \tag{10.1.17}$$

Also,

$$\frac{\partial^2 V}{\partial\,\boldsymbol{\theta}\,\partial\,\boldsymbol{\theta}^T} = \boldsymbol{\Psi}^T\,\boldsymbol{\Psi}, \tag{10.1.18}$$

has to be positive-definite such that the loss function $V$ has a minimum. Both requirements are satisfied if and only if

$$\det\,[\boldsymbol{\Psi}^T\,\boldsymbol{\Psi}] = \det\,\mathbf{P}^{-1} > 0. \tag{10.1.19}$$

This condition also includes that the input signal is persistently exciting the process and that the process is *stable*.

From parameter-estimation methods, it is usually required that the estimate is not biased for a finite number of data samples $N$

$$E\,\{\boldsymbol{\theta}\,(N)\} = \boldsymbol{\theta}_0, \tag{10.1.20}$$

($\boldsymbol{\theta}_0$ denotes the true parameters) and is consistent in the quadratic mean

$$\lim_{N\to\infty} E\left\{\hat{\boldsymbol{\theta}}(N)\right\} = \boldsymbol{\theta}_0, \tag{10.1.21}$$

$$\lim_{N\to\infty} E\left[\hat{\boldsymbol{\theta}}(N) - \boldsymbol{\theta}_0\right]\left[\boldsymbol{\theta}(N) - \boldsymbol{\theta}_0\right]^T = \mathbf{0}. \tag{10.1.22}$$

For the least squares method, (10.1.21) becomes, by substituting (10.1.10) into (10.1.15)

$$\begin{aligned} E\left\{\hat{\boldsymbol{\theta}}(N)\right\} &= \boldsymbol{\theta}_0 + E\left\{[\boldsymbol{\Psi}^T\,\boldsymbol{\Psi}]^{-1}\boldsymbol{\Psi}^T\mathbf{e}\right\} \\ &= \boldsymbol{\theta}_0 + \mathbf{b}. \end{aligned} \tag{10.1.23}$$

In order to have a vanishing bias (systematic estimation error) $\mathbf{b}$, $\boldsymbol{\Psi}^T$ and $\mathbf{e}$ must be uncorrelated. Consequently, $e(k)$ must not be correlated and $E\{e(k)\} = 0$. The estimation is unbiased if the disturbance signal $n(k)$ is generated by the disturbance filter

$$G_v(z) = \frac{n(z)}{v(z)} = \frac{1}{A(z^{-1})}, \tag{10.1.24}$$

where $v(k)$ is discrete white noise. Since this filter does not exist in practice, the least squares estimation, in general, yields biased estimates. These systematic estimation errors are the larger the greater the variance $\sigma_n^2$ of the disturbance signal is compared to the output signal $\sigma_{yu}^2$.

Because of the biased estimates for the least squares algorithm, this method can only be used for processes with no or only small disturbance signals. A big advantage of the least squares algorithm, however, is that the parameter vector $\hat{\boldsymbol{\theta}}$ can be determined in one batch calculation and no iterative methods are necessary. This is possible since the employed error measure is linear in the parameters.

**(b) Output Error Methods**

Instead of the equation error the output error

$$e'(k) = y(k) - y_M\left(\hat{\boldsymbol{\theta}}, k\right), \tag{10.1.25}$$

can be used, where

$$y_M\left(\hat{\boldsymbol{\theta}}, z\right) = \frac{\hat{B}(z^{-1})}{\hat{A}(z^{-1})}\,u(z), \tag{10.1.26}$$

is the model equation output; see Fig. 10.2b. But then no direct calculation of the parameter estimates $\hat{\boldsymbol{\theta}}$ is possible because $e'(k)$ is nonlinear in the parameters. Therefore, the loss function (10.1.13) is minimized by a numerical optimization method, e.g. downhill-simplex. The computational effort is then larger, and online real-time application, in general, not possible. However, relative precise parameter estimates may be obtained.

### (c) Recursive Least Squares (RLS) Methods, Discrete Time

Writing the nonrecursive estimation equations for $\hat{\boldsymbol{\theta}}(k+1)$ and $\hat{\boldsymbol{\theta}}(k)$ and subtracting one from the other, results in the recursive parameter-estimation algorithm

$$\begin{array}{cccccc} \hat{\boldsymbol{\theta}}(k+1) = & \hat{\boldsymbol{\theta}}(k) & + \quad \boldsymbol{\gamma}(k) & [y(k+1) & - & {}^{T}(k+1)\hat{\boldsymbol{\theta}}(k)] \\ \text{new} & \text{old} & \text{correction} & \text{new} & - & \text{one - step - ahead} \\ \text{estimate} & \text{estimate} & \text{vector} & \text{measurement} & & \text{prediction of the new} \\ & & & & & \text{measurement} \end{array} \quad (10.1.27)$$

The correcting vector is given by

$$\begin{aligned} \boldsymbol{\gamma}(k) &= \mathbf{P}(k+1)\ \boldsymbol{\psi}(k+1) \\ &= \frac{1}{\boldsymbol{\psi}^T(k+1)\ \mathbf{P}(k)\ \boldsymbol{\psi}(k+1) + 1}\ \mathbf{P}(k)\ \boldsymbol{\psi}(k+1), \end{aligned} \quad (10.1.28)$$

and

$$\mathbf{P}(k+1) = [\mathbf{I} - \boldsymbol{\gamma}(k)\ \boldsymbol{\psi}^T(k+1)]\ \mathbf{P}(k). \quad (10.1.29)$$

To start the recursive algorithm, one sets

$$\begin{aligned} \hat{\boldsymbol{\theta}}(0) &= \mathbf{0}, \\ \mathbf{P}(0) &= \alpha \mathbf{I}. \end{aligned} \quad (10.1.30)$$

with $\alpha$ large ($\alpha = 100, \ldots, 1000$). The expectation of the matrix $\mathbf{P}$ is proportional to the covariance matrix of the parameter estimates

$$E\,\{\mathbf{P}(k+1)\} = \frac{1}{\sigma_e^2}\ \text{cov}\,[\Delta\boldsymbol{\theta}(k)], \quad (10.1.31)$$

with

$$\sigma_e^2 = E\left\{\mathbf{e}^T\mathbf{e}\right\}, \quad (10.1.32)$$

and the parameter error

$$\Delta\boldsymbol{\theta}(k) = \hat{\boldsymbol{\theta}}(k) - \boldsymbol{\theta}_0. \quad (10.1.33)$$

Hence, the recursive algorithm contains the variances of the parameter estimates (diagonal elements of covariance matrix). (10.1.27) can also be written as

$$\hat{\boldsymbol{\theta}}(k+1) = \hat{\boldsymbol{\theta}}(k) + \boldsymbol{\gamma}(k)\ e(k+1). \quad (10.1.34)$$

To improve the numerical properties of the basic RLS algorithms, modified versions are recommended like discrete square root filtering.

As for process parameter estimation, the variations of $u(k)$ and $y(k)$ of the measured signals $U(k)$ and $Y(k)$ have to be used. The DC (direct current or steady state) values $U_{00}$ and $Y_{00}$ either have also to be estimated or have to be removed; see Isermann and Münchhof (2011).

In order to obtain unbiased parameter estimates, modified versions of the basic LS method were developed, like Extended Least Squares (ELS), instrumental variables or maximum likelihood method; see e.g. Isermann and Münchhof (2011).

### 10.1.2 Method of Least Squares Parameter Estimation (LS), Continuous Time

Parameter-estimation methods for dynamic processes were first considered for process models in discrete time as applied for digital control systems. For other applications, e.g. the validation of theoretical models or for fault diagnosis, however, parameter-estimation methods for models with continuous-time signals are needed.

A stable process with lumped parameters is considered, which can be described by the linear, time-invariant differential equation

$$\begin{aligned} &a_n y_u^{(n)}(t) + a_{n-1}\, y_u^{(n-1)}(t) + \ldots + a_1\, y_u^{(1)}(t) + y_u(t) \\ &= b_m\, u^{(m)}(t) + b_{m-1}\, u^{(m-1)}(t) + \ldots + b_1\, u^{(1)}(t) + b_0\, u(t) \quad m < n. \end{aligned} \tag{10.1.35}$$

It is assumed that the derivatives of the output signal

$$y^{(j)}(t) = d^{\,j}\, y(t)/dt^{\,j}, \quad j = 1, 2, \ldots, n, \tag{10.1.36}$$

and of the input signal for $j = 1, 2, \ldots, m$ exist. $u(t)$ and $y(t)$ are the deviations

$$\begin{aligned} u(t) &= U(t) - U_{00}, \\ y(t) &= Y(t) - Y_{00} \end{aligned} . \tag{10.1.37}$$

of the absolute signals $U(t)$ and $Y(t)$ from the operating point described by $U_{00}$ and $Y_{00}$. The transfer function corresponding to (10.1.35) is

$$\begin{aligned} G_P(s) &= \frac{y_u(s)}{u(s)} = \frac{B(s)}{A(s)} \\ &= \frac{b_0 + b_1\, s + \ldots + b_{m-1}\, s^{m-1} + b_m\, s^m}{1 + a_1\, s + \ldots + a_{n-1}\, s^{n-1} + a_n\, s^n}. \end{aligned} \tag{10.1.38}$$

The measurable signal $y(t)$ contains an additional disturbance signal $n(t)$

$$y(t) = y_u(t) + n(t). \tag{10.1.39}$$

Substituting (10.1.39) into (10.1.35) and introducing an equation error $e(t)$ yields

$$y(t) = \boldsymbol{\psi}^T(t)\, \boldsymbol{\theta} + e(t), \tag{10.1.40}$$

with

$$\psi^T(t) = \left[-y^{(1)}(t) \ldots - y^{(n)}(t) \mid u(t) \ldots u^{(m)}(t)\right], \tag{10.1.41}$$

$$\Theta = [a_1 \ldots a_n \quad b_0 \ldots b_m]^T \,. \tag{10.1.42}$$

The input and output signals are measured at discrete-time samples $t = k\ T_0,\ k = 0, 1, 2, \ldots, N$ with sampling time $T_0$ and the derivatives are generated. Based on this, $N + 1$ equations can be written down

$$y(k) = \psi^T(k)\ \hat{\theta} + e(k). \tag{10.1.43}$$

This system of equations can be written in matrix notation as

$$\mathbf{y} = \mathbf{\Psi}\ \hat{\theta} + \mathbf{e}. \tag{10.1.44}$$

Minimizing the loss function

$$V = \mathbf{e}^T(N)\ \mathbf{e}(N) = \sum_{k=0}^{N} e^2(k), \tag{10.1.45}$$

yields with $dV/d\hat{\boldsymbol{\theta}} = \mathbf{0}$ as previously shown in Sect. 10.1.1 the vector of parameter estimates for the least squares method

$$\hat{\theta}(N) = \left[\mathbf{\Psi}^T\ \mathbf{\Psi}\right]^{-1}\ \mathbf{\Psi}^T\ y. \tag{10.1.46}$$

The existence of a unique solution requires that the matrix $\mathbf{\Psi}^T \mathbf{\Psi}$ is positive-definite. After dividing this matrix by the measurement time, the elements of the resulting matrix are the estimates of the correlation functions $\mathbf{\Phi}(\tau)$ of the derivatives of the signals for $\tau = 0$ with no time shift. It can be seen that the form is very similar to the least squares method for models with discrete-time signals. Hence, a lot of the derivations can be directly transferred, such as the recursive formulation and the numerically improved versions. However, particular problems arise concerning the convergence and the evaluation of the needed derivatives of the signals.

A *convergence analysis* shows that the estimates for continuous signals are also biased if the error signal $e(k)$ is not statistically independent. Hence, the estimates in general are biased for disturbed processes.

If the needed derivatives of the signals are directly measurable (e.g. as for vehicle applications), these values can be written in the data matrix $\mathbf{\Psi}$ and the correlation functions in the matrix $[\mathbf{\Psi}^T\ \mathbf{\Psi}]/(N+1)$ can be directly calculated. However, if the derivatives are not measurable, the derivatives have to be evaluated from the sampled signals $u(t)$ and $y(t)$. For this, there basically exist the following methods. The *numerical differentiation* in combination with interpolation approaches (splines,

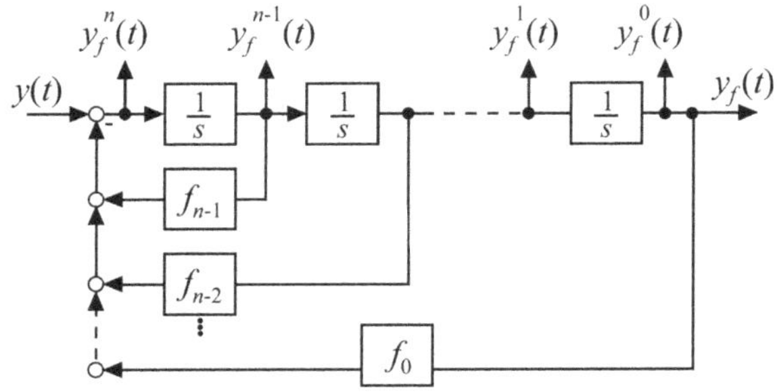


**Fig. 10.3** State variable filter

Newton's method) is usually not able to suppress noise due to disturbance signals. State variable filters (SVF), see Fig. 10.3,

$$F(s) = \frac{y_f(s)}{y(s)} = \frac{1}{f_0 + f_1 s + \ldots + f_{n-1} s^{n-1} + s^n}, \tag{10.1.47}$$

have proven to yield good results. The state variable filter is a low-pass filter that provides the derivatives as well as filters the disturbance signals. With the state variable filter, the input signal $u(t)$ and the output signal $y(t)$ is filtered. The choice of the filter parameters $f_i$ is relatively free. The design of a Butterworth filter is recommended. A further possibility is the application of finite impulse response filters (FIR), where the derivations of the impulse response of a low-pass filter are convoluted with the signal.

For large signal-to-disturbance ratios, this least squares method has been shown to yield good results. For larger disturbance signals, consistent parameter-estimation methods should be employed such as *the instrumental variables method*.

Since these parameter-estimation methods are based on discrete-time estimation methods, a lot of results and estimation methods for discrete-time models can be transferred to models for continuous time.

For the identification of *nonlinear processes*, the method of least square parameter estimation can be directly applied if the equation error is linear in the parameters like for static polynomial models or special dynamic nonlinear dynamic models, like a Hammerstein or Volterra model. A future possibility is the application of neural network models like Radial Basis Functions or Multilayer Perceptrons. However, the determined parameters have in general no transparent relation to physical parameters. However, this improves with local linear model networks like LOLIMOT. More details on these identification methods are given, e.g. in Isermann (2006) and Isermann and Münchhof (2011).

## 10.2 State Variable Estimation

State observers and state estimators adjust the internal state variables $\mathbf{x}(t)$ of dynamic models based on measured input signals $u(t)$ and output signals $y(t)$ and assume that the parameters $\theta$ of the models are known.

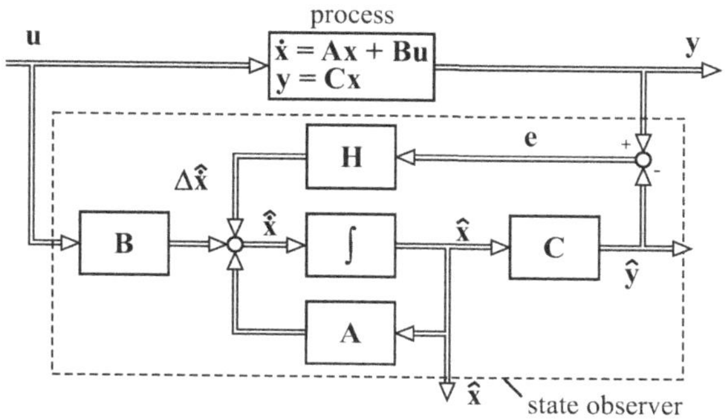


**Fig. 10.4** Process and state observer

In the case of drive dynamic modeling and control state, observers are required to obtain access to nonmeasurable variables, for example, the vehicle slip angle, rolling and pitch angles, road gradients, lateral velocity, and vehicle position.

### 10.2.1 State Observer, Continuous Time

A linear time-invariant process is considered which can be described by the state-space model

$$\dot{\mathbf{x}}(t) = \mathbf{A}\,\mathbf{x}(t) + \mathbf{B}\,\mathbf{u}(t), \tag{10.2.1}$$

$$\mathbf{y}(t) = \mathbf{C}\,\mathbf{x}(t). \tag{10.2.2}$$

Here $p$ input signals $u(t)$ and $r$ output signals $\mathbf{y}(t)$ are assumed as for multi-variable processes. With the assumption that the structure and the parameters of the model are known, a state observer is used to reconstruct the unmeasurable state variables based on measured inputs and outputs

$$\hat{\dot{\mathbf{x}}}(t) = \mathbf{A}\,\hat{\mathbf{x}}(t) + \mathbf{B}\,\mathbf{u}(t) + \mathbf{H}\,\mathbf{e}(t), \tag{10.2.3}$$

$$\mathbf{e}(t) = \mathbf{y}(t) - \mathbf{C}\,\hat{\mathbf{x}}(t); \tag{10.2.4}$$

compare Fig. 10.4. $\mathbf{e}(t)$ is an output error which acts through the observer matrix $\mathbf{H}$ on the reconstructed state derivatives $\hat{\dot{\mathbf{x}}}(t)$. Inserting (10.2.4) in (10.2.3) yields the implementation form of the state observer

$$\begin{aligned}\hat{\dot{\mathbf{x}}}(t) &= \mathbf{A}\,\hat{\mathbf{x}}(t) + \mathbf{B}\,\mathbf{u}(t) + \mathbf{H}\left[\mathbf{y}(t) - \mathbf{C}\hat{\mathbf{x}}(t)\right] \\ &= [\mathbf{A} - \mathbf{H}\,\mathbf{C}]\;\hat{\mathbf{x}}(t) + \mathbf{B}\,\mathbf{u}(t) + \mathbf{H}\,\mathbf{y}(t).\end{aligned} \tag{10.2.5}$$

where it is assumed that the system is observable.

The state error

$$\tilde{\dot{\mathbf{x}}}(t) = \dot{\mathbf{x}}(t) - \hat{\dot{\mathbf{x}}}(t), \tag{10.2.6}$$

between the real process states and the observed states becomes under the assumption that process and model parameters are identical and by introducing (10.2.1) and (10.2.5)

$$\dot{\tilde{\mathbf{x}}}(t) = [\mathbf{A} - \mathbf{H}\ \mathbf{C}]\ \tilde{\mathbf{x}}(t). \tag{10.2.7}$$

Hence, the state error is independent on the input $\mathbf{u}$ and vanishes asymptotically

$$\lim_{t \to \infty} \tilde{\mathbf{x}}(t) = \mathbf{0}, \tag{10.2.8}$$

for any initial state deviation $[\mathbf{x}(0) - \hat{\mathbf{x}}(0)]$ if the observer is stable.

For the state error follows from (10.2.7) after Laplace transformation

$$(s\mathbf{I} - \mathbf{A} + \mathbf{HC})\, \tilde{x}(s) = 0, \tag{10.2.9}$$

and the characteristic equation becomes

$$\begin{aligned} \det\ (s\mathbf{I} - \mathbf{A} + H\mathbf{C}) &= (s - s_1)\,(s - s_2)\ldots(s - s_n) \\ &= \gamma_m + \gamma_{m-1}s + \ldots + s^n = 0. \end{aligned} \tag{10.2.10}$$

In order to obtain a stable observer the poles

$$s_j = \delta_j \pm i\omega_j \quad j = 1, \ldots, n,$$

have to be placed in the left half of the s-plane, hence for the real part follows

$$\delta_j < 0.$$

This can be reached by proper design of the observer feedback matrix $\mathbf{H}$ by pole placement; see e.g. Föllinger (1994), Chen and Patton (1999).

### 10.2.2 Nonlinear State Observer, Continuous Time

As the dynamic models of vehicles are frequently nonlinear, the linear observers cannot be applied directly. Therefore, they have to be extended to nonlinear model equations in the general form of a nonlinear state-space model

$$\dot{\mathbf{x}} = f(\mathbf{x}, \mathbf{u}), \tag{10.2.11}$$

$$\mathbf{y} = \mathbf{h}(\mathbf{x}, \mathbf{u}). \tag{10.2.12}$$

A nonlinear state observer which is close to a linear Luenberger observer is obtained after linearisation of the process equation around the operation point according to a Taylor approximation, Föllinger (1994), Adamy (2018)

The observer equation for the nonlinear models then are, leaning on the linear observer (10.2.5)

$$\dot{\hat{\mathbf{x}}} = \mathbf{f}(\hat{\mathbf{x}}, \mathbf{u}) + \mathbf{L}(\hat{\mathbf{x}}, \mathbf{u})(\mathbf{y} - \hat{\mathbf{y}}), \tag{10.2.13}$$

$$\hat{\mathbf{y}} = \mathbf{h}(\hat{\mathbf{x}}, \mathbf{u}). \tag{10.2.14}$$

After Taylor approximation around $\mathbf{x} = \hat{\mathbf{x}}$ and neglecting the higher order terms one obtains for the process equation (10.2.11) and (10.2.12)

$$\mathbf{f}(\mathbf{x}, \mathbf{u}) = f(\hat{\mathbf{x}}, \mathbf{u}) + \frac{\partial \mathbf{f}(\hat{\mathbf{x}}, \mathbf{u})}{\partial \mathbf{x}}(\mathbf{x} - \hat{\mathbf{x}}), \tag{10.2.15}$$

$$\mathbf{h}(\mathbf{x}, \mathbf{u}) = \mathbf{h}(\hat{\mathbf{x}}, \mathbf{u}) + \frac{\partial \mathbf{h}(\hat{\mathbf{x}}, \mathbf{u})}{\partial \mathbf{x}}(\mathbf{x} - \hat{\mathbf{x}}), \tag{10.2.16}$$

where $\partial \mathbf{f}/\partial \mathbf{x}$ and $\partial \mathbf{h}/\partial \mathbf{x}$ are Jacobi matrices, of the form

$$\frac{\partial \mathbf{f}}{\partial \mathbf{x}} = \begin{bmatrix} \frac{\partial \mathbf{f}_1}{\partial \mathbf{x}_1} & \frac{\partial \mathbf{f}_1}{\partial \mathbf{x}_2} & \cdots & \frac{\partial \mathbf{f}_1}{\partial \mathbf{x}_n} \\ \vdots & \vdots & & \vdots \\ \frac{\partial \mathbf{f}_n}{\partial \mathbf{x}_1} & \frac{\partial \mathbf{f}_n}{\partial \mathbf{x}_2} & \cdots & \frac{\partial \mathbf{f}_n}{\partial \mathbf{x}_n}, \end{bmatrix} \tag{10.2.17}$$

for

$$\mathbf{f}(\mathbf{x}, \mathbf{u}) = \begin{bmatrix} \mathbf{f}_1(\mathbf{x}, \mathbf{u}) \\ \mathbf{f}_2(\mathbf{x}, \mathbf{u}) \\ \vdots \\ \mathbf{f}_n(\mathbf{x}, \mathbf{u}). \end{bmatrix} \tag{10.2.18}$$

The process state equation then becomes with the linearisation of (10.2.15)

$$\dot{\mathbf{x}} = \mathbf{f}(\hat{\mathbf{x}}, \mathbf{u}) + \frac{\partial \mathbf{f}(\hat{\mathbf{x}}, \mathbf{u})}{\partial \mathbf{x}}(\mathbf{x} - \hat{\mathbf{x}}), \tag{10.2.19}$$

$$\mathbf{y} = \mathbf{h}(\mathbf{x}, \mathbf{u}) = \mathbf{h}(\hat{\mathbf{x}}, \mathbf{u}) + \frac{\partial \mathbf{h}(\hat{\mathbf{x}}, \mathbf{u})}{\partial \mathbf{x}}(\mathbf{x} - \hat{\mathbf{x}}), \tag{10.2.20}$$

and for the observer follows from (10.2.13) and (10.2.18)

$$\begin{aligned} \dot{\hat{\mathbf{x}}} &= \mathbf{f}(\hat{\mathbf{x}}, \mathbf{u}) + \mathbf{L}(\hat{\mathbf{x}}, \mathbf{u})(\mathbf{y} - \mathbf{h}(\hat{\mathbf{x}}, \mathbf{u})), \\ &= \mathbf{f}(\hat{\mathbf{x}}, \mathbf{u}) + \mathbf{L}(\hat{\mathbf{x}}, \mathbf{u})\frac{\partial \mathbf{h}(\hat{\mathbf{x}}, \mathbf{u})}{\partial \mathbf{x}}(\mathbf{x} - \hat{\mathbf{x}}), \end{aligned} \tag{10.2.21}$$

$$\hat{\mathbf{y}} = \mathbf{h}(\hat{\mathbf{x}}, \mathbf{u}). \tag{10.2.22}$$

The equation describing the state error is analogous to (10.2.7)

$$\dot{\tilde{\mathbf{x}}} = \dot{\mathbf{x}} - \dot{\hat{\mathbf{x}}} = \left[ \frac{\partial \mathbf{f}(\hat{\mathbf{x}}, \mathbf{u})}{\partial \mathbf{x}} - \mathbf{L}(\hat{\mathbf{x}}, \mathbf{u}) \frac{\partial \mathbf{h}(\hat{\mathbf{x}}, \mathbf{u})}{\partial \mathbf{x}} \right] (\mathbf{x} - \hat{\mathbf{x}}), \tag{10.2.23}$$

$$\dot{\tilde{\mathbf{x}}} = \mathbf{F}(\hat{\mathbf{x}}, \mathbf{u}) \tilde{\mathbf{x}}. \tag{10.2.24}$$

Similar to (10.2.7) the coefficients of the feedback matrix $\mathbf{L}$ have to selected such that the observer is stable, e.g. by pole placement design with stable poles.

After further considerations with a desired diagonal matrix

$$\mathbf{G} = \begin{bmatrix} \lambda_1 & 0 & 0 \\ 0 & \lambda_2 & 0 \\ 0 & 0 & \lambda_3 \end{bmatrix}$$

one obtains for a process with three state variables and two outputs

$$\mathbf{L}(\hat{\mathbf{x}}, \mathbf{u}) = \begin{pmatrix} \frac{\partial f_1}{\partial x_1} - \lambda_1 & \frac{\partial f_1}{\partial x_3} \\ \frac{\partial f_2}{\partial x_1} & \frac{\partial f_2}{\partial x_3} \\ \frac{\partial f_3}{\partial x_1} & \frac{\partial f_3}{\partial x_3} - \lambda_3 \\ . & \end{pmatrix} \tag{10.2.25}$$

with selected stable poles $\lambda_1$ and $\lambda_3$; see Daiss (1996), Börner (2004), Schorn (2007), Kiencke and Nielsen (2005), and Adamy (2018). Figure 10.5 shows the signal flow for the nonlinear observer illustrating the higher complexity compared to the linear case, Fig. 10.4.

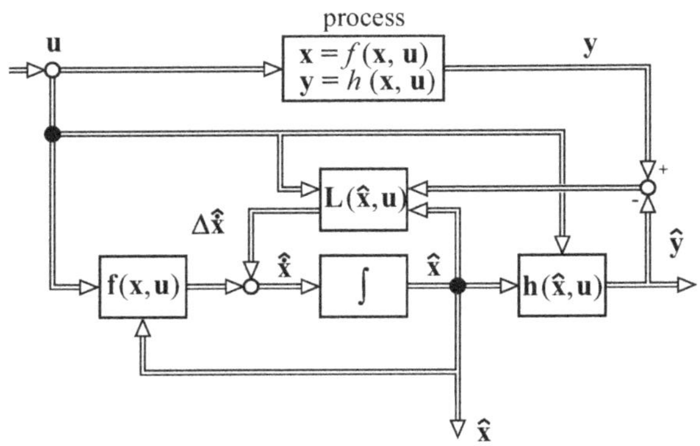


**Fig. 10.5** Nonlinear state observer

### 10.2.3 State Estimation (Kalman Filter), Discrete Time

For the linear multi-input multi-output process with *discrete-time signals* and without stochastic disturbances

$$\mathbf{x}(k+1) = \mathbf{A}\,\mathbf{x}(k) + \mathbf{B}\,\mathbf{u}(k), \tag{10.2.26}$$

$$\mathbf{y}(k) = \mathbf{C}\,\mathbf{x}(k), \tag{10.2.27}$$

the state observer equation becomes, corresponding to (10.2.3), (10.2.4)

$$\hat{\mathbf{x}}(k+1) = \mathbf{A}\,\hat{\mathbf{x}}(k) + \mathbf{B}\,\mathbf{u}(k) + \mathbf{H}\left[\mathbf{y}(k) - \mathbf{C}\,\hat{\mathbf{x}}(k)\right]. \tag{10.2.28}$$

The equation error as an output error then is

$$\mathbf{e}(k) = \mathbf{y}(k) - \mathbf{C}\,\hat{\mathbf{x}}(k), \tag{10.2.29}$$

and the state error equation according to (10.2.7) becomes

$$\tilde{\mathbf{x}}(k+1) = [\mathbf{A} - \mathbf{H}\,\mathbf{C}]\,\tilde{\mathbf{x}}(k). \tag{10.2.30}$$

If no disturbances act on the process, the observer converges to the true state variables if the eigenvalues $\mathbf{A} - \mathbf{H}\,\mathbf{C}$ are asymptotically stable. The speed of convergence can be made fast by a large influence of the observer gain $\mathbf{H}$. However, under the influence of *stochastic disturbances* the state reconstruction with these observers is not optimal. The state reconstruction must then simultaneously follow the true state variables and reject the noise effects which then leads to an estimation problem. The following shorted derivation follows Isermann (2006) and Isermann and Münchhof (2011).

The process is now supplemented by stochastic noise $\mathbf{v}(k)$ of the state and $\mathbf{n}(k)$ at the output

$$\mathbf{x}(k+1) = \mathbf{A}\,\mathbf{x}(k) + \mathbf{B}\,\mathbf{u}(k) + \mathbf{v}(k), \tag{10.2.31}$$

$$\mathbf{y}(k) = \mathbf{C}\,\mathbf{x}(k) + \mathbf{n}(k). \tag{10.2.32}$$

The process matrices $\mathbf{A}$, $\mathbf{B}$ and $\mathbf{C}$ are assumed to be known. The initial state $\mathbf{x}(0)$ is not known, but probabilistic information is known about $\mathbf{x}(0)$ and also about $\mathbf{v}(k)$ and $\mathbf{n}(k)$. These stochastic variables are assumed to be statistically independent, have a normal (Gaussian) distribution with the mean values

$$E\left\{\mathbf{x}(0)\right\} = \mathbf{x}_0\,; \quad E\left\{\mathbf{v}(k)\right\} = \mathbf{0}\,; \quad E\left\{\mathbf{n}(k)\right\} = \mathbf{0}, \tag{10.2.33}$$

and the covariance matrices

$$\begin{aligned} &E\left\{(\mathbf{x}(0) - \mathbf{x}_0)(\mathbf{x}(0) - \mathbf{x}_0)^T\right\} = \mathbf{X}_0, \\ &E\left\{\mathbf{v}(k)\,\mathbf{v}^T(k)\right\} = \mathbf{Q}, \\ &E\left\{\mathbf{n}(k)\,\mathbf{n}^T(k)\right\} = \mathbf{R}. \end{aligned} \tag{10.2.34}$$

Furthermore, it is assumed that $\mathbf{Q}$ and $\mathbf{R}$ are known as a measure about the size of the noises.

As the state-estimation error cannot converge to zero, a best estimate has to be found for the state vector $\mathbf{x}(k)$ based on the measured input variables $\mathbf{u}(k)$ and output variables $\mathbf{y}(k)$. A *least squares estimation* then requires

$$\min \ ||\mathbf{x}(k) - \hat{\mathbf{x}}(k|j)||^2. \tag{10.2.35}$$

Two different time instances are used here. $k$ means the present time and $j$ the used time instant of the measurements. The state estimation can then be given different names, Tomizuka (1998);

$k > j$ prediction problem,
$k = j$ filtering problem,
$k < j$ smoothing problem.

The filtering problem and *one-step ahead prediction* is further considered. The used measurements of the output are

$$\mathbf{Y}_j = \{\mathbf{y}(0),\ \mathbf{y}(1), \ldots, \mathbf{y}(j)\}\,. \tag{10.2.36}$$

Following notations are used:

Optimal state estimates:

$$\hat{\mathbf{x}}(k|j) = E\left\{\mathbf{x}(k)|\mathbf{Y}_j\right\}. \tag{10.2.37}$$

State-estimation error:

$$\tilde{\mathbf{x}}(k|j) = \mathbf{x}(k) - \hat{\mathbf{x}}(k|j). \tag{10.2.38}$$

Covariance matrices of the one-step ahead estimation error

$$\mathbf{P}^{-}(k+1) = E\left\{\tilde{\mathbf{x}}(k+1|k)\,\tilde{\mathbf{x}}^T(k+1|k)\right\}, \tag{10.2.39}$$

$$\mathbf{P}(k+1) = E\left\{\tilde{\mathbf{x}}(k+1|k+1)\,\tilde{\mathbf{x}}^T(k+1|k+1)\right\}. \tag{10.2.40}$$

For time instant $k+1$ the state variable $\mathbf{x}(k+1)$ can be predicted by using the state model (10.2.29) with the information at time $k$

$$\hat{\mathbf{x}}(k+1|k) = \mathbf{A}\,\hat{\mathbf{x}}(k|k) + \mathbf{B}\,\mathbf{u}(k) + \bar{\mathbf{v}}, \tag{10.2.41}$$

as the exact $\mathbf{v}(k)$ is unknown.

With the assumption $E\{\mathbf{v}(k)\} = \bar{\mathbf{v}} = \mathbf{0}$ it yields

$$\hat{\mathbf{x}}(k+1|k) = \mathbf{A}\,\hat{\mathbf{x}}(k|k) + \mathbf{B}\,\mathbf{u}(k). \tag{10.2.42}$$

At time $k+1$ also the measurement of the output $\mathbf{y}(k+1)$ is available. It holds

$$\mathbf{y}(k+1) = \mathbf{C}\,\mathbf{x}(k+1) + \mathbf{n}(k+1). \qquad (10.2.43)$$

However, $\mathbf{x}(k+1)$ is unknown. The prediction $\hat{\mathbf{x}}(k+1|k)$ is disturbed by the noise $\mathbf{v}(k)$ and the measurable output $\mathbf{y}(k+1)$ by $\mathbf{n}(k+1)$. It is assumed the $\mathbf{v}(k)$ and $\mathbf{n}(k)$ are statistically independent.

If both, $\hat{\mathbf{x}}(k+1|k)$ and $\mathbf{x}(k+1)$, would be known, one could calculate a *weighted mean as an estimate.*

$$\begin{aligned}\hat{\mathbf{x}}(k+1|k+1) &= (\mathbf{I} - \mathbf{K}')\,\hat{\mathbf{x}}(k+1|k) + \mathbf{K}'\,\mathbf{x}(k+1)\\ &= \hat{\mathbf{x}}(k+1|k) + \mathbf{K}'[\mathbf{x}(k+1) - \hat{\mathbf{x}}(k+1|k)],\end{aligned} \qquad (10.2.44)$$

where $\mathbf{K}'$ is a $(m \times m)$ weighting matrix which is to be chosen such that the covariance of the estimation error $\mathbf{P}(k+1)$ becomes a minimum. Now, instead of $\mathbf{x}(k+1)$, the measurable output vector $\mathbf{y}(k+1) = \mathbf{C}\,\mathbf{x}(k+1)$ using the process model (10.2.32), is taken. Then with $\mathbf{K}' = \mathbf{K}\,\mathbf{C}$ holds

$$\begin{aligned}\hat{\mathbf{x}}(k+1|k+1) &= \hat{\mathbf{x}}(k+1|k) + \mathbf{K}\mathbf{C}[\mathbf{x}(k+1) - \hat{\mathbf{x}}(k+1|k)]\\ &= \hat{\mathbf{x}}(k+1|k) + \mathbf{K}[\mathbf{y}(k+1) - \mathbf{C}\hat{\mathbf{x}}(k+1|k)]\\ &= [\mathbf{I} - \mathbf{K}\mathbf{C}]\,\hat{\mathbf{x}}(k+1|k) + \mathbf{K}\,\mathbf{y}(k+1).\end{aligned} \qquad (10.2.45)$$

This equation contains

$\hat{\mathbf{x}}(k+1|k)$ : the model prediction of $\mathbf{x}(k+1)$ based on the last estimate $\hat{\mathbf{x}}(k|k)$, (10.2.42), .
$\mathbf{y}(k+1)$ : the new measurement

A recursive estimation algorithm then follows from (10.2.43)

$$\hat{\mathbf{x}}(k+1|k+1) = \hat{\mathbf{x}}(k+1|k) + \mathbf{K}(k+1)[\mathbf{y}(k+1) - \mathbf{C}\,\hat{\mathbf{x}}(k+1|k)], \qquad (10.2.46)$$

where the correction matrix $\mathbf{K}(k+1)$ has to be chosen as to minimize the covariance matrix of the estimation error. As this variance changes with time, also $\mathbf{K}(k+1)$ must be time-variant. The error of the prediction is

$$\tilde{\mathbf{x}}(k+1|k) = \hat{\mathbf{x}}(k+1|k) - E\left\{\hat{\mathbf{x}}(k+1|k)\right\},$$

and the error in the measurement, see (10.2.32)

$$\tilde{\mathbf{y}}(k+1) = \mathbf{y}(k+1) - E\left\{\mathbf{y}(k+1)\right\} = \mathbf{n}(k).$$

The corresponding covariance matrices are

$$\begin{aligned}\mathbf{P}^-(k+1) &= E\left\{\tilde{\mathbf{x}}(k+1|k)\,\tilde{\mathbf{x}}^T(k+1|k)\right\}\\ \mathbf{Y} = E\left\{\tilde{\mathbf{y}}(k+1)\,\tilde{\mathbf{y}}^T(k+1)\right\} &= E\left\{\mathbf{n}(k)\,\mathbf{n}^T(k)\right\} = \mathbf{R}.\end{aligned}$$

The further derivations of the minimization of the covariance matrix of the estimation error is shown in AGARD (1970) by forming the covariance matrix into a complete square; see also Isermann (1991). This leads to a sequence of calculations:

*prediction*:

$$\hat{\mathbf{x}}(k+1|k) = \mathbf{A}\,\hat{\mathbf{x}}(k|k) + \mathbf{B}\,\mathbf{u}(k), \qquad (10.2.47)$$

$$\mathbf{P}^{-}(k+1) = \mathbf{A}\,\mathbf{P}(k)\,\mathbf{A}^{T} + \mathbf{Q}. \qquad (10.2.48)$$

*correction*

$$\mathbf{K}(k+1) = \mathbf{P}^{-}(k+1)\,\mathbf{C}^{T}[\mathbf{C}\,\mathbf{P}^{-}(k+1)\,\mathbf{C}^{T} + \mathbf{R}]^{-1}, \qquad (10.2.49)$$

$$\begin{aligned}\hat{\mathbf{x}}(k+1|k+1) &= \hat{\mathbf{x}}(k+1|k) + \mathbf{K}(k+1)[\mathbf{y}(k+1)\\ &\quad -\mathbf{C}\,\hat{\mathbf{x}}(k+1|k)],\end{aligned} \qquad (10.2.50)$$

$$\mathbf{P}(k+1) = [\mathbf{I} - \mathbf{K}(k+1)\mathbf{C}]\mathbf{P}^{-}(k+1). \qquad (10.2.51)$$

If the prediction (10.2.47) is inserted in (10.2.48), it follows

$$\underset{\text{new estimate}}{\hat{\mathbf{x}}(k+1|k+1)} = \underset{\text{old estimate}}{\mathbf{A}\,\hat{\mathbf{x}}(k|k)} + \mathbf{B}\,\mathbf{u}(k)$$

$$+ \underset{\substack{\text{correction}\\ \text{matrix}}}{\mathbf{K}(k+1)}\;\underset{\substack{\text{new measure-}\\ \text{ment}}}{[\mathbf{y}(k+1)}\; - \underset{\substack{\text{predicted measurement}\\ \text{based on the process}\\ \text{model and old estimate}}}{\mathbf{C}(\mathbf{A}\,\hat{\mathbf{x}}(k|k) + \mathbf{B}\,\mathbf{u}(k))]}\,. \qquad (10.2.52)$$

which is identical to the recursive state estimation (10.2.46), now with determined gain $\mathbf{K}(k+1)$. These filtering equations are called the *Kalman filter*.

The state estimation is a recursive estimation of the state $\hat{\mathbf{x}}(k+1|k+1)$ based on a predicted state $\hat{\mathbf{x}}(k+1|k)$ by the process model and a correction based on the new measurement $y(k+1)$ and a correction matrix which minimizes the covariance of the estimation error $\tilde{x}(k+1|k)$. A comparison with the state observer (10.2.28) shows that the observer only uses past information $\hat{\mathbf{x}}(k)$ and $\mathbf{y}(k)$ and not predicted values $\hat{x}(k+1|k)$ and the new measurement $y(k+1)$.

The correction matrix or gain $\mathbf{K}(k+1)$ depends on the covariance matrices $\mathbf{Q}$ of the state noise $\mathbf{v}(k)$ and $\mathbf{R}$ of the output noise. It can be computed in advance for stationary noise, since it does not depend on measured signals.

If the process matrices $\mathbf{A}$, $\mathbf{B}$ and $\mathbf{C}$ and the noise covariance matrices do not depend on time, the Kalman filter gain $\mathbf{K}(k+1)$ approaches asymptotically a steady-state value $\bar{\mathbf{K}}$. The steady-state estimation error covariance matrix $\mathbf{P}^{-}$ follows from the elimination of $\mathbf{P}(k)$ from (10.2.48) using (10.2.51) and (10.2.49), leading to

$$\mathbf{P}^{-}(k+1) = \mathbf{A}\,\mathbf{P}^{-}(k)\,\mathbf{A}^{T} - \mathbf{A}\,\mathbf{P}^{-}(k)\mathbf{C}^{T}[\mathbf{C}\,\mathbf{P}^{-}(k)\,\mathbf{C}^{T} + \mathbf{R}]^{-1}\,\mathbf{C}\,\mathbf{P}^{-}(k)\mathbf{A}^{T} + \mathbf{Q}, \qquad (10.2.53)$$

which is a Riccati equation. (One recognizes the duality to the optimal state control). Its asymptotic solution results in the steady- state matrix $\mathbf{P}^-$. Then, the *steady-state Kalman filter gain* becomes

$$\bar{\mathbf{K}} = \mathbf{P}^- \mathbf{C}^T [\mathbf{C}\,\mathbf{P}^-\,\mathbf{C}^T + \mathbf{R}]^{-1}. \tag{10.2.54}$$

The calculations then reduce to

*prediction*:

$$\hat{\mathbf{x}}(k+1|k) = \mathbf{A}\,\hat{\mathbf{x}}(k|k) + \mathbf{B}\,\mathbf{u}(k), \tag{10.2.55}$$

*correction*:

$$\underset{\text{new estimate}}{\hat{\mathbf{x}}(k+1|k+1)} = \underset{\text{old estimate}}{\hat{\mathbf{x}}(k+1|k)} + \underset{\substack{\text{correction}\\\text{matrix}}}{\bar{\mathbf{K}}}\ \underset{\substack{\text{new}\\\text{measurement}}}{[\mathbf{y}(k+1)} - \underset{\substack{\text{predicted}\\\text{measurement with}\\\text{process model and}\\\text{old estimate.}}}{\mathbf{C}\,\hat{\mathbf{x}}(k+1|k)]}. \tag{10.2.56}$$

To obtain a *comparison to a state observer* the previous correction in the form

$$\hat{\mathbf{x}}(k|k) = \hat{\mathbf{x}}(k|k-1) + \bar{\mathbf{K}}[\mathbf{y}(k) - \mathbf{C}\,\hat{\mathbf{x}}(k|k-1)], \tag{10.2.57}$$

is inserted in the prediction (10.2.53), leading to

$$\hat{\mathbf{x}}(k+1|k) = \mathbf{A}\,\hat{\mathbf{x}}(k|k-1) + \mathbf{B}\,\mathbf{u}(k) + \mathbf{A}\,\bar{\mathbf{K}}[\mathbf{y}(k) - \mathbf{C}\,\hat{\mathbf{x}}(k|k-1)]. \tag{10.2.58}$$

A comparison with the observer (10.2.28) shows that if the observer gain is chosen as

$$\mathbf{H} = \mathbf{A}\,\bar{\mathbf{K}}, \tag{10.2.59}$$

the observer equation corresponds to the Kalman filter.

Figure 10.6 shows a signal flow diagram of the Kalman filter by using (10.2.58) and (10.2.55). (This scheme is depicted for the case that at time instant $k$ the input signal $u(k)$ and the output signal $y(k)$ is available, as for an observer. If, however, the next output sample $y(k+1)$ is used for estimating $\hat{\mathbf{x}}(k+1|k+1)$ the correction follows (10.2.56)).

In the original work of Kalman (1960), the recursive state estimator was derived by applying the orthogonality condition between the estimation errors and the measurements

$$E\left\{\tilde{\mathbf{x}}(i)\,\mathbf{Y}^T\,(j)\right\} = \mathbf{0} \ \text{ for } \ j < i. \tag{10.2.60}$$

An alternative derivation of the Kalman filter follows using the conditional expectation of a least squares estimate

$$E\left\{\mathbf{x}(k)|\mathbf{Y}_j\right\}; \tag{10.2.61}$$

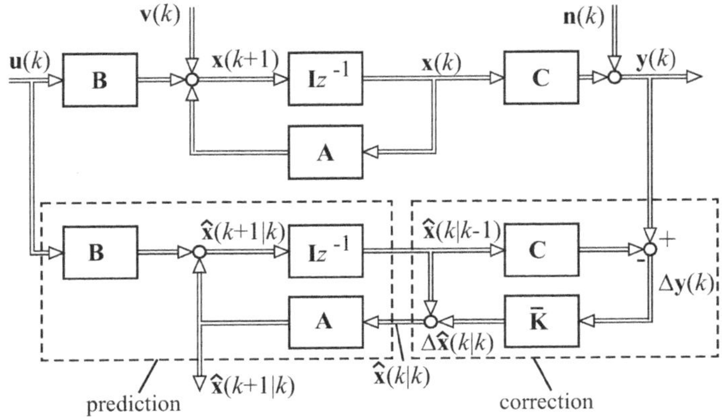


**Fig. 10.6** Signal flow diagram of a Kalman filter with the prediction according to (10.2.55) and the correction according (10.2.57)

see (10.2.35) and (10.2.36). Hence, the estimate is the vertical projection of $\mathbf{x}(k)$ on $\mathbf{Y}_j$

$$\hat{\mathbf{x}}(k|j) = E\left\{\mathbf{x}(k)|\mathbf{Y}_j\right\}, \tag{10.2.62}$$

with the estimation error

$$\tilde{\mathbf{x}}(k|j) = \mathbf{x}(k) - \hat{\mathbf{x}}(k|j);$$

see e.g. Tomizuka (1998). Other references for the Kalman filter are Brown and Hwang (1992), Gelb (1974), Åström and Wittenmark (1984), Lewis (1986), Welch and Bishop (2004), and Grewal and Andrews (2015).

Figure 10.7 summarizes the prediction and correlation equations for implementing the Kalman filter in the case of linear processes. The Kalman filter equations determine the process state vector $\hat{\mathbf{x}}(k+1)$ according to a prediction-correction principle in two steps. Based on the known process model, a one-step ahead prediction is made for the state $\hat{\mathbf{x}}(k+1|k)$ and the covariance of the state-estimation error $\mathbf{P}(k+1|k)$, based on the values at time step $k$.

The *correction step* begins by updating the Kalman gain as a weighting matrix by using the predicted state- estimation covariance matrix $\mathbf{P}(k+1|k)$. Then, an output error $\Delta y(k+1)$ is calculated as a difference of the new process output measurements $y(k+1)$ and its prediction $\mathbf{C}\hat{\mathbf{x}}(k+1|k)$. A new state estimate $\hat{\mathbf{x}}(k+1|k+1)$ is obtained by correcting the state prediction $\hat{\mathbf{x}}(k+1|k)$ with the output error $\Delta y(k+1)$ multiplied with the updated Kalman gain $\mathbf{K}(k+1)$. Then, the covariance matrix $\mathbf{P}(k+1|k+1)$ is corrected using the corrected Kalman gain $\mathbf{K}(k+1)$ and a new prediction step begins.

In principle, the Kalman filter is similar to that of a state observers. But it is especially suitable for processes with relatively large state variable and output noise. The Kalman filter was derived for the process model (10.2.26) and (10.2.27).

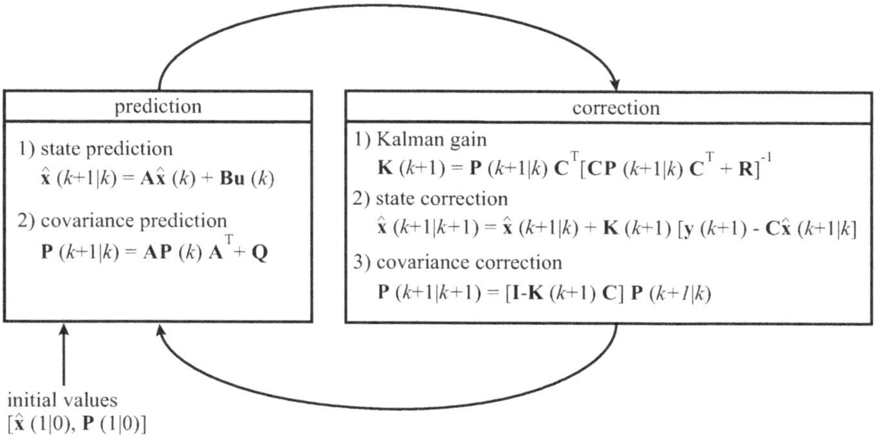


**Fig. 10.7** Kalman filter equations for the process equation (10.2.26) and (10.2.27)

## 10.2.4 Extended Kalman Filter

**(a) Nonlinear Processes**

In many applications, one is confronted with nonlinear system models of the form

$$\mathbf{x}(k+1) = \mathbf{f}_k(\mathbf{x}(k), \mathbf{u}(k)) + \mathbf{v}(k), \tag{10.2.63}$$

$$\mathbf{y}(k) = \mathbf{g}_k(\mathbf{x}(k)) + \mathbf{n}(k)\ , \tag{10.2.64}$$

where the index $k$ in $\mathbf{f}_k$ and $\mathbf{g}_k$ indicate that also the functions themselves can be time-varying.

For processes of this form, the *Extended Kalman Filter* (EKF) has been used in many applications. In a few first publications, this filter was called the Kalman-Schmidt-Filter; see Grewal and Andrews (2015). In the EKF, the update equation for the states is based on the "true" nonlinear model, whereas the update for the error covariance matrix $\mathbf{P}(k)$ is based on a first order Taylor series expansion of (10.2.63) and (10.2.62). The prediction step for the states is hence given as

$$\hat{\mathbf{x}}(k+1|k) = \mathbf{f}_k\big(\hat{\mathbf{x}}(k), \mathbf{u}(k)\big). \tag{10.2.65}$$

The update of the covariance matrix requires the calculation of the Jacobian matrices in each update step. The Jacobian matrices are given as

$$\mathbf{F}(k) = \left.\frac{\partial \mathbf{f}_k(\mathbf{x}, \mathbf{u})}{\partial \mathbf{x}}\right|_{\mathbf{x}=\hat{\mathbf{x}}(k),\ \mathbf{u}=\mathbf{u}(k)} = \begin{bmatrix} \frac{\partial f_1}{\partial x_1} & \frac{\partial f_1}{\partial x_2} & \cdots \\ \frac{\partial f_2}{\partial x_1} & \frac{\partial f_2}{\partial x_2} & \cdots \\ \vdots & \vdots & \end{bmatrix}, \tag{10.2.66}$$

$$\mathbf{G}(k+1) = \left.\frac{\partial \mathbf{g}_{k+1}(\mathbf{x})}{\partial \mathbf{x}}\right|_{\mathbf{x}=\hat{\mathbf{x}}(k+1|k)} = \begin{bmatrix} \frac{\partial g_1}{\partial x_1} & \frac{\partial g_1}{\partial x_2} & \cdots \\ \frac{\partial g_2}{\partial x_1} & \frac{\partial g_2}{\partial x_2} & \cdots \\ \vdots & \vdots & \end{bmatrix}. \tag{10.2.67}$$

Then, the update equation for $\mathbf{P}(k+1)$ and the calculation $\mathbf{K}(k+1)$ are given as

$$\mathbf{P}^{-}(k+1) = \mathbf{F}(k)\mathbf{P}(k)\mathbf{F}^{\mathrm{T}}(k) + \mathbf{Q}(k), \tag{10.2.68}$$

$$\mathbf{K}(k+1) = \mathbf{P}^{-}(k+1)\mathbf{G}(k+1)\big[\mathbf{G}(k+1)\mathbf{P}^{-}(k+1)\mathbf{G}^{\mathrm{T}}(k+1) + \mathbf{R}(k+1)\big]^{-1}, \tag{10.2.69}$$

and

$$\mathbf{P}(k+1) = \big[\mathbf{I} - \mathbf{K}(k+1)\mathbf{G}(k+1)\big]\mathbf{P}^{-}(k+1). \tag{10.2.70}$$

The state estimates are corrected using the true nonlinear relation as

$$\hat{\mathbf{x}}(k+1|k+1) = \hat{\mathbf{x}}(k+1|k) + \mathbf{K}(k+1)\big[\mathbf{y}(k+1) - \mathbf{g}_{k+1}\big(\hat{\mathbf{x}}(k+1|k)\big)\big]. \tag{10.2.71}$$

While the derivation of the EKF seems quite simple, it must be stressed at this point that the EKF does *not provide optimal estimates*. While the random variables were remaining Gaussian at all times for the Kalman filter, the distribution of the random variables will change after going through the nonlinear transformations in the EKF. Furthermore, one should be aware that the filter can quickly diverge due to the linearization around false operating points, if, for example, the initial conditions are chosen wrongly. While these points seem to be severe drawbacks of the EKF, it is still used in many applications, the most prominent being navigation systems and GPS devices.

The final extended Kalman filter calculations result in the sequence:

*prediction:*

$$\hat{\mathbf{x}}(k+1|k) = \mathbf{f}_k\big(\hat{\mathbf{x}}(k), \mathbf{u}(k)\big), \tag{10.2.72}$$

$$\mathbf{F}(k) = \left.\frac{\partial \mathbf{f}_k(\mathbf{x}, \mathbf{u})}{\partial \mathbf{x}}\right|_{\mathbf{x}=\hat{\mathbf{x}}(k),\ \mathbf{u}=\mathbf{u}(k)}, \tag{10.2.73}$$

$$\mathbf{P}^{-}(k+1) = \mathbf{F}(k)\mathbf{P}(k)\mathbf{F}^{\mathrm{T}}(k) + \mathbf{V}(k)\mathbf{M}(k)\mathbf{V}^{\mathrm{T}}(k), \tag{10.2.74}$$

*correction:*

$$\mathbf{G}(k+1) = \left.\frac{\partial \mathbf{g}_{k+1}(\mathbf{x})}{\partial \mathbf{x}}\right|_{\mathbf{x}=\hat{\mathbf{x}}(k+1|k)}, \tag{10.2.75}$$

$$\mathbf{K}(k+1) = \mathbf{P}^{-}(k+1)\mathbf{G}(k+1)\big[\mathbf{G}(k+1)\mathbf{P}^{-}(k+1)\mathbf{G}^{\mathrm{T}}(k+1) + \mathbf{R}(k+1)\big]^{-1}, \tag{10.2.76}$$

$$\hat{\mathbf{x}}(k+1|k+1) = \hat{\mathbf{x}}(k+1|k) + \mathbf{K}(k+1)\mathbf{y}(k+1) - \mathbf{g}_{k+1}\big(\hat{\mathbf{x}}(k+1|k)\big), \tag{10.2.77}$$

$$\mathbf{P}(k+1) = \big[\mathbf{I} - \mathbf{K}(k+1)\mathbf{G}(k+1)\big]\mathbf{P}^{-}(k+1). \tag{10.2.78}$$

**(b) Extended Kalman Filter for Parameter Estimation**
The extended Kalman filter can also be used for parameter estimation. Here, the state vector $\mathbf{x}(k)$ is augmented with a parameter vector $\boldsymbol{\theta}$, leading to the state-space system

$$\begin{pmatrix} \hat{\mathbf{x}}(k+1) \\ \hat{\boldsymbol{\theta}}(k+1) \end{pmatrix} = \begin{pmatrix} \mathbf{f}\big(\hat{\mathbf{x}}(k), \boldsymbol{\theta}(k), \mathbf{u}(k)\big) \\ \boldsymbol{\theta}(k) \end{pmatrix} + \begin{pmatrix} \mathbf{v}(k) \\ \boldsymbol{\xi}(k), \end{pmatrix} \tag{10.2.79}$$

$$\mathbf{y}(k) = \mathbf{g}\big(\hat{\mathbf{x}}(k)\big) + \mathbf{n}(k), \tag{10.2.80}$$

where in comparison with (10.2.63), the parameter vector $\boldsymbol{\theta}(k)$ has been introduced and obeys the dynamics

$$\boldsymbol{\theta}(k+1) = \boldsymbol{\theta}(k) + \boldsymbol{\xi}(k); \tag{10.2.81}$$

see Grewal and Andrews (2015). One can see that the parameters are modeled as constant quantities. However, the model includes a stochastic disturbance, i.e. the parameters are modeled as being disturbed by a white noise in the sense of a random walk.

### 10.2.5 Determination of Derivatives

If the required derivatives of the signals for state estimation or parameter estimation are directly measurable (e.g. as for vehicle applications), these values can be entered directly. On the contrary, if the derivatives are not measurable, the derivatives have to be determined from the sampled signals $u(t)$ and $y(t)$. For this, one basically has a choice between numerical differentiation and state variable filtering.

**(a) Numerical Differentiation**
The *numerical differentiation* in combination with interpolation approaches (e.g. splines) is usually not able to suppress noise due to disturbance signals for higher derivatives which limits this technique to the application of second or third order derivatives as a maximum. Often however, only first order derivatives can be determined reliably.

One can use in general the following methods of determining the derivatives:

$$\text{Forward Differential Quotient: } \hat{\dot{x}}(k) = \frac{x(k+1) - x(k)}{T_0}, \tag{10.2.82}$$

$$\text{Backward Differential Quotient: } \hat{\dot{x}}(k) = \frac{x(k) - x(k-1)}{T_0}, \tag{10.2.83}$$

$$\text{Central Differential Quotient: } \hat{\dot{x}}(k) = \frac{x(k+1) - x(k-1)}{2T_0}. \tag{10.2.84}$$

All of them have distinct advantages and disadvantages: While the forward and the central differential quotient depend on future values, the backward differential quotient introduces a delay of half a sample step.

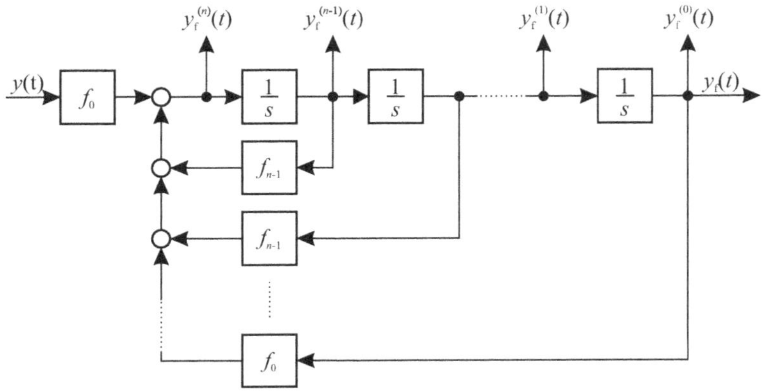


**Fig. 10.8** State variable filter for filtering and determining the derivatives of a signal simultaneously

**Table 10.1** Polynomials $B(s)$ of the normalized Butterworth filter

| Order $n$ | Polynomial $B(s)$ of the Butterworth Filter |
|---|---|
| $n = 1$ | $s + 1$ |
| $n = 2$ | $s^2 + 1.4142s + 1$ |
| $n = 3$ | $s^3 + 2s^2 + 2s + 1$ |
| $n = 4$ | $s^4 + 2.6131s^3 + 3.4142s^2 + 2.6131s + 1$ |
| $n = 5$ | $s^5 + 3.2361s^4 + 5.2361s^3 + 5.2361s^2 + 3.2361s + 1$ |
| $n = 6$ | $s^6 + 3.8637s^5 + 7.4641s^4 + 9.1416s^3 + 7.4641s^2 + 3.8637s + 1$ |

### (b) State Variable Filters

The principle of state variable filters is to use a low-pass filter that dampens out the higher frequent noise and transform it into a state-space representation, such that the states are the derivatives of the filter output, i.e. the filtered signal. They are designed to suppress any noise above the cut-off frequency, i.e. for $f > f_c$.

State variable filters (SVF), see Fig. 10.8, with a transfer function

$$G_F(s) = \frac{y_F(s)}{y(s)} = \frac{K}{B(s)} = \frac{f_0}{f_0 + f_1 s + \ldots + f_{n-1}s^{n-1} + s^n}, \tag{10.2.85}$$

have proven to yield good results in the identification of continuous-time systems. A state variable filter is an analog filter with a certain topology that is subsequently discretized for the realization on a digital computer. The input signal $u(t)$ and the output signal $y(t)$ of a process must both be filtered with the same state variable filter. The choice of the filter parameters $f_i$ is relatively free. For example, the parameters of a Butterworth filter can be used; see Table 10.1. For the discretization, different approximation techniques can be employed. For more details see Isermann and Münchhof (2011).

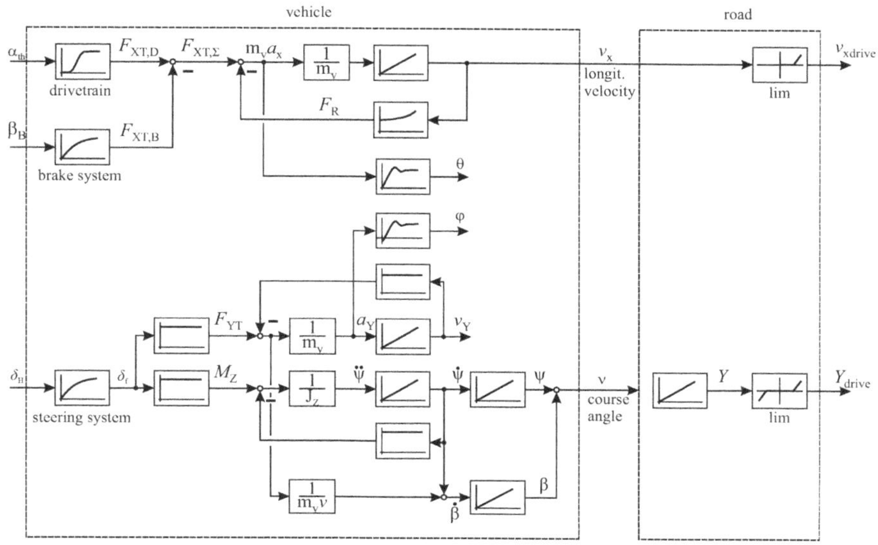


**Fig. 10.9** Main inputs and outputs of a driving vehicle (simplified vehicle models).

## 10.3 Driving Maneuvers

Driving maneuvers are used to test the stationary and dynamic behavior of the real vehicle. The main inputs from the driver are the positions of the gas pedal $\alpha_{th}$, brake pedal $\beta_B$, and steering angle $\delta_H$ and the main outputs are the longitudinal velocity $v_X$ and course angle $\nu$. The inputs also influence the roll angle $\varphi$ and the pitch angle $\theta$; see Fig. 10.9. The time-dependent input test signals have to excite the stationary and dynamic behavior of the vehicle represented by the outputs and intermediate variables. In order to change the inputs for the stationary behavior, their size (amplitude) has to be adapted to the gains of the vehicle (accelerator gain, braking gain, and steering gain). With regard to the excitation of dynamics, the size of the inputs, as well as the excited frequencies, have to be properly selected and their form has to be suitably designed with regard to the type of dynamic transfer function, e.g. a proportional, integral or double integral behavior. Therefore, different types of test maneuvers are applied.

Standardized driving maneuvers have been worked out in technical committees and are available for example as follows:

| | |
|---|---|
| DIN/OSI 4138: | Steady-state circular test procedure ($R < 80\,\text{m}$; $v$ changed in steps. Measurements of $\delta_H$, $a_Y$, $\dot{\psi}$, $\beta$, $\varphi$, $M_H$ |
| DIN (OSI 7401): | Lateral transient response test methods (steering angle: step, sinusoidal; time response, frequency response) |
| DIN/OSI 7975: | Braking in a turn ($R = 30\ldots50\,\text{m}$; $v = 44$ to $57\,\text{km/h}$; $a_y = 5\text{m/s}^2$, $R = 100\,\text{m}$; $v = 72\,\text{km/h}$; $a_y = 4\,\text{m/s}^2$. Determination of $\dot{\psi}(t)$, $a_Y(t)$, $\beta(t)$, $a_X(t)$) |

DIN/OSI 3888: Double lane change. Dimensions of the test track are standardized. $v_{\text{entry}} = 80\,\text{km/h}$; $\alpha_{\text{th}} \approx$ const. or variable. Maximum of $v_{\text{entry}}$ can be determined.

One distinguishes *open-loop tests*, where the input signals are applied to the vehicle either by the driver or automated by a robot and *closed-loop tests*, where the driver or an autopilot after applying a test input has to control certain driving states in a feedback loop, like cornering with constant radius or driving with constant velocity, or driving through a way betweens cones.

Some typical driving maneuvers which are used for testing the lateral dynamics are depicted in Fig. 10.10; see e.g. Ersoy and Gies (2017). Because of the limited speed to change the steering wheel angle, the inputs are actually ramp functions with a speed of about 3.5 to 10 rad/s. Step inputs can be realized with the accelerator pedal or braking pedal within 300 ms.

Figure 10.10a shows a *step input* (idealized) $\delta_{\text{H}}$ of the steering wheel to generate a transient function of the yaw rate $\dot{\psi}$ with constant speed $v_{\text{X}}$. The yaw angle then increases (for slow speeds) linearly and the vehicle corners continuously on a circle with radius $R$.

The *lane change*, Fig. 10.10b, is a maneuver which represents an evasion maneuver with constant velocity before an object. To support the driver and because of repeatability the driveable track is limited by cones. The maneuver is driven for certain constant velocity $v_{\text{X}}$ and begins with a ramp of amplitude $\delta_{\text{H}0}$, e.g. to the left and after holding time a ramp of double amplitude $2\delta_0$ to the right and then back with a ramp size $\delta_0$ to the left. This signal form is required because of the integrating behavior of the vehicle with regard to the yaw angle $\psi$ and the lateral position $y_{\text{E}}$ and to reach a straight line parallel to the initial driving line.

A *double lane change* like in Fig. 10.10c represents an overtaking or evasive maneuver to the left and then to the right. It is a severe test drive for a transient road-holding ability in closed-loop control by the driver. The initial entry speed is $v = 80\,\text{km/h}$. The throttle position should be a steady as possible or can be any.

A *slalom maneuver* excites the lateral dynamics for a larger time period. Slalom maneuvers can be realized with constant amplitude, with increasing amplitude, Fig. 10.10d, or with constant amplitude and increasing frequency, Fig. 10.10e.

A further maneuver is a *fast manual step input* $\delta_{\text{H}0}$ of the steering wheel and then taking off the hands. This allows to observe the dynamic eigenbehavior and returning behavior of the steering system with the vehicle dynamics. The result is usually a more or less damped oscillation to a steady state, usually in neutral position, see e.g. Mitschke and Wallentowitz (2014), shown for such behavior during cornering.

The analytical and subjective *evaluation* of the lateral behavior can be based on the transient response of the yaw rate after a fast ramp input of the steering wheel angle. According to the second order behavior, the transient function shows a delayed reaction with an overshoot; see also the simulations in Sects. 7.2 and 7.3

$$\Delta\dot{\psi}_{\text{OS}} = \dot{\psi}_{\text{max}} - \dot{\psi}_{\text{stat}}; \tag{10.3.1}$$

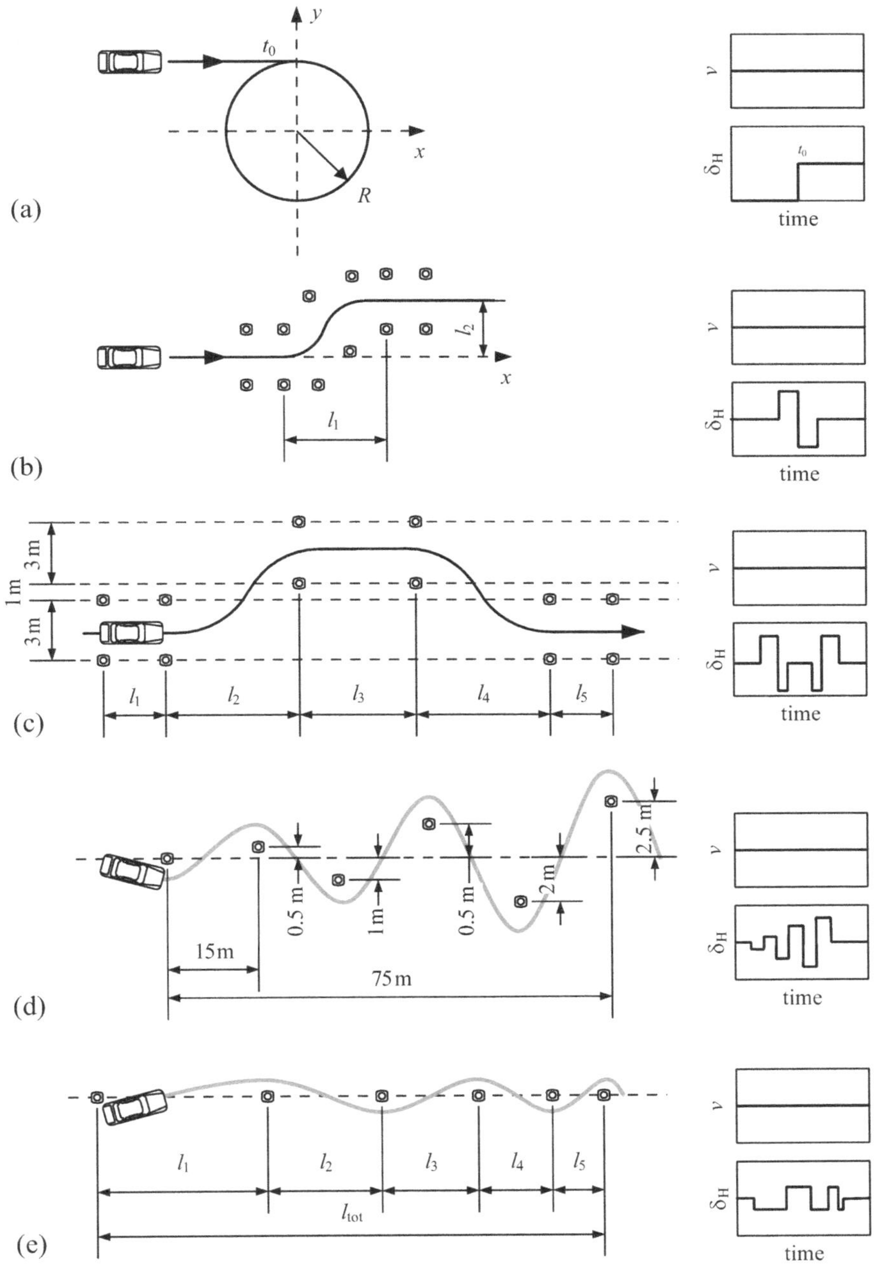


**Fig. 10.10** Some driving maneuvers for testing the lateral behavior, Börner (2004). **a** Steering wheel step. **b** Lane change. **c** Double lane change (ISO 3888: $l_1 = 25\,\text{m}$, $l_2 = 30\,\text{m}$, $l_3 = 25\,\text{m}$, $l_4 = 25\,\text{m}$, $l_5 = 15\,\text{m}$; $b_1$, $b_2$, $b_3$: dependent on vehicle width). **d** slalom maneuver with increasing amplitude. **e** slalom maneuver with increasing frequency

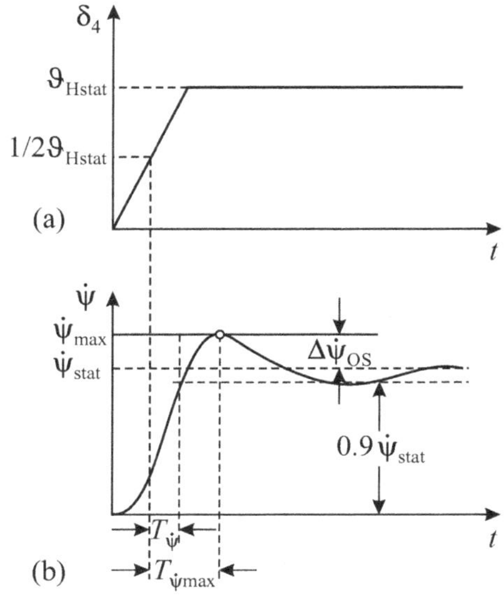


**Fig. 10.11** Some characteristic values of a transient function of the yaw rate after a ramp input of the steering wheel angle, ISO 7401, Mitschke and Wallentowitz (2014)

see Fig. 10.11. Some further characteristic values, defined in ISO 7401 and Mitschke and Wallentowitz (2014) are:

$$
\begin{aligned}
K_{st,\dot{\psi}} &= \dot{\Psi}_{stat}/\delta_{Hstat} \text{ yaw rate steering gain,}\\
T_{\dot{\psi}} &= \text{peak response time, delay until reaching a first } \dot{\psi}_{max},\\
T_{\dot{\psi}max} &= \text{peak response time, delay until reaching a first } \dot{\psi}_{max}. \qquad (10.3.2)
\end{aligned}
$$

Typical values for passenger cars are, Mitschke and Wallentowitz (2014):

$$
\begin{aligned}
K_{st,\dot{\psi}} &= 0.2\ldots0.3 \text{ 1/s},\\
T_{\dot{\psi}max} &= 0.33\ldots0.5 \text{ s},\\
\Delta\dot{\psi}_{OS} &= 0.13\ldots0.27.
\end{aligned}
$$

A discussion on the subjective evaluation of the lateral steering behavior is given in Mitschke and Wallentowitz (2014) and Ersoy and Gies (2017).

The *frequency content* of input test signals is analyzed; e.g. in Isermann and Münchhof (2011). The amplitude densities of non-periodic test signals as shown in Fig. 10.12 is obtained by a Fourier transform

$$
\mathcal{F}\{u(t)\} = u(i\omega) = \int_{-\infty}^{\infty} u(t)e^{-i\omega t}dt. \qquad (10.3.3)
$$

Figure 10.13 depicts the amplitude densities $|u^*(i\omega)| = |u(i\omega)|/u_0 T$ for various non-periodic test signals. A step function excites mainly the low frequencies. Rectangular and trapezoidal test signals show higher amplitude densities at middle frequencies, but smaller ones at lower frequencies and zeros at higher frequencies.

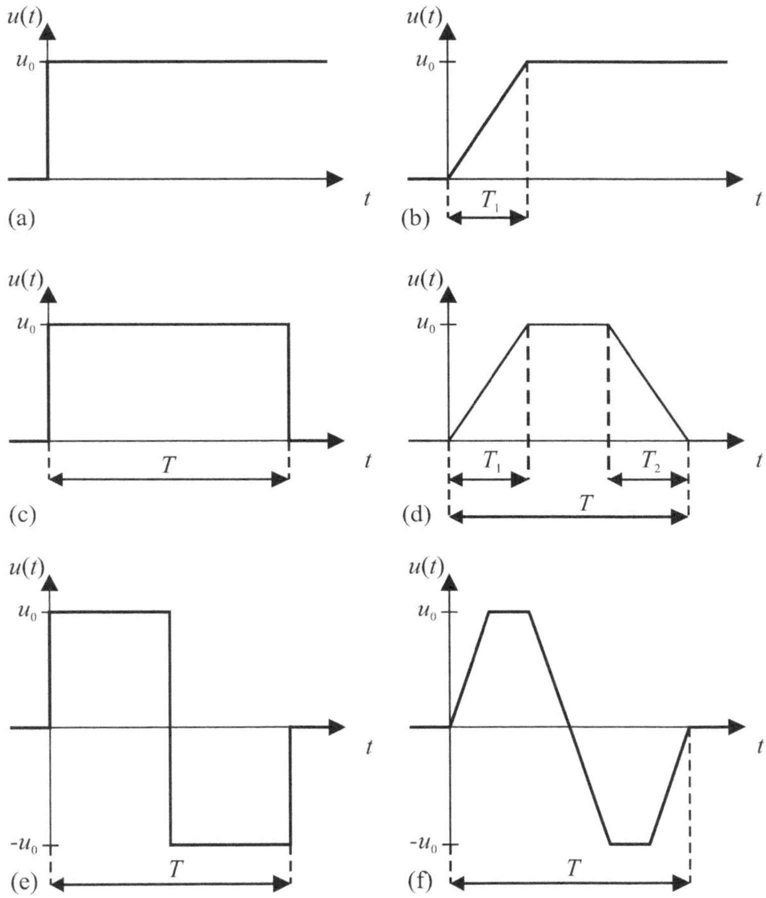


**Fig. 10.12** Simple non-periodic test signals. **a** Step function. **b** Ramp function. **c** Rectangular pulse. **d** Trapezoidal pulse. **e** Double-rectangular pulse. **f** Double trapezoidal pulse

These non-periodic test signals allow to determine a frequency response by the ratio of Fourier transforms of the input and output signal

$$G(i\omega) = \frac{y(i\omega)}{u(i\omega)} = \frac{\int_0^\infty y(t)e^{-i\omega t}dt}{\int_0^\infty u(t)e^{-i\omega t}dt}. \tag{10.3.4}$$

A direct way to obtain a frequency response, which is defined for linear processes, is to determine for *sinusoidal inputs* and *outputs* in an oscillating stationary state the amplitude ratio and phase angle for different frequencies $\omega_\nu$

$$\begin{aligned} G(i\omega_\nu) &= \frac{y_0(\omega_\nu)}{u_0\omega_\nu} \\ \varphi(i\omega_\nu) &= -\omega_\nu t_\varphi. \end{aligned} \tag{10.3.5}$$

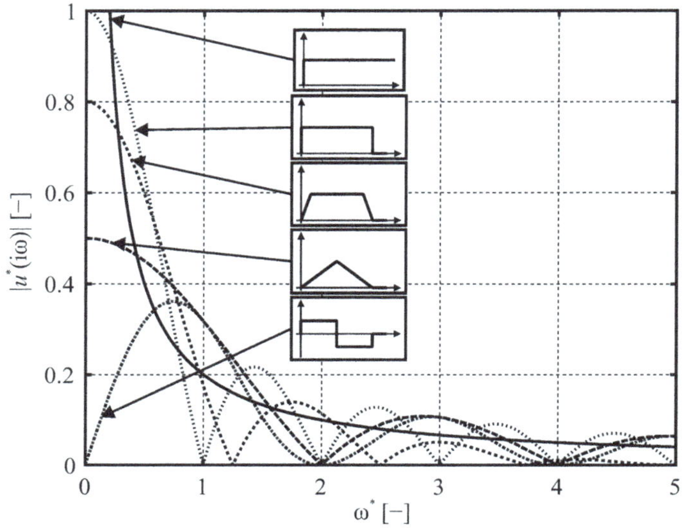


**Fig. 10.13** Referred amplitude density of various non-periodic test signals, Isermann and Münchhof (2011)

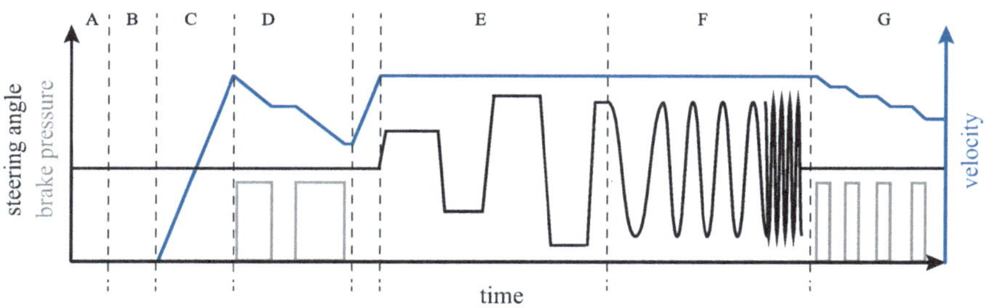


**Fig. 10.14** Identification driving maneuver with a following-up of different test signals and phases for determination of basic vehicle models. **A** Weighing; **B** stand still; **C** slow acceleration; **D** braking maneuver; **E** stationary steering maneuvers; **F** dynamic steering maneuvers (slalom); **G** dynamic braking maneuvers

The evaluation of the measured signals can be made directly from the measured signals or by orthogonal correlation. However, for driving automobiles the required measurement time may be too long.

An alternative is to use *sweep sine test signals*, also called chirp signals, with a continuously varying frequency; e.g.

$$y(t) = y_0 \sin(2\pi f(t)t), \tag{10.3.6}$$

with a linear frequency transition

$$f(t) = at + b.$$

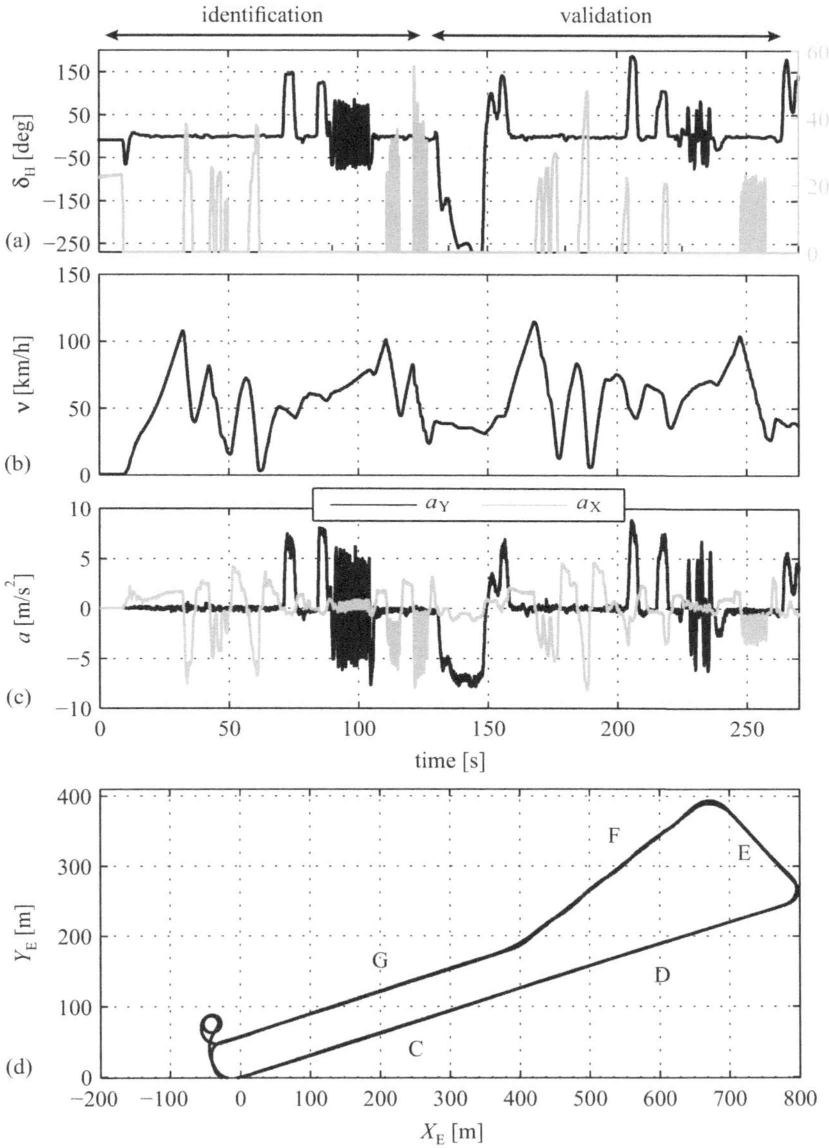


**Fig. 10.15** Measured signals during an identification driving maneuver of Fig. 10.13, Bauer (2015)

Figure 10.14F shows an example. This test signal is more suitable for automobiles. For more details on the identification of process dynamics; see Isermann and Münchhof (2011).

In order to identify different models of vehicles various test signals can be applied sequentially, as shown in Fig. 10.14 and Bauer (2015). Phase A is for preparing the

**Fig. 10.16** A passenger car driving a lane change

maneuvers and determination of the mass and coordinates of the center of gravity. Then the sensor and the data acquisition are checked in phase B. Phase C is a slow acceleration with about $\dot{v}_x \leq 1\,\text{m/s}^2$ to determine the dynamic roll radii of the wheels. Then moderate braking maneuvers are applied in phase D to determine dynamic longitudinal models. After reaching a certain velocity circular drives are performed in phase E with different radii, also with different velocities. In phase F slalom maneuvers with (approximately) sinusoidal steering inputs and changing frequency are driven to excite the lateral and roll dynamics. The excitation of the longitudinal dynamics is performed in phase G with strong braking pulses, also with different starting velocities.

Figure 10.15 depicts as an example the resulting measured signals and a graph of the test field, see Fig. 10.16, with a second time period for validation measurements.

In the following chapters on parameter and state estimation, the described driving test maneuvers are applied to different vehicles and obtained results are shown.

## References

Adamy J (2018) Nichtlineare Systeme und Regelungen, 3rd edn. Springer, Wiesbaden

AGARD (1970) Theory and applications of Kalman filtering, vol AGARDOgraph 139. Zentralstelle für Luftfahrtdokumentation und ESRO/ELDO, München and Neuilly-Seine

Åström K, Wittenmark B (1984) Computer controlled systems - theory and design. Prentice-Hall, Englewood Cliffs

Bauer M (2015) Methoden zur modellbasierten Fahrdynamikanalyse und Bewertung von Fahrdynamikregelsystemen. Dissertation Technische Universität Darmstadt. Fortschr.-Ber. VDI Reihe 12, 792. VDI Verlag, Düsseldorf

Börner M (2004) Adaptive Querdynamikmodelle für Personenkraftfahrzeuge - Fahrzustandserkennung und Sensorfehlertoleranz. Dissertation Technische Universität Darmstadt. Fortschr.-Ber. VDI Reihe 12, Nr. 563. VDI Verlag, Düsseldorf

Brown R, Hwang Y (1992) Introduction to random signals and applied Kalman filtering, 2nd edn. Wiley, Hoboken

Chen J, Patton R (1999) Robust model-based fault diagnosis for dynamic systems. Kluwer, Boston

Daiss A (1996) Beobachtung fahrdynamischer Zustände und Verbesserung einer ABS-und Fahrdynamikregelung. Diss. Universität Karlsruhe, Fortschr.-Ber. VDI Reihe 12, 283. VDI Verlag, Düsseldorf

Ersoy M, Gies S (eds) (2017) Fahrwerkhandbuch: Grundlagen, Fahrdynamik, Komponenten, Elektronische Systeme, Fahrerassistenz, Autonomes Fahren, Perspektiven, 5th edn. ATZ/MTZ Fachbuch, Springer Vieweg

Föllinger O (1994) Regelungstechnik, 8th edn. Hüthig Buch Verlag, Heidelberg

Gelb A (ed) (1974) Applied optimal estimation. MIT Press, Cambridge

Grewal M, Andrews A (2015) Kalman filtering: theory and practice with MATLAB, 4th edn. Wiley, Hoboken, New Jersey

Isermann R (1991) Digital control systems, vol 2, 2nd edn. Springer, Berlin

Isermann R (2006) Fault-diagnosis systems - an introduction from fault detection to fault tolerance. Springer, Heidelberg

Isermann R, Münchhof M (2011) Identification of dynamic systems. Springer, Berlin, Heidelberg

Kalman R (1960) A new approach to linear filtering and prediction problems. Trans ASME, Series D, J Basic Eng 82(March):35–45

Kiencke U, Nielsen L (2005) Automotive control systems: for engine, driveline, and vehicle, 2nd edn. Springer, Berlin

Lewis R (1986) Optimal estimation with an introduction to stochastic control theory. Wiley, Hoboken

Mitschke M, Wallentowitz H (2014) Dynamik der Kraftfahrzeuge, 5th edn. Springer, Berlin

Schorn M (2007) Quer- und Längsregelung eines Personenkraftwagens für ein Fahrerassistenzsystem zur Unfallvermeidung. Diss. Universität Darmstadt, Fortschr.-Ber. VDI Reihe 12, 651. VDI Verlag, Düsseldorf

Tomizuka M (1998) Advanced control systems II, Class Notes for ME233. University of California at Berkeley, Dept of Mechanical Engineering

Welch G, Bishop G (2004) Introduction to Kalman filter, update monday, april, 5th edn. University of North Carolina, Chapel Hill, Department of Computer Science

# Parameter Estimation (Identification) of Vehicle Dynamics

# 11

The application of parameter-estimation methods to determine the unknown vehicle model parameters is treated for a selection of dynamic and static vehicle or vehicle component models. Herewith, the methods of Sect. 10.1 are used, either in continuous or discrete time, in non-recursive or recursive form. In general, one tries to use series on-board sensors only depending on the standard equipment of the vehicle, (Robert Bosch GmbH 2018). However, in some cases also additional sensors are applied which are used for research vehicles. Compare the discussion of available sensors and measurement systems discussed in the introduction to Chap. 10.

## 11.1 Vehicle Mass and Resistance Parameters

The most direct method to determine a (constant) vehicle mass $m$ is to weight the vehicle as a whole or with a wheel load scale for the front and rear axle, which leads to

$$m = (G_{\mathrm{f}} + G_{\mathrm{r}})/g = m_{\mathrm{f}} + m_{\mathrm{r}}. \tag{11.1.1}$$

Another possibility is to use the travel ways of the wheel suspensions if measurements are available. Using a linear suspension model, according to Sect. 8.2.1, it holds for the *steady state* of one suspension on an even ground, carrying the corresponding body mass $m_{\mathrm{Bij}}$

$$F_{\mathrm{Bij}} = m_{\mathrm{Bij}} g = \bar{F}_{\mathrm{Bij}} = c_{\mathrm{Bij}} z_{\mathrm{WBij}}, \tag{11.1.2}$$

where

$$\bar{z}_{\mathrm{WB}} = \bar{z}_{\mathrm{W}} - \bar{z}_{\mathrm{B}}, \tag{11.1.3}$$



R. Isermann, *Automotive Control*, ATZ/MTZ-Fachbuch,
https://doi.org/10.1007/978-3-642-39440-9_11

is the deflection of the spring. For the body mass then follows

$$m_{\mathrm{B}} = \sum m_{\mathrm{Bij}} = \sum c_{\mathrm{Bij}} z_{\mathrm{WBij}}, \tag{11.1.4}$$

with $i = f, r$ and $j = r, l$.

To obtain the total vehicle mass, the mass of the wheels have to be added. A better accuracy may be obtained by using nonlinear suspension models considered in Sect. 8.2.2, where dry friction is also taken into account, creating hysteresis effects.

The vehicle mass can also be estimated from the *longitudinal acceleration* $a_{\mathrm{X}}$. The force balance for the longitudinal behavior for an engine torque input are, on an inclined road, yields; see (6.5.1)–(6.5.10) and (6.6.1), (6.6.2)

$$m a_{\mathrm{X}}(t) = F_{\mathrm{XT}}(t) - F_{\mathrm{X,R}}(t) - F_{\mathrm{X,A}}(t) - F_{\mathrm{X,C}}(t), \tag{11.1.5}$$

$$F_{\mathrm{XT}}(t) = \sum_i F_{\mathrm{XT,i}}(t) = \frac{i_{\mathrm{tot}} \eta_{\mathrm{tot}}}{r_{\mathrm{dyn}}} M_{\mathrm{eng}}(t). \tag{11.1.6}$$

Inserting the equations for the rolling resistance $F_{\mathrm{X,R}}$, the aero dynamic drag $F_{\mathrm{X,A}}$ and the climbing resistance $F_{\mathrm{X,C}} = mg\lambda$ leads to the *nonlinear acceleration model*

$$m a_{\mathrm{X}} = -\left(f_{\mathrm{R0}} + f_{\mathrm{R1}} v + f_{\mathrm{R4}} v^4\right) mg - \frac{c_{\mathrm{air}}}{2} v^2 - mg\lambda + f_{\mathrm{eng}} M_{\mathrm{eng}}, \tag{11.1.7}$$

$$f_{\mathrm{eng}} = \frac{i_{\mathrm{tot}} \eta_{\mathrm{eng}}}{r_{\mathrm{dyn}}}; \quad c_{\mathrm{air}} = c_{\mathrm{WX}} A_{\mathrm{V}} \rho_{\mathrm{air}},$$

or in vector form

$$\mathbf{y} = \boldsymbol{\psi}^{\mathrm{T}} \boldsymbol{\theta}, \tag{11.1.8}$$

with

$$\begin{aligned} \mathbf{y} &= a_{\mathrm{X}}, \\ \boldsymbol{\psi}^{\mathrm{T}} &= \left[ -1 \; -v \; -v^2 \; -v^4 \; -1 \; M_{\mathrm{eng}} \right], \\ \boldsymbol{\theta} &= \left[ f_{\mathrm{R0}} g \; f_{\mathrm{R1}} g \; \tfrac{c_{\mathrm{air}}}{2m} \; f_{\mathrm{R4}} g \; \lambda g \; \tfrac{f_{\mathrm{eng}}}{m} \right]. \end{aligned} \tag{11.1.9}$$

The estimation of the parameter vector $\boldsymbol{\theta}$ can be performed with the method of least squares (LS) for continuous time, see Sect. 10.1.2, after determination of the inputs $a_{\mathrm{X}}(t)$, $v(t)$, and $M_{\mathrm{eng}}(t)$ over a certain time period with discrete time samples. The acceleration $a_{\mathrm{X}}(t)$ in the x-direction of the vehicle can be determined from the measured speeds of the non-driven wheels $v_{\mathrm{X}} = r_{\mathrm{dyn}} \omega_{\mathrm{W}}(t)$ and generation of the first derivative by using a state variable filter; see Sect. 10.2.5. The method can be applied off-line or online.

Assuming that the air drag constant $c_{\mathrm{air}}$ and $f_{\mathrm{eng}}$ are known, the unknown mass $m$ and the rolling resistance parameters can be determined theoretically from the estimated parameters in $\boldsymbol{\theta}$ for an acceleration of the vehicle. In the case of braking, (6.7.1) and (6.7.11) can be used with

$$F_{\mathrm{XT}}(t) = -\sum_i F_{\mathrm{XT,B,i}}(t) = -c_\beta \Delta\beta(t). \tag{11.1.10}$$

Then $c_\beta$ has to be known for mass estimation. A similar mass estimation method was applied in Daiss (1996) for the case that the engine torque is known (small acceleration and closed lock-up clutch). The vehicle mass could be estimated after averaging of the estimates over 6 s driving with an error of 1%.

The vehicle mass $m$ can also be directly calculated with (11.1.5) if the driving forces are separately calculated as shown in Kiencke and Nielsen (2005).

The application of the described parameter-estimation method, however, by using directly the calculated longitudinal tire forces $F_{\mathrm{XT}}(t)$ was described in Halfmann and Holzmann (2003) for accelerations between 70 and 120 km/h. The vehicle mass could be estimated with an accuracy of about 1%.

The mass estimation method with the nonlinear acceleration model can be simplified for the case of a *decelerating free rolling vehicle* on an even road with an disengaged engine, $F_{\mathrm{XT}} = 0$, Halfmann and Holzmann (2003). Then (11.1.9) simplifies to

$$\begin{aligned} \mathbf{u}^{\mathrm{T}} &= \left[ -1 \;\; v \;\; -v^2 \;\; -v^4 \right], \\ \boldsymbol{\theta}^{\mathrm{T}} &= \left[ f_{\mathrm{R}0} g \;\; f_{\mathrm{R}1} g \;\; \tfrac{c_{\mathrm{air}}}{2m} \;\; f_{\mathrm{R}4} g \right]. \end{aligned} \tag{11.1.11}$$

Figure 11.1a shows the velocity of a free rolling vehicle after disengagement of the engine at 190 km/h. Using the measured and state variable filtered acceleration $a_{\mathrm{X}}(t)$, an off-line estimation of the parameters $\boldsymbol{\theta}$ with the method of least squares led to the following parameters

$$\hat{m} = 1659\,\mathrm{kg},$$

$$\hat{f}_{\mathrm{R0}} = 0.691 \cdot 10^{-2}; \quad f_{\mathrm{R1}} = 0.457 \cdot 10^{-2} \frac{s}{m},$$

$$\hat{f}_{\mathrm{R4}} = 0.1 \cdot 10^{-4} \frac{s^4}{m^4}.$$

The real mass was 1623 kg and the estimation error about 2%.

Figure 11.1.1b compares the measured and the vehicle model (11.1.11) simulated acceleration with a plausible good agreement.

If the mass $m$ is known, then the air drag coefficient $c_{\mathrm{air}}$ can be estimated.

If only *small changes* $\Delta v$ of the vehicle velocity are considered for an even road, the linear first order differential equation (6.6.6) of the vehicle with a stiff powertrain can be used. Writing this equation in the form of

$$\Delta v(t) = -T_{\mathrm{acc}} \dot{v}(t) + K_{\mathrm{acc}} \Delta\alpha(t), \tag{11.1.12}$$

one obtains the notation

$$y(t) = \boldsymbol{\Psi}^{\mathrm{T}}(t)\boldsymbol{\theta}, \tag{11.1.13}$$

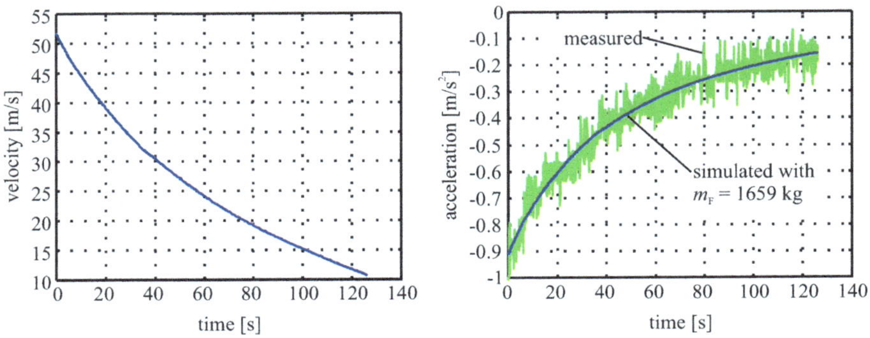


**Fig. 11.1** **a** Measured velocity for a free rolling vehicle. **b** Measured and simulated acceleration with the estimated parameters. (Opel Vectra V6, 125 kW, $c_W = 0.29$, 1993)

with

$$\psi^T = [-\dot{v}(t) \quad \Delta\alpha(t)],$$
$$\theta^T = [T_{acc} \quad K_{acc}.].$$

The input and output signals are measured for the discrete times $t = kT_0$ with $k = 0, 1, 2, \ldots$ and $T_0$ the sampling time. $\dot{v}(t)$ can be determined from $\Delta v(t)$ with a state variable filter. The application of the LS parameter-estimation method of Sect. 10.1.2 then allows to determine $\hat{T}_{acc}$ and $\hat{K}_{acc}$. The vehicle mass (including rotational masses (6.5.23) can then be calculated with

$$\hat{m}_{tot} = \frac{\hat{T}_{acc}}{\hat{K}_{acc}} \frac{i_{tot}\eta_{tot}}{r_{dyn}} c_{eng,\alpha}. \tag{11.1.14}$$

The engine constant $c_{eng,\alpha} = \partial M_{eng}/\partial\alpha$ is a function of the operating point $\bar{M}_{eng}$ and $\bar{n}_{eng}$ and follows from the lookup table of the engine torque; see Sect. 6.2.

In the case of braking with $\Delta\beta(t)$ and small changes of $v(t)$, (6.7.13) can be used and similarly the total vehicle mass can be determined after parameter estimation of $\hat{T}_{brake}$ and $\hat{K}_{brake}$ by

$$\hat{m}_{tot} = \frac{\hat{T}_{acc}}{\hat{K}_{acc}} \left(-c_{\beta F}\right). \tag{11.1.15}$$

The brake constant $c_{\beta F}$ follows from (6.7.11) and has to be known.

A further simplification is obtained if only the *initial acceleration* after, e.g. a step input of $\Delta\alpha(t = 0)$ is used. Then (6.6.7) leads to

$$\left.\frac{dv}{dt}\right|_{t=0} = \frac{K_{acc}}{T_{acc}} \Delta\alpha(0) = \frac{i_{tot}\eta_{tot}}{r_{dyn}} c_{eng,\alpha} \frac{1}{\hat{m}_{tot}} \Delta\alpha(0),$$

and

$$\hat{m}_{tot} = \frac{i_{tot}\eta_{tot}}{r_{dyn}} \Delta\alpha(0) \left.\frac{dv}{dt}\right|_{t=0}. \tag{11.1.16}$$

These methods can be applied off-line or online.

Another possibility for an *indirect estimation* of the vehicle mass is to compare the measured acceleration for a simulation with a basic mass $m_{\text{base}}$ and the real vehicle with another mass $m_{\text{load}}$, Halfmann and Holzmann (2003). Then is holds for an even road

$$\begin{aligned} a_{\text{X,base}} &= \frac{1}{m_{\text{base}}} \left(F_{\text{XT}} - F_{\text{X,R}} - F_{\text{X,A}}\right)_{\text{sim}}, \\ a_{\text{X,load}} &= \frac{1}{m_{\text{load}}} \left(F_{\text{XT}} - F_{\text{X,R}} - F_{\text{X,A}}\right)_{\text{meas}}. \end{aligned} \tag{11.1.17}$$

For small differences of the accelerations follows

$$\frac{a_{\text{X,base}}}{a_{\text{X,load}}} = \frac{m_{\text{load}} \left[\Delta F_{\text{XT}}(t) - \Delta F_{\text{X,R}}(t) - \Delta F_{\text{X,A}}(t)\right]_{\text{sim}}}{m_{\text{base}} \left[\Delta F_{\text{XT}}(t) - \Delta F_{\text{X,R}}(t) - \Delta F_{\text{X,A}}(t)\right]_{\text{meas}}} \approx \frac{m_{\text{load}}}{m_{\text{base}}} \cdot c(t). \tag{11.1.18}$$

For similar driving maneuvers with small to medium accelerations, it holds $c(t) \approx 1$ and it follows

$$\frac{a_{\text{X,base}}}{a_{\text{X,load}}} \approx \frac{m_{\text{load}}}{m_{\text{base}}} = \frac{m_{\text{base}} + \Delta m}{m_{\text{base}}} = K_{\text{a}}(\Delta m). \tag{11.1.19}$$

If $K_{\text{a}}(\Delta m)$ can be estimated with appropriate driving maneuvers, it is

$$\Delta m = \left(\hat{K}_{\text{a}} - 1\right) m_{\text{base}}. \tag{11.1.20}$$

The simulated and the measured accelerations show approximately a linear relationship with the slope $K_{\text{a}}$. However, this method, which uses only acceleration measurements, needs a longitudinal dynamic vehicle simulation. In Halfmann and Holzmann (2003), it was shown that, after a selection of suitable driving maneuvers and filtered acceleration measurements, the vehicle mass could be determined with a maximal error of 60 kg or 4%.

A general property of the considered methods for the determination of the vehicle mass by measuring the longitudinal acceleration and the velocity is that they require a selection of suitable driving maneuvers, where the respective assumptions are met.

## 11.2 Center of Gravity Coordinates

As the dynamic forces on the vehicle are assumed to act in the center of gravity (CG), the coordinates $h_{\text{CG}}$, $l_{\text{f}}$ and $l_{\text{r}}$ have to be determined for the individual vehicle. A direct way it to weigh the tire load forces at the contact points by scales. Firstly, the wheel respective loads $F_{\text{WZ,r}} = m_{\text{r}} g$ are weighted; see Fig. 11.2a. A torque balance around the front axle then leads to

$$l_{\text{f}} = \frac{m_{\text{r}}}{m_{\text{r}} + m_{\text{f}}} l \text{ and } l_{\text{f}} = l - l_{\text{f}}. \tag{11.2.1}$$

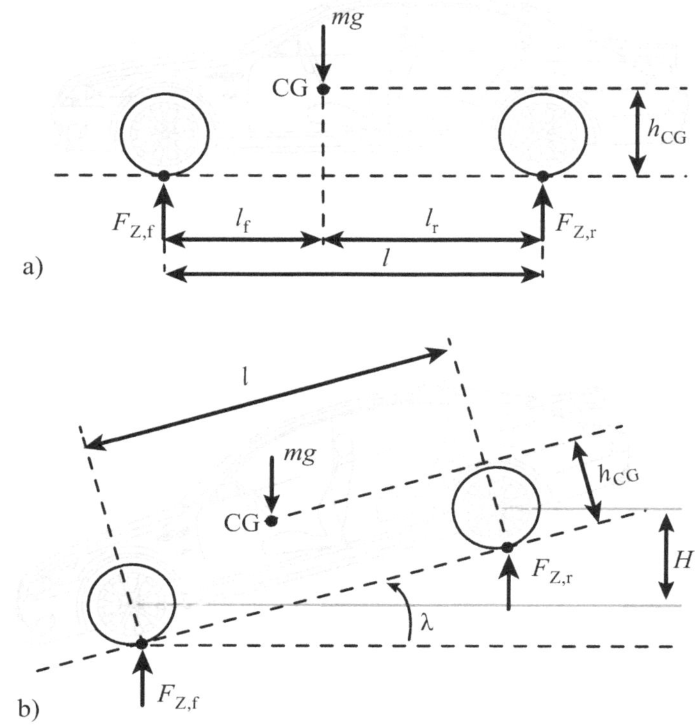


**Fig. 11.2** Determination of the center of gravity coordinates through weighting of the tire connect forces **a** even position. **b** Inclined position Determination of the center of gravity coordinates through weighting of the tire connect forces **a** even position. **b** Inclined position

To determine the height $h_{CG}$ of the CG the vehicle can be lifted at the rear axle or brought to an inclined ground with gradient $\lambda$; see Fig. 11.2b. If the vertical forces of the rear axle $F_{Z,r} = m_r g$ can be measured, a torque balance on the front axle results in

$$l_f \text{mg} \cos\lambda - l m_r g \cos\lambda - h_{CG}\ \text{mg} \sin\lambda = 0,$$

and the height of CG becomes

$$h_{CG} = \frac{1}{\tan\lambda}\left(l_f - \frac{m_r}{m} l\right). \qquad (11.2.2)$$

## 11.3 Dynamic Rolling Tire Radius

For driving on a straight road without braking, it holds for the non-driven wheels

$$v_{\omega,\text{fl}} = r_{\text{dyn,fl}}\omega_{\text{fl}} \quad v_{\omega,\text{fr}} = r_{\text{dyn,fr}}\omega_{\text{fr}}.$$

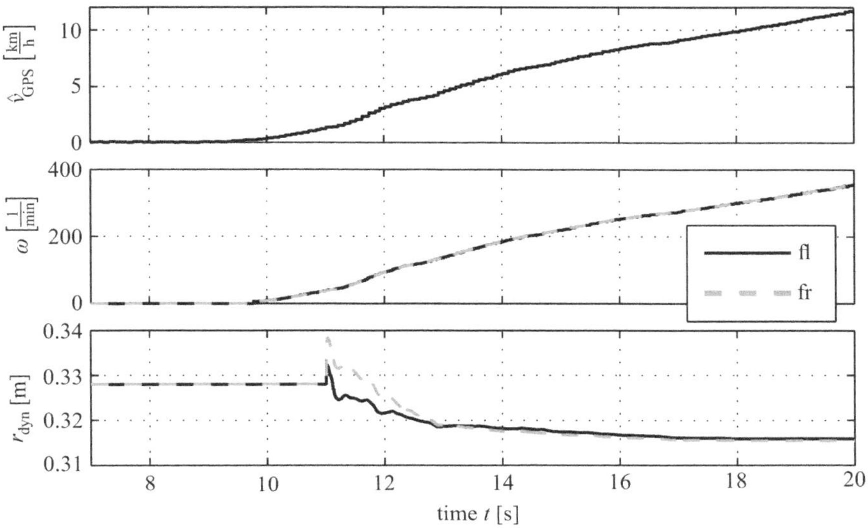


**Fig. 11.3** Identification of the dynamic tire radius $r_{dyn,fl}$ and $r_{dyn,fr}$ based on measurement of $v_{ref}$ (GPS) and $w_{fl}$, $w_{fr}$. Tire: 225/50/R17 (outer radius $r_{out} = 328.4$ mm, static radius for $v = 0$: $r_{stat} = 310$ mm), Bauer (2015)

If the velocity of the center of gravity respectively the vehicle can be measured with a GPS system or a Correvit sensor and delivers $v_{ref}$, then one can state the equation, Bauer (2015)

$$\mathbf{y} = \boldsymbol{\psi}^T \boldsymbol{\theta}, \tag{11.3.1}$$

with

$$\begin{aligned} \mathbf{y}^T &= [v_{ref}\ v_{ref}], \\ \boldsymbol{\psi}^T &= [\omega_{fl}\ \omega_{fr}], \\ \boldsymbol{\Theta}^T &= [r_{dyn,fl}\ r_{dyn,fr}]. \end{aligned} \tag{11.3.2}$$

Applying the parameter-estimation method of least squares enables to obtain an estimation of $\hat{r}_{dyn,fl}$ and $\hat{r}_{dyn,fr}$.

Figure 11.3 depicts the time history of the described parameter-estimation method for increasing the speed from standstill. The estimation begins for $v \geq 2m/s$. Above $10s$, the estimate converges to $r_{dyn} = 315, 4$ mm, which is 96% of the outer tire radius for zero load.

## 11.4 Road Gradients

The longitudinal and lateral road gradient angles $\lambda$ and $\eta$ have a significant influence on vehicle models. They change continuously during driving and should be available.

### 11.4.1 Longitudinal Road Gradient

According to Fig. 4.3 or 7.26, for a vehicle on a road with gradient angle $\lambda$, and therefore, the slope $p = \tan\lambda$ [%], it holds for the measured acceleration

$$a_{X,\mathrm{sens}}(t) = \dot{v}_X(t) + \dot{\theta}(t)v_Z(t) - \dot{\psi}(t)v_Y(t) - g\sin(\lambda(t) - \theta(t)) + n_a(t), \tag{11.4.1}$$

where $\theta$ is the pitch angle, $\psi$ is the yaw angle (compare (7.3.4)) and $n_a$ is a noise term.

The longitudinal velocity $v_X(t)$ can be determined from the speed of the non-driven wheels

$$v_X = r_{\mathrm{dyn}}(\omega_{\mathrm{w,r}} + \omega_{\mathrm{w,l}})/2. \tag{11.4.2}$$

$\dot{v}_X$ follows by applying a state variable filter. The lateral velocity $v_Y$ is either reconstructed by a measured $\dot{v}_Y$ or it is determined with a one-track model according to Eq. (7.2.9)

$$v_Y = v_X \sin\hat{\beta}, \tag{11.4.3}$$

where the slip angle $\hat{\beta}(\delta, v_X)$ is estimated with one of the methods in Sect. 12.3, e.g. via Eq. (12.3.8). Usually the product $\dot{\theta}v_Z$ is small and can be neglected. Then the road gradient can be calculated from

$$\lambda(t) = \theta(t) + \frac{1}{g}\arcsin(\dot{v}_X(t) - a_{X,\mathrm{sens}}(t) + \dot{\psi}(t)v_X(t)\sin\hat{\beta}(t)) + n_{\mathrm{a}}(t). \tag{11.4.4}$$

An application of this method is shown in Halbe (2008).

For straight driving on a smooth road with $v_Y = 0$, $v_Z = 0$ and $\theta = 0$ (11.4.1) simplifies to

$$a_{X,sens}(t) = \dot{v}_X(t) - g\sin\lambda. \tag{11.4.5}$$

The road gradient can then be determined by

$$\begin{aligned} \lambda(t) &= \frac{1}{g}\arcsin(\dot{v}_X(t) - a_{X,\mathrm{sens}}(t)), \\ &\approx \frac{1}{g}(\dot{v}_X(t) - a_{X,\mathrm{sens}}(t)); \end{aligned} \tag{11.4.6}$$

see also Kiencke and Nielsen (2005).

Considering the force balance equation of an inclined road (11.1.7) for the longitudinal acceleration $a_X(t)$, the road gradient is already part of the estimated parameter vector $\hat{\boldsymbol{\theta}}$ and can, therefore, be determined together with the mass estimation; see (11.1.9). Based on a similar equation, a linear state observer is described in Kiencke and Nielsen (2005) and Halfmann and Holzmann (2003) and good results have been obtained after knowledge of the wheel forces.

### 11.4.2 Lateral Road Gradient

For a vehicle on a lateral inclined road with gradient angle $\eta$ following equation can be stated by looking at Figs. 7.26 and 7.27

$$a_{Y,\mathrm{sens}}(t) = \dot{v}_Y(t) + \dot{\psi}(t)v_X(t) - \dot{\varphi}(t)v_Z(t) + g\sin(\eta(t) + \varphi(t)), \quad (11.4.7)$$

$\dot{v}_Y$ is either directly measured or can be determined with (11.4.3) and its derivative $\dot{\varphi}(t)$. If it can be assumed that $\dot{\varphi}(t) = 0$, then the lateral gradient angle can be determined from

$$\eta(t) = -\varphi(t) + \arcsin\frac{1}{g}(a_{Ysens}(t) - \dot{v}_Y(t) - \dot{\psi}(t)v_X(t)). \quad (11.4.8)$$

For straight driving with $\dot{\psi} = 0$, it is

$$\eta(t) = -\varphi(t) + \arcsin\frac{1}{g}(a_{Y,\mathrm{sens}}(t) - \dot{v}_Y(t)), \quad (11.4.9)$$

and with negligible lateral acceleration $\dot{v}_Y = 0$

$$\eta(t) = -\varphi(t) + \arcsin\frac{1}{g}a_{Y,\mathrm{sens}}(t). \quad (11.4.10)$$

## 11.5 Understeer Gradient

The analysis of the linearized one-track model in Sect. 7.2.5 shows for stationary cornering with radius $R_P$ for the steering wheel angle

$$\delta_H = i_S\frac{l}{R_P} + i_S\frac{l}{v_{ch}^2}a_Y, \quad (11.5.1)$$

or with $\delta_f = \delta_H/i_S$; see (7.2.20)

$$\delta_f = \frac{l}{R_P} + \frac{l}{v_{ch}^2}a_Y. \quad (11.5.2)$$

If the characteristic velocity $v_{ch}$ can be considered as a constant parameter, then the steering wheel angle changes linearly with the lateral acceleration

$$\delta_f = \delta_A + \mathrm{SG}a_y, \quad (11.5.3)$$

where $\delta_A$ is the dynamic reference steer angle (slow velocity); see (7.2.1)

$$\delta_A = \frac{l}{R_P}. \quad (11.5.4)$$

Hence, the understeer gradient

$$\mathrm{SG} = \frac{d\delta_{\mathrm{f}}}{da_{\mathrm{Y}}} = \frac{l}{v_{\mathrm{ch}}^2}, \tag{11.5.5}$$

is the gradient of the relation (11.5.3), see Fig. 7.13, in the linear range for about $a_{\mathrm{Y}} < 4.5\ \mathrm{m/s}^2$. Out of the linear range (11.5.3) may be approximated by a nonlinear term

$$\delta_{\mathrm{f}}(a_{\mathrm{y}}) = c_0 + c_1 a_{\mathrm{Y}} + c_2 a_{\mathrm{Y}}^n, \tag{11.5.6}$$

with $c_0 = \delta_{\mathrm{A}} = \frac{l}{R_{\mathrm{P}}}$, $c_1 = \mathrm{SG}$ and e.g. $n = 2$.

The experimental determination of $\delta_{\mathrm{f}} = f(a_{\mathrm{Y}}^2)$ can be made by driving with constant radius $R_{\mathrm{P}}$ and increasing velocity such that different measured values of $\delta_{\mathrm{f}}(t) = \delta_{\mathrm{H}}(t)/i_{\mathrm{S}}$ and $a_{\mathrm{Y}} = v^2(t)/R_{\mathrm{P}}$ are obtained. As $c_0$ is known, it is brought on the left side

$$\begin{aligned} y(t) &= \boldsymbol{\psi}(t)\boldsymbol{\theta}, \\ y(t) &= \delta_{\mathrm{f}}(t) - c_0, \\ \boldsymbol{\psi}^T(t) &= \left[a_{\mathrm{Y}}(t) \quad a_{\mathrm{Y}}^n(t),\right] \\ \boldsymbol{\theta}^T &= [c_1 \quad c_2.] \end{aligned} \tag{11.5.7}$$

Least squares estimation with measurements at $t = kT_0$, $k = 0, 1, 2 \ldots N$ then yields

$$\hat{\boldsymbol{\theta}}^T = \left[\hat{c}_1 \quad \hat{c}_2,\right]$$

with $\mathrm{SG} = \hat{c}_1$

The standard ISO/DIN 4138 requires that the measurements have to be taken for steps of $0.5\ \mathrm{m/s}^2$ in $a_{\mathrm{y}}$ with a duration of $\geq 3$ s.

## 11.6 Tire Model Parameters

As treated in Chap. 5, the tire models describe the longitudinal tire forces $F_{\mathrm{XT}}$ in dependence on the longitudinal slip $S_{\mathrm{X}}$ and the lateral tire force $F_{\mathrm{YT}}$ independence on the side slip angle $\alpha$ or side slip $S_{\mathrm{Y}}$ through nonlinear algebraic approximation equations.

A linear behavior can only be assumed for small longitudinal and lateral forces. In the following, it is shown how the usually unknown parameters of different tire models can be estimated based on drive dynamic sensors of driving vehicles. Figure 11.4 gives a survey of the considered cases.

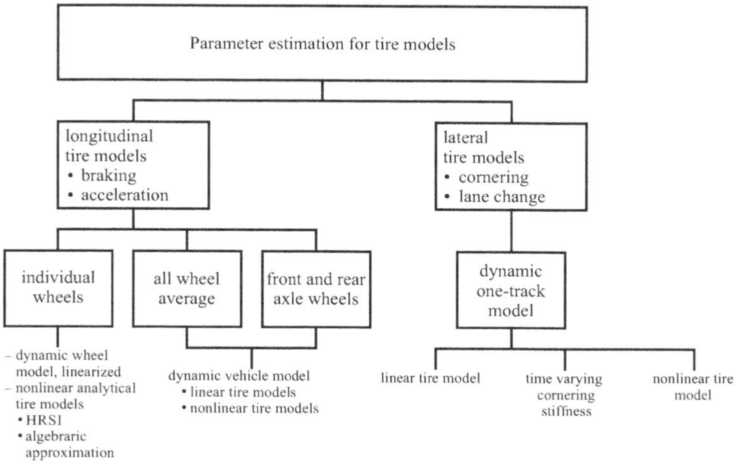


| | | | | | | |
|---|---|---|---|---|---|---|
| estimated parameters: | $\hat{\mu}_{X,ij}$ | $\overline{\mu}_X$ | $\hat{\mu}_{X,f}, \hat{\mu}_{X,r}$ | $\hat{c}_{\alpha f}, \hat{c}_{\alpha r}$ <br> $\hat{\mu}_{Yf}, \hat{\mu}_{Yr}$ | $\hat{k}_{\alpha f}, \hat{k}_{\alpha r}$ <br> $\hat{\mu}_{Yf}, \hat{\mu}_{Yr}$ | $\hat{c}_{\alpha f}, \hat{c}_{\alpha r}$ <br> $\hat{\mu}_{Yf}, \hat{\mu}_{Yr}, \mu_{Y,max}$ |
| measurements: | $M_D, M_B,$ <br> $\omega_W, a_X$ | $a_X$ | $a_X$ | $\delta_f, a_Y, \dot{\psi}$ | $\delta_f, a_Y, \dot{\psi}, (\beta)$ | $\delta_f, a_Y, a_X, \dot{\psi}$ |

**Fig. 11.4** Survey for the estimation of longitudinal and lateral tire paramters with different tire and vehicle models

### 11.6.1 Longitudinal Tire Model Parameters: Friction Coefficient Estimation

The friction coefficient $\mu_X$ in longitudinal direction is according to (5.1.3) defined by the relation

$$\mu_X(S_X) = \frac{F_{XT}}{F_{ZT}}. \tag{11.6.1}$$

In order to estimate $\mu_X$, the forces $F_{XT}$ and $F_{ZT}$ have to be known. The longitudinal $F_{XT}$ has to be measured, e.g. by rim force measurement equipment for research cars or it has to be reconstructed from on-board sensors.

**(a) Dynamic Wheel Model (Individual Wheels)**

To obtain $\mu_X(t)$ for the individual wheel, the dynamic behavior of a wheel for accelerations is considered; see also Kiencke (1993). Equation (5.5.11) yields

$$J_W^* \frac{d\omega_W(t)}{dt} - r_{dyn}\mu_X S_X(t) F_Z(t) + d'_{ds}\omega_W(t) = -M_{WB}(t) + M_{ds}(t). \tag{11.6.2}$$

It is assumed that $d'_{ds}$ is a coefficient for the friction torque of the drive shaft and wheel bearings and includes the linear part of the rolling resistance (6.5.4). The

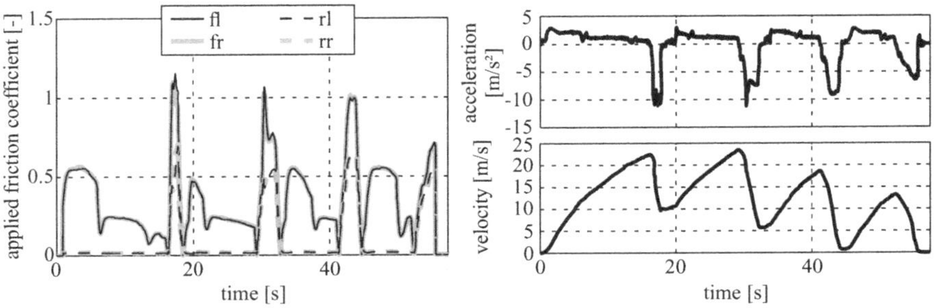


**Fig. 11.5** Estimation of the longitudinal friction coefficients with dynamic wheel models by driving with changing throttle and brake pedal input. Front driven VW Golf IV (2005), $m = 1505\text{kg}$, dry asphalt, Schorn (2007)

determination of the braking torque $M_{\mathrm{WB}}$ is based on the measured braking pressure $p_{\mathrm{Wcyl}}$ calculated with (5.5.4) and (5.5.6) or on the braking pedal position $\beta$ according to Eq. (6.7.7). $M_{\mathrm{ds}} = i_{\mathrm{tot}}\eta_{\mathrm{tot}}M_{\mathrm{eng}}/2$ is the wheel torque from the drivetrain, see e.g. (11.1.7). Then it follows

$$\mu_{\mathrm{X}}(S_{\mathrm{X}}(t)) = \frac{1}{r_{\mathrm{dyn}}F_{\mathrm{ZT}}(t)}(J_{\mathrm{W}}^{*}\frac{d\omega_{\mathrm{W}}(t)}{dt} + d_{\mathrm{ds}}^{'}\omega_{\mathrm{W}}(t) + M_{\mathrm{WB}}(t) - M_{\mathrm{ds}}(t)). \tag{11.6.3}$$

The direct use of this equation results in too noisy values. Therefore, Schorn (2007) has proposed to perform a recursive least squares parameter estimation with forgetting factor. For the front left wheel, (11.6.3) becomes

$$\begin{aligned} y(t) &= \psi^T(t)\boldsymbol{\theta}, \\ y(t) &= -M_{\mathrm{dsfl}}(t) + M_{\mathrm{WBfl}}(t) + d_{\mathrm{ds}}^{'}\omega_{\mathrm{Wfl}}(t) + J_{\mathrm{W}}^{*}\frac{d\omega_{\mathrm{Wfl}}(t)}{dt}, \\ \psi^T(t) &= \left[r_{\mathrm{dyn}}\ F_{\mathrm{Zfl}}(t)\right], \\ \boldsymbol{\theta} &= [\hat{\mu_{\mathrm{X}}}(t)]. \end{aligned} \tag{11.6.4}$$

The vertical tire force $F_{\mathrm{Zf}}$ is determined with (6.5.39). For parameter estimation, (11.6.4) is discretized with discrete time $t = kT_0$. This estimation then provides the actually applied friction coefficient $\hat{\mu}_{\mathrm{X}}(t)$ for each wheel.

Figure 11.5 depicts an example for a front driven car. The applied friction coefficient time history is obtained for a positive acceleration with two wheels and for braking with four wheels. A similar approach is published in Kiencke (1993) and it is shown that by recursive parameter estimation, the actual $\mu_{\mathrm{X}}(t)$ value is obtained after about 0.1s braking time.

**(b) Dynamic Vehicle Model (Average of All Wheels)**

If an overall friction coefficient of two or all the wheels is of interest, one can use the equations applied for mass estimation. It holds for *braking*; see (11.1.5) and (11.6.1).

$$ma_{\mathrm{X}}(t) = -r_{\mathrm{dyn}}\mu_{\mathrm{X}}(S_{\mathrm{X}}(t))F_{\mathrm{Z}} - F_{\mathrm{X,R}}(t) - F_{\mathrm{X,A}}(t) - F_{\mathrm{X,C}}(t), \tag{11.6.5}$$

which leads to

$$\mu_X(S_X(t)) = \frac{1}{r_{dyn} F_{ZT}}(-ma_X(t) - F_{X,R}(t) - F_{X,A}(t) - F_{X,C}(t)). \quad (11.6.6)$$

If it is assumed that the vehicle tire force $F_Z$ is constant and the parameters for the resistance force from the mass estimation (11.1.9) are known, then a parameter-estimation method as for (11.6.3) can be applied. However, this gives only an average approximate friction coefficient for all braking or driving wheels. Therefore, the method described next is much better suited.

**(c) Determination of Analytical Tire Friction Model Parameters (Individual Wheels)**

It is assumed that the friction coefficients $\mu_X$ in dependence on the slip values are experimentally determined by measurements or calculations of $F_{XT}$ and $F_Z$ and that an analytical relation $\mu_X(S_X)$ has to be found. The approximation of the friction coefficient $\mu_X(S_X)$ according to the *HRSI - model* (5.1.4) with two straight lines

$$\begin{aligned} \mu_X &= c_{\mu_0} S_X; & S_X &< S_{X,max}, \\ \mu_X &= \mu_{X_0} - c_{\mu_1} S_X; & S_X &> S_{X,max}. \end{aligned} \quad (11.6.7)$$

enables to estimate the unknown parameters by stating, Semmler (2006),

$$y(k) = \psi^T(k)\boldsymbol{\theta}, \quad (11.6.8)$$

with

$$y(k) = \mu_X(k) = F_{XT}(k)/F_{ZT},$$

$$\left.\begin{aligned} \psi^T(k) &= S_X, \\ \boldsymbol{\theta} &= c_{\mu_0}, \end{aligned}\right\} S_X < S_{X,max},$$

$$\left.\begin{aligned} \psi^T(k) &= \left[1 \;\; -S_X\right], \\ \boldsymbol{\theta}^T &= \left[\mu_{X_0} \;\; c_{\mu_1}\right], \end{aligned}\right\} S_X > S_{X,max}.$$

and $k = t/T_0$ the discrete time. The wheel braking forces can be calculated with (6.7.7) and (6.7.8) by measurement of $p_{mcy'}(\beta)$ and $F_{ZT}$ is determined with (6.5.39), (6.5.40), and measurement of $a_X$.

The slip values are determined with (5.1.1) and $v_X$ according to Sect. 12.2. The unknown parameters $c_{\mu_0}$, $\mu_{X_0}$ and $c_{\mu_1}$ are estimated by least squares parameter estimation.

Figure 11.6 depicts an example of some measurements with 25 samples with sampling time $T_0 = 7$ ms for increasing slip during a brake maneuver. The value $S_{X,max}$ which corresponds to $\mu_{X,max}$ is obtained by applying the LS estimation for certain time segments and then the change of the gradient is determined by

$$\text{sign}(c_{\mu_0}(k)) \neq \text{sign}(c_{\mu_0}(k-1)); \quad (11.6.9)$$

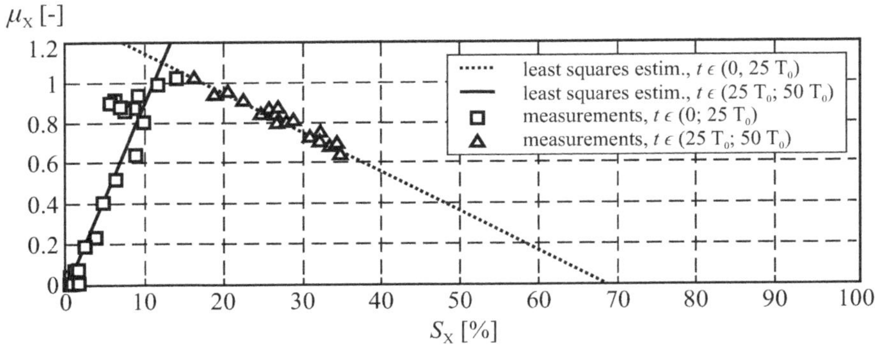


**Fig. 11.6** Least squares estimation for the friction coefficient $\mu_X$ of the HRSI-model. Semmler (2006)

see Semmler (2006).

However, because of the strongly disturbed signal during braking the determination of $S_{X,max}$ and $\mu_{X,max}$ is not reliable with this relative simple HRSI-model.

Therefore, tire models which cover the whole $\mu(S)$-curve are to be preferred also with regard to determine the important $\mu_{max}$-value.

An *algebraic tire friction approximation function*, proposed by Kiencke (1993), Daiss (1996), and Kiencke and Nielsen (2005) is

$$\mu_X(S_X) = c_{\mu 0} \frac{1}{1 + c_1 S_X + c_1 S_X^2} S_X, \tag{11.6.10}$$

which can be brought into a form which is linear in the unknown parameters

$$\begin{aligned} \frac{S_X}{\mu_X(S_X)} &= \frac{1}{c_{\mu 0}}(1 + c_1 S_X + c_2 S_X^2), \\ \frac{1}{\mu_X(S_X)} &= \frac{1}{c_{\mu 0} S_X} + \frac{c_1}{c_{\mu 0}} + \frac{c_2}{c_{\mu 0}} S_X. \end{aligned} \tag{11.6.11}$$

A LS parameter estimation is then based on

$$\begin{aligned} y &= \boldsymbol{\psi}^T \boldsymbol{\theta}, \\ y &= \frac{1}{\mu_X(S_X)}, \\ \boldsymbol{\psi}^T &= \left[ \frac{1}{S_X} \; 1 \; S_X \right], \\ \boldsymbol{\theta}^T &= \left[ \frac{1}{c_{\mu 0}} \; \frac{c_1}{c_{\mu 0}} \; \frac{c_2}{c_{\mu 0}}. \right] \end{aligned} \tag{11.6.12}$$

Hence, if several values of $\mu_X(kT_0)$ are known for varying slip $S_X(kT_0)$, $k = 1, 2, \ldots N$, for example estimated with (11.6.4) or (11.6.8), the parameters $c_{\mu 0}$, $c_1$ and $c_2$ can be determined with the method of least squares.

The initial gradient of (11.6.10) is

$$d\mu_X/dS_X\Big|_{S_X\to 0} = c_{\mu_0}, \tag{11.6.13}$$

and the maximum friction coefficient is

$$\mu_{max} = \frac{c_{\mu_0}\sqrt{1/c_2}}{2 + c_1\sqrt{1/c_2}}. \tag{11.6.14}$$

With the assumption that $c_{\mu_0} = 30$ is known, experimental results for the estimation of parameters $c_1$ and $c_2$ are shown in Kiencke and Nielsen (2005). The parameters converge after 200 ms during braking. However, sufficient excitation of braking force is required, close to the maximum of $\mu_X$; see also Börner (2004).

### 11.6.2 Lateral Tire Model Parameters: Cornering Stiffness Estimation

The *friction coefficient* in lateral direction is according to (5.2.7)

$$\mu_Y(S_Y) = \frac{F_{YT}}{F_{ZT}}, \tag{11.6.15}$$

and requires for its determination the knowledge of the lateral tire force $F_{YT}$ and the vertical tire force $F_{ZT}$. $F_{YT}$ can either be directly measured with the rim force equipment for research cars, see Fig. 5.18, or it has to be reconstructed based on lateral vehicle models and on-board sensors. Instead of the friction coefficients, one can use for small side slip angles the *cornering stiffness*; see (5.2.4)

$$c_\alpha = \frac{dF_{YT}(\alpha)}{d\alpha}\Big|_{\alpha=0} = \frac{d\mu_Y(\alpha)}{d\alpha}\Big|_{\alpha=0} F_{ZT}, \tag{11.6.16}$$

and thus; see Fig. 11.7

$$F_{YT} = c_\alpha(F_{ZT})\alpha, \tag{11.6.17}$$

with

$$c_\alpha = c_{\alpha_1} F_{ZT}, \tag{11.6.18}$$

for small vertical forces and

$$c_\alpha = \left(c_{\alpha_1} - c_{\alpha_2}\frac{F_Z}{F_{Z,\mathrm{rate}}}\right) F_{ZT}, \tag{11.6.19}$$

for large $F_{ZT}$; see (5.2.4)–(5.2.6). The lateral slip $S_Y$ and the sideslip angle $\alpha$ are related by; see (5.2.2)

$$S_Y = \frac{v_{YT}}{v_{XT}} = \tan\alpha. \tag{11.6.20}$$

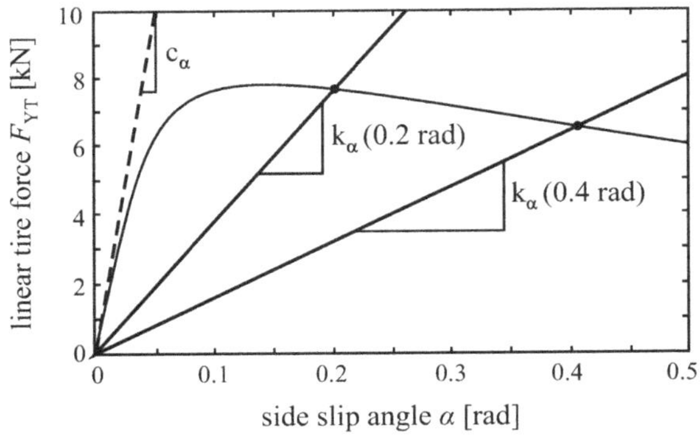


**Fig. 11.7** Lateral tire force in dependence on the side slip angle and approximation by a constant cornering stiffness coefficient $c_\alpha$ for small $\alpha$ and operating point depending cornering stiffness factor $k_\alpha$ for larger $\alpha$, Börner (2004), Halbe (2008)

For small side slip angles, it holds, therefore

$$S_Y \approx \alpha. \tag{11.6.21}$$

According to (7.2.52) for the one-track model the cornering stiffness of the front axis $c_{\alpha f}$ and the rear axis $c_{\alpha r}$ determine the characteristic velocity $v_{ch}^2$, and therefore, also the understeer gradient (7.2.74)

$$\mathrm{SG} = \frac{l}{v_{ch}^2} = \frac{m(c_{\alpha r} l_r - c_{\alpha f} l_f)}{c_{\alpha f} c_{\alpha r}}, \tag{11.6.22}$$

If SG$= d\delta_f/da_Y$ is known, which is a basic steering characterstic. Then $c_{\alpha r}$ can be calculated from known $c_{\alpha f}$ due to

$$c_{\alpha r} = \frac{m c_{\alpha f} l_f}{m l_r - \mathrm{SG} c_{\alpha f} l}. \tag{11.6.23}$$

As the understeer gradient SG directly can be determined by cornering with different lateral accelerations $a_Y$, see Sect. 11.5, only $c_{\alpha f}$ has to be identified if the one-track model can be applied.

In the following, the determination of lateral tire model parameters is considered with use of sensors which are available for development and for series vehicles with usually installed on-board sensors.

**(a) Linearized One-Track Model, Linear Tire Models, and Dynamic Steering Input**

The linearized one-track model, (7.2.33)–(7.2.36), is valid with the assumption of small steering angles $\delta_f$, small side slip angles $\beta$, and linear tire models

$$F_{YT,i} = c_{\alpha,i} \alpha_i \quad i = \mathrm{f,r}. \tag{11.6.24}$$

The goal is now to determine the cornering stiffness values $c_{\alpha f}$ and $c_{\alpha r}$ for the front and rear axis. For constant velocity $\dot{v} = 0$, the lateral behavior is described using (7.2.45), (7.2.46), and $d_A = 0$ by:

(a) Lateral force balance (y - direction)

$$a_Y = \frac{c_{\alpha f}}{m}\left(\delta_f - \beta - \frac{l_f}{v}\dot{\psi}\right) + \frac{c_{\alpha r}}{m}\left(-\beta + \frac{l_r}{v}\dot{\psi}\right). \quad (11.6.25)$$

(b) Torque balance around the vertical axis

$$\ddot{\psi} = c_{\alpha f}\frac{l_f}{J_z}\left(\delta_f - \beta - \frac{l_f}{v}\dot{\psi}\right) - c_{\alpha r}\frac{l_r}{J_z}\left(-\beta + \frac{l_r}{v}\dot{\psi}\right). \quad (11.6.26)$$

It is assumed that $\delta_f$, $\dot{\psi}$ and $a_Y$ are directly measured and that the ground velocity $v$ is determined by wheel speed measurements, see Sect. 12.2 and that the side slip angle $\beta$ is measured, e.g. with a correlation measurement technique (e.g. CORREVIT).

Further assumptions are that $l_f$, $l_r$, and $J_Z$ are known and that $\ddot{\psi}$ is determined from the measured $\dot{\psi}$ by a state variable filter. Wesemeier (2012) has shown how the cornering stiffness can be estimated by applying LS parameter estimation to both Eqs. (11.6.25) and (11.6.26). The equation system is brought into the form

$$\mathbf{y}(t) = \boldsymbol{\psi}(t)\boldsymbol{\theta}, \quad (11.6.27)$$

with the measurements

$$\mathbf{y}^T(t) = \begin{bmatrix} a_Y(t) & \ddot{\psi}(t), \end{bmatrix}$$

the regressors

$$\boldsymbol{\psi}(t) = \begin{bmatrix} \chi_{11}(t) & \chi_{12}(t) \\ \chi_{21}(t) & \chi_{22}(t), \end{bmatrix}$$

$$\chi_{11}(t) = \frac{1}{m}\left(\delta_f(t) - \beta(t) - \frac{l_f}{v(t)}\dot{\psi}(t)\right),$$
$$\chi_{12}(t) = \frac{1}{m}\left(-\beta(t) - \frac{l_r}{v(t)}\dot{\psi}(t)\right),$$
$$\chi_{21}(t) = \frac{l_f}{J_Z}\left(\delta_f(t) - \beta(t) - \frac{l_f}{v(t)}\dot{\psi}(t)\right),$$
$$\chi_{22}(t) = \frac{l_r}{J_Z}\left(-\beta(t) - \frac{l_r}{v(t)}\dot{\psi}(t)\right),$$

and the parameter vector

$$\boldsymbol{\theta}^T = \begin{bmatrix} c_{\alpha f} & c_{\alpha r}. \end{bmatrix}$$

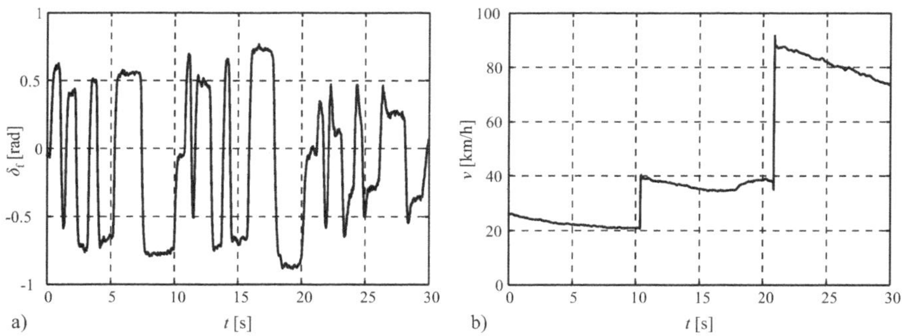


**Fig. 11.8** Driving maneuver with a series of pulses of the steering wheel angle and different velocities to estimate the cornering stiffness, Opel Omega A 2.0i, Wesemeier (2012)

The measured signals are then sampled for continuous time $t = kT_0$, with $k = 1, 2, 3, ...N$ and $T_0$ the sampling time. Building an equation system according to (10.1.12) resp. (10.1.44) leads then after defining an equation error as in (10.1.10) to the least squares estimate (10.1.46)

$$\boldsymbol{\theta}(N) = [\boldsymbol{\Psi}^T \ \boldsymbol{\Psi}]^{-1}\boldsymbol{\Psi}^T \mathbf{y}. \tag{11.6.28}$$

Figure 11.8 depicts the steering wheel inputs in form of rectangular/trapezoidal pulses of different length with amplitude of about ±45° for the velocities $v = 30.5$ and 80 km/h. A state variable filter is applied for the measurements. The parameter estimates for the off-line evaluation over the complete time interval are

$$\hat{c}_{\alpha f} = 63900 \text{ N/rad} \; ; \quad \hat{c}_{\alpha r} = 97990 \text{ N/rad} \; ;$$
$$\hat{c}_{\alpha f} l_\text{f} / \hat{c}_{\alpha r} l_\text{r} = 0.592; \quad v_\text{ch} = 76.6 \text{ km/h}.$$

The evaluation of these results by comparing of measurements of a slalom maneuver with a one-track model simulation and the estimates of the cornering stiffness has shown good results for $\dot{\psi}$ and $a_\text{Y}$, but less for $\beta$, especially if $a_\text{Y} > 5\frac{m}{s^2}$ (where the one-track model is not valid), Wesemeier (2012). The use of only one of the equations (11.6.25) or (11.6.26) did not give reasonable results. The lateral friction coefficient is obtained with (5.2.9) by

$$\mu_{\text{Y},i} = c_{\alpha,i}\alpha_i \quad i = \text{f,r},$$

where $\alpha_i$ is determined with (7.2.25) and (7.2.26).

**(b) Linearized One-Track Model, Linear Tire Models, and Quasi-stationary Cornering**

Cornering is an excitation of lateral vehicle dynamics, which is a part of normal driving. The dynamic one-track models then lead to algebraic equations according (7.2.57)–(7.2.59), with the gains

$$\frac{\dot{\psi}}{\delta_\text{H}} = \frac{1}{i_\text{S} l} \frac{v}{1 + (\frac{v}{v_\text{ch}})^2} = \frac{b_1 v}{1 + a_2 v^2} = K_{\dot{\psi}}, \tag{11.6.29}$$

$$\frac{a_Y}{\delta_H} = \frac{1}{i_S l} \frac{v^2}{1 + (\frac{v}{v_{ch}})^2} = \frac{b_1 v^2}{1 + a_2 v^2} = K_{ay}, \qquad (11.6.30)$$

$$\frac{\beta}{\delta_H} = \frac{l_r}{i_S l} \frac{1 - \frac{m l_f}{c_{\alpha r} l_r l}}{1 + (\frac{v}{v_{ch}})^2} = \frac{b_0 + b_2 v^2}{1 + a_2 v^2} = K_\beta, \qquad (11.6.31)$$

with

$$\begin{aligned} a_2 &= \frac{l}{v_{ch}^2} = \frac{m(c_{\alpha r} l_r - c_{\alpha f} l_f)}{c_{\alpha f} c_{\alpha r} l}, \\ b_0 &= \frac{l_r}{i_S l}, \\ b_1 &= \frac{1}{i_S l}, \\ b_2 &= \frac{m l_r}{c_{\alpha r} l^2 i_S}. \end{aligned} \qquad (11.6.32)$$

These models can now be used to estimate $c_{\alpha f}$ and $c_{\alpha r}$ if the other parameters are known. The three gain equations then are:

$$\begin{aligned} \dot{\psi} &= K_{\dot{\psi}} \delta_H, \\ a_y &= K_{ay} \delta_H, \\ \beta &= K_\beta \delta_H. \end{aligned} \qquad (11.6.33)$$

As two gains $K_{ay} = K_{\dot{\psi}} v$ are linearly dependent, only two parameters can be determined from (11.6.33). If then the parameters $m, l_f, l_r$, and $i_S$ are known, $c_{\alpha f}$ and $c_{\alpha r}$ can be estimated.

Wesemeier (2012) has shown that it is possible to estimate the parameters

$$\boldsymbol{\theta}^T = [a_2 \;\; b_0 \;\; b_2], \qquad (11.6.34)$$

with the method of least squares, if the slip angle $\beta$ can be measured and then the following parameters are determined

$$\begin{aligned} \hat{l}_r &= b_0 l, \\ \hat{c}_{\alpha r} &= \frac{m \hat{l}_r}{b_2 l}, \\ \hat{c}_{\alpha f} &= \frac{m c_{\alpha r} \hat{l}_r}{a_2 c_{\alpha r} l + m l_f}, \\ l_f &= l - \hat{l}_r. \end{aligned} \qquad (11.6.35)$$

The measurements of $\dot{\psi}$, $a_y$, and $\beta$ are taken for cornering with constant radius and increasing and decreasing velocity; see Fig. 11.9. The estimates are

$$c_{\alpha f} = 66100 \text{ N/rad} \quad ; \quad c_{\alpha r} = 96400 \text{ N/rad}; \quad l_r = 1.42 \text{ m},$$

and thus values close to the measurements according to Fig. 11.8.

If the measurement of the slip angle $\beta$ is not available, then only (11.6.29) and (11.6.30) can be used. This leads to

$$\begin{aligned} \dot{\psi} - b_1 v \delta_H &= -a_2 v^2 \dot{\psi}, \\ a_Y - b_1 v^2 \delta_H &= -a_2 v^2 a_Y. \end{aligned} \tag{11.6.36}$$

Hence, only one parameter $a_2$ can be estimated, and therefore, either $c_{\alpha f}$ for given $c_{\alpha r}$, or vice versa, can be determined.

One further assumption is to use a specified ratio

$$\kappa_{c\alpha} = \frac{c_{\alpha f}}{c_{\alpha r}}, \tag{11.6.37}$$

as both cornering stiffness values may change simultaneously and $\kappa_{c\alpha}$ is usually in the range of 0.5....0.9, Mitschke and Wallentowitz (2004). However, according

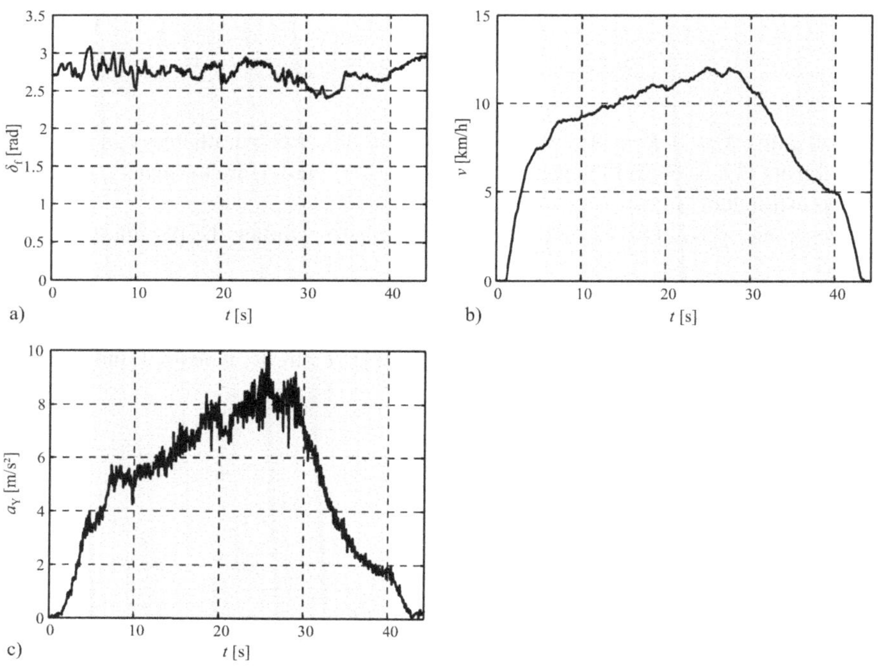


**Fig. 11.9** Measurements for cornering with constant radius and increasing and decreasing velocity, Wesemeier (2012)

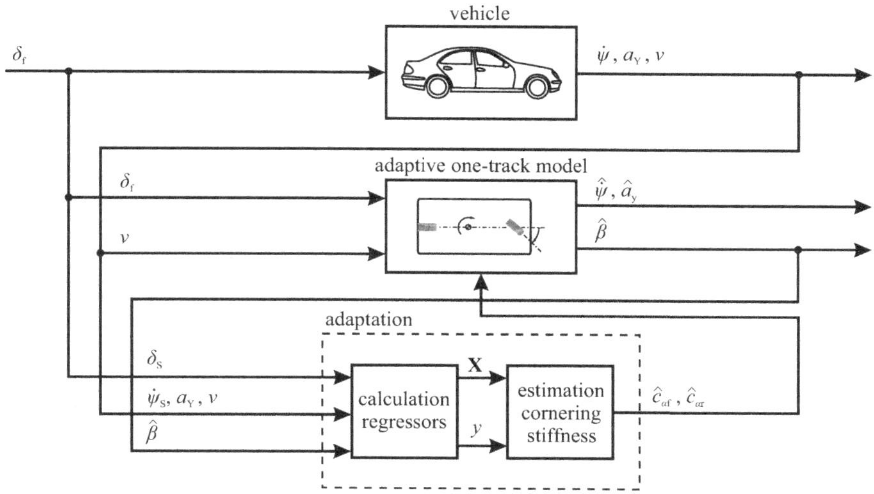


**Fig. 11.10** Signal flow for using an adaptive one-track model to calculate the slip angle $\hat{\beta}$ and to estimate the cornering stiffness $\hat{c}_{\alpha f}$ and $\hat{c}_{\alpha r}$, Wesemeier (2012)

to (11.6.23) $\kappa$ depends depends on $c_{\alpha f}$. This equation can be used to specify $\kappa_{c\alpha}$ approximately. Based on (11.6.36) the parameter $\hat{a}_2$ is then estimated with the LS-method. Then for given $\kappa_{c\alpha}$ it follows from (11.6.35)

$$\hat{c}_{\alpha r} = m\frac{l_r - \kappa_{c\alpha} l_f}{a_2 l} \quad \text{and} \quad \hat{c}_{\alpha f} = \kappa_{c\alpha}\hat{c}_{\alpha r}. \tag{11.6.38}$$

**(c) Linearized One-Track Model, Linear Tire Model, and Parallel Adaptive Model**

The disadvantage of the slip angle measurement is avoided by a sequentially operating one-track simulation model or an observer as shown in Fig. 11.10.

The adaptive model obtains after each sampling the estimated cornering stiffness and estimates the slip angle $\hat{\beta}(t_k)$, $t_k = kT_0$. This can be made for a general dynamic one-track model as described in section (a) and for cornering according to section (b). Depending on the kind of excitation, the cornering stiffness values determined for quasi-stationary cornering and for dynamic slalom like excitation are blended via a fuzzy-logic membership weighting. For details; see Wesemeier (2012).

**(d) One-Track Model and Time-Varying Cornering Stiffness**

In order to apply the cornering stiffness estimation for large lateral accelerations $a_Y > 4$ m/s$^2$ Börner (2004) has extended the *cornering stiffness coefficient* by

$$k_\alpha(\alpha, t) = \frac{F_{YT}(\alpha, t)}{\alpha(t)}, \tag{11.6.39}$$

describing the secant to the present operating point of the $F_Y(\alpha)$-curve; see Fig. 11.7. This time-varying cornering stiffness factor describes only the lateral force for operating point depending $\alpha$.

Using the force balance equation for the lateral direction (7.2.23) and the torque balance equation around the vertical axis (7.2.24) of the *one-track model* in yaw rate/slip angle representation, it follows for the lateral tire force of one axle after solving for $F_{YT,f}$ and $F_{YT,r}$, Börner (2004)

$$F_{YT,f} = \frac{1}{l \cos \delta_f}(J_Z\ddot{\psi} + l_r m \ddot{y} \cos \beta + l_r m \dot{v} \sin \beta), \tag{11.6.40}$$

$$F_{YT,r} = \frac{1}{l}(-J_Z\ddot{\psi} + l_f m \ddot{y} \cos \beta + l_f m \dot{v} \sin \beta). \tag{11.6.41}$$

For small $\beta$ these equations can be simplified by $\beta \approx 0$

$$\begin{aligned} F_{YT,f} &= \frac{1}{l}(J_Z\ddot{\psi} + l_r m \ddot{y}), \\ F_{YT,r} &= \frac{1}{l}(-J_Z\ddot{\psi} + l_f m \ddot{y}). \end{aligned} \tag{11.6.42}$$

Another possibility is to determine the lateral tire force from the lateral force and torque balance with $a_Y = v^2/R$

$$F_{YT,f} = \frac{l_r}{l} m a_Y; \qquad F_{YT,r} = \frac{l_f}{l} m a_Y, \tag{11.6.43}$$

which follows for $v = const$, $\ddot{\psi} = 0$, $\beta \approx 0$, $\delta_f \approx 0$ for stationary cornering with low speed and large radius; see also Mitschke and Wallentowitz (2004).

The side slip angles are calculated with (7.2.21) and the slip angle $\beta$ is determined with the one-track model (7.2.34), see section 12.3.2, with the assumption that $J_Z$ is known.

Figure 11.11 summarizes the different calculation steps. The cornering stiffness factor for the front and rear axle wheels are determined by a recursive LS parameter estimation with a forgetting factor; see (10.1.27)

$$k_{\alpha i}(k) = k_{\alpha i}(k-1) + \gamma(k)[F_{YT,i}(k) - \alpha_i(k) k_{\alpha i}(k-1)]; \qquad i = \text{f,r}, \tag{11.6.44}$$

in order to reduce noise effects.

The experimental results for a circular drive are depicted in Fig. 11.12. Within few seconds after turning the steering wheel, the cornering stiffness factor of the front axis converges to $k_{\alpha f} \approx 50000$ N/rad. Best results were obtained with the

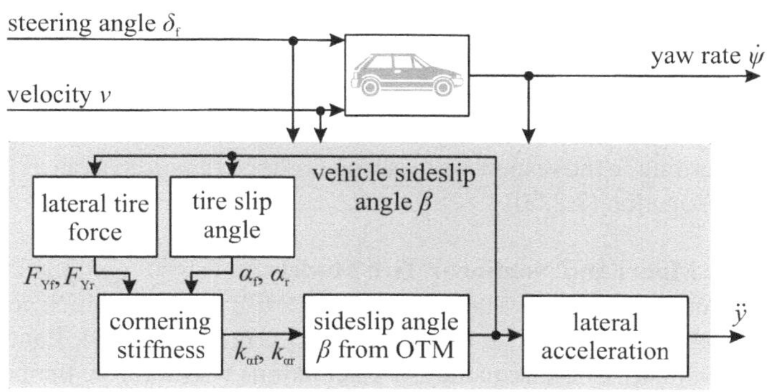


**Fig. 11.11** Schematic for the estimation of the cornering stiffness factors $k_{\alpha f}$, $k_{\alpha r}$ by using a one-track model (OTM) and measurement of $\delta_f$, $v$, and $\dot{\psi}$, Börner (2004)

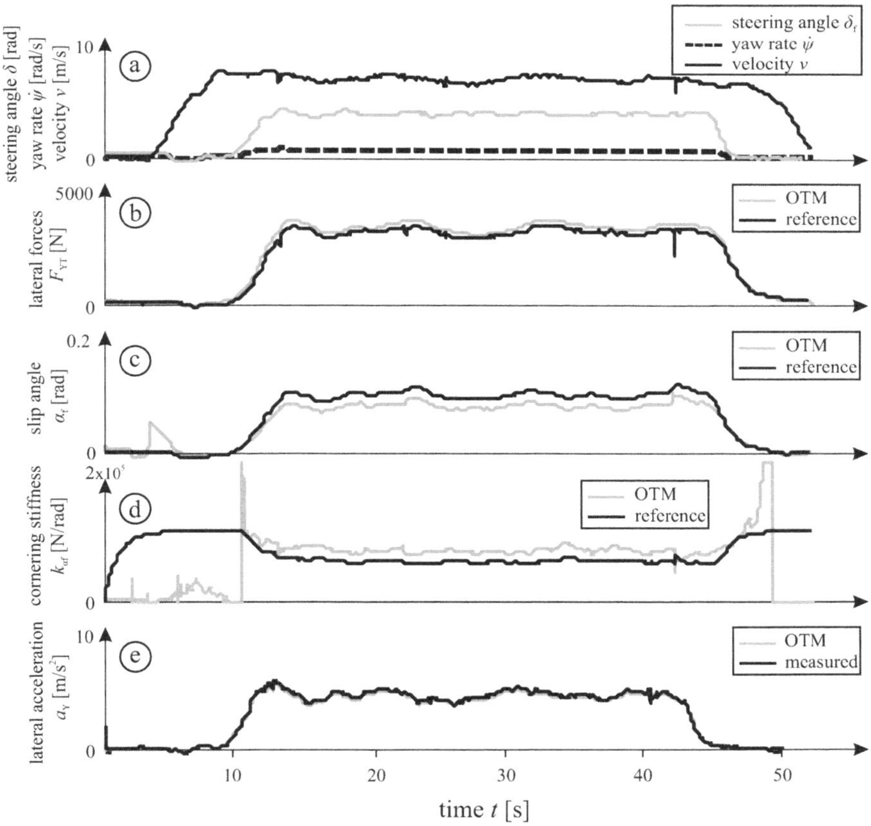


**Fig. 11.12** Experimental results for the estimation of the cornering stiffness factor $k_{\alpha f}$ of the front tires during a circular drive with $v \approx 30$ km/h and radius $R \approx 11.5$ m. Reference values are determined with a two-track model, Opel Omega A 2.0i (1993), Börner (2004)

determination of the lateral tire force with (11.6.40) and (11.6.41) compared to the simplified versions (11.6.42) and (11.6.43).

The measured signals are transient functions after a ramp wise change of the steering wheel. They show the short delayed reactions of $F_{\mathrm{YT}}$, $\alpha_{\mathrm{f}}$, and $\ddot{y}$ with a small overshoot, according to the response of a second order dynamic system as expressed by the transfer function (7.2.50).

**(e) One-Track Model and Nonlinear Tire Models**

Tire friction models for the longitudinal and lateral tire behavior which are suitable for parameter estimation are (5.1.11), (5.1.12), (5.2.11) or (5.2.13), Bauer (2015). Figure 11.13 summarizes the sequence of calculations to determine the parameters of longitudinal and lateral tire models. The parameter estimation for the *longitudinal tire models* was already described in Sect. 11.6.1 c).

In case of *lateral friction models*, a one-track model is considered with the lateral tire forces $F_{\mathrm{YT,i}}(i = f, r)$ of a centered wheel of the one-track model at the front and rear axle. Based on force and torque balance equations, the lateral tire forces are calculated with (11.6.42) and measurement of $\dot{\psi}$ and $a_{\mathrm{Y}}$, assuming small slip angles $\beta$. $\ddot{\psi}$ is determined with a state variable filter (SVF). An alternative is (11.6.43) for stationary cornering. The vertical forces $F_{\mathrm{ZT,i}}$ follow from (7.3.22) (or for small $a_{\mathrm{Y}}$ from (7.3.66). Thereby, it is assumed that for each axle holds

$$F_{\mathrm{ZT,i}} = F_{\mathrm{ZT,ir}} + F_{\mathrm{ZT,il}},$$

and that $a_{\mathrm{X}}$ and $a_{\mathrm{Y}}$ are measured. The side slip for the front and rear axle is

$$S_{\mathrm{Y,i}} = \frac{v_{\mathrm{YT,i}}}{v_{\mathrm{XT,i}}} = \tan\alpha_{\mathrm{i}}. \tag{11.6.45}$$

according to (5.2.2) requires the determination of the tire velocities $v_{\mathrm{YT,i}}$ and $v_{\mathrm{XT,i}}$ or the tire side slip angles $\alpha_{\mathrm{i}}$. These tire velocities are calculated with (7.3.19), where the chassis velocities $v_{\mathrm{X}}$ and $v_{\mathrm{Y}}$ follow from an extended Kalman filter; see Fig. 11.13.

The actual lateral friction coefficient $\mu_{\mathrm{YT},i}$ is then determined according to the definition (5.2.7)

$$\mu_{\mathrm{YT},i} = \frac{F_{\mathrm{YT},i}}{F_{\mathrm{ZT},i}} \quad i = f, r \tag{11.6.46}$$

For the lateral friction model (5.2.13), now one pair of calculated actual values $\mu_{\mathrm{YT},i}(k)$ and $S_{\mathrm{Y},i}(k)$ is known. Several determined values for $k = 1, 2, 3..., N$ for small side slip are used to estimate the initial slope $c_{\alpha 1}$ of the tire characteristic

$$c_{\alpha 1,i} = \left.\frac{\mu_{\mathrm{YT},i}}{S_{\mathrm{Y},i}}\right|_{S_{\mathrm{Y}}=0} = \left.\frac{F_{\mathrm{YT},i}}{F_{\mathrm{ZT},i}S_{\mathrm{Y},i}}\right|_{S_{\mathrm{Y}}=0}, \tag{11.6.47}$$

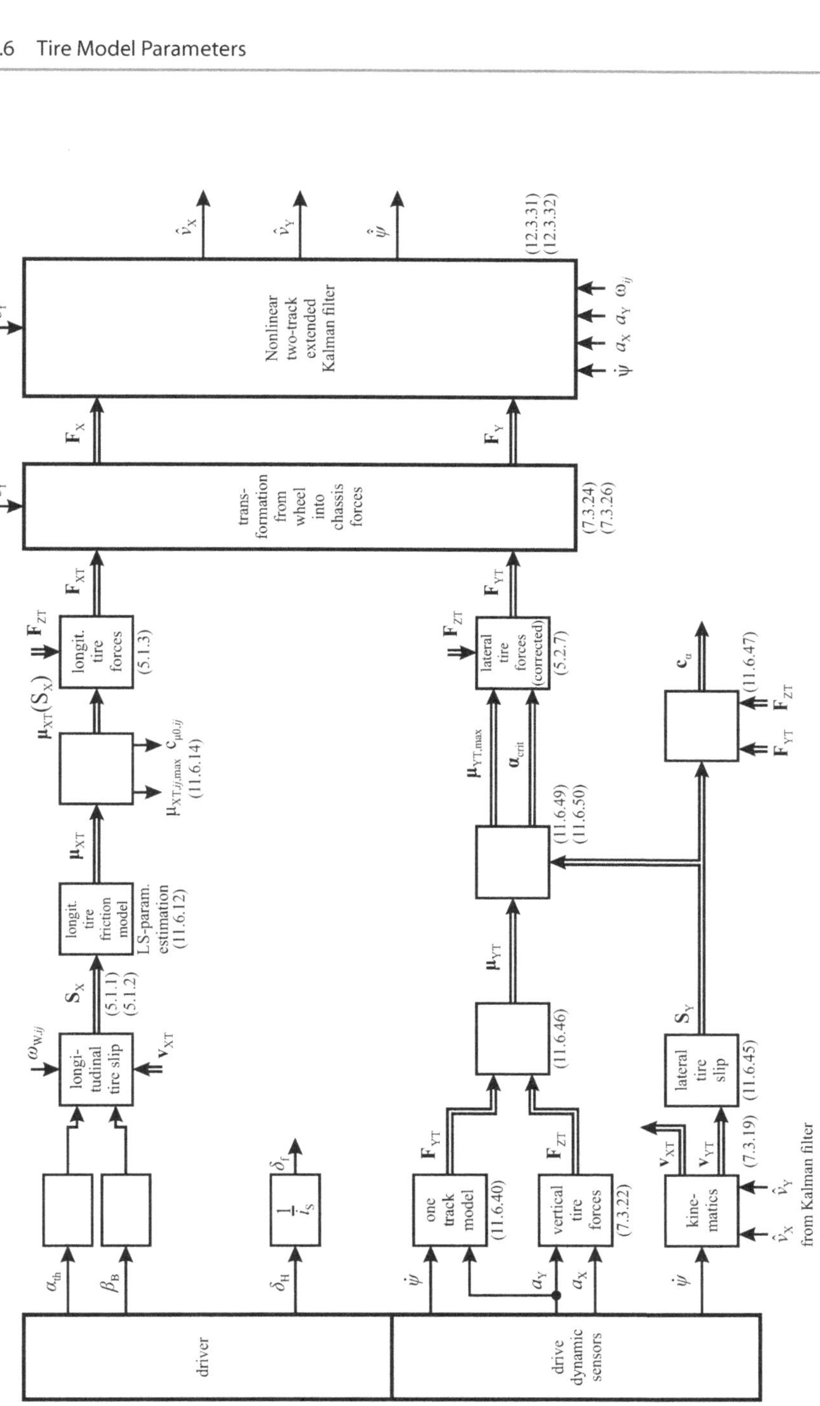


**Fig. 11.13** Signal flow schematic for the estimation of the parameters of the nonlinear lateral tire model (5.2.13), Bauer (2015)

according to (5.2.14), for example, by averaging. Then also an averaged value of

$$c_{\alpha,i} = c_{\alpha 1,i} F_{ZT,i} \tag{11.6.48}$$

is known.

Equation (5.2.13) can then be used to solve for $\mu_{YT,i,max}$ for the front and rear axle wheels; see Bauer (2015)

$$\mu_{YT,max} = -\frac{1}{2}\frac{c_{\alpha_1} S_Y - \frac{2S_Y}{\alpha_{crit,ref}}}{1-\frac{c_{\alpha 1} S_Y}{\mu_{YT}}} + \sqrt{\left(\frac{1}{2}\frac{c_{\alpha_1} S_Y - \frac{2S_Y}{\alpha_{crit,ref}}}{1-\frac{c_{\alpha 1} S_Y}{\mu_{YT}}}\right)^2 - \frac{(\frac{S_Y}{\alpha_{crit,ref}})^2}{1-\frac{c_{\alpha 1} S_Y}{\mu_{YT}}}}. \tag{11.6.49}$$

This equation is used to determine several values $\mu_{YT,i,max}(k)$ based on several values $\mu_{YT,i}(k)$ and $S_{Y,i}(k), k = 0, 1, 2, \ldots, N$, for larger slip values. An average then yields $\hat{\mu}_{YT,i,max}$. The corresponding side slip angle $\alpha_{crit}$ for $\mu_{YT,max}$ follows from

$$\alpha_{crit} = \frac{\mu_{YT,max}}{\mu_{YT,max,ref}} \alpha_{crit,ref}. \tag{11.6.50}$$

The reference values $\mu_{YT,max,ref}$ and $\alpha_{crit,ref}$ are taken from a similar tire with known values; see e.g. the measurements of Fig. 5.18.

Using the known parameters, the lateral forces $\mathbf{F}_{YT}^T = \left[F_{YT,f} \quad F_{YT,r}\right]$ can be calculated by; see (5.2.7)

$$F_{YT,i} = \mu_{YT,i}(S_{Y,i}) F_{ZT,i}, \tag{11.6.51}$$

and used as a correction of the value from (11.6.42).

The tire forces $\mathbf{F}_{YT}$ and $\mathbf{F}_{XT}$ are then transformed into chassis forces $\mathbf{F}_X$ and $\mathbf{F}_Y$, by using (7.3.23) and (7.3.24).

Finally, the velocities of the center of the gravity $v_X$ and $v_Y$, which are required for the determination of the tire velocities with (7.3.19) are estimated with the Extended Kalman filter of Sect. 12.3.3 (b) with (12.3.31) and (12.3.32), which is based on a non linear two-track model.

The final results for the parameters of the lateral tire models of the front and rear axle are then the initial gradients $c_{\alpha 1,f}$, $c_{\alpha 1,r}$, according to (11.6.47) and $\hat{\mu}_{YT,max,f}$ and $\hat{\mu}_{YT,max,r}$, according to (11.6.49) of the nonlinear lateral tire model (5.2.13).

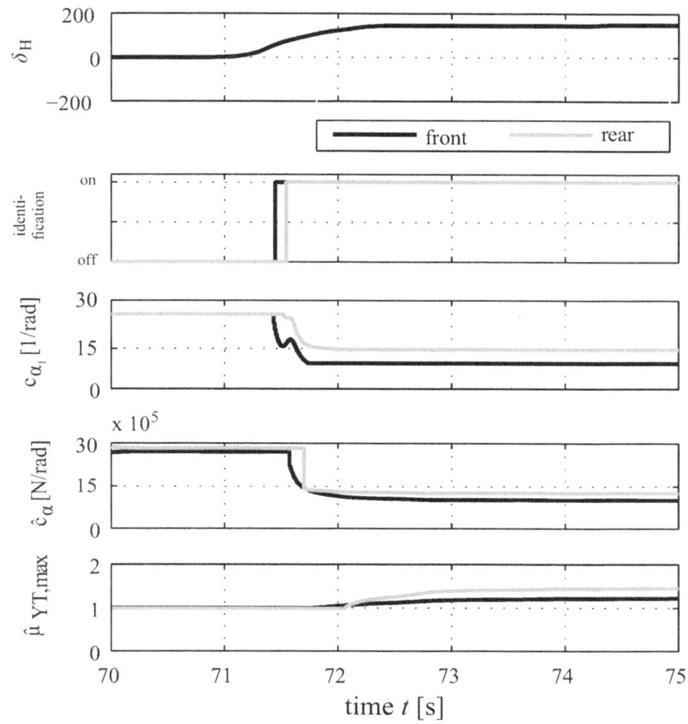


**Fig. 11.14** Experimental results for the estimation of the cornering stiffness $c_\alpha$ and the maximal friction coefficient $\mu_{\max}$ of the tires of front and rear axle after a change of the steering wheel, Bauer (2015), BMW 540i (E60), $v \approx 50$ km/h

Figure 11.13 contains also the signal flow for the determination of the longitudinal tire friction model of Sect. 11.6.1 (d) with friction model (11.6.10).

The application of this estimation procedure is shown in Fig. 11.14 for a maneuver with steering angle $\delta_{\mathrm{H}} \geq 20°$. The estimated cornering stiffness converge after start of the parameter estimation in about 0.3 s. The results of the cornering stiffnesses of the one-track model are

$$\begin{aligned} \text{front}: c_{\alpha 1,f} &= 115000 \text{ N/rad} \ ; \ \hat{\mu}_{\mathrm{YT,max,f}} = 1.09, \\ \text{rear}: c_{\alpha 1,r} &= 138000 \text{ N/rad} \ ; \ \hat{\mu}_{\mathrm{YT,max,r}} = 1.2. \end{aligned}$$

A further possibility is that the lateral tire model parameters are estimated together with the time-varying state variables. This is shown in Sect. 12.6.

## 11.7 Mass Moments of Inertia

Modeling the rotational behavior of vehicles requires the knowledge of the mass moments of inertia:

$J_Z$: moment of inertia for rotation around the vehicle axis (yawing),

$J_X$: moment of inertia for rotation around the longitudinal axis (rolling),

$J_Y$: moment of inertia for rotation around the lateral axis (pitching).

The yaw motion includes the total vehicle mass $m$ whereas, for the roll and pitch motion the sprung mass $m_S$, respectively, the body mass $m_{Bo} = m - m_{un}$ without the unsprung mass $m_{un}$ (axles and wheels) is relevant.

The moment of inertia of the planar body with an axle through the center of gravity is defined by

$$J = \int r^2 dm. \tag{11.7.1}$$

If the mass is assumed to be concentrated with a distance $i$ from the axis of rotation like on a thin wall of a cylinder, it holds

$$J = mi^2, \tag{11.7.2}$$

where $i$ is called the *radius of gyration*.

As the calculation of the moments of inertia of a complete vehicle is rather complicated, a first approach to determine those parameters is to use approximate calculation formulae. Thus, based on the evaluation for different passenger cars, radii of gyration $i$ have been obtained, leading to estimates of the three mass moments of inertia:

$$\begin{aligned} J_Z &\approx m i_Z^2, \\ J_X &\approx m_{Bo} i_X^2, \\ J_Y &\approx m_{Bo} i_Y^2. \end{aligned} \tag{11.7.3}$$

Table 11.1 lists some values for middle class passenger cars, Preukschat (1988), Börner (2004), and Kiencke and Nielsen (2005). The unsprung mass is e.g. 80...105

**Table 11.1** Radii of the gyration for middle class passenger cars, Preukschat (1988)

| Load | $i_X$ (m) | $i_Y$ (m) | $i_Z$ (m) |
|---|---|---|---|
| Empty | 0.65 | 1.21 | 1.20 |
| Two people | 0.64 | 1.13 | 1.15 |
| Four people | 0.60 | 1.10 | 1.14 |
| Maximal mass | 0.56 | 1.13 | 1.18 |

**Table 11.2** Examples of geometrical data, masses and mass moments of inertia for different passenger cars due to Wesemeier (2012), Bauer (2015), Goertz (2007), Schorn (2007)

| | VW Golf IV (2000) | Mercedes A-class W168 | Opel Omega A 2.0 i (1993) | BMW 540 i E60 | BMW 550 i E10 |
|---|---|---|---|---|---|
| $m$ (kg) | 1505 | 1315 | 1450 | 1823 | 2005 |
| $l_f$ (m) | 0.975 | | 1.30 | 1.46 | 1.42 |
| $l_r$ (m) | 1.55 | | 1.43 | 1.41 | 1.55 |
| $h_{CG}$ (m) | 0.55 | 0.68 | 0.58 | 0.54 | 0.54 |
| $J_X$ (kgm$^2$) | 411 | 424 | 618 | 510 | 730 |
| $J_Y$ (kgm$^2$) | 1853 | 1227 | 1928 | 2080 | 3860 |
| $J_Z$ (kgm$^2$) | | 1505 | 1920 | 2809 | 4060 |
| $h_{RA}$ (m) | | | 0.07 | | |

kg for the front and 110...130 kg for the rear part. Some values for different cars are shown in Table 11.2. They are taken from Börner (2004), Bauer (2015), Bechtloff (2018), Goertz (2007), and Schorn (2007).

Preukschat (1988) has developed another estimation formula

$$J_Z = A_Z m l l_V + \Delta m l_{\Delta m}^2, \tag{11.7.4}$$

with $A_Z = 0.1269$ for passenger cars, $l$ the wheel base, $l_V$ the vehicle length, $\Delta m$ the additional mass in the trunk with its distance $l_{\Delta m}$ from the CG; see also Börner (2004).

The experimental determination of the mass moments of inertia requires the excitation of the vehicle with applied forces or torques. This can be realized either with special test rigs or with dynamic driving maneuvers. Test rigs allow vehicle motions on all three axis and use force measurements; see e.g. Goertz (2007).

In the following an example is described which determines the mass moment $J_Z$ by driving maneuvers, like lane changes or slalom maneuvers.

The experimental estimation of the yaw mass moment of inertia $J_Z$ can be based on the torque balance around the vertical vehicle axis which follows from (7.3.40) for the simplified two-track model in the yaw rate/slip representation, taking into account lateral tire forces $F_{YT,ij}$ and longitudinal tire forces $F_{XT,ij}$.

$$\begin{aligned} J_Z\ddot{\psi} &= l_f((F_{YT,fl} + F_{YT,fr})\cos\delta_f + (F_{XT,fl} + F_{XT,fr})\sin\delta_f, \\ &- l_r(F_{YT,rl} + F_{YT,rr}), \\ &+ \frac{b_f}{2}\Big((F_{XT,fr} - F_{XT,fl})\cos\delta_f + (F_{XT,fr} - F_{YT,fl}\sin\delta_f)\Big), \\ &+ \frac{b_r}{2}(F_{XT,rr} - F_{XT,rl}). \end{aligned} \tag{11.7.5}$$

The required longitudinal and lateral tire forces have to be determined with appropriate equations

$$F_{\mathrm{XT},ij} = \mu_{\mathrm{XT},ij} F_{\mathrm{ZT},ij}, \tag{11.7.6}$$

$$F_{\mathrm{YT},ij} = \mu_{\mathrm{YT},ij} F_{\mathrm{ZT},ij}, \tag{11.7.7}$$

treated in Chap. 5, and using the estimated tire model parameters from Sect. 11.6.

The torque balance equation can be considerably simplified by using a one-track model (7.2.13) without longitudinal tire forces from braking or acceleration and $F_{\mathrm{YT,A}} = 0$

$$J_{\mathrm{Z}}\ddot{\psi} = l_{\mathrm{f}} F_{\mathrm{YT,f}} \cos\delta_{\mathrm{f}} - l_{\mathrm{r}} F_{\mathrm{YT,r}}. \tag{11.7.8}$$

The tire forces may be determined with (11.6.42).

For parameter estimation (11.7.5) or (11.7.8) have to be brought into the form

$$y(t) = \boldsymbol{\psi}^T(t)\boldsymbol{\theta}, \tag{11.7.9}$$

with

$$\boldsymbol{\psi}^T(t) = \ddot{\psi}(t),$$
$$\boldsymbol{\theta} = J_{\mathrm{Z}},$$

and $y(t)$ the remaining part of the equations with measured and calculated variables.

The experimental application of this estimation procedure with (11.7.5) is shown in Kiencke and Nielsen (2005). $\hat{J}_{\mathrm{Z}}$ converges after about 10 s.

A much simpler approach is obtained by using (11.6.43), which is originally valid for stationary cornering. Introduction into the one-track torque balance (11.7.8) yields

$$J_{\mathrm{Z}}\ddot{\psi} = \frac{l_{\mathrm{f}} l_{\mathrm{r}}}{l} m\ddot{y}(\cos\delta_{\mathrm{f}} - 1), \tag{11.7.10}$$

which then can be used for parameter-estimation analog to (11.7.9).

## 11.8 Roll and Pitch Dynamic Parameters

### 11.8.1 Roll Dynamic Parameters

The roll dynamic behavior can with various simplifications be described by a linear second order differential equation; see (9.1.15)

$$J_{\mathrm{X}}\ddot{\varphi}(t) + d_{\mathrm{rol}}\dot{\varphi}(t) + c_{\mathrm{rol}}\varphi = -K_{\mathrm{rol}} a_{\mathrm{Y}}(t) - K_{\mathrm{rol}} g\eta, \tag{11.8.1}$$

with $J_{\mathrm{X}}$ the mass moment of inertia around the (longitudinal) roll axis, $c_{\mathrm{rol}}$ and $d_{\mathrm{rol}}$ the torsional spring rate and torsional damping rate, and $K_{\mathrm{rol}}$ a torque constant; see (9.1.16). If the roll angle $\varphi$ and roll angle velocity $\dot{\varphi}$ can be measured, the roll angle acceleration $\ddot{\varphi}$ can be determined with a state variable filter, Würtenberger (1997). If only the roll angle velocity $\dot{\varphi}$ can be measured, $\varphi$ can be determined with a Kalman filter, Halbe (2008) and $\ddot{\varphi}$ by a state variable filter. The excitation of the roll dynamics requires a lateral acceleration $a_Y(t)$ which can, for example, be generated by a lane change or a slalom maneuver.

For parameter estimation with the method of least squares estimation, (11.8.1) is written as

$$y(t) = \boldsymbol{\psi}^T \boldsymbol{\theta}, \tag{11.8.2}$$

with

$$\begin{aligned} y(t) &= -a_Y(t), \\ \boldsymbol{\psi}^T(t) &= [\ddot{\varphi}(t) \;\; \dot{\varphi}(t) \;\; \varphi(t) \;\; 1], \\ \boldsymbol{\theta}^T &= \left[\frac{J_X}{K_{rol}} \;\; \frac{d_{rol}}{K_{rol}} \;\; \frac{c_{rol}}{K_{rol}} \;\; \frac{c_0}{K_{rol}}\right], \\ &= \left[J_X^* \;\; d_{rol}^* \;\; c_{rol}^* \;\; c_0^*\right], \end{aligned}$$

where $c_0 = g\eta$ is an offset parameter taking into account a lateral road gradient $\eta$. $K_{rol}$ is assumed to be known; see (9.1.16). Figure 11.15 depicts measurements of a slalom maneuver and Fig. 11.16 shows the estimated parameters, Würtenberger (1997). $c_{rol}$ and $d_{rol}$ show a fast convergence. The estimation of $J_X$ is more stochastic and needs more convergence time though the maximal lateral acceleration is about 5 m/s$^2$.

Other experimental results are shown in Fig. 11.17, $\varphi$, $\dot{\varphi}$, and $a_Y$ are measured and $\ddot{\varphi}$ is determined with a state variable filter. $c_{rol}^*$ converges during stationary cornering and $J_X^*$ and $d_{rol}^*$ after about 20 s during the slalom maneuver. Hence, the determination of the mass moment of inertia $J_X$ and the roll damping $d_{rol}^*$ needs a strong and persistent excitation of the lateral vehicle dynamics.

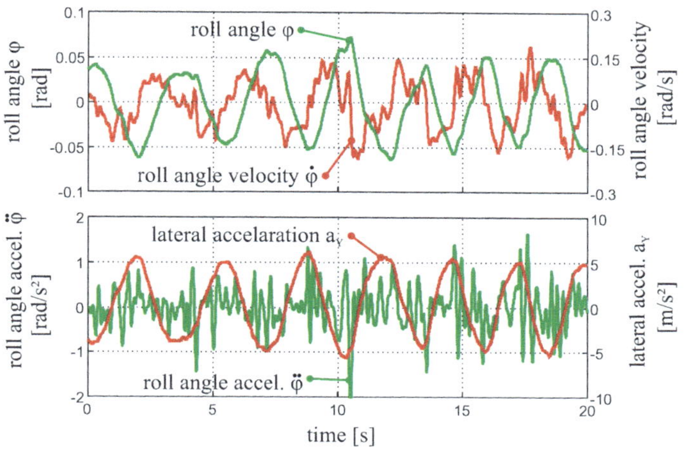


**Fig. 11.15** Measured signals during a slalom maneuver after state variable filtering, Würtenberger (1997)

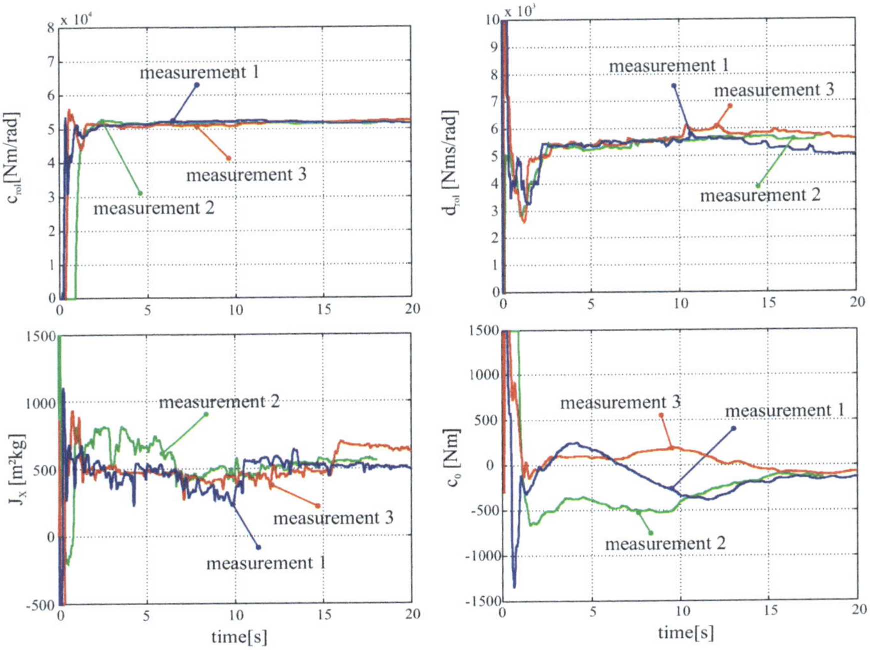


**Fig. 11.16** Parameter estimates for the roll behavior (3 measurements), Opel Vectra 85 kW (1996), Würtenberger (1997)

### 11.8.2 Pitch Dynamic Parameters

As for roll dynamics a second order system model describes the motion around the pitch axis for a longitudinal acceleration; see (9.2.4)

$$J_Y\ddot{\theta}(t) + d_{\text{pit}}\dot{\theta}(t) + c_{\text{pit}}\theta(t) = -K_{\text{pit}}a_X(t) - K_{\text{pit}}g\lambda, \tag{11.8.3}$$

with $J_Y$ the mass moment of inertia around the (lateral) pitch axis, $c_{\text{pit}}$ and $d_{\text{pit}}$ the torsional spring rate and damping rate and $K_{\text{pit}}$ a torque constant. $\eta$ takes into account a longitudinal road gradient and serves as an offset parameter.

The pitch angle is usually not measured, but could be reconstructed if the suspension deflections are measured. Otherwise, it may be reconstructed from the measurement of $\dot{\theta}$ with an additional inertial measurement unit (IMU).

The parameter estimation follows then the procedure as for the roll behavior; see Eq. (11.8.2). However, it is generally difficult to obtain enough excitation from acceleration during driving to estimate the pitch dynamic parameters.

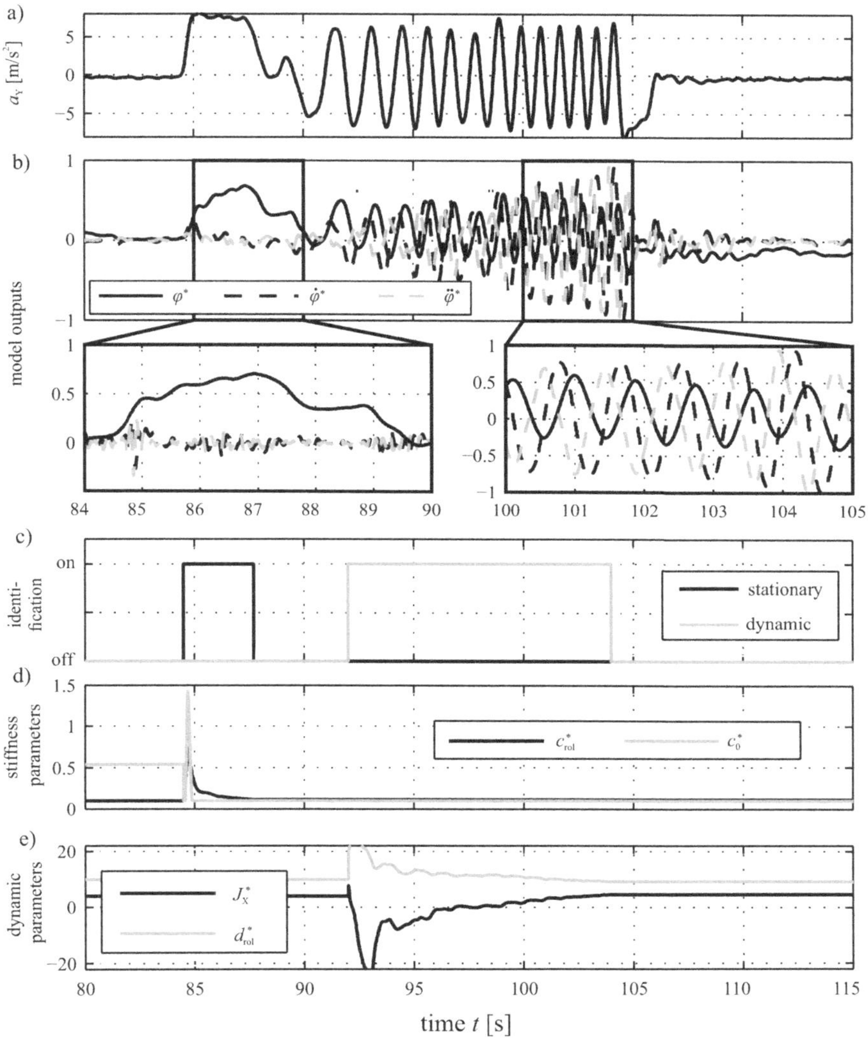


**Fig. 11.17** Measured signals for stationary cornering and slalom maneuver with variable frequency and resulting parameter estimates, BMW 540i (E60), $v \approx 70$ km/h: **a** Input: lateral acceleration **b** Normalized model outputs **c** Identification status **d** Stiffness parameters **e** Mass moment of inertia $J^*_X$ and damping coefficient $d^*_{rol}$. Parameter estimates: $c_{rol} = 101395$ Nm/rad, $d_{rol} = 13781$ Nms/rad, $\eta = 0.46°$, $J_X = 1598$ kgm$^2$, Bauer (2015)

## References


Bauer M (2015) Methoden zur modellbasierten Fahrdynamikanalyse und Bewertung von Fahrdynamikregelsystemen. Dissertation Technische Universität Darmstadt. Fortschr.-Ber. VDI Reihe 12, 792. VDI Verlag, Düsseldorf

Bechtloff J (2018) Schätzung des Schwimmwinkels und fahrdynamischer Parameter zur Verbesserung modellbasierter Fahrdynamikregelungen. Dissertation Technische Universität Darmstadt. Fortschr.-Ber. VDI Reihe 12, 809. VDI Verlag, Düsseldorf

Börner M (2004) Adaptive Querdynamikmodelle für Personenkraftfahrzeuge - Fahrzustandserkennung und Sensorfehlertoleranz. Dissertation Technische Universität Darmstadt. Fortschr.-Ber. VDI Reihe 12, Nr. 563. VDI Verlag, Düsseldorf

Daiss A (1996) Beobachtung fahrdynamischer Zustände und Verbesserung einer ABS-und Fahrdynamikregelung. Diss. Universität Karlsruhe, Fortschr.-Ber. VDI Reihe 12, 283. VDI Verlag, Düsseldorf

Goertz H (2007) Idenfikation von Fahrzeugträgheitsparametern in fahrversuch und am prüfstand. PhD thesis, RWTH Aachen

Halbe I (2008) Modellgestützte Sensoreninformationsplattform für die Quer- und Längsdynamik von Kraftfahrzeugen: Anwendungen zur Fehlerdiagnose und Fehlertoleranz. Diss. TU Darmstadt, Fortschr.-Ber. VDI Reihe 12, 680. VDI Verlag, Düsseldorf

Halfmann C, Holzmann H (2003) Adaptive Modelle für die Kraftfahrzeugdynamik VDI-Buch. Springer, Berlin

Kiencke (1993) Realtime estimation of adhesion characteristic between Tyre and Road. IFAC 12th Trienn World Congr 26(2):15–18

Kiencke U, Nielsen L (2005) Automotive control systems: for engine, driveline, and vehicle, 2nd edn. Springer, Berlin

Mitschke M, Wallentowitz H (2004) Dynamik der Kraftfahrzeuge, 4th edn. Springer, Berlin

Preukschat A (1988) Antriebsarten. In: Reimpell J (ed) Fahrwerktechnik, 2nd edn. Vogel Buchverlag

GmbH Robert Bosch (ed) (2018) Automotive handbook, 10th edn. Wiley, Chichester, England

Schorn M (2007) Quer- und Längsregelung eines Personenkraftwagens für ein Fahrerassistenzsystem zur Unfallvermeidung. Diss. Universität Darmstadt, Fortschr.-Ber. VDI Reihe 12, 651. VDI Verlag, Düsseldorf

Semmler S (2006) Regelung der Fahrzeugbremsdynamik mit kontinuierlich einstellbaren Radbremsen. Diss. Universität Darmstadt, Fortschr.-Ber. VDI Reihe 12, 632. VDI Verlag, Düsseldorf

Wesemeier D (2012) Modellbasierte Methoden zur Schätzung nicht messbarer Größen der Fahrzeugquerdynamik und des Reifenluftdrucks. Diss. Universität Darmstadt, Fortschr.-Ber. VDI Reihe 12, 747. VDI Verlag, Düsseldorf

Würtenberger M (1997) Modellgestützte Verfahren zur Überwachung des Fahrzustandes eines Pkw. Diss. Universität Darmstadt, Fortschr.-Ber. VDI Reihe 12, 314, VDI Verlag, Düsseldorf

# State Estimation of Vehicles 12

The determination of not or not precisely measurable driving state variables as velocity over ground, slip angle, yaw, rolling, and pitch angle can be reconstructed by state estimation methods with kinematic or dynamic vehicle models. The estimation of some parameters may also be integrated in a state estimation procedure.

## 12.1 State Estimation of the Vehicle Position

The determination of the vehicle position within an inertial coordinate system is a basic requirement for driver assistance and automatic driving systems. It requires calculations by using the vehicle's translatory and rotatory degrees of freedom. In the case of a vehicle on an even earth surface, the position is determined by the coordinates $X$ and $Y$ within an earth fixed coordinate system ($X_E$, $Y_E$) and the yaw angle $\psi$ (orientation). The translatory position may be given in general for the center of gravity or the rear axle in the case of slowly moving vehicles, like mobile robots or parking vehicles, Strunz (1993), Borenstein et al. (1996), Kochem (2005).

The determination of the position is divided into external and internal methods, compare Fig. 12.1. *External methods* use sensors to indicate the surrounding either by artificial or natural reference objects. *Active artificial reference objects* are sending signals, like radio signals, lighthouse beams or from satellites of the global navigation satellite system (GNSS) or GPS, with higher precision, of the differential version system (DGNSS) or DGPS. They require a corresponding receiver in the vehicle.

*Passive artificial reference objects* are landmarks, as reflectors or barcode shields, which are detected by an active sensor system in the vehicle as, for example, laser scanners or cameras

*Natural reference objects* are using the natural surrounding, like edges, columns, or walls and require a mathematical picture within the surrounding of the vehicle.



R. Isermann, *Automotive Control*, ATZ/MTZ-Fachbuch,
https://doi.org/10.1007/978-3-642-39440-9_12

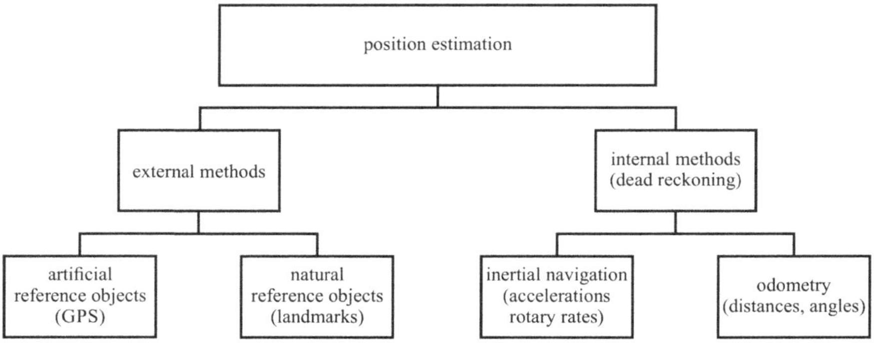


**Fig. 12.1** Survey of different methods for the position estimation of vehicles

Sensors on-board the vehicle are, for example, cameras, laser scanners of ultrasonic sensors.

*Internal methods* use on-board motion sensors and determine the driven distance and vectors with regard to a starting position through integration to obtain the actual position. The corresponding methods are called *dead reckoning* and can be divided into *inertial navigation* and *odometry*.

*Inertial navigation* uses in general three acceleration sensors and three gyros. The acceleration sensors have to be integrated twice to obtain the position and the angle changes of the gyros deliver the orientation within a Cartesian space. However, for a motion in a plane, only two accelerometers and one gyro are sufficient.

*Odometry* uses velocity measurements and determines the position by integration over a time period. For example, the angular velocities of the wheels can be used and with the knowledge of the wheels' circumferences the traveled distances from a starting point can be determined, and by using a kinematic vehicle model and the steering angle also the change of the orientation. This can, for example, be used in parking garages or tunnels, if GPS-based positioning is not possible.

The advantage of these internal methods or dead reckoning is that they do not require outside information and are not affected by weather conditions. Their disadvantage is that the current position is calculated from a previous position by integration of (noisy) sensor data. Therefore, errors are cumulative and sum up with time thus generating drift. Hence, inertial navigation methods in general have to be corrected after some time by using more precise localization methods, as, for example, GPS. This can be realized with a Kalman filter to result in a combination of internal and external position state estimation.

In the following, two odometry methods are described. An example for the position estimation with inertial navigation in combination with other state estimates is shown in Sect. 12.6.

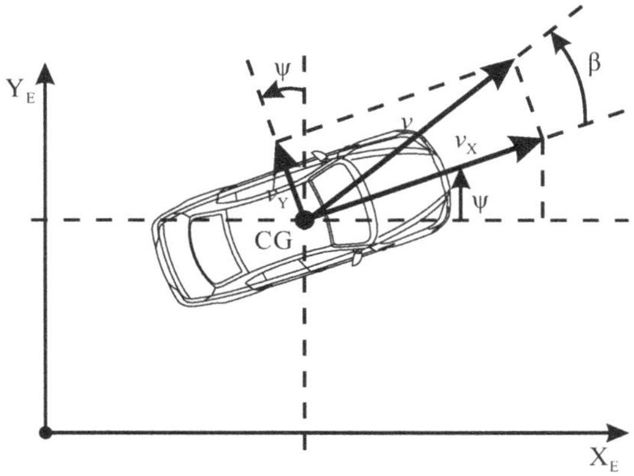


**Fig. 12.2** Motion of a vehicle in the earth fixed coordinate system

### 12.1.1 Odometric Position Estimation for an Earth Fixed Coordinate System

The determination of the position of a vehicle on a straight road with on-board drive dynamic sensors may be based on the measurement of the vehicle speed $v$, the yaw rate $\dot{\psi}$, and the estimated vehicle slip angle $\beta$. The vehicle orientation then is with a known initial yaw angle $\psi_0$ given by

$$\psi(t) = \psi_0 + \int_{t_0}^{t} \dot{\psi}(\tau)\,\mathrm{d}\tau, \tag{12.1.1}$$

at least for a limited time period.

If the vehicle drives with a velocity of the center of gravity (ground velocity) $\mathbf{v}^{\mathrm{T}}(t) = [v_X(t)\ v_Y(t)]$ in the vehicle fixed coordinate system $(X_{\mathrm{V}}, Y_{\mathrm{V}})$ and an orientation (yaw angle) $\psi$ with regard to the earth fixed coordinate system $(X_{\mathrm{E}}, Y_{\mathrm{E}})$ it holds for the earth fixed velocities, see Fig. 12.2,

$$\begin{aligned} v_{X_{\mathrm{E}}}(t) &= v_X(t)\cos\psi - v_Y(t)\sin\psi, \\ v_{Y_{\mathrm{E}}}(t) &= v_X(t)\sin\psi + v_Y(t)\cos\psi. \end{aligned} \tag{12.1.2}$$

The position of the center of gravity is, starting with a position $(X_{\mathrm{E0}}, Y_{\mathrm{E0}})$,

$$\begin{aligned} X_{\mathrm{E}}(t) &= X_{\mathrm{E0}} + \int_0^t \left(v_X(\tau)\cos\psi(\tau) - v_Y(\tau)\sin\psi(\tau)\right)\mathrm{d}\tau, \\ Y_{\mathrm{E}}(t) &= Y_{\mathrm{E0}} + \int_0^t \left(v_X(\tau)\sin\psi(\tau) + v_Y(\tau)\cos\psi(\tau)\right)\mathrm{d}\tau, \end{aligned} \tag{12.1.3}$$

while the components of the ground velocity can be obtained from the vehicle slip angle $\hat{\beta}$ by using (7.3.68) and the ground velocity $\hat{v}$, determined, as in Sect. 12.2, by

$$v_X = \hat{v}\cos\hat{\beta}; \quad v_Y = \hat{v}\sin\hat{\beta}; \quad (\hat{v} = \hat{v}_{\mathrm{CG}}) \tag{12.1.4}$$

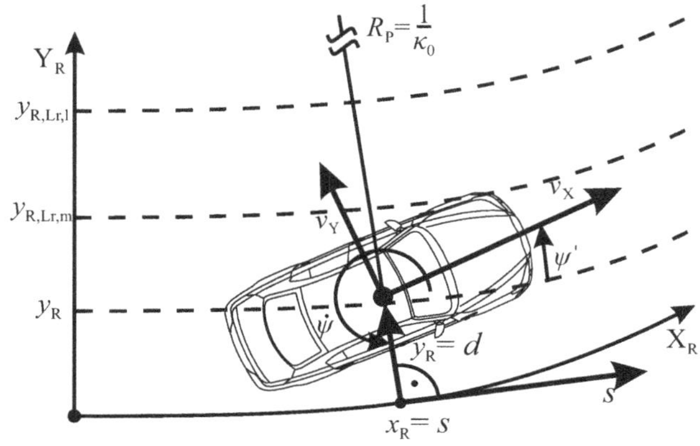


**Fig. 12.3** Motion of a vehicle in a road fixed coordinate system, a Frenet coordinate system

The slip angle estimate follows from the state estimation methods of Sect. 12.3.

The application of this method has shown that after a round trip of 2 km with velocities of up to 27 m s$^{-1}$ and lateral acceleration of up to 6 m s$^{-2}$, the difference between start point and estimation of the endpoint differed about 0.6 m and the orientation $\hat{\psi}$ by 0.7°, Schorn (2007).

### 12.1.2 Odometric Position Estimation for a Bent Road

During driving on a bent road, the position and orientation relative to a moving road coordinate system ($X_R$, $Y_R$) is of interest, see Fig. 12.3, where the $X_R$-axis follows the right border of a bent road and $Y_R$ is orthogonal to the $X_R$-axis, Eidehall et al. (2007), Schmitt and Isermann (2009). This corresponds to a *Frenet coordinate system* $[s, d]$ which is a moving reference frame for describing a curve locally. $d$ is perpendicular to $s$ and $s$ is a tangent to $X_R$. The velocities in this coordinate system are then, Schmitt (2012),

$$\begin{aligned} v_{X_R}(t) &= v_X(t)\cos\psi' - v_Y(t)\sin\psi', \\ v_{Y_R}(t) &= v_X(t)\sin\psi' + v_Y(t)\cos\psi', \end{aligned} \tag{12.1.5}$$

where $\psi'$ is a *relative yaw angle* with regard to a tangent to the road curve and, therefore,

$$\begin{aligned} \dot{\psi}'(t) &= \dot{\psi}(t) - \kappa_0 v_{XR} \\ &= \dot{\psi}(t) - \kappa_0\left(v_X(t)\cos\psi' - v_Y(t)\sin\psi'\right), \end{aligned} \tag{12.1.6}$$

where $\kappa_0 v$ describes the yaw rate due to the curvature $\kappa_0$ of the road.

Figure 12.3 depicts also the lateral positions $y_{R,Lr,m}$ and $y_{R,Lr,l}$ of the middle and left marking of the right lane. Hence, following state vector describes the vehicle state and its position and orientation on the right lane, Schmitt (2012),

$$\mathbf{x}_{\text{veh}}^{\text{T}} = \left[ v_{\text{X}} v_{\text{Y}} \dot{\psi} x_{\text{R}} y_{\text{R}} \psi' y_{\text{R,Lr,m}} \;\; y_{\text{R,Lr,1}} \kappa_0 \right]. \tag{12.1.7}$$

This state vector will, for example, be required for lane keeping control.

## 12.2 State Estimation of the Ground Velocity with Kinematic Vehicle Models

The use of basic kinematic or geometrical relations has the advantage that no dynamic vehicle models are required. The velocity $v$ of the center of gravity, or velocity over ground, in the earth fixed coordinate system is required for several vehicle control systems and especially for the ABS-system to determine the slip values of the tires. The most common way to determine $v$ is to use the angular speed sensors $\omega_{ij}$ of the wheels. However, during driving the driven wheels and during braking all wheels show slips and a direct calculation of the velocity from the circumferential velocity $v_{\omega,i} = \omega_{\text{W}i} r_i$ is not exactly possible. However, various approaches exist to use the information from the wheel angular velocities.

### 12.2.1 Use of the Wheel Angular Velocities

Taking the yaw rate $\dot{\psi}$ during cornering into account and neglecting the wheel slip, one obtains for the wheel circumferential velocities, Bauer (2015).

$$\begin{aligned}
v_{\omega,\text{fl}} &= \frac{\omega_{\text{fl}} r_{\text{dyn,fl}}}{\cos(\delta_{\text{f}} - \beta)} + \dot{\psi}\left(\frac{b_{\text{f}}}{2} - l_{\text{f}}\beta\right), \\
v_{\omega,\text{fr}} &= \frac{\omega_{\text{fr}} r_{\text{dyn,fr}}}{\cos(\delta_{\text{f}} - \beta)} - \dot{\psi}\left(\frac{b_{\text{f}}}{2} + l_{\text{f}}\beta\right), \\
v_{\omega,\text{rl}} &= \frac{\omega_{\text{rl}} r_{\text{dyn,rl}}}{\cos(\beta)} + \dot{\psi}\left(\frac{b_{\text{r}}}{2} + l_{\text{r}}\beta\right), \\
v_{\omega,\text{rr}} &= \omega_{\text{rr}} r_{\text{dyn,rr}} - \dot{\psi}\left(\frac{b_{\text{r}}}{2} - l_{\text{r}}\beta\right).
\end{aligned} \tag{12.2.1}$$

During normal driving situations with higher speed the slip angle can be neglected, Daiss (1996), by setting $\beta$ to zero. According to the select-low principle, Burckhardt (1993), the velocity $v = v_\omega$ is taken from the wheel with the smallest slip, or from a maximum of all $v_{\omega_{ij}}$, assuming there is the smallest slip:

$$v_\omega = \max\left[v_{\omega_{ij}}\right]. \tag{12.2.2}$$

### 12.2.2 Use of the Wheel Angular Velocities and the Acceleration

The acceleration of the center of gravity $a_{\mathrm{CG}}$ measured by an accelerometer or an inertial measurement unit (IMU) may be used to determine the velocity by integration and then to combine it with the velocity from the wheel velocities.

The relation between the acceleration and the velocity is given by

$$a = \dot{v} = \frac{\mathrm{d}v}{\mathrm{d}t}. \tag{12.2.3}$$

Hence, the velocity can be obtained by an integration of the acceleration:

$$v(t) = \int a(t)\,\mathrm{d}t. \tag{12.2.4}$$

After discretization of (12.2.3) with the sample time $T_0$,

$$\dot{v}(k) = \frac{\Delta v(k)}{T_0} = \frac{1}{T_0}\left(v(k) - v(k-1)\right), \tag{12.2.5}$$

it holds for the velocity

$$v_{\mathrm{CG}}(k) = v_{\mathrm{CG}}(k-1) + T_0 a_{\mathrm{CG}}(k). \tag{12.2.6}$$

This leads to a state-space process representation:

$$\begin{gathered}\begin{bmatrix}\hat{v}_{\mathrm{CG}}(k)\\ \hat{a}_{\mathrm{CG}}(k)\end{bmatrix} = \begin{bmatrix}1 & T_0\\ 0 & 1\end{bmatrix}\begin{bmatrix}\hat{v}_{\mathrm{CG}}(k-1)\\ \hat{a}_{\mathrm{CG}}(k-1)\end{bmatrix} + \begin{bmatrix}\nu_v(k)\\ \nu_{\mathrm{a}}(k)\end{bmatrix}\\ \hat{\mathbf{x}}(k) = \mathbf{A}\hat{\mathbf{x}}(k-1) + \boldsymbol{\nu}(k).\end{gathered} \tag{12.2.7}$$

This equation can be considered as a process model without input signal and with stochastic disturbances $\nu(k)$, where $\hat{v}_{\mathrm{CG}}(k)$ and $\hat{a}_{\mathrm{CG}}(k)$ are predicted values based on the last values $\hat{v}_{\mathrm{CG}}(k-1)$ and $\hat{a}_{\mathrm{CG}}(k-1)$.

The measurement model for a new measurement at time $k$ gives the output signals

$$\begin{gathered}\begin{bmatrix}v_\omega(k)\\ a_{\mathrm{CG}}(k)\end{bmatrix} = \begin{bmatrix}1 & 0\\ 0 & 1\end{bmatrix}\begin{bmatrix}\hat{v}_{\mathrm{CG}}(k)\\ \hat{a}_{\mathrm{CG}}(k)\end{bmatrix} + \begin{bmatrix}n_v(k)\\ n_a(k)\end{bmatrix}\\ \mathbf{y}(k) = \mathbf{C}\hat{\mathbf{x}}(k) + \mathbf{n}(k),\end{gathered} \tag{12.2.8}$$

where $v_\omega(k)$ is the velocity determined by the wheel velocities (11.4.2) and (12.2.1) and $a_{\mathrm{CG}}(k)$ is the measured acceleration.

A Kalman filter now combines the measurements $v_\omega(k)$ and $a_{\mathrm{CG}}(k)$, see Fig. 10.7,

$$\hat{\mathbf{x}}(k) = \hat{\mathbf{x}}(k-1) + \mathbf{K}\left[\mathbf{y}(k) - \mathbf{C}\hat{\mathbf{x}}(k)\right], \tag{12.2.9}$$

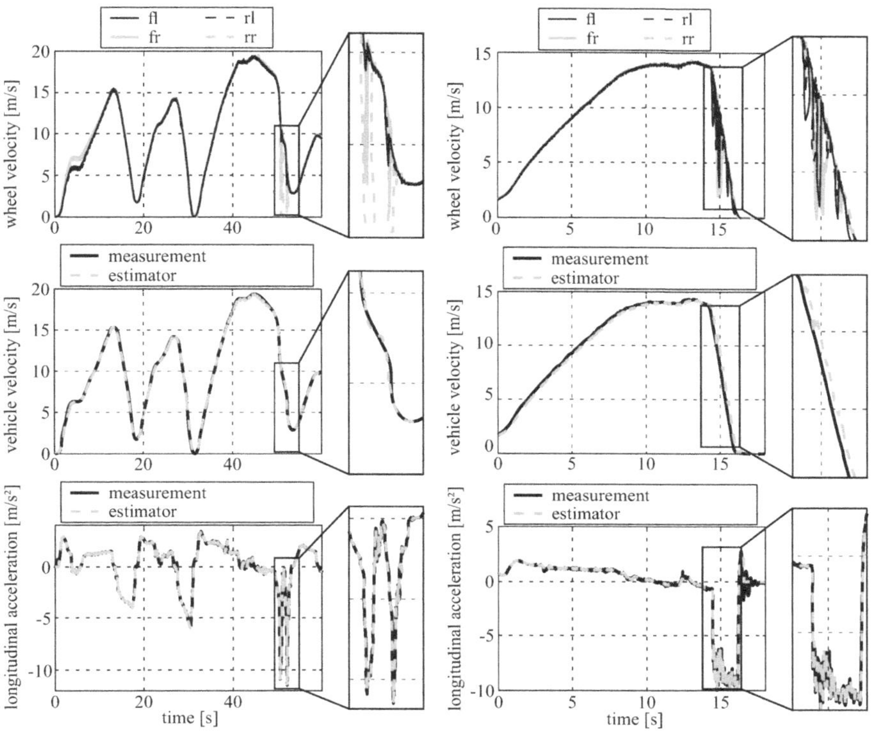


(a) dry asphalt

(b) wet asphalt

**Fig. 12.4** Estimation of the vehicle velocity $\hat{v}_{\mathrm{CG}}$ and longitudinal acceleration $\hat{a}_{\mathrm{CG}}$ by Kalman filtering with a classification of the covariance matrix **R** in several cases. Front-driven VW Golf IV, 74 kW (2000). **a** Acceleration and braking on dry asphalt, **b** acceleration and full braking on wet asphalt, Schorn (2007)

with the gain (10.2.52)

$$\mathbf{K} = \mathbf{P}^{-}\mathbf{C}^{\mathrm{T}}\left[\mathbf{C}\mathbf{P}^{-}\mathbf{C}^{\mathrm{T}} + \mathbf{N}\right]^{-1}, \tag{12.2.10}$$

and the covariances of the state variables

$$\mathbf{Q} = \begin{bmatrix} \sigma_v^2 & 0 \\ 0 & \sigma_a^2 \end{bmatrix}, \tag{12.2.11}$$

and of the measured outputs

$$\mathbf{R} = \begin{bmatrix} \sigma_{v_\omega}^2 & 0 \\ 0 & \sigma_{a_{\mathrm{CG}}}^2 \end{bmatrix}. \tag{12.2.12}$$

The state estimate $\hat{\mathbf{x}}(k)$ thus determines an estimate of the velocity $v_{\mathrm{CG}}$ of the center of gravity as a weighted combination of the measured wheel circumferential velocity $v_\omega(k)$ and the measured acceleration $a_{\mathrm{CG}}(k)$.

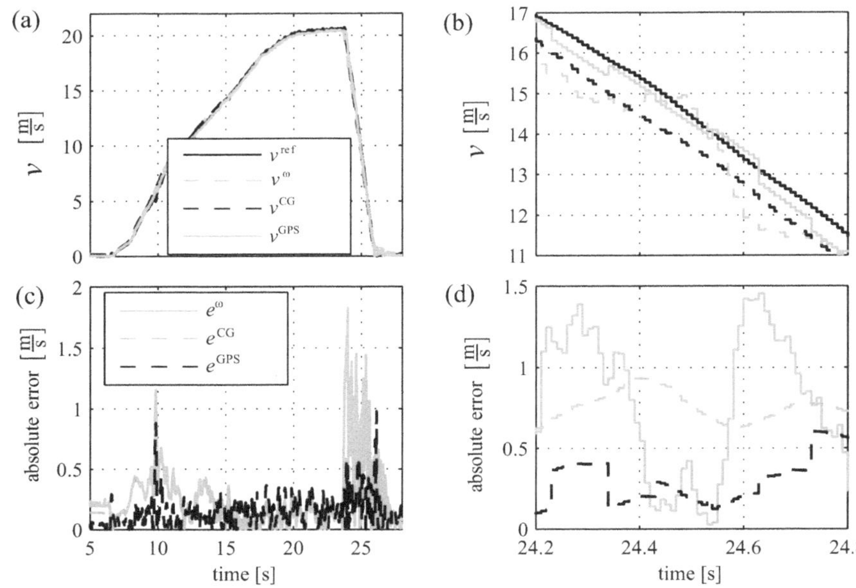


**Fig. 12.5** Determination of the velocity $\hat{v}_{CG}$ of the center of gravity based on wheel velocities and acceleration measurement and use of a Kalman filter, Bauer (2015). ABS-braking on a wet road, BMW 540i (E60). **a** velocities, **b** zoom of **a**, **c** error of estimates $e_i$, **d** zoom of **c**. $v_{ref}$: direct measurement with a CORREVIT-sensor; $\hat{v}^{\omega}$: velocity from wheel velocities; $\hat{v}^{CG}$: velocity as combination of $\hat{v}_{\omega}$ and $a_{CG}$; $\hat{v}^{GPS}$: fusion with GPS-measurement. $T_0 = 10$ ms

The variances **R** of the measurements can be adapted to the driving situation. For small slip values $\sigma^2_{v_\omega}$ they can be smaller and for large ones (strong braking) larger, Halbe (2008), Daiss (1996). Schorn (2007) distinguishes several cases for the adaptation of **R**, i.e. positive and negative acceleration, cornering, and locking of wheels, and obtains good results for a front-driven car, also for full braking on a wet road with active ABS; see Fig. 12.4.

Figure 12.5 shows measured results for an ABS-braking on a wet road, Bauer (2015). The largest error is obtained by using only the wheel velocity sensors ($R^2 = 0.88$), smaller errors by the combination with the acceleration measurement ($R^2 = 0.93$) and the smallest errors by additional combination with a GPS-sensor ($R^2 = 0.99$); see Sect. 12.6.

## 12.3 State Estimation for the Lateral Vehicle Behavior

For controlling the lateral vehicle behavior with regard to lane following, lane change, and especially for electronic stability control (ESC) and emergency swerving maneuvers, the lateral state variables side slip angle $\beta(t)$ or the longitudinal and lateral velocity components of the center of gravity $v_X(t)$ and $v_Y(t)$ are required. As they cannot be measured directly, they have to be calculated for special driving cases as stationary cornering or estimated by state observers or Kalman filters for general dynamic cornering. The basic models for these estimation procedures are one-track and two-track models. The available on-board measurements are usually steering angle $\delta_{\mathrm{f}}(t)$, the yaw rate $\dot{\psi}(t)$, the (estimated) ground velocity $v(t)$, and for advanced instrumentation $a_{\mathrm{Y}}(t)$. In addition to the estimated variables $\hat{\beta}(t)$ or $\hat{v}_{\mathrm{X}}(t)$ and $\hat{v}_{\mathrm{Y}}(t)$ also improved variables as $\hat{\dot{\psi}}(t)$ are obtained. Figure 12.6 depicts an overview of the various possibilities.

The use of *state observers* for the reconstruction of vehicle state variables and especially the vehicle slip angle $\beta$ are, for example, treated in Daiss (1996), Kiencke and Nielsen (2005), Börner (2004), Vietinghoff and Hiemer (2005), Baffet et al. (2007), Bechtloff et al. (2015).

Approaches of vehicle state estimates with *Kalman filters* are, for example, published by Ray (1995), Samadi et al. (2001), Arndt et al. (2004), Baffet et al. (2007), Doumiati et al. (2011).

The following treatment of vehicle state reconstruction and state estimation methods is mainly based on developments by Würtenberger (1997), Börner (2004), Wesemeier and Isermann (2007), Schorn (2007), Halbe (2008), Bauer (2015), Bechtloff (2018) partially also with tire model adaptation.

### 12.3.1 Slip Angle Estimation for Special Driving Maneuvers with Kinematic Models

The vehicle slip angle $\beta$ is an important driving variable, because it shows the direction of the center of gravity velocity $v$ relative to the vehicle longitudinal axis $X_{\mathrm{V}}$; see Figs. 7.2 and 7.3. It can usually not directly be measured and has to be calculated from measurable variables. For some special cases, the slip angle can be directly determined. However, in general it has to be reconstructed with state estimation methods by using dynamic vehicle models.

Assuming a constant momentary pole MP, a constant radius $R$, and slow velocity, e.g. for parking, it holds for a one-track model according to Fig. 7.2:

$$\sin\beta = \frac{l_{\mathrm{r}}}{R_{\mathrm{r}}} = \frac{l_{\mathrm{r}}}{R\cos\beta}, \tag{12.3.1}$$

and

$$\sin\delta_{\mathrm{f}} = \frac{l}{R_{\mathrm{f}}} = \frac{l\cos\delta_{\mathrm{f}}}{R\cos\beta}. \tag{12.3.2}$$

Dividing both equations yields

$$\frac{\sin\beta}{\sin\delta_{\mathrm{f}}} = \frac{l_{\mathrm{r}}}{l\cos\delta_{\mathrm{f}}} \tag{12.3.3}$$

and

$$\sin\beta = \frac{l_{\mathrm{r}}}{l}\tan\delta_{\mathrm{f}}, \tag{12.3.4}$$

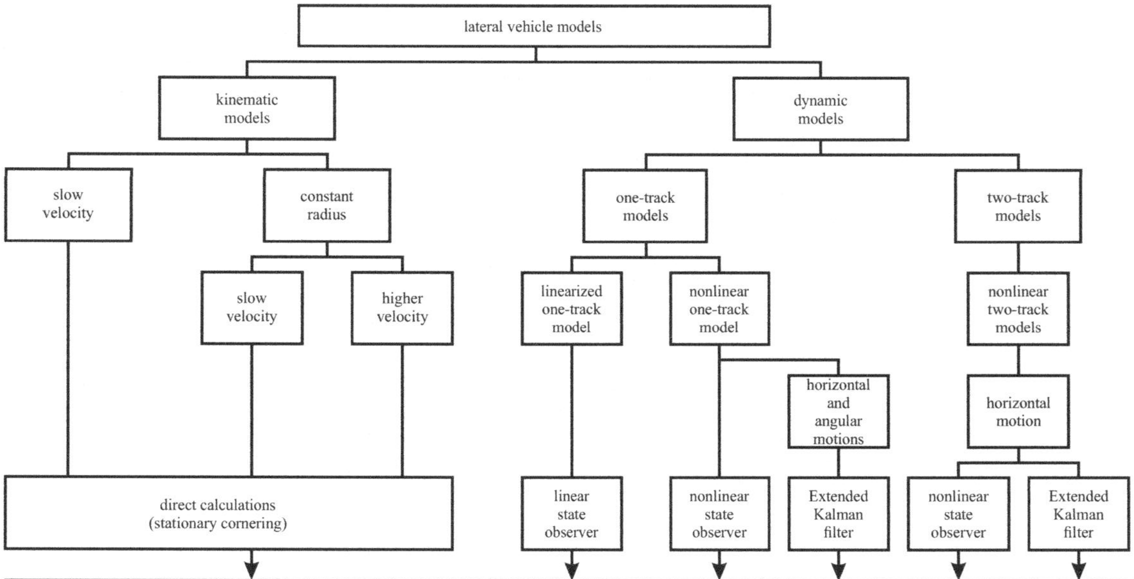


| | | | | | | | | |
|---|---|---|---|---|---|---|---|---|
| measured variables | | $\delta_f$ or $a_Y,\dot{\psi},v$ | $\delta_f,v$ | $\delta_f,\dot{\psi}$ | | $\delta_f,\delta_r,a_X,a_Y,\dot{\psi},\omega_{ij}$ | $\delta_f,\dot{\psi},\omega_{ij}$ | $\delta_f,\dot{a}_X,a_Y,\psi$ |
| state variable estimate | yaw-rate/ velocity representation | $\hat{\beta}$ | $\hat{\beta}$ | | | | | $\hat{v}_X,\hat{v}_Y,\hat{\dot{\psi}},\hat{\beta}$ |
| | yaw-rate/ slip angle representation | | | $\hat{\beta},\hat{\dot{\psi}}$ | | $\hat{v},\hat{\beta},\hat{\mu}_{max},\hat{\varphi},\hat{\theta},\hat{c}_{\alpha i}$, $\hat{SG},\Delta\hat{\mu}$ | $\hat{v}_X,\hat{\beta},\hat{\dot{\psi}}$ | |
| section | | (12.3.7) | (12.3.8) | (12.3.9), (12.3.10) | | Sect. 12.5 | (12.3.15) | Sect. 12.3.3, 12.5, 12.6 |

**Fig. 12.6** Overview scheme for the estimation of vehicle state variables with different lateral vehicle models (treated in Sects. 12.3 and 12.5), without GPS signals

or

$$\beta = \arcsin\left(\frac{l_{\mathrm{r}}}{l}\tan\delta_{\mathrm{f}}\right) \tag{12.3.5}$$

which can be simplified for small angles to

$$\beta \approx \frac{l_{\mathrm{r}}}{l}\delta_{\mathrm{f}}, \tag{12.3.6}$$

see, for example, Kochem (2005).

A further relation for slow cornering follows from (7.1.7) and (7.1.8):

$$\begin{aligned} \cos\beta &= \frac{a_{\mathrm{Y}}}{a_{\mathrm{c}}} = \frac{a_{\mathrm{Y}}}{v\dot{\psi}} \\ \beta &= \arccos\left(\frac{a_{\mathrm{Y}}}{v\dot{\psi}}\right). \end{aligned} \tag{12.3.7}$$

Thus, $\beta$ can be determined for these special cases, if $\delta_{\mathrm{f}}$ or $a_{\mathrm{Y}}$, $v$, and $\dot{\psi}$ are available as measurements. However, both methods are only applicable for slow velocities, if no tire slip occurs and the measurements for (12.3.7) are not too noisy with regard to the small measured variables.

For higher velocities, the lateral tire forces generate side slip angles $\alpha_{\mathrm{f}}$ and $\alpha_{\mathrm{r}}$; see Fig. 7.3. The *one-track model* (7.2.59) for *stationary cornering* then yields with $i_{\mathrm{S}} = \delta_{\mathrm{H}}/\delta_{\mathrm{f}}$

$$\beta = \delta_{\mathrm{f}}\frac{l_{\mathrm{r}}}{l}\frac{1 - \frac{ml_{\mathrm{f}}}{c_{\alpha,\mathrm{r}}l_{\mathrm{r}}l}v^2}{1 + \left(\frac{v}{v_{\mathrm{ch}}}\right)^2}, \tag{12.3.8}$$

with the characteristic velocity $v_{\mathrm{ch}}$ due to (7.2.52). This relation assumes a linear lateral tire force characteristic with the tire stiffnesses $c_{\alpha\mathrm{f}}$ and $c_{\alpha\mathrm{r}}$, see (7.2.18), which have to be known.

## 12.3.2 Slip Angle Estimation with State Observers (General Dynamic Driving Maneuvers)

**(a) Linear State Observer (One-Track Model)**

Assuming dynamic driving maneuvers with small steering angles $\delta_{\mathrm{f}}$ and small side slip angles $\alpha$, the *linearized one-track model* of Sect. 7.2.2 can be used. The dynamic state variable model is according to (7.2.34)

$$\begin{bmatrix} \dot{\beta}(t) \\ \ddot{\psi}(t) \end{bmatrix} = \mathbf{A}\begin{bmatrix} \beta(t) \\ \dot{\psi}(t) \end{bmatrix} + \mathbf{b}\,\delta_{\mathrm{f}}(t), \tag{12.3.9}$$

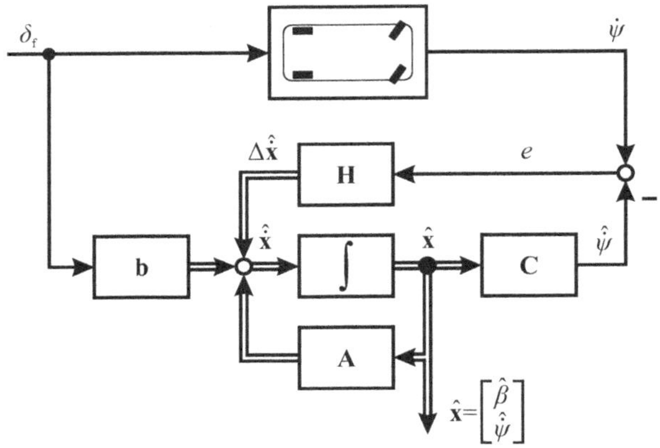


**Fig. 12.7** Linear slip angle state observer with a one-track model

with the measured output

$$y(t) = \dot{\psi}(t) = \mathbf{C}\begin{bmatrix}\beta(t)\\ \dot{\psi}(t)\end{bmatrix}. \tag{12.3.10}$$

The signal flow between the steering angle $\delta_f$ as input and $\dot{\psi}$, respectively $\psi$, as output in Fig. 7.5 shows that the slip angle $\beta$ influences as feedback the side slip angles of the front and the rear tire and, therefore, the lateral forces. Hence, the slip angle $\beta(t)$ is an interrelated variable of the one-track model, and it has to be reconstructed by a state observer or estimated with a Kalman filter.

Figure 12.7 depicts the signal flow of a linear state observer. The feedback matrix **H** has to be selected, for example, by pole placement of the eigenvalues of (10.2.10), so that the observer is stable.

The reconstructed state variables of this linear one-track model observer are then $\hat{\beta}(t)$ and $\hat{\dot{\psi}}(t)$. However, **A** and **b** require that the following parameters are known: $m$, $J_Z$, $c_{\alpha,f}$, and $c_{\alpha,r}$. The estimation of these parameters is described in Chap 11. Different methods to estimate the cornering stiffnesses $c_{\alpha,f}$ and $c_{\alpha,r}$ are treated in Sect. 11.6.2. The determination of these stiffnesses within the linear tire force characteristic or of $k_{\alpha,f}$ and $k_{\alpha,r}$ for the (simplified) nonlinear case applies the least squares parameter estimation according to (11.6.28) via the determination of the lateral axle forces with (11.6.40) and (11.6.41). The herewith required $\beta$ may be obtained in a sequential, delayed procedure from the observer.

An application of this combined one-track observer and recursive parameter estimation is described in Schorn (2007). Results for the slip angle estimation with a linear state observer are presented in Daiss (1996), with better results by an adaption of the cornering stiffnesses. Important are good tire parameters. However, several publications have shown that significantly better estimates are obtained with nonlinear observers and two-track models.

**(b) Nonlinear State Observer (Two-Track Model, Horizontal Motion)**

A more precise estimation of the side slip angle $\beta(t)$ can be obtained by using a *two-track model*, where the lateral and longitudinal forces of all wheels are taken into account. This is also required for ESC-control functions with individual wheel braking. A first used two-track model is the *yaw rate/slip angle representation*, (7.3.39)–(7.3.41); see also Fig. 7.25. As it is nonlinear, a nonlinear state observer has to be designed, Daiss (1996), Schorn (2007). The required state variable representation of the process model is

$$\dot{\mathbf{x}}(t) = \begin{bmatrix} \dot{v}(t) \\ \dot{\beta}(t) \\ \ddot{\psi}(t) \end{bmatrix} = \begin{bmatrix} f_1(\mathbf{x}, \mathbf{u}) \\ f_2(\mathbf{x}, \mathbf{u}) \\ f_3(\mathbf{x}, \mathbf{u}) \end{bmatrix} = \mathbf{f}(\mathbf{x}, \mathbf{u}), \tag{12.3.11}$$

with the output measurements

$$\mathbf{y}(t) = \begin{bmatrix} v(t) \\ \dot{\psi}(t) \end{bmatrix} = \mathbf{h}(\mathbf{x}), \tag{12.3.12}$$

with the state variable vector

$$\mathbf{x}(t) = \begin{bmatrix} v_\mathrm{X}(t) & \beta(t) & \dot{\psi}(t) \end{bmatrix}^\mathrm{T}, \tag{12.3.13}$$

and the input vector

$$\mathbf{u}(t) = \begin{bmatrix} \delta_\mathrm{f}(t) & F_{\mathrm{XT,fl}} & F_{\mathrm{XT,fr}} & F_{\mathrm{XT,rl}} & F_{\mathrm{XT,rr}} \end{bmatrix}^\mathrm{T}. \tag{12.3.14}$$

The considered yaw rate/slip angle equations of the two-track model are (7.3.39)–(7.3.41), now abbreviated as

$$\dot{v} = f_1(\mathbf{F}_{\mathrm{XT}}, \mathbf{F}_{\mathrm{YT}}, \mathbf{F}_{\mathrm{XA}}, \mathbf{F}_{\mathrm{XR}}, \delta_\mathrm{f}, \beta),$$

$$\dot{\beta} = f_2(\mathbf{F}_{\mathrm{XT}}, \mathbf{F}_{\mathrm{YT}}, \mathbf{F}_{\mathrm{YA}}, \delta_\mathrm{f}, \beta, \dot{\psi}),$$

and

$$\ddot{\psi} = f_3(\mathbf{F}_{\mathrm{XT}}, \mathbf{F}_{\mathrm{YT}}, \delta_\mathrm{f}).$$

They already contain the transformation from the tire forces to vehicle forces.

The state observer for the nonlinear two-track model (7.3.35)–(7.3.38) has to be designed for the state variables $\mathbf{x}(t)$ of (12.3.13), the input variables $\mathbf{u}(t)$ of (12.3.14), and the measured output variables $\mathbf{y}(t)$ of (12.3.12).

The observer design follows Sect. 10.2.2 in the form

$$\begin{aligned} \dot{\hat{x}} &= \mathbf{f}(\hat{\mathbf{x}}, \mathbf{u}) + \mathbf{L}(\hat{\mathbf{x}}, \mathbf{u})(\mathbf{y} - \hat{\mathbf{y}}) \\ \hat{\mathbf{y}} &= \mathbf{h}(\hat{\mathbf{x}}, \mathbf{u}), \end{aligned} \tag{12.3.15}$$

compare Fig. 12.8. It is or 3rd order and the poles $\lambda_1$ and $\lambda_3$ of the feedback matrix $\mathbf{L}$, see (10.2.23), have to be selected such that the observer is stable; see Börner (2004), Schorn (2007), Adamy (2018).

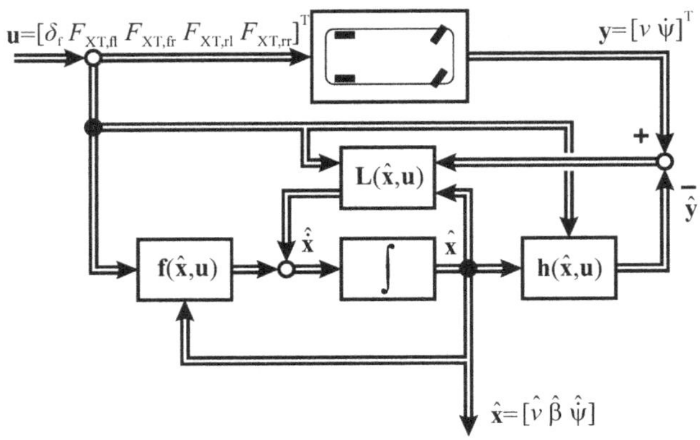


**Fig. 12.8** Nonlinear slip angle state observer with a two-track model in yaw rate/slip angle representation

A signal-flow scheme of used models and calculations for the nonlinear state observer is depicted in Fig. 12.9. It shows the various steps from the required on-board measurements through the parameter estimation of the tire models until the final state estimates $\hat{\mathbf{x}}(t)$ and follows Börner (2004) and Schorn (2007).

The used two-track model according to Sect. 7.3.2b) and Fig. 7.25 requires the longitudinal and lateral forces of the tires $\mathbf{F}_{\mathrm{XT}}$ and $\mathbf{F}_{\mathrm{YT}}$, which are treated separately, and of the chassis $\mathbf{F}_{\mathrm{X}}$ and $\mathbf{F}_{\mathrm{Y}}$.

The determination of the longitudinal forces $\mathbf{F}_{\mathrm{XT}}$ with (5.1.3),

$$F_{\mathrm{XT}} = \mu_{\mathrm{X}}(S_{\mathrm{X}})F_{\mathrm{ZT}},$$

needs as inputs the wheel braking torques $\mathbf{M}_{\mathrm{B}}$ or the wheel driving torques $\mathbf{M}_{\mathrm{D}}$, see Sect. 5.5 and 6.5, and the friction coefficients ¯$_{\mathrm{X}}(S_{\mathrm{X}})$, see Sect. 5.1, with the vertical tire forces $\mathbf{F}_{\mathrm{ZT}}$ from (7.3.22).

To calculate the lateral tire forces $\mathbf{F}_{\mathrm{YT}}$, (5.2.3) can be used with (5.2.5) for small slip angles $\alpha$,

$$F_{\mathrm{YT}} = c_{\alpha}(F_{\mathrm{ZT}})\alpha,$$

or (5.2.11) with (5.2.7) for larger $\alpha$. The side slip angles follow from (7.2.25) and (7.2.26).

As the estimation of the vehicle slip angle $\beta$ is especially required for dynamic steering wheel inputs $\delta_{\mathrm{H}}(t)$, respectively, cornering with $\delta_{\mathrm{H}} = \text{const.}$, the models for the lateral tire forces should be adaptive to the actual tire/road condition. One possibility is to apply the cornering stiffness parameter estimation of Sect. 11.6.2a in order to obtain an estimate of $\hat{c}_{\alpha\mathrm{f}}$ and $\hat{c}_{\alpha\mathrm{r}}$ for the front and rear axle wheels by using a one-track model.

The lateral forces are distributed to the left and right wheel according to the individual vertical tire forces, as for the front wheels, Schorn (2007),

$$\begin{aligned} F_{\mathrm{YT,fl}} &= F_{\mathrm{YT,f}} \frac{F_{\mathrm{ZT,fl}}}{F_{\mathrm{ZT,fl}} + F_{\mathrm{ZT,fr}}}, \\ F_{\mathrm{YT,fr}} &= F_{\mathrm{YT,f}} \frac{F_{\mathrm{ZT,fr}}}{F_{\mathrm{ZT,fl}} + F_{\mathrm{ZT,fr}}}. \end{aligned} \tag{12.3.16}$$

The determination of the side slip angles $\alpha_i$ and the cornering stiffness adaptation require the vehicle slip angle $\beta$. Therefore, the finally estimated $\hat{\beta}$ introduces feedback signal paths in Fig. 12.9.

Finally, the two-track model in yaw rate/slip angle representation (7.3.39)–(7.3.41) can be used for the nonlinear observer.

Experimental results with this nonlinear observer and parameter estimation of the cornering stiffnesses have been obtained by Schorn (2007) for a front-driven car. Figures 12.10 and 12.11 depict the vehicle slip angle estimates $\hat{\beta}(t)$ for a slalom maneuver and a double lane change. The agreement with the direct measurement of $\beta(t)$ by using an optical Correvit-sensor as reference is quite good. Additionally, the side slip angle of the rear axle wheel was determined by (7.2.26) of the linearized one-track model

$$\hat{\alpha}_{\mathrm{r}} = -\hat{\beta} + \frac{l_{\mathrm{r}}}{v}\dot{\psi}, \tag{12.3.17}$$

and shows also a good agreement.

In order to circumvent the lateral tire force models $F_{\mathrm{YT}}(\alpha)$, the lateral forces of the vehicle can alternatively directly be determined via the force and torque balance equations (7.2.12) and (7.2.13) for a one-track model. With the assumption of small vehicle slip angles $\beta$, it holds, see (11.6.42),

$$\begin{aligned} F_{\mathrm{Y,f}} &= \frac{1}{l}(J_{\mathrm{Z}}\ddot{\psi} + l_{\mathrm{r}} m a_{\mathrm{Y}}), \\ F_{\mathrm{Y,r}} &= \frac{1}{l}(-J_{\mathrm{Z}}\ddot{\psi} + l_{\mathrm{f}} m a_{\mathrm{Y}}). \end{aligned} \tag{12.3.18}$$

In addition to the measurement of $\delta_{\mathrm{f}}(t)$, $v(t)$, and $\dot{\psi}(t)$, (12.3.9) and (12.3.10), now the lateral acceleration $a_{\mathrm{Y}}$ has to be measured.

### 12.3.3 Slip Angle Estimation with Kalman Filters

State observers are designed under the assumption that no disturbances from measurements and from the process occur. Kalman filters (KF) for state estimation include as well process noise as measurement noise in their design. In addition, they use a state prediction by one sampling interval. To operate with nonlinear process models extended Kalman filters (EKF) can be applied. In this section, different approaches for slip angle estimation are described operating with nonlinear two-track

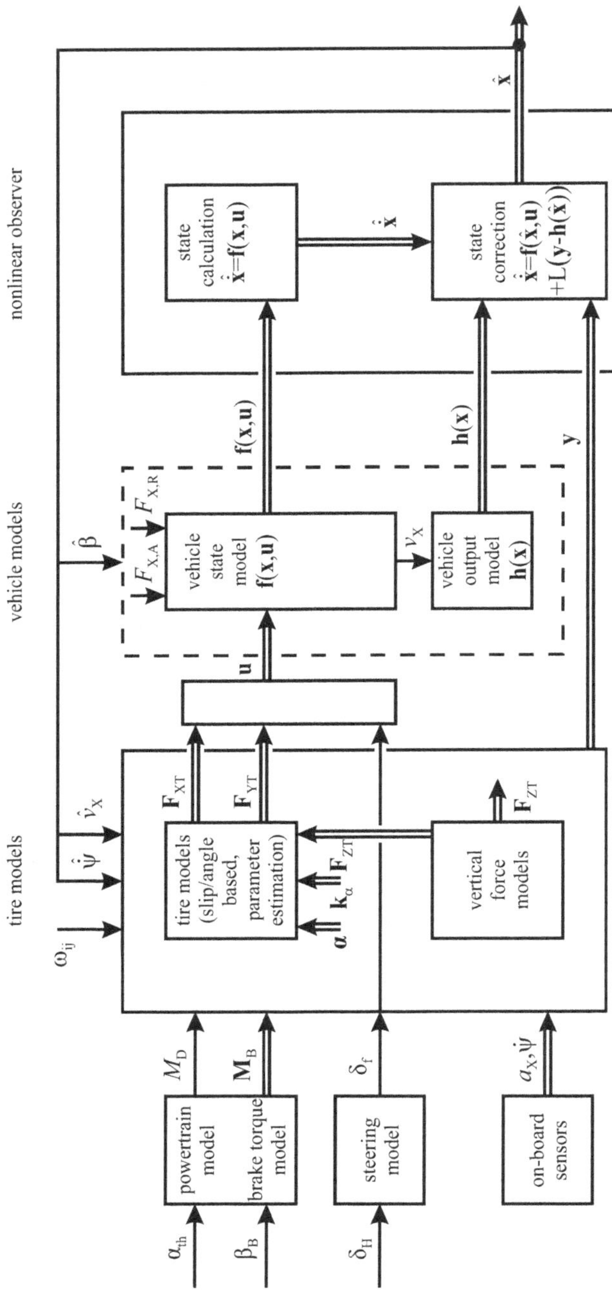


**Fig. 12.9** Signal flow diagram for a nonlinear vehicle slip angle state observer using a two-track model in yaw rate/slip angle representation and a one-track model for the adaptive tire cornering stiffnesses, continuous-time representation, Schorn (2007)

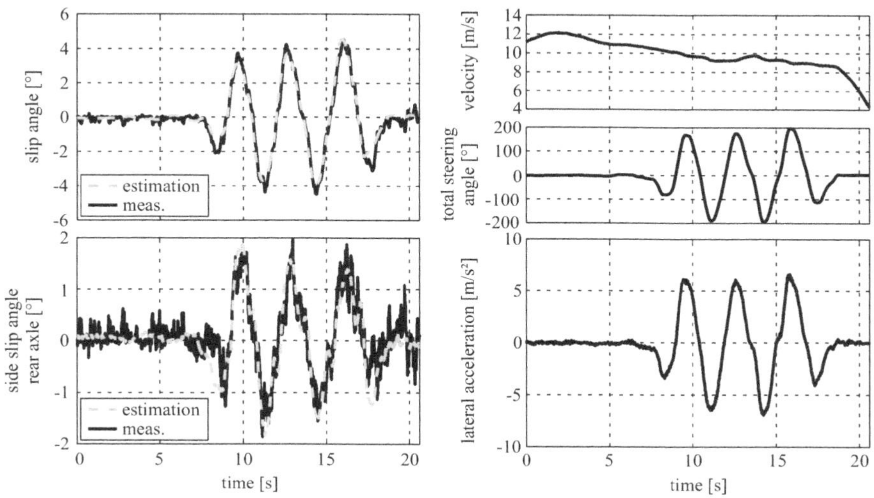


**Fig. 12.10** Experimental results with a nonlinear slip angle observer with parameter estimation of the cornering stiffness $k_{\alpha,i}$ for a slalom course, Golf IV, Schorn (2007)

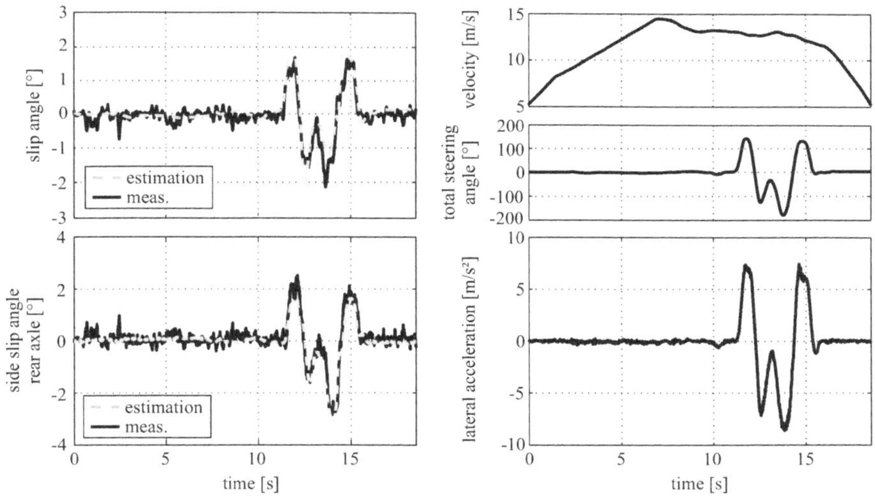


**Fig. 12.11** Experimental results of a nonlinear slip angle observer with parameter estimation of the cornering stiffness $k_{\alpha,i}$ for a double lane change, Golf IV, Schorn (2007)

models. Section 12.5 shows an extended Kalman filter for a one-track model, and Sect. 12.6 a method with the inclusion of 3D-GPS-measurements.

**(a) Extended Kalman Filter with Yaw Rate/Velocity Two-Track Model and Variable Longitudinal Velocity**
The basic recursive equations for the Kalman filter (KF) and the extended Kalman filter (EKF) are described in Sect. 10.2. As the classical Kalman filter operates with discrete-time models the *nonlinear two-track model*, now in the yaw rate/velocity representation for an even road, (7.3.12)–(7.3.14) and Fig. 7.22, has to be discretized.

Assuming small sampling times $T_0$ and, therefore, the discrete time $k = t/T_0$, it holds for $t = k + 1$

$$\dot{v}(t) = \frac{\mathrm{d}v(t)}{\mathrm{d}t} \approx \frac{\Delta v(k+1)}{T_0} = \frac{v(k+1) - v(k)}{T_0}. \tag{12.3.19}$$

The two-track model for horizontal motion then becomes

$$\begin{aligned} f_1(k) = v_\mathrm{X}(k+1) = v_\mathrm{X}(k) + \frac{T_0}{m}\Big(&F_{\mathrm{X,fl}} + F_{\mathrm{X,fr}} + F_{\mathrm{X,rl}} \\ &+ F_{\mathrm{X,rr}} - F_{\mathrm{XA}} - F_{\mathrm{XR}}\Big)_k + T_0\dot{\psi}(k)v_\mathrm{Y}(k), \end{aligned} \tag{12.3.20}$$

$$\begin{aligned} f_2(k) = v_\mathrm{Y}(k+1) = v_\mathrm{Y}(k) + \frac{T_0}{m}\Big(&F_{\mathrm{Y,fl}} + F_{\mathrm{Y,fr}} + F_{\mathrm{Y,rl}} \\ &+ F_{\mathrm{Y,rr}} - F_{\mathrm{YA}}\Big)_k - T_0\dot{\psi}(k)v_\mathrm{X}(k), \end{aligned} \tag{12.3.21}$$

$$\begin{aligned} f_3(k) = \dot{\psi}(k+1) = \dot{\psi}(k) + \frac{T_0}{J_\mathrm{Z}}\Big[&(F_{\mathrm{X,fr}} - F_{\mathrm{X,fl}})\frac{b_\mathrm{f}}{2} + (F_{\mathrm{X,rr}} - F_{\mathrm{X,rl}})\frac{b_\mathrm{r}}{2} \\ &+ (F_{\mathrm{Y,fl}} + F_{\mathrm{Y,fr}})J_\mathrm{f} - (F_{\mathrm{Y,rl}} + F_{\mathrm{Y,rr}})l_\mathrm{r}\Big]_k, \end{aligned} \tag{12.3.22}$$

see Schorn (2007), Bauer (2015).

This nonlinear process model is now the basis for the *state prediction* of an extended Kalman filter according to (10.2.77)

$$\hat{\mathbf{x}}(k+1|k) = f_k(\mathbf{x}(k), \mathbf{u}(k)) + \mathbf{v}(k), \tag{12.3.23}$$

with the output equation (10.2.64)

$$\mathbf{y} = \mathbf{g}_\mathrm{k}(x(k)) + \mathbf{n}(k), \tag{12.3.24}$$

where $\mathbf{v}(k)$ and $\mathbf{n}(k)$ are independent white noise signals with (known) covariance matrices $\mathbf{Q}$ and $\mathbf{R}$; see (10.2.32). The state vector is

$$\mathbf{x}(k) = \left[v_\mathrm{X}(k) \;\; v_\mathrm{Y}(k) \;\; \dot{\psi}(k)\right]^\mathrm{T}, \tag{12.3.25}$$

and the input vector

$$\mathbf{u}(k) = [\mathbf{F}_\mathrm{X} \ \ \mathbf{F}_\mathrm{Y} \ \ \mathbf{F}_\mathrm{XA} \ \ \mathbf{F}_\mathrm{XR}]_k^\mathrm{T} , \tag{12.3.26}$$

with the longitudinal and lateral wheel forces, acting on the chassis,

$$\begin{aligned} \mathbf{F}_\mathrm{X} &= \left[F_\mathrm{X,fl} \ \ F_\mathrm{X,fr} \ \ F_\mathrm{X,rl} \ \ F_\mathrm{X,rr}\right] , \\ \mathbf{F}_\mathrm{Y} &= \left[F_\mathrm{Y,fl} \ \ F_\mathrm{Y,fr} \ \ F_\mathrm{Y,rl} \ \ F_\mathrm{Y,rr}\right] . \end{aligned} \tag{12.3.27}$$

Hence, the two-track model (12.3.20)–(12.3.22) provides the state prediction vector $\hat{\mathbf{x}}(k+1|k)$ for the extended Kalman filter, compare (10.2.65), (12.3.23), and Fig. 10.6.

The process state function matrix is

$$\mathbf{f}_k = [f_1(k) \ \ f_2(k) \ \ f_3(k)]^\mathrm{T} . \tag{12.3.28}$$

As measurements for the two-track model (7.3.12)–(7.3.14), the on-board signals

$$\mathbf{y}(k) = \left[a_\mathrm{X}(k) \ \ a_\mathrm{Y}(k) \ \ \dot{\psi}(k)\right]^\mathrm{T} \tag{12.3.29}$$

are available.

The predicted process model outputs then become with (12.3.24), for a next sample step

$$\mathbf{y}(k+1) = \mathbf{g}_{k+1}(\hat{\mathbf{x}}(k+1) + \mathbf{n}(k+1)), \tag{12.3.30}$$

$$\begin{aligned} \mathbf{g}_{k+1} = \begin{bmatrix} g_1 \\ g_2 \\ g_3 \end{bmatrix}_{k+1} &= \begin{bmatrix} \hat{a}_\mathrm{X}(k+1) \\ \hat{a}_\mathrm{Y}(k+1) \\ \hat{\dot{\psi}}(k+1) \end{bmatrix}_\mathrm{model} \\ [0.25cm] &= \begin{bmatrix} \left(\frac{1}{m}\sum F_{\mathrm{X},ij} - F_\mathrm{XA} - F_\mathrm{XR}\right)_{k+1} \\ \left(\frac{1}{m}\sum F_{\mathrm{Y},ij}\right)_{k+1} \\ \dot{\psi}(k+1) \end{bmatrix}_\mathrm{model} . \end{aligned} \tag{12.3.31}$$

Herewith, the lateral air resistance $F_\mathrm{YA}$ and the gyroscopic terms are neglected. As the small differences of the tire forces in (12.3.22) result in rather noisy estimates of $\dot{\psi}$, the measurement for $\dot{\psi}(k+1)$ is directly used and not corrected.

The *state correction* equation of the EKF is finally according to (10.2.71)

$$\hat{\mathbf{x}}(k+1|k+1) = \hat{\mathbf{x}}(k+1|k) + \mathbf{K}(k+1)\left[\mathbf{y}(k+1) - \mathbf{g}_{k+1}(\hat{x}(k+1|k))\right] , \tag{12.3.32}$$

where $\mathbf{y}(k+1)$ are the new measurements for $(k+1)$ and $\mathbf{g}_{k+1}$ the process model output due to (12.3.31).

The EKF requires the determination of the Jacobi matrices $\mathbf{F}(k)$ and $\mathbf{G}(k+1)$ according to (10.2.66) and (10.2.67) to update the covariance matrix $\mathbf{P}^-(k+1)$ of the state estimation error and the gain matrix $\mathbf{K}$ with (10.2.69).

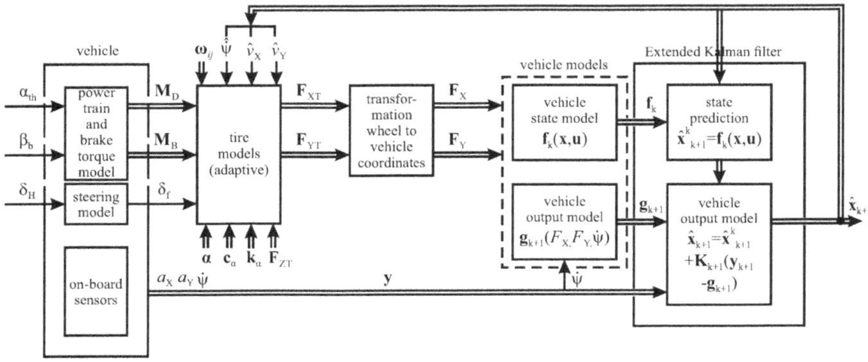


**Fig. 12.12** Signal flow for the vehicle state estimation with a two-track model and an extended Kalman filter with $\hat{\mathbf{x}} = [\hat{v}_X \ \ \hat{v}_Y \ \ \dot{\psi}]^T$ and $\hat{\mathbf{y}} = [a_X \ \ a_Y \ \ \dot{\psi}]$

The vehicle slip angle $\beta$ follows finally due to (7.1.1)

$$\hat{\beta} = \arctan \frac{\hat{v}_Y}{\hat{v}_X}. \tag{12.3.33}$$

The inputs $\mathbf{u}(k)$ of the EKF are the wheel forces $\mathbf{F}_X$ and $\mathbf{F}_Y$ and the resistance forces; see (12.3.26). The wheel forces results from the tire forces $\mathbf{F}_{XT}$ and $\mathbf{F}_{YT}$. The longitudinal forces $\mathbf{F}_{XT}$ are determined with the longitudinal vehicle models treated in chapter 6. The calculation of the lateral forces can be performed as follows, Schorn (2007).

*Version 1* uses a linear lateral tire force model (5.2.3)

$$F_{YT,ij} = c_{\alpha,ij}(F_{ZT,ij})\alpha_{ij}, \tag{12.3.34}$$

with the cornering stiffness $c_\alpha = c_{\alpha_1} F_{ZT}$, $c_{\alpha_1}$ known, the vertical wheel forces $F_{ZT,ij}(a_X, a_Y)$ according to (7.3.22), and $\alpha_{ij}(\delta_f, v_Y, v_X, \dot{\psi})$ due to (7.3.21) for all wheels $ij$.

*Version 2* uses an adaptive cornering stiffness coefficient $k_{\alpha,ij}(\alpha_{ij})$, (11.6.39), equal for the wheels of the front and rear axle by inserting the $k_{\alpha,f}$ and $k_{\alpha,r}$ in the state vector $\hat{\mathbf{x}}$ of the EKF for estimation.

*Version 3* determines the adaptive $k_{\alpha,ij}(\alpha_{ij})$ with a separate recursive least squares parameter estimation; see (11.6.44).

The tire forces $\mathbf{F}_{XT}$ and $\mathbf{F}_{YT}$ are then transformed from the wheel coordinates to the vehicle coordinates by (7.3.38).

Figure 12.12 summarizes the main steps for the state estimation with an extended Kalman filter, providing estimates for $\hat{v}_X$, $\hat{v}_Y$, and $\hat{\beta}$ based on measurements of $\delta_f$, $a_X$, $a_Y$, $\dot{\psi}$ for the state estimation.

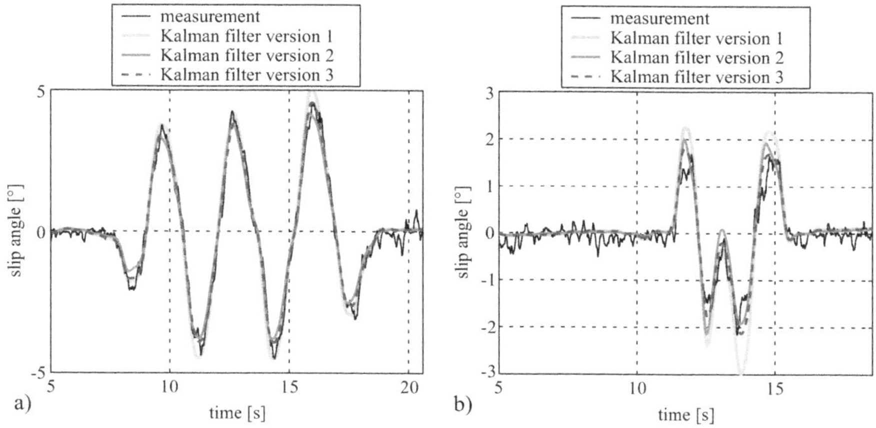


**Fig. 12.13** Experimental results for the slip angle estimation $\hat{\beta}$ with an extended Kalman filter with different constant cornering stiffness parameters. **a** Slalom course, **b** double lane change. Version 1: $c_{\alpha,i}$ = const.; version 2: adaptive $k_{\alpha,i}$; version 3: RLS estimation of $k_\alpha$. VW Golf IV, Schorn (2007)

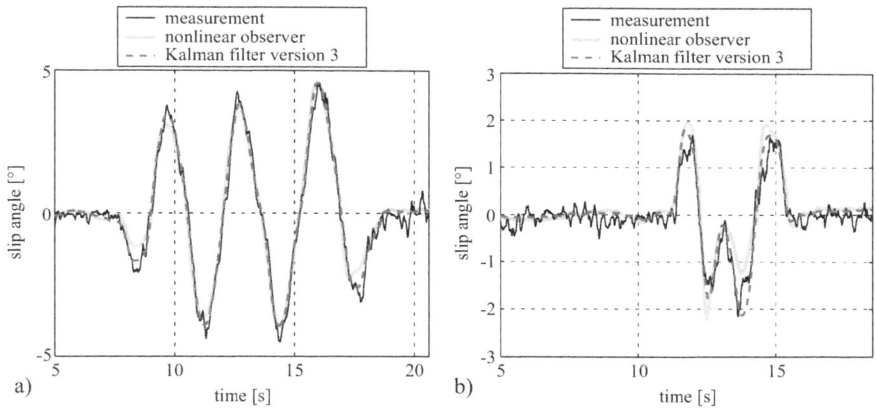


**Fig. 12.14** Experimental results for the slip angle estimation. Comparison of the estimates for a nonlinear state observer and an extended Kalman filter (version 3) for **a** a slalom course and **b** a double lane change. VW Golf IV, Schorn (2007)

The extended Kalman filter then consists of the prediction and correction equations, stated in (10.2.70)–(10.2.76). The covariance matrices **Q** and **R** can just be specified as unity matrices or properly selected by probing.

Figure 12.13 illustrates the resulting slip angle estimates for a slalom course and a double lane change with a front-driven car. Three versions of EKF estimates are compared to each other for the cases: $c_{\alpha_i}$ = const., $k_{\alpha_i}$ time-variant adaptive and $k_{\alpha_i}$ as recursive least squares parameter estimates. The agreement with the direct measurements by an optical CORREVIT-sensor system is for all versions good and the best for version 3.

A comparison of slip angles estimates from the nonlinear observer with the extended Kalman filter, version 3, shows that better results could be obtained with the Kalman filter; see Fig. 12.14.

**(b) Extended Kalman Filter with Simplified Two-Track Model and Constant Velocity**

A simplified version for the estimation of the vehicle slip angle is obtained, if a constant longitudinal velocity $v_X = \text{const.}$ and constant longitudinal tire forces $F_{XT,ij}$ are assumed. Furthermore, gyroscopic forces are neglected. As measurements $\delta_f$ as input and $a_Y$ and $\dot{\psi}$ as outputs are used, it follows from (7.3.13) and (7.3.27) by including the transformation of the tire forces to the chassis forces and neglecting small terms

$$\begin{aligned} a_Y &= \frac{1}{m}\Big[(F_{YT,fl} + F_{YT,fr})\cos\delta_f + (F_{YT,rl} + F_{XT,rr})\Big] \\ &= \frac{1}{m}\sum F_Y, \end{aligned} \tag{12.3.35}$$

and from (7.2.14) and (7.3.28)

$$\begin{aligned} \ddot{\psi} &= \frac{1}{J_Z}\Big[(F_{YT,fl} + F_{YT,fr})l_f\cos\delta_f - (F_{YT,rl} + F_{YT,rr})l_r \\ &\quad + \frac{b_f}{2}(F_{YT,fr} - F_{YT,fl})\sin\delta_f\Big] = \frac{1}{J_Z}\sum M_Z. \end{aligned} \tag{12.3.36}$$

Further, (7.2.8) yields

$$\dot{\beta} = \frac{\dot{v}_Y}{v} - \dot{\psi} = \frac{a_Y}{v} - \dot{\psi}. \tag{12.3.37}$$

Discretization with sampling time $T_0$ then gives

$$a_Y = \frac{\mathrm{d}v_Y}{\mathrm{d}t} \approx \frac{\Delta v_Y(k+1)}{T_0} = \frac{v_Y(k+1) - v_Y(k)}{T_0}, \tag{12.3.38}$$

$$\ddot{\psi} = \frac{\mathrm{d}\dot{\psi}}{\mathrm{d}t} \approx \frac{\Delta\dot{\psi}(k+1)}{T_0} = \frac{\dot{\psi}(k+1) - \dot{\psi}(k)}{T_0}, \tag{12.3.39}$$

$$\dot{\beta} = \frac{\mathrm{d}\beta}{\mathrm{d}t} \approx \frac{\Delta\beta(k+1)}{T_0} = \frac{\beta(k+1) - \beta(k)}{T_0}. \tag{12.3.40}$$

As $\dot{\beta}$ can directly be calculated with (12.3.37) from the measured variables $a_Y$ and $\dot{\psi}$, it is not included in the state vector. Thus, the state vector is

$$\mathbf{x}^T(k) = \big[v_Y(k) \ \ \dot{\psi}(k)\big], \tag{12.3.41}$$

as (12.3.25), but without $v_X(k)$ and (12.3.38) and (12.3.39) are represented in the discrete-time state equation model

$$\begin{aligned} \mathbf{x}(k+1) &= f_k(\mathbf{x}(k), \mathbf{u}(k)) + \mathbf{v}(k), \\ \mathbf{y}(k) &= \mathbf{g}_k(\mathbf{x}(k)) + \mathbf{n}(k), \end{aligned} \tag{12.3.42}$$

with

$$f_1(k) = v_Y(k+1) = v_Y(k) + \frac{T_0}{m} \sum F_Y(k), \tag{12.3.43}$$

$$f_2(k) = \dot{\psi}(k+1) = \dot{\psi}(k) + \frac{T_0}{J_Z} \sum M_Z(k). \tag{12.3.44}$$

(12.3.37) and (12.3.38) yield

$$f_3(k) = \beta(k+1) = \beta(k) + \frac{1}{v}\left(v_Y(k+1) - v_Y(k)\right) - T_0\dot{\psi}(k) \tag{12.3.45}$$

The *state prediction* for the extended Kalman filter according to (12.3.23) uses

$$\mathbf{f}_k = [f_1(k) \;\; f_2(k)]^{\mathrm{T}}, \tag{12.3.46}$$

$$\mathbf{y}(k) = \left[a_{Y(k)} \;\; \dot{\psi}(k)\right]^{\mathrm{T}}, \tag{12.3.47}$$

and the *state correction* is made according (12.3.32) with

$$\mathbf{g}_{k+1} = \begin{bmatrix} g_1 \\ g_2 \end{bmatrix}_{k+1} = \begin{bmatrix} a_Y(k+1) \\ \dot{\psi}(k+1) \end{bmatrix}_{\text{model}} = \begin{bmatrix} \frac{1}{m}\sum F_Y(k+1) \\ \frac{1}{J_Z}\sum M_Z(k+1) \end{bmatrix} \tag{12.3.48}$$

and the measurements

$$\mathbf{y}(k+1) = \left[a_Y(k+1) \;\; \dot{\psi}(k+1)\right]. \tag{12.3.49}$$

The state estimates $\hat{\mathbf{x}}(k+1)$ result from the *state correction* equation (12.3.32) and $\hat{\beta}(k+1)$ results via (12.3.45) from

$$\hat{\beta}(k+1) = \hat{\beta}(k) + \left(\frac{\hat{v}_Y(k+1) - \hat{v}_Y(k)}{v} - T_0\dot{\psi}(k+1)\right). \tag{12.3.50}$$

The lateral tire forces $F_{YT,ij}$ are obtained with linear or nonlinear models as in version 1 to 3 of the last section. Reif et al. (2007) have used the Paceika-model for a similar vehicle slip angle estimation.

## 12.4 State Estimation of the Roll Angle and Pitch Angle

The roll and pitch dynamic behavior can with simplifying assumptions be described by linear second-order differential equations, as shown in Chap. 9. In the following, state estimation methods are described which allow to determine the roll angle $\varphi$ and pitch angle $\theta$, if they cannot be measured directly; see also Ryu and Gerdes (2004).

### 12.4.1 State Estimation of the Roll Angle

The roll dynamic behavior is for small $\varphi$ described by

$$J_X\ddot{\varphi}(t) + d_{rol}\dot{\varphi}(t) + c_{rol}\varphi(t) = -K_{rol}a_Y(t) - K_{rol}g\eta \tag{12.4.1}$$

see (9.1.15) and (11.8.1). The parameters can be estimated if $K_{rol}$ is determined with (9.1.16); see also Sect. 11.8.1.

For *stationary cornering* with $a_Y =$ const. and no lateral gradient, i.e. $\eta = 0$, follows in a steady state

$$\hat{\varphi} = \frac{-K_{rol}}{c_{rol}}a_Y \tag{12.4.2}$$

and the roll angle can directly be determined. In *dynamic driving situations*, (12.4.1) has to be used, which becomes in state-space form with $\eta = 0$

$$\begin{aligned} \begin{bmatrix} \dot{\varphi} \\ \ddot{\varphi} \end{bmatrix} &= \begin{bmatrix} 0 & 1 \\ -\frac{c_{rol}}{J_X} & -\frac{d_{rol}}{J_X} \end{bmatrix} \begin{bmatrix} \varphi \\ \dot{\varphi} \end{bmatrix} + \begin{bmatrix} 0 \\ -\frac{K_{rol}}{J_X} \end{bmatrix} a_Y, \\ \dot{\varphi} &= [0\ 1] \begin{bmatrix} \varphi \\ \dot{\varphi} \end{bmatrix} \end{aligned} \tag{12.4.3}$$

The goal is now to estimate the roll angle $\varphi$ based on measurements of the roll rate $\dot{\varphi}$ and the lateral acceleration $a_Y$.

As for state estimation, a process model in discrete time $k = t/T_0$, where $T_0$ is the sampling time, is preferred. The differential equation (12.4.1) can for small $T_0$ directly be discretized, leading to

$$a_0\varphi(k) + a_1\varphi(k-1) + a_2\varphi(k-2) = b_0a_Y(k), \tag{12.4.4}$$

with, see, for example, Isermann (1989), p. 21,

$$\begin{aligned} a_0 &= \frac{J_X}{T_0^2} + \frac{d_{rol}}{T_0} + 1, \\ a_1 &= -\frac{d_{rol}}{T_0} - \frac{2J_X}{T_0^2}, \\ a_2 &= \frac{J_X}{T_0^2}, \\ b_0 &= -K_{rol}. \end{aligned} \tag{12.4.5}$$

In order to write (12.4.4) in a form with a prediction for time $k+1$, one obtains

$$\varphi(k+1) + a_1'\varphi(k) + a_2'\varphi(k-1) = b_0'a_Y(k+1) \tag{12.4.6}$$

with $a'_1 = a_1/a_0$, $a'_2 = a_2/a_0$, and $b'_0 = b_0/a_0$. A corresponding state model then becomes

$$\begin{bmatrix} \varphi(k) \\ \varphi(k+1) \end{bmatrix} = \begin{bmatrix} 0 & 1 \\ -a'_2 & -a'_1 \end{bmatrix} \begin{bmatrix} \varphi(k-1) \\ \varphi(k) \end{bmatrix} + \begin{bmatrix} 0 \\ b'_0 \end{bmatrix} a_{\mathrm{Y}}(k+1) \tag{12.4.7}$$

with the measurable variable

$$\dot{\varphi}(t) = \frac{\Delta\varphi(k)}{T_0} \approx \frac{1}{T_0}\left(\varphi(k) - \varphi(k-1)\right) = \begin{bmatrix} -\frac{1}{T_0} & \frac{1}{T_0} \end{bmatrix} \begin{bmatrix} \varphi(k-1) \\ \varphi(k) \end{bmatrix}. \tag{12.4.8}$$

The last two equations represent the state variable model

$$\begin{aligned} \mathbf{x}(k+1) &= \mathbf{A}\mathbf{x}(k) + \mathbf{b}\mathbf{u}(k+1), \\ y(k) &= \mathbf{c}^{\mathrm{T}}\mathbf{x}(k), \end{aligned} \tag{12.4.9}$$

and the Kalman filter equations (10.2.53) and (10.2.54) become

*prediction:*

$$\hat{\mathbf{x}}(k+1|k) = \mathbf{A}\hat{\mathbf{x}}(k|k) + \mathbf{B}u(k+1). \tag{12.4.10}$$

*correction:*

$$\hat{\mathbf{x}}(k+1|k+1) = \hat{\mathbf{x}}(k+1|k) + \bar{\mathbf{K}}\left[y(k+1) - \mathbf{c}^{\mathrm{T}}\hat{\mathbf{x}}(k+1|k)\right] \tag{12.4.11}$$

with

$$\mathbf{x}^{\mathrm{T}}(k|k) = \left[\varphi(k-1)\ \varphi(k)\right],$$

and the measurements

$$\begin{aligned} u(k+1) &= a_{\mathrm{Y}}(k+1) \\ y(k+1) &= \dot{\varphi}(k+1). \end{aligned}$$

The Kalman filter then provides an estimate of the roll angle $\hat{\varphi}(k+1)$ from the correction $\hat{\mathbf{x}}(k+1|k+1)$.

By appropriate selection of the covariance matrices $\mathbf{R}$ and $\mathbf{Q}$, a weighting of the influence of the measurements $\dot{\varphi}$ and $a_{\mathrm{Y}}$ can be made, depending on the noise variance of the sensors, see, for example, Halbe (2008).

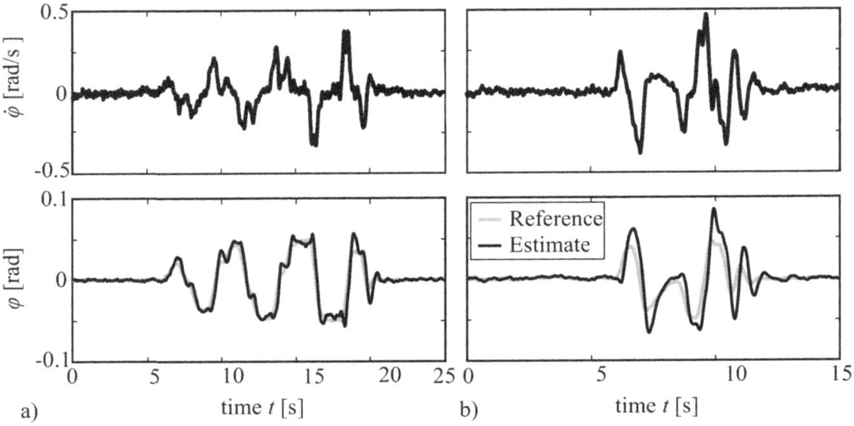


**Fig. 12.15** Measured roll rate and estimated roll angle with a Kalman filter and measurement of the lateral acceleration $a_Y$ and roll rate $\dot{\varphi}$. Opel Omega A2.0i (1993), Halbe (2008). **a** Slalom drive with $v = 30$ km h$^{-1}$, **b** double lane change with $v = 40$ km h$^{-}1$

Figure 12.15 depicts driving results for the roll angle estimation. The reference measurement is based on the travelways of the suspensions; see (9.1.13). The agreement of estimated and measured roll angle is good.

The estimation of the roll angle with a Kalman filter based on measurements of $\dot{\varphi}$ from an IMU and $\varphi$ from a 3D-GPS system is illustrated in Bauer (2015).

## 12.4.2 State Estimation of the Pitch Angle

The simplified pitch model is according to (9.2.4) and (11.8.3)

$$J_Y\ddot{\theta}(t) + d_{\text{pit}}\dot{\theta}(t) + c_{\text{pit}}\theta(t) = -K_{\text{pit}}a_X(t) - K_{\text{pit}}g\lambda, \qquad (12.4.12)$$

and has the same structure as the roll dynamic model (12.4.1). The parameters can be estimated as described in Sect. 11.8.2. $K_{\text{pit}}$ is determined with (9.2.4).

For estimating the pitch angle $\theta$, measurements of the pitch rate $\dot{\theta}$ and the longitudinal acceleration $a_X$ are required. The state estimation of the pitch angle $\hat{\theta}(t)$ with a Kalman filter then follows a procedure as for the roll angle, (12.4.3)–(12.4.11).

However, because the pitch angle $\theta$ usually is small compared to noise effects in the measurements, especially for the required values of $a_X$ for braking, it is difficult to obtain good estimates.

Experimental results for the pitch angle estimation for strong braking maneuvers are shown in Bauer (2015).

## 12.5 Expanded Vehicle State Estimation with an Extended Kalman Filter, a Nonlinear One-Track Model and Front and Rear Wheel Steering

The state estimation methods in Sect. 12.3 are mainly designed for the determination of the slip angle $\beta$ and include also the velocities $v_Y$ and/or $v_X$ and the yaw rate $\dot{\psi}$; see Fig. 12.6. The parameters of the tire models are separately estimated by parameter estimation methods; see Figs. 12.9 and 12.12. However, by expanding the state vector, the parameters of the tire models can also be included in an overall estimation with an extended Kalman filter, as developed by Bechtloff et al. (2015), Bechtloff et al. (2016), Bechtloff (2018).

In order to reduce the computational effort for the two-track model, the nonlinear one-track model (7.3.71) is applied.

**(a) Vehicle and Tire Models**

Neglection of road grades $\lambda$ and $\eta$, but inclusion of the roll angle $\varphi$ and the pitch angle $\theta$ results in the following equations for the center of gravity in the *yaw rate/slip angle representation* (7.3.71)

$$\dot{v} = \left(\frac{F_X}{m} + g\sin\theta\right)\cos\beta + \left(\frac{F_Y}{m} - g\cos\theta\sin\varphi\right)\sin\beta, \tag{12.5.1}$$

$$\dot{\beta} = \left(\frac{F_Y}{m} - g\cos\theta\sin\varphi\right)\frac{1}{v}\cos\beta - \left(\frac{F_X}{m} + g\sin\theta\right)\frac{1}{v}\sin\beta - \dot{\psi}, \tag{12.5.2}$$

$$\ddot{\psi} = \frac{1}{J_Z}M_Z, \tag{12.5.3}$$

where $F_X$, $F_Y$, and $M_Z$ are due to (7.3.48), (7.3.49), and (7.3.51). As (7.3.51) shows, the longitudinal tire forces $F_{X,ij}$ of all four wheels are included, as required for individual wheel braking and driving.

The resulting longitudinal and lateral forces are with $a_X = F_X/m$ and $a_Y = F_Y/m$ replaced by the measurable accelerations as they can be used as model inputs, yielding

$$\dot{v} = (a_X + g\sin\theta)\cos\beta + (a_Y - g\cos\theta\sin\varphi)\sin\beta, \tag{12.5.4}$$

$$\dot{\beta} = (a_Y - g\cos\theta\sin\varphi)\frac{1}{v}\cos\beta - (a_X + g\sin\theta)\frac{1}{v}\sin\beta - \dot{\psi}. \tag{12.5.5}$$

The continuous-time nonlinear vehicle model is as for the nonlinear observer (10.2.13) formulated by

$$\dot{\mathbf{x}}(t) = \mathbf{f}(\mathbf{x}(t), \mathbf{u}(t)) + \boldsymbol{\nu}(t), \tag{12.5.6}$$

$$\mathbf{y}(t) = \mathbf{g}(\mathbf{x}(t), \mathbf{u}(t)) + \mathbf{n}(t), \tag{12.5.7}$$

with an expanded state variable vector

$$\mathbf{x} = \Big[\underbrace{v\beta\mu_{\max}\varphi\theta}_{\mathbf{x}_{\text{fast}}} \;\; \underbrace{c_{\alpha,\text{f}}SG\Delta\mu}_{\mathbf{x}_{\text{slow}}}\Big]^{\text{T}}, \tag{12.5.8}$$

and the input vector

$$\mathbf{u} = \Big[\dot{\psi}a_{\text{X}}a_{\text{Y}}\delta_{\text{f}}\delta_{\text{r}} \;\; \omega_{\text{fl}}\omega_{\text{fr}}\omega_{\text{rl}}\omega_{\text{rr}} \;\; \varphi_{\text{K}}\dot{\varphi}_{\text{K}}\Big], \tag{12.5.9}$$

according to Bechtloff (2018).

The process noise $\boldsymbol{\nu}(t)$ and the output noise $n(t)$ are assumed to be statistically independent, have zero mean with covariance matrices **Q** and **R**, as (10.2.31), (10.2.32).

The input vector contains the measured front and rear wheel steering angles, $\delta_{\text{f}}$ and $\delta_{\text{r}}$, the measured yaw rate $\dot{\psi}$ and roll rate $\dot{\varphi}_{\text{K}}$, the accelerations $a_{\text{X}}$ and $a_{\text{Y}}$, and the angular wheel velocities $\omega_{ij}$. The roll angle $\varphi_{\text{K}}$ is determined with the roll angle model (9.1.15) or a first-order dynamic model as in Bechtloff (2018).

The state vector is divided in fast variables $\mathbf{x}_{\text{fast}}(t)$ and slow variables or quasi constant parameters $\mathbf{x}_{\text{slow}}$. The inclusion of the roll angle $\varphi$ and the pitch angle $\theta$ allows to consider high dynamic situations and also banked corners. The cornering stiffness of the front axle and the understeer gradient $SG$ are also estimated. The cornering stiffness of the rear axle is calculated with (7.2.52) and (7.2.74) by

$$c_{\alpha\text{r}} = \frac{mc_{\alpha\text{f}}l_{\text{f}}}{ml_{\text{r}} - SGc_{\alpha\text{f}}l^2}. \tag{12.5.10}$$

Instead of estimating the maximal friction coefficients $\mu_{\text{YT,f,max}}$ and $\mu_{\text{YT,r,max}}$, see (5.2.13), their difference

$$\Delta\mu_{\text{Y}} = \Delta\mu_{\text{YT,r,max}} - \Delta\mu_{\text{YT,f,max}} \tag{12.5.11}$$

is used.

**(b) State Prediction**

After discretization of (12.5.6), analogously to (12.3.23), the state prediction of the EKF is

$$\mathbf{x}(k+1|k) = \mathbf{f}_k(\mathbf{x}(k), \mathbf{u}(k)) + \boldsymbol{\nu}(k). \tag{12.5.12}$$

The *prediction model* in continuous time (12.5.6) then becomes with the one-track model (12.5.1) and (12.5.2)

$$\dot{\mathbf{x}} = \begin{bmatrix} \dot{v}_X \\ \dot{\beta} \\ \dot{\mu}_{max} \\ \dot{\varphi} \\ \dot{\theta} \\ \dot{c}_{\alpha f} \\ \dot{SG} \\ \Delta\dot{\mu}_Y \end{bmatrix} = \begin{bmatrix} (a_X + g\sin\theta)\cos\beta + (a_Y - g\cos\theta\sin\varphi)\sin\beta \\ (a_Y - g\cos\theta\sin\varphi)\frac{1}{v}\cos\beta - (a_X + g\sin\theta)\frac{1}{v}\sin\beta - \dot{\psi} \\ \frac{1}{T_\mu}(1-\mu_{max}) \\ \dot{\varphi}_K + \frac{1}{T_\varphi}(\varphi_K - \varphi) \\ -\frac{1}{T_\theta}\theta \\ 0 \\ 0 \\ 0 \end{bmatrix} + \begin{bmatrix} \nu_{aX} \\ \frac{\nu_{aY}}{v_X} + \nu_{\dot{\psi}} \\ \nu_{\mu_{max}} \\ \nu_\varphi \\ \nu_\theta \\ \nu_{c_{\alpha f}} \\ \nu_{SG} \\ \nu_{\Delta\mu_Y} \end{bmatrix}. \tag{12.5.13}$$

The derivatives of $\mu_{max}$, $\varphi$, and $\theta$ are modeled as Gauss-Markov processes of 1st order and the derivatives of the parameters $c_{\alpha,f}$, $SG$, and $\Delta\mu$ as random-walks, as usually for the parameter estimation with a Kalman filter; see Grewal and Andrews (2015). $\dot{\varphi}_K$ and $\varphi_K$ follow from the roll model (12.4.1) with $a_Y$ as input, see also Fig. 12.16.

**(c) State Correction**

The state correction equation of the EKF is as (10.2.69) and (12.3.30)

$$\hat{\mathbf{x}}(k+1|k+1) = \hat{\mathbf{x}}(k+1|k) + \mathbf{K}(k+1)\left(\mathbf{y}(k+1) - \mathbf{g}_{k+1}(\hat{x}(k+1|k))\right). \tag{12.5.14}$$

The *vehicle model output* $\mathbf{y}$, which corresponds to the prediction model (12.5.13) is defined as

$$\mathbf{y} = \mathbf{g}(\mathbf{x}, u)$$

$$\mathbf{y} = \begin{bmatrix} v_{Tm,fl} \\ v_{Tm,fr} \\ F_{XT,f} \\ F_{XT,r} \\ F_{YT,f} \\ F_{YT,r} \\ \beta \end{bmatrix} = \begin{bmatrix} v\cos(\delta_f - \beta) + l_f\dot{\psi}\sin\delta_f \\ v\cos(\delta_f - \beta) + l_f\dot{\psi}\sin\delta_f \\ F_{XT,f}(F_{Zf}, S_{Xf}, \alpha_f, \mu_{X,max}, c_{SX,f}, S_{X,crit,f}, c_{Xf}) \\ F_{XT,r}(F_{Zr}, S_{Xr}, \alpha_r, \mu_{X,max}, c_{SX,r}, S_{X,crit,r}, c_{Xr}) \\ F_{YT,f}(F_{Zf}, S_{Xf}, \alpha_f, \mu_{Y,max}, c_{c_{\alpha f}}, \alpha_{crit,f}, c_{Yf}) \\ F_{YT,r}(F_{Zr}, S_{Xr}, \alpha_r, \mu_{Y,max}, c_{c_{\alpha r}}, \alpha_{crit,r}, c_{Yr}) \\ \beta \end{bmatrix}. \tag{12.5.15}$$

Hence, the vehicle model output is expressed by the tire velocity of the middle front wheel and the four slip and side-slip angle based force models. The wheel velocity

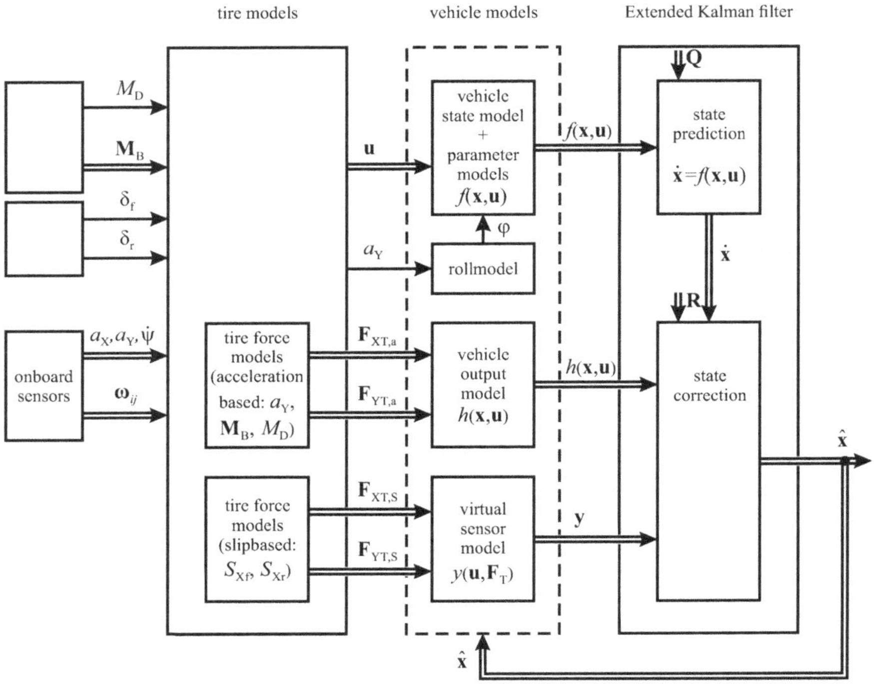


**Fig. 12.16** Signal flow for a vehicle state estimation with a nonlinear one-track model, virtual sensor models, front and rear wheel steering, and an extended Kalman filter in continuous-time representation, Bechtloff (2018)

of the middle front wheel is, see Bechtloff (2018),

$$y_1 = y_2 = v_{\mathrm{Tm}} = v \cos(\delta_{\mathrm{f}} - \beta) + l_{\mathrm{f}} \dot{\psi} \sin \delta_{\mathrm{f}}. \tag{12.5.16}$$

The tire forces per axle follow from the Pacejka-formula (5.1.7) in longitudinal direction

$$y_3 = F_{\mathrm{XT,f}}(F_{\mathrm{Zf}}, S_{\mathrm{Xf}}, \alpha_{\mathrm{f}}, \mu_{\mathrm{X,max}}, c_{\mathrm{SX}}, S_{\mathrm{X,crit}}, c_{\mathrm{Xf}}), \tag{12.5.17}$$

$$y_4 = F_{\mathrm{XT,r}}(F_{\mathrm{Zr}}, S_{\mathrm{Xr}}, \alpha_{\mathrm{r}}, \mu_{\mathrm{X,max}}, c_{\mathrm{SX}}, S_{\mathrm{X,crit}}, c_{\mathrm{Xr}}), \tag{12.5.18}$$

and in lateral direction, Table 5.1,

$$y_5 = F_{\mathrm{YT,f}}(F'_{\mathrm{Zf}}, \alpha_{\mathrm{f}}, S_{\mathrm{Yf}}, \mu_{\mathrm{Y,max,f}}, c_{\alpha,\mathrm{f}}, \alpha_{\mathrm{crit,f}}, c_{\mathrm{Y,f}}), \tag{12.5.19}$$

$$y_6 = F_{\mathrm{YT,r}}(F'_{\mathrm{Zr}}, \alpha_{\mathrm{r}}, S_{\mathrm{Xr}}, \mu_{\mathrm{Y,max,r}}, c_{\alpha,\mathrm{r}}, \alpha_{\mathrm{crit,r}}, c_{\mathrm{Y,r}}). \tag{12.5.20}$$

The vertical forces for each axle are, (6.5.50), (6.5.51),

$$F'_{\mathrm{Zf}} = m\frac{l_{\mathrm{r}}}{l}\left(\frac{g}{2} + \frac{h_{\mathrm{CG}}}{b_{\mathrm{f}}}a_{\mathrm{Y}}\right), \qquad (12.5.21)$$

$$F'_{\mathrm{Zf}} = m\frac{l_{\mathrm{r}}}{l}\left(\frac{g}{2} - \frac{h_{\mathrm{CG}}}{b_{\mathrm{f}}}a_{\mathrm{Y}}\right). \qquad (12.5.22)$$

$c_{\alpha\mathrm{f}}$ is obtained by parameter estimation and $c_{\alpha\mathrm{r}}$ follows from (12.5.10) after parameter estimation of the understeer gradient $SG$. For the maximal friction coefficients holds

$$\begin{aligned} \mu_{\mathrm{X,max,f}} &= \mu_{\mathrm{X,max,f0}}\ \mu_{\mathrm{max}}, \\ \mu_{\mathrm{X,max,r}} &= \mu_{\mathrm{X,max,r0}}\ \mu_{\mathrm{max}}, \end{aligned} \qquad (12.5.23)$$

$$\begin{aligned} \mu_{\mathrm{Y,max,f}} &= \mu_{\mathrm{Y,max,f0}}\ \mu_{\mathrm{max}}, \\ \mu_{\mathrm{Y,max,r}} &= (\mu_{\mathrm{Y,max,f0}} + \Delta\mu_{\mathrm{Y}})\mu_{\mathrm{max}}, \end{aligned} \qquad (12.5.24)$$

and for the critical side-slip angle

$$\alpha_{\mathrm{crit,f}} = \alpha_{\mathrm{crit,f0}}\frac{c_{\alpha\mathrm{f0}}}{c_{\alpha\mathrm{f}}}; \quad \alpha_{\mathrm{crit,r}} = \alpha_{\mathrm{crit,r0}}\frac{c_{\alpha\mathrm{r0}}}{c_{\alpha\mathrm{r}}}. \qquad (12.5.25)$$

$\mu_{\mathrm{X,max,f0}}$, $\mu_{\mathrm{Y,max,f0}}$, and $\alpha_{\mathrm{crit,f0}}$ have to be known from the basic parameters of the used tire as well as a starting value

$$\Delta\mu_{\mathrm{Y}}(t = 0) = \mu_{\mathrm{Y,max,r0}} - \mu_{\mathrm{Y,max,f0}}. \qquad (12.5.26)$$

The coefficients $c_{\alpha\mathrm{f}}$ and $\Delta\mu$ are estimated with the vehicle model (12.5.13). $c_{\alpha\mathrm{r}}$ follows from (12.5.10) and estimated $SG$.

The *measured outputs*

$$\mathbf{y} = \mathbf{g}(\mathbf{x}, \mathbf{u}) \qquad (12.5.27)$$

have to be based on measurable outputs of the vehicle. Because the variables of the output vector $\mathbf{y}$ of (12.5.15) are not directly measurable, they have to be reconstructed from measurable quantities, leading to a so-called *virtual sensor model* according to

Bechtloff (2018):

$$\mathbf{y} = \begin{bmatrix} v_{\mathrm{Tm,fl}} \\ v_{\mathrm{Tm,fr}} \\ F_{\mathrm{XT,f}} \\ F_{\mathrm{XT,r}} \\ F_{\mathrm{YT,f}} \\ F_{\mathrm{YT,r}} \\ \beta \end{bmatrix}$$

$$= \begin{bmatrix} \omega_{\mathrm{fl}} r_{\mathrm{dyn,f}} + \dot{\psi}\frac{b_{\mathrm{f}}}{2}\cos\delta_{\mathrm{f}} \\ \omega_{\mathrm{fr}} r_{\mathrm{dyn,f}} - \dot{\psi}\frac{b_{\mathrm{r}}}{2}\cos\delta_{\mathrm{f}} \\ (M_{\mathrm{B,fl}} + M_{\mathrm{B,fr}}) r_{\mathrm{dyn,f}} \\ (M_{\mathrm{B,rl}} + M_{\mathrm{B,rr}} + M_{\mathrm{D}}) r_{\mathrm{dyn,r}} \\ \frac{1}{\cos\delta_{\mathrm{f}}}\left[\frac{1}{l}\left(m l_{\mathrm{r}} a_{\mathrm{YT}} + J_{\mathrm{Z}}\dot{\psi} - M_{\mathrm{Z}}\right)\right] - \left(F_{\mathrm{XT,fl}} + F_{\mathrm{XT,fr}}\right)\sin\delta_f \\ \frac{1}{\cos\delta_{\mathrm{r}}}\left[\frac{1}{l}\left(m l_{\mathrm{f}} a_{\mathrm{YT}} - J_{\mathrm{Z}}\dot{\psi} + M_{\mathrm{Z}}\right)\right] - \left(F_{\mathrm{XT,rl}} + F_{\mathrm{XT,rr}}\right)\sin\delta_r \\ \beta_{\mathrm{kin}} \end{bmatrix}. \tag{12.5.28}$$

The vehicle slip angle $\beta_{\mathrm{kin}}$ is determined by using a one-track model

$$\beta_{\mathrm{kin}} = \arctan\left[\frac{l_{\mathrm{f}}}{l}\tan\delta_{\mathrm{f}} + \frac{l_{\mathrm{r}}}{l}\tan\delta_{\mathrm{r}}\right]. \tag{12.5.28a}$$

The state correction according to (12.5.14) then can be performed with $\mathbf{y}(k+1)$ from (12.5.15) and $g(k+1)$ from (12.5.28) after discretization. The determination of the Kalman gain $\mathbf{K}(k+1)$ follows (10.2.67) by appropriate selection of covariance matrices $\mathbf{Q}$ and $\mathbf{R}$; see Bechtloff (2018).

**(d) Experimental Results**

The described state and parameter estimation with an extended Kalman filter (EKF) and an unscented Kalman filter (UKF) was applied in many driving situations.

Figure 12.17 shows the estimation of the slip angle for a double lane change without ESC. The maximal error of the slip angle for lateral accelerations until 10 m $s^{-1}$ is about 1.2° or 7%. Figure 12.18 further depicts the estimates of the slip angle for high speeds and a drift maneuver with strong accelerations and decelerations until $a_{\mathrm{X}} = -10\ \mathrm{ms}^{-2}$ and until $a_{\mathrm{Y}} = 10\ \mathrm{ms}^{-2}$ by strong braking on a handling test field. The vehicle was driven until drive dynamic limits including banked corners. Therefore, several drift situations occurred. The largest error of the slip angle is about $\Delta\beta \approx 3°$ or 13%.

The state estimation for many other driving experiments, like ABS-braking, stationary cornering to determine $\mu_{\mathrm{max}}$, and steep curves gave also good results for several estimates like $\hat{v}$, $\hat{\varphi}$, $\hat{\mu}_{\mathrm{max}}$, $F_{\mathrm{YT,r}}$, $F_{\mathrm{YT,f}}$. Hence, the state estimation with the applied one-track model and the state vector (12.5.8) is applicable also for high dynamic driving situations.

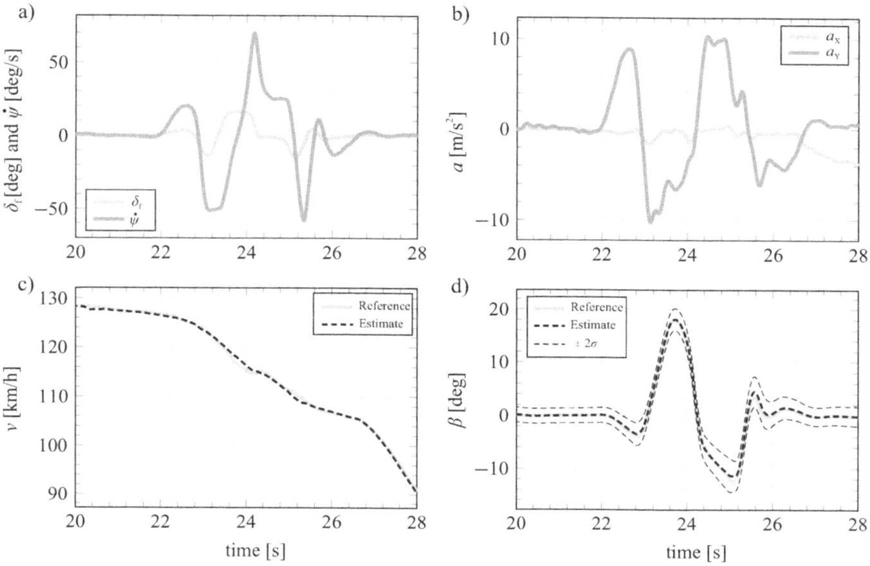


**Fig. 12.17** State estimation of the slip angle for a double lane change, BMW 550i (E60), Bechtloff (2018). **a** steering angle and yaw rate, **b** accelerations, **c** velocity, **d** slip angle

## 12.6 Vehicle State Estimation with Additional 3D-GPS Measurements and an Extended Kalman Filter

The usually implemented on-board sensors and also additionally installed inertial measurement units do not allow to measure all important drive dynamic variables. This situation can be improved by the use of signals from a global positioning system (GPS). A GPS with one antenna provides the velocity vector

$$\mathbf{v}_{\mathrm{E}}^{\mathrm{GPS}} = \left[ v_{\mathrm{X,E}}^{\mathrm{GPS}} \;\; v_{\mathrm{Y,E}}^{\mathrm{GPS}} \;\; v_{\mathrm{Z,E}}^{\mathrm{GPS}} \right]^{\mathrm{T}},$$

and the position vector

$$\mathbf{p}_{\mathrm{E}}^{\mathrm{GPS}} = \left[ p_{\mathrm{X,E}}^{\mathrm{GPS}} \;\; p_{\mathrm{Y,E}}^{\mathrm{GPS}} \;\; p_{\mathrm{Z,E}}^{\mathrm{GPS}} \right]^{\mathrm{T}},$$

in the earth fixed coordinate system E, if at least 4 satellites are available.

A 3-antenna system also delivers the position angles

$$\Psi = \left[ \varphi^{\mathrm{GPS}} \;\; \theta^{\mathrm{GPS}} \;\; \psi^{\mathrm{GPS}} \right]^{\mathrm{T}},$$

relative to the earth coordinate system.

Figure 12.19 depicts the installation of 3 antennas on the used vehicle for drive dynamic measurements. This 3D-GPS system is from the manufacturer Sepentrio (2009) and provides measurements with sampling time $T_0 = 100$ ms. The 3D-GPS

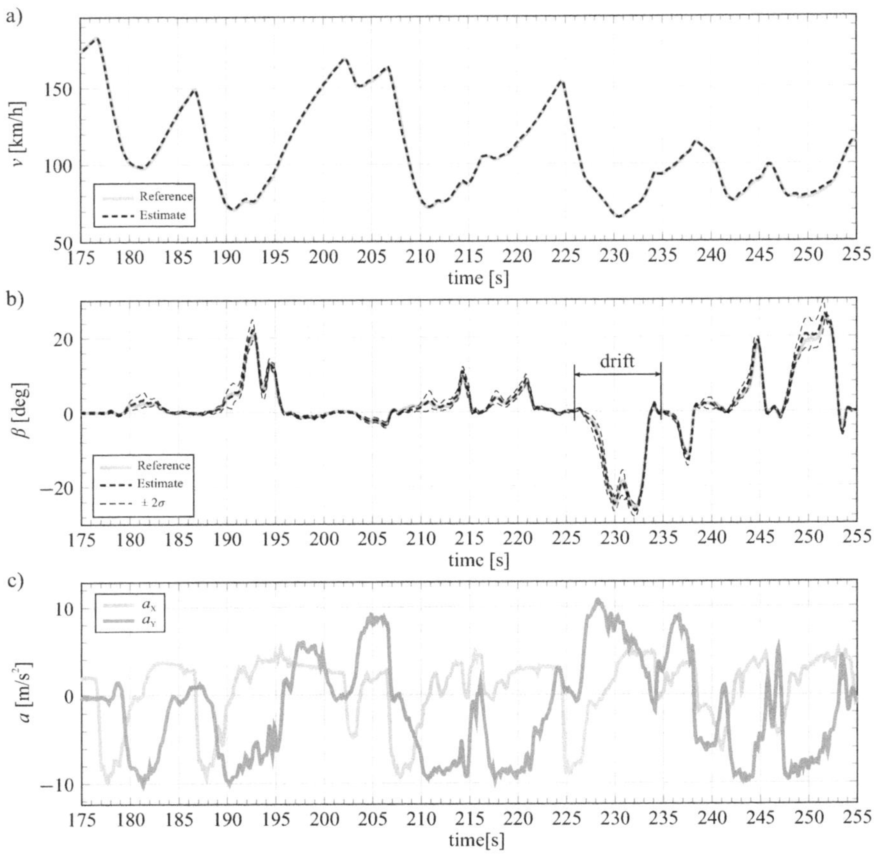


**Fig. 12.18** State estimation of the slip angle for high speeds and drift maneuvers on the handling course of the Automotive Testing Area in Papenburg (ATP), Germany, BMW 550i (E60), Bechtloff (2018). **a** velocity, **b** slip angle, **c** accelerations

delivers the velocity and position vector and its variances in an earth fixed coordinate system. Its origin is fixed after initialization. For more details, see, for example, Hofmann-Wellenhof et al. (2001), Kaplan and Hegarty (2006), Parkinson and Spilker (1996).

This 3D-GPS allows not only to provide position measurements with an accuracy in the range of some cm, but also the yaw angle, roll angle, and pitch angle of the vehicle, which are usually not available in production cars. Especially in combination with Kalman filters by sensor fusion with the measurements of an IMU, precise estimates of vehicle state variables are possible.

Thus, 3D-GPS is a valuable additional source of measurements for prototype and research vehicles to develop dynamic vehicle models and control systems. In the following, some examples are given.

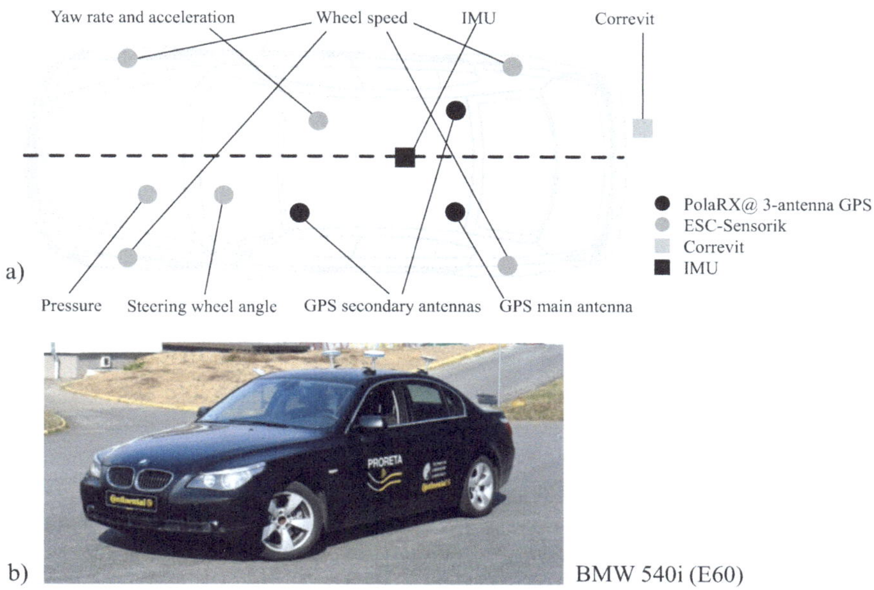


**Fig. 12.19** Installation of a 3-antenna GPS system on a BMW 540i (E60). **a** Schematic illustration, **b** photo

### 12.6.1 Roll Angle and Yaw Angle Estimation

The inertial measurement unit (IMU) on board the vehicle measures angle rates and the 3D-GPS system absolute values of the angles of the driving vehicle. In order to improve the accuracy of the relatively small and noisy signal values of the IMU, the goal is to design a fusion of both measurements. This can be performed with a Kalman filter and will be described for the roll rate $\dot{\varphi}$ and transferred to the yaw rate $\dot{\psi}$, according to Bauer (2015). See also Ryu and Gerdes (2004).

The measured signal of the roll rate sensor is modeled by

$$\dot{\varphi}^{\text{IMU}} = s_{\dot{\varphi}}\dot{\varphi} + b_{\dot{\varphi}} + \nu_{\dot{\varphi}}, \tag{12.6.1}$$

where $\dot{\varphi}$ is the real roll rate, $s_{\dot{\varphi}}$ the sensor gain (sensitivity), $b_{\dot{\varphi}}$ an offset, and $\nu_{\dot{\varphi}}$ sensor noise. The real roll rate then becomes

$$\dot{\varphi} = \frac{1}{s_{\dot{\varphi}}}\dot{\varphi}^{\text{IMU}} - \frac{1}{s_{\dot{\varphi}}}b_{\dot{\varphi}} - \frac{1}{s_{\dot{\varphi}}}\nu_{\dot{\varphi}}. \tag{12.6.2}$$

Discretization leads to

$$\dot{\varphi}(t) = \frac{\mathrm{d}\varphi(t)}{\mathrm{d}t} \approx \frac{\Delta\varphi(k+1)}{T_0} = \frac{\varphi(k+1) - \varphi(k)}{T_0}, \tag{12.6.3}$$

and one obtains a one-step-prediction for the real roll rate

$$\varphi(k+1) = \varphi(k) + \frac{T_0}{s_{\dot\varphi}}\dot\varphi^{\mathrm{IMU}}(k) - \frac{T_0}{s_{\dot\varphi}} b_{\dot\varphi}(k) - \frac{T_0}{s_{\dot\varphi}}\nu_{\dot\varphi}(k). \tag{12.6.4}$$

This *sensor model* is now turned into a state variable representation

$$\begin{bmatrix} \varphi(k+1) \\ \frac{1}{s_{\dot\varphi}}(k+1) \\ \frac{b_{\dot\varphi}}{s_{\dot\varphi}}(k+1) \end{bmatrix} = \begin{bmatrix} 1 & T_0\dot\varphi^{\mathrm{IMU}}(k) & -T_0 \\ 0 & 1 & 0 \\ 0 & 0 & 0 \end{bmatrix} \begin{bmatrix} \varphi(k) \\ \frac{1}{s_{\dot\varphi}}(k) \\ \frac{b_{\dot\varphi}}{s_{\dot\varphi}}(k) \end{bmatrix} + \begin{bmatrix} \frac{T_0}{s_{\dot\varphi}}\nu_{\dot\varphi}(k) \\ \nu_{\frac{1}{s_{\dot\varphi}}}(k) \\ \nu_{\frac{b_{\dot\varphi}}{s_{\dot\varphi}}}(k) \end{bmatrix}, \tag{12.6.5}$$

$$\hat{\mathbf{x}}(k+1|k) = \mathbf{A}\hat{\mathbf{x}}(k|k) + \boldsymbol{\nu}(k),$$

which is used for the *prediction part* of the Kalman filter. The GPS-measurement for the roll angle $\varphi^{\mathrm{GPS}}(k)$ relates to the state variables by

$$\varphi^{\mathrm{GPS}}(k) = \begin{bmatrix} 1 & 0 & 0 \end{bmatrix} \begin{bmatrix} \varphi(k) \\ \frac{1}{s_{\dot\varphi}}(k) \\ \frac{b_{\dot\varphi}}{s_{\dot\varphi}}(k) \end{bmatrix} + n_\varphi(k), \tag{12.6.6}$$

$$y(k) = \mathbf{c}^{\mathrm{T}}\mathbf{x}(k) + n_\varphi(k),$$

and is used in the *correction part* of the Kalman filter.

The noise signals $\nu$ and $n$ are assumed as white, zero mean, Gaussian random variables with known or assumed covariance matrices $\mathbf{Q}$ and $\mathbf{R}$.

The Kalman filter then follows from (10.2.54)

$$\hat{\mathbf{x}}(k+1|k+1) = \hat{\mathbf{x}}(k+1|k) + \bar{\mathbf{K}}\left[y(k+1) - \mathbf{c}^{\mathrm{T}}\hat{\mathbf{x}}(k+1|k)\right], \tag{12.6.7}$$

with a gain matrix $\bar{\mathbf{K}}$ according to (10.2.52). The estimated values for $s_{\dot\varphi}$ and $b_{\dot\varphi}$ are then used to correct the measured $\dot\varphi^{\mathrm{IMU}}$. Figure 12.20 summarizes the roll rate estimation through sensor fusion. The application of this roll angle estimation method is depicted in Fig. 12.21 for a sampling time $T_0 = 0.2$ s (5 Hz) and a sweep sine test signal until 1 Hz for the steering angle. The lateral acceleration reaches about 5 $\mathrm{ms}^{-1}$ for a speed around 45 km $\mathrm{h}^{-1}$. Figure 12.21c and f show a good agreement of the estimated roll angle with the reference values from the 3D-GPS. Figure 12.21g and h depict the offset and the gain deviation of the IMU roll rate sensor. The final gain deviation with $s_{\dot\varphi} = 1.05$, i.e. 5 %, agrees with the sensor specification. Taking these deviations into account, the roll rate measurement can be corrected according to Fig. 12.20 and results as shown in Fig. 12.21d. The same procedure as for the roll angle was applied for the *yaw angle estimation* and Fig. 12.22 shows the results. As reference signal for a validation, the yaw rate measurement from the ESC-unit is used, noted by "ref". The corrected yaw rate $\hat{\dot\psi}^{\mathrm{IMU}}_{\mathrm{corr}}$ shows a good agreement with

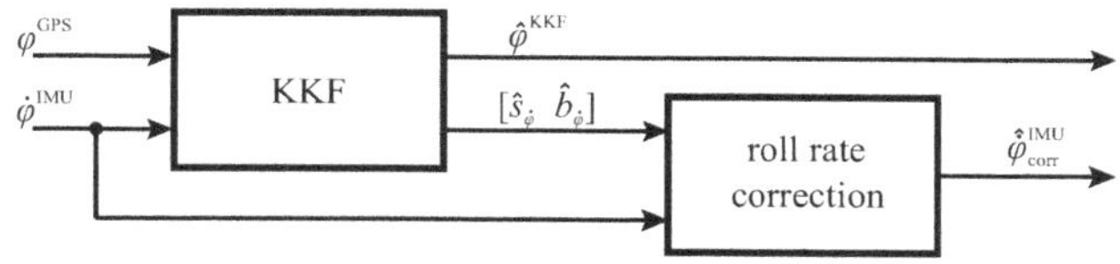


**Fig. 12.20** Signal flow for the roll rate estimation by a fusion of the measured roll rate of the IMU and the roll angle from the 3D-GPS with a kinematic Kalman filter (KKF), Bauer (2015)

the reference values, Fig. 12.21c and d, taking into account the estimated sensor deviations in (g) and (h). Finally, the estimated yaw angle in (c) and (d) indicates a good agreement with the reference signal.

Summarizing, the sensor fusion of the roll rate and yaw rate measurements from the IMU with 3D-GPS measurements of the roll angle and yaw angle by using a kinematic Kalman filter gives relative precise estimates of the roll angle and the yaw angle and allows to estimate the offset and gain of on-board roll rate and yaw rate sensors.

### 12.6.2 Vehicle State Estimation with a Two-Track Model

In order to include more state variables as in (12.3.25) for the state estimation with an extended Kalman filter, the nonlinear two-track model in the yaw rate/velocity representation (7.3.12)–(7.3.14) is used and discretized with sampling time $T_0$. The state vector is an extension of (12.3.25) and consists of

$$\mathbf{x}(k) = \left[v_X(k) \;\; v_Y(k) \;\; a_Y(k) \;\; \dot{\psi}(k) \;\; \psi(k) \;\; X_E(k) \;\; Y_E(k)\right]^T . \qquad (12.6.8)$$

It contains additionally the lateral acceleration $a_Y(k)$, the yaw angle $\psi(k)$ and the vehicle position $X_E(k)$ and $Y_E(k)$ in order to allow for corrections of the Kalman filter with the respective measurements, Bauer (2015). Hence, compared to the process model (12.3.20)–(12.3.22) equations for the lateral acceleration $a_Y(k)$, the yaw angle $\psi(k)$, and the odometry position estimation, (12.1.3), have to be introduced. This leads for driving on a plane surface to

– *Process model*

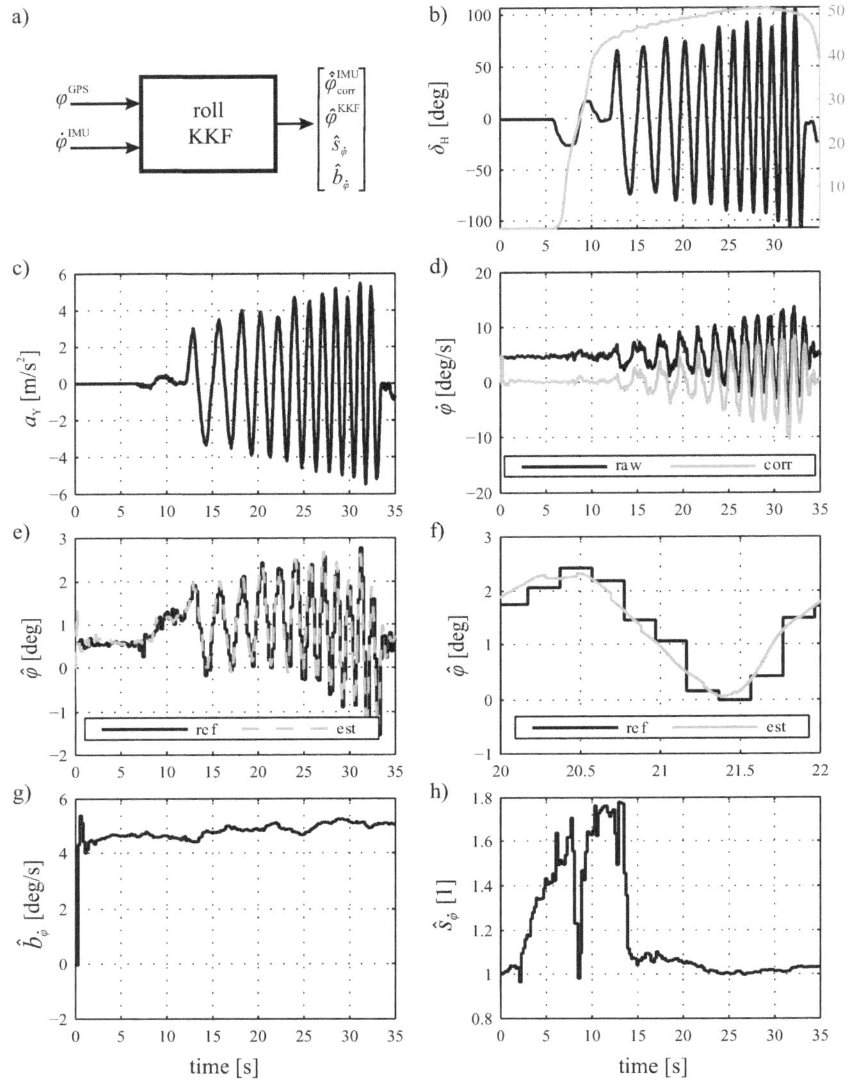


**Fig. 12.21** Results of the roll angle estimation, Bauer (2015), for a driving maneuver with a sweep sine slalom maneuver on an even and dry test field with varying frequency. **a** Inputs and outputs of the kinematic Kalman filter, **b** steering angle and velocity, **c** lateral acceleration, **d** measured (raw) and corrected roll rate, **e** roll angle estimate, **f** zoom of the roll angle estimate, **g** offset estimate, **h** sensor gain estimate. Reference data are GPS measurements of the roll angle for the same test drive which were not used for estimation. Vehicle: BMW 540i (E60)

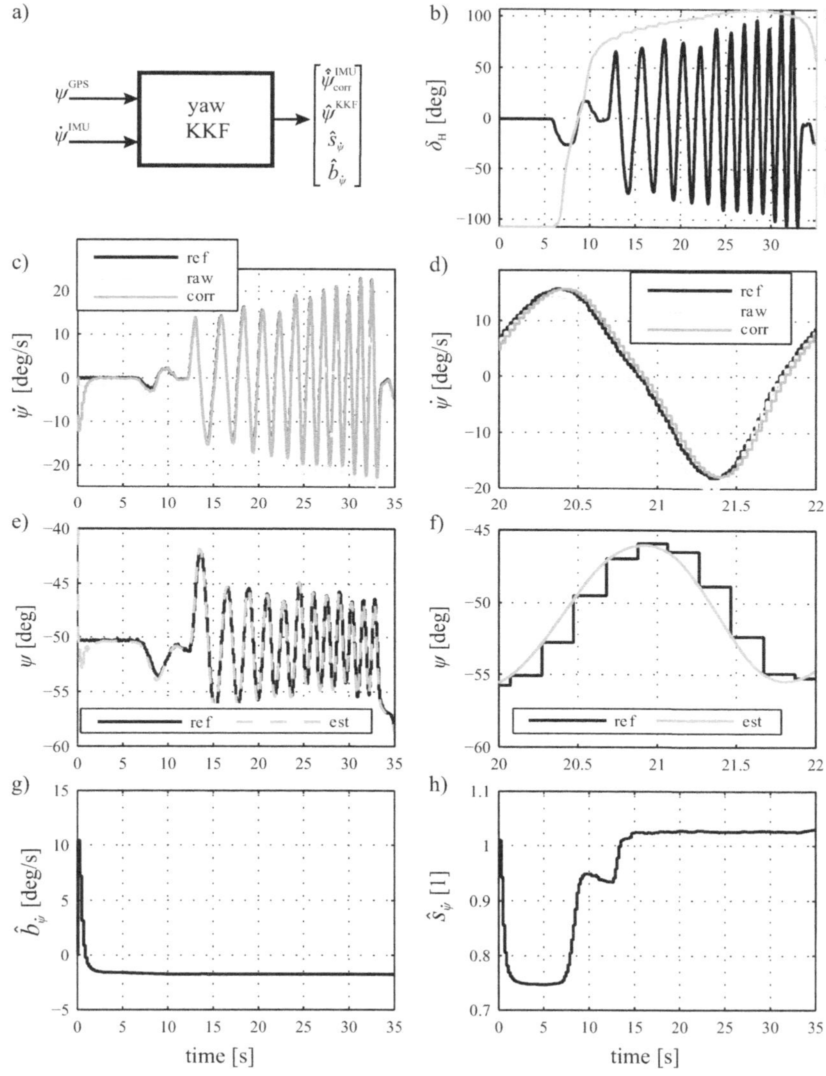


**Fig. 12.22** Results of the yaw angle estimation, Bauer (2015), for a driving maneuver with a sweep sine slalom maneuver on an even and dry test field with varying frequency. $\dot{\psi}^{\mathrm{IMU}}$ is the raw signal of the IMU and $\dot{\psi}^{\mathrm{IMU}}_{\mathrm{corr}}$ the corrected signal. **a** Inputs and outputs of the kinematic Kalman filter, **b** steering angle and velocity, **c** yaw rate, **d** zoom of **c**, **e** yaw angle, **f** zoom of **e**, **g** offset estimate, **h** gain estimate. Vehicle: BMW 540i (E60)

$$
\begin{aligned}
\mathbf{x}(k+1) &= f_k(\mathbf{x}(k), \mathbf{u}(k)) + \boldsymbol{\nu}(k), \\
v_X(k+1) &= v_X(k) + T_0\Bigg[\frac{1}{m}\sum_{ij} F_{X,ij} - F_{XA}(v_X(k)) \\
&\quad - F_{XR}(v_X(k)) + \dot{\psi}(k)v_Y(k)\Bigg] + \nu_{v_X}(k), \\
v_Y(k+1) &= v_Y(k) + T_0\left[a_Y(k) - \dot{\psi}(k)v_X(k)\right] + \nu_{v_Y}(k), \\
a_Y(k+1) &= a_Y(k) + \frac{T_0}{m}\sum_{ij} F_{Y,ij}(k) + \nu_{a_Y}(k), \\
\dot{\psi}(k+1) &= \dot{\psi}(k) + \frac{T_0}{J_Z}\Bigg[\left(F_{X,fr} - F_{X,fl}\right)\frac{b_f}{2} + \left(F_{X,rr} - F_{X,rl}\right)\frac{b_r}{2} \\
&\quad + \left(F_{Y,fl} + F_{Y,fr}\right)l_f - \left(F_{Y,rl} + F_{Y,rr}\right)l_r\Bigg]_k + \nu_{\dot{\psi}}(k), \\
\psi(k+1) &= \psi(k) + T_0\dot{\psi}(k) + \nu_{\psi}(k), \\
X_E(k+1) &= X_E(k) + T_0\big[v_X(k)\cos\psi(k) \\
&\quad - v_Y(k)\sin\psi(k)\big] + \nu_{X_E}(k), \\
Y_E(k+1) &= Y_E(k) + T_0\big[v_X(k)\sin\psi(k) \\
&\quad + v_Y(k)\cos\psi(k)\big] + \nu_{Y_E}(k).
\end{aligned}
\tag{12.6.9}
$$

This vehicle model is then used for the *state prediction* as (12.3.23). The state function matrix is

$$
\begin{aligned}
\mathbf{f}_k &= [f_1(k) \;\; f_2(k) \;\; \ldots \;\; f_7(k)] \\
&= [v_X(k+1) \;\; v_Y(k+1) \;\; \ldots \;\; Y_E(k+1)].
\end{aligned}
\tag{12.6.10}
$$

– *Process output model*

The measured output variables of the on-board IMU and the 3D-GPS

$$
\mathbf{y}(k+1) = \left[v_X^{\text{GPS}} \;\; a_Y^{\text{IMU}} \;\; \dot{\psi}^{\text{IMU}} \;\; \psi^{\text{GPS}} \;\; X_E^{\text{GPS}} \;\; Y_E^{\text{GPS}}\right]_{k+1}
\tag{12.6.11}
$$

are inserted in the state correction (12.3.32) together with the predicted states

$$
\mathbf{g}_{k+1}(\hat{x}(k+1|k)) = \left[\hat{v}_X \;\; \hat{v}_Y \;\; \hat{a}_Y \;\; \hat{\dot{\psi}} \;\; \hat{\psi} \;\; \hat{X}_E \;\; \hat{Y}_E\right]_{k+1}
\tag{12.6.12}
$$

and deliver the new state estimate $\hat{\mathbf{x}}(k+1|k+1)$.

The structure of the Jacobi matrix $\mathbf{G}$ of the EKF is relatively simple, as all state variables except $v_Y$ are measured, Bauer (2015).

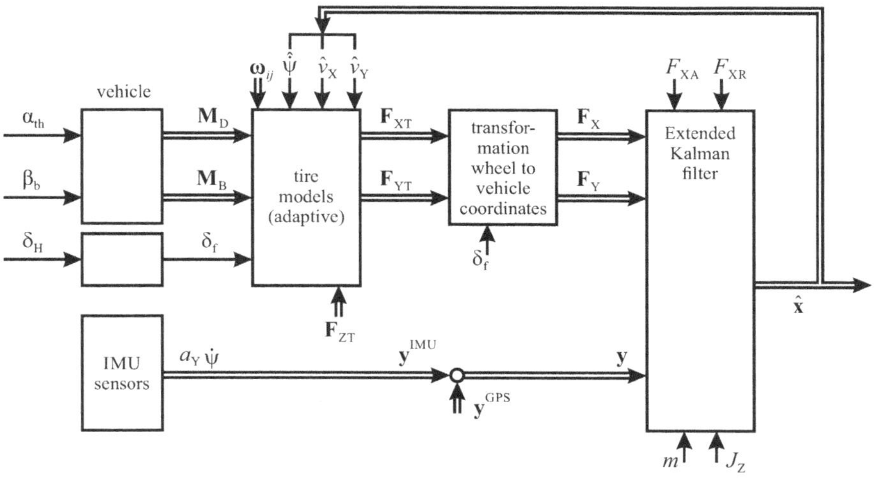


**Fig. 12.23** Signal flow for vehicle state estimation with a two-track model and an extended Kalman filter including parameter estimation of the tire models and the use of measurements of a 3D-GPS system with $\hat{\mathbf{x}} = \left[\hat{v}_X \ \hat{v}_Y \ \hat{a}_Y \ \hat{\dot{\psi}} \ \hat{\psi} \ \hat{X}_E \ \hat{Y}_E\right]^T$ and $\mathbf{y} = \left[v_X^{GPS} \ a_Y^{IMU} \ \dot{\psi}^{IMU} \ \psi^{GPS} \ X_E^{GPS} \ Y_E^{GPS}\right]$

The signal flow of the EKF is depicted in Fig. 12.23. The tire models are obtained with parameter estimation for combined longitudinal and lateral friction coefficients $\mu_{res,ij}$ as described in Sect. 11.6.2.

It shows that the tire model adaptation needs the state estimates $\hat{v}_X$, $\hat{v}_Y$, and $\hat{\dot{\psi}}$ determined by the Kalman filter. This introduces a feedback in the signal flow for an online real time application.

The results of the application of the drive dynamic extended Kalman filter for a double lane change are shown in Fig. 12.24. The agreement of the estimated variables with the reference measurements from the 3D-GPS system and an optical CORREVIT-sensor is very good. These results show that the additional use of the 3D-GPS system allows to obtain very good estimates $\hat{v}_X$, $\hat{v}_Y$, $\hat{\dot{\psi}}$, and $\hat{a}_Y$ for drive dynamics and $\hat{X}_E$, $\hat{Y}_E$, and $\hat{\psi}$ for the position and orientation of the vehicle. The range of the validity function is about $0.89 \leq R^2 \leq 0.99$.

A comparison of the use of 1D and 3D-GPS signals shows better results especially for the roll and pitch angle with the 3D-GPS system, Bechtloff et al. (2014). Further results are shown in Bauer et al. (2012). A fast state estimation and parameter estimation with onboard sensors and a 3D-GPS system is illustrated by using a nonlinear two-track model in Bechtloff et al. (2014).

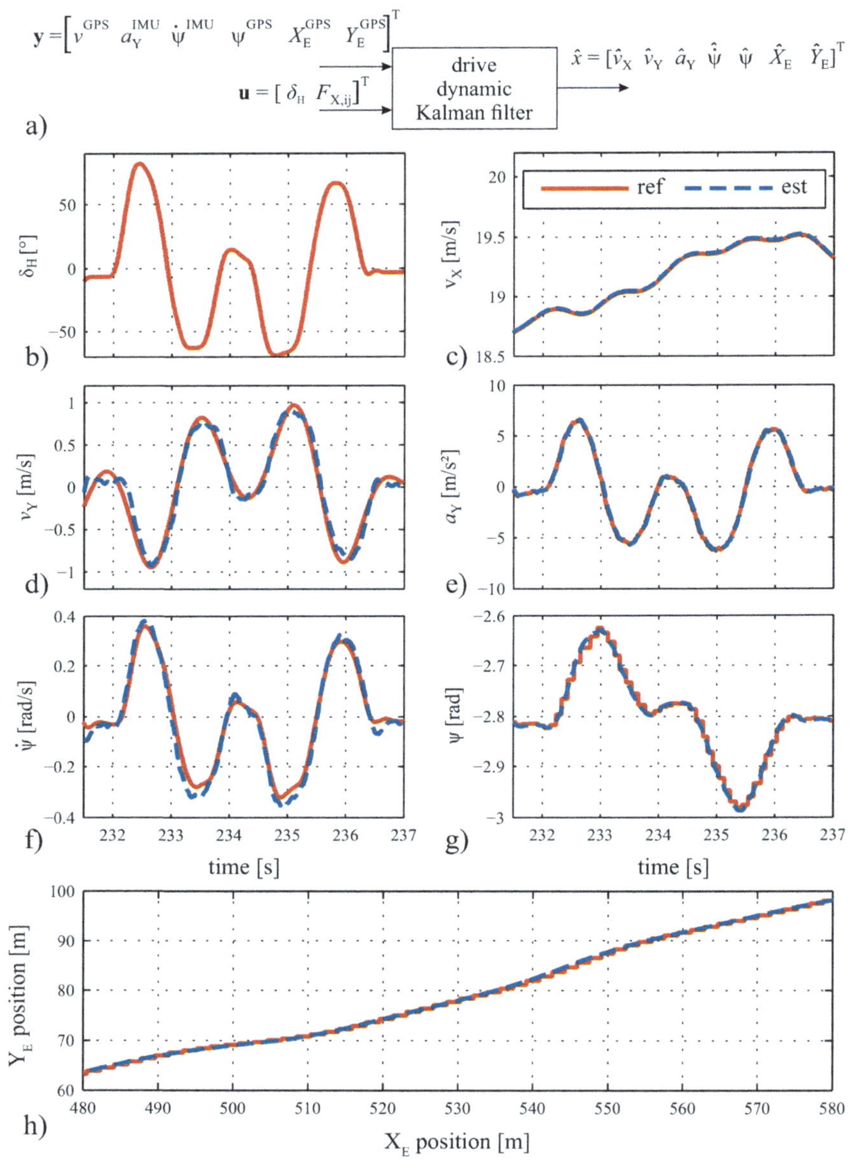


**Fig. 12.24** Results for the application of an extended Kalman filter for a double lane change, Bauer (2015), comparing reference values from GPS and CORREVIT-sensor (solid line) to estimated variables (dashed line). **a** Input and output signals, **b** steering angle, **c** longitudinal velocity, **d** lateral velocity, **e** lateral acceleration, **f** yaw rate, **g** yaw angle, **h** driven trajectory of BMW 540i (E60)

# References

Adamy J (2018) Nichtlineare Systeme und Regelungen, 3rd edn. Springer Vieweg, Wiesbaden

Arndt M, Ding E, Massel T (2004) Fehlertolerante Überwachung des Wankratensensors. Automatisierungstechnische Praxis - atp 46(7)

Baffet G, Charara A, Dherbomez G (2007) An observer of tire-road forces and friction for active security vehicle systems. IEEE/ASME Trans Mechatron 12(6):651–661

Bauer M (2015) Methoden zur modellbasierten Fahrdynamikanalyse und Bewertung von Fahrdynamikregelsystemen. Dissertation Technische Universität Darmstadt. Fortschr.-Ber. VDI Reihe 12, 792. VDI Verlag, Düsseldorf

Bauer M, Ackermann C, Isermann R (2012) Vehicle state estimation with fusion of driving dynamic sensors and GPS data to develop driving dynamic control systems. In: FISITA World Automotive Congress, Peking, China

Bechtloff J (2018) Schätzung des Schwimmwinkels und fahrdynamischer Parameter zur Verbesserung modellbasierter Fahrdynamikregelungen. Dissertation Technische Universität Darmstadt. Fortschr.-Ber. VDI Reihe 12, 809. VDI Verlag, Düsseldorf

Bechtloff J, Bauer M, Ackermann C, Isermann R (2014) Fast identification of a detailed two-track model with on-board sensors and GPS. In: 5th International Munich chassis technology symposium, Munich

Bechtloff J, Ackermann C, Isermann R (2015) Adaptive state observers for driving dynamics – Online estimation of tire parameters under real conditions. In: ATZ live conference, Chassis Tech, Munich

Bechtloff J, König L, Isermann R (2016) Cornering stiffness and side slip angle estimation for integrated vehicle dynamics control. In: IFAC symposium on advanced automatic control (ACC), Norrköping, Sweden

Borenstein J, Everett HR, Feng L (1996) Where am I?. Sensors and methods for mobile robot positioning. Technical report, University of Michigan, Mobile Robots Laboratory

Börner M (2004) Adaptive Querdynamikmodelle für Personenkraftfahrzeuge - Fahrzustandserkennung und Sensorfehlertoleranz. Dissertation Technische Universität Darmstadt. Fortschr.-Ber. VDI Reihe 12, Nr. 563. VDI Verlag, Düsseldorf

Burckhardt M (1993) Fahrwerktechnik: Radschlupf-Regelsysteme. Vogel-Verlag, Würzburg

Daiss A (1996) Beobachtung fahrdynamischer Zustände und Verbesserung einer ABS-und Fahrdynamikregelung. Diss. Universität Karlsruhe, Fortschr.-Ber. VDI Reihe 12, 283. VDI Verlag, Düsseldorf

Doumiati M, Victorino A, Charara A, Lechner D (2011) Estimation of road profile for vehicle dynamics motion: experimental validation. In: Proceedings of the 2011 IEEE American control conference, IEEE, San Francisco, CA, USA, pp 5237–5242

Eidehall A, Pohl J, Gustafsson F (2007) Joint road geometry estimation and vehicle tracking. Control Eng Pract 15(12):1484–1494

Grewal M, Andrews A (2015) Kalman filtering: theory and practice with MATLAB, 4th edn. Wiley, Hoboken

Halbe I (2008) Modellgestützte Sensoreninformationsplattform für die Quer- und Längsdynamik von Kraftfahrzeugen: Anwendungen zur Fehlerdiagnose und Fehlertoleranz. Diss. TU Darmstadt, Fortschr.-Ber. VDI Reihe 12, 680. VDI Verlag, Düsseldorf

Hofmann-Wellenhof B, Lichtenegger H, Collins J (2001) Global positioning system: theory and practice, 5th edn. Springer, Wien

Isermann R (1989) Digital control systems, 2nd edn. Springer, Berlin

Kaplan E, Hegarty C (2006) GPS: principles and applications. Artech House Publishers, Norwood

Kiencke U, Nielsen L (2005) Automotive control systems: for engine, driveline, and vehicle, 2nd edn. Springer, Berlin

Kochem M (2005) Ein Fahrerassistenzsystem zur Unterstützung des rückwertigen Parkvorgangs für Pkw. Dissertation Technische Universität Darmstadt. Fortschr.-Ber. VDI Reihe 12, 590. VDI Verlag, Düsseldorf

Parkinson B, Spilker J (1996) Progress in astronautics and aeronautics: the global positioning system, vols 163, 164. American Institute of Aeronautics and Astronautics, Washington
Ray LR (1995) Nonlinear state and tire force estimation for advanced vehicle control. IEEE Trans Control Syst Technol 3(1):117–124
Reif K, Renner K, Seeger M (2007) Vehicle state estimation on the basis of a nonlinear two-track model. Automobiltechnische Zeitschrift - ATZ 109(7–8):33–36
Ryu J, Gerdes J (2004) Estimation of vehicle roll and road bank angle. Proc American Control Confer 3:2110–2115
Samadi B, Kazemi R, Nikravesh KY, Kabganian M (2001) Real-time estimation of vehicle state and tire-road friction forces. In: Proceedings of the 2001 IEEE American control conference, IEEE, Arlington, VA, USA, vol 5, pp 3318–3323
Schmitt K (2012) Situationsanalyse für ein Fahrerassistenzsystem zur Vermeidung von Überholunfällen auf Landstraßen. Diss. Univ. Darmstadt, Fortschritt Bericht VDI, Reihe 12 Nr.763
Schmitt K, Isermann R (2009) Vehicle state estimation in curved road coordinates for a driver assistance system for overtaking situations. In: 21st IAVSD symposium on dynamics of vehicles on roads and tracks, Stockholm, Sweden
Schorn M (2007) Quer- und Längsregelung eines Personenkraftwagens für ein Fahrerassistenzsystem zur Unfallvermeidung. Diss. Universität Darmstadt, Fortschr.-Ber. VDI Reihe 12, 651. VDI Verlag, Düsseldorf
Strunz U (1993) Umgebungsmodellierung und sensorgestützte Navigation für mobile Roboter. Dissertation. RWTH Aachen, Aachen
Vietinghoff A, Hiemer M (2005) Nonlinear observer design for lateral vehicle dynamics. IFAC Proc Vol 38(1):988–993
Wesemeier D, Isermann R (2007) Identification of vehicle parameters using stationary driving maneuvers. IFAC Proc Vol 40(10):33–40
Würtenberger M (1997) Modellgestützte Verfahren zur Überwachung des Fahrzustandes eines Pkw. Diss. Universität Darmstadt, Fortschr.-Ber. VDI Reihe 12, 314, VDI Verlag, Düsseldorf

# Part III
# Dynamic Control of Chassis Components

# Braking Control

# 13

In the frame of active vehicle safety, the optimal function of brake systems has one of the highest priorities. In the long history of the development of mechanical, hydraulic, and pneumatic brake systems, the improvements of the braking performance by electronic control since the introduction of ABS-brake control around 1978 play a dominant role for automotive control.

This chapter considers hydraulic brake systems with pneumatic and electrical brake boosters and their mathematical models. The principles of anti-lock-control (ABS) are briefly described and measurements for ABS control with switching valves are shown. Then an electro-hydraulic brake system (EHB) with proportionally acting valves is considered, and it is illustrated in which braking results can be obtained with wheel-slip based control. The chapter concludes with the description and modeling of an electromechanical disk brake. A rich literature with books and published articles exists for these braking systems. Some comprehensive books are Mitschke and Wallentowitz (2004); Isermann (2006); Savaresi and Tanelli (2010); Robert Bosch GmbH (2018); Ulsoy et al (2012); Winner et al (2016); Breuer and Bill (2017); Ersoy and Gies (2017).

## 13.1 Hydraulic Brake System

Service brake systems for passenger cars and light utility vehicles consist usually of two independent hydraulic circuits. They are configured either *diagonal*, i.e. the right front wheel and the left rear wheel belong to one circuit, or *parallel*, where the two front wheels and the two rear wheels form each one circuit. The scheme in Fig. 13.1 shows that the brake pedal acts on the brake booster, which amplifies the pedal force, e.g. by a vacuum-operated booster with a factor 4–10, and acts on


R. Isermann, *Automotive Control*, ATZ/MTZ-Fachbuch,
https://doi.org/10.1007/978-3-642-39440-9_13

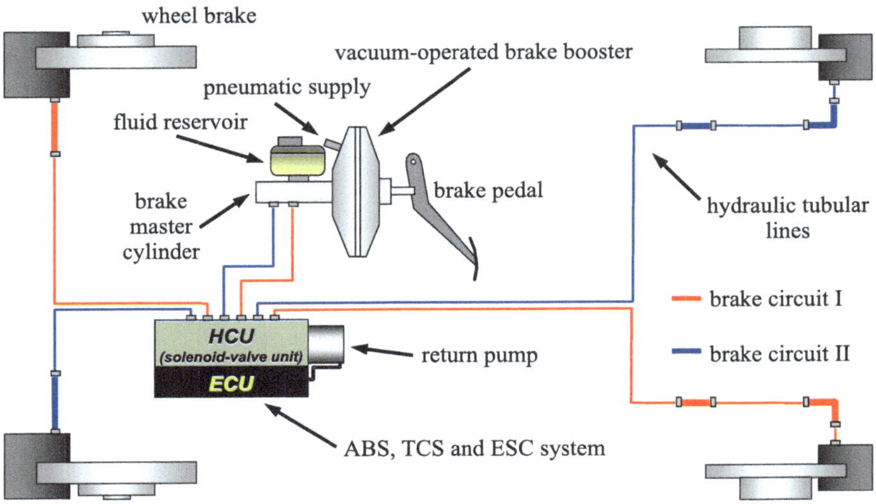


**Fig. 13.1** Schematic of a hydraulic brake system with two independent brake circuits

the brake master cylinder. There, the force is converted into hydraulic pressure with up to 120...180 bar.

A hydraulic unit carries the solenoid valves and the ECU for the brake control system (ABS, ESC). The pressurized brake fluid transmits the brake pressure through brake lines and flexible brake hoses to the wheel brake cylinders. The pistons of the brake cylinder then generate the force to press the brake pads against the disc or drum brake.

The *brake master cylinder*, or dual master cylinder, for the two braking circuits is illustrated in Fig. 13.2. The primary piston generates the pressure for circuit I, the primary circuit. The brake fluid pressure acts on the secondary piston and pressurizes the secondary circuit II. If the brake circuit I has a leak, the push rod acts directly on the second piston, such that braking with about half power is still possible. If the brake circuit II is leaky, the primary piston moves the secondary piston to its final position and then pressurizes the primary circuit allowing also for braking with about half of the power. In both cases, the driver recognizes this by an increased pedal displacement.

Brake force boosters use an auxiliary energy to amplify the pedal force, either stemming from air vacuum or hydraulic pressure. Vacuum brake boosters use the (variable) vacuum in the intake manifold of gasoline engines or from special vacuum pumps in the case of diesel engines. A cut-away drawing of a vacuum brake booster is shown in Fig. 13.3. The foot force exerted by the driver is supplied to the vacuum brake booster via a linkage. The vacuum brake booster is divided into two chambers by means of the diaphragm. The vacuum chamber is always kept at a pressure substantially lower than the atmospheric pressure whereas the pressure in the working chamber is controlled by the pneumatic valves located inside the vacuum brake booster. These valves allow to open or shut a flow path from the working

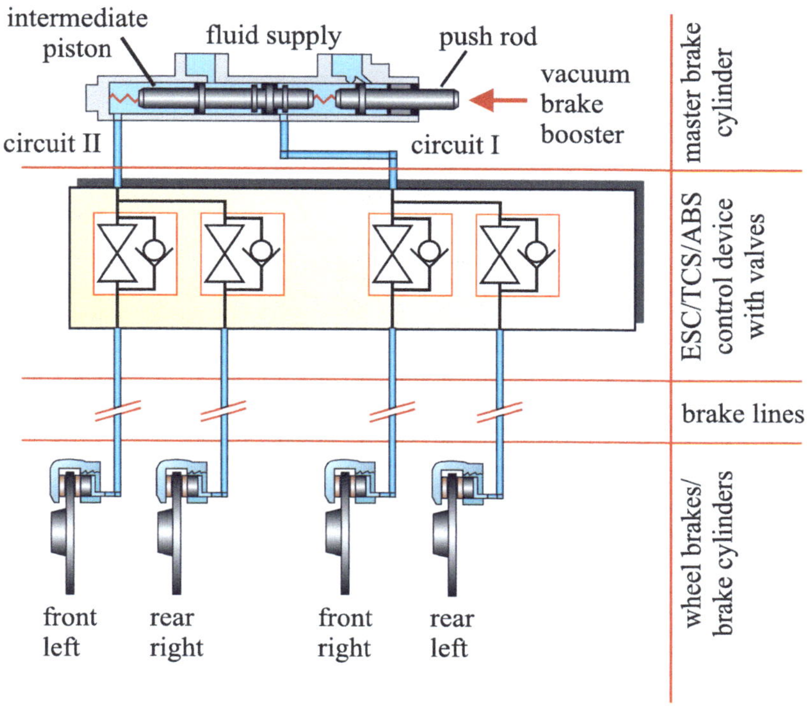


**Fig. 13.2** Schematic of the brake master cylinder and two hydraulic circuits

chamber to the vacuum chamber or from the working chamber to the surroundings, respectively. Opening and closing of the valves is controlled by the reaction washer, which is basically an elastic rubber disc. The air sucked in from the surroundings is traversing an air filter and thereby cleaned. The pressure difference between the two chambers of the vacuum brake booster exerts a force onto the membrane, which is transmitted to the master brake cylinder via a push rod. For a more detailed description of the vacuum brake booster, the reader is referred to Burckhardt (1991); Breuer and Bill (2017). Active brake force boosters have an integrated solenoid drive and manipulate the duct between the working chamber and the atmosphere and allow therefore electrical controlled braking for ACC.

Figure 13.4 depicts a signal flow scheme of a hydraulic brake system for a passenger car. The brake pedal, with position $\beta_B$, controlled by the driver's foot, transfers via a linkage the foot force $F_F$ to the push rod of the master piston. The movement of the linkage also acts on the brake booster, which generates, in addition to the foot force, a brake booster force $F_{BB}$ such that the force on the push rod becomes

$$F_{PR}(t) = F_F(t) + F_{BB}(t). \quad (13.1.1)$$

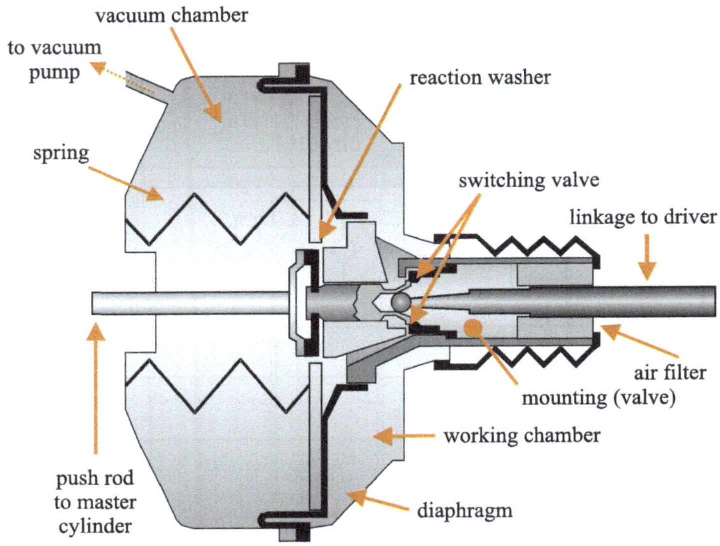


**Fig. 13.3** Cut-away scheme of a vacuum brake booster, (Münchhof et al 2003)

The generation of the booster force is, regarding the dynamics within the pneumatic

acting membrane actuator, delayed by a small time constant $T_{BB}$.

The movement of the master piston $z_{MPI}$ generates, according to the compressibility of the brake fluid volume in the brake lines, brake hoses and the brake caliper of brake circuit I, a brake pressure $p_{II}$ in the master cylinder after traveling through clearances between the brake pads and the brake disk. This pressure acts then back to the push rod.

The movement of the push rod in the brake master cylinder is also transferred to the secondary piston for the brake circuit II and generates the brake pressure $p_{II}$. The brake pressures $p_{BI}$ and $p_{BII}$ in the brake calipers are dynamically delayed by a small time constant of about $T_p \approx 25$ ms and lead via the brake forces on the brake disk or brake drum to the wheel brake torques $M_{BI}$ and $M_{BII}$.

## 13.2 Models of a Hydraulic Brake Circuit

### 13.2.1 Pneumatic Brake Booster

Figure 13.5 shows a simplified scheme of a vacuum brake booster, consisting of a vacuum chamber and a working chamber which are separated by a diaphragm. Both the vacuum chamber and the working chamber have the low pressure $p_{VC}$ for the case of no braking. In a braking situation, the pedal travel $z_{BP}$ opens a valve to the surrounding air with pressure $p_a$, such that an increased pressure $p_{WC}$ acts on the diaphragm with area $A_D$ for the amplification of the foot force.

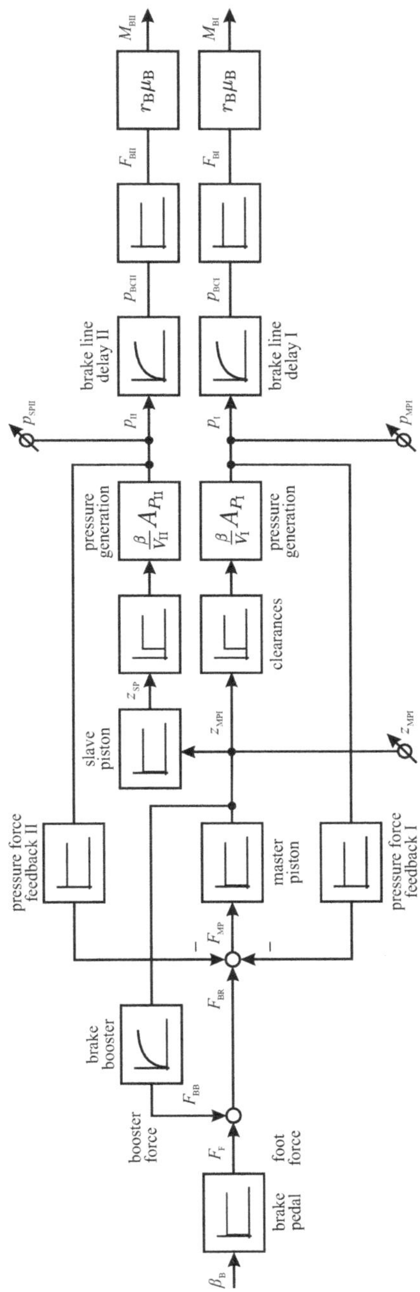


: measurements in production cars

**Fig. 13.4** Signal flow scheme for the hydraulic brake system of a passenger car

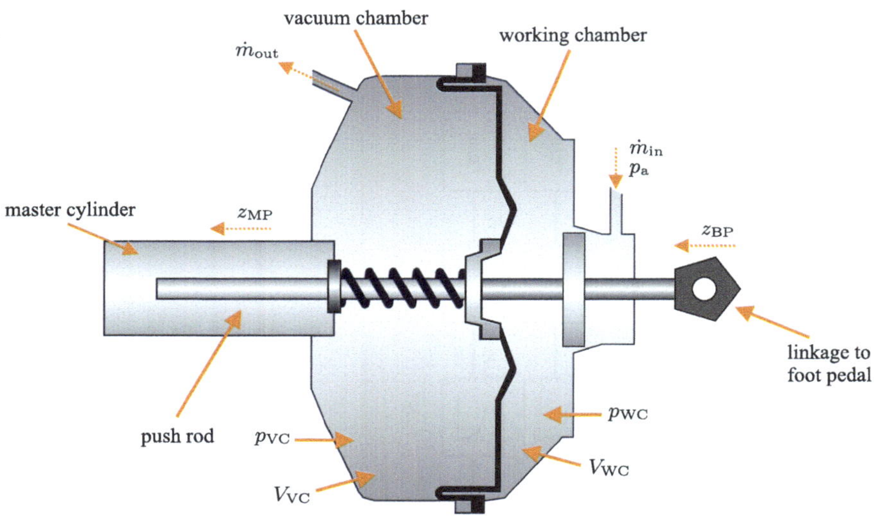


**Fig. 13.5** Simplified representation of a vacuum brake booster as a pneumatic valve-accumulator element

It is assumed that the vacuum pressure $p_{\mathrm{VC}}$ is constant. Then the airflow through the valve can be described by

$$\dot{m}_{\mathrm{in}} = A_{\mathrm{V}}(z_{\mathrm{BP}})\,\psi\left(\frac{p_{\mathrm{WC}}}{p_{\mathrm{a}}}\right) p_{\mathrm{a}}\sqrt{\frac{2}{RT_{\mathrm{a}}}}, \tag{13.2.1}$$

where $\psi$ is the outflow function of a compressible gas through a contraction (nozzle) and $p_{\mathrm{a}}$ and $T_{\mathrm{a}}$ the pressure and temperature of the ambient air. For overcritical pressure ratios ($\frac{p_{\mathrm{WC}}}{p_{\mathrm{a}}} < 0.53$) the outflow function for air is $\psi = 0.484$, while for undercritical pressure ratios ($\frac{p_{\mathrm{WC}}}{p_{\mathrm{a}}} \geq 0.53$) $\psi \leq 0.484$ holds.

Another possibility is to use the pressure loss for turbulent flow of orifices in the case of an undercritical pressure ratio. Then it holds

$$\Delta\dot{m}_{\mathrm{in}} = A_{\mathrm{V}}(z_{\mathrm{BP}})\sqrt{\frac{2\rho_{\mathrm{a}}}{\zeta}}\sqrt{p_{\mathrm{a}} - p_{\mathrm{WC}}}, \tag{13.2.2}$$

see, e.g. Backé (1992); Isermann (2003). Here, $\zeta$ is a loss coefficient and depends on the geometry of the valve; see, e.g. Merrit (1967). Linearization of this equation around a certain operation point $(\bar{A}_{\mathrm{V}}, \bar{p}_{\mathrm{WC}})$ leads to

$$\begin{aligned}\Delta\dot{m}_{\mathrm{in}} &= \frac{\partial\dot{m}_{\mathrm{in}}}{\partial p_{\mathrm{WC}}}\Delta p_{\mathrm{WC}} + \frac{\partial\dot{m}_{\mathrm{in}}}{\partial A_{\mathrm{V}}}\Delta A_{\mathrm{V}} \\ &= -\bar{A}_{\mathrm{V}}\sqrt{\frac{\rho_{\mathrm{a}}}{2\zeta}}\frac{1}{\sqrt{\bar{p}_{\mathrm{a}} - \bar{p}_{\mathrm{WC}}}}\Delta p_{\mathrm{WC}} + \sqrt{\frac{2\rho_{\mathrm{a}}}{\zeta}}\sqrt{\bar{p}_{\mathrm{a}} - \bar{p}_{\mathrm{WC}}}\,\Delta A_{\mathrm{V}} \\ &= -c_1\Delta p_{\mathrm{WC}} + c_2\Delta A_{\mathrm{V}}.\end{aligned} \tag{13.2.3}$$

The mass flow balance of the working chamber with volume $V_{WC} = m_{WC}/\rho_{WC}$ is

$$\Delta \dot{m}_{in}(t) = \frac{d}{dt} m_{WC}(t) = V_{WC} \frac{d}{dt} \rho_{WC}(t). \tag{13.2.4}$$

Assuming an isothermal gas state change of the form

$$p_{WC} = \rho_{WC} R T_{WC} \text{ or } \rho_{WC} = \frac{1}{R T_{WC}} p_{WC}, \tag{13.2.5}$$

results with (13.2.3) and (13.2.4) in a first-order lag function

$$T_{BB} \frac{d\, p_{WC}(t)}{dt} + \Delta p_{WC}(t) = K_{BB} \Delta A_V, \tag{13.2.6}$$

with the gain and time constant

$$K_{BB} = \frac{\Delta p_{WC}}{\Delta A_V} = \frac{c_2}{c_1} = \frac{2}{A_V} (\bar{p}_a - \bar{p}_{WC}) \tag{13.2.7}$$

and

$$T_{BB} = \frac{V_{WC}}{c_1 R T_{WC}}, \tag{13.2.8}$$

respectively.

The stationary booster force to the push rod is

$$\bar{F}_{BB}(z_{BP}) = A_D [\bar{p}_{WC}(z_{BP}) - \bar{p}_{VC}] - c_{VCS} z_{BP}, \tag{13.2.9}$$

where $c_{VCS}$ is the spring constant of the return spring in the vacuum chamber. For small changes and when neglecting the spring force, it is

$$\Delta F_{BB} = A_D \Delta p_{WC} = A_D K_{BB} \Delta A_V. \tag{13.2.10}$$

Hence, the gain $K_{BB}$ of the brake booster pressure is for small changes of the valve area

$$A_V = f(z_{BP}) \tag{13.2.11}$$

due to (13.2.7) the larger, the smaller the pressure $\bar{p}_{WC}$ in the working chamber and the smaller $A_V$. On the other hand, the time constant $T_{BB}$ is the larger, the larger the volume $V_{WC}$, the smaller $p_{WC}$, and the smaller $A_V$.

Therefore, the static and dynamic behavior of the brake booster depends mainly on the operating point $(\bar{A}_V, \bar{p}_{WC})$ and the pedal-valve characteristic $A_V(z_{BP})$. Usually, the travelway of the push rod to the master cylinder and the travelway of the foot pedal are the same, $z_{MP} = z_{BP}$. For modeling of a vacuum brake booster see also Münchhof et al (2003).

### 13.2.2 Brake Circuit

A hydraulic brake circuit consists of the brake master cylinder with the primary and secondary piston, the brake lines and hoses, the cylinders of the caliper, and the brake fluid. To model the dynamic behavior, the brake circuit is first considered as a mass-spring-damper model and then, after neglecting the mass, as a resistance-storage model.

**(a) Mass-spring-damper Model** Figure 13.6 depicts a simplified scheme of a hydraulic circuit. The brake master cylinder moves with a displacement $z_1 = z_{\mathrm{MP}}$ and forces the brake fluid with displacement $z_2 = (A_1/A_l)\, z_1$ through the brake line to the wheel brake cylinder of the caliper and generates there a displacement $z_4 = z_{\mathrm{CP}}$. Taking the moved mass $m_l = A_l \rho l$, a laminar flow resistance coefficient $d_l$ and a stiffness coefficient $c_l$ into account and assuming $z_3 = 0$, the application of the law of momentum yields

$$m_l \ddot{z}_2(t) + d_l \dot{z}_2(z) + c_l z_2(t) = F_1(t). \tag{13.2.12}$$

Introducing $V_2(t) = A_l z_2(t)$ and $F_1 = p_1(t) A_l$ leads to

$$m_l \ddot{V}_2(t) + d_l \dot{V}_2(t) + c_l V_2(t) = A_l^2 p_1(t). \tag{13.2.13}$$

This equation describes the dynamic change of the volume $V_2$ into the brake line after introducing a pressure $p_1$ at the brake line entrance.

The compressibility $\kappa$ of a closed hydraulic fluid mass is defined with the bulk modulus $\beta$

$$\beta = \frac{1}{\kappa} = -V \left( \frac{\partial p}{\partial V} \right)_{T=\mathrm{const.}} . \tag{13.2.14}$$

If an additional fluid volume $\mathrm{d}V$ is pressed into the fluid column of length $l$ it holds

$$\frac{\mathrm{d}V}{V} = \frac{\mathrm{d}z}{l} = \frac{1}{\beta}\mathrm{d}p = \frac{1}{\beta}\frac{\mathrm{d}F}{A_l}, \tag{13.2.15}$$

and its stiffness becomes

$$c_l = \frac{\mathrm{d}F}{\mathrm{d}z} = \beta \frac{A_l}{l} = \beta \frac{A_l^2}{V_l}. \tag{13.2.16}$$

The pressure $p_2$ depends now on the compressibility of the brake line and the wheel brake cylinder with the storage volume $V_{\mathrm{st}}$.

It follows from (13.2.15) that

$$\Delta p_2 = \frac{\beta}{V_{\mathrm{st}}} \Delta V_2, \tag{13.2.17}$$

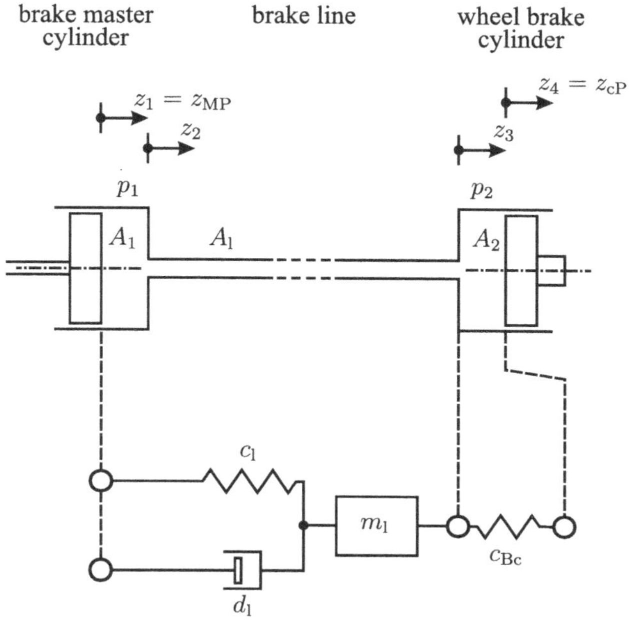


**Fig. 13.6** Scheme of a hydraulic brake circuit and replacement with mass-spring-damper elements

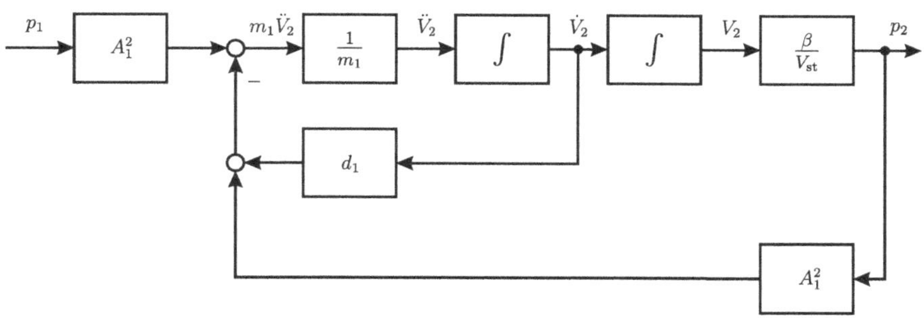


**Fig. 13.7** Signal flow for a hydraulic brake circuit to one wheel with linear pressure storage behavior

where $\Delta V_2$ is the volume flowing into the brake line. If the bulk modulus is assumed to be constant over a wide pressure range then

$$p_2 = \frac{\beta}{V_{st}} V_2 \tag{13.2.18}$$

can be applied. Based on (13.2.14) and (13.2.19), the signal flow scheme of Fig. 13.7 can be drawn.

Replacing $V_2$ in (13.2.13) by $p_2$ with (13.2.18) gives

$$m_l \frac{V_{st}}{\beta} \ddot{p}_2(t) + d_l \frac{V_{st}}{\beta} \dot{p}_2(t) + A_l^2 p_2(t) = A_l^2 p_1(t), \tag{13.2.19}$$

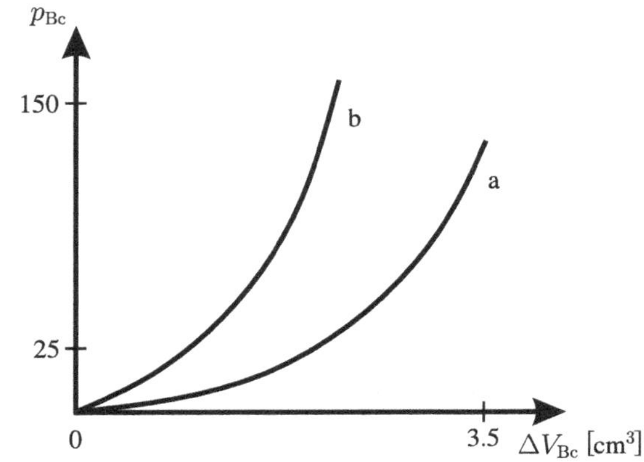


**Fig. 13.8** Storage characteristics of a brake circuit. Measured brake cylinder pressure in dependence on brake fluid volume change. **a** Front brake, **b** Rear brake, (Straky 2003)

which corresponds to the transfer function

$$G_{pp}(s) = \frac{p_2(s)}{p_1(s)} = \frac{K_{pp}}{a_2 s^2 + a_1 s + 1}, \tag{13.2.20}$$

where

$$\begin{aligned} a_2 &= \frac{m_l V_{st}}{\beta A_l^2} = \frac{\rho l V_{st}}{\beta A_l}, \\ a_1 &= \frac{d_l V_{st}}{\beta A_l^2}, \\ K_{pp} &= 1. \end{aligned} \tag{13.2.21}$$

Hence, a second-order system for the pressure $p_2 = p_{CP}$ is obtained.

However, because of the elasticities of the brake pads and the caliper a nonlinear storage characteristic results, which can be approximated by

$$p_2 = k_{BC0} \Delta V_2 + k_{BC1} \Delta V_2^2, \tag{13.2.22}$$

compare Fig. 13.8. Then (13.2.18) has to be replaced by (13.2.22). The brake fluid volume for the front wheel is larger than that of the rear wheel because of the larger caliper cylinder area of the front brake. The hydraulic capacity of the brake system was estimated for a Continental MK 20 brake with VW-Golf chassis to $\bar{k}_{BC0} = \Delta V_2 / \Delta p_2 = 0.974 \times 10^{-7}\ \mathrm{cm^3\ Pa^{-1}}$ for the brake master cylinder.

The inlet valves in the ABS control module and their non-return valves have a significant smaller orifice as the brake lines and are therefore additional resistances. They are modeled by

$$\Delta p_V(t) = k_V \dot{V}^2 \text{ with } k_V = \begin{cases} k_+ & \dot{V} \geq 0 \\ k_- & \text{otherwise} \end{cases}, \tag{13.2.23}$$

with $k$ as flow direction dependent constants, (Straky 2003). Then the signal flow in Fig. 13.9 with a corresponding feedback path results.

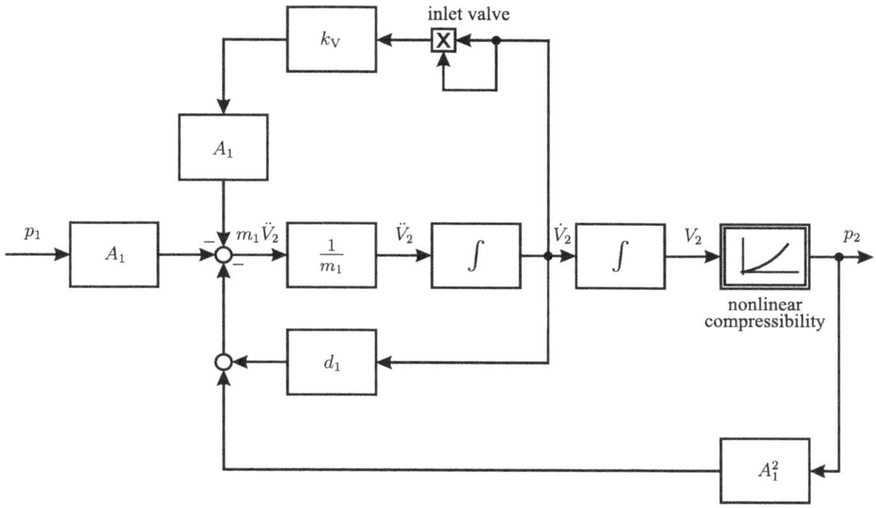


**Fig. 13.9** Signal flow for a hydraulic brake circuit to one wheel with nonlinear pressure storage behavior and inlet valve resistance

If the motion energy can be neglected compared to the storage energy, the fluid mass $m_l$ can be set to zero. Then (13.2.19) simplifies to

$$d_l \frac{V_{\mathrm{st}}}{\beta} \dot{p}_2(t) + A_l^2 p_2(t) = A_l^2 p_1(t), \tag{13.2.24}$$

or

$$G'_{\mathrm{pp}}(s) = \frac{p_2(s)}{p_1(s)} = \frac{K'_{\mathrm{pp}}}{T'_{\mathrm{p}}s + 1}, \tag{13.2.25}$$

with $K'_{\mathrm{pp}} = 1$ and the time constant

$$T'_{\mathrm{p}} = a_1 = \frac{d_l V_{\mathrm{st}}}{\beta A_l^2}. \tag{13.2.26}$$

This time constant was experimentally parameter estimated to be about 20...25 ms for a Continental MK 20 brake system and a VW-Golf chassis, Straky (2003).

**(b) Resistance-storage-model**

Another approach to model the behavior of the pressure in the caliper brake cylinder is to replace the brake line by a resistance element with laminar or turbulent flow and the compressibility of the brake fluid by a storage element, as depicted in Fig. 13.10. Using this lumped parameter approximation, the acceleration forces for the fluid are neglected.

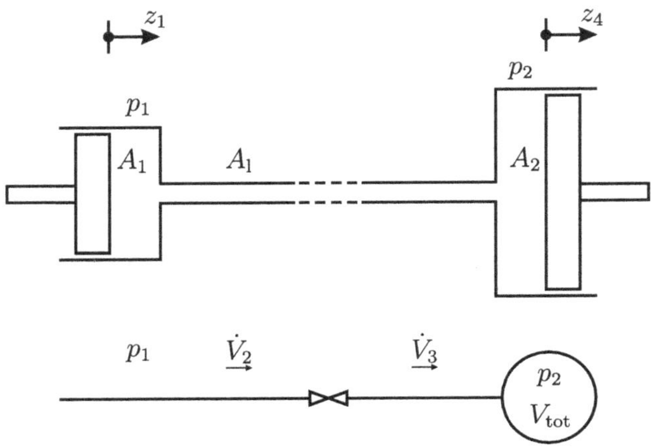


**Fig. 13.10** Scheme of a hydraulic brake circuit and replacement by a resistance-storage element with lumped parameters

Assuming a laminar flow for the brake line, it holds according to the law of Hagen-Poisseuille that

$$p_1(t) - p_2(t) = k_l \dot{V}_2, \tag{13.2.27}$$

$$k_l = \frac{128 \cdot \nu_f \rho_f l}{\pi d^4}, \tag{13.2.28}$$

with $\nu_f$ being the kinematic viscosity, $\rho_f$ the fluid density, $l$ the pipe length, and $d$ the pipe diameter. For turbulent flow, a quadratic law is valid

$$p_1(t) - p_2(t) = k_t \dot{V}_2^2, \tag{13.2.29}$$

$$k_t = k_{t1} = \frac{f l}{2 d A_l^2}, \tag{13.2.30}$$

with $f$ being a friction factor depending on the surface roughness and Reynolds number and $A_l$ being the cross section area of the pipe. In the case of flow restrictions in the form of orifices, it holds

$$k_t = k_{t2} = \zeta \frac{\rho_f}{2 A_l^2}, \tag{13.2.31}$$

where $\zeta$ is a loss coefficient depending on the geometry; see Merrit (1967); Backé (1992); Isermann (2003).

The storage behavior follows from (13.2.19)

$$\frac{\mathrm{d}p_2}{\mathrm{d}t} = \frac{\beta}{V_{st}} \dot{V}_2 = k_{BC} \dot{V}_2, \tag{13.2.32}$$

where $V_{st}$ is the volume of the brake fluid in one circuit. Insertion into (13.2.27) leads to the linear first-order differential equation

$$T_p'' \frac{\mathrm{d}p_2(t)}{\mathrm{d}t} + p_2(t) = p_1(t), \qquad (13.2.33)$$

$$T_p'' = \frac{k_l}{k_{BC}} = k_l \frac{V_{st}}{\beta}, \qquad (13.2.34)$$

which corresponds to (13.2.24), (13.2.26).

In the case of turbulent flow and especially with orifice flow restrictions in the brake control device, the brake circuit should be modeled according to the signal flow in Fig. 13.9, as shown by Straky (2003).

The transfer behavior from a small brake pedal change $\Delta\beta$ to a change of all braking forces $\Delta F_{XT,B,i}$ can now be determined as follows. For the brake booster holds (13.2.6)

$$\frac{\Delta p_{wc}(s)}{\Delta A_v(s)} = \frac{K_{BB}}{1 + T_{BB}s} \qquad (13.2.35)$$

which yields the brake booster force with $\Delta A_v = c_{fiA}\Delta\beta$

$$\frac{\Delta F_{BB}(s)}{\Delta\beta(s)} = \frac{K_{BB}A_D c_{fiA}}{1 + T_{BB}s}. \qquad (13.2.36)$$

The dynamics of the hydraulic brake circuit is with (13.2.25) and $p_2 = p_{wcyl}$ and $p_1 = F_{BB}/A_1$

$$\frac{\Delta p_{wcyl}(s)}{\Delta F_{BB}(s)} = \frac{K'_{pp}}{A_1\left(1 + T_p' s\right)} \qquad (13.2.37)$$

The sum of the front and rear wheel forces becomes with (6.7.7) and (6.7.8)

$$F_{XT,B,\Sigma} = \sum_i F_{XT,B,i} = 2(c_{B,f} + c_{B,r})p_{wcyl} = c_B\, p_{wcyl} \qquad (13.2.38)$$

and after linearization with (6.7.6)

$$\Delta F_{XT,B,\Sigma} = c_B \Delta p_{wcyl}. \qquad (13.2.39)$$

The series connection of the brake booster and the hydraulic brake then results with (13.2.36), (13.2.38), and (13.2.39) in the transfer function of the hydraulic brake system

$$\frac{\Delta F_{XT,B,\Sigma}(s)}{\Delta\beta(s)} = \frac{K_{Bf}}{(1 + T_{BB}s)(1 + T_p' s)} \qquad (13.2.40)$$

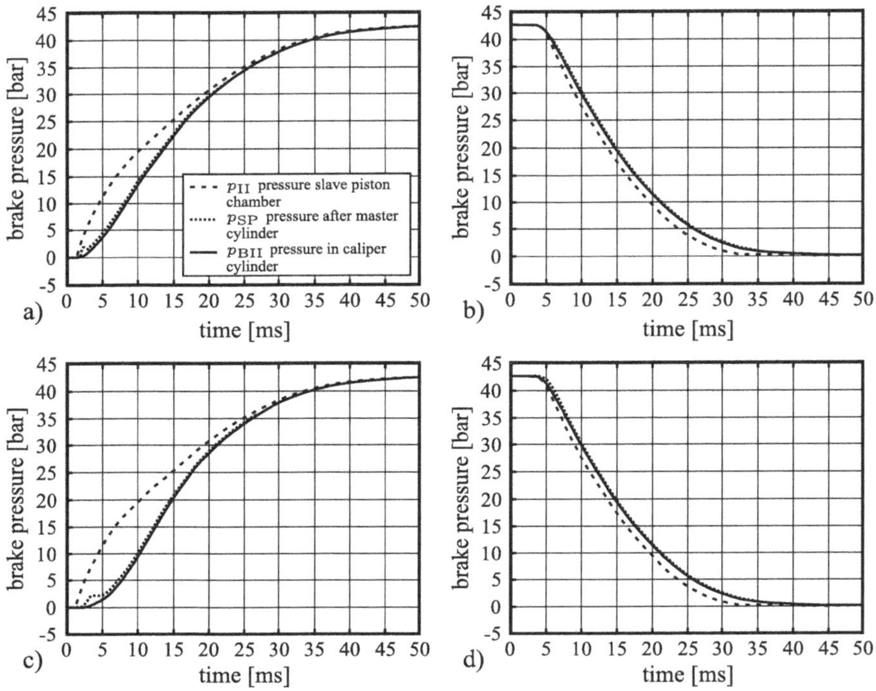


**Fig. 13.11** Dynamic behavior of various pressures in the hydraulic brake system Continental MK 20 built as test bench according to dimensions of a VW Golf V6, (Straky 2003). **a** Pressure increase front left, **b** pressure decrease front left, **c** pressure increase rear right, **d** pressure decrease rear right

with the gain

$$K_{\mathrm{Bf}} = \frac{K_{\mathrm{BB}} K'_{\mathrm{pp}} A_{\mathrm{D}} c_{\mathrm{fiA}} c_{\mathrm{B}}}{A_1}. \tag{13.2.41}$$

The overall brake system can therefore be described by a second-order delay with two time constants.

Figure 13.11 shows the measured dynamic behavior of the pressure of a secondary brake circuit on a brake system test bench with a configuration as a compact car. The transient functions $p_i(t)$ are obtained from a fast ramp function of the linkage to the brake booster/master piston. The transient pressure courses for the *pressure increase* (braking situation) show:

- The measured transient functions in Fig. 13.11 can be evaluated with the method of characteristic parameter determination; see Isermann and Münchhof (2011). For the front brake, Fig. 13.11a, the delay time is $T_{\mathrm{D}} \approx 3$ ms and the settling time $T_{\mathrm{S}} \approx 20$ ms which gives $T_{\mathrm{D}}/T_{\mathrm{S}} = 0.15$. A second order low pass model with equal time constants $T$ then yields $T_i \approx T_{\mathrm{S}}/2.72 = 7.4$ ms. The sum of two time constants is about $T_{\Sigma} \approx 14.8$ ms.

- In the case of rear brakes it holds $T_D \approx 5$ ms and $T_S \approx 20$ ms, $T_D/T_S = 0.25$. This yields for $n = 3$, $T_i = 20/3.7 = 5.4$ ms, $T_\Sigma = 16.2$ ms. For $n = 2$ one obtains $T_i \approx 7.35$ ms, $T_\Sigma = 14.7$ ms.
- The pressures $p_I$ after the master cylinder (beginning of the brake line) and in the caliper brake cylinder $p_{BCII}$ differ only very little (compare with Fig. 13.4). However, they lag with about 3 to 5 ms behind the pressure $p_{II}$ in the slave piston chamber. This is mainly due to the pressure loss through the valve orifices in the brake control device; see Figs. 13.2 and 13.9
- The pressure $p_{BII}$ in the caliper cylinder is more delayed for the rear wheel brake than for the front wheel brake to allow a dynamically delayed brake force increase for the rear wheel to support vehicle stability.
- The time constants of the model (13.2.40) are $T_{BB} \approx 7$ ms and $T'_P \approx 8$ ms.

The transient pressure courses for the *pressure decrease* (release situation) indicates:

- A delay of 4 to 5 ms due to clearance effects
- The behavior of the pressure $p_{BII}$ in the caliper brake cylinder shows a first order lag behavior with a time constant of about 17 to 20 ms and a settling time of about 35 ms
- The pressures $p_{SP}$ and $p_{BII}$ are almost the same
- The caliper brake cylinder pressure $p_{BII}$ lags only about 1 to 2 ms after the pressure $p_{II}$ in the slave piston chamber or $p_{SP}$. This is because the non-return valves in the brake control device are opened for the back flow.

The measurements illustrate the real behavior of the considered brake system and underline the theoretically developed models treated in the last sections, summarized in Fig. 13.4.

## 13.3 Anti-lock Control with Switching Valves (ABS)

In order to avoid locking wheels during strong braking, anti-lock braking systems (ABS) are installed. They ensure that the wheels rotate with a certain angular speed close to optimal slip with the goals to keep the vehicle's steerability and stability and to shorten braking distances, (Burckhardt 1993).

A hydraulic-electronic control unit (HECU) consists of a central hydraulics block with valves, an electric driven pump (90 W to 220 W), see Fig. 13.12, and a coil carrier with an electronic control unit (2 micro-controllers, ~ 20 MHz, ROM with 256 kB to 1 MB). The coil carrier is connected to the hydraulics unit with a magnetic connector.

As shown in Fig. 13.12, an inlet valve and an outlet valve (both 2/2 way) within the hydraulic block enable to modulate the brake pressure for each wheel. The inlet valve is normally open and the outlet valve is closed.

Based on the wheel-speed sensor, the wheel's acceleration is determined. If this acceleration shows during braking a sharp negative value, a deceleration, the braking

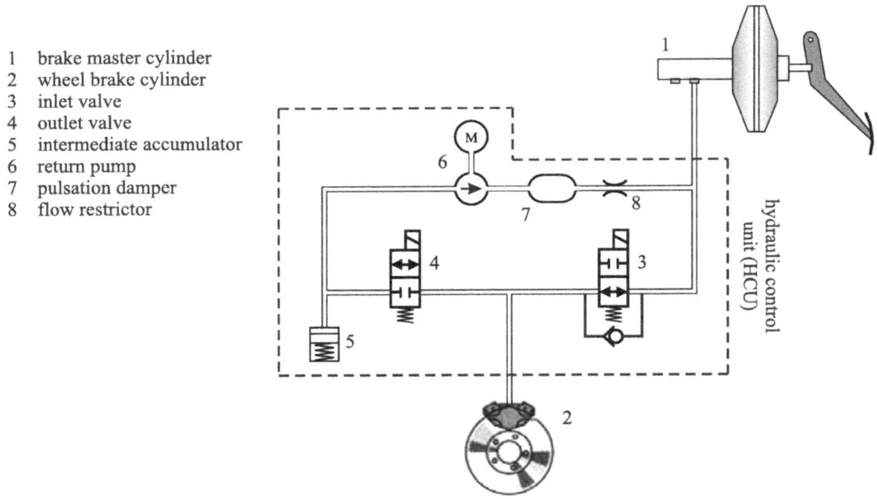


**Fig. 13.12** Hydraulic brake system with magnetic 2/2-way valves for ABS-braking

pressure in the master cylinder has to be reduced in order to avoid locking of the wheel, though the braking pedal is applied by the driver to strong braking. Therefore, first the inlet valve is closed and the brake pressure remains constant. If the deceleration continues the outlet valve is opened, the braking pressure gets smaller, and the wheel results in smaller slip, thus avoiding complete locking of the wheel.

The brake fluid leaving the brake circuit is stored in an intermediate accumulator and then pumped back by the return pump to the brake master cylinder; see Fig. 13.12.

In the following, an example for a control cycle of an ABS is described, leaning on Semmler (2006a); Robert Bosch GmbH (2011); van Zanten and Kost (2016); Breuer and Bill (2017).

The measurements for the ABS control system are usually:

- $\omega_W$ angular wheel speed for each wheel
- $v_X$ vehicle ground speed

Based on these measurements, the following variables are calculated in the ABS-ECU:

- $\dot{\omega}_X = \begin{cases} \mathrm{d}\omega_X/\mathrm{d}t > 0 & \text{wheel speed acceleration} \\ \mathrm{d}\omega_X/\mathrm{d}t < 0 & \text{wheel speed deceleration} \end{cases}$
- $S_X = \frac{v_X - \omega_W r_{dyn}}{v_X}$ tire slip for each wheel; see (5.1.1).

Figure 13.13 depicts a signal flow for a braking wheel with the angular speed acceleration $\dot{\omega}_X$ as output, based on Fig. 5.23. It shows that $\dot{\omega}_W$ is a direct output after an application of a braking torque $M_B$, however, with feedback including an integrator and the $\mu$-slip characteristic. The feedback is negative, if $\mathrm{d}\mu_X/\mathrm{d}S_X > 0$, leading to

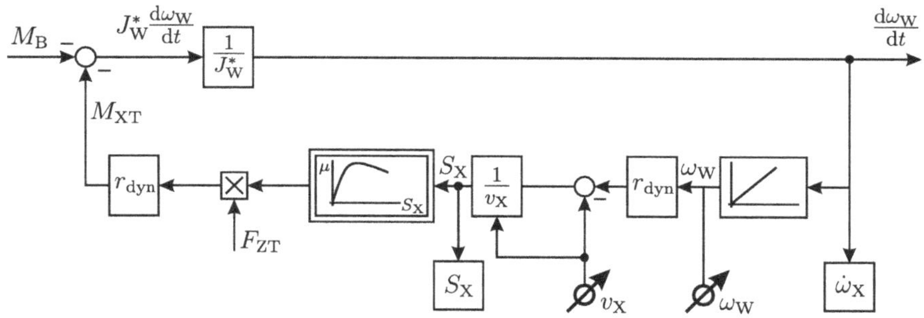


**Fig. 13.13** Signal flow chart for a braking wheel with braking torque $M_B$ as input and angular wheel acceleration $\dot{\omega}_W$ as output. Measurements for ABS control: $\omega_W$ and $v_X$, calculated variables: $\dot{\omega}_X$ and $S_X$

stable behavior and is positive for $d\mu_X/dS_X < 0$, which results in monotonic unstable behavior and, therefore, locking.

After linearization and assuming small changes, one obtains from (5.5.15)

$$G_{\dot{\omega}MB}(s) = \frac{\Delta\dot{\omega}_W}{\Delta M_B} = \frac{K_{WB}s}{1 + T_{WB}s}. \tag{13.3.1}$$

Hence, a differential behavior with a time delay results ($DT_1$-behavior). This means that the wheel acceleration indicates a fast pulse-like behavior, which is difficult to control with a continuous controller.

A typical ABS control cycle for one wheel with switching valves is depicted in Fig. 13.14 after a stepwise brake command with the brake pedal. Then different phases for switching the valve solenoids follow sequentially.

*Phase 1*: The brake pressure $p_B$ applied by the driver increases, the wheel speed decreases, and the deceleration $\dot{\omega}_W(t) < 0$ increases.

*Phase 2*: If $\dot{\omega}_W < -a$, the wheel deceleration transgresses the threshold $(-a)$, the inlet valve (Fig. 13.12) is closed, such that the brake pressure is hold at $p_B = \text{const.}$. The wheel slip further increases and moves from the stable range to the unstable range of the $\mu$-slip characteristic. The reference speed $v_{Xref}$ is determined, corresponding to an a priori defined value for maximal vehicle deceleration.

*Phase 3*: If the wheel speed is smaller than the reference speed, i.e. $v_W < v_{Xref}$, and the tire slip becomes larger than a slip-switching threshold $S_{X1}$, i.e. $S_X(t) < S_{X1}$, the outlet valve is opened while the inlet valve is closed, such that the braking pressure falls. The tire slip is therefore reduced to bring it in the stable range of the $\mu$-slip characteristic.

*Phase 4*: If the wheel speed deceleration reaches the threshold $(-a)$, i.e. $\dot{\omega}_W > -a$, the outlet valve is closed, the braking pressure is hold, $p_B = \text{const.}$, the wheel

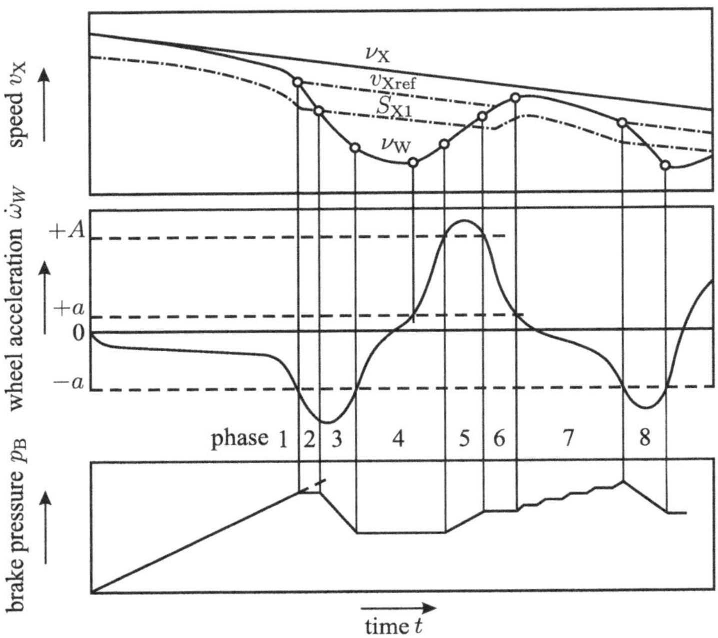


**Fig. 13.14** Phases of the ABS control functions, (Robert Bosch GmbH 2004)

deceleration becomes smaller, the wheel accelerates, $\dot{\omega}_W > 0$, and the wheel speed $\omega_W$ increases.

*Phase 5*: If the wheel speed acceleration exceeds the threshold $(+A)$, i.e. $\dot{\omega}_W > +A$, the inlet valve is opened, the braking pressure $p_B$ increases, and the wheel acceleration $\dot{\omega}_W$ decreases.

*Phase 6*: If the wheel acceleration $\dot{\omega}_W < +A$ the inlet valve is closed, the braking pressure $p_B$ stays constant, and the wheel acceleration $\dot{\omega}_W$ decreases.

*Phase 7*: If the wheel acceleration decreases under the threshold $(+a)$, i.e. $\dot{\omega}_W < +a$, the inlet valve is opened with pulse signals, such that the braking pressure increases slowly and the wheel deceleration increases. It is assumed that the slip is now in the stable range and near to the maximal value.

*Phase 8*: If the wheel deceleration exceeds the threshold $(-a)$, i.e. $\dot{\omega}_W < -a$, then the outlet valve is opened and the inlet valve is closed, the braking pressure drops, and the wheel deceleration becomes smaller, such that wheel speed $\omega_W$ increases again. This corresponds to phase 3 and the ABS control cycle repeats from phase 4 to phase 8.

In phase 2 of the first control cycle, a short pressure holding phase is necessary at the beginning, in order to avoid a locking wheel for slow brake pressure rise and small slip values, e.g. on ice. This means that the second and following control cycles begin with phase 4 and end with phase 8.

Fig. 13.15 gives an overview of the sequentially arising phases together with the switching conditions based on limit values. It shows how the brake pressure is increased, decreased, and hold constant by switching the inlet and outlet valves depending on exceeding the different limit values of the wheel acceleration and wheel-road slip (the influence of the reference speed is omitted).

Measurements over the first 0.6 s of an initial ABS cycle are illustrated in Fig. 13.16. One recognizes the decrease of the wheel circumferential velocity from 75 km $h^{-1}$ to 50 km $h^{-1}$ in the first 200 ms and a maximal slip of $S_X = 0.21$, which is reduced to smaller values.

Figure 13.17 depicts the measurements of the ground speed $v_X$ and the wheel speeds $v_{W,fl}$ and $v_{W,rl}$ for a full braking. The wheel speed oscillations show the control cycles of the ABS-system with larger values within the first second, an average difference velocity of about 5 m s, and an average deceleration of 9.8 m $s^{-2}$, (Semmler 2006b).

An extension of the basic hydraulic system for ABS-braking of Fig. 13.12 is depicted in Fig. 13.18 for 4 wheels of a passenger car. The ABS control unit also provides functions for brake force distribution between the front and rear wheels. In order to prevent the rear wheels to lock during increasing braking pressure, the brake pressure to the rear wheel is limited by closing the inlet valves if increased slip at the rear axle is detected. Here, a "select-low principle" is applied, which means that the rear wheel which first shows higher slip determines the brake pressure limitation.

Figure 13.18 contains, compared to Fig. 13.12, two additional magnetic valves, a suction valve, and a change-over valve to allow a *traction control system* (TCS) for the case that one or two driven wheels are spinning compared to the non-driven wheels. Then, first the engine torque is reduced. If then one of the driven wheels continues to spin because of different frictions on both sides, the brake pressure is applied to this wheel and the drive torque is transferred to the other non-spinning wheel through the differential. In this case, the suction valve is opened, the change-over valve is closed, and the activated return pump increases the pressure to the brake of the spinning wheel through the inlet valve; see Sect. 17.1.

## 13.4 Electromechanical Brake Booster

The conventional vacuum brake booster requires a vacuum source, either from the manifold of gasoline engines or an extra vacuum pump for diesel engines. Additionally, the required packaging space is relatively large. With the goal of an integrated mechatronic brake force generator, an electromechanical brake booster has been developed, which uses the dual master cylinder and an electrical motor with gear to amplify the pedal force and is used since 2013 for series production. The advantages are a faster brake pressure generation, better use of electrical inputs for driver assis-

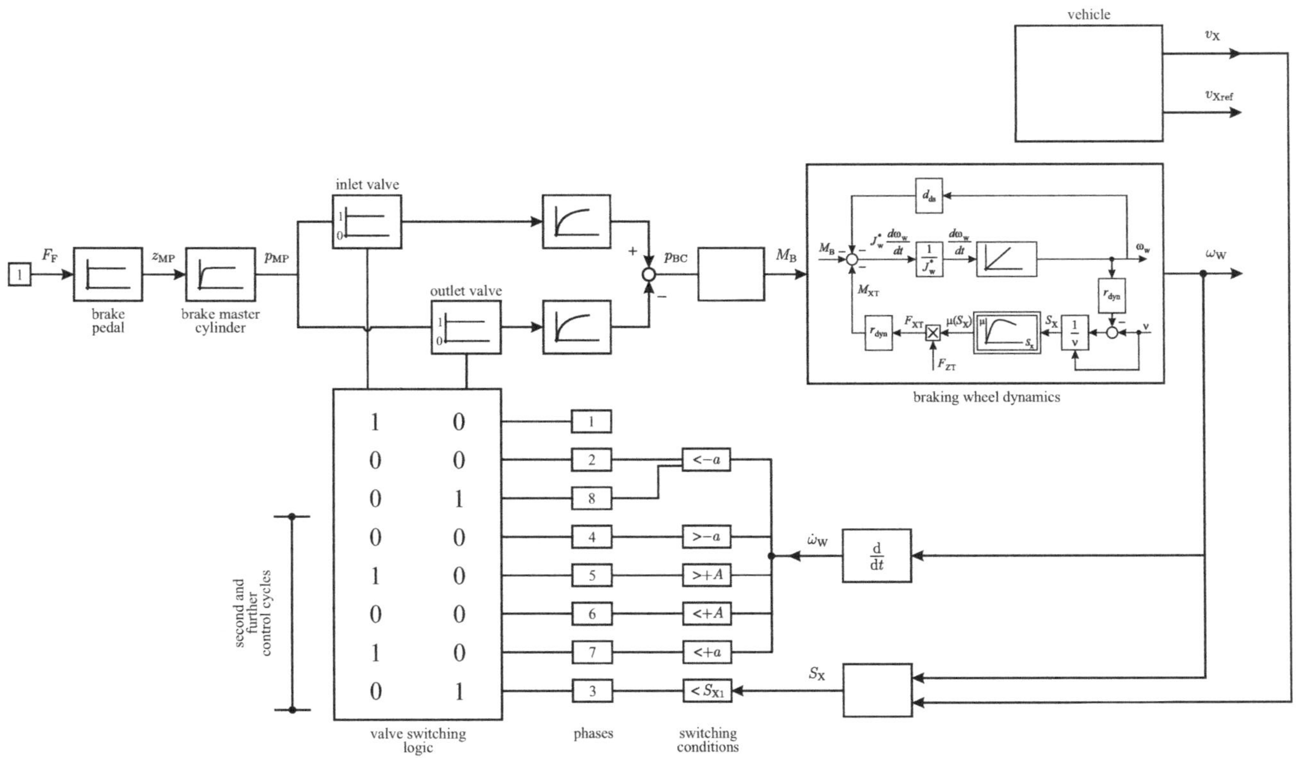


**Fig. 13.15** Signal flow for the phases of ABS control cycles. First cycle: 1 to 3. Second and following phases: 4 to 8

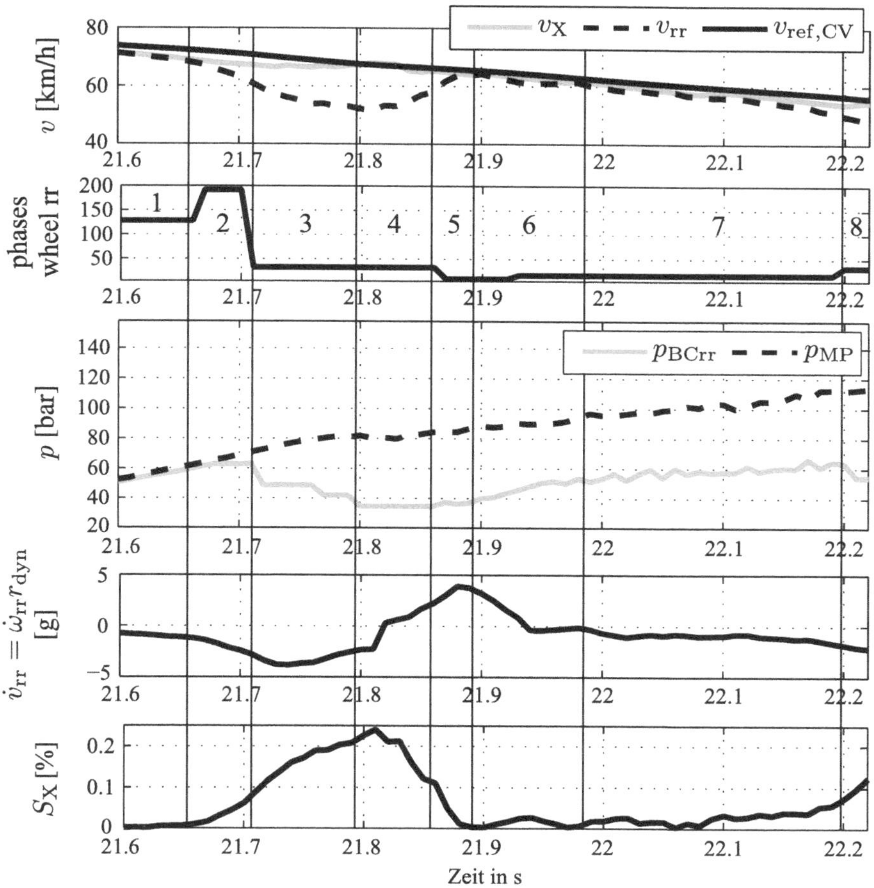


**Fig. 13.16** Measurements for an ABS cycle of the rear right wheel (rr) during a full braking with $a_X = 8.8 \text{ m s}^{-2}$, (Bauer 2015). $v_X$: ground velocity (ABS), $v_{ref,CV}$: reference ground velocity (Correvit-sensor), $v_{rr}$: wheel circumferential velocity, $p_{MPI}$: pressure in master cylinder, $p_{BCrr}$: pressure in caliper cylinder

tance systems, redundancy for automatic driving, and use for electrical and hybrid drives.

The generation of the supporting brake force is performed, e.g. by a permanently excited synchronous motor (PMSM) followed by a two-step reduction gear and a worm gear, with a power of about 300 W, (Kunz et al 2018).

A schematic illustration of the signal flow is depicted in Fig. 13.19. The generation of the supporting braking force is based on the measurement of the travelways of the pedal push rod and of the electromotorically driven rack. The control variable is the difference of the two travelways with the goal to bring it to zero. An alternative is to measure directly the difference of the two travelways, (Kunz et al 2018).

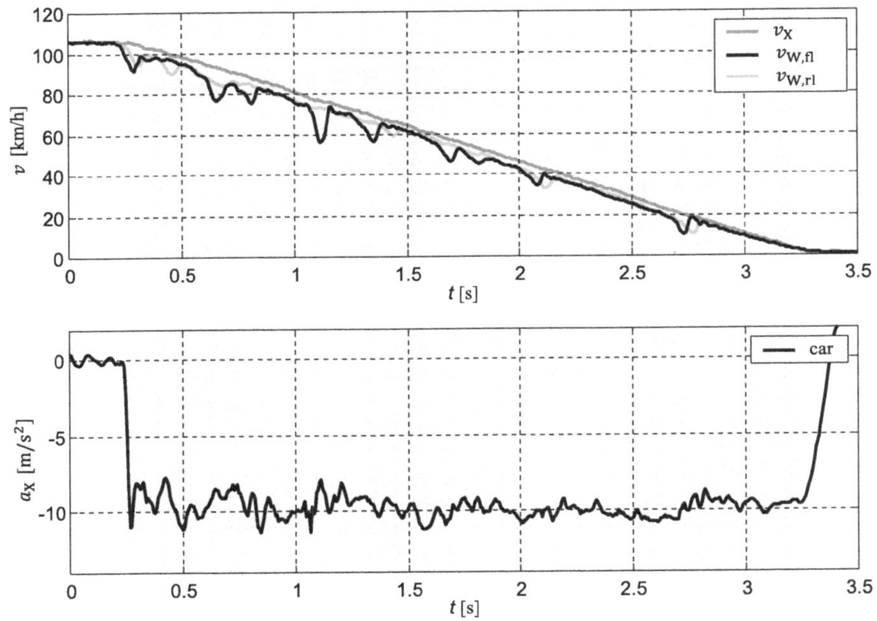


**Fig. 13.17** Measurements for the front left and rear left wheel during a full braking on dry asphalt, VW Golf IV, 1.6 l, (Semmler 2006b)

This arrangement is more precise and faster than a vacuum booster. Because of the electrically actuated braking force, it allows the direct contribution and blending for recuperative braking with generators or for electrical drives and direct use for automatic driving.

## 13.5 Electro-Hydraulic Brake System (EHB)

The electro-hydraulic brake system is a brake-by-wire system, where the brake forces for each wheel are individually controlled electronically by *continuously operating valves*. This allows to directly integrate ABS, TCS, and ESC functions in one compact braking unit. The EHB consists of the following components, see Fig. 13.20:

- Actuation unit with dual master cylinder, pedal travel force simulator, backup travel sensor, and brake fluid reservoir
- Hydraulic control unit (HCU) with motor, three piston pumps, high-pressure accumulator, 8 continuously operating electromechanical valves, 2 isolation valves, 2 balancing valves, and 6 pressure sensors
- Electronic control unit (ECU) with two microprocessors

The brake pedal force simulator generates a suitable pedal travel/force characteristic with damping. The pedal travel is measured by 2 separate angle-position sensors.

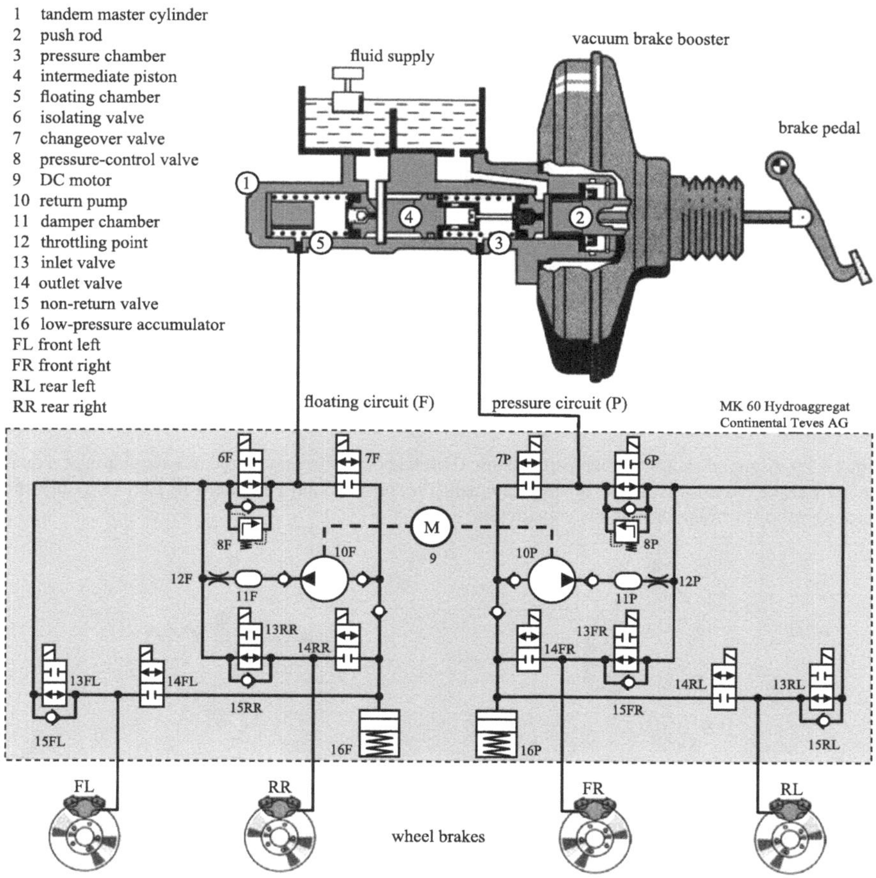


**Fig. 13.18** Hydraulic brake system for antilock braking of a passenger car in a diagonal configuration, (Bauer 2015)

Additionally, the brake pressure of the brake master cylinder is measured, such that a threefold redundancy of the drivers brake command exists.

In normal driving mode, the electrically driven hydraulic three piston pump charges the high-pressure metal diaphragm accumulator leading to a pressure between 90 and 180 bar, which is measured by an accumulator pressure sensor. The isolating valves are closed if the brake pedal is activated and separate the pressure in the master cylinder from the brake calipers. The braking demand from the driver, measured by the pedal travel sensors, is then transferred to the ECU, which controls the brake pressures individually for each wheel "by wire", Jonner et al (1996).

As the braking pressure does not have to be different for the wheels of one axle in many situations, the balancing valves connecting the brakes of one axle can stay open and, thus, increase lifetime and allow for diagnosis functions.

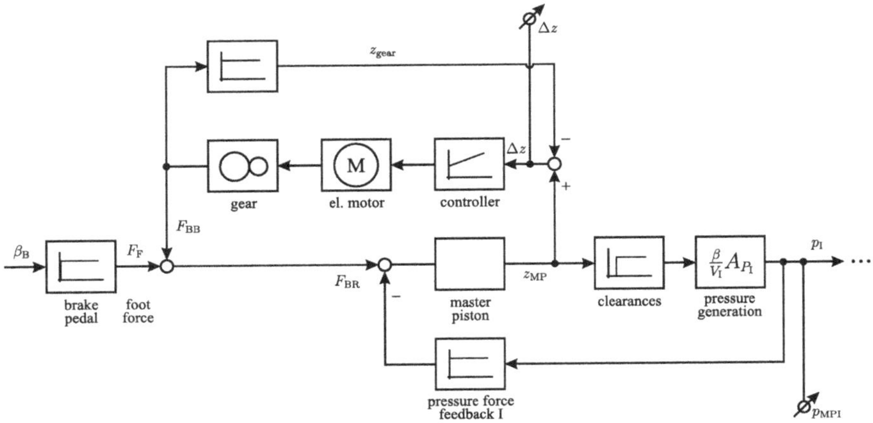


**Fig. 13.19** Signal flow for an electromechanical brake booster acting on the master piston of a dual master cylinder with difference position control. Please note that compared to Fig. 13.4 only the master piston for brake circuit I is considered

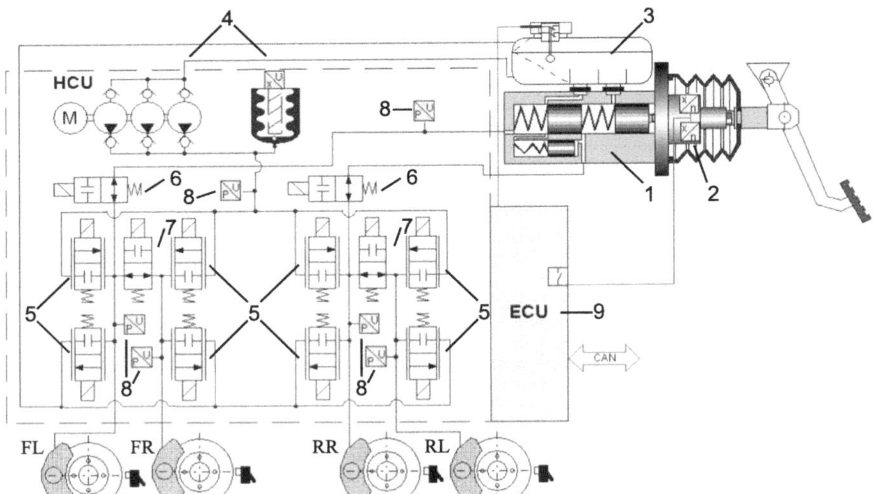


**Fig. 13.20** Electro-hydraulic brake system (EHB) for a passenger car with dual master cylinder, pedal force simulator, and high-pressure brake fluid pump, (Remfrey et al 2016; Semmler 2006b). 1: dual master cylinder with pedal feel simulator; 2: travelway sensors; 3: brake fluid accumulator; 4: HCU with motor-pump accumulator; 5: valve block with eight continuous control valves; 6: two isolating valves; 7: two balancing valves; 8: six pressure sensors; 9: ECU

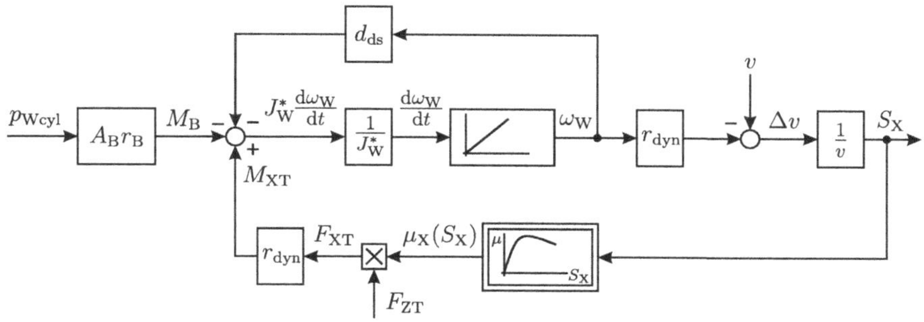


**Fig. 13.21** Signal flow for a braked wheel with braking pressure as input and braking slip as output

The software structure of the ECU is characterized by a modular structure with different levels from basic functions through yaw control, traction control (TCS), and anti-lock control (ABS) to electronic stability control (ESC), see Remfrey et al (2016).

In the case of electrical or electronic faults, the pedal simulator is shut off and the direct connections from the master cylinder to the brake circuits allow to brake with pedal force only. The EHB was introduced into series production in the years 2001 to 2004, (Remfrey et al 2016). However, because of its complexity, high costs, and maintenance efforts, the EHB concept was after some years not continued by some manufacturers.

## 13.6 EHB Slip Control with Proportional Valves

The individual brake pressure control with the *proportionally acting valves* of the EHB allows to control directly the wheel slip during braking. If then a suitable value for the desired slip is known, a certain brake pressure can be generated and a locking of the wheel avoided. This was demonstrated by Semmler (2006b). In the following, a simplified slip control is considered.

The torque balance equation of one wheel is according to (5.5.9) and Fig. 5.22 given by

$$J_W^* \frac{d\omega_W(t)}{dt} = M_{ds,out}(t) + M_{XT}(t) - M_B(t) - M_{ds,F}(t). \tag{13.6.1}$$

Assuming no drive torque, i.e. $M_{ds,out} = 0$, and for the road/tire torque through slip $S_X$, (5.5.2), and (5.5.5)

$$M_{XT} = r_{dyn}\mu_X(S_X)F_{ZT}, \tag{13.6.2}$$

one obtains

$$J_W^* \frac{d\omega_W(t)}{dt} - r_{dyn}\mu_X(S_X(t))F_{ZT}(t) + d_{ds}\omega_W(t) = -M_B(t). \tag{13.6.3}$$

A corresponding signal flow is depicted in Fig. 13.21. The relation $\mu_X(S_X)$ introduces a nonlinearity in the feedback of the block diagram. In order to simplify the derivations, the linearized approximation (5.1.5) for the $\mu$-slip characteristic is used, i.e.

$$\mu_X = c_{\mu 0} S_X, \tag{13.6.4}$$

and the wheel/drive friction is neglected, so that $d_{ds} = 0$. This yields

$$J_W^* \frac{d\omega_W(t)}{dt} - r_{dyn} c_{\mu 0} S_X(t) F_{ZT}(t) = -M_B(t). \tag{13.6.5}$$

Introduction of (5.5.1)

$$S_X = \frac{\Delta v}{v} = 1 - \frac{\omega_W r_{dyn}}{v} \text{ or } \omega_W = \frac{v}{r_{dyn}}(1 - S_X), \tag{13.6.6}$$

and

$$\dot{\omega}_W = -\frac{v}{r_{dyn}} \dot{S}_X, \tag{13.6.7}$$

with $\dot{v} \approx 0$ gives

$$\underbrace{J_W^* \frac{v}{r_{dyn}}}_{=a_1} \dot{S}_X(t) + \underbrace{r_{dyn} c_{\mu 0} F_{ZT}(t)}_{=a_0} S_X(t) = M_B(t), \tag{13.6.8}$$

or

$$a_1 \dot{S}_X(t) + a_0 S_X(t) = M_B(t). \tag{13.6.9}$$

Hence, a first-order delay results with transfer function

$$G_{MS}(s) = \frac{S_X(s)}{M_B(s)} = \frac{1}{a_1 s + a_0}, \tag{13.6.10}$$

with time varying parameters. The brake torque follows (6.7.5)

$$M_B = c_{WB} p_{Wcyl}, \tag{13.6.11}$$

and the slip transfer function becomes

$$G_{pS}(s) = \frac{S_X(s)}{p_{Wcyl}(s)} = \frac{c_{WB}}{a_1 s + a_0}. \tag{13.6.12}$$

Applying a PD-slip controller

$$G_c(s) = \frac{\Delta p_{Wcyl}(s)}{\Delta S_X(s)} = (K_P + K_D s), \tag{13.6.13}$$

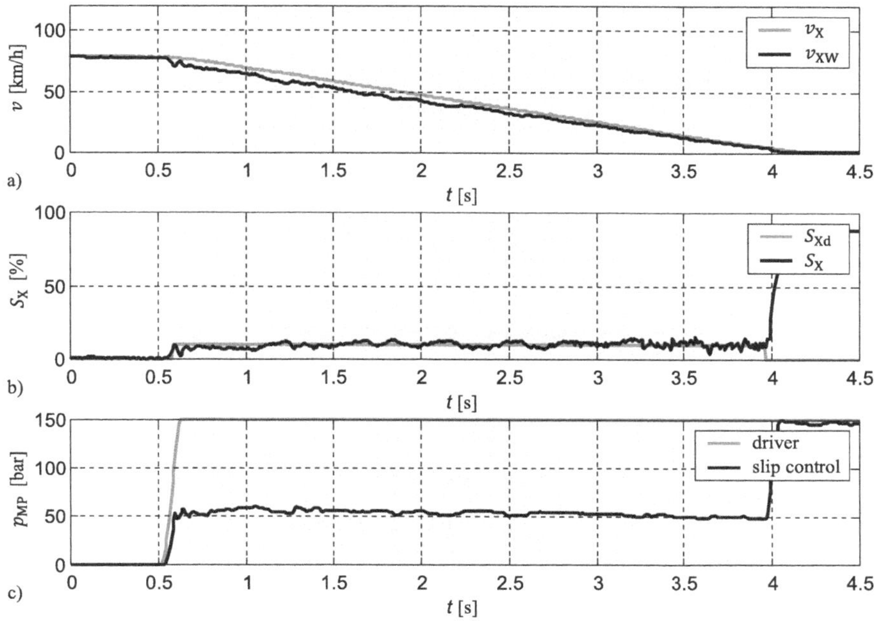


**Fig. 13.22** Slip control of the front wheels on dry asphalt for a desired slip of $S_{Xd} = 10\,\%$, VW Golf IV, 1.6 l, (Semmler 2006b). **a** vehicle and braked wheel velocities; **b** reference and real slip; **c** brake pressure in the master brake cylinder for the driver input and the reduced value by the slip control

results in a closed-loop transfer function with the brake pressure as manipulated variable for a desired slip $S_{Xd}$

$$G_{pS(s)} = \frac{S_X(s)}{S_{Xd}(s)} = \frac{G_{pS}(s)G_c(s)}{1 + G_{pS}(s)G_c(s)} = K_S. \qquad (13.6.14)$$

A proportional behavior can be obtained if the controller parameters are properly selected, as, for example, $K_P = a_0$ and $K_D = a_1$.

As the parameters $a_0$ and $a_1$ of the simplified slip model depend on $v$, $F_{ZT}$, and the linear approximation parameter $c_{\mu 0}$, which becomes smaller with increasing slip $S_X$, the controller parameters have to be adapted.

The estimation of the slip parameter $c_{\mu 0}$ can be obtained with least squares parameter estimation as described in Sect. 11.6.1 (d) either with a linear HRSI-model (11.6.15) or with a nonlinear algebraic model (11.6.18). Semmler (2006a) has developed a slip controller which compensates the time varying variables.

To find optimal desired slip values the sign of the parameter $c_{\mu 0}$ of the linear approximation (13.6.4) is observed with increasing slip. The change of the sign then indicates passing the critical slip $S_{X,crit}$ and, therefore, a too large slip value. Another approach is to observe oscillations of the slip values around $S_{X,crit}$.

Figure 13.22 and 13.23 depict brake maneuvers with desired slip values of 10 % for dry asphalt and 17 % on wet asphalt. The front wheels are slip controlled for the

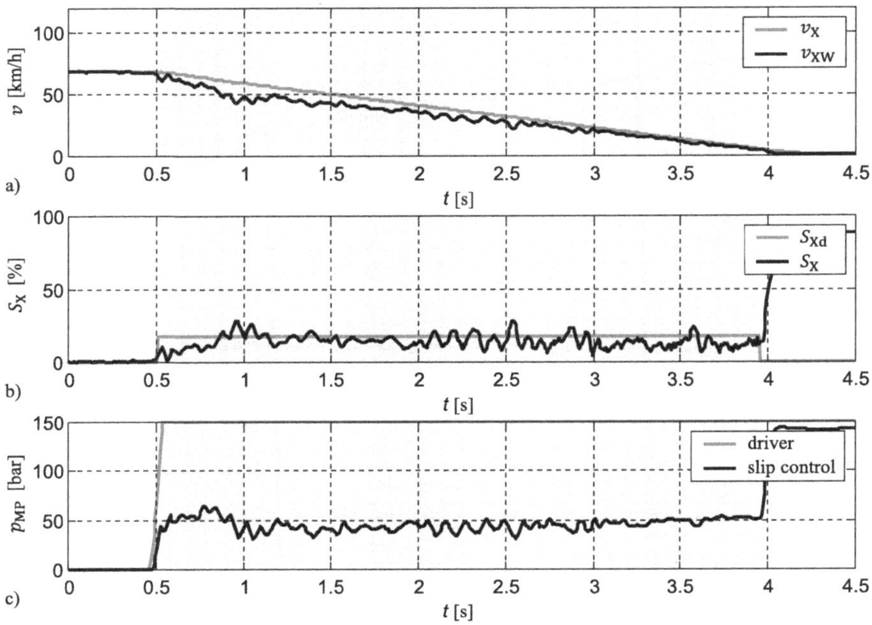


**Fig. 13.23** Slip control of the front wheels on wet asphalt for a desired slip of $S_{Xd} = 17\,\%$, VW Golf IV, 1.6 l, (Semmler 2006b). **a** vehicle and braked wheel velocities; **b** reference and real slip; **c** brake pressure in master brake cylinder for the driver input and the reduced value by the slip control

desired slip values. The rear wheels are rolling free and are used to determine the vehicle velocity. The slip controller is activated after $\dot{\omega}_W$ and $S_X$ pass a threshold and shut off for low velocity.

Further experiments have shown also precise slip control for snow with $S_{X,d} = 44\,\%$ and on ice with $S_{X,d} = 70\,\%$ without locking wheels.

Figure 13.24 depicts an ABS-braking with slip controlled wheels, where the optimal desired slip $S_{X,d}$ is adapted via a frequency analysis of the wheel slip. The real slip oscillates with about 10 Hz around the critical slip $S_{X,crit}$ for maximal $\mu_X$-values. On average, the resulting deceleration is -10.2 m s$^{-2}$. A comparison with the ABS control with switching valves of Fig. 13.17 shows that the resulting maximal slip is with $S_{X,max} = 0.11$ somewhat higher than $S_{X,max} = 0.08$ for the conventional ABS-braking and the resulting shortage of the brake distance is about 2 to 3 % smaller. For more details see Semmler (2006a, b).

## 13.7 Electromechanical Brake (EMB)

### 13.7.1 Introduction

The number of registered cars being equipped with modern electronic driver-assisting brake control systems such as anti-lock braking (ABS), traction control system

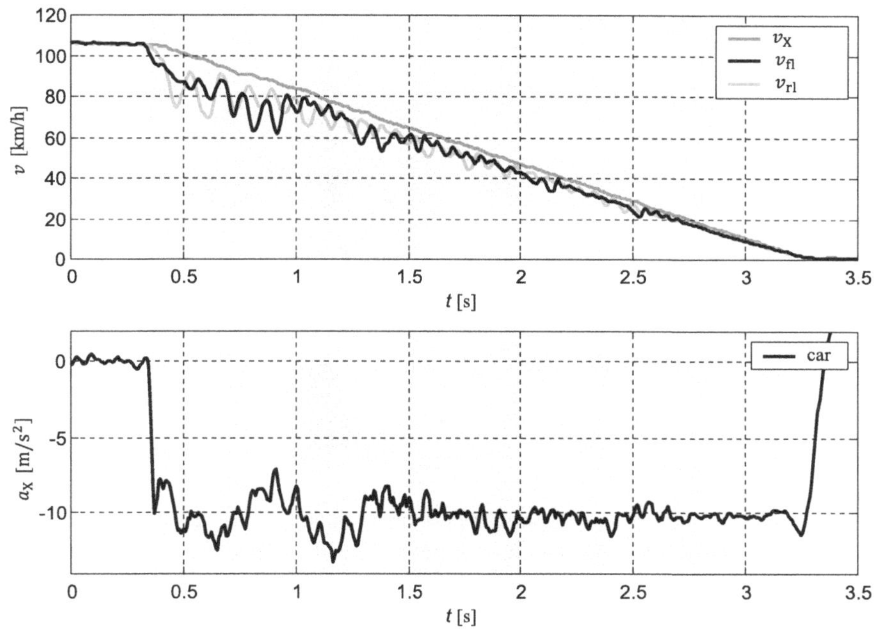


**Fig. 13.24** Slip controlled ABS-braking on dry asphalt with determination of the desired optimal slip with frequency analysis of the wheel velocity, VW Golf IV, 1.6 l, (Semmler 2006b)

(TCS), and electronic stability control (ESC) have increased steadily. However, to embed such functionality in conventional hydraulic brake systems, a large number of electro-hydraulic components are required. In recent years, the automotive industry and their suppliers have therefore started to develop *brake-by-wire systems* as typical mechatronic solutions. There are two concepts favored, the electro-hydraulic system (EHB) and the fully electromechanical system (EMB). The electro-hydraulic brake-by-wire system still uses brake fluid and conventional brake actuators but proportional valves. This system came on the market in 2001.

However, with regard to the disadvantages of the electro-hydraulic system (brake fluid, brake lines, proportional valves, *etc.*), the fully electromechanical brake system is a promising concept and is therefore also investigated. Figure 13.25 shows an electromechanical brake actuator, Rieth (1999). An electromechanically actuated brake system provides an ideal basis for converting electrical command signals into clamping forces, or rather peripheral forces, at the brakes. Standard and advanced braking functions can then be realized with uniform hardware and software. The software modules of the control unit and the sensor equipment determine the functionality of the brake-by-wire system.

The reduction of vehicle hardware and entire system weight are not the only motivational factors contributing to the development of a fully electromechanical brake-by-wire system. The system is environmentally friendly due to the lack of brake fluid and requires little maintenance (only pads and discs). Its decoupled brake pedal

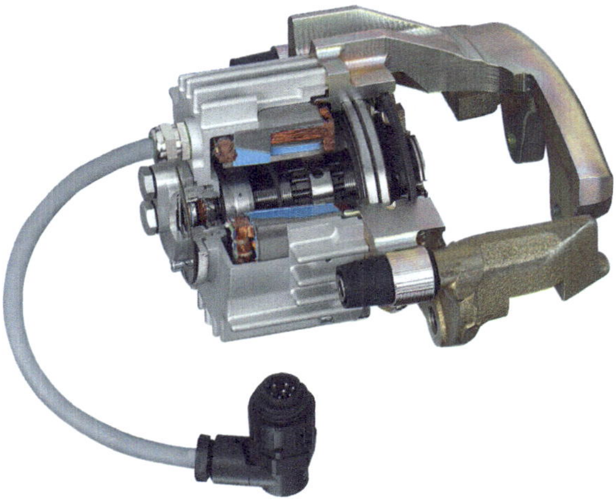

**Fig. 13.25** Electromechanical disc brake (Continental Teves), (Rieth 1999)

can be mounted in a crash-compatible and space-saving manner in the passenger compartment. There are no restraints to the design of the pedal characteristics, so ergonomic and safety aspects can be directly included. This "plug and play" concept with a minimized number of parts also reduces production and logistics costs.

However, brake-by-wire systems also entail disadvantages. One of these is the varying efficiency of the brake actuator due to changing rough environmental conditions and wear, see Fig. 13.26. The transmission characteristics of each subsystem, such as the electromechanical actuator, the gear unit, the friction brake, and the tire/road/vehicle system, is subject to large parameter variations. Therefore, the individual wheel brakes need to be operated in a closed control loop. For disc brakes, an obvious physical quantity to control is the clamping force. Yet, a force sensor is difficult to integrate in the electric brake. Therefore, it can be tried to estimate their force based on easier measurable variables. The following treatment is taken from Schwarz (1999); Isermann (2003).

## 13.7.2 Electromechanical Brake Module

A wheel module consists of an electromechanical brake with servo-amplifier and microcontroller unit, see Fig. 13.27, Schwarz (1999). With the aim of using components that are suitable for large volume production, a standard 16-bit microcontroller was chosen as the programming platform for the digital control algorithms. The microcontroller performs the clamping force and position control as well as the clearance management, the communication, and the supervision of the wheel module.

To drive the DC brushless motor of the wheel brake, a compact and powerful servo-amplifier is used. It processes the motor rotor position, which is provided by a resolver, for electronic commutation. Additionally, the servo-amplifier is equipped with two analog controllers, a motor current controller, and a rotor velocity controller.

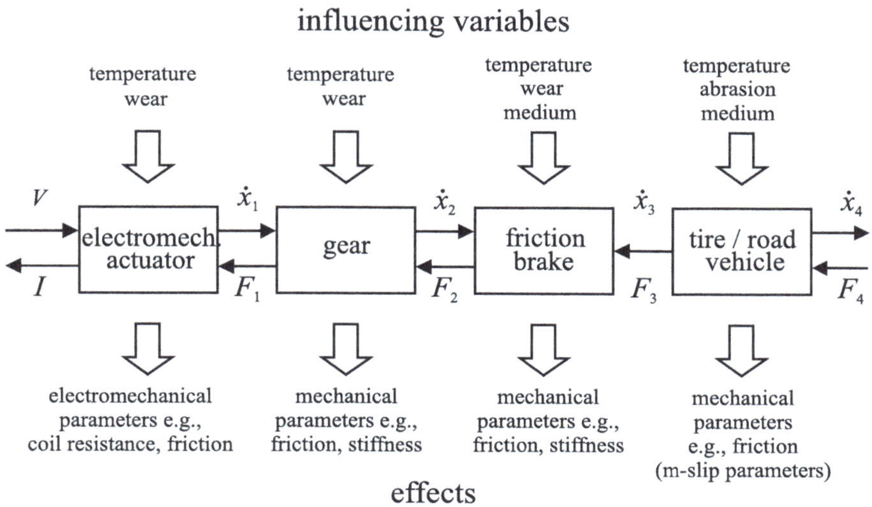


**Fig. 13.26** Two-port model representation of a brake-by-wire actuator with influencing variables and their effects

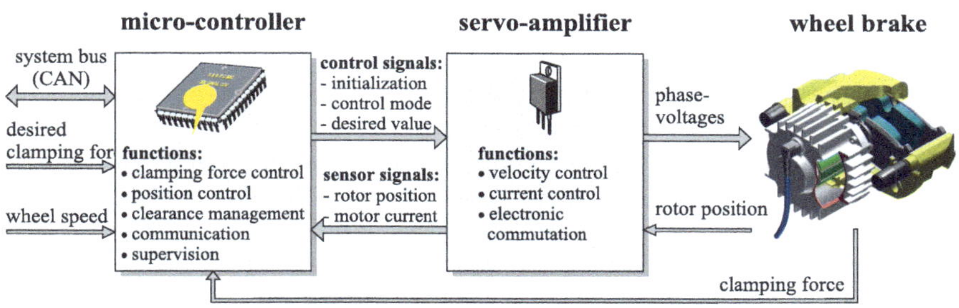


**Fig. 13.27** Block diagram of a wheel module

The design of an "intelligent disc brake" consisting of an electromechanical converter, gear, friction brake, and sensors is mainly driven by the demand for minimized space and lightweight construction. Based on a reliable and well-proven caliper, a very compact electric brake was developed. Figure 13.28 depicts the third-generation actuator with an additional planetary gear, a clamping-force sensor, and other modifications.

Instead of the piston in hydraulic brakes, the inner brake pad is actuated by a spindle. The nut of the planetary roller gear (PRG) is driven by the rotor of a brushless torque motor via a planetary gear (PG). By integrating the coil of the servo-motor directly into the brake housing and by supporting the gear unit with one central bearing only, a compact design was made possible. However, the central bearing has to absorb both radial and axial forces. A resolver measures the position of the rotor for electronic commutation.

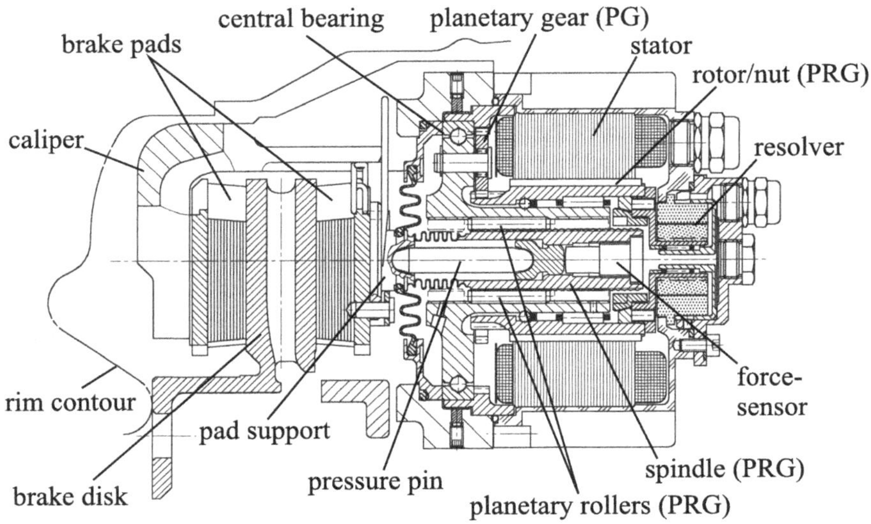


**Fig. 13.28** Sectional drawing of the EMB (Continental Teves), (Schwarz 1999)

### 13.7.3 EMB-brake Model

Figure 13.29 shows a block diagram of the EMB, (Schwarz 1999). The microcontroller block contains the force and position controller, both having proportional behavior. The analog controller comprises the velocity and the current controller of the servo-amplifier. The angular velocity controller is a PI controller and the current controller is a PI controller with a series-connected first-order lag ($PT_1$) element. They are cascaded with the velocity controller in an outer loop.

The next block in the model structure is the motor, Fig. 13.29. For simplification, the DC brushless motor was modeled as a standard DC motor with its output torque $M_m$ being proportional to the motor current $I_m$:

$$M_m = I_m \Psi. \tag{13.7.1}$$

The planetary gear (PG), Fig. 13.30, is modeled as two series-connected gear units. The first gear unit is the sun-planet wheel unit, the second is the planet-carrier wheel unit. Each of the gear units has friction torque due to bearing friction $M_{FB1}$ and $M_{FB2}$ and due to tooth friction $M_{FT}$. The stiffness and damping of one gear unit are lumped together as $c_G$ and $d_G$. For the calculation of the inertias and friction losses, special attention has to be paid to the speed ratios of the components. With the support of the component supplier, all parameters could either be evaluated or measured. Figure 13.31 depicts the two-mass model of one gear unit.

One of the most challenging tasks in the modeling process was the design of the planetary roller gear (PRG) subsystem. The PRG was also modeled as a two-mass oscillator with friction due to bearing mounting as well as for the nut-roller and roller-spindle contact. Since the thread-face contact contour of a planetary roller gear

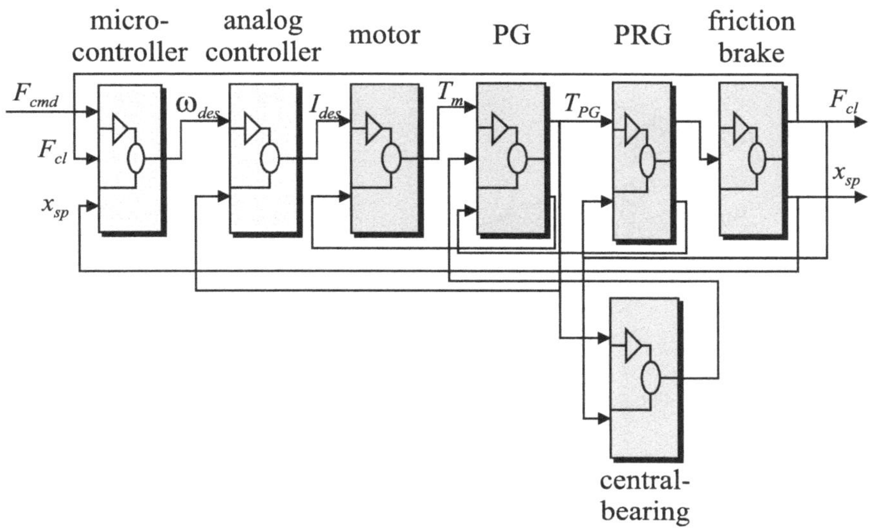


**Fig. 13.29** Matlab®/Simulink™ block diagram of EMB-brake control, (Schwarz 1999)

**Fig. 13.30** Principle and components of the planetary gear

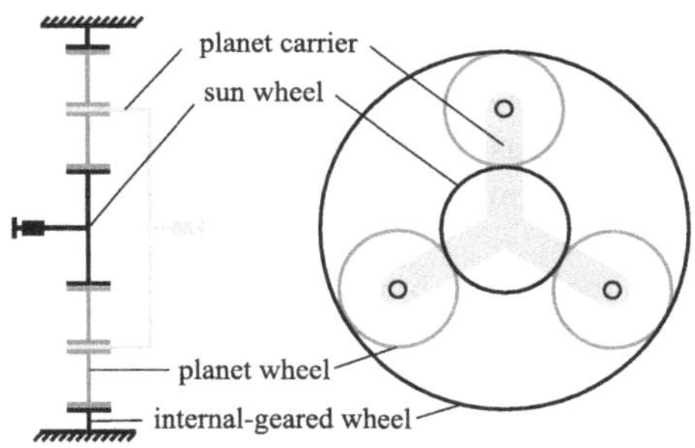


**Fig. 13.31** Two-mass model of one gear unit

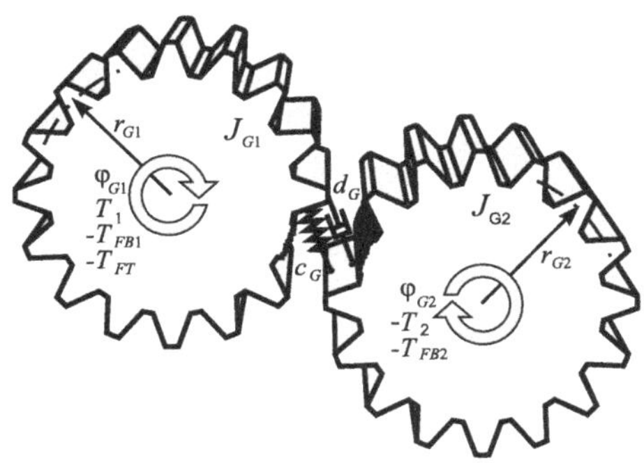

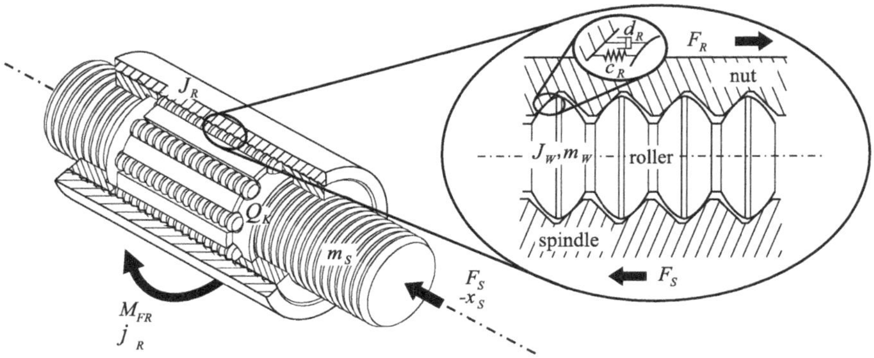


**Fig. 13.32** Planetary roller gear, (Schwarz 1999)

is identical to the contact contour of an angular ball bearing, the models could be derived from such bearing models. The final PRG subsystem includes, see Fig. 13.32,

- stiffness of the nut, spindle, and rollers (lumped together as one stiffness and one damping)
- Coulomb friction of the PRG with axial load dependency $M_{\mathrm{FR}}$
- dead zone of the PRG
- moments of inertia of the nut $J_{\mathrm{R}}$, the rollers $J_{\mathrm{W}}$, and the roller cage $J_{\mathrm{K}}$ as well as the mass of the spindle $m_{\mathrm{S}}$ and the rollers $m_{\mathrm{W}}$ (lumped together as one inertia and one mass).

In the friction brake subsystem, Fig. 13.33, the static and dynamic behavior of the caliper, the brake pads, and the brake disc are modeled. The stiffness of the caliper and the brake pads were determined with a specifically designed test unit.

Two states of the friction brake have to be distinguished:

- the brake pads do not touch the brake disc: clearance (air gap) has to be taken into account ($x_{\mathrm{AG}} > 0$)
- the brake pads touch the brake disc: no clearance ($x_{\mathrm{AG}} = 0$).

For each state, the mass $m_{\mathrm{B}}$, the translatory friction force $F_{\mathrm{FB}}$, the stiffness $c_{\mathrm{B}}$, and the damping $d_{\mathrm{B}}$ must be calculated individually. Based on the value of the clearance $x_{\mathrm{AG}}$, the model is switched between the two states.

A bearing test bench served to identify the parameters of the central bearing subsystem. The moment of friction could be measured for different axial forces and rotational velocities. By means of artificial excitation signals and measurements, the parameters could be estimated by parameter estimation methods, (Schwarz 1999).

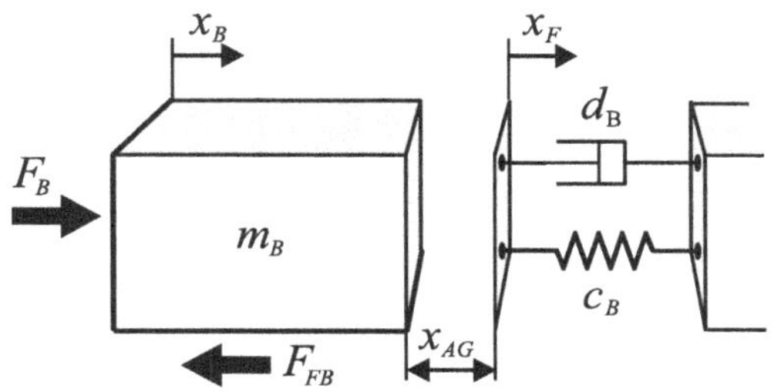


**Fig. 13.33** Model of a friction brake

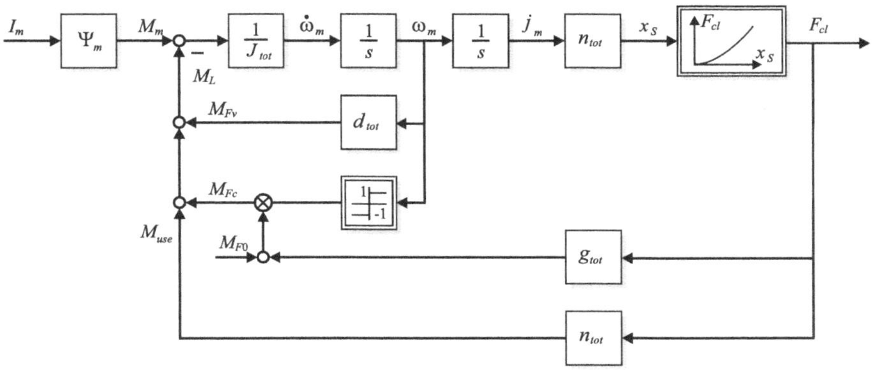


**Fig. 13.34** Block diagram of the simplified electromechanical brake model with the clamping force as output

## 13.7.4 Simplified EMB-brake Model

For the development of a reduced model, the different physical parameters were lumped together, see Fig. 13.34. $J_{tot}$ includes all inertias as well as all translatory accelerated masses. The parameter $d_{tot}$ comprises the viscous friction torques of the individual brake components. With the parameter $g_{tot}$, the load-dependent friction is considered. The gear ratio $n_{tot}$ results from the series connection of the planetary gear ratio and the planetary roller gear ratio. A torque balance for the lumped inertias results in

$$\begin{aligned} J_{tot}\dot{\omega}_m &= M_m - M_{F\nu} - M_{Fc} - M_{use} \\ &= M_m - d_{tot}\omega_m - \text{sign}(\omega_m)(M_{F0} + g_{tot}F_{cl}) - \nu_{tot}F_{cl}. \end{aligned} \quad (13.7.2)$$

The clamping force $F_{cl}$ is modeled as a characteristic curve, depending on the spindle travel $x_S$ with the nonlinear equation

$$F_{cl} = \alpha_B\sqrt{\beta_B + x_S^2} + \chi_B \quad (\beta_B, \chi_B : \text{offset parameters}). \quad (13.7.3)$$

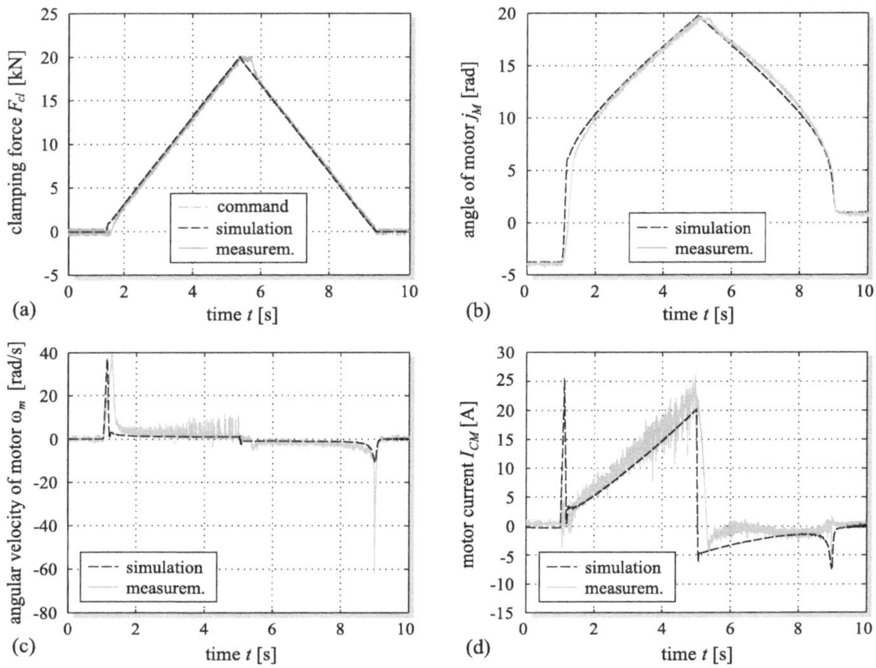


**Fig. 13.35** Simulation and measurement results: **a** clamping force; **b** angle of motor shaft; **c** rotational velocity of motor shaft; **d** motor current, (Schwarz 1999)

### 13.7.5 Simulation and Measurement

In order to validate the brake model mentioned above, the same input signal is used both for simulation and measurement. Figure 13.35 shows the results, Schwarz (1999).

The output signals of the brake model correspond with the output signals obtained from the physical brake. Minor differences in clamping force, angle of the motor shaft, angular velocity of the motor shaft, and motor current between model and brake can be traced back due to the fact that the variation of the friction parameters due to the transverse forces are neglected here.

By using these models, a complete brake control system could be developed and tested on special test rigs and vehicles.

## References

Backé W (1992) Grundlagen der Ölhydraulik. Lecture notes, 9th edn. RWTH Aachen, Aachen

Bauer M (2015) Methoden zur modellbasierten Fahrdynamikanalyse und Bewertung von Fahrdynamikregelsystemen. Dissertation Technische Universität Darmstadt. Fortschr.-Ber. VDI Reihe 12, 792. VDI Verlag, Düsseldorf

Breuer B, Bill K (eds) (2017) Bremsenhandbuch: Grundlagen, Komponenten, Systeme, Fahrdynamik, 5th edn. Springer, Berlin

Burckhardt M (1991) Fahrwerktechnik: Bremsdynamik und Pkw-Bremsanlagen. Vogel-Verlag, Würzburg

Burckhardt M (1993) Fahrwerktechnik: Radschlupf-Regelsysteme. Vogel-Verlag, Würzburg

Ersoy M, Gies S (eds) (2017) Fahrwerkhandbuch: Grundlagen, Fahrdynamik, Komponenten, Elektronische Systeme, Fahrerassistenz, Autonomes Fahren, Perspektiven, 5th edn. ATZ/MTZ Fachbuch, Springer Vieweg, Berlin

Isermann R (2003) Mechatronic systems—fundamentals. Springer, London

Isermann R (ed) (2006) Fahrdynamik-Regelung. Vieweg-Verlag, Wiesbaden

Isermann R, Münchhof M (2011) Identification of dynamic systems. Springer, Berlin, Heidelberg

Jonner W, Winner H, Dreilich L, Schunck E (1996) Electrohydraulic brake system—the first approach. In: SAE Technical paper Series, Warrendale, PA, 960991

Kunz A, Kunz M, Vollert H, Förster M (2018) Elektromechanischer Bremskraftverstärker für alle Antriebskonzepte und automatisiertes Fahren. Automobiltechnische Zeitschrift 4:60–65

Merrit H (1967) Hydraulic control systems. Wiley, New York

Mitschke M, Wallentowitz H (2004) Dynamik der Kraftfahrzeuge, 4th edn. Springer, Berlin

Münchhof M, Straky H, Isermann R (2003) Model-based supervision of a vacuum brake booster. In: 5th IFAC international symposium on fault detection, supervision and safety of technical processes. SAFEPROCESS 2003, Washington DC, USA, pp 429–434

Remfrey J, Gruber S, Ocvirk N (2016) Hydraulic brake systems for passenger vehicles. In: Winner H, Hakuli S, Lotz F, Singer C (eds) Handbook of driver assistance systems. Springer International Publishing AG, Cham, Switzerland, pp 1–23

Rieth P (1999) Brake-by-wire. Bremstechnologie im Wandel. In: Fahrzeugtechnisches Seminar, TU Darmstadt, Germany

Robert Bosch GmbH (2004) Sicherheits- und Komfortsysteme. Vieweg-Verlag, Wiesbaden

GmbH Robert Bosch (ed) (2011) Automotive handbook, 8th edn. Bentley publishers, Cambridge

GmbH Robert Bosch (ed) (2018) Automotive Handbook, 10th edn. Wiley, Chichester, England

Savaresi SM, Tanelli M (2010) Active braking control systems design for vehicles. Springer, London

Schwarz R (1999) Rekonstruktion der Bremskraft bei Fahrzeugen mit elektromechanisch betätigten Radbremsen. Dissertation Technische Universität Darmstadt. Fortschr.-Ber. VDI Reihe 12, 393. VDI Verlag, Düsseldorf

Semmler S (2006a) Bremsregelungen für mechatronische Bremsen. In: Isermann R (ed) Fahrdynamik-Regelung. Vieweg, Wiesbaden, pp 137–168

Semmler S (2006b) Regelung der Fahrzeugbremsdynamik mit kontinuierlich einstellbaren Radbremsen. Diss. Universität Darmstadt, Fortschr.-Ber. VDI Reihe 12, 632. VDI Verlag, Düsseldorf

Straky H (2003) Modellgestützter Funktionsentwurf für Kfz-Stellglieder; Regelung der elektromechanischen Ventiltriebaktorik und Fehlerdiagnose der Bremssystemhydraulik. Dissertation Technische Universität Darmstadt. Fortschr.-Ber. VDI Reihe 12, 546. VDI Verlag, Düsseldorf

Ulsoy A, Peng H, Çakmakci M (2012) Automotive control systems. Cambridge University Press, New York

Winner H, Hakuli S, Lotz F, Singer C (eds) (2016) Handbook of driver assistance systems. Springer International Publishing AG, Cham, Switzerland

van Zanten A, Kost F (2016) Brake-based asstistance functions. In: Winner H, Hakuli S, Lotz F, Singer C (eds) Chapter 39 inHandbook of driver assistance systems, Springer International Publishing AG, Cham, Switzerland, pp 919–967

# 14 Steering Control Systems

The steering system translates the rotation of the steering wheel through the steering shaft, a steering gear, tie rods steering arms, and the wheel carrier to a turn of the steered wheels; see Fig. 14.1. The required steering forces are generated by muscular forces of the driver and are transferred to the wheels without power assistance for very small cars. However, for larger cars a power assistance is required which acts with hydraulic or electrical auxiliary energy in addition to the drivers muscular forces. Many requirements have to be taken into account in designing the steering system; see, e.g. Heissing and Ersoy (2011b), Pfeffer and Harrer (2013), Ersoy and Gies (2017), Bosch GmbH (2018). On the one side, the steering system has to move light with direct orientation to the wheel steering angles, without any play. On the other side, it has to give haptic feedback to the driver about the wheel/road condition, but has to damp irregularities of the road. In the following, after considering pure mechanical steering, hydraulic and electrical power steering systems are considered and models for their behavior are derived, as required for lateral vehicle control.

## 14.1 Mechanical Steering Systems

### 14.1.1 Types of Steering Systems

For passenger cars, two main steering systems have proven to fulfill the general requirements. *Rack and pinion steering* consists of a pinion connected to steering shaft, a rack connected via a tie rod, and a steering arm to the turnable front wheels; see Fig. 14.1. This type is usually used for small cars. Its steering ratio is given by the ratio of steering wheel (or pinion) revolutions to the rack travel way. It may be constant or may be varied by appropriate design of the tooth distances, e.g. from



R. Isermann, *Automotive Control*, ATZ/MTZ-Fachbuch,
https://doi.org/10.1007/978-3-642-39440-9_14

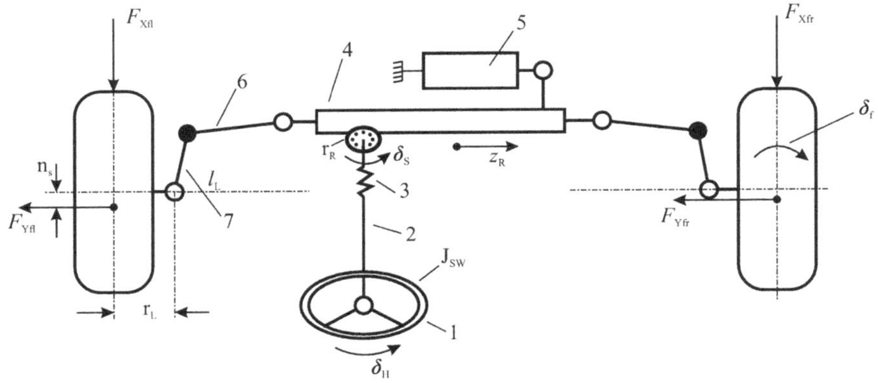


**Fig. 14.1** Power-assisted rack and pinion steering 1. steering wheel 2. steering column 3. column stiffness $c_{col}$ 4. Rack and pinion steering gear 5. power steering actuator 6. tie rod 7. steering lever. $n_S$ caster offset, $F_{YF}$ lateral tire forces, $r_R$ pinion radius, $\delta_H$ steering wheel angle, $\delta_S$ steering shaft angle, $l_L$ steering lever length, $\delta_f$ wheel steering angle, $z_R$ travel way of the rack, $r_L$ radius of turning wheel (lateral)

indirect ratio in the center (for straight driving) and more direct ratio for medium and large steering angles (e.g. for parking).

*Recirculating-ball steerings* have a tooth segment at the steering shaft and transfer the shaft torque via a nut and recirculating balls to a steering worm connected to the tie rods. Because of the low friction this type is a comfortable design and is or was used for larger passenger cars and especially for commercial, heavy-duty vehicles.

Rack and pinion steering systems dominate now for passenger cars. Different versions can be distinguished. The steering rack can be located for or after the front axle center line, and the pinion may be built above or below the rack.

Due to increasing weight of passenger cars and comfort requirements, *power-assisted steering systems* are now standard down to compact cars. The manual steering force by the driver is then amplified by hydraulic or electric servo systems.

### 14.1.2 Stationary and Dynamic Behavior of Mechanical Steering Systems

In order to model various steering systems, first a pure mechanical system is considered, as now only used in very light weight vehicles. The schematic representation in Fig. 14.2 illustrates that this mechanical steering in the case of a pinion-rack type consists of eight main components, the steering wheel and steering shaft (column), the pinion-rack gear, the steering rack, the tie rod and steering lever, the wheel carrier with steering axis (kingpin), and the wheels with suspensions. The mechanical steering system then represents a multi-mass-spring-damper system. Different models have been published with, e.g. 2 to 5 inertias and masses, with regard to various applications; see, e.g. Pfeffer and Harrer (2013), Braess and Seiffert (2007), Fankem et al. (2014). However, in view of using the steering system as an actuation device for

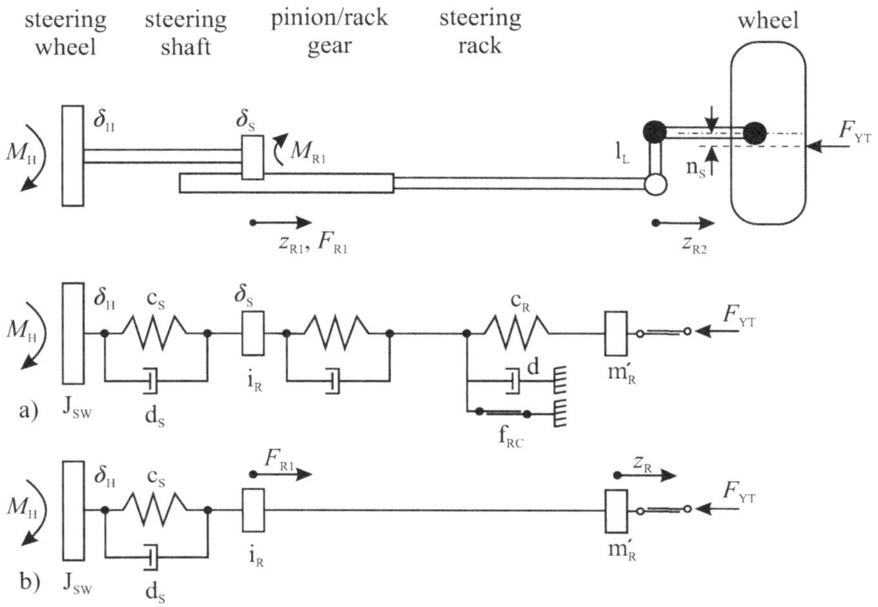


**Fig. 14.2** Schematic for modeling of a pinion-rack steering system without power assistance: **a** with 3 spring-damper elements; **b** with 1 spring-damper element

lateral vehicle dynamics, some simplifications can be made. As shown in Fig. 14.2a the steering wheel and steering shaft are considered as a torsional spring-damper system, the pinion-rack gear as a transformer with elasticity and damping, and the steering rack and linkage to the wheels as a further spring-damper system with a representative mass. The lateral force of the tire generates during cornering an aligning torque which acts on the rack and comprises a dry friction. Hence, the whole steering consists basically of three coupled spring-damper systems with two masses and with dry friction. Models of general multi-mass torsional oscillating system without and with gear are treated in Isermann (2005). The resulting models are used for the development of models for steering systems.

Figure 14.2b depicts a simplified schematic with two masses and one representative spring-damper system, which will be used later.

The mechanical steering system can also be represented by a serial connection of *two-port systems* with *energy flows* at its terminals; see Fig. 14.3. According to, e.g. Isermann (2005) p. 52, one can then define potential differences $e(t)$, which are forces $F(t)$ or torques $M(t)$ and flows $f(t)$, which are transitional velocities or angular velocities. Their product is then a power

$$P(t) = f(t)e(t). \tag{14.1.1}$$

The two-port systems then consist of a power source, which is the driver and transformers like the steering wheel and shaft, the pinion-rack gear, the rack, the tie rod,

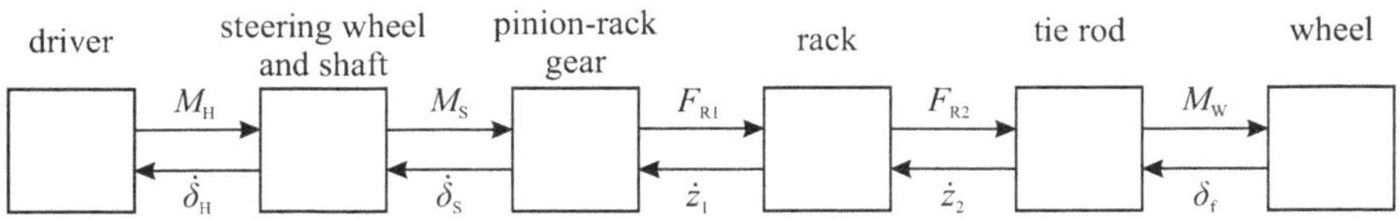


**Fig. 14.3** Two-port representation of a mechanical steering system with power variables at the terminals

and a sink, the turned wheels. The causality follows by the manual torque $M_H$ of the driver, which is an applied variable as input to the steering system. However, also the steering angle can be used as applied variable.

**(a) Kinematic Relations for Mechanical Steering Systems**

The kinematic relation from the steering wheel to the steering angle of the front wheels without considering elasticity is defined as the *steering ratio*, ISO 8855,

$$i_S = \frac{\Delta\delta_H}{\Delta\delta_f} = \frac{\text{change of steering wheel angle}}{\text{change of mean wheel steering angle}} \tag{14.1.2}$$

where $\Delta\delta_f$ is an average of the left and right wheel steering angle in the absence of tire forces and torques, but including static tire vertical load,

$$\Delta\delta_f = \frac{1}{2}(\Delta\delta_{fl} + \Delta\delta_{fr}). \tag{14.1.3}$$

This overall steering ratio is not constant, but changes in general with the steering wheel angle.

For a pinion-rack steering, the ratio between the steering shaft angle and the rack position with $r_R$ the pinion radius is the gear ratio

$$i_G = \frac{\Delta\delta_S}{\Delta z_R} = \frac{1}{r_R} \tag{14.1.4}$$

and the ratio between the rack position and the steering angle with $l_L$ the steering lever is the *steering linkage ratio*

$$i_T = \frac{\Delta z_R}{\Delta\delta_f} = l_L(\delta_f). \tag{14.1.5}$$

Hence it follows for the *overall steering ratio* with $\Delta\delta_S = \Delta\delta_H$

$$i_S = i_G i_T = \frac{\Delta\delta_H}{\Delta\delta_f} = \frac{l_L}{r_R} \tag{14.1.6}$$

for small changes around the center point.

The gear ratio $i_G(\delta_S)$ changes usually with $\delta_H$ in order to obtain a more direct steering behavior for larger steering wheel angles and the linkage gear ratio changes because of the nonlinear kinematics $\delta_f(z_R)$ of the linkages.

The effective steering ratio has to include elasticities of the steering system, which arise especially in the steering shaft and its different joints, the tie rod and its joints and all elastic bearings, if tire forces act on the steering. The so-called *dynamic overall steering ratio* is than

$$i_{\mathrm{S,dyn}} = \frac{\Delta\delta_{\mathrm{H}} + \Delta\delta_{\mathrm{H,elast}}}{\Delta\delta_{\mathrm{f}}}. \tag{14.1.7}$$

The additional steering wheel angle $\Delta\delta_{\mathrm{H,elast}}$ depends on the elasticity or compliance with respect to a tire force or torque, ISO 8855. It can be measured by fixing the front wheels and applying a torque at the steering wheel, see, e.g. Reimpell et al. (2001) and Pfeffer and Harrer (2013) and results in a hysteresis characteristic $\delta_{\mathrm{H}}(M_{\mathrm{H}})$.

The *overall stiffness* of a mechanical steering system is defined as

$$c_{\mathrm{S}} = \frac{\Delta M_{\mathrm{H}}}{\Delta\delta_{\mathrm{H}}} \tag{14.1.8}$$

and depends on the stiffness of the steering shaft $c_{\mathrm{shaft}}$, of the gear $c_{\mathrm{gear}}$ and of the linkage $c_{\mathrm{link}}$ between the rack and the steering axis (kingpin axis: axis about which the wheel and hub assembly rotates). Because of the series connection it holds

$$\frac{1}{c_{\mathrm{S}}} = \frac{1}{c_{\mathrm{shaft}}} + \frac{1}{c_{\mathrm{gear}}} i_{\mathrm{G}}^2 + \frac{1}{c_{\mathrm{link}}} i_{\mathrm{G}}^2 i_{\mathrm{T}}^2. \tag{14.1.9}$$

Some published values in Pfeffer and Harrer (2013) are $c_{\mathrm{S}} = 14000\ldots22000$ Nm/rad and in Mitschke and Wallentowitz (2004) $c_{\mathrm{S}} = 5000\ldots20000$ Nm/rad.

### (b) Steering Wheel Torque As Applied Input Variable

For modeling the dynamic behavior of a mechanical pinion-rack steering system, the simplified schematic in Fig. 14.2b is considered. The steering system is thus represented by an elastic steering shaft with linear damping, the steering wheel with inertia and it is assumed that the pinion-rack gear, the rack and the linkage to the front wheels are rigid but have viscous and dry friction. This means that the elasticity respective stiffness is representatively concentrated in the steering shaft in order to avoid too complicated equation systems, but to model the main effects.

The dynamic behavior of the steering wheel and shaft follows after applying the principle of angular momentum or a torque balance, assuming linear behavior to

$$J_{\mathrm{SW}}\ddot{\delta}_{\mathrm{H}}(t) = M_{\mathrm{H}}(t) - c_{\mathrm{S}}(\delta_{\mathrm{H}}(t) - \delta_{\mathrm{S}}(t)) - d_{\mathrm{S}}(\dot{\delta}_{\mathrm{H}}(t) - \dot{\delta}_{\mathrm{S}}(t)) \tag{14.1.10}$$

where $J_{\mathrm{SW}}$ is the ratio of inertia of the steering wheel, $M_{\mathrm{H}}$ the manual torque by the driver, $\delta_{\mathrm{H}}$ the steering wheel angle, $\delta_{\mathrm{S}}$ the shaft angle at the input to the pinion-rack gear, $c_{\mathrm{S}}$ and $d_{\mathrm{S}}$ the stiffness and viscous damping coefficient of the shaft.

Following torque components can be distinguished

$$J_{\mathrm{SW}}\ddot{\delta}_{\mathrm{H}} = M_{\mathrm{H}} - M_{\mathrm{S1}} + M_{\mathrm{R1}} \tag{14.1.11}$$

with the torque from the shaft to the steering wheel

$$M_{\mathrm{S1}} = c_{\mathrm{S}}\delta_{\mathrm{H}} + d_{\mathrm{S}}\dot{\delta}_{\mathrm{H}} \tag{14.1.12}$$

and the torque from the pinion to the shaft

$$M_{\mathrm{R1}} = c_{\mathrm{S}}d_{\mathrm{S}} + d_{\mathrm{S}}\dot{\delta}_{\mathrm{S}}. \tag{14.1.13}$$

The force on the rack becomes

$$\begin{aligned} F_{\mathrm{R1}} &= \frac{1}{r_{\mathrm{R}}}(M_{\mathrm{S1}} - M_{\mathrm{R1}}) \\ &= \frac{1}{r_{\mathrm{R}}}(c_{\mathrm{S}}(\delta_{\mathrm{H}} - \delta_{\mathrm{S}}) + d_{\mathrm{S}}(\dot{\delta}_{\mathrm{H}} - \dot{\delta}_{\mathrm{S}})). \end{aligned} \tag{14.1.14}$$

The rack is now considered as a rigid component without elasticity but with friction, thus that $z_{\mathrm{R1}} = z_{\mathrm{R2}} = z_{\mathrm{R}}$ can be assumed. Its mass $m_{\mathrm{R}}$ is enlarged by the masses of the components of the linkages and the turned wheels, such that for a representative mass holds

$$m'_{\mathrm{R}} = m_{\mathrm{R}} + 2m_{\mathrm{TR}} + \frac{2}{l_{\mathrm{L}}^2}J_{\mathrm{WZ}} \tag{14.1.15}$$

where $m_{\mathrm{TR}}$ is the mass of the tie rods and $J_{\mathrm{WZ}}$ the ratio of inertia of the front wheels turning around its vertical steering or kingpin axis.

A force balance of the rack and the coupled masses then becomes

$$m'_{\mathrm{R}}(t)\ddot{z}_{\mathrm{R}}(t) = F_{\mathrm{R}}(t) - d'_{\mathrm{R}}\dot{z}_{\mathrm{R}}(t) - f'_{\mathrm{RC}}\,\mathrm{sign}\,\dot{z}_{\mathrm{R}}(t) \tag{14.1.16}$$

$$F_{\mathrm{R}}(t) = F_{\mathrm{R1}}(t) - F_{\mathrm{y}}(t) \tag{14.1.16a}$$

with the rack force $F_{\mathrm{R}}$ and the force from the pinion due to (14.1.14).

$F_{\mathrm{Y}}$ is the sum of the tire forces on the rack, and $d'_{\mathrm{R}}$ and $f'_{\mathrm{RC}}$ are viscous and dry friction coefficients. For the pinion-rack gear holds $z_{\mathrm{R}} = r_{\mathrm{R}}\delta_{\mathrm{S}}$; see (14.1.4).

– *Effects of tire forces*

The *lateral tire forces* $F_{\mathrm{YT,i}}$ generate a torque at each wheel $i$, compare Sect. 5.2,

$$M_{\mathrm{ZTYi}} = F_{\mathrm{YTfi}}\ n_{\mathrm{S}} \tag{14.1.17}$$

with

$$n_{\mathrm{S}} = n_{\mathrm{T}} + n_{\varnothing} \tag{14.1.18}$$

where $n_{\mathrm{T}}$ is the caster offset of the tire which depends on the vertical force and velocity and $n_{\varnothing}$ a constructive offset (spindle trial), ($\approx 0.02$m) due to the inclined vertical steering axis, Pfeffer and Harrer (2013) and ISO 8855, which acts as a *return torque* (aligning torque). The resulting torque is then for 2 front wheels

$$M_{\mathrm{ZTY}\Sigma} = (F_{\mathrm{YTfl}} + F_{\mathrm{YTfr}})n_{\mathrm{S}} = F_{\mathrm{YTf}\Sigma}n_{\mathrm{S}}. \tag{14.1.19}$$

For equal lateral forces of both wheels it holds

$$M_{\mathrm{ZTY}\Sigma} = 2F_{\mathrm{YTf}}n_{\mathrm{S}}. \tag{14.1.20}$$

The aligning torque of a tire during cornering with small side slip angles $\alpha_{\mathrm{f}} = 0°...20°$ is due to (5.2.10).

$$M_{\mathrm{ZTYi}}(F_{\mathrm{Z}}) = c_{\alpha}(F_{\mathrm{Zi}})n_{\mathrm{S}}(F_{\mathrm{Zi}})\alpha_{\mathrm{i}}. \tag{14.1.21}$$

As well the side slip coefficient $c_{\alpha}$ as the caster offset $n_{\mathrm{T}}$ depend strongly on the vehicle wheel forces $F_{\mathrm{Z}}$ and velocities $v_{\mathrm{x}}$; see Fig. 5.6. However, as during cornering the vehicle force on one side increases and on the other side decreases it is assumed that both changes compensate each other and $F_{\mathrm{Z}} = const$ is taken. The lateral tire force during cornering becomes

$$F_{\mathrm{YTfi}} = \frac{M_{\mathrm{ZTYi}}}{n_{\mathrm{S}}} = c_{\alpha}(F_{\mathrm{Zi}})\alpha_{\mathrm{fi}}. \tag{14.1.22}$$

For small side slip angles it holds, see (7.2.25)

$$\alpha_{\mathrm{f}} = \delta_{\mathrm{f}} - \beta - \frac{l_{\mathrm{f}}}{v_{\mathrm{x}}}\dot{\psi} \tag{14.1.23}$$

and using a one-track model the yaw rate becomes for stationary cornering, see (7.2.57)

$$\dot{\psi} = \delta_{\mathrm{f}}\frac{1}{l}\frac{v_{\mathrm{x}}}{1 + v_{\mathrm{x}}^2/v_{\mathrm{ch}}^2}. \tag{14.1.24}$$

Therefore one obtains for constant velocity

$$\begin{aligned} \dot{\psi} &= c_{\dot{\psi}}\delta_{\mathrm{f}} \\ c_{\dot{\psi}} &= \frac{1}{l}\frac{v_{\mathrm{x}}}{1 + v_{\mathrm{x}}^2/v_{\mathrm{ch}}^2} \end{aligned} \tag{14.1.25}$$

As the slip angle $\beta$ for stationary cornering with not very large lateral acceleration is relatively small, (14.1.23) yields simplified

$$\alpha_{\mathrm{f}} \approx (1 - \frac{l_{\mathrm{f}}}{v_{\mathrm{x}}}c_{\dot{\psi}})\delta_{\mathrm{f}} = c_{\delta}(v_{\mathrm{x}})\delta_{\mathrm{f}} \tag{14.1.26}$$

and the side slip angle can be assumed as being proportional to the steering angle, and with (14.1.22) follows

$$F_{\mathrm{YTfi}} = c_\alpha(F_{\mathrm{Zi}})c_\delta\delta_{\mathrm{f}}. \tag{14.1.27}$$

An alternative to determine the lateral tire force during cornering with a radius $R_{\mathrm{P}}$ is to use the lateral acceleration $a_{\mathrm{Y}}$

$$F_{\mathrm{YTf}} = m_{\mathrm{f}}a_{\mathrm{Y}} = m\frac{l_{\mathrm{r}}}{l}a_{\mathrm{Y}} = m\frac{l_{\mathrm{r}}}{l}\frac{v_{\mathrm{X}}^2}{R_{\mathrm{P}}} \tag{14.1.28}$$

where $m_{\mathrm{f}}$ is the partial front mass and $a_{\mathrm{Y}}$ is the lateral acceleration; see section 14.2.1.

The use of (14.1.27) requires that $c_\alpha$ , $v_{\mathrm{ch}}^2$ are known. For (14.1.28) $m_{\mathrm{f}}$ has to be known and $a_{\mathrm{Y}}$ is usually measured.

The *longitudinal tire forces* $F_{\mathrm{Xfi}}$ generate for each front wheel a torque

$$M_{\mathrm{ZTXi}} = F_{\mathrm{XTfi}}\ r_{\mathrm{L}} \tag{14.1.29}$$

where $r_{\mathrm{L}}$ is the turning radius of the wheel around its vertical axis.

The resulting torque for 2 front wheels is then

$$M_{\mathrm{ZTX\Sigma}} = (F_{\mathrm{XTfl}} - F_{\mathrm{XTfr}})r_{\mathrm{L}} = \Delta F_{\mathrm{XTf}}\ r_{\mathrm{L}}. \tag{14.1.30}$$

For symmetric data of the wheel suspensions and on even road without left and right road disturbances this torque is $M_{\mathrm{TX\Sigma}} = 0$. However, for uneven roads, road disturbances and different vertical tire forces, e.g. through cornering the torque is $M_{\mathrm{ZTX\Sigma}} \neq 0$.

The resulting torque from the lateral and longitudinal tire forces is then for the 2 front wheels

$$\begin{aligned} M_{\mathrm{ZTf}} &= M_{\mathrm{ZTY\Sigma}} + M_{\mathrm{ZTX\Sigma}} \\ &= (F_{\mathrm{YTfl}} + F_{\mathrm{YTfr}})n_{\mathrm{S}} + (F_{\mathrm{XTfl}} - F_{\mathrm{XTfr}})r_{\mathrm{L}}. \end{aligned} \tag{14.1.31}$$

If $\Delta F_{\mathrm{XTf}}$ can be neglected (14.1.19) gives for two front wheels

$$M_{\mathrm{ZTf}} = 2F_{\mathrm{YTf\Sigma}}n_{\mathrm{S}}. \tag{14.1.32}$$

The torque of the steering wheel because of these tire forces follows to $M_{\mathrm{H}} = \frac{1}{i_{\mathrm{S}}}M_{\mathrm{ZTf}}$

For *small vehicle speed* and for *parking* where the lateral forces vanish a tire turning friction torque $M_{\mathrm{ZT,c}}$ has to be taken into account, also called as boring torque. For modeling it is assumed that this torque consists of a dry friction and a viscous friction part

$$M_{\mathrm{ZTf}} = f_{\mathrm{ZT,c}}^{M}\,\mathrm{sign}\,\dot{\delta}_{\mathrm{f}} + f_{\mathrm{ZT,v}}^{M}\dot{\delta}_{\mathrm{f}}. \tag{14.1.33}$$

A model to describe the steering torque to turn a non-rolling tire is given in Mitschke and Wallentowitz (2004) and Bastow et al. (2004), resulting in a friction parameter

$$f^M_{ZT,c} = \mu_f \frac{F_Z^{\frac{3}{2}}}{\sqrt{p_T}} \tag{14.1.34}$$

with $\mu_f$ a friction coefficient, $F_Z$ the vertical load and $p_T$ the tire pressure in [N/m$^2$]. As friction coefficient on concrete or tarmac $\mu_f \approx 1$ can be used. The friction torque leads to a force on the rack

$$F_{YRF} = \frac{1}{l_L} M_{ZTf} = \frac{1}{l_L}(f^M_{ZT,c} \operatorname{sign} \dot{\delta}_f + f^M_{ZT,v}\dot{\delta}_f). \tag{14.1.35}$$

Hence, the tire forces on the rack through the aligning torque and through the 2 tires turning friction torque becomes

$$F_{YR} = \frac{1}{l_L}(2c_\alpha c_\delta n_S \delta_f + 2 f^M_{ZT,v}\dot{\delta}_f + 2 f^M_{ZT,c} \ \operatorname{sign} \dot{\delta}_f). \tag{14.1.36}$$

Because of the inclined steering axis (kingpin axis) of the front wheels arises a *static returning torque*, which depends on the steering angle $\delta_f$, the kingpin inclination angle $\sigma$, the castor angle $\tau$, the camber angle $\epsilon_v$, the scrub radius $r_k$ (see ISO 8855) and Mitschke (1972), Pfeffer and Harrer (2013)

$$M_{Z,st} = m_f g k_{Mst}(\sigma, \tau, \epsilon_v, r_k)\delta_f = f^M_{ZT,st}\delta_f. \tag{14.1.37}$$

Introduction in (14.1.16) yields

$$\begin{aligned} m'_R \ddot{z}_R =& F_{R1} - d'_R \dot{z}_R - f'_{RC} \operatorname{sign} \dot{z}_R - \frac{2}{l_L} c_\alpha c_\delta n_S \delta_f - \frac{2}{l_L} f^M_{ZT,v}\dot{\delta}_f \\ & - \frac{2}{l_L} f^M_{ZT,c} \operatorname{sign} \dot{\delta}_f - \frac{2}{l_L} f^M_{ZT,st}\delta_f. \end{aligned} \tag{14.1.38}$$

The dry friction terms can be joined together

$$f''_{RC} = f'_{RC} + \frac{2}{l_L} f^M_{ZT,c}. \tag{14.1.39}$$

Figure 14.4 represents a signal flow scheme based on (14.1.10), (14.1.16), and (14.1.39). It shows a linear torsional inertia spring-damper system of the steering wheel and shaft, the pinion-rack gear as a static transformer and a nonlinear translational mass-damper system of the rack with coupled linkages and wheels.

For higher velocities it can be assumed that $f^M_{ZT,c} \approx 0$, $f^M_{ZT,v} \approx 0$, and $f^M_{ZT,st} \approx 0$. Then (14.1.38) with (14.1.14) leads to

$$\begin{aligned} m'_R \ddot{z}_R = \frac{c_S}{r_R}(\delta_H - \frac{z_R}{r_R}) + \frac{d_S}{r_R}(\dot{\delta}_H - \frac{\dot{z}_R}{r_R}) \\ -d'_R \dot{z}_R - f'_{RC} \operatorname{sign} \dot{z}_R - \frac{2}{l_L} c_\alpha c_\delta n_S \delta_f \end{aligned} \quad (14.1.40)$$

or with (14.1.28)

$$\begin{aligned} m'_R \ddot{z}_R = \frac{c_S}{r_R}(\delta_H - \frac{z_R}{r_R}) + \frac{d_S}{r_R}(\dot{\delta}_H - \frac{\dot{z}_R}{r_R}) \\ -d'_R \dot{z}_R - f'_{RC} \operatorname{sign} \dot{z}_R - m_f \frac{n_S}{l_L} a_Y. \end{aligned} \quad (14.1.41)$$

As Fig. 14.4 illustrates the overall model represents a series connection of two second-order systems with several feedbacks. The signal flow scheme allows to analyze the main effects of various design parameters and to simulate the stationary and dynamic behavior provided the parameters are known, by taking into account several simplifying assumptions. Table 14.2 gives ranges of some steering parameters.

The signal flow scheme shows the superposition of the different forces on the rack. Thus, the rack force $F_R$ is according to (14.1.14) determined by the steering shaft-pinion output torque $M_{S1}$, the torque $M_{R1}$from the rack with $\delta_S = z_R/r_R$ and the transformed lateral tire forces $F_Y$ from the wheels. In the case of power steering, the power assist force $F_{PA}$ has to be added. The resulting torque $M_{R1}$ as well as $M_{S1}$ are fed back to the steering shaft and act against the manual torque $M_H$ and thus influence significantly the so-called "steering feel".

Using the equations (14.1.10) and (14.1.40) with the kinematic relation

$$\delta_S = z_R/r_R \quad (14.1.42)$$

leads to the equation system

$$\ddot{\delta}_H = \frac{1}{J_{SW}}[M_H - c_S\delta_H - d_S\dot{\delta}_H + \frac{c_S}{r_R} z_R + \frac{d_S}{r_R} \dot{z}_R] \quad (14.1.43)$$

$$\begin{aligned} \ddot{z}_R = \frac{1}{m'_R}[\frac{c_S}{r_R}\delta_H + \frac{d_S}{r_R}\dot{\delta}_H - \frac{c_S}{r_R^2} z_R + \frac{d_S}{r_R^2}\dot{z}_R - d'_R \dot{z}_R \\ - f'_{RC} \operatorname{sign} \dot{z}_R - \frac{2}{l_L} c_\alpha c_\delta n_S \delta_f] \end{aligned} \quad (14.1.44)$$

which can be brought into a linear state-space representation if $f'_{RC} = 0$ is set

$$\dot{\mathbf{x}}(t) = \mathbf{A}\mathbf{x}(t) + \mathbf{b}u(t)$$

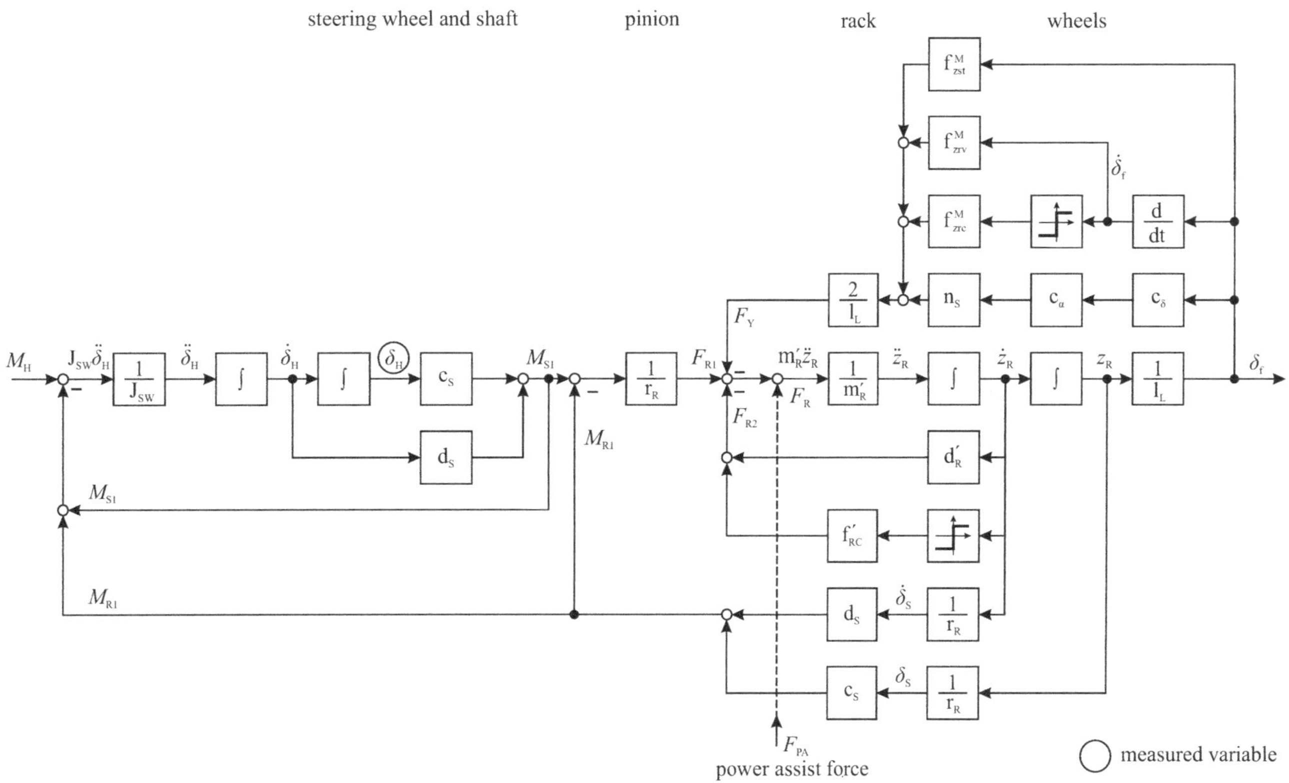


**Fig. 14.4** Signal flow scheme of a rack and pinion mechanical steering system for an applied steering wheel torque during cornering with steering angle $\delta_f$

$$\begin{bmatrix} \dot{\delta}_H \\ \ddot{\delta}_H \\ \dot{z}_R \\ \ddot{z}_R \\ \dot{\delta}_f \end{bmatrix} = \begin{bmatrix} 0 & 1 & 0 & 0 & 0 \\ -\frac{c_S}{J_{SW}} & -\frac{\delta_S}{J_{SW}} & \frac{c_S}{J_{SW} r_R} & \frac{d_S}{J_{SW} r_R} & 0 \\ 0 & 0 & 0 & 1 & 0 \\ \frac{c_S}{m_R^{'} r_R} & \frac{d_S}{m_R^{'} r_R} & -\frac{c_S}{m_R^{'} r_R^2} & (-\frac{d_S}{m_R^{'} r_R^2} + \frac{d_R^{'}}{m_R^{'}}) & -\frac{2c_\alpha c_\delta n_S}{m_R^{'} l_L} \\ 0 & 0 & 0 & \frac{1}{l_L} & 0 \end{bmatrix} \begin{bmatrix} \delta_H \\ \dot{\delta}_H \\ z_R \\ \dot{z}_R \\ \delta_f \end{bmatrix} +$$

$$\begin{bmatrix} 0 \\ \frac{1}{J_{SW}} \\ 0 \\ 0 \\ 0 \end{bmatrix} M_H$$

$$y(t) = \mathbf{c}^T \mathbf{x}(t)$$

$$\delta_f = [0\ 0\ 0\ 0\ 1] \begin{bmatrix} \delta_H \\ \dot{\delta}_H \\ z_R \\ \dot{z}_R \\ \delta_f \end{bmatrix} \tag{14.1.45}$$

– *Stationary cornering with constant velocities*

Introduction of $z_R = l_L \delta_f$ in (14.1.40) results in

$$\begin{aligned} m_R^{'} l_L \ddot{\delta}_f = & \frac{c_S}{r_R} \delta_H + \frac{d_S}{r_R} \dot{\delta}_H - \frac{c_S}{r_R^2} l_L \delta_f - l_L (\frac{d_S}{r_R^2} + d_R^{'}) \dot{\delta}_f \\ & - f_{RC}^{'} \operatorname{sign} \dot{\delta}_f - \frac{2}{l_L} f_{ZT,v}^M \dot{\delta}_f - \frac{2}{l_L} f_{ZT,c}^M \operatorname{sign} \dot{\delta}_f \\ & - \frac{2}{l_L} c_\alpha c_\delta n_S \delta_f . \end{aligned} \tag{14.1.46}$$

In the stationary case, one obtains by setting the derivatives of $\delta_f$ and $\delta_H$, $f_{ZT,v}^M$ and $f_{ZT,st}^M$ to zero, but taking the rack and tire friction $f_{RC}^{''}$ (14.1.40) into account

$$\begin{aligned} \delta_H = & (\frac{l_L}{r_R} + \frac{2 r_R n_S}{c_S l_L} c_\alpha c_\delta) \delta_f + \frac{r_R}{c_S} f_{RC}^{''} \operatorname{sign} \dot{\delta}_f \\ & + \frac{2 r_R}{c_S l_L} f_{ZT,c}^M \operatorname{sign} \dot{\delta}_f \\ = & (i_S + \frac{2 r_R n_S}{c_S l_L} c_\alpha c_\delta) \delta_f + \frac{r_R}{c_S} f_{RC}^{''} \operatorname{sign} \dot{\delta}_f . \end{aligned} \tag{14.1.47}$$

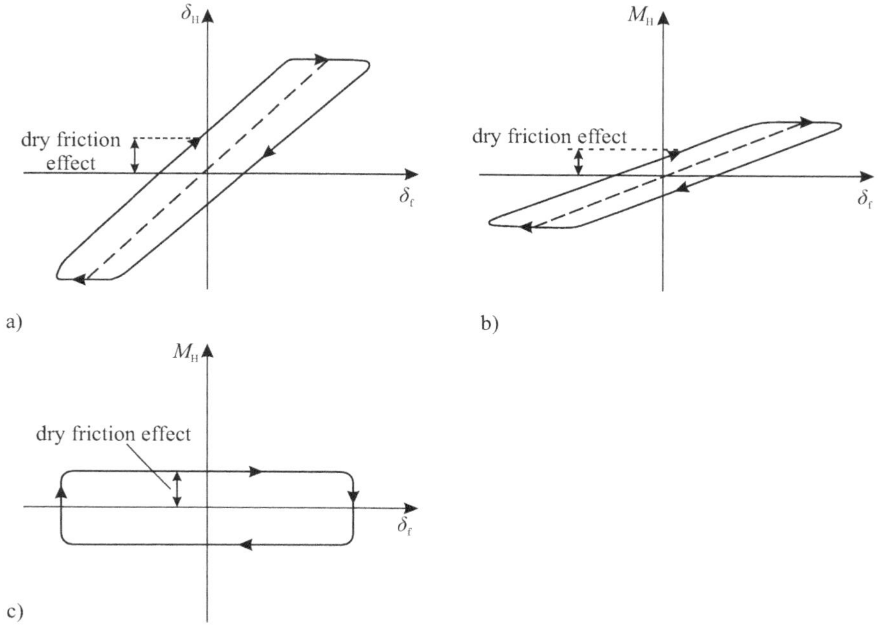


**Fig. 14.5** Steering wheel angle and torque characteristics **a** $\delta_H = f(\delta_f)$ ; $v_x$ = const.; **b** $M_H = f(\delta_f)$ ; $v_x$ = const. **c** $M_H = f(\delta_f)$ for parking

Hence, the steering wheel angle is proportional to the steering angle with the steering ratio $i_S = l_L/r_R$ plus a term depending on the side slip coefficient $c_\alpha$ and the side slip angle coefficients $c_\delta(v_x)$; see (14.1.26).

If (14.1.28) is used to determine the lateral tire forces during cornering (14.1.47) changes to

$$\delta_H = (i_S + \frac{r_R n_S}{c_S l_L}\frac{m l_r}{l} a_Y)\delta_f + \frac{r_R}{c_S} f_{RC}'' \operatorname{sign} \dot{\delta}_f. \tag{14.1.48}$$

The additional term to $i_S$ depends now on the lateral acceleration $a_Y$.

The steering wheel angle has an offset, depending on the dry friction coefficient $f_{RC}'$ of the rack and $f_{RT,c}^M$ of the turning wheel; see (14.1.33). This leads to a hysteresis characteristic, as shown in Fig. 14.5a.

The required torque of the steering wheel follows from (14.1.10) for stationary cornering

$$M_H = c_S\delta_H - c_S\delta_S = c_S\delta_H - c_S\frac{l_L}{r_R}\delta_f. \tag{14.1.49}$$

Insertion of (14.1.47) gives

$$M_H = 2\frac{n_S}{i_S} c_\alpha c_\delta \delta_f + r_R f_{RC}'' \operatorname{sign} \dot{\delta}_f \tag{14.1.50}$$

or with (14.1.48)

$$M_{\mathrm{H}} = \frac{n_{\mathrm{S}}}{i_{\mathrm{S}}} \frac{m l_{\mathrm{r}}}{l} a_{\mathrm{Y}} \delta_{\mathrm{f}} + r_{\mathrm{R}} f_{\mathrm{RC}}'' \operatorname{sign} \dot{\delta}_{\mathrm{f}}. \tag{14.1.51}$$

A corresponding characteristic is depicted in Fig. 14.5b. The steering wheel torque is proportional to the steering angle and inverse proportional to the steering ratio $i_{\mathrm{S}}$. The dry friction term acts additionally, such that a hysteresis characteristic results.

For *low velocity* it can be assumed, that the lateral tire forces (14.1.27) through $c_\alpha \alpha_{\mathrm{f}}$ are small compared to the dry friction term, as $\alpha_{\mathrm{f}} \approx 0$. Then it can be set $c_\alpha = 0$ and (14.1.47) simplifies to

$$\delta_{\mathrm{H}} = i_{\mathrm{S}} \delta_{\mathrm{f}} + \frac{r_{\mathrm{R}}}{c_{\mathrm{S}}} f_{\mathrm{RC}}'' \operatorname{sign} \dot{\delta}_{\mathrm{f}}. \tag{14.1.52}$$

The required torque at the steering wheel yields with (14.1.50)

$$M_{\mathrm{H}} = r_{\mathrm{R}} f_{\mathrm{RC}}'' \operatorname{sign} \dot{\delta}_{\mathrm{f}}, \tag{14.1.53}$$

and gives a hysteresis characteristic (almost) independent of $\delta_{\mathrm{f}}$, depending on the dry friction of the tire and the rack.

For *parking* the turning tire friction torque is larger than the dry friction of the rack such that with $f_{\mathrm{RC}}' = 0$ in (14.1.39) and (14.1.51) the required torque becomes

$$M_{\mathrm{H}} = \frac{2}{i_{\mathrm{S}}} f_{\mathrm{ZT,c}}^{M} \operatorname{sign} \dot{\delta}_{\mathrm{f}}, \tag{14.1.54}$$

see the characteristic in Fig. 14.5c.

It is noted that in modeling the mechanical steering system the stiffness, respectively, elasticity has been concentrated in the steering shaft and assumed as linear. In reality, the elasticities are distributed over the whole steering system and are partially nonlinear. Measured steering characteristics show therefore hysteresis curves with nonlinear parts and rounded corners.

– *The driver as a controller*

The manual input torque $M_{\mathrm{H}}$ is generated by the *driver* as a *steering angle controller*. Assuming that the driver reacts simplified as a P-controller it holds

$$M_{\mathrm{H}}(t) = K_{\mathrm{Dr}}(\delta_{\mathrm{Hd}}(t) - \delta_{\mathrm{H}}(t)) \tag{14.1.55}$$

where $\delta_{\mathrm{Hd}}$ is a desired value or setpoint. Laplace transformation of (14.1.10) for the linear case then yields

$$\delta_{\mathrm{H}}(s) = \frac{K_{\mathrm{Dr}} \delta_{\mathrm{Hd}}(s) - M_{\mathrm{R1}}(s)}{J_{\mathrm{SW}} s^2 + d_{\mathrm{S}} s + (c_{\mathrm{S}} + K_{\mathrm{Dr}})}. \tag{14.1.56}$$

The steering angle $\delta_{\mathrm{H}}$ depends on the desired value $\delta_{\mathrm{Hd}}$ and the torque feedback $M_{\mathrm{R1}}$ from the rack which acts as a disturbance, as depicted in Fig. 14.6.

Hence, the steering angle control system with the driver as a controller is a second-order system with a natural angular frequency

$$\omega_{0H} = \sqrt{\frac{c_S + K_{Dr}}{J_{SW}}} \tag{14.1.57}$$

which is in the range of the natural angular frequency of the free oscillating steering wheel with shaft; see the following Example 14.1.1.

### (c) Steering Wheel Angle As Applied Variable

The natural input for steering a vehicle is the steering wheel angle $\delta_H$. (As illustrated in Fig. 14.6 the steering wheel torque $M_H$ is the manipulated variable by a human driver). Therefore, $\delta_H$ is now considered as an applied variable.

For a given steering wheel angle, the torque from the steering shaft to the pinion is

$$M_{S1} = c_S(\delta_H - \delta_S) + \delta_S(\dot{\delta}_H - \dot{\delta}_S) \tag{14.1.58}$$

where $\delta_S$ is the shaft angle at the pinion. This gives a rack force

$$F_{R1} = \frac{1}{r_R} M_{S1}. \tag{14.1.59}$$

It is now assumed that the overall stiffness and the viscous friction of the steering system is lumped into the shaft parameters $c_S$ and $d_S$ and that $m_R^{'}$ is a representative mass as (14.1.15) and $f_{RC}^{'}$ a corresponding dry friction coefficient. For stationary cornering with constant higher velocity then (14.1.46) is used with the assumption $f_{ZT,c}^{M} = 0$ and $f_{ZT,v}^{M} = 0$ resulting in

$$\begin{aligned} m_R^{'} l_L \ddot{\delta}_f + l_L\left(\frac{d_S}{r_R^2} + d_R^{'}\right)\dot{\delta}_f + \left(\frac{c_S}{r_R^2} l_L + 2\frac{c_\alpha c_\delta n_S}{l_L}\right)\delta_f \\ + f_{RC}^{'} \operatorname{sign} \dot{\delta}_f = \frac{c_S}{r_R}\delta_H + \frac{d_S}{r_R}\dot{\delta}_H \end{aligned} \tag{14.1.60}$$

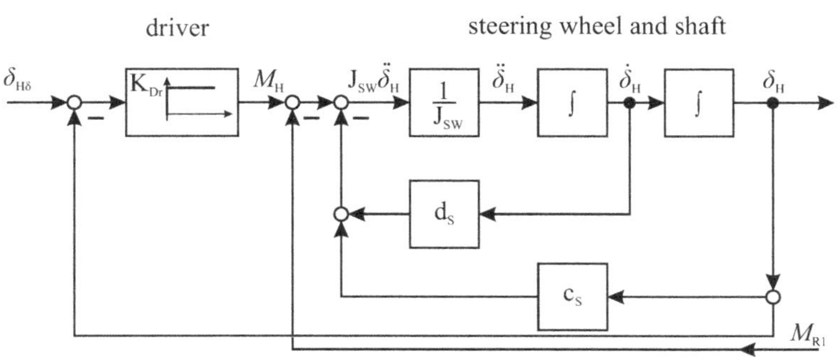


**Fig. 14.6** Signal flow scheme of the steering wheel shaft with the driver as a steering angle controller

or

$$a_2\ddot{\delta}_f + a_1\dot{\delta}_f + a_0\delta_f = b_0\delta_H + b_1\dot{\delta}_H - c_1 \operatorname{sign}\dot{\delta}_f \tag{14.1.61}$$

with

$$a_0 = (\frac{c_S}{r_R^2} l_L + 2\frac{c_\alpha c_\delta n_S}{l_L}) \qquad a_1 = l_L(\frac{d_S}{r_R^2} + d_R^{'})$$

$$a_2 = m_R^{'} l_L \qquad b_0 = \frac{c_S}{r_R}$$

$$b_1 = \frac{d_S}{r_R} \qquad c_1 = f_{RC}^{'}$$

Laplace transformation after omitting the dry friction term yields a transfer function with the steering wheel angle as input and the steering angle of the front wheels as output

$$G_{\delta_F\delta_H} = \frac{\delta_F(s)}{\delta_H(s)} = \frac{b_0 + b_0 s}{a_2 s^2 + a_1 s + a_0}. \tag{14.1.62}$$

Hence, the front wheel steering angle $\delta_f$ results after a change of the steering wheel angle $\delta_H$ as a second-order oscillating system with a derivative (lead) term, or a linear system with two poles and one zero. Its natural frequency is

$$\omega_{0\delta} = \sqrt{\frac{a_0}{a_2}} = \sqrt{\frac{1}{m_R^{'} l_L}\left(\frac{c_S}{r_R^2} l_L + 2\frac{c_\alpha c_\delta n_S}{l_L}\right)} \tag{14.1.63}$$

and its damping ratio

$$\zeta_\delta = \frac{a_1}{2}\frac{1}{\sqrt{a_0 a_2}} = \frac{l_L(\frac{d_S}{r_R^2} + d_R^{'})}{2\sqrt{m_R^{'}(c_S(\frac{l_L}{r_R})^2 + 2c_\alpha c_\delta n_S)}}. \tag{14.1.64}$$

Hence, the natural frequency increases with smaller $m_R^{'}$ and larger stiffness $c_S$ and larger side slip coefficient $c_\alpha$. The damping ratio increases with larger damping factors of the steering shaft $d_S$ and of the (enlarged) rack $d_R^{'}$.

If the effects of the lateral tire force $c_\alpha c_\delta$ is assumed to be relatively small, then the characteristic oscillation factors become simpler

$$\omega_{0\delta} = \frac{1}{r_R}\sqrt{\frac{c_S}{m_R^{'}}} \tag{14.1.65}$$

$$\zeta_\delta = \frac{r_R(\frac{d_S}{r_R^2} + d_R^{'})}{2\sqrt{m_R^{'} c_S}}. \tag{14.1.66}$$

However, these characteristic values are only valid for the linearized system without the dry friction coefficient $f'_{RC}$ of the rack, the linkages and the steering axis and without the friction coefficient $f^M_{ZT,c}$ and $f^M_{ZT,v}$ of the tire due to (14.1.33). Therefore, the damping factor is considerably larger as expressed by (14.1.64) and (14.1.66). As a rough approximation $d'_R$ can be increased to cope with the neglected dry frictions as shown in the next example.

**Example 14.1.1** For a pinion-rack mechanical steering system following parameters are used, see, e.g. Table 14.2,

$$c_S = 10000 \text{ Nm/rad}; \quad m'_R = 1500 \text{ kg}; \quad l_L = 0.12 \text{ m}$$
$$r_R = 0.015 \text{ m}; \quad d_S = 0.12 \text{ Nms/rad}; \quad d'_R = 6.10^5 \text{ Ns/m}$$

($d'_R$ is approximated by using: rack force $F_R$/ rack velocity $\dot{z}_R$ = 12000N/ 0.02 m/s = $6.10^5$ Nm/s, thus imitating the influence of the turning tire dry friction). With these assumptions it follows from (14.1.65) and (14.1.66)

$$\omega_{0\delta} = 172 \text{ 1/s} = 27.4 \text{ Hz}$$
$$\zeta_\delta = 1.16$$

Hence, a well-damped oscillation results with a natural frequency which is larger than the range of lateral vehicle dynamics. □

The steering gain follows from (14.1.62)

$$K_{\delta HF} = \frac{\delta_F(0)}{\delta_H(0)} = \frac{1}{\frac{l_L}{r_R} + 2(\frac{r_R}{l_L})\frac{c_\alpha c_\delta n_S}{c_S}}. \tag{14.1.67}$$

For a large shaft stiffness $c_S$ or small $c_\alpha$, i.e. $c_\alpha / c_S \to 0$ one obtains the basic kinematic relation

$$K_{\delta HF} = \frac{r_R}{l_L} = \frac{1}{i_S}. \tag{14.1.68}$$

Figure 14.7 depicts a signal flow scheme for the considered steering system, which is based on (14.1.60) and Fig. 14.4.

### 14.1.3 Frequency Ranges of Interest

With regard to the dynamic behavior of steering systems, one can distinguish mainly 4 frequency ranges; see Table 14.1. The driver's input through the steering wheel as well as the resulting lateral dynamics are in a low-frequency range of about $f < 3$

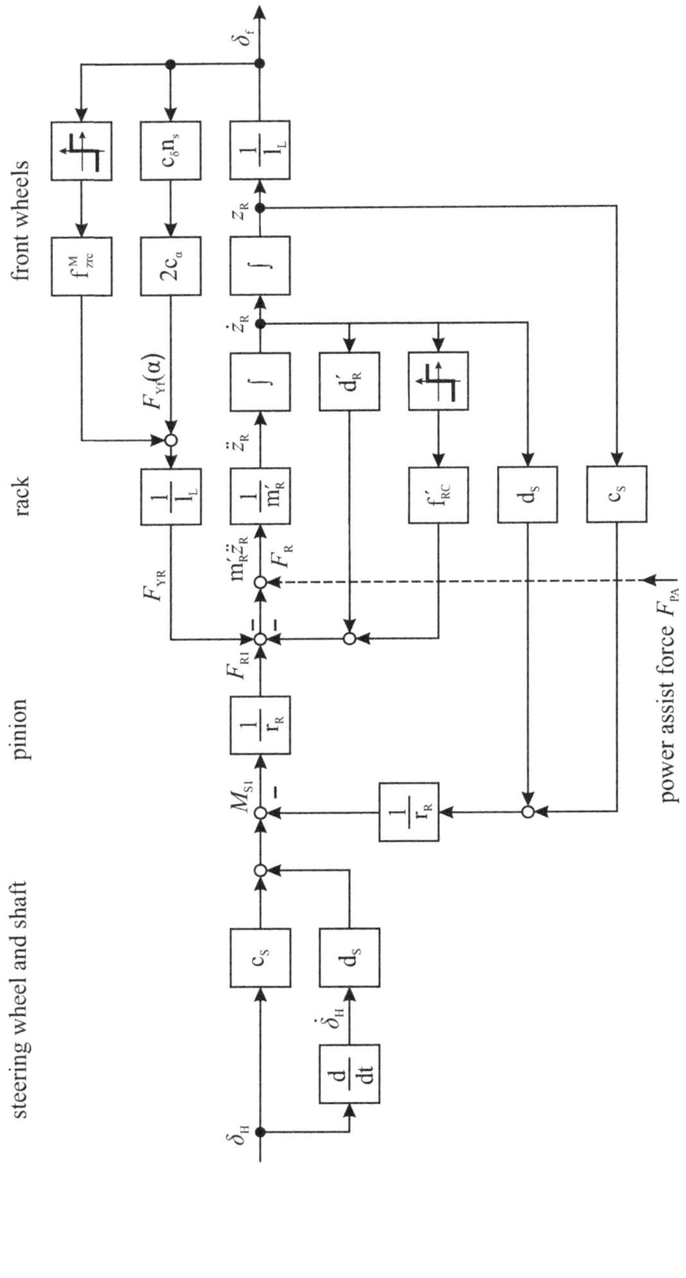


**Fig. 14.7** Signal flow scheme of a rack and pinion mechanical steering system for an applied steering wheel angle during cornering with constant velocity

Hz. The driver can usually compensate drive dynamic oscillations within 0–0.4 Hz. An upper limit of driver's capabilities is about 2 Hz, Heissing and Ersoy (2011b). The frequency responses of passenger cars for sinusoidal steering wheel input $\delta_f$ and, e.g. the yaw rate $\dot{\psi}$ as output show resonance frequencies in the range of 0.25–1.5 Hz, Mitschke and Wallentowitz (2004); see Fig. 7.8. Experiments with removed driver's hand from the steering wheel after cornering, so-called "free control", result in yaw rate oscillations of about 1.7 Hz, Mitschke and Wallentowitz (2004). Hence, the *lateral dynamics of passenger cars* arise in a range of about 0–2 Hz and depend on load and velocity.

The *haptic feedback* in the form of torque feedback of the steering wheel from the tire side forces and from the kind of road contact are observed in a medium frequency range of about 3–20 Hz. These frequencies are an important information on the lateral driving condition and part of "steering feel", Harrer (2008). Frequencies above 20 Hz result from *road irregularities, imbalances* of the tires and have to be considered as impairing feedback. The natural frequency of the *steering column* is usually around 40 Hz, such that, in the case of an overcritical design, the 2nd harmonics of 4- or 6-cylinder engines in idle running in a range of 20...35 Hz, and imbalances of tires of $f < 25$ Hz do not excite vertical and lateral steering wheel vibrations, Pfeffer and Harrer (2013). If necessary, absorbers in the steering wheel with eigenfrequencies around 32 Hz may damp unwanted vibrations. However, the natural frequency of the *steering column with free swinging steering wheel* lies in a range of 1–3 Hz; see Fankem et al. (2014).

Hence, the steering system as a precise and fast actuation system for the lateral driving control of passenger cars has to transfer frequency inputs in a range of 0–3 Hz and feedback information from the road and tires in a range of 3–20 Hz.

## 14.2 Power-Assisted Steering Systems

Except very small cars electrical power-assisted steering systems (EPS) are standard for small to larger passenger cars and hydraulic power-assisted steering systems (HPS) for large passenger cars, and light weight and heavy commercial vehicles. These power-assisted steering systems amplify the driver's muscular forces by adding additional steering forces resulting from hydraulic or electrical auxiliary energy. Some development steps for power steering systems are shown in Fig. 14.8.

Figure 14.9 depicts a signal flow scheme of a power steering system for the case of a rack and pinion steering. In addition to the mechanical steering system a torque sensor in the steering column measures the manual torque $M_H$ of the driver via a torsion bar generating a small twist angle $\Delta\delta_M$. This value then directly changes an orifice valve in the case of an hydraulic actuator to influence hydraulic flows or is an electrical output of a torque sensor in the case of electrical power steering. Based on this torque measurement the hydraulic or electrical power actuator is feed forward controlled via a calibrated characteristic curve $u_{PA} = f(\Delta\delta_M)$ which determines the actuator force dependent on the steering angle $\delta_H$ and the vehicle speed $v_x$. For both, hydraulic and electric power steering systems many different constructions

**Table 14.1** Frequency ranges of front steering systems

| | Low freq. range $f < 3\text{Hz}$ | Medium freq. range $3\text{Hz} < f < 20\text{Hz}$ | Higher freq. range $20\text{Hz} < f < 40\text{Hz}$ | High freq. range $f > 40\text{Hz}$ |
|---|---|---|---|---|
| Driving input for normal driving<br>– Parking<br>– Rural road<br>– Lane change | $\longleftarrow --\longrightarrow$ | | | |
| Lateral driving<br>– Dynamics<br>– Yaw rate<br>– lateral accel | $\longleftarrow --\longrightarrow$ | | | |
| Steering feedback<br>– Haptic inform<br>– Driving condition<br>– Road contact<br>– Side forces | | $\longleftarrow ---\longrightarrow$ | | |
| Steering wheel and shaft | $\longleftarrow --\longrightarrow$ | $\longleftarrow ---\longrightarrow$ | $\longleftarrow ---\longrightarrow$ | |
| Disturbances<br>– Tire imbalances<br>– Road regularities | | | $\longleftarrow ---\longrightarrow$ | |
| Drivetrain and chassis vibrations<br>– Engine<br>– Road regularities | | | $\longleftarrow ----$ - | $----\longrightarrow$ |

exist; see, e.g. Pfeffer and Harrer (2013), Heissing and Ersoy (2011a), Bosch GmbH (2018), Ersoy and Gies (2017). In the following models for typical basic designs are considered.

### 14.2.1 Kinematic Relations for Power Steering

The kinematic relations without considering elasticities and forces for a rack and pinion steerings are described with (14.1.2)–(14.1.9).

In the case of a *power steering system* with a support of the rack force $F_{\text{PA}}$ the torque at the vertical steering axis (kingpin axis) becomes

$$M_{\text{W}} = M_{\text{H}} i_{\text{S}} + F_{\text{PA}} l_{\text{L}} = M_{\text{H}} i_{\text{S}} + M_{\text{PA}} \tag{14.2.1}$$

with $i_{\text{S}} = \delta_{\text{H}}/\delta_{\text{f}}$ the wheel steering ratio, and the corresponding torque by the power assistance is for a required steering torque $M_{\text{W}}$

$$M_{\text{PA}} = F_{\text{PA}} l_{\text{L}} = M_{\text{W}} - M_{\text{H}} i_{\text{S}}. \tag{14.2.2}$$

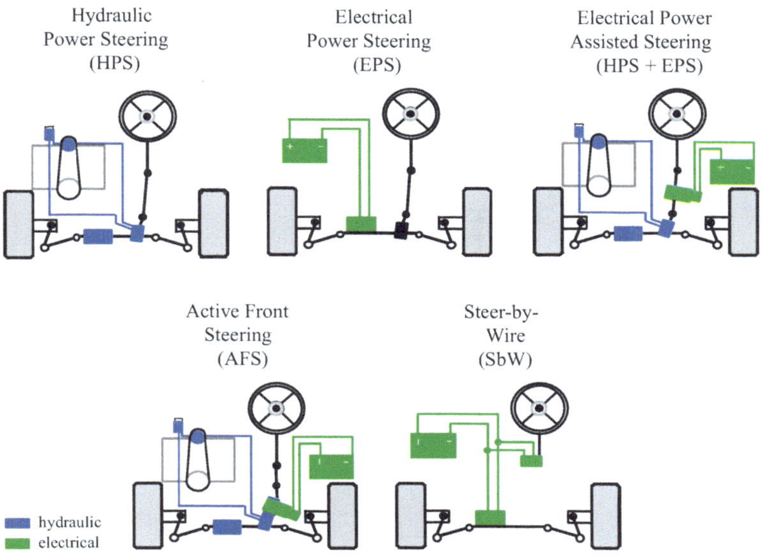


**Fig. 14.8** On the development of different power steering systems

The lateral tire force is for stationary cornering with a path radius $R_P$ and the vehicle front mass $m_f$

$$F_{Yf} = F_{YT} = m_f a_y = m_f \frac{v_x^2}{R_P}, \tag{14.2.3}$$

which leads to a torque at the wheels

$$M_W = F_{Yf} n_S = m_f n_S a_y \tag{14.2.4}$$

and the resulting steering wheel torque becomes without power steering

$$\begin{aligned} M_H &= m_f \frac{n_S}{i_S} a_y = C_A \frac{v_X^2}{R_P} \\ C_A &= m_f \frac{n_S}{i_S}. \end{aligned} \tag{14.2.5}$$

Hence, it is proportional to the lateral acceleration $a_y$ or to $v_X^2$. $C_A$ is called a torque return factor, Pfeffer and Harrer (2013).

A steering gain is defined as the ratio of the required steering torque at the front wheel to the manually activated torque of the rack, Mitschke and Wallentowitz (2004),

$$K_S = \frac{M_W}{i_S M_H} = \frac{i_S M_H + M_{PA}}{i_S M_H} = 1 + \frac{M_{PA}}{i_S M_H}. \tag{14.2.6}$$

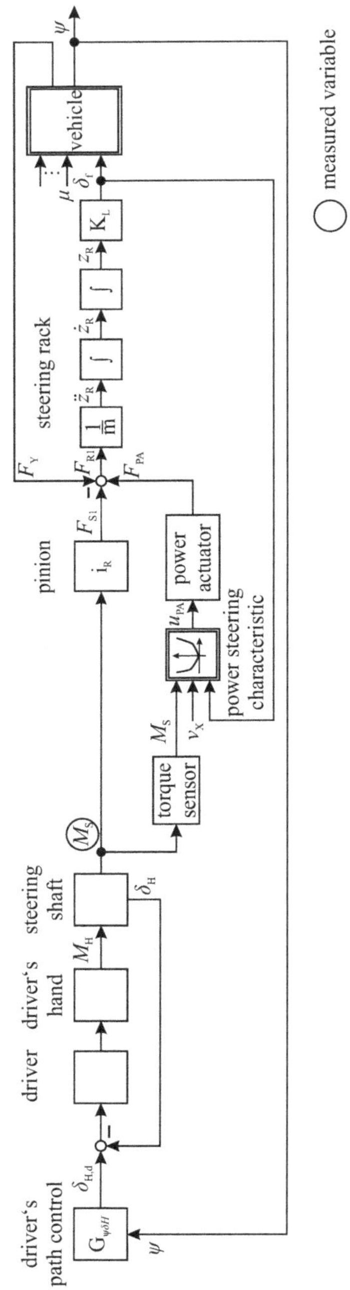


**Fig. 14.9** Signal flow scheme for a power-assisted pinion and rack steering

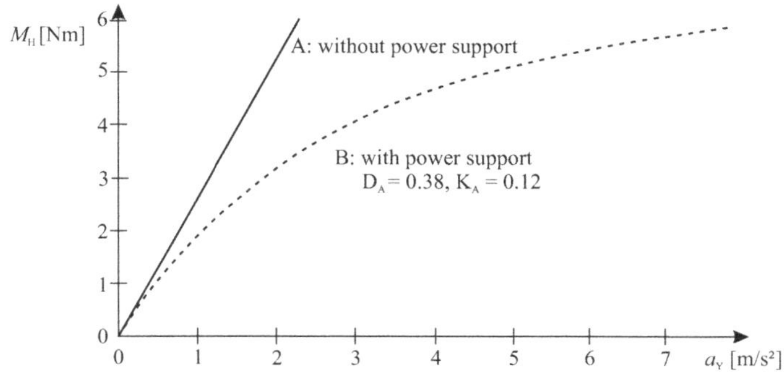


**Fig. 14.10** Steering torque characteristics in dependence on the steering wheel type, Pfeffer and Harrer (2013)

Hence, it is $K_S = 1$ without power steering. The generated torque from the power assist is then,

$$M_{PA} = i_S M_H (K_S - 1). \tag{14.2.7}$$

The evaluation of cars with harmonic steering behavior has shown, that the steering gain is approximately linearly depending on the lateral acceleration, Pfeffer and Harrer (2013),

$$K_S(a_y) = C_A(D_A + K_A a_y) = \frac{m_f n_S}{i_S}(D_A + K_A a_y) \tag{14.2.8}$$

and the steering wheel torque becomes then with (14.2.4), (14.2.5), and (14.2.7)

$$M_H = \frac{M_W}{i_S K_S} = \frac{a_y}{(D_A + K_A a_y)} = \frac{1}{\frac{D_A}{a_y} + K_A} \tag{14.2.9}$$

Without power steering, i.e. $K_S = 1$, it is

$$D_A = 1/C_A = \frac{i_S}{m_f n_S}$$

$$K_A = 0$$

The steering wheel torque is then, as (14.2.5), proportional to $a_y$. With power steering and $K_A > 0$ the steering wheel torque shows a degressive behavior; see Fig. 14.10.

The wheel torque which has to be generated from the power steering actuator follows then from (14.2.7), (14.2.8) and (14.2.9) to

$$M_{PA} = \frac{a_y(m_f n_S(D_A + K_A a_y) - i_S)}{D_A + K_A a_y} \tag{14.2.10}$$

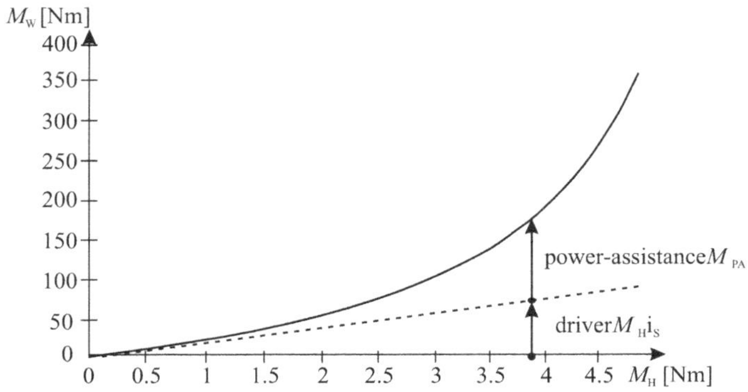


**Fig. 14.11** Steering torque characteristics at the front wheels $M_W = M_H i_S + M_{PA}$, according to (14.2.1) and (14.2.11), Pfeffer and Harrer (2013)

or by replacing $a_y$ with (14.2.9)

$$a_y = \frac{D_A M_H}{1 - K_A M_H}$$

then results in

$$\begin{aligned} M_{PA}(M_H) &= \frac{i_S K_A M_H^2}{1 - K_A M_H} \\ &= F_{PA}(M_H) l_L. \end{aligned} \tag{14.2.11}$$

Figure 14.11 depicts the steering torque $M_W$ due to (14.2.1) at the front wheels in dependence on the steering wheel torque $M_H$. The part from the driver is then due to (14.2.1) $M_H i_S$ and increases linearly with $M_H$ and the part from the power assistance is $M_{PA}$ and increases progressively with $M_H$; see (14.2.11) and Pfeffer and Harrer (2013).

## 14.3 Hydraulic Power Steering (HPS)

### 14.3.1 Basic Designs of HPS

Hydraulic power steering systems for passengers cars are used since about 1945. Mainly two types of steering gears, the rack and pinion steering or the recirculating ball steering are applied, the first one for most of the passenger cars and the second one for heavier, usually commercial vehicles.

In the case of a conventional hydraulic power steering (HPS) with a rack and pinion gear, the hydraulic system consists of an engine driven displacement pump, an oil tank, a steering valve and a hydraulic cylinder. The cylinder and the control valve are usually integrated with the rack and pinion housing. Typical pumps are sliding vane, radial piston, or rolling vane pumps. A schematic of an hydraulic power steering system is depicted in Fig. 14.12

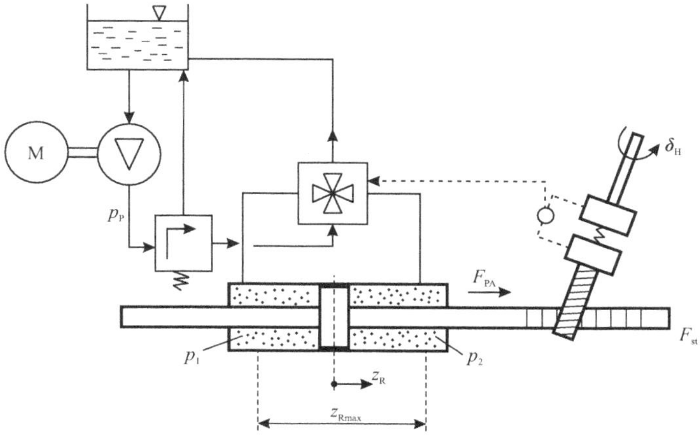


**Fig. 14.12** Schematic of an hydraulic power steering system

The oil pump generates a pressure and a permanent circulation oil flow. A control valve changes the oil difference pressure on the piston in the hydraulic cylinder according to the torque on the steering wheel. Therefore an elastic torque sensor in the form of a torsion is connected to the control valve which opens the high-pressure side of the pump to one side of the cylinder and the other side back to the pump. The travel of the control valve is a movement of a rotary slide, forming a displacement of control edges as chamfers and facets and thus determines the opening area of the oil flow. If no torque acts on the steering column the valves let the oil flowing back to the oil tank with only a small change in pressure. This is called the "open-center" principle.

The flow principle of the control valve is depicted in Fig. 14.13a. It consists of a fixed outer sleeve and an inner rotary shaft. Both parts have several axial slots which are opened or closed by turning the inner shaft, thus resulting in at least 4 changeable flow resistances, as depicted in Fig. 14.13b. In neutral position, the oil flow streams from the inflow to the outflow with small resistance.

If a steering torque turns the inner shaft several slots are increased or decreased, according to the schematic in Fig. 14.13b. In the case of a turn to the left, for example, the slots 1 and 3 are increased and 2 and 4 are decreased, thus increasing the pressure in the left part of the cylinder and reducing it in the right part. Hence, the flow resistances of several orifices are changed simultaneously, according to an electric Wheatstone bridge, Pfeffer and Harrer (2013). See also, Bosch GmbH (2018).

The resulting difference pressure of the hydraulic power steering cylinder then results in a nonlinear characteristic as in Fig. 14.14. Because of friction effects, a small hysteresis is observed in the control valve. The characteristic curve depends on the design of the orifices of the control edges. A maximum pressure is limited by a pressure relief valve, normally integrated with the pump.

In order to damp force impulses from the road through the wheels to the steering wheels special *damping valves* are required. They are for example built in at the connections of the fluid hoses to the cylinder and generate a volume flow-dependent

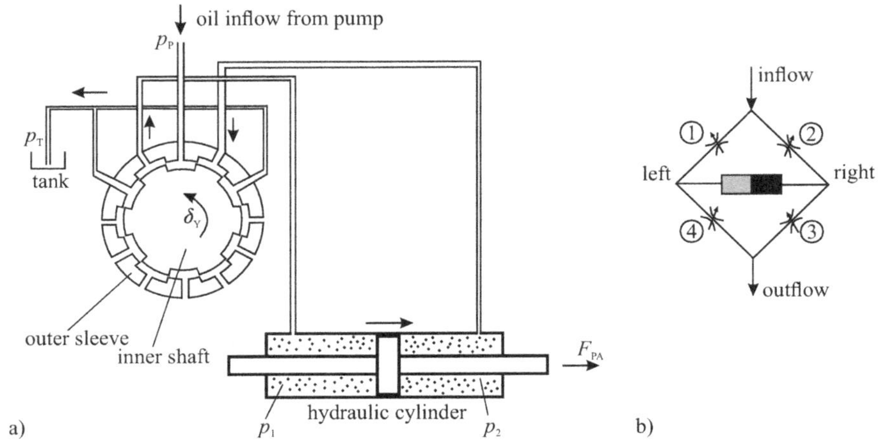


**Fig. 14.13** Schematic of the control valve for an hydraulic power steering system. **a** Hydraulic control with axial slots, due to Kageyama (2014). **b** Function principle of the flow resistances according to a Wheatstone bridge, due to Pfeffer and Harrer (2013)

resistance for the out flowing fluid combined with a pressure relief function for the in flowing fluid; see, e.g. Pfeffer and Harrer (2013).

*Parametrization hydraulic power steering* allows to change the pressure-torque characteristic depending on the vehicle speed, see Fig. 14.15 and Barthenheier (2004). An electro-hydraulic converter modifies the feedback pressure of the cylinder with an ECU and the vehicle speed as input. Thus, when parking and large steering torques at the wheel are required the characteristic is steep in order to need only small steering torques by the driver. With increasing speed the power assistance is reduced compared to the steering wheel torque, because smaller steering torque at the wheels are required and thus precise steering is possible by the driver.

The design and modeling approaches for HPS have a long tradition, see, for example, Edge and Darling (1986), Milliken and Milliken (2002), Bootz et al. (2003), and Pfeffer (2006).

## 14.3.2 Dynamic Models of HPS

The delivered theoretical volume flow $\dot{V}_p$ of a positive displacement pump, as, e.g. a sliding vane pump, is

$$\dot{V}_{p,th} = V_{p,th}\omega_p \tag{14.3.1}$$

where $V_{p,th}$ is the theoretical (ideal) displacement per turn. Because of volume flow losses through internal leaks and non ideal charging the effective volume flow becomes.

$$\dot{V}_{p,eff} = \dot{V}_{p,th}\eta_{vol} = V_{p,th}\eta_{vol}\omega_p \tag{14.3.2}$$

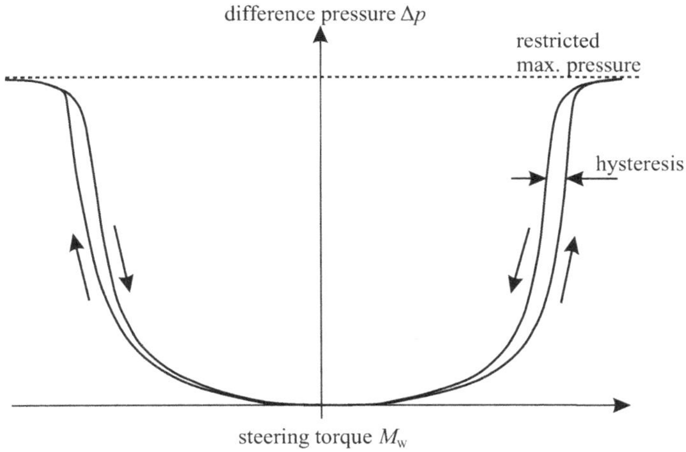


**Fig. 14.14** Pressure vs. steering wheel support torque characteristic for a hydraulic power steering, Barthenheier (2004), ZF-Lenksysteme (2010)

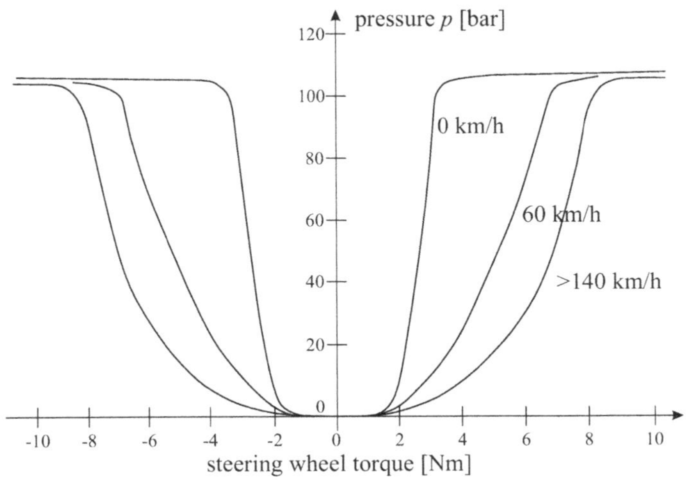


**Fig. 14.15** Difference pressure of the hydraulic power steering cylinder vs. steering wheel torque for different vehicle velocities

where $\eta_{vol}$ is a volumetric efficiency.

Hence, the volume flow of displacement pumps is proportional to the angular velocity $\omega_p$ which depends for an engine driven pump from the engine speed $\omega_{eng}$ or $n_{eng} = 2\pi\omega_{eng}$.

As the maximum oil flow for power steering is required for parking and low vehicle speed, the linear increase is realized until, e.g. idle speed $n_{idle}$; see Fig. 14.16. However, for the higher vehicle resp. engine velocity the required steering forces

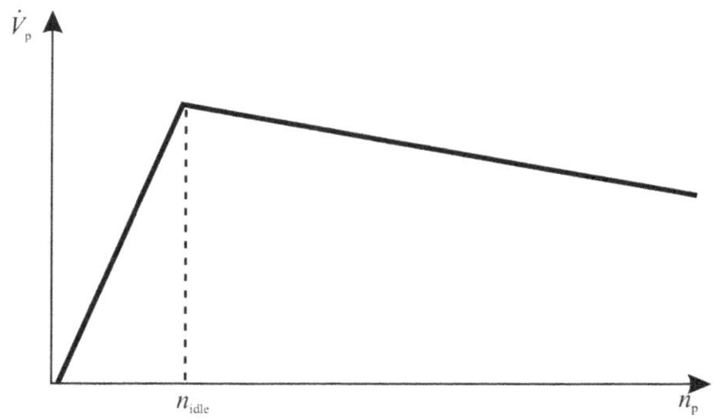


**Fig. 14.16** Volume flow characteristic of a sliding vane pump with flow limiting valve

decrease and therefore usually a flow limiting valve is built in thus that the volume flow decreases with increasing velocity, as shown in Fig. 14.16.

To model the stationary and dynamic behavior of the hydraulic power steering actuator, it is required to describe the behavior between the angle of the inner shaft of the control valve and the actuation force $F_{PA}$ of the hydraulic actuator. This corresponds to the models of an hydraulic servo axis, as described in Münchhof (2006) and Isermann (2005), consisting of a proportional hydraulic 4/3 valve, two cylinder chambers, a piston, and a shaft. The actuation force is; see Fig. 14.12.

$$f_{PA} = A_P(p_1 - p_2) - f_c \operatorname{sign} \dot{z}_R \tag{14.3.3}$$

where $A_P$ is the effective area of the piston and $f_c$ a dry friction coefficient. For the volume flow of an orifice holds

$$\begin{aligned} \dot{V}_1 &= A'_V \sqrt{\frac{2}{\rho}} \sqrt{\Delta p_V} \operatorname{sign}(\Delta p) \\ A'_V &= \alpha_D A_V \end{aligned} \tag{14.3.4}$$

where $\alpha_D$ is a contraction parameter, $A_V$ the opening area of the valve, $\rho$ the oil density, and $\Delta p_V$ the pressure difference which is, e.g.

$$\Delta p_{V1} = p_P - p_1 \tag{14.3.5}$$

for the valve of the left cylinder and

$$\Delta p_{V2} = p_2 - p_T \tag{14.3.6}$$

for the right cylinder in Fig. 14.12. As the valve then opens to the tank, it is assumed that $\Delta p_{V2} \approx 0$. Therefore in the following only one cylinder is considered.

To obtain the difference pressure $\Delta p_{V1}$ at the valve in dependence on the twist angle $\delta_V$ as measure for the steering wheel torque one obtains with (14.3.4).

$$\Delta p_V = \frac{1}{A'^2_V} \frac{\rho}{2} \dot{V}^2 = R_V \dot{V}^2 \tag{14.3.7}$$

where $R_V$ is the valve resistance. Introducing a relative valve opening area

$$u_A = \frac{A'_V}{A'_{V_{max}}} \tag{14.3.8}$$

leads to

$$\Delta p_V = \frac{R_{min}}{u_A^2}\dot{V}^2 = \frac{R_{min}}{u_A^2}\dot{V}|\dot{V}| \tag{14.3.9}$$

with the minimal resistance for the fully opened valve

$$R_{min} = \frac{\rho}{2A'^2_{V_{max}}}. \tag{14.3.10}$$

The effective opening area of the valve depends now on a designed characteristic in dependence on the twist angle $\delta_V$ between the inner rotary shaft and the outer sleeve

$$A'_V = f(\delta_V). \tag{14.3.11}$$

The volume flow through the valve follows from (14.3.9) for the pressurized chamber 1

$$\dot{V}_{V1} = \frac{1}{\sqrt{R_{min}}}\sqrt{\Delta p_{V1}}u_A = \frac{1}{\sqrt{R_{min}}}\sqrt{p_P - p_1}u_A. \tag{14.3.12}$$

Hence, the volume flow into chamber 1 depends for constant pump pressure $\bar{p}_P =$ const. on $p_1(t)$and $u_A(t)$.

To obtain the resulting power assistance force $F_{PA}$ of the hydraulic cylinder, the mass balance equation for the chamber 1 is considered. It holds

$$\dot{m}_{V1}(t) = \dot{m}_{cyl}(t)$$

$$\begin{aligned} \dot{V}_{V1}(t)\rho_V(t) &= \frac{d}{dt}[V_{cyl}(t)\rho_{cyl} + \rho_{cyl}(t)V_{cyl}] \\ &= A_P\dot{z}_R\rho_1 + \dot{\rho}_1(t)(A_P(z_{R_{max}} - z_{R1})). \end{aligned} \tag{14.3.13}$$

The density can be replaced by the pressure in using the definition of the bulk modulus

$$\beta = -V\left(\frac{\partial p}{\partial V}\right)\bigg|_{T=\text{const}}. \tag{14.3.14}$$

This leads with

$$\frac{\partial p}{\partial \rho} = \left(\frac{\partial p}{\partial V}\right)\left(\frac{\partial V}{\partial \rho}\right) = -\frac{V}{\rho}\frac{\partial p}{\partial V} = \frac{\beta}{\rho}$$

to

$$d\rho = \frac{\rho}{\beta}dp \tag{14.3.15}$$

Equation (14.3.13) then becomes

$$A_P(z_{R_{max}} - z_{R1})\frac{\rho_1}{\beta}\dot{p}_1(t) = \dot{V}_{V1}(t)\rho_V(t) - A_P\rho_1(t)\dot{z}_R(t)$$

or

$$\frac{A_P}{\beta}\frac{\rho_1(t)}{\rho_V}(z_{R_{max}} - z_{R1}(t))\dot{p}_1(t) = \dot{V}_{V1}(t) - A_P\frac{\rho_1(t)}{\rho_V}\dot{z}_R(t) \tag{14.3.16}$$

and with (14.3.12) and the effective cylinder length

$$\Delta z_{cyl} = (z_{R_{max}} - z_{R1}) \tag{14.3.17}$$

$$\frac{A_P}{\beta}\frac{\rho_1(t)}{\rho_V}\Delta z_{cyl}(t)\dot{p}_1(t) = \frac{1}{\sqrt{R_{min}}}(\bar{p}_P - p_1(t))^{\frac{1}{2}}u_A(t) - A_P\frac{\rho_1(t)}{\rho_V}\dot{z}_R(t). \tag{14.3.18}$$

The actuator force follows from (14.3.3)

$$\begin{aligned} F_{PA}(t) &= A_P(p_1(t) - p_2(t)) - f_C \operatorname{sign} \dot{z}_R \\ &\approx A_P p_1(t) - f_C \operatorname{sign} \dot{z}_R. \end{aligned} \tag{14.3.19}$$

Based on the considerations for the design of the power assist torque $M_{PA}$ according to (14.2.11) the required pressure difference in the cylinder can be calculated with

$$\Delta p_{cyl}(M_H) = p_1 - p_2 = \frac{M_{PA}(M_H)}{A_{cyl}l_L} = \frac{F_{PA}(M_H)}{A_{cyl}}. \tag{14.3.20}$$

A signal flow scheme of this nonlinear differential equation of first order for the pressure $p_1$ in chamber 1 is depicted in Fig. 14.17. It shows a proportional behavior with time delay and the influence of the piston position in the cylinder length $\Delta z_{cyl}$ and the piston speed $\dot{z}_R$.

For small changes around an operating point ($\dot{V}_{V1,0}$, $p_{P,0}$, $p_{1,0}$, $z_{R,0}$) the valve equation (14.3.12) is now expressed as a Taylor-series expansion

$$\dot{V}_{V1} = \dot{V}_{V1,0} + \frac{\partial \dot{V}_{V1}}{\partial p_1}\Delta p_1 + \frac{\partial \dot{V}_{V1}}{\partial u_A}\Delta u_A \tag{14.3.21}$$

leading to the flow gain

$$K_A = \frac{\partial \dot{V}_{V1}}{\partial u_A} = \frac{1}{\sqrt{R_{min}}}\sqrt{p_{P,0} - p_{1,0}} \tag{14.3.22}$$

and a flow-pressure coefficient

$$K_B = \frac{\partial \dot{V}_{V1}}{\partial p_1} = \frac{1}{\sqrt{R_{min}}\sqrt{p_{P,0} - p_{1,0}}}, \tag{14.3.23}$$

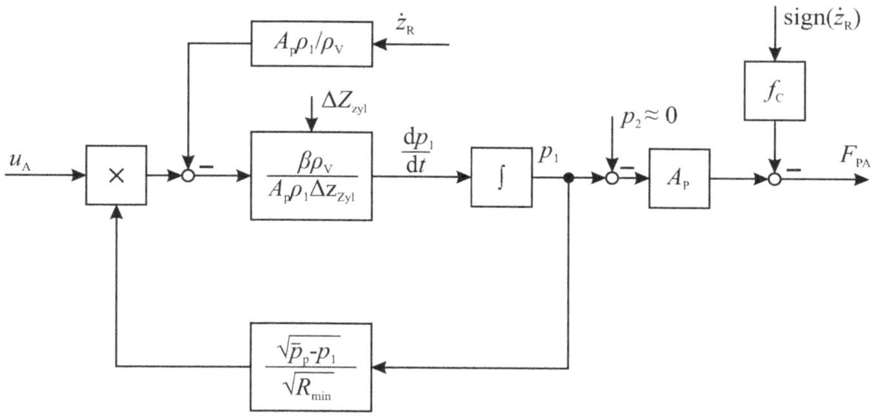


**Fig. 14.17** Nonlinear model of the hydraulic power assistance actuator Fig. 14.12 with the relative valve area $u_A$ as input and the cylinder chamber 1 pressure $p_1$ and actuation force $F_{PA}$ as output (only one chamber considered)

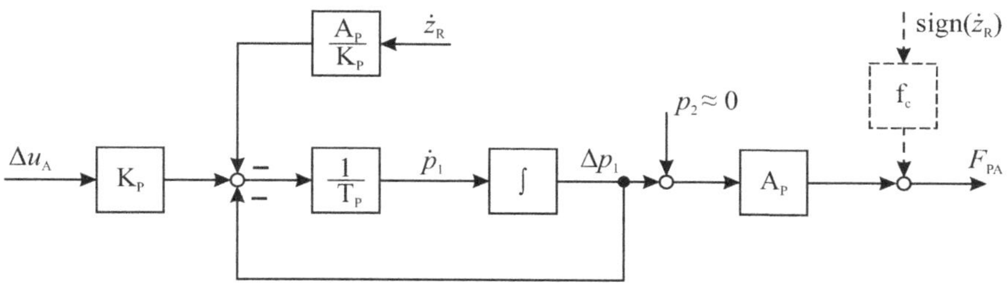


**Fig. 14.18** Linearized model of the hydraulic power assistance actuator

Equation (14.3.18) then results with the assumption of $\rho_1(t)/\rho_V \approx 1$ in

$$T_P \frac{dp_1(t)}{dt} + \Delta p_1(t) = \frac{K_A}{K_B}\Delta u_A(t) - \frac{A_P}{K_P}\dot{z}_R(t) \tag{14.3.24}$$

with the time constant and gain

$$T_P = \frac{A_P \Delta z_{cyl,0}}{\beta K_B} \; ; \; K_P = \frac{K_A}{K_B} = (p_{P,0} - p_{1,0}) = \Delta p_{P1}. \tag{14.3.25}$$

Hence, the pressure $p_1$ and also the hydraulic cylinder force $F_{PA} = A_P p_1$ behave for small changes of the valve as a linear first order differential equation; see Fig. 14.18. The gain is proportional to the pressure difference $\Delta p_{P1}$ between the pump pressure $p_{P,0}$ and the cylinder pressure $p_{1,0}$ and the (small) time constant increases with the momentaneous cylinder length $\Delta z_{cyl}$ and the difference pressure $\Delta p_{P1}$.

Figure 14.19 depicts the overall signal flow of the considered hydraulic power steering system.

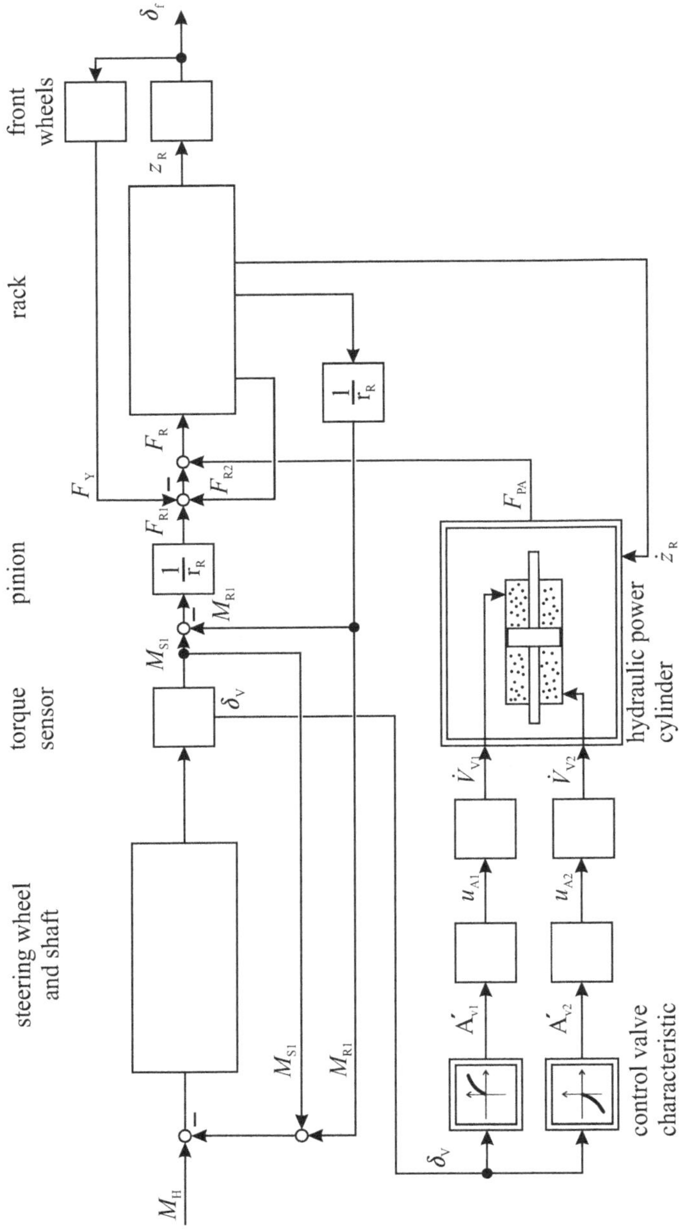


**Fig. 14.19** Schematic signal flow of the rack and pinion system with hydraulic power support, based on Figs. 14.3 and 14.9

It shows that the power assistance from $F_{PA}$ depends on the torque measurement (twist angle $\delta_V$) and is added to the manual generated rack force $F_S$, such that the total force becomes

$$F_R(t) = F_{R1}(t) + F_{PA}(t) - F_Y(t). \tag{14.3.26}$$

The power assistance force depends also on $z_R(t)$ and $\dot{z}_R(t)$ as feedback from the coupled rack system; see Fig. 14.4. The rack dynamics result in a torque component $M_{R1}$ on the steering wheel and thus gives a feedback from the tire forces $F_Y(t)$ and the power assistance force $F_{PA}$ to the driver, as part of the so-called "steering feel". This is treated in more detail in section 14.4.

## 14.4 Electrical Power Steering (EPS)

An electrical power steering system (EPS) uses an electromotor powered by the electrical board net to support the manual torque by the driver through a mechanical transmission. Though the energy density of electrical system is less than that for hydraulic systems and therefore more package space may be required around the actuator space, they offer many advantages and have therefore penetrated passenger cars since about 1988 from the A-segment to heavy vehicles, with electrical power from 150W to 1000W. This required partially a 24V power supply. The upcoming 48V voltage level will further increase their application rate.

The advantages of EPS compared to HPS are mainly:

- reduction of fuel consumption
- elimination of engine-driven oil pump, flexible hoses, hydraulic cylinder, and many seals
- complete system modules delivered as a single assembly, ready and tested units
- an EPS electronic control unit (ECU) together with a steering angle sensor allows a precise position and torque control
- ECU enables the use of additional inputs from chassis sensors and other ECU's via bus systems
- enhanced steering functions like electrical damping, active return of the steering wheel to straight line driving, compensation of disturbance torques from road or sidewind by control software
- addition of driver assistance functions like automated parking, lane keeping and trailer back-up assist, and for automated driving.

Examples for EPS systems are shown in Fig. 14.20. A torque sensor in the steering shaft as a measure of the driver's manual torque is transmitted to a specific ECU which determines together with other measurements like the vehicle's speed the power steering assisting torque of the electric motor; see Fig. 14.31.

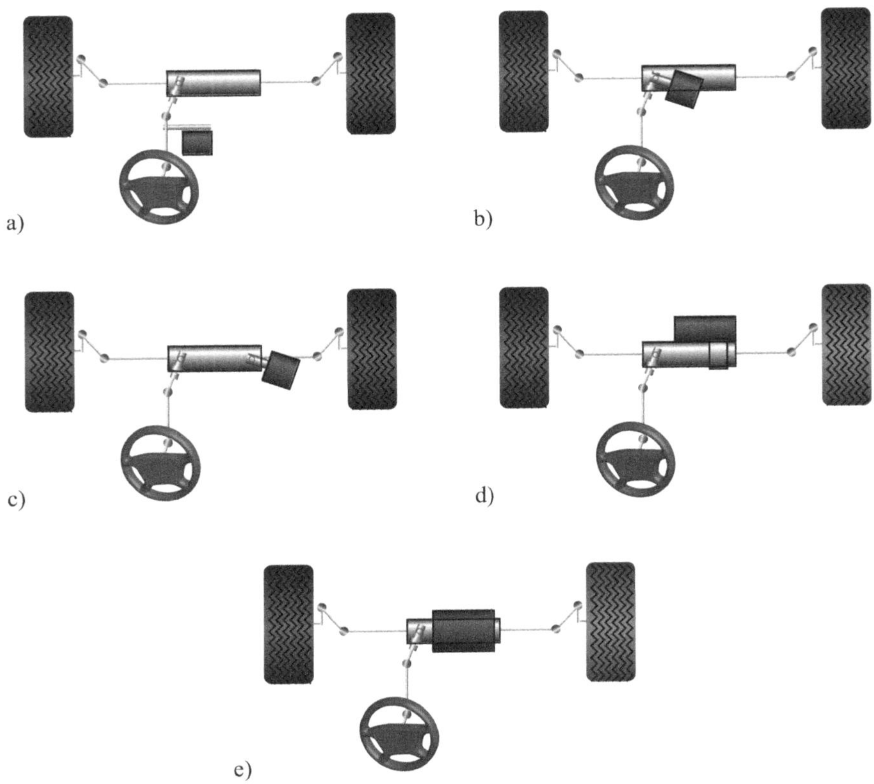

**Fig. 14.20** Different types of electrical power steering systems: **a** column-type **b** pinion-type **c** dual pinion-type **d** axis-parallel-type **e** rack-centered axis-parallel type, Kessler (2021)

### 14.4.1 Basic Designs of EPS Systems

Depending on the arrangement of the electrical motor at the column or at the rack with different gears following EPS-type can be distinguished;see, e.g. Reimann et al. (2016), Pfeffer and Harrer (2013) and Fig. 14.20.

*Column-type EPS*, consisting of the torque sensor, the electric motor, and a reduction gear, is arranged directly at the steering column inside the vehicle. The ECU is installed separately or attached to the motor or torque sensor. The reduction gear can be a worm-gear, allowing sufficient back-turning capability. This type is usually applied for light vehicles and uses the existing rack and pinion steering system.

*Pinion-type EPS* (single pinion) may also use the existing rack and pinion system. The electric motor with ECU is connected to the pinion by a worm-gear. This enables more direct connection to the rack, but undergoes harder environmental influences (temperature, splash water).

*Dual pinion-type EPS* acts with a second pinion on an additional rack, e.g. by a worm gear and allows more freedom for the integration in the vehicle. It also contains a gear set and the ECU.

*Axis parallel-type EPS* (APA) have the electric motor arranged parallel to the rack and transmit the rotary movement to a straight rack movement by a ball screw which is driven by a tooth belt and the electric motor. The transfer of forces from the ball nut to the rack is performed by a chain of hardened steel balls with ball screw threads. The gear set operates without backlash. The advantages are high precision and suitability also for heavier vehicles.

*Axis parallel-rack-type EPS* (RC: rack concentric) have the ball nut of the ball screw without additional gear directly operating on the rack. The electrical motor has then a hollow shaft where the rack ball screw passes through. This very direct, stiff design requires a high torque motor and precise control.

The generated rack forces range from about 6000N for the column-type to 15000N for the axis-parallel-type EPS, Runge et al (2009) with electrical motors power from 200W to 1000W.

## 14.4.2 Components of EPS Systems

### (a) Electric Motors and Gears

The required torque/speed characteristic for the electric motors has to fulfill a high torque at relatively low speed for parking maneuvers, small torque and low speed for normal driving and medium torque and high speed during evasion maneuvers. Hence, a torque/speed characteristic as shown in Fig. 14.21 with a part for $M_{\text{m,max}} = \text{const}$ for low speed and $P_{\text{el,max}} = M_{\text{m,max}}\omega_{\text{M}}$ for medium and high speed meets these goals, Pfeffer and Harrer (2013).

In addition the electric motor has to posses a steady torque generation without ripples, high power density and efficiency, small braking torque in case of short-cuts or current-off situation, robust design, maintenance-free for the whole lifetime, etc. Suitable types are direct current (DC) motors, asynchronous and synchronous alternating current (AC) or induction motors; see, e.g. Binder (2017).

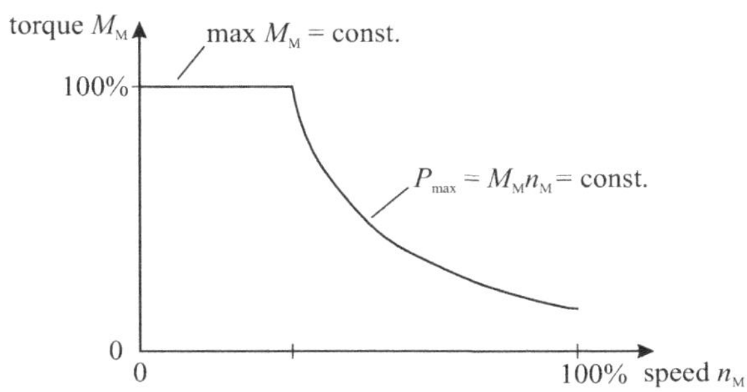


**Fig. 14.21** Required torque/speed characteristics for EPS-motors

*DC motors* with brush commutation, stator windings, and commutated rotor coils have the advantage that their speed can be directly be controlled by the DC power supply voltage. However, because of the mechanical commutation and therefore restricted lifetime and electrical disturbances (sparking) they are not suited for EPS. Better properties show *brushless DC motors* with a permanent magnet rotor and several stator windings which are electronically switched depending on the rotor position.

*Alternating current motors* consist of at least three stator windings which are supplied by a three-phase voltage system. In the case of an *asynchronous motor* the rotor is designed as a squirrel cage consisting of bars in axial direction and is short-circuited at the front and backside via end-rings. Hence, this represents a robust design, needs permanent magnets, and has the advantage that no generator operation with disturbing braking torque results for a failure of the stator excitation. The speed can be controlled by a field-oriented control; see, e.g. Leonhard (1996), Binder (2017), or Isermann (2005).

*Synchronous motors* have also stator windings supplied by a three-phase sinusoidal voltage system. The rotor has axially permanent magnets consisting of rare earth magnetic materials such as Sm-Co. It rotates synchronously with the stator magnetic field and by changing the stator voltage frequency the speed of the rotor can be controlled, also with a field-oriented control system. These permanently excited synchronous motors (PMSM) are most frequently used for middle and high-segment passenger cars. Asynchronous motors need a rotor angular speed sensor and synchronous motors for the measurement of the rotor angle.

The *EPS-gear* has the task to transmit the rotary motion of the electric motor into a translatoric motion of the rack, with an appropriate transfer of the motor torque to the rack force and a suitable speed ratio in the range of 15...30. Further requirements are static and dynamic robustness, no self-inhibition, high efficiency, small or no backlash, maintenance-free over lifetime. The dominating types are worm drives, recirculation ball, and belt transmissions.

Worm drives have a relatively large sliding area and allow a continuous and noise-free transmission. However, they have a higher friction. Recirculating ball gears have a direct force transfer through the steel balls to the spindle of the rack. They have a high mechanical load capacity, large gear ratios, and less friction. Belt transmissions consist of a toothed driving and driven pulley and allow a transmission ratio of 2...4. They need a certain pretension and have no slip. They are in axis parallel configuration frequently combined with recirculation ball gears. Figure 14.22 depicts an example.

**(b) Sensors**

The *torque sensor* for measurement of the manual torque is built in the steering shaft. One distinguishes measurement principles with or without torsion bar. Mostly applied is the measurement of the twist angle of a torsion bar. As a torsion bar influences the stiffness of the steering system, its stiffness should be relatively large. Usual designs have a torsion stiffness of 2 to 2.5 Nm/° and measurement range of $\pm$ 8 to $\pm$ 10 Nm. As this measurement plays a central, highly safety-relevant role it must be very reliable, robust, and fault-tolerant.

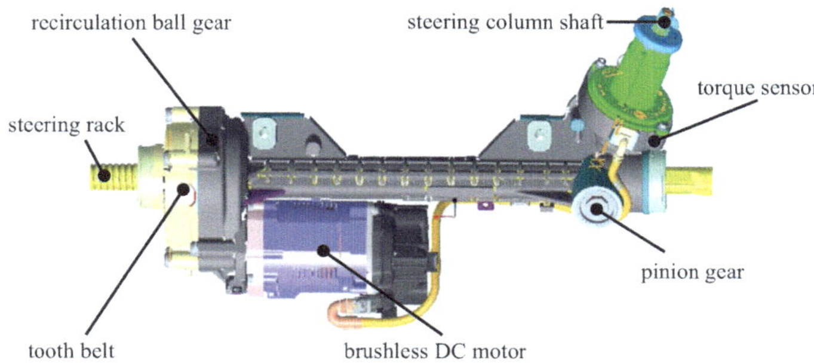


**Fig. 14.22** Axis-parallel EPS with PMSM electric motor, tooth belt, recirculation ball gear, rack spindle. Source: Servo electric, ZF-Lenksysteme (2010), $I_{\max} = 80\text{A}$, $P_{\text{el},\max} = 960\text{W}$, $F_{\text{rack},\max} = 12000\text{N}$

The twist angle between the two ends of a torsion bar can be measured based on potentiometric, inductive, magnetic, or optical principles; see, e.g. Pfeffer and Harrer (2013). Inductive torque sensors use, e.g. the change of an air gap of magnetic, toothed rings which leads to a change of the impedance of a surrounding coil supplied with an alternating current.

Magnetic sensors operate in a static magnetic field provided by a permanent magnet. One end of the torsion bar is, e.g. provided with a multiple pole magnetic ring, the other end with a magnetoresistive sensor, where the electrical resistance of an isotropic element is changed by the magnetic field; see Bosch GmbH (2018).

The measurement of the *steering angle* requires a measurement principle for the full steering angle range of about $\pm 2 \times 360° = \pm 720°$. This is, e.g. realized by a dual configuration of rotation sensors using the Hall effect. Two associated permanent magnets on small tooth wheels are driven by a larger diameter tooth wheel attached to the steering shaft. The two smaller gear wheels differ by one tooth. Therefore, their difference angle is a measure for the absolute angular position $\delta_H$. This arrangement allows to obtain an accuracy of about $1°$, Bosch GmbH (2018).

### (c) Electric Control Unit (ECU)

The ECU for an EPS is usually directly connected to the electromotor actuator in the case of pinion-and rack-type EPS. For column-type EPS it can also be separated. The direct connection with plugin connection has advantages like short cable losses, integration of the rotor angle measurement, and easy assembly. However, it must be extremely well be protected against water, oil, salt and high temperatures.

The principle design of an EPS-ECU is depicted in Fig. 14.23. The microcontroller receives sensor measurements of the steering shaft torque, the steering shaft angle, and, e.g. the rotor position of the EPS-motor in order to calculate the required assistance torque of the electromotor. It may also contain a speed control in the case of DC motors or a field-oriented control of the rotor position for synchronous motors.

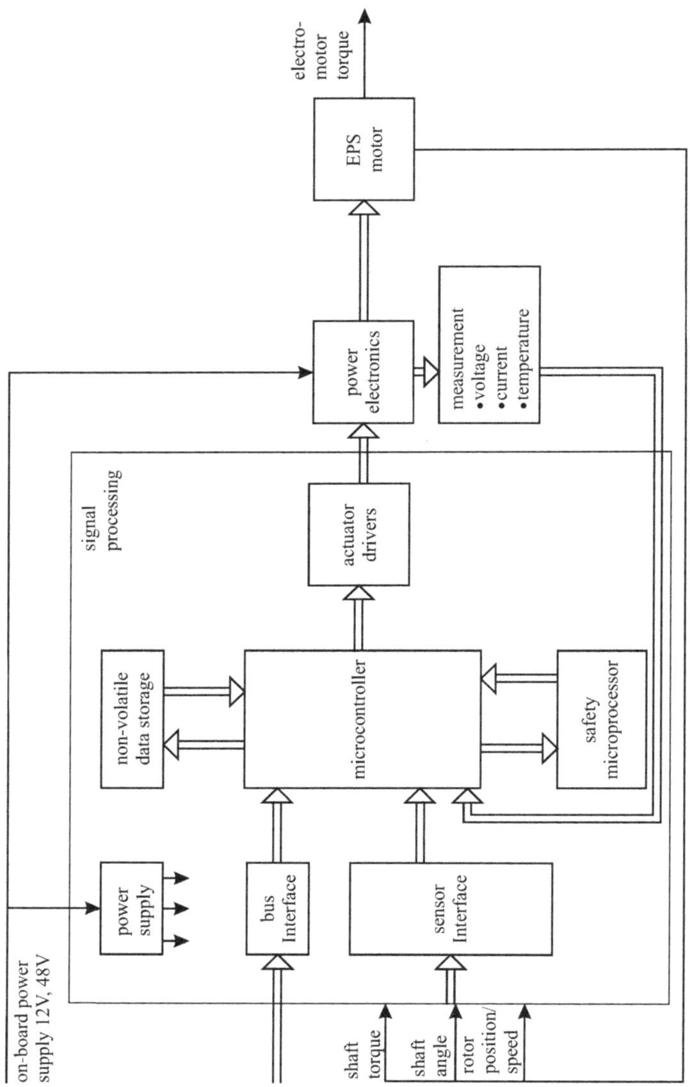


**Fig. 14.23** Simplified block diagram of an electronic control unit (ECU) for an EPS-steering, leaning on Pfeffer and Harrer (2013)

The microcomputer is also connected with a vehicle bus system (CAN or Flexray) in order to obtain measurements of the vehicle speed and signals from other driver assistance systems.

The microcontroller shows the principle design with CPU, different storages like RAM and ROM and A/D-conversion; see, e.g. Borgeest (2014), Bosch GmbH (2018), Isermann (2005). The programming language may be C, and use of MISRA-rules, OSEX-VDX standards, and AUTOSAR-architecture; see Sect. 2.3.

The outputs of the microcontroller are calculated manipulated variables for the electromotor, like current and motor position. For example, pulse-width modulated (PWM) signals are given to the power electronics, where the DC-voltage of the board net (12V or 48V) is controlled by switching to a certain mean value of a current. This is reached by multiple-quadrant converters with MOSFET or IGBT-transistors; see, e.g. Bose (1997), Erickson (1997), Pressman (1997), Isermann (2005).

Because of the very high safety requirement of EPS systems, an integrated monitoring and supervision system with fault-detection plays an important role. Therefore a fault-management system is implemented as software, making also use of a second safety microprocessor, which supervises the main microcomputer. In the case of faults and failures the EPS-motor is switched off, allowing then conventional manual steering.

### 14.4.3 Dynamic Models of Electrical Power Steering Systems (EPS)

For the design of the electrical power assistance system a relation between the electromotor command, the stator current $I_S$ and the power assistance force $F_{PA}$ is needed.

#### (a) Steering Shaft, Rack, and Wheel Suspension Models

As an example an *axis-parallel-type EPS* (APA) with a dual pinion-rack system is considered. Figure 14.24 depicts a schematic representation of the involved components. The permanent excited synchronous motor (PMSM) acts with torque $M_M$ and rotor angle $\varphi_M$ via a toothwheel with radius $r_M$ and a tooth belt on the tooth wheel with radius $r_{BS}$ of a ball screw nut; see Fig. 14.25. The gear ratio of the belt drive follows from $\delta_M = \delta_{BS} r_{BS}$ to

$$i_B = \frac{\text{input angle}}{\text{output angle}} = \frac{\varphi_M}{\delta_{BS}} = \frac{r_{BS}}{r_M} = \frac{M_{BS}}{M_M}. \tag{14.4.1}$$

The rack is in the considered case a screw spindle with radius $r_{RS}$ and gradient angle $\epsilon$. This gives $\tan\epsilon = r_{BS}/r_{RS}$. As for the spindle travel way holds $z_{R1} = \delta_{BS} r_{BS} tan\epsilon$ the recirculation ball screw gear ratio becomes

$$i_{BS} = \frac{\delta_{BS}}{z_{R1}} = \frac{1}{r_{BS} tan\epsilon}. \tag{14.4.2}$$

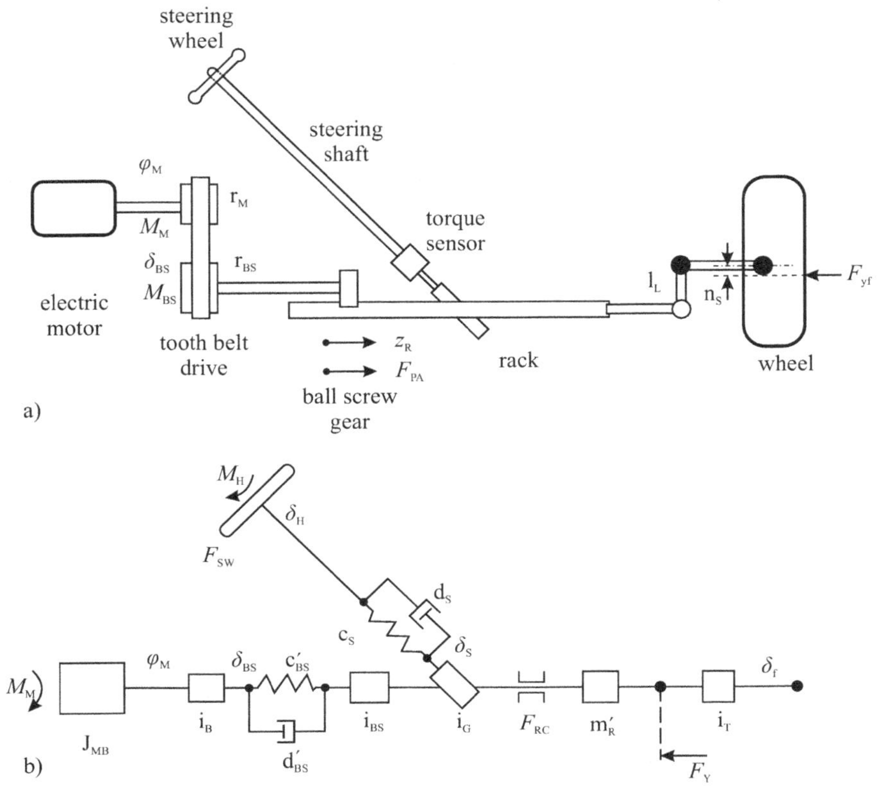


**Fig. 14.24** Schematic representation of a dual pinion axis-parallel-type EPS: **a** main components; **b** simplified representation with two lumped mass-spring-damper elements in the steering shaft and in the ball screw gear and a representative mass $m'_R$

The overall ratio of the belt drive from the motor angle to the rack traveling is then

$$i_{BD} = \frac{\varphi_M}{z_{R1}} = i_B i_{BS}. \tag{14.4.3}$$

Dynamic models of belt drives are treated, e.g. in Isermann (2005).

Experiments on a test stand have shown that the tooth belt can be in this case assumed as stiff; see Beck (2010). The ratio of inertia of the electric motor and the belt screw drive can therefore be lumped together and the enlarged inertia for the acceleration $\ddot{\delta}_M$ of the electric motor becomes

$$J_{MB} = J_M + \frac{1}{i_B^2} J_{MB}. \tag{14.4.4}$$

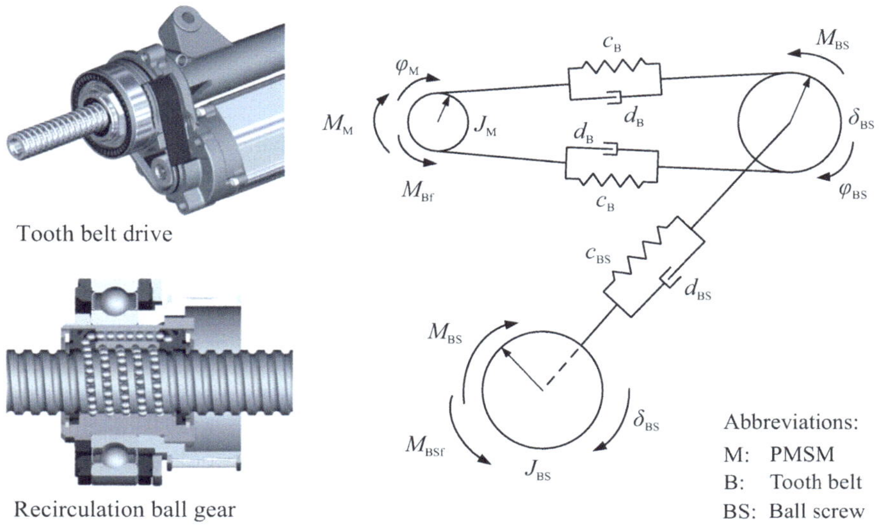


**Fig. 14.25** **a** Tooth belt drive and recirculation ball gear, ZF-Lenksysteme (2010). **b** Schematic representation of the torsional mass spring-damper element of the belt and ball screw gear

The torque balance for the electric motor with the coupled stiff tooth belt then yields

$$J_{MB}\ddot{\varphi}_M = M_M(t) - \frac{1}{i_B} M_{BS}(t) \tag{14.4.5}$$

where $M_{BS}$ is a load torque from the ball screw belt pulley. The load torque $M_{BS}$ of the ball screw depends on the rack force $F_{PA}$ which is transmitted by the EPS drive. For the screw spindle the mechanical work without considering friction losses is

$$M_{BS}\delta_{BS} = F_{PA}z_{R1} \tag{14.4.6}$$

and with (14.4.2) one obtains

$$M_{BS} = \frac{1}{i_{BS}} F_{PA}. \tag{14.4.7}$$

For dynamic modeling the EPS-system from the motor to the turning front wheels, it is now assumed that the motor and the rack spindle with the wheel are rigid components and that the elasticity or stiffness of the whole system is lumped into the belt drive and ball screw gear with a stiffness factor $c'_{BS}$ and a viscous friction factor $d'_{BS}$.

A torque balance of a first system consisting of the electrical motor and the belt and ball screw drive then yields.

$$J'_{MB}\ddot{\delta}_{BS} = i_B M_M - c'_{BS}(\delta_{BS} - i_{BS}z_R) - d'_{BS}(\dot{\delta}_{BS} - i_{BS}\dot{z}_R)$$

or

$$J'_{\text{MB}}\ddot{\delta}_{\text{BS}} + d'_{\text{BS}}\dot{\delta}_{\text{BS}} + c'_{\text{BS}}\delta_{\text{BS}} = i_{\text{B}}M_{\text{M}} + d'_{\text{BS}}i_{\text{BS}}\dot{z}_{\text{R}} + c'_{\text{BS}}i_{\text{BS}}z_{\text{R}}. \tag{14.4.8}$$

This equation takes into account that the output of the ball screw gear is coupled with the rack due to (14.4.2). For the lumped ratio of inertia holds now

$$J'_{\text{MB}} = J_{\text{BS}} + i_{\text{B}}^2 J_{\text{M}}. \tag{14.4.9}$$

The power assist force of the EPS-drive spring-damper system is

$$F_{\text{PA}} = i_{\text{BS}}(c'_{\text{BS}}(\delta_{\text{BS}} - i_{\text{BS}}z_{\text{R}}) + d'_{\text{BS}}(\dot{\delta}_{\text{BS}} - i_{\text{B}}\dot{z}_{\text{R}})). \tag{14.4.10}$$

A second dynamic system consists of the rack, linkage to the wheels, and turning wheels. Its mass is lumped into a representative mass $m'_{\text{R}}$ according to (14.1.15). The friction of the spindle rack is taken into account by a viscous friction coefficient $d'_{\text{R}}$ and dry friction coefficient $f'_{\text{RC}}$.

The dynamic behavior of the rack then results from a force balance as (14.1.16)

$$m'_{\text{R}}\ddot{z}_{\text{R}} = F_{\text{PA}} - d'_{\text{R}}\dot{z}_{\text{R}} - f'_{\text{RC}}\,\text{sign}\,\dot{z}_{\text{R}} - F_{\text{Y}} \tag{14.4.11}$$

with the load force from the front tires for cornering with higher velocity, see (14.1.36), with $f^{M}_{\text{ZT},v} \approx 0$ and $f^{M}_{\text{ZT,c}} \approx 0$

$$F_{\text{Y}} = \frac{2}{l_{\text{L}}} c_{\alpha} c_{\delta} n_{\text{S}} \delta_{\text{f}} \tag{14.4.12}$$

and for small velocity, see (14.1.36) with $f^{M}_{\text{ZT},v} \approx 0$

$$F_{\text{Y}} = \frac{2}{l_{\text{L}}}(c_{\alpha} c_{\delta} n_{\text{S}} \delta_{\text{f}} + f_{\text{ZT,c}}\,\text{sign}\,\dot{\delta}_{\text{f}}). \tag{14.4.13}$$

A signal flow scheme on the basis of these equations is depicted in Fig. 14.26. It shows that the ball screw gear with the representative lumped parameters $c'_{\text{BS}}$ and $d'_{\text{BS}}$ is located between the motor with the elastic belt drive and the rigid spindle rack with representative mass $m'_{\text{R}}$ and friction parameters $d'_{\text{R}}$ and $f'_{\text{RC}}$.The turning front wheels generate during cornering with steering angle $\delta_{\text{f}}$ the lateral load force $F_{\text{Y}}$ on the rack which results in a feedback path of the rack. Hence, the EPS-system consists of two serially coupled second-order dynamic systems with several feedback paths.

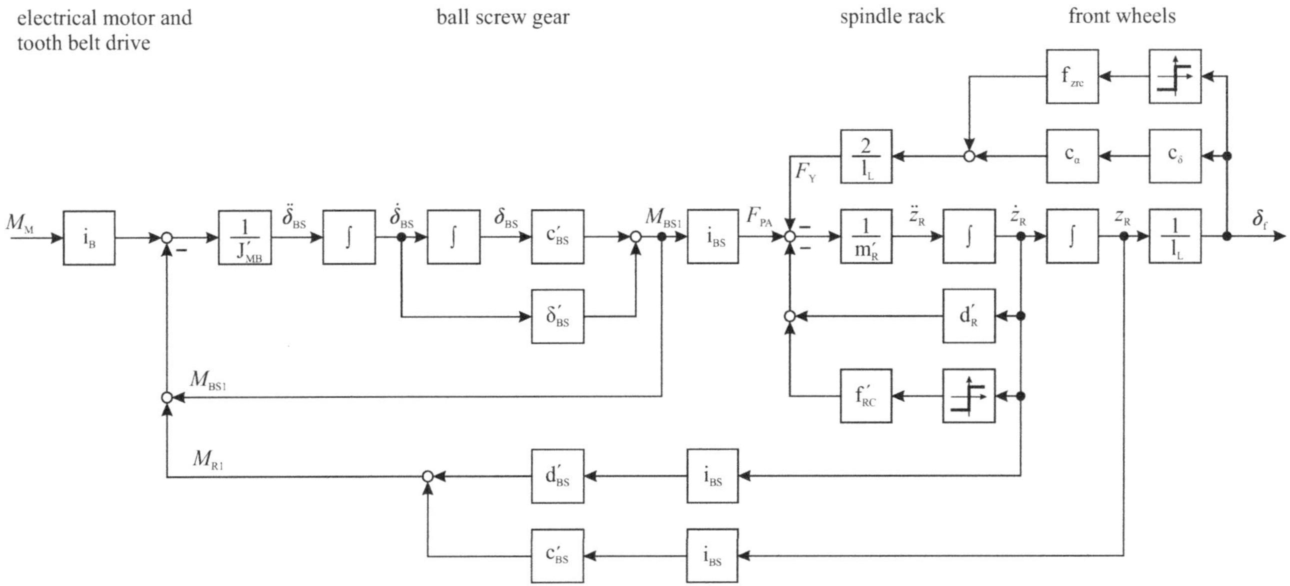


**Fig. 14.26** Signal flow scheme of an axial-parallel-type EPS-system with electrical motor, tooth belt drive, ball screw gear, spindle rack, and front wheels for cornering with steering angle $\delta_f$

### (b) Fixed Rack Spindle

On steering test benches, it is possible to fix the rack in order to make dynamic investigations. Then it follows with $z_R = 0$, $\dot{z}_R = 0$ from (14.4.8) for the transfer function, assuming small changes

$$G_M(s) = \frac{\Delta\delta_{BS}(s)}{\Delta M_M(s)} = \frac{i_B}{J'_{MS}s^2 + d_{BS}s + c_{BS}}. \tag{14.4.14}$$

This second-order oscillating system has a natural angular frequency

$$\omega_{oBS} = \sqrt{\frac{c'_{BS}}{J'_{MS}}}, \tag{14.4.15}$$

a damping ratio

$$\zeta = \frac{d'_{BS}}{2}\frac{1}{\sqrt{c'_{BS}J_{MB}}'} \tag{14.4.16}$$

and a resonance frequency

$$\omega_{res,BS} = \omega_{oBS}\sqrt{1-2\zeta^2} = \sqrt{\frac{c'_{BS}}{J'_{MB}} - \frac{d'^2_{BS}}{2J'^2_{MB}}}, \tag{14.4.17}$$

see, e.g. Isermann (2005).

Figure 14.27 depicts the results of an experimental determination of a frequency response by least squares parameter estimation and excitation of the electric motor torque with a pseudo-random binary-signal (PRBS), Beck (2010), Kessler (2021). The investigated axis-parallel EPS, shown in Fig. 14.22, was mounted on a special test bench and the rack spindle could be blocked for this experiment.

The used z-transfer function for parameter estimation is

$$G'_M(z) = \frac{\Delta\varphi_M(z)}{\Delta M_M(z)} = \frac{i_B\Delta\delta_{BS}(z)}{\Delta M_M(z)} = \frac{b_1z^{-1} + b_2z^{-2}}{1 + a_1z^{-1} + a_2z^{-2}}. \tag{14.4.18}$$

Figure 14.27a shows the PRBS excitation signal of the electric motor torque $M_M$ and a comparison of the measured and with the estimated model calculated output, which is in this case the rotor angle of the electric motor. The maximal error is $\Delta\delta_M \approx 0.05\,\text{rad} = 2.9°$ and can be considered as a good agreement. The resulting frequency response in Fig. 14.27c indicates also a good agreement of measured and with the estimated model simulated values. It illustrates that the EPS drive with fixed rack spindle can be described well by a second-order dynamic model (14.4.14). Its resonance frequency according to (14.4.17) is $\omega_{res,\delta M} \approx 80\ 1/\text{s} \approx 13\,\text{Hz}$.

Frequency response measurements with a sinus-sweep input of the rack force and freely oscillating steering wheel for a similar EPS-steering system on a test bench

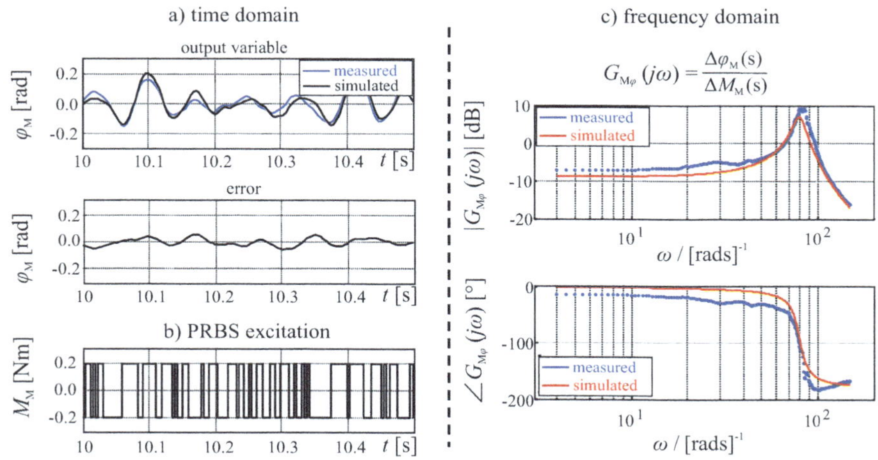


**Fig. 14.27** Parameter estimation of an EPS-unit (Fig. 14.22) with fixed rack spindle on a test bench, Beck (2010), Kessler (2021). **a** rotor angle **b** electric motor PRBS excitation signal **c** frequency response $G_{M^{\cdot}}(s) = \Delta\varphi_M(s)/\Delta M_M(s)$

in Fankem et al. (2014) show a natural frequency of 7.5Hz for the motor, belt drive and ball gear and about 1Hz for the steering wheel and shaft. Frequency response measurements are also published in Düsterloh et al. (2018).

Assuming that the rack spindle is stiff compared to the recirculation ball screw gear it follows from (14.4.10) with $z_R = 0$ and $\dot{z}_R = 0$ for the rack force

$$\begin{aligned} F_{PA}(t) &= i_{BS} M_{BS}(t) \\ &= i_{BS}(c'_{BS}\delta_{BS}(t) + d'_{BS}\dot{\delta}_{BS}(t)), \end{aligned}$$

or with Laplace-transform for small changes

$$\Delta F_{PA}(s) = i_{BS}(c'_{BS} + d'_{BS}s)\Delta\delta_{BS}(s). \tag{14.4.19}$$

Using (14.4.14), it follows for the rack force with the ball screw drive and motor dynamics

$$G_{MFPA}(s) = \frac{\Delta F_{PA}}{\Delta M_M(s)} = \frac{i_B i_{BS}(c'_{BS} + d'_{BS}s)}{J'_{MB}s^2 + d'_{BS}s + c'_{BS}}. \tag{14.4.20}$$

The transfer function for a power assist force is therefore a second-order oscillating system with a derivative term, a so-called $PDT_2$ element and contains a pole pair and a zero and has a natural frequency (14.4.15).

These discussions show that the models for the EPS-drive-train are relevant for frequencies of about $f > 8$Hz. Therefore they are mainly required for the design of the EPS-system with regard to oscillations and its control.

If for the lateral vehicle dynamics the lower frequency part $f < 8\text{Hz}$ of the transfer function (14.4.20) is of interest and if it is well damped it may be approximated by a first-order system.

$$G_{\mathrm{MFPA}}(s) = \frac{\Delta F_{\mathrm{PA}}(s)}{\Delta M_{\mathrm{M}}(s)} \approx \frac{i_{\mathrm{B}} i_{\mathrm{BS}}}{(1 + T_{\mathrm{BS}} s)} \tag{14.4.21}$$

with a time constant taken from the cut-off (corner) frequency of (14.4.15)

$$T_{\mathrm{BS}} = \frac{1}{\omega_{\mathrm{oBS}}} = \sqrt{\frac{J'_{\mathrm{MB}}}{c'_{\mathrm{BS}}}}. \tag{14.4.22}$$

Using $\omega_{\mathrm{oBS}} \approx \omega_{\mathrm{res},\delta\mathrm{M}}$ it holds $T_{\mathrm{BS}} = 1/80 = 125$ ms for the EPS-system of the experiments of the Fig. 14.27.

Table 14.2 presents some values of parameters for EPS-system, taken from literature.

**(c) Electric Motor and Power Assist Force Models**

The torque of a permanently excited synchronous motor (PMSM) is, see, e.g. Vas (1990), Leonhard (1996), Isermann (2005), Schröder (1995), Binder (2017).

$$M_{\mathrm{M}} = \frac{3}{2} \mathrm{p} \Psi_{\mathrm{Rd}} I_{\mathrm{sq}} = \Psi' I_{\mathrm{sq}} \tag{14.4.23}$$

where p is the number of pole pairs, $\Psi_{\mathrm{Rd}} = \Psi_{\mathrm{R}}$ the rotor flux and $I_{\mathrm{sq}}$ the q-current component of a field-oriented control. The dynamic behavior of the q-current after a change of the q-voltage is

$$G_{\mathrm{UI}}(s) = \frac{\Delta I_{\mathrm{sq}}(s)}{\Delta U_{\mathrm{sq}}(s)} = \frac{K_{\mathrm{s}}}{1 + T_{\mathrm{s}} s} \tag{14.4.24}$$

with $K_{\mathrm{s}} = 1/R_{\mathrm{s}}$ and $T_{\mathrm{s}} = L_{\mathrm{s}}/R_{\mathrm{s}}$ where $R_{\mathrm{s}}$ and $L_{\mathrm{s}}$ are the resistance and inductance of the stator. Using a PI-controller with the input voltage $U_{\mathrm{sq}}$ as manipulated variable and gain $K_{\mathrm{c}}$ for q-current controller and pole-zero cancelation results in

$$G_{\mathrm{II}}(s) = \frac{\Delta I_{\mathrm{sq}}(s)}{\Delta I_{\mathrm{sq,d}}(s)} = \frac{1}{1 + T_{\mathrm{q}} s} \tag{14.4.25}$$

with $I_{\mathrm{sq,d}}$ a desired value and $T_{\mathrm{q}} = L_{\mathrm{s}}/K_{\mathrm{c}}$; see, e.g. Isermann (2005).

Hence, the closed-loop current control loop and therefore the torque behavior can be described by the first-order lag behavior (14.4.25) and for the torque model one obtains with (14.4.23)

$$G_{\mathrm{IM}}(s) = \frac{\Delta M_{\mathrm{M}}(s)}{\Delta I_{\mathrm{sq,d}}(s)} = \frac{K_{\mathrm{M,PMSM}}}{1 + T_{\mathrm{q}} s} \tag{14.4.26}$$

**Table 14.2** Parameters of EPS systems according to Düsterloh et al. (2018), Fankem et al. (2014), Fankem (2017), Mitschke and Wallentowitz (2004)

| Parameter | | Dimension | Value |
|---|---|---|---|
| Steering wheel inertia | $J_{SW}$ | $kgm^2$ | 0.033...0.05 |
| Steering shaft stiffness | $c_S$ | Nm/° | (4.4)...(4.8) |
| | | Nm/rad | 252.1...275 |
| Steering shaft damping | $d_S$ | Nm/° | 0.0008...0.0018 |
| | | Nm/rad | 0.045...0.103 |
| Torsion bar damping | $d_M$ | Nm/° | 0.0055...0.0135 |
| | | Nm/rad | 0.315...0.77 |
| Steering shaft friction | $f_S$ | Nm | 0.2 |
| Steering system stiffness | $c_{st}$ | Nm/° | 90...350 |
| Torque sensor stiffness | $c_{TS}$ | Nm/rad | 75 |
| Torque sensor damping | $d_{TS}$ | Nm/rad | $1.6 \times 10^{-4}$ |
| Rack-pinion ratio | $i_G$ | rad/m | 117 |
| Rack mass | $m_R$ | kg | 3...3.6 |
| Equivalent rack mass | $m_R^{'}$ | kg | 630...2230 |
| Rack dry friction | $f_R$ | N | 150...300 |
| Rack viscous friction | $d_R^{'}$ | Ns/m | 12000 |
| Pinion ratio | $i_P$ | mm/rev | 52...57 |
| Ball screw drive ratio | $i_{BS}$ | rev/mm | 0.11...0.15 |
| | $i_{BS}$ | rad/m | 897 |
| Belt ratio | $i_B$ | | 1...2.85 |
| Belt and ball screw drive | $i_{BS}$ | rad/m | 1514 |
| Belt stiffness | $c_B$ | Nm/rad | 200 |
| Belt damping | $d_B$ | Nms/rad | 0.001 |
| Ball screw drive inertia | $J_{BS}$ | $kgm^2$ | 0.00035...0.0019 |
| Ball screw stiffness | $c_{BS}$ | N/m | $3 \times 10^8$ |
| Ball screw damping | $d_{BS}$ | Ns/m | $1.6 \times 10^4$ |
| Electric motor inertia | $J_M$ | $kgm^2$ | 0.00011...0.00058 |
| Electric motor time constant | $T_M$ | ms | 18 |

with $K_{M,PMSM} = \frac{3}{2}p\Psi_{Rd} = \Psi^{'}$.

In the case of a *DC-motor* with brushes it holds for the torque

$$M_M = \Psi I_A = K_{M,DC} I_A \tag{14.4.27}$$

where $\Psi$ is the magnetic flux and $I_A$ the armature current. For *squirrel cage induction motors* with field-oriented control the torque is

$$M_M = \frac{3}{2}p\frac{M}{L_R}\Psi_{Rd} I_{sq} = K_{M,IM} I_{sq} \tag{14.4.28}$$

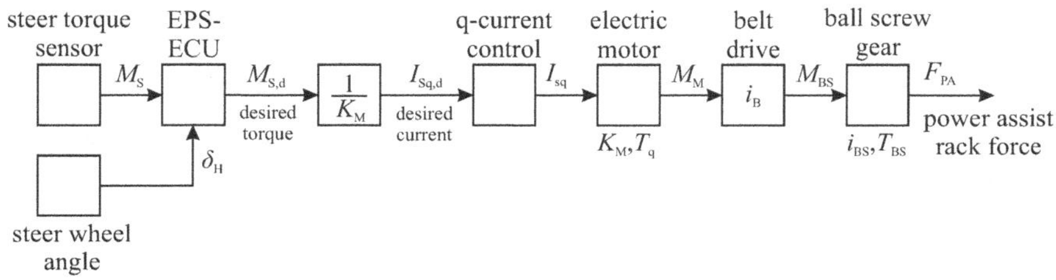


**Fig. 14.28** Simplified signal flow for the generation of the rack power assistance force of an ESP actuator system

where M is the mutual inductance between stator and rotor, $L_R$ the self -inductance of the rotor and $\Psi_{Rd}$ the rotor flux; see, e.g. Vas (1990), Schröder (1995), Leonhard (1996), Isermann (2005), Binder (2017).

The torque of the electrical motors can therefore for all three motor types be described by a first-order transfer function according to (14.4.26) with the appropriate gain factor and time constant.

Combining (14.4.21) and (14.4.26) leads to a transfer function of the power assist rack force

$$\begin{aligned} G_{\mathrm{IFPA}}(s) &= \frac{\Delta F_{\mathrm{PA}}(s)}{\Delta I_{\mathrm{sq,d}}(s)} = G_{\mathrm{IM}}(s) G_{\mathrm{MFPA}}(s) \\ &= \frac{K_{\mathrm{M}} i_{\mathrm{B}} i_{\mathrm{BS}}}{(1 + T_{\mathrm{q}} s)(1 + T_{\mathrm{BS}} s)} \end{aligned} \tag{14.4.29}$$

which includes the dynamics of the electrical motor, the belt drive and recirculating ball screw gear.

Figure 14.28 depicts a corresponding signal flow. Hence, the rack force results after a change of the q-current setpoint approximately delayed by a second-order low pass system with small time constants. The motor time constant is about $T_q \approx$ 20ms and the belt drive ball screw drive time constant is due to Fig. 14.27c and (14.4.22) estimated to $T_{BS} \approx$ 100ms, and the sum of time constants about $T_{qBS\Sigma} = T_q + T_{BS} \approx$ 120ms or a cut-off frequency $\omega_{c,qBS\Sigma} \approx 52$ 1/s.

A further simplification leads to a first-order system with a sum time constant $T_{qBS\Sigma}$

$$G_{\mathrm{IFPA}}(s) = \frac{\Delta F_{\mathrm{PA}}(s)}{\Delta I_{\mathrm{sq,d}}(s)} \approx \frac{K_{\mathrm{M}} i_{\mathrm{B}} i_{\mathrm{BS}}}{(1 + T_{\mathrm{qBS}\Sigma} s)}. \tag{14.4.30}$$

If a low frequency behavior with $f <$ 5Hz is of interest this time constant may be neglected and a proportional behavior for the EPS-generated rack force can be assumed

$$\begin{aligned} \Delta F_{\mathrm{PA}}(t) &= K_{\mathrm{M}} i_{\mathrm{B}} i_{\mathrm{BS}} \eta_{\mathrm{BS}} \Delta I_{\mathrm{sq,d}}(t) \\ &= K_{\mathrm{IF}} \Delta I_{\mathrm{sq,d}}(t). \end{aligned} \tag{14.4.31}$$

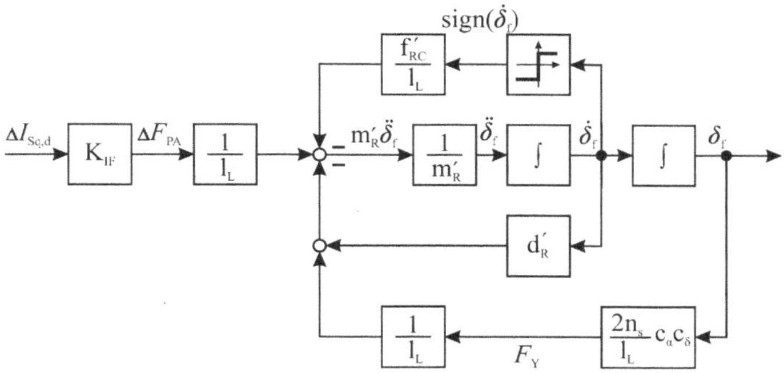


**Fig. 14.29** Signal flow of the rack-suspension system with the EPS-motor current setpoint as input

where an efficiency factor $\eta_{BS}$ takes into account friction losses in the bearings, belt drive and ball screw gear.

### (d) EPS-Drive and Rack-Suspension System

The force balance of the rack with the representative mass $m'_R$ due to (14.1.15) and the power assist force $F_{PA}$ is according to (14.1.16), by neglecting a torque from the steering wheel and the rack force $F_{R1}$ and thus the coupled steering shaft and steering wheel

$$m'_R \ddot{z}_R = F_{PA} - d'_R \dot{z}_R - f'_{RC} \operatorname{sign} \dot{z}_R - F_Y. \tag{14.4.32}$$

Inserting $z_R = l_L \delta_f$ and the feedback force from the tires during cornering $F_Y = 2F_{YTfi}$ due to (14.1.27) leads to

$$m'_R \ddot{\delta}_f + d'_R \dot{\delta}_f + \frac{2n_S}{l_L^2} c_\alpha c_\delta \delta_f + \frac{f'_{RC}}{l_L} \operatorname{sign} \dot{\delta}_f = \frac{F_{PA}}{l_L}. \tag{14.4.33}$$

Figure 14.29 depicts a corresponding signal flow. Hence, the EPS-motor drives a second-order oscillating system with a dry friction of the rack and suspension.

Neglecting the dry friction by setting $f'_{RC} = 0$, results in a transfer function for small changes

$$G_{F\delta}(s) = \frac{\Delta\delta_f(s)}{\Delta F_{PA}(s)} = \frac{b'_0}{a'_2 s^2 + a'_1 s + 1} \tag{14.4.34}$$

with

$$a'_2 = \frac{m'_R l_L^2}{2n_S c_\alpha c_\delta} \quad b'_0 = \frac{l_L}{2n_S c_\alpha c_\delta}$$
$$a'_1 = \frac{d'_R l_L^2}{2n_S c_\alpha c_\delta}$$

Hence, a linear second-order system results with gain

$$K_{F\delta} = b_0^{'} = \frac{l_L}{2n_S c_\alpha c_\delta}. \quad (14.4.35)$$

and natural angular frequency

$$\omega_{oF\delta} = \sqrt{\frac{1}{a_2^{'}}} = \sqrt{\frac{2n_S c_\alpha c_\delta}{m_R^{'} l_L^2}}. \quad (14.4.36)$$

with the overall gain

$$K_{I\delta} = \frac{\Delta\delta_f}{\Delta I_{sq,d}} = K_{IF} K_{F\delta} = K_M i_B i_S \eta_{BS} K_{F\delta}.$$

As (14.4.33) and Fig. 14.29 shows, the steering angle $\delta_f$ obtains with regard to the power assist force $F_{PA}$ a proportional acting behavior because of the feedback of the lateral force $F_Y$ of the cornering front tires.

**(e) Overall EPS-Model for Manual Steering**

The EPS-drive consisting of the power electronics, the electromotor, belt drive, and ball screw gear, which generates the power assist force $F_{PA}$ on the steering rack, see Fig. 14.28, acts parallel to the manually actuated mechanical steering system; see Fig. 14.9. The setpoint of the desired motor torque $M_{M,d}$ is given by the programmed torque support characteristic in the EPS-ECU; see Fig. 14.10 as an example.

The generated power assist force $F_{PA}$ then acts on the steering rack as indicated in Fig. 14.4 and the rack force (14.1.16a) becomes

$$F_R = F_{R1}(\delta_H, \delta_S) - F_Y(\delta_f) + F_{PA}(M_{M,d}). \quad (14.4.37)$$

The power assist rack force thus excites the dynamics of the complete EPS-steering system and effects also the feedback torque $M_{R1}$ from the rack to the steering wheel angle $\delta_H$, which is part of the "steering feel".

Figure 14.30 represents an overall signal flow of an EPS-system in a condensed form which results from Figs. 14.4 to 14.28.

The mechanical torque and the force flow of the steering shaft and the pinion-rack gear is now paralleled by an electromechanical torque and force flow from the electrical power assist. Because the measured torque generates an additional rack force $F_{PA}$ to the manually generated rack force $F_{R1}$ the manual torque $M_H$ is reduced for generating a certain steering angle $\delta_f$ according to the programmed torque support characteristics.

Figure 14.31 depicts another more condensed schematic of an EPS-system. It again shows two serially coupled mass-spring-damper systems of second order and the parallel acting electrical power assist motor and belt and ball screw drive. The driver controls the steering wheel angle with the manual torque $M_H$ as manipulated

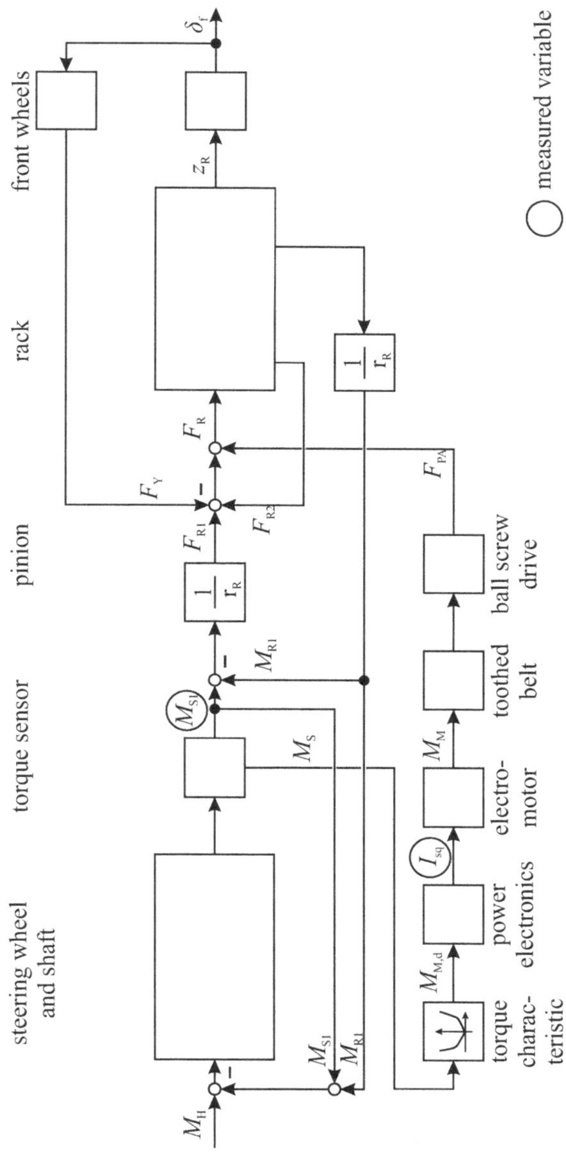


**Fig. 14.30** Block chart of a rack and pinion EPS

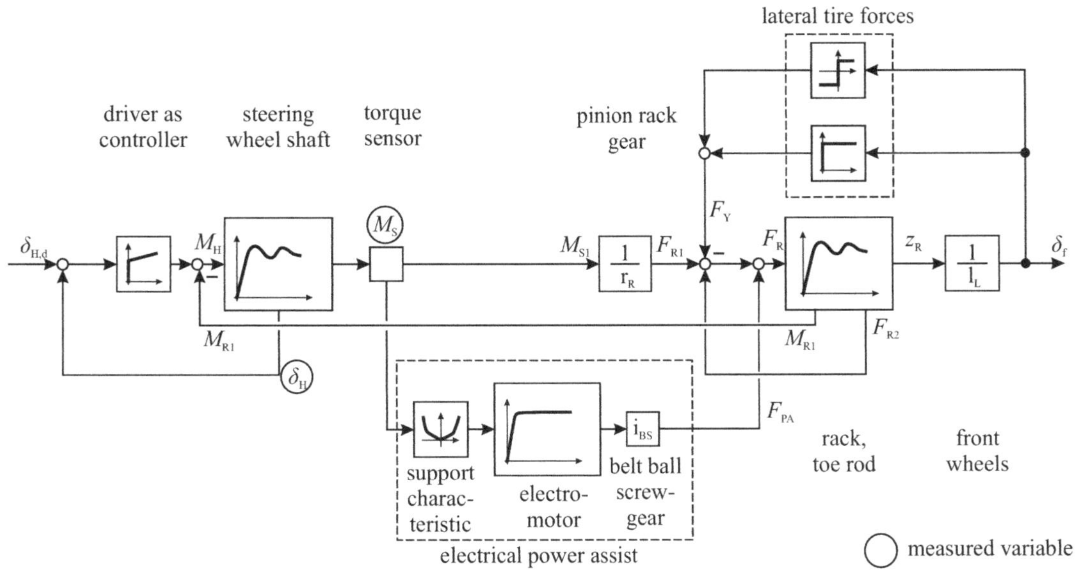


**Fig. 14.31** Schematic overall signal flow scheme of a rack and pinion electrical power steering

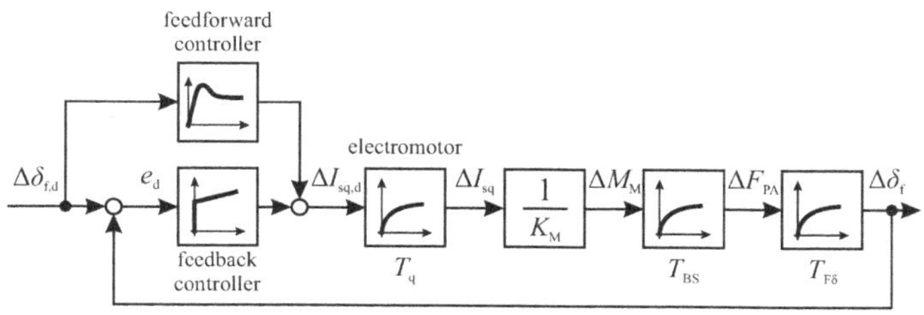


**Fig. 14.32** Signal flow of a linearized steering angle control loop for an EPS steering system

variable and the torque sensor in the steering shaft acts as input to determine the desired motor torque due to the support characteristic $M_M(M_S, v_X)$.

### (f) Steering Angle Control with an EPS System

Automatic lateral vehicle control for lane keeping control (LKC), automatic vehicle control (AC), or drive-by-wire control uses a controller which acts directly on the steering actuator. The output of the steering angle controller then manipulates the desired value $I_{sq,d}$ of the current control loop of the electromotor, which generates the torque $M_M$ acting on the belt drive/gear and rack-suspension system; see Fig. 14.32.

The second-order rack-suspension system (14.4.34) can be considered as well damped and can therefore for low frequencies approximated by a first-order system

$$\tilde{G}_{F\delta}(s) = \frac{\Delta\delta_f(s)}{\Delta F_{PA}(s)} = \frac{K_{F\delta}}{(1 + T_{F\delta}s)} \tag{14.4.38}$$

where $T_{F\delta}$ follows from the undamped natural angular frequency (14.4.36) (cut-off frequency) to

$$T_{F\delta} = \frac{1}{\omega_{0F\delta}} = \sqrt{\frac{m_R' l_L^2}{2 n_S c_\alpha c_\delta}}. \tag{14.4.39}$$

Together with the simplifying assumptions for the belt drive, recirculating ball screw gear and the electrical motor one obtains with (14.4.30), (14.4.31)

$$G_{I\delta}(s) = \frac{\Delta\delta_f(s)}{\Delta I_{sq,d}(s)} = \frac{K_{IF}}{\left(1 + T_{qBS\Sigma}s\right)} \cdot \frac{K_{F\delta}}{(1 + T_{F\delta}s)} \tag{14.4.40}$$

with the gain

$$K_{I\delta} = K_{IF}K_{F\delta} = K_M i_B i_{BS} \eta_{BS} K_{F\delta}. \tag{14.4.41}$$

If the steering angle controller is designed as an internal model controller (IMC); see Appendix A.1.3.2. Example A.2, it holds

$$G_{\delta I}(s) = \frac{\Delta I_{sq,d}}{e_\delta(s)} = \frac{1}{2K_{I\delta}T_r} \frac{\left(1 + T_{qBS\Sigma}s\right)(1 + T_{F\delta}s)}{s\left(1 + \frac{T_r}{2}s\right)} \tag{14.4.42}$$

where $T_r$ is the time constant of a second-order realizability filter (tuning parameter). Selecting $T_r = 2T_{qBS\Sigma}$ leads to a feedback PI-controller

$$G_{\delta I}(s) = \frac{\Delta I_{sq,d}(s)}{e_\delta(s)} = K_{\delta I}\left(1 + \frac{1}{T_{\delta I}s}\right) \tag{14.4.43}$$

with

$$\begin{aligned} K_{\delta I} &= \frac{T_{F\delta}}{4K_{I\delta}T_{qBS\Sigma}} \\ T_{\delta I} &= T_{F\delta}. \end{aligned}$$

The closed-loop transfer function then results in

$$G_{\delta\delta}(s) = \frac{\Delta\delta_f(s)}{\Delta\delta_{F,\delta}(s)} = \frac{1}{4T_{qBS\Sigma}^2 s^2 + 4T_{qBS\Sigma}s + 1}. \tag{14.4.44}$$

For a well-damped controller parameter tuning the steering angle control loop may be approximated by a first-order delay

$$G_{\delta\delta}(s) = \frac{\Delta\delta_f(s)}{\Delta\delta_{F,d}(s)} = \frac{K_{\delta\delta}}{1 + T_{\delta\delta}s} \tag{14.4.45}$$

with

$$\begin{aligned} K_{\delta\delta} &= 1 \\ T_{\delta\delta} &= 2T_{qBS\Sigma}. \end{aligned}$$

The closed-loop behavior of the EPS steering control system behaves then approximately as a first-order delay and its time constant is for $T_{qBS\Sigma} \approx 120\,\text{ms}$, see (14.4.30), $T_{\delta\delta} \approx 240\,\text{ms}$.

The steering angle control may be added by a feedforward control with the reference variable $\Delta\delta_{f,d}$ as input; see Fig. 14.32. According to Appendix A.1.2, Example A.1 one obtains

$$G_{\delta I,ff}(s) = \frac{\Delta I_{sq,d}(s)}{\Delta\delta_{f,d}(s)} = \frac{\left(1 + T_{qBS\Sigma}s\right)(1 + T_{F\delta}s)}{K_{I\delta}\,(1 + T_r s)^2}. \tag{14.4.46}$$

If the time constant of the realizability filter is selected as $T_r = T_{qBS\Sigma}$ it yields

$$G_{\delta I,ff}(s) = \frac{(1 + T_{F\delta}s)}{K_{I\delta}\left(1 + T_{qBS\Sigma}s\right)}. \tag{14.4.47}$$

Thus the feedforward controller acts as a PD element with a time lag, which means a speed adding by-pass element parallel to the feedback controller.

The IMC controller design gives a direct relation of the controller parameters to the constructive data of a steering system; see (14.4.43). For the final implementation test experiments have to be made, resulting eventually in controller parameter modifications with regard to, e.g. steering comfort. Also special measures may be introduced, e.g. for friction compensation; see Isermann (2005).

### 14.4.4 Fault-Tolerant EPS-Structures

In the case of faults and even failures of the electrical power steering system, the driver can take over the required steering torque for small and medium cars. However, for larger cars and light commercial vehicles the EPS should be fail-operational with regard to failures.

For *automatic driving* the driver is out of the loop. Automatic closed-loop systems cover usually smaller additive and parametric faults of the controlled process. However, if the faults become larger, either sluggish, less damped, or unstable behavior may result. The not engaged driver may then not perform the right steering command in time. Therefore, it will be required that EPS systems in the case of automatic controlled driving must have *fail-operational functions* in the case of certain faults, i.e. have to be fault-tolerant.

For most components of the steering actuator system hardware redundancy is required in form of dynamic redundancy with hot or cold standby; see Sect. 22.4, Isermann (2011). This means that a selection of components has to be doubled. Figure 14.33 depicts redundancy structures with different degrees of redundancy. At first the torque sensor can be duplicated or an analytical redundancy concept can be programmed, Schöttler (2007). In a next step, the torque sensor, the ECU, and inverter are duplicated. The arrangement can be cold or hot standby. In the case of a fault in one channel the other channel stays active and the faulty channel is switched off. Case C in Fig. 14.33 has a further redundance in the windings by a multi-phase configuration, Hayashi (2013), Yoneki et al (2013). A serial connection of two motors is shown in Fig. 14.33D. Figure 14.33E depicts a duplication of the complete EPS actuator. In this parallel arrangement also the gear is duplicated, Beck and Isermann (2010), Kessler (2015). The degree of redundancy increases from case A to E, however on cost of hardware extent, installation space, cost, and weight. The selection of the redundancies also depends on fault-statistics for the different components. Case C seems to be a reasonable compromise, however requires a special motor design. In the case of a winding fault the power is reduced. In the cases D and E, the power can be distributed differently to both motors in the normal operating range.

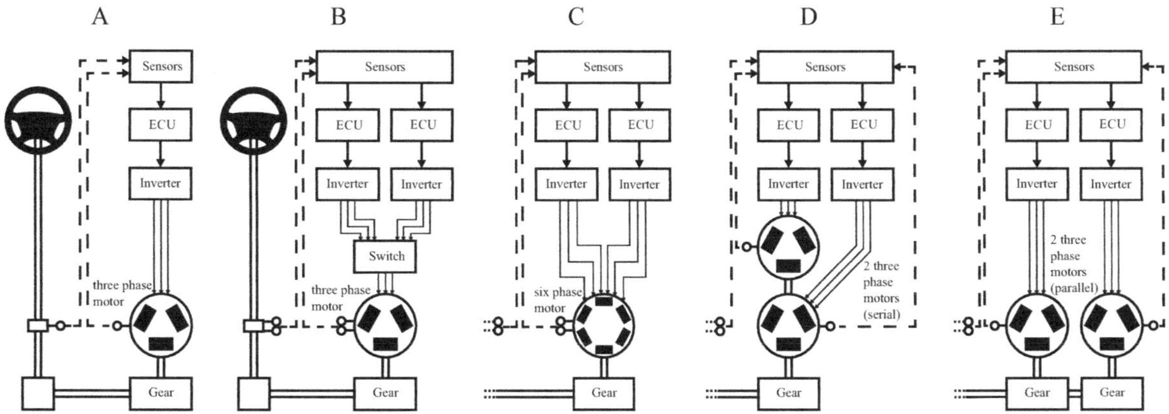


| faults | redundancy | redundancy | redundancy | redundancy | redundancy |
|---|---|---|---|---|---|
| Sensors | - | √ | √ | √ | √ |
| ECU | - | √ | √ | √ | √ |
| Inverter | - | √ | √ | √ | √ |
| Motor | - | - | √ | √ | √ |
| Gear | - | - | - | - | (√) |

**Fig. 14.33** Redundancy structures for EPS systems, Isermann (2017)

## References


Barthenheier T (2004) Potential einer Fahrertyp- und fahrsituationsabhängigen Lenkradmomentgestaltung. Dissertation, TU Darmstadt

Bastow D, Howard G, Whitehead JP (2004) Car suspension and handling. SAE international, Warrendale(USA)

Beck M (2010) Modelling of a duplex electric power steering prototype. In: 4th, iQPC-conference on advanced steering systems, Wiesbaden, Germany

Beck M, Isermann R (2010) Fehlertoleranzstrategien für mechatronische Systeme - Abschlussbericht. Arbeitsgemeinschaft industrieller Forschungsvereinigungen "Otto von Guericke" e.V. (BMWi), Frankfurt

Binder A (2017) Elektrische Maschinen und Antriebe, 2nd edn. Springer Verlag, Heidelberg

Bootz A, Brander B, Stoffel B (2003) Efficiency analysis of hydraulic power steering systems. 8th Scandinavian conference on fluid power

Borgeest K (2014) Elektronik in der Fahrzeugtechnik, 4th edn. Springer Vieweg, Wiesbaden

Bose B (ed) (1997) Power Electronics and Variable Frequency Drives - Technology and Applications. IEEE Press, New York

Braess H, Seiffert H (eds) (2007) Vieweg Handbuch Kraftfahrzeugtechnik, 5th edn. Vieweg, Wiesbaden

Düsterloh D, Uselmann A, Scherhaufer J, Bittner C, Schramm D (2018) Objectification of the feedback behavior of the suspension and steering system. In: 9th international Munich chassis symposium 2018, Springer, pp 505–526

Edge K, Darling J (1986) Cylinder pressure transient in hydraulic oil pumps with sliding plate valves. Proc Inst Mech Eng, Part B: Manag Eng Manufact 200(1):45–54

Erickson R (1997) Fundamentals of power electronics. Chapman and Hall, London

Ersoy M, Gies S (eds) (2017) Fahrwerkhandbuch: Grundlagen, Fahrdynamik, Komponenten, Elektronische Systeme, Fahrerassistenz, Autonomes Fahren, Perspektiven, 5th edn. ATZ/MTZ Fachbuch, Springer Vieweg

Fankem S (2017) Steering feel generation in steer-by-wire vehicles and driving simulators, Dissertation Technische Universität Kaiserslautern. Verlag Dr. Hut, Muenchen

Fankem S, Weiskircher T, Müller S (2014) Model-based rack force estimation for electric power steering. IFAC World Congress, Cape Town, South Africa, 24-29, Aug 2014

Harrer M (2008) Characterisation of steering feel. University of Bath (United Kingdom), Diss

Hayashi J (2013) Road map of the motor for an electric power steering system. In: 4. ATZ-Konferenz chassis.tech plus, München, Germany

Heissing B, Ersoy M (2011a) Chassis handbook, Berlin: Vieweg-Teubner Verlag. Springer, Wiesbaden

Heissing B, Ersoy M (eds) (2011b) Fahrwerkhandbuch: Grundlagen, Fahrdynamik, Komponenten, Systeme, Mechatronik. Perspektiven, ATZ/MTZ Fachbuch, Vieweg

Isermann R (2005) Mechatronic systems - fundamentals, 2nd edn. Springer, London

Isermann R (2011) Fault-diagnosis applications. Springer, Heidelberg

Isermann R (2017) Combustion engine diagnosis. Springer, Berlin

Kageyama I (2014) Steering systems. In: Road and off-road vehicle dynamics handbook

Kessler P (2015) Einstellbare Verkopplungsregelung für ein Duplex-Aktorsystem einer achsparallelen elektrischen Servolenkung (EPS). In: Bäker B, Unger A (eds) Diagnose in mechatronischen Fahrzeugsystemen IX. TUDpress, Dresden, pp 217–231

Kessler P (2021) Fehlertolerante mechatronische Aktorsysteme mit einer elektrischen Duplex-Servolenkung. Dissertation Technische Universität Darmstadt, Institute of Automatic Control (IAT)

Leonhard W (1996) Control of electrical drives, 2nd edn. Springer, Berlin

Milliken WF, Milliken DL (2002) Chassis design: principles and analysis. SAE International, Warrendale

Mitschke M, Wallentowitz H (2004) Dynamik der Kraftfahrzeuge, 4th edn. Springer, Berlin

Mitschke M (1972) Dynamik der Kraftfahrzeuge. Springer-Verlag, Berlin

Münchhof M (2006) Fault management for a smart hydraulic servo axis. In: Proceedings of the Actuator (2006) Messe Bremen GmbH. Bremen
Pfeffer P (2006) Interaction of vehicle and steering system regarding on-centre handling. Dissertation, University of Barth, UK
Pfeffer P, Harrer M (2013) Lenkungshandbuch, 2nd edn. Vieweg Teubner Verlag, Springer Wiesbaden
Pressman A (1997) Switching power supply design, 2nd edn. McGraw-Hill, New York
Reimann G, Brenner P, Büring H (2016) Steering actuator systems. In: Winner et al (eds) Handbook of driver assistance systems. Springer-Vieweg International Publishing, Zürich
Reimpell J, Stoll H, Betzler J (2001) The automotive chassis, 2nd edn. Reed Educational and Professional Publishing Ltd, Oxford
Bosch GmbH R (ed) (2018) Automotive handbook, 10th edn. Wiley, Chichester
Runge W, Gaedke A, Heger M, Vähning A et al (2009) Elektrisch Lenken. Automob Z 111(10):738–744
Schöttler F (2007) Functional safety in electrical power steering systems. In: Proceedings of IQPC conference on steering systems, Frankfurt, Germany
Schröder D (1995) Elektrische Antriebe, vol 1: Grundlagen. Springer, Berlin
Vas P (ed) (1990) Vector control of AC machines. Clarendon Press, Oxford
Yoneki S, Hitozumi E, Collerais B (2013) Fail-operational eps by distributed architecture. 4. ATZ-Konferenz chassis.tech plus, München, Germany, pp 421–442
ZF-Lenksysteme (2010) Produktioninformation

# 15 Suspension Control Systems

The basic behavior of suspension systems, their relevance for driving comfort and safety, their components, and linear and nonlinear mathematical models for passive suspensions were treated in Chapter 8. It was shown that ride quality and driving safety are conflicting properties and that an adaptation to changing road and load conditions can be reached with active suspensions. Therefore, this chapter considers semi-active as well as active hydraulic suspensions. A further topic is the indirect tire pressure monitoring based on wheel velocity and suspension sensors.

## 15.1 Classification of Suspension Systems

Suspension systems are classified into passive, semi-active, and active systems. Typical features are the required energy and the characteristic frequencies of the actuator. This fact is visualized in Fig. 15.1. Some properties of suspension systems are summarized in Table 15.1.

*Passive systems* consist of invariable components with fixed parameters. Extensions with extra spring-, damper-, or mass systems in order to absorb further energy are possible. But all systems are tuned for only one fixed frequency or a set of frequencies and vehicle mass and tire pressure.

If the system or its parameters are *slowly variable*, an adaptation of the suspension is possible. Examples for these systems are load-leveling systems. Here, the initial deflection of the spring is varied by changing the air volume of air springs or the oil volume of hydraulic cylinders, which are installed in series to the actual spring. This can be achieved by manual manipulation or by electromechanic / electro-hydraulic / electro-pneumatic systems. However, the manipulation frequency is very low.

Increasing the manipulation frequency of the slowly variable passive or adaptive systems leads to *semi-active systems*. They do not generate forces out of their own. Therefore, they require less control energy. Here, the mode of operation is the variation of the system's parameters. For instance, the mechanism of semi-active sus-



R. Isermann, *Automotive Control*, ATZ/MTZ-Fachbuch,
https://doi.org/10.1007/978-3-642-39440-9_15

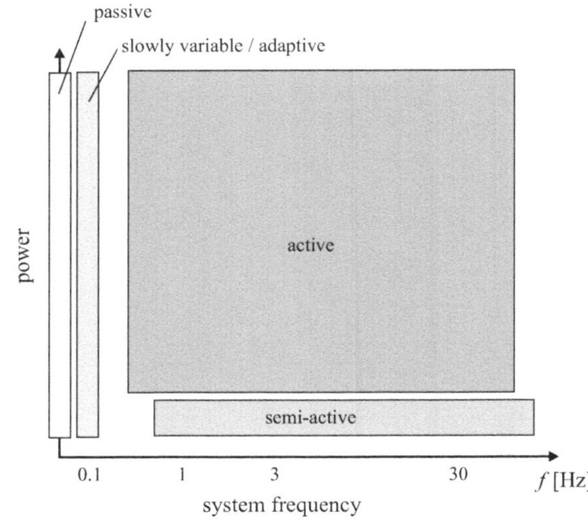


**Fig. 15.1** Power vs. frequency for various suspension systems; see Streiter (1996)

pension systems is the adaptation of the stiffness of the spring and/or the damping to the actual demands. Therefore, external energy is only required to adjust parameter changes, but not to influence the motion of the sprung or unsprung mass directly. One distinguishes between fast and slow semi-active systems. A slow system adapts to various car settings, e.g. mass or driving modes, whereas a fast system is able to adapt directly to road disturbances.

*Active systems* provide active forces with usually higher control frequencies, intervening directly in the system dynamics. As a result, the power demand is relatively high and increasing quadratically with the frequency. Active suspension systems provide an extra force in addition to possible existing passive systems. This extra force is used for compensating road disturbances and roll and pitch motions of the body. Depending on the system's frequency, partially and fully active systems can be distinguished. Furthermore, there are non-, partially- and fully loaded suspension systems, depending on whether the controlled system contributes to carrying the body. *Parameter-adaptive* means the control engineering approach how to adjust the free parameters of semi-active or active systems based on measured suspension signals. For the sake of completeness, Fig. 15.2 gives a short survey of active suspension systems. Various control strategies exist for these active systems. For example, Streiter (1996) describes the control of a hydropneumatic system, in which a hydraulic cylinder is implemented in series to a conventional steel spring. Such systems have gone into production, e.g. Mercedes CL-Class, Pyper et al. (2003) and Streiter (2008); see Sect. 15.4.

**Table 15.1** Classification of suspension systems, Bußhardt (1995). Natural frequencies: $f_B$ body and $f_W$ wheel

| System | System representation | Force range | Operation range (frequency) | Demand of • actuators • sensors | Max. energy demand | Improvements compared to passive systems | |
|---|---|---|---|---|---|---|---|
| | | | | | | Comfort | Safety |
| Passive | 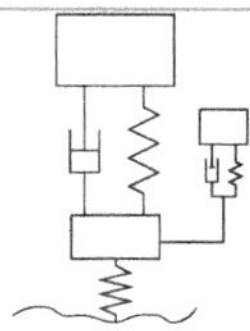 | 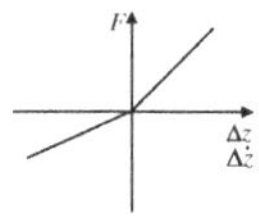 | – | – | – | – | – |
| Slowly variable adaptive | 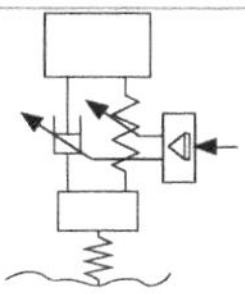 | 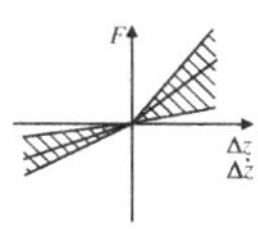 | $< f_B$ | • 4 – 8<br>• ≥ 1 | c. 50 W | 15 – 20% | 10 – 25% |
| Semi-active | 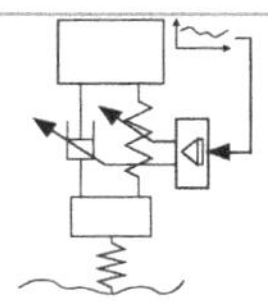 | 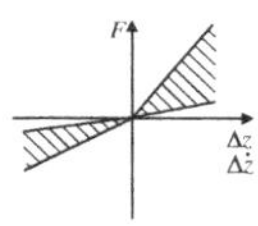 | $f_B - f_W$ | • 4 – 8<br>• ≥ 8 | c. 50 W | 20 – 30% | 10 – 25% |
| Active partially loaded | 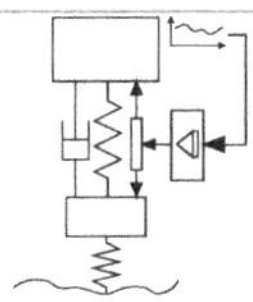 | 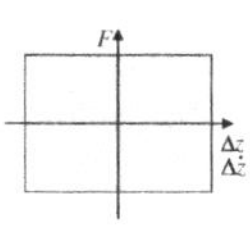 | $0 - f_B$ | • 4 – 8<br>• ≥ 12 | 1 – 2 kW | > 30% | – |
| Active fully loaded | 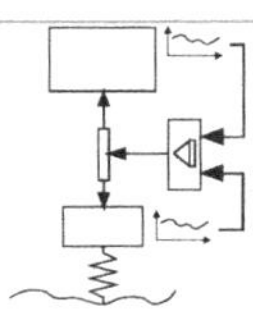 | 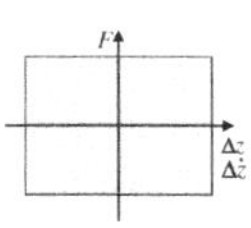 | $0 - f_W$ | • 4 – 8<br>• ≥ 12 | 1 – 2 kW | > 30% | ≥ 25% |

Figure 8.4 summarizes the possible ranges of suspension characteristics in the conflict diagram for passive suspensions. Compared to passive systems, semi-active and active systems expand their operating ranges in the direction of improved driving comfort and safety.

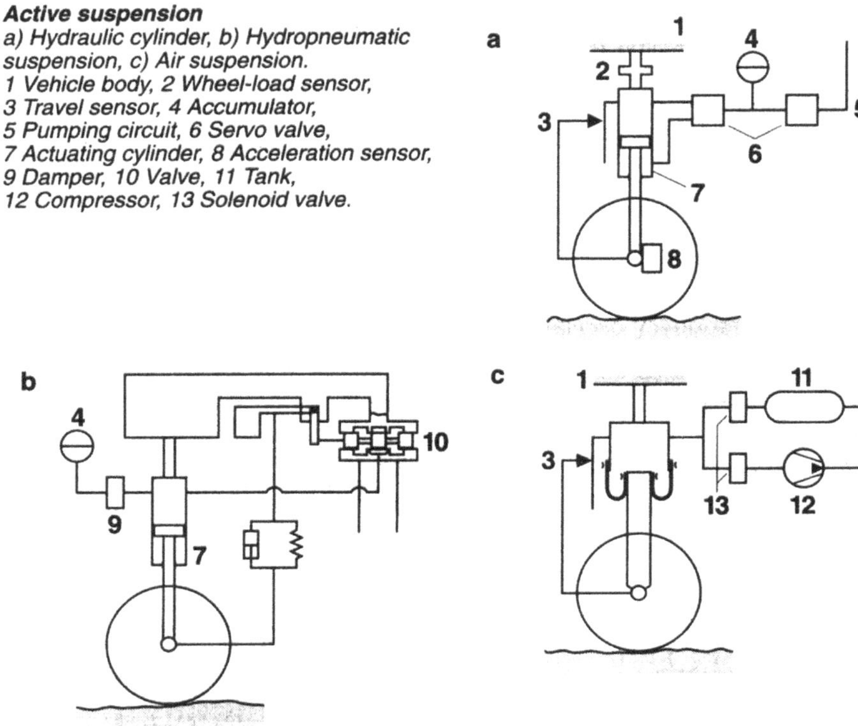


**Fig. 15.2** Active suspension systems; Robert Bosch GmbH (1996)

## 15.2 Semi-active Suspensions

The basic principles of semi-active suspension systems are briefly presented in order to give some insight into the control functions.

### 15.2.1 Semi-active Dampers

The common function of conventional semi-active shock absorbers is the variation of the throttle cross-sectional area either in fixed steps or continuously in order to obtain multiple performance curves from a single shock absorber. In the continuous case, a continuous valve controls the flow resistance and in the discrete case, various throttles or throttle systems are switched on or off. The *active throttle* can either be integrated into the damper or into an external bypass, which often is constructed in the form of a ring channel. The throttle's adjustment is achieved using an electric motor or by means of magnet valves. The latter provide short switching times and a continuous adjustment; see Reimpell et al. (2001) and Causemann (2008). The fundamental function is demonstrated in Fig. 15.3.

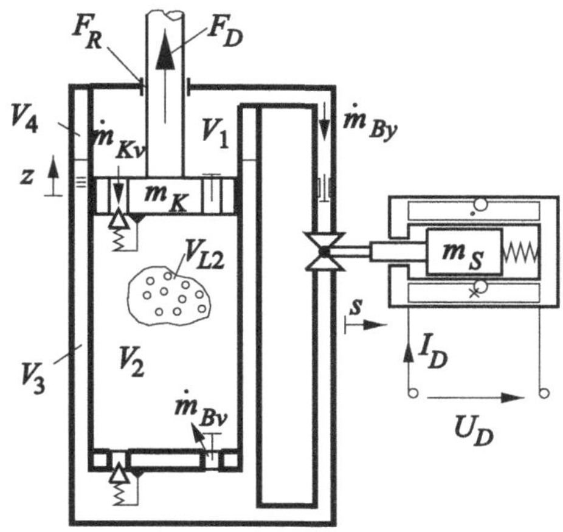


**Fig. 15.3** Schematic of an orifice changing semi-active damper with a magnet valve

During the pull cycle, the oil of the upper chamber ($V_1$) flows through the bypass and the spring-weighted piston valve. The rectifier valve in the piston bottom compensates for a possible additional deficiency of oil in chamber two $V_2$. Therefore, the pressure losses in the bypass and at the spring-weighted piston valve determine the damping forces. For the duration of the push cycle, the volume being taken up by the piston generates an oil stream through the bottom spring-weighted valve and the bypass. Consequently, these valves determine the damping forces for the push part of the cycle. On account of the bypass's role in both the push and pull cycles, the variation of the bypass's resistance influences the damper characteristics. As illustrated in Fig. 15.3, an electromagnetic valve controls the bypass's cross-sectional area, but other actuators such as electric motors are also common. The air volume $V_{L2}$ is the effect of aging or damages and results in fading effects of the damping. Figure 15.4 illustrates the performance characteristics. The manipulated variable is the current applied to an electromagnetic valve.

An extension to this principle is the *variation of the working volume*. An additional chamber with a controllable pressure is connected through additional valves. In the case of high pressures in the piston, these valves open depending on the pressure in the additional chamber. This supplementary influences the damper characteristics. An important property for the potential of a semi-active suspension is the time constant or rather response time of the actuators. The response time determines the achievable performance of suspension controllers to react directly to road excitations.

Some years ago, a new development has arisen in the form of dampers taking advantage of the electro-rheological or magneto-rheological property of liquids. Both principles are based on the alteration of the damping medium's viscosity depending on an applied electric or magnetic field. Commonly, these fields are applied in the duct connecting the upper and lower chambers of a damper; for details, see, e.g. Choi et al. (1998), Choi et al. (2001), Fees (2001), and Moosheimer and Waller (1999). One of the salient properties of electro-rheological fluid is that it has a fast response to an electric field and hence a wide control bandwidth; see Choi et al. (2001). Magneto-rheological shock absorbers have many attractive features such as

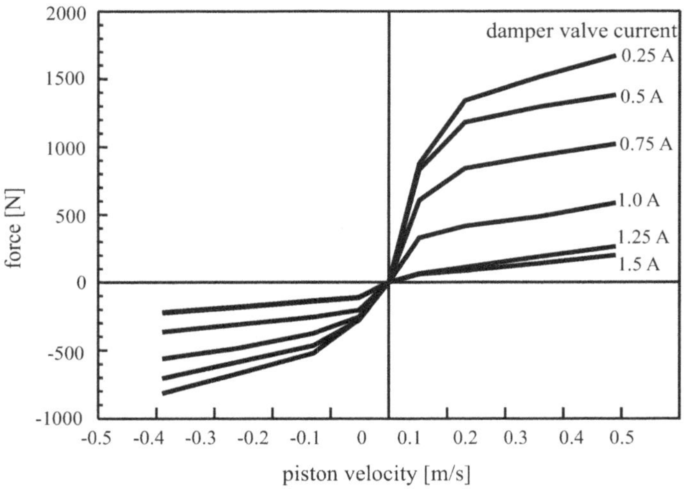


**Fig. 15.4** Family of characteristics of a continuous variable damper, manufacturer's specification

high yield strength and stable hysteric behavior over a broad temperature range; see Yokoyama et al. (2001). In order to achieve the damping forces required for a semi-active electro-rheological vehicle suspension system, a high electric field, about $E = 3..5$ kV/mm, has to be applied; see Kim and Choi (1999). Obviously, this requirement and the long-term stability of the fluid properties are presently obstacles to the application in common vehicle systems.

## 15.2.2 Load-Leveling System

A comfortable suspension is achieved by means of soft springs. However, this increases spring displacement. In order to maintain the vehicle-body height at an acceptable level and have soft springs, air springs or hydropneumatic springs are employed. The advantages of these systems are a velocity-dependent reduction of the ride height resulting in lower aerodynamic resistance and fuel savings, an increase in vehicle height on poor road surfaces, and enhanced stability in curves achieved through lateral blocking of the suspension elements on a single axle; see Robert Bosch GmbH (2011). A serial connection of a conventional steel spring with a hydraulic piston is also possible, for example, the active body control system of DaimlerChrysler; see Sect. 15.4.3. However, these systems are in principle active suspension systems, which include a leveling function as part of their design. Beyond this, self-pumping dampers exist being able to regulate their level automatically. This is achieved by taking energy out of the body's motion. The damper uses this energy to pump itself to a constant mean level.

### 15.2.3 Semi-active Spring

Load-leveling systems can generally be used in a semi-active sense if they provide the required manipulation dynamics. The common semi-active spring is based on a system containing an air spring or a hydropneumatic system. The stiffness of an air spring can be approximated according to Reimpell et al. (2001) by the following equation:

$$c = n\frac{pA^2}{V}. \tag{15.2.1}$$

As a result, the pneumatic systems are based on the alteration of the air volume $V$ by addition or removal of air resulting in a change of the stiffness. This leads to a continuously adjustable stiffness. Unfortunately, this procedure leads directly to a changing initial deflection. In order to prevent a varying initial deflection, hydropneumatic systems can be implemented, which replaces the removed air with oil or *vice versa*; see Bußhardt (1995). Switchable air volumes are another possibility. This attempts to alter the total volume of air, and thereby the stiffness without significant change of the mean deflection. The common systems, compare Reimpell and Stoll (1989), consist of two connected volumes. By cutting off or opening the pneumatic connection between these volumes, two different stiffness values are achievable. Often, the air or hydropneumatic spring is arranged in parallel to a conventional steel spring, in order to reduce the forces and stress on these systems. Also, variable steel spring systems exist. Their principle of operation is based upon the alteration of the point of the force application. This is attained either by the direct shifting of the point of application at one spring or bypassing several springs in a system of serially arranged springs, e.g. Venhovens et al. (1993). However, the dynamic response of these mechanical solutions is small.

## 15.3 Control of Semi-active Suspensions

The automatic control of semi-active suspensions is characterized by manipulation of the semi-active damper and/or the semi-active spring based on measurements of the body and wheel movements. These semi-active suspensions modulate the damper or spring by electronically controlled actuators and just superimpose additional forces compared to the basic passive system. However, they do not deliver external power to the suspension and are therefore not critical with regard to dynamic instability. Figure 15.5 depicts the range of adjustable force-velocity characteristics of a semi-active damper. The individual characteristics can be adjusted by manipulation of the current of an electromechanic bypass valve as shown in Fig. 15.3. There are mainly the following principles to adjust the damper characteristic:

- manual adjustment of discrete fixed characteristics (e.g. soft or hard damping);
- automatic adjustment of characteristics based on vehicle movement measurements.

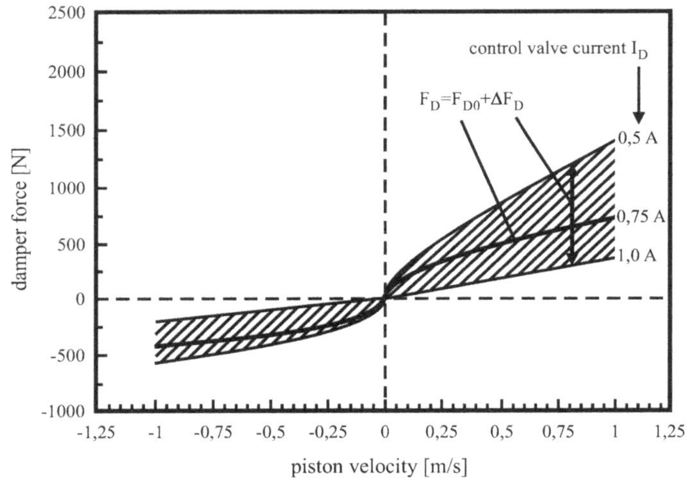


**Fig. 15.5** Adjustable force-velocity characteristics of a semi-active damper

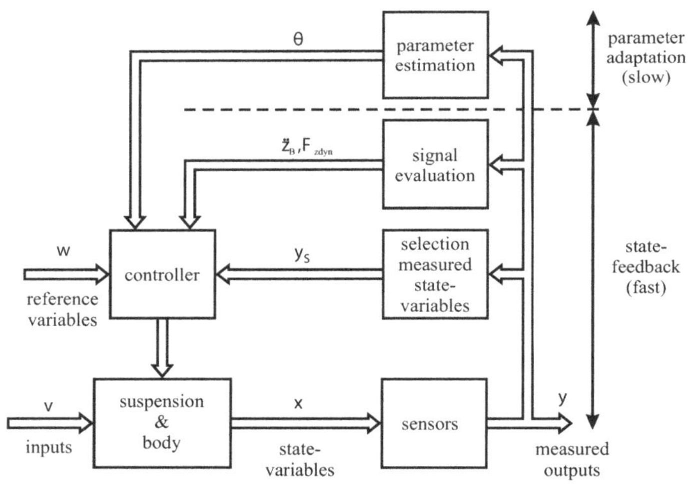


**Fig. 15.6** Control configuration for a semi-active suspension

In the following, the last case is further considered for continuously variable semi-active shock absorbers.

Two principles for the control of semi-active dampers can be distinguished. The first possibility is to estimate the parameters or characteristics. Based on measured signals of the suspension, the actual damping coefficient $d_B$ or the body mass $m_B$ can be determined by recursive parameter estimation, as shown in Table 8.4, and the damping coefficient or the damping characteristic can be adjusted by feedback control; see Fig. 15.6. This is called *parameter-adaptive semi-active suspension.*

The second possibility is to use the measured suspension signals directly to change the parameter settings of the suspension. This leads to *state feedback controlled semi-active suspensions*; see Fig. 15.6.

The state variables of the simplified suspension model, Fig. 8.1, are

$$\mathbf{x}^T = [z_W \;\; \dot{z}_W \;\; z_B \;\; \dot{z}_B \;\; r], \tag{15.3.1}$$

which are, however, usually not directly measurable. The implementation of sensors may provide

- suspension deflection $(z_W - z_B)$,
- body acceleration $\ddot{z}_B$, and
- wheel acceleration $\ddot{z}_W$,

such that for example the output vector becomes

$$\mathbf{y}^T = \left[z_W \; -z_B \; \ddot{z}_B \; \ddot{z}_W\right], \tag{15.3.2}$$

where a selection of the variables can be made. This allows then to determine $(\dot{z}_W - \dot{z}_B)$ through numerical differentiation and $\dot{z}_B$ and $\dot{z}_W$ through numerical integration. Usually the suspension deflection $(z_W - z_B)$ is not measured, because of costs. However, mass produced airbag sensors can be used to measure the acceleration $\ddot{z}_B$ and $\ddot{z}_W$. The state variables (15.3.1) as absolute values may be reconstructed by using the Kalman filters; see Bußhardt (1995).

### 15.3.1 Parameter-Adaptive Semi-active Dampers

The characteristic of this method is the successive parameter estimation and damping control in a loop, see Fig. 15.7. In the first step, the unknown coefficients are estimated as shown in Sect. 8.3.1. The actual damping $d_{\mathrm{Bact}}$ is compared with a reference value $d_{\mathrm{Bref}}$, which either is given externally or is calculated internally by the estimated values. The shock absorber is adjusted by an electrical voltage $U_d$, respectively, and a current $I_d$. This method is applicable for relatively slowly changing conditions like damping fluid temperature, tire pressure, or changing load. The concept is characterized by the feedback control of an *estimated coefficient*, based on *measured variables*. It adapts the real (estimated) damping coefficient $d_{\mathrm{Bact}}$ to the desired set point $d_{\mathrm{Bref}}$. The damping coefficient is estimated based on measurements of the suspension deflection $(z_W - z_B)$ and the body acceleration $\ddot{z}_B$; see Table 8.4. As an example, Fig. 15.8 shows a result of a test rig experiment. The body mass $m_B$ was changed from 260 kg to 340 kg. The damping factor $D$ of the body mass was set to

$$D_{\mathrm{ref}} = \frac{d_B}{2\sqrt{c_B m_B}} = 0.7, \tag{15.3.3}$$

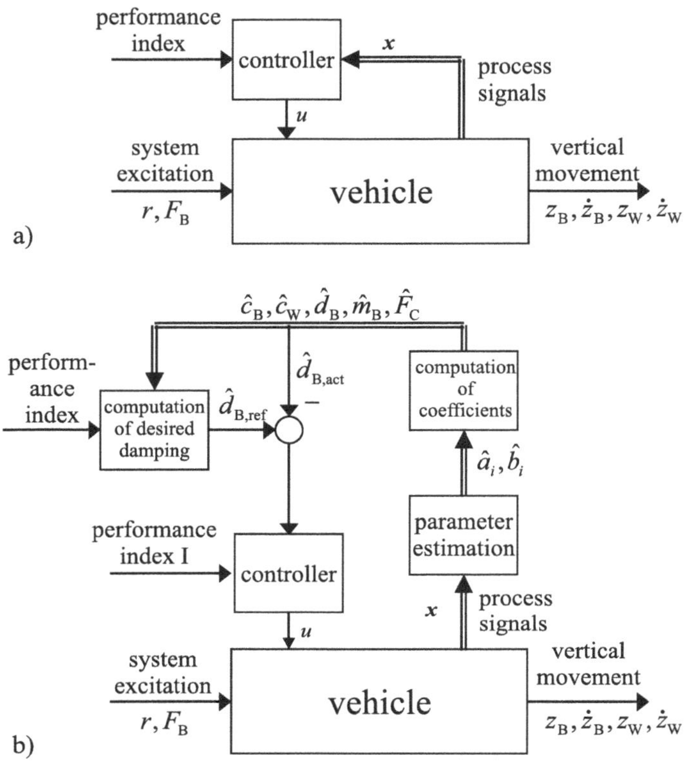


**Fig. 15.7** Suspension control principles **a** state variable feedback **b** parameter adaptation (parameter feedback)

assuming a linearized damper and neglecting the motion of the wheel. The damping coefficient $d_B(t)$ is adapted automatically to the higher load by a PI controller with sampling time $T_0 = 5$ ms, so that the damping factor remains approximately constant. The behavior of the estimated damping for both body masses after a stepchange on the wheel side can be seen in Fig. 15.8.

The results of the parameter estimation can also be used to select a damper characteristic out of discrete sets, as in Fig. 8.9.

### 15.3.2 State Feedback Controlled Semi-active Suspensions

The measured signals $\mathbf{y}$ of the suspension can be directly used to manipulate the semi-active suspension, as shown in Fig. 15.7a. The following description uses first the measured suspension deflection $(z_W - z_B)$.

To describe the principle, a linear model of the suspension is assumed. By changing the damper actuator, an additional force $\Delta F_D$ is supposed to be introduced between body and wheel mass; see Fig. 15.5. This results in the following momen-

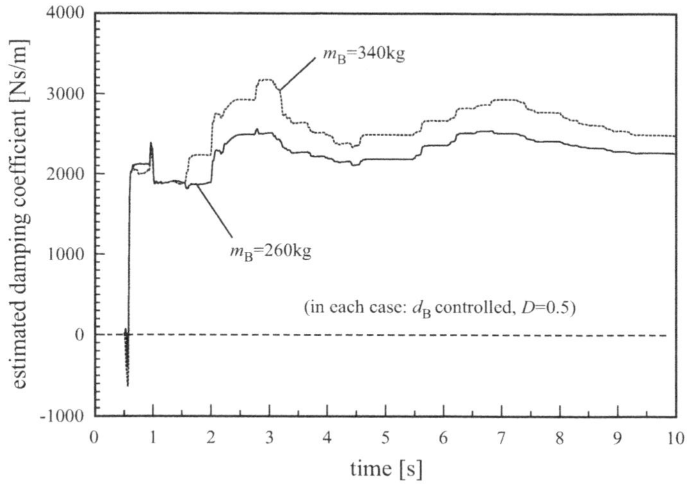


**Fig. 15.8** Estimated damping coefficient $d_B(t)$ for a desired damping factor $D_{ref} = 0.5$ and changing body mass, after an excitation on the wheel side in form of a step function $r = 0.027$ m (see test rig Fig. 8.8); see Bußhardt (1995)

tum balances for the body and the wheel

$$m_B \ddot{z}_B = c_B (z_W - z_B) + d_B (\dot{z}_W - \dot{z}_B) + \Delta F_D - F_B, \tag{15.3.4}$$

$$m_W \ddot{z}_W = c_W (r - z_W) - c_B (z_W - z_B) - d_B (\dot{z}_W - \dot{z}_B) - \Delta F_D; \tag{15.3.5}$$

compare (8.1.3) and (8.1.4). These equations can be directly transformed to a state-space form

$$\dot{\mathbf{x}}(t) = \mathbf{A}\mathbf{x}(t) + \mathbf{B}\mathbf{u}(t) + \mathbf{F}v(t), \tag{15.3.6}$$

$$\mathbf{y}(t) = \mathbf{C}\mathbf{x}(t) + \mathbf{D}\mathbf{u}(t) + \mathbf{G}v(t), \tag{15.3.7}$$

with the state variables of suspension deflections

$$\mathbf{x}^T = [(z_W - z_B) \quad (\dot{z}_W - \dot{z}_B)] \tag{15.3.8}$$

and

$$\begin{aligned} \mathbf{u}^T &= \Delta F_u, \\ \mathbf{y}^T &= \left[\ddot{z}_B \quad F_{zdyn}\right], \\ \mathbf{v}^T &= [v - z_W \quad F_B], \end{aligned} \tag{15.3.9}$$

with

$$\mathbf{A} = \begin{bmatrix} 0 & 1 \\ -c_{\mathrm{B}}\left(\frac{1}{m_{\mathrm{W}}} + \frac{1}{m_{\mathrm{B}}}\right) & -d_{\mathrm{B}}\left(\frac{1}{m_{\mathrm{W}}} + \frac{1}{m_{\mathrm{B}}}\right) \end{bmatrix},$$
$$\mathbf{B} = \begin{bmatrix} 0 \\ -\left(\frac{1}{m_{\mathrm{W}}} + \frac{1}{m_{\mathrm{B}}}\right) \end{bmatrix},$$
$$\mathbf{C} = \begin{bmatrix} \frac{c_{\mathrm{B}}}{m_{\mathrm{B}}} & \frac{d_{\mathrm{B}}}{m_{\mathrm{B}}} \\ 0 & 0 \end{bmatrix},$$
$$\mathbf{D} = \begin{bmatrix} \frac{1}{m_{\mathrm{B}}} \\ 0 \end{bmatrix}. \tag{15.3.10}$$

The used outputs are $\ddot{z}_{\mathrm{B}}$ and $F_{\mathrm{zdyn}}$ because they will be used for the control performance criterion. In case of semi-active systems, the additional force $\Delta F_{\mathrm{u}}$ has to be generated by the adjustment of the suspension parameters $c_{\mathrm{B}}$ and $d_{\mathrm{B}}$. For the linearized system, the spring and damper forces are as follows:

$$F_{\mathrm{S}}(k) = c_{\mathrm{B}}(k)\left(z_{\mathrm{W}}(k) - z_{\mathrm{B}}(k)\right), \tag{15.3.11}$$

$$F_{\mathrm{D}}(k) = d_{\mathrm{B}}(k)\left(\dot{z}_{\mathrm{W}}(k) - \dot{z}_{\mathrm{B}}(k)\right). \tag{15.3.12}$$

By dividing the parameters $c_{\mathrm{B}}$ and $d_{\mathrm{B}}$ in a constant value ($c_{\mathrm{B0}}$ and $d_{\mathrm{B0}}$) and a changing part, these equations can be converted to

$$c_{\mathrm{B}}(k) = \frac{F_{\mathrm{S}}(k)}{z_{\mathrm{W}}(k) - z_{\mathrm{B}}(k)} = c_{\mathrm{B0}} + \frac{\Delta F_{\mathrm{Su}}(k)}{z_{\mathrm{W}}(k) - z_{\mathrm{B}}(k)}, \tag{15.3.13}$$

$$d_{\mathrm{B}}(k) = \frac{F_{\mathrm{D}}(k)}{\dot{z}_{\mathrm{W}}(k) - \dot{z}_{\mathrm{B}}(k)} = d_{\mathrm{B0}} + \frac{\Delta F_{\mathrm{Du}}(k)}{\left(\dot{z}_{\mathrm{W}}(k) - \dot{z}_{\mathrm{B}}(k)\right)}. \tag{15.3.14}$$

Hence, the extra force $F_{\mathrm{u}}$ is divided into the additive parts $\Delta F_{\mathrm{Su}}$ and $F_{\mathrm{Du}}$ for the spring and damper, respectively.

The state-space model is then transformed into discrete-time form

$$\mathbf{x}(k+1) = \mathbf{A}\mathbf{x}(k) + \mathbf{B}\mathbf{u}(k) + \mathbf{F}\mathbf{v}(k), \tag{15.3.15}$$

$$\mathbf{y}(k+1) = \mathbf{C}\mathbf{x}(k) + \mathbf{D}\mathbf{u}(k) + \mathbf{G}\mathbf{v}(k), \tag{15.3.16}$$

and a state feedback controller

$$\mathbf{u}(k+1) = -\mathbf{k}^T\mathbf{x}(k), \tag{15.3.17}$$

or a minimum-variance controller is designed by minimization of the performance criterion,

$$I(k+1) = E\left\{q_0\left(\frac{\ddot{z}_{\mathrm{B}}(k)}{g}\right)^2 + \left(\frac{F_{\mathrm{z,dyn}}}{F_{\mathrm{z,stat}}}\right)^2\right\}. \tag{15.3.18}$$

Simulation results have then shown that for a given road profile as excitation $r(k)$ compared to a passive suspension, the mean body acceleration $E\{\ddot{z}_B\}$ for wheel side excitation can be improved by 23% with the variable damper, 30% with the variable spring, and 46% for the combined variable spring and damper; see Bußhardt (1995).

However, the experimental results at the test rig have shown that with a semi-active damper and a sampling time of $T_0 = 5$ ms, only an improvement of 10% could be obtained, i.e. only half of the improvement results from simulations. The reason is the time delay of the magnetic valve of the damper, the time delay of a required low pass filter, and the computation time which sum up to about $T_\Sigma \approx 30$ ms. Simulations have shown that the time delay should be $T_\Sigma \leq 10$ ms and the manipulation range of the damping factor should be increased to result in significant improvements. In contrast to the slowly reacting parameter estimation-based adaptive dampers in Fig. 15.7b, the state variable feedback controlled damper according to Fig. 15.7a changes the damping characteristic very fast. However, this requires a fast magnetic valve actuator. Another state-space-based nonlinear control of a semi-active suspension is considered in Dessort and Chucholowski (2017).

Commercially available semi-active dampers use as inputs for control mainly sensors for body acceleration in the front and rear part of the vehicle and sensors for wheel acceleration at each wheel; see Kutsche and Rappelt (2006) and Causemann (2008). This allows to obtain measures for comfort and safety. Modified skyhook algorithms are applied. The continuous manipulation of the damping forces leads to a reduction of the body movements, improvements for rolling and pitching, reduction of vibrations $< 3$ Hz, and reduction of the braking distance for rough roads.

## 15.4 Active Suspensions

The parametrization of passive suspension systems has shown that trade-offs between ride quality and driving safety have to be made. Improvements in one direction are mostly only obtained at the expense of deterioration in other directions. The behavior can be improved by adding controlled actuators in the suspension. The semi-active suspensions lead to better individually tuned passive suspensions or better adaptation to the actual excitations from the road or from driving. However, active suspensions, which supply additional forces by using an external energy source, result in more significant improvements; compare Figs. 15.1 and 15.2. The goal of active suspensions is to improve the ride comfort and driving safety by optimally dampened vibrations, oscillations, and roll and pitch movements.

### 15.4.1 Active Suspension Principles

Active suspension systems are usually designed with hydraulic, hydrodynamic, or pneumatic actuators. Figure 15.9 shows some principles in addition to Fig. 15.2. The *hydraulic systems* contain a cylinder as shown in Fig. 15.2a either parallel or serial to steel springs which carry the static load. Figure 15.9 depicts a version with a serial

| system | hydraulic system | hydro-pneumatic system | airspring system | active stabiliser |
|---|---|---|---|---|
| scheme | 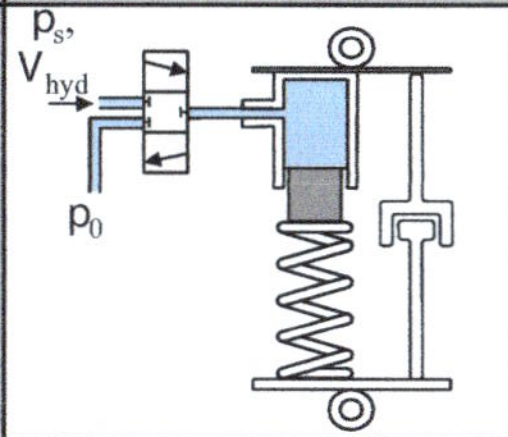 | 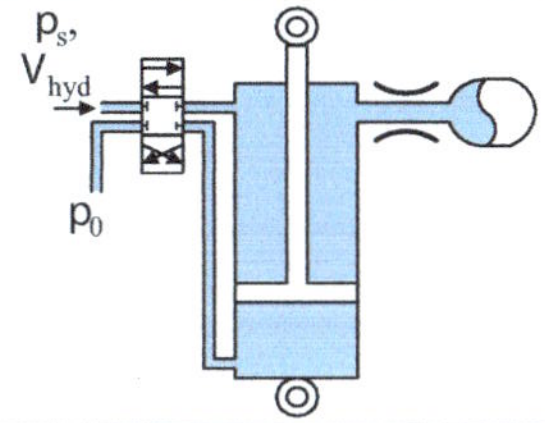 | 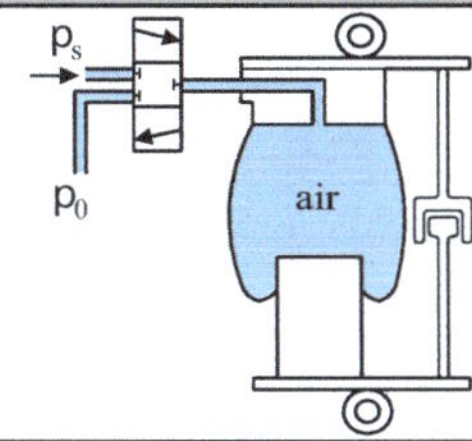 | 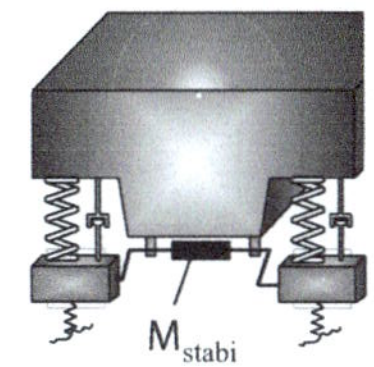 |
| principle | • spring mounted plunger and cylinder<br>• active lift control | • hydrodynamic spring and hydraulic cylinder<br>• active lift control | • air bellow spring<br>• active lift control | • additional torque in stabiliser |
| advantages | • robust hydraulics<br>• conventional suspension components<br>• individual corner control | • robust hydraulics<br>• combination with airspring<br>• individual corner control | • robust pneumatics<br>• no steel spring<br>• individual corner control | • robust hydraulics |
| dis-advantages | • hydraulic oil pressure supply<br>• many parts<br>• expensive | • air pressure supply<br>• many parts | • air pressure supply<br>• many parts | • hydraulic oil pressure supply<br>• only for roll control |

**Fig. 15.9** Some principles of active suspension systems

spring-mounted point adjustment. These hydraulic systems require an hydraulic pump and an accumulator for fast actuation and damping of oil pressure oscillations. The *hydropneumatic system* has in addition to the hydraulic cylinder a hydropneumatic spring, where gas and fluid are separated by a membrane. Pumping in and out of oil is used for low-frequency control with magnetic valves and the hydrodynamic spring for higher frequency control. Historically, the hydropneumatic suspension was developed by Citroën and used first in 1954 for rear suspensions. A nitrogen reservoir with pressure until 75 bar within a suspension sphere is separated from hydraulic oil through a rubber/steel membrane. An engine-powered hydraulic pump pressurizes the oil to an accumulator sphere and the suspension cylinders. The suspension works by means of a piston pressuring the fluid with about $150 - -180$ bar into the sphere, where a valve in the entrance to the sphere causes a resistance for damping. Some advantages of this kind of hydropneumatic system are the progressive spring rate, such that it adapts continuously to the load, and an automatic level control of the body, however on the cost of complexity.

*Active air suspensions* use bellows with control valves which are connected to an air accumulator and an air compressor; see, e.g. Folchert (2006) and Hilgers et al. (2009). The manipulated air pressure in the bellows allows then directly to add additional forces.

The air supply system can be open to the atmosphere or closed. The stiffness of an air spring is, see, e.g. Reimpell and Stoll (1989) and Isermann (2005),

$$c = n p_{\mathrm{i}}(z) \frac{A}{z_{\mathrm{BW}}}. \tag{15.4.1}$$

It is inverse proportional to the deflection $z_{\mathrm{BW}}$ and therefore progressive and proportional to the pressure $p_{\mathrm{i}}$. With changing load, the pressure increases and also the stiffness. Therefore, the oscillating behavior does not change for varying load because the natural frequency $\omega_0 = \sqrt{c/m_{\mathrm{veh}}}$ becomes independent on the vehicle mass $m_{\mathrm{veh}}$. In addition, the stiffness can be varied by changing the pressure. If combined with a level control of the body, the air spring operates, different from steel springs, in a constant middle operating position, and the travel ways to stop buffers are not reduced. The damping can be realized in a conventional way with hydraulic shock absorbers. However, a damper system can be integrated with two bellows, where the airflow between both chambers is manipulated by a valve; see Heissing and Ersoy (2011) and Sommer (2003). By switching to an additional air volume, the stiffness can be varied to a larger extent, and a selection from comfortable to sporty modes becomes possible; see Hilgers et al. (2009). Another type of active suspension is *active stabilizers* which can be used to control and damp the rolling behavior. The stabilizer actuators are usually hydraulic.

### 15.4.2 On Active Suspension Control

Historically, there are two theoretical control principles, the skyhook and the groundhook strategy. The *skyhook control strategy* aims to increase the driving comfort by keeping the vehicle body calm, independent of the driving state and road surface. The goal is to decouple the movement of the body from the excitation of the road. Therefore, the idea is to connect the body with an ideal damper to the "sky". The required damping force then becomes, Karnopp (1979) and Karnopp (1987),

$$F_{\mathrm{DB}} = d_{\mathrm{sky}} \dot{z}_{\mathrm{B}} . \tag{15.4.2}$$

This skyhook strategy can also be used for the design of semi-active suspensions. However, the dynamic tire forces are not taken into account, which are relevant for driving safety. The *groundhook control* aims to improve the safety by reducing the vertical dynamic wheel forces. The idea is therefore that the wheel is connected by a damper to the road. The damping forces are then

$$F_{\mathrm{DW}} = d_{\mathrm{ground}} \left( \dot{z}_{\mathrm{W}} - \dot{r} \right) . \tag{15.4.3}$$

However, as both aspects, comfort and safety, have to be considered, optimized suspension control tries to optimize a performance criterion for stochastic road disturbances

$$J = E \left\{ q_0 \left( \frac{\ddot{z}_{\mathrm{B}}(t)}{g} \right)^2 + \left( \frac{F_{\mathrm{z,dyn}}(t)}{F_{\mathrm{z,stat}}} \right)^2 \right\} , \tag{15.4.4}$$

where it is assumed for the design that the disturbances have a zero mean value. There exist many publications on the control of active suspensions. For example, the design of linear quadratic (LQ) state feedback is considered by Salman et al. (1988), Yue et al. (1989), Krtolica and Hrovat (1990), and Ursu et al. (1997). Nonlinear robust control is treated by Alleyne and Hedrick (1995) and Roukieh and Titli (1993). $H^\infty$ controller designs can be found in Hayakawa et al. (1999) and Wang et al. (2001). See also Rajamani (2012). Also, preview information is taken into account, as in Oya (1998), Yue et al. (1989), Karlsson et al. (2001), Dessort and Chucholowski (2017), and Strom and Lohmann (2017).

### 15.4.3 Active Hydraulic Suspension

As an example, the active body control system (ABC) of Mercedes is considered; see Streiter (1996) and Streiter (2008). Figure 15.10 shows a test rig for this fully loaded hydraulic system which is of a type with spring-mounted point adjustment; see Robert Bosch GmbH (2011). It consists of a hydraulic plunger, which is connected in series with a steel spring. The damper connects directly the body and the wheel. The plunger is connected to a motor-pump unit, which provides a controlled hydraulic flow from/to a hydraulic accumulator. The test rig replaces a quarter of a real car with suspension arms, a real tire, and an appropriate body mass. A realistic road excitation is simulated by the vertical movement of the base plate. The test rig is equipped with a variety of sensors, whereas the presented algorithms are based on a sensor setup that can be suitable for automotive production. The sensor setup complies sensors for

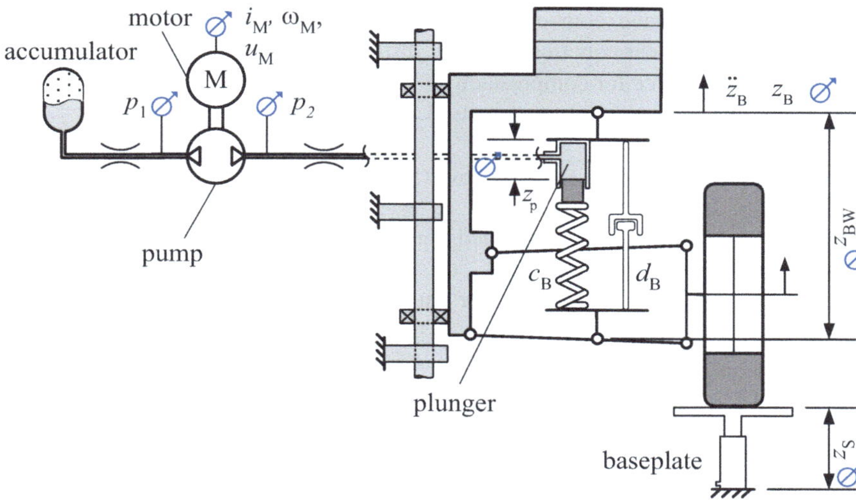


**Fig. 15.10** Schematic of the active hydraulic suspension (ABC) on a test rig; see Fischer (2006)

the suspension deflection $z_{WB}$, the body acceleration $\ddot{z}_B$, the plunger position $z_P$, the hydraulic pressure at both sides of the pump ($p_1$, $p_2$), the motor speed $\omega_M$, voltage $u_M$, and current $i_M$. In the following, the modeling of the different components of the electro-hydraulic suspension system such as hydraulic accumulator, hydraulic lines, motor, pump, plunger, and suspension system is performed according to Fischer (2006) and Isermann (2003).

Using the momentum balance with forces, the differential equations for the body and the wheel mass can be derived

$$m_B\ddot{z}_B(t) = F_{hyd}(t) + F_D(t), \tag{15.4.5}$$

$$m_W\ddot{z}_W(t) = F_{zydn}(t) + F_S(t), \tag{15.4.6}$$

where $F_{hyd}$ describes the plunger force and $F_D$ the damper force to the body. The steel spring and the wheel can be approximated by a linear spring in their operating point, leading to the following forces:

$$F_S(t) = c_B\left(z_{WB}(t) + z_P(t)\right), \tag{15.4.7}$$

$$F_{zdyn}(t) = c_W\left(z_S(t) - z_W(t)\right). \tag{15.4.8}$$

The hydraulic plunger force $F_{hyd}$ results with the plunger's effective cross-sectional area $A_P$. Neglecting friction and mass effects, the forces $F_{hyd}$ and $F_S$ are identical

$$F_{hyd}(t) = F_S(t) = A_p p_P(t). \tag{15.4.9}$$

The strongly degressive course of the dampers characteristic curve is usually described by a nonlinear equation with two different square root characteristics and an equal Coulomb force for compression (+) and decompression (−) of the damper (Bußhardt 1995; Majjad 1997), compare (8.2.16),

$$F_{\mathrm{D}}(t) = d_{\mathrm{B}+,-}\sqrt{|\dot{z}_{\mathrm{WB}}(t)|}\,\mathrm{sign}\,(\dot{z}_{\mathrm{WB}}(t)) + F_{\mathrm{C}}\,\mathrm{sign}\,(\dot{z}_{\mathrm{WB}}(t))\,. \tag{15.4.10}$$

The suspension arms are not modeled separately, as their transmission ratio is included in the physical parameters $A_{\mathrm{p}}$, $c_{\mathrm{B}}$, and $d_{\mathrm{B}}$; compare Würtenberger (1997).

The motor is modeled as a DC motor with

$$u_{\mathrm{M}}(t) = R_{\mathrm{M}} i_{\mathrm{M}}(t) + L_{\mathrm{M}} \dot{i}_{\mathrm{M}}(t) + \psi\omega_{\mathrm{M}}, \tag{15.4.11}$$

$$J\dot{\omega}(t) = \psi_{\mathrm{M}} i_{\mathrm{M}}(t) - M_{\mathrm{Load}}(t), \tag{15.4.12}$$

where $R_{\mathrm{M}}$ is the armature resistance, $L_{\mathrm{M}}$ the armature inductance, $\psi$ the motor back electro-motive force constant, $\psi_{\mathrm{M}}$ the motor torque constant, $J$ the moment of inertia of the motor, and $\omega$ the angular speed.

The motor drives a positive displacement pump that produces the hydraulic volume flow $\dot{V}$ and generates the load torque $M_{\mathrm{Load}}$

$$M_{\mathrm{Load}}(t) = V_{\mathrm{Z0}}\,(p_2(t) - p_1(t)) + M_{\mathrm{R2}}\,\mathrm{sign}\,(\omega(t))\,\omega^2(t), \tag{15.4.13}$$

$$\dot{V}(t) = V_{\mathrm{Z0}}\omega_{\mathrm{M}}(t) - k_{\mathrm{L}}\,(p_2(t) - p_1(t))\,, \tag{15.4.14}$$

where $V_{\mathrm{Z0}}$ is the basic volume of the pumpe, $M_{\mathrm{R2}}$ its friction torque, and $k_{\mathrm{L}}$ an inner pump leakage coefficient.

The hydraulic lines are modeled considering a laminar and an inductive hydraulic resistance resulting in the following equation for the pipe pressure loss:

$$\Delta p(t) = R\dot{V}(t) + L\ddot{V}(t). \tag{15.4.15}$$

The hydraulic accumulator is modeled by an air spring with effective stiffness $c_{\mathrm{accu}}$

$$\dot{p}_{\mathrm{accu}} = -c_{\mathrm{accu}}\dot{V}_{\mathrm{accu}}(t). \tag{15.4.16}$$

As result of the serial arrangement of plunger and body spring, the elasticity of the hydraulic liquid in the plunger is neglected as well as friction and gap losses. Hence, the following equations for the plunger can be obtained:

$$\dot{z}_{\mathrm{P}}(t) = \frac{1}{A_{\mathrm{p}}}\dot{V}(t), \tag{15.4.17}$$

$$p_{\mathrm{P}}(t) = \frac{c_{\mathrm{B}}}{A_{\mathrm{p}}}\cdot(z_{\mathrm{WB}}(t) + z_{\mathrm{P}}(t))\,. \tag{15.4.18}$$

**Table 15.2** Parameter estimation equations of the test rig

| Co mponent | Estimation equations | Parameters | |
|---|---|---|---|
| | | Estimated | known |
| Plunger (Volume) | $\dot{z}_{\mathrm{P}}(t) = \frac{V_{Z0}}{A}\omega(t) - \frac{k_{\mathrm{L}}}{A}(p_2(t) - p_1(t))$ | $V_{Z0}, \omega_0, k_{\mathrm{L}}$ | $A_{\mathrm{p}}$ |
| Plunger (Force) | $p_2(t) = \frac{c_{\mathrm{B,PF}}}{A}(z_{\mathrm{WB}} + z_{\mathrm{P}})(t) + p_{2,0} + R_{\mathrm{PF}} V_{Z0}\omega(t) + L_{\mathrm{P}} V_{Z0}\dot{\omega}(t)$ | $c_{\mathrm{B,PF}}, R_{\mathrm{PF}}, L_{\mathrm{P}}, p_{2,0\mathrm{PF}}$ | $A, V_{Z0}$ |
| Suspension (hydraulic) | $\ddot{z}_{\mathrm{B}}(t) = \frac{A_{\mathrm{p}}}{m_{\mathrm{B}}} p_2(t) - \frac{A_{\mathrm{p}} R_{\mathrm{p2}} V_{Z0}}{m_{\mathrm{B}}}\omega(t) + \frac{d_{\mathrm{Bi,p2}}}{m_{\mathrm{B}}}\dot{z}_{\mathrm{WB}}(t) + \frac{F_{\mathrm{C,p2}}}{m_{\mathrm{B}}}\operatorname{sign}(\dot{z}_{\mathrm{WB}}(t))$ | $A, R_{\mathrm{p2}}, d_{\mathrm{B+,p2}}, d_{\mathrm{B-,p2}}, F_{\mathrm{C,p2}}$ | $m_{\mathrm{B}}$ |
| Suspension (mechanic) | $\ddot{z}_{\mathrm{B}}(t) = \frac{c_{\mathrm{B,zP}}}{m_{\mathrm{B}}}(z_{\mathrm{WB}}(t) + z_{\mathrm{P}}(t)) + \frac{d_{\mathrm{B,i,zP}}}{m_{\mathrm{B}}}\dot{z}_{\mathrm{WB}}(t) + \frac{F_{\mathrm{C,zP}}}{m_{\mathrm{B}}}\operatorname{sign}(\dot{z}_{\mathrm{WB}}(t))$ | $c_{\mathrm{B,zP}}, d_{\mathrm{B+,zP}}, d_{\mathrm{B-,zP}}, F_{\mathrm{C,zP}}$ | $m_{\mathrm{B}}$ |
| Motor | $u_{\mathrm{M}}(t) = R_{\mathrm{M}} i_{\mathrm{M}}(t) + \psi\omega_{\mathrm{M}}(t)$ | $R_{\mathrm{M}}, \psi$ | |
| Pump | $i(t) = \frac{V_{Z0}}{\psi_{\mathrm{M}}}(p_2(t) - p_1(t)) + \frac{J}{\psi_{\mathrm{M}}}\dot{\omega}(t) + \frac{M_{\mathrm{R2}}}{\psi_{\mathrm{M}}}\operatorname{sign}(\omega(t))\,\omega^2(t)$ | $\psi_{\mathrm{M}}, c_{\mathrm{S,M}}, M_{\mathrm{R2}}, J, p_{10,\mathrm{M}}$ | $V_{Z0}$ |
| Hydraulic accumulator | $p_1(t) = -c_{\mathrm{S,S}} V_{Z0}\int\omega(t)dt + p_{1,0} - R_{\mathrm{S}} V_{Z0}\omega(t) - L_{\mathrm{S}} V_{Z0}\omega(t)$ | $c_{\mathrm{S,S}}, p_{10,\mathrm{S}}, R_{\mathrm{S}}, L_{\mathrm{S}}$ | $V_{Z0}$ |

The models of all components are combined and used for substituting nonmeasured signals. The resulting equations are summarized in Table 15.2. As these equations only contain measured signals, parameter estimation can be applied to estimate the coefficients. In order to calculate particular physical parameters, some parameters have to be known a priori. As noted in the last column of the table, by successive calculation of the physical parameters of the various equations, all physical parameters can be obtained with only some parameters known a priori. Details are described in Fischer (2006). The dynamic models of this active hydraulic suspension can be used for the design of the control system and can be used for model-based fault diagnosis.

The parameters of the active hydraulic suspension have been estimated on the test rig of Fig. 15.10. The base plate was excited with a measured road profile for $v_X = 15$ m/s. The power density spectrum is depicted in Fig. 15.12 and shows resonance frequencies around 1 to 2 Hz for the body and around 12 Hz for the wheel. The estimated parameters with the DSFI-method (discrete square root filtering information form) are exemplarily shown in Fig. 15.13 for the mechanical part of the suspension; see Table 15.2, fourth row. They converge within a few seconds. For more details, see Fischer et al. (2004) and Fischer (2006).

The models of Table 15.2 have been used to develop a model-based fault diagnosis system for the considered active suspension system; see Fischer (2006).

The control system for the active hydraulic suspension system on the test rig is depicted in Fig. 15.11. It consists of a cascaded control with a q-current controller of the brushless DC motor applying field-oriented control in the lowest level and a superimposed motor speed controller, which obtains its reference value from the body controller. The body controller is based on a state-space controller. It uses the state variables suspension deflection $z_{WB}$, body velocity $\dot{z}_B$, and body acceleration $\ddot{z}_B$.

The combination of the ABC system with a laser-aided road profile sensing allows a chassis-control with preview; see Streiter (2008). It was demonstrated first time with the research vehicle F700 from Mercedes called PRE-SCAN and has demonstrated further progress in ride comfort.

## 15.5 Tire Pressure Monitoring with Wheel and Suspension Sensors

A survey conducted by Michelin in 2006 has shown that only 6.5% of the 20.300 inspected cars had the required tire pressure at all four wheels. More than 39.5% of the cars had at least one extremely underinflated tire ($<$1.5 bar); see Bridgestone (2007). It is well known that it is dangerous to drive with underinflated tires. First of all, the risk of an accident increases due to the worse vehicle dynamic properties and the increased probability of a tire burst. Because of the increasing deformation, the tire heats up and its structure destabilizes; see Normann (2000). Furthermore, tire deflation increases fuel consumption and tire wear.

There are different approaches to monitor the tire pressure. *Directly measuring systems* use *pressure sensors* for tire pressure measurement; see Normann (2000), Mate and Zittlau (2006), and Wagner (2004). However, battery for power supply and wireless data transmission increase the complexity and costs of such systems. Due to extreme environmental conditions like a large operating range of temperature and high accelerations, the sensor unit in the tire has to be very robust. Therefore, alternatives like indirectly measuring systems are desired. *Indirectly measuring systems* estimate the tire pressure using sensor signals of the wheel or the suspension which may be already available for other vehicle dynamic control systems and are influenced by the tire pressure. Thus, the wheel speed $\omega$ and the vertical wheel acceleration $\ddot{z}_W$ can be used for indirect tire pressure monitoring.

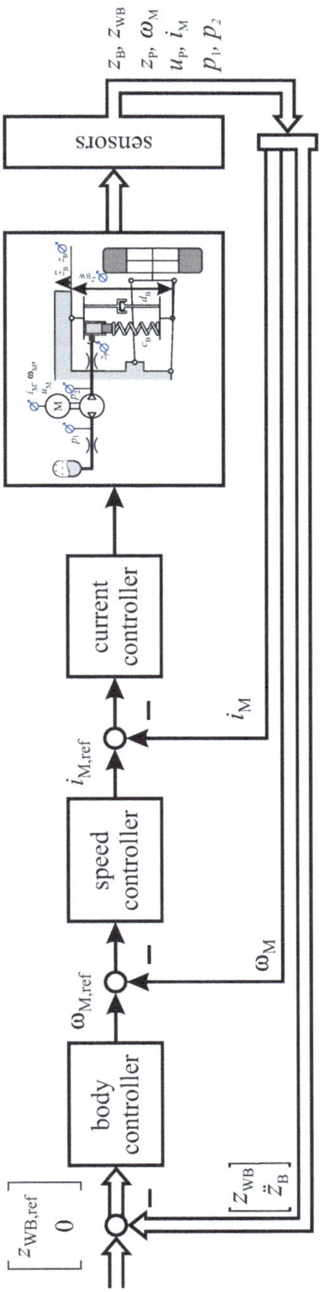


**Fig. 15.11** Signal flow of the control system of the active hydraulic suspension on the test rig

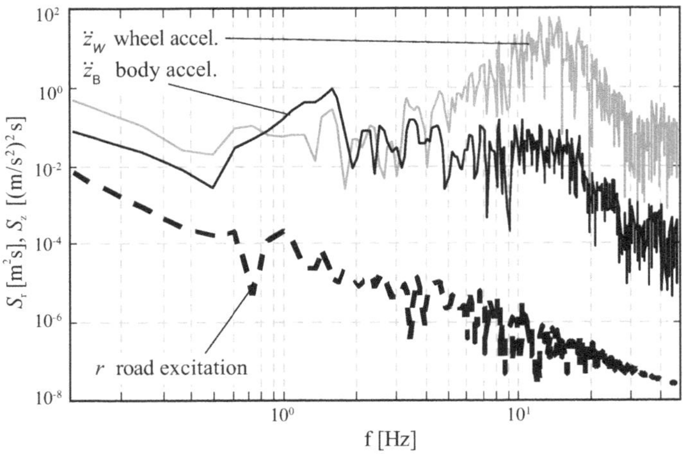


**Fig. 15.12** Excitation of the test rig; power density spectrum of measured road excitation, wheel and body acceleration for $v_X = 15$ m/s

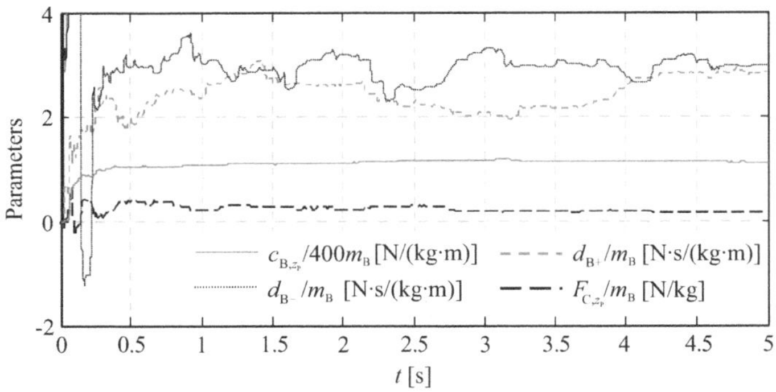


**Fig. 15.13** Parameter estimation results of Eq. (15.4.5) and Table 15.2

### 15.5.1 Comparison of Wheel's Speeds

One option for indirect tire pressure monitoring is to compare the wheel speeds. As shown in Fig. 15.14a, the vertical force $F_Z$ of a tire is supported by the tire wall with its stiffness $c_W$ and by the pressure $p$ inside the tire. A loss in tire pressure will result in an increase of the contact area $A_c$ and an increasing compression of the tire wall. Both result in a decrease of the dynamic tire radius $r_{dyn}$. Therefore, the wheel speed increases to achieve the same velocity; see Persson et al. (2002) and Mayer et al. (1996). However, with increasing speed, the tire radius increases again because of centrifugal and speed-dependent stiffness effects. This compensates for a part of the tire radius reduction.

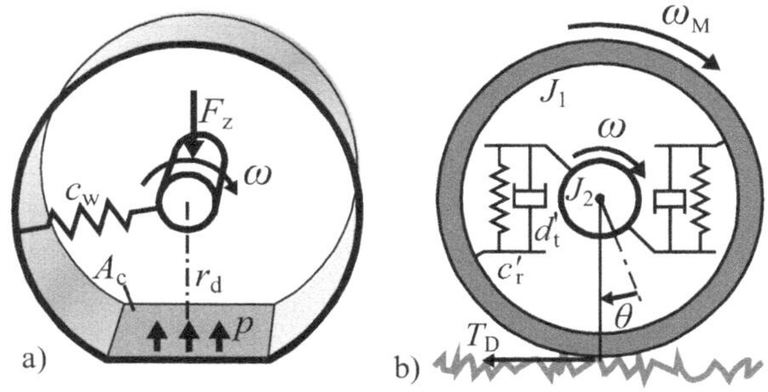


**Fig. 15.14** Influence of tire pressure on **a** wheel radius; **b** torsional wheel oscillations

As the exact velocity of a vehicle is unknown, the wheel speeds can only be compared among one another. Therefore, an equal loss of pressure in all tires cannot be detected. To evaluate the differences among the wheel speeds, *different residuals* can be composed. One option to detect differences among the wheel speeds is by using the relative wheel speed difference

$$r_{\mathrm{rel},i} = \frac{\omega_i - \bar{\omega}}{\bar{\omega}} \qquad \textit{with } i \in \{\mathrm{fl}, \mathrm{fr}, \mathrm{rl}, \mathrm{rr}\}, \tag{15.5.1}$$

where $\bar{\omega}$ denotes the mean value of all wheel speeds

$$\bar{\omega} = \frac{\omega_{\mathrm{fl}} + \omega_{\mathrm{fr}} + \omega_{\mathrm{rl}} + \omega_{\mathrm{rr}}}{4}. \tag{15.5.2}$$

Another option to evaluate the wheel speed differences is to compute a diagonal wheel speed difference

$$r_{\mathrm{diag}} = \frac{(\omega_{\mathrm{fl}} + \omega_{\mathrm{rr}}) - (\omega_{\mathrm{fr}} + \omega_{\mathrm{rl}})}{\omega_{\mathrm{fl}} + \omega_{\mathrm{fr}} + \omega_{\mathrm{rl}} + \omega_{\mathrm{rr}}}. \tag{15.5.3}$$

This residual already contains some compensation of disturbing influences by its way of computation. As only one residual is computed, it fails to locate the deflated tire.

In Mayer et al. (1996), wheel speed ratios of two neighboring wheels

$$r_{\mathrm{rat}} = \frac{\omega_i}{\omega_j} - 1 \qquad \textit{with } i, j \in \{\mathrm{fl}, \mathrm{fr}, \mathrm{rl}, \mathrm{rr}; \mathrm{fl}, \mathrm{fr}, \mathrm{rl}, \mathrm{rr}\} \tag{15.5.4}$$

are introduced to detect changes among the wheel speeds.

Figures 15.15 and 15.16 show the application of the residuals (15.5.1) to (15.5.4) to measured data of a test vehicle Opel Omega 2.0i performing a straight driving maneuvre. Figure 15.15 shows the results from a test drive where all tires were inflated with the recommended tire pressure of 2.0 bar. The wheel speed ratios $r_{\mathrm{rat,fl,rl}}$ and $r_{\mathrm{rat,fr,rr}}$ are a little smaller than 0. These ratios are composed of a front wheel speed in the numerator and a rear wheel speed in the denominator. As the rear wheels are the driven wheels, they are subject to a higher wheel slip. Therefore, the rear

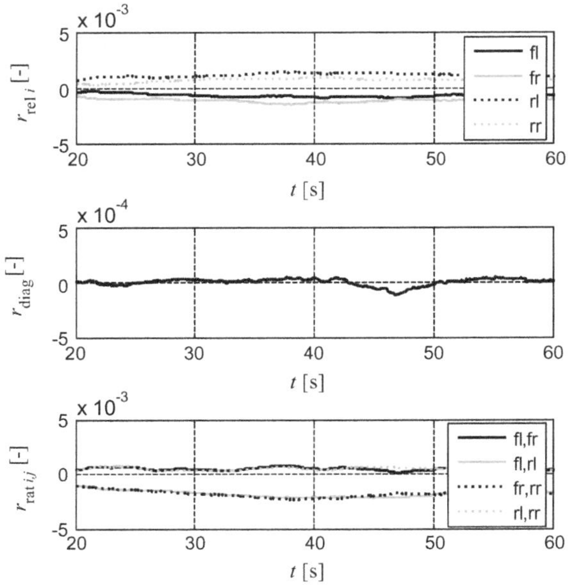


**Fig. 15.15** Residuals for the evaluation of wheel speed differences, straight driving, $p_i = 2.0$ bar with $i \in \{\text{fl, fr, rl, rr}\}$

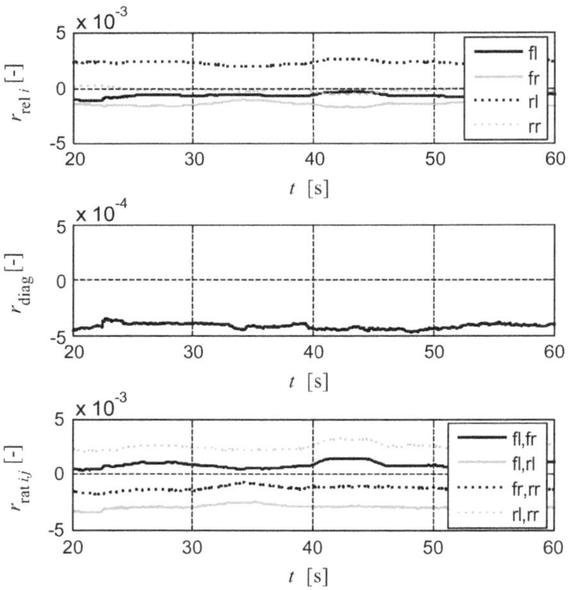


**Fig. 15.16** Residuals for the evaluation of wheel speed differences, straight driving, $p_i = 2.0$ bar with $i \in \{\text{fl, fr, rl, rr}\}$, $p_{rl} = 1.5$ bar

wheel speeds are a little larger than the ones at the front wheels and cause the small deflection of these residuals.

Figure 15.16 shows the results from a test drive where the pressure of the left rear tire was reduced to 1.5 bar. A comparison of these results with that from Fig. 15.15 shows that all three criteria react to the deflation with at least one residual. The relative difference of the left rear tire $r_{\mathrm{rel,rl}}$ is larger than the remaining ones. This indicates an increasing $\omega_{\mathrm{rl}}$ of the left rear tire. This influences also the corresponding wheel speed ratios $r_{\mathrm{rat,fl,rl}}$ and $r_{\mathrm{rat,fr,rr}}$ in an increasing and decreasing manner, respectively. Both methods, eqs. (15.5.1) and (15.5.4), allow a location of the inflated tire. This is not possible using the diagonal difference $r_{\mathrm{diag}}$. In the case of a slalom maneuver, all residuals are affected by cornering. Therefore, a *curve compensation* by taking the lateral speed differences into account was developed. This allows to detect a deflated tire also during cornering; see Wesemeier (2012).

### 15.5.2 Torsional Wheel Speed Oscillations

At the mantle of a tire, a disturbance torque $T_{\mathrm{D}}$ is generated by variations in the street height and the friction coefficient. This and the elasto-kinematic suspension causes oscillations which are transferred from the wheel speed at the mantle $\omega_{\mathrm{M}}$ to the wheel speed $\omega$ of the wheel rim; see Persson et al. (2002) and Prokhorov (2005). As shown in Fig. 15.14b, the tire possesses a torsional stiffness $c'_{\mathrm{t}}$ and a damping factor $d'_{\mathrm{t}}$. Balance equations of torques at the mantle lead to

$$J_1 \dot{\omega}_{\mathrm{M}} = T_{\mathrm{D}} - c'_{\mathrm{t}}\theta - d'_{\mathrm{t}}\dot{\theta}, \tag{15.5.5}$$

and at the wheel rim

$$J_2 \dot{\omega}_{\mathrm{M}} = c'_{\mathrm{t}}\theta + d'_{\mathrm{t}}\dot{\theta}, \tag{15.5.6}$$

where $J_1$ and $J_2$ are the moments of inertia of the mantle and the rim, respectively. The twist angle $\theta$ between mantle and rim results from $\dot{\theta} = \omega_{\mathrm{M}} - \omega$. Applying the Laplace Transformation and inserting Equation (15.5.6) in (15.5.5) yield the transfer function

$$G(s) = \frac{\omega(s)}{T_{\mathrm{D}}(s)} = \frac{d'_{\mathrm{t}}s + c'_{\mathrm{t}}}{J_1 J_2 s^3 + (J_1 + J_2)d'_{\mathrm{t}}s^2 + (J_1 + J_2)c'_{\mathrm{t}}s}. \tag{15.5.7}$$

As the torsional tire stiffness $c'_{\mathrm{t}}$ is dependent on the tire pressure, changes of it can be detected by applying spectral analysis methods to the wheel speed $\omega$.

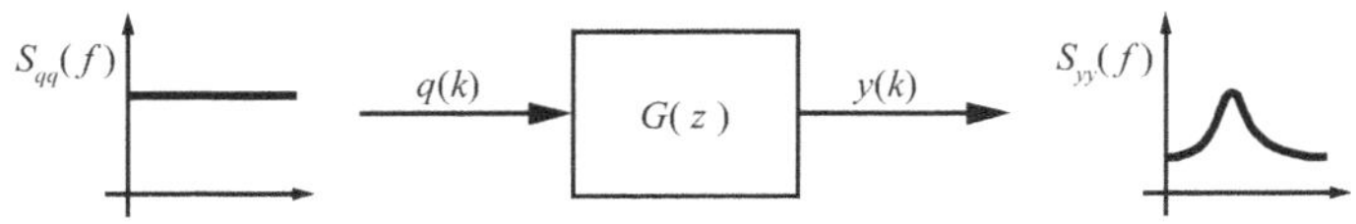


**Fig. 15.17** Assumed composition of signals for parametric spectral estimation

**(a) Parametric Spectral Analysis**

To detect changes of the tire pressure from changes in a signal's spectrum, a time-dependent analysis in the frequency domain has to be performed. This can be achieved by methods of parametric spectral analysis. Using parametric spectral analysis, a signal $y(k)$ is assumed to be composed as an output of a linear shaping filter $G(z)$ excited by white noise $q(k)$, e.g. Isermann (2005), as shown in Fig. 15.17.

Following the Wiener–Lee relation, the power spectral density (PSD) $S_{\mathrm{yy}}(f)$ of the output of the filter $G(z)$ is obtained from $S_{\mathrm{qq}}(f)$ of the input noise by

$$S_{\mathrm{yy}}(f) = S_{\mathrm{qq}}(f)\,|G(z)|^2 = \sigma_{\mathrm{q}}^2\,|G(z)|^2\,, \tag{15.5.8}$$

where $\sigma_{\mathrm{q}}^2$ is the variance of the white noise $q(k)$. For determination of the PSD $S_{\mathrm{yy}}(f)$, the parameters of the shaping filter $G(z)$ and the variance $\sigma_{\mathrm{q}}^2$ have to be estimated. Locations of spectral peaks are given by the resonance frequencies of the shaping filter. Following Kammeyer and Kroschel (2006), the filter $G(z)$ is modeled in autoregressive form with the discrete difference equation

$$y(k) = q(k) - a_1 y(k-1) - \ldots - a_n y(k-n) \tag{15.5.9}$$

in order to receive a linear parameter estimation problem. The model parameters $a_1, \ldots, a_n$ can be estimated using least squares parameter estimation (LS). Another option for the identification of a dynamic system is the total least squares parameter estimation (TLS) where input and output of a system are assumed to be corrupted by noise; see Isermann and Münchhof (2011). Both methods result in good results. For a detailed description of the parametric spectral estimation methods, see Wesemeier (2012).

**(b) Power Spectral Density (PSD) of the Wheel Speed**

The method of spectral analysis is now applied to the measured wheel speeds from the test vehicle. As frequency components influenced by the tire pressure are expected between 40 Hz and 50 Hz (Persson et al. 2002; Prokhorov 2005), a band pass filter is applied to focus the analysis to a smaller frequency range. Figure 15.18 shows the PSD of a rear tire which was inflated with the recommended pressure of 2.0 bar. In Fig. 15.19, the tire pressure was reduced to 1.5 bar. The PSD from a fast Fourier transformation (FFT) has many peaks such that the frequency of the resonance maximum is hard to detect. The PSDs from the parametric spectral analysis are much smoother. The PSD using LS spectral analysis matches the general trend of the FFT quite well. Although the maximum of the PSD using TLS is far too large, the resonance frequencies $f_{\mathrm{res}}$ where the maxima are located are almost the same for all

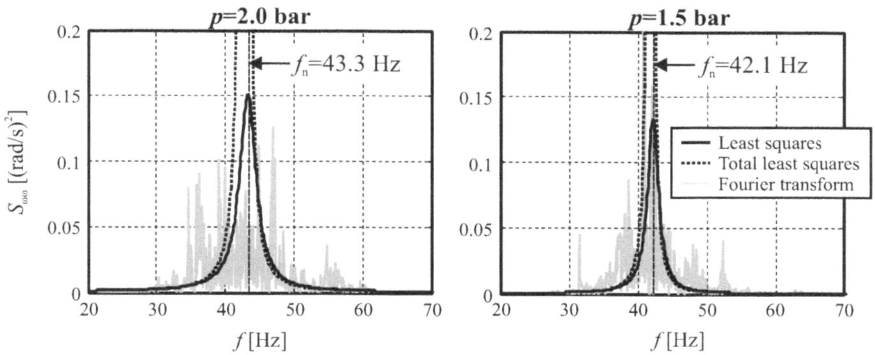


**Fig. 15.18** Estimated PSDs of the wheel speed, $p = 2.0$ bar and $p = 1.5$ bar

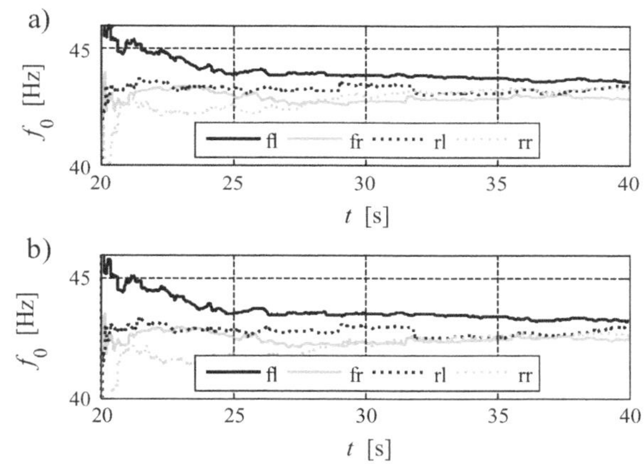


**Fig. 15.19** Time-dependent spectral analysis of the wheel speeds of a straight driving maneuver, $p_i = 2.0$ bar with $i$ =fl,fr,rl,rr. **a** LS spectral analysis, **b** TLS spectral analysis

three methods. In the LS spectrum, the resonance frequency is given by 43.3 Hz. In Fig. 15.18, the resonance frequency of the deflated tire decreases to 42.1 Hz, i.e. by a difference of 1.2 Hz or 2.7%.

**(c) Time-Dependent Spectral Analysis of the Wheel Speeds**

Applying recursive parameter estimation methods, a time-dependent spectral analysis can be performed. A new set of resonance frequencies is then calculated in each time sample. Instead of the resonance frequency $f_{\text{res}}$, the natural frequency $f_0$ can be used; see Wesemeier and Isermann (2008). Figures 15.19 and 15.20 show the results of the spectral analysis of a straight driving maneuver. In (a), the result using least squares is shown while (b) shows the result using total least squares. In Fig. 15.19, all tires are inflated with the correct tire pressure of 2.0 bar. After some larger variations resulting from insufficient data for a reliable parameter estimation, all estimated natural frequencies $f_0$ converge around 43 Hz for both estimation methods. When the pressure of the right rear tire is reduced to 1.5 bar, $f_0$ of the rear tires is smaller than that of one of the front tires; see Fig. 15.20. The smallest $f_0$ is estimated for

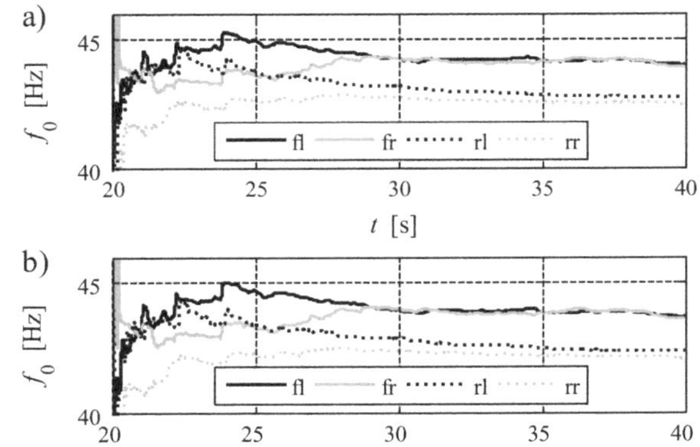


**Fig. 15.20** Time-dependent spectral analysis of the wheel speeds of a straight driving maneuver, $p_i = 2.0$ bar with $i$ =fl,fr,rl; $p_{\mathrm{rr}} = 1.5$ bar. **a** LS spectral analysis, **b** TLS spectral analysis

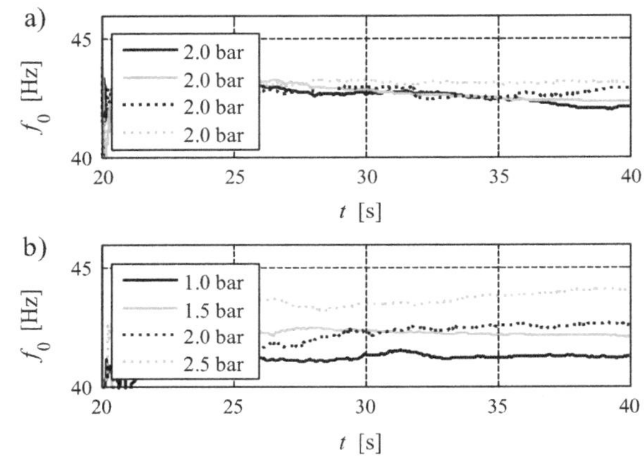


**Fig. 15.21** Natural frequencies of the wheel speeds of the left (**a**) and right (**b**) rear tire using total least squares and different tire pressures at the right rear tire

the deflated right rear tire. As the rear axle is the driven axle featuring a differential gear, the left rear tire is also influenced by the tire deflation.

Figure 15.21 shows the natural frequencies of the rear tires using total least squares spectral analysis for several test drives. The right rear tire in b) was inflated with different tire pressures while the pressure of the left rear tires in a) was kept constant. A correlation between tire pressure and natural frequency is observable. The lower the pressure of the right rear tire, the smaller is its natural frequency. $f_0$ of the left rear tire is less affected by the deflation of the right rear tire.

### 15.5.3 Vertical Wheel Acceleration

Other signals from the suspension system are also influenced by the tire pressure; see Börner et al. (2002), Weispfenning (1996), and Börner et al. (2002). These are, e.g. the spring deflection $z_{WB}$, vertical body acceleration $\ddot{z}_B$, or vertical wheel acceleration $\ddot{z}_W$. Figure 15.22 shows the schematic of a vehicle suspension system as a quarter car. Excited by variations in road height $z_{\mathrm{T}}$, the vertical displacement $z_{\mathrm{W}}$ of the wheel

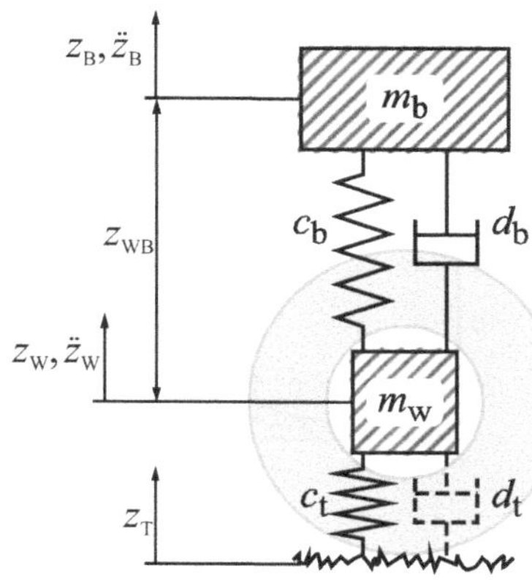


**Fig. 15.22** Quarter car model

with its mass $m_W$ and the vertical displacement $z_B$ of the vehicle body with its mass $m_b$ are subject to oscillations. The wheel is connected to the vehicles body by the suspension system represented by the spring stiffness $c_b$ and the shock absorbers damping factor $d_b$. The wheel's elastic properties can be expressed by the vertical tire stiffness $c_t$. The tires damping factor $d_t$ is very small such that it can be neglected. Following first principles, the following equations concerning the balance of forces can be derived

$$m_b\ddot{z}_B = d_b(\dot{z}_W - \dot{z}_B) + c_b(z_W - z_B), \qquad (15.5.10)$$

$$m_W\ddot{z}_W = -d_t(\dot{z}_W - \dot{z}_B) - c_b(z_W - z_B) + c_t(z_T - z_W). \qquad (15.5.11)$$

As the body motion $z_B, \dot{z}_B$ is much slower than the wheel motion $z_W, \dot{z}_W$, it can be neglected to derive a transfer function from the road height $z_T$ to the displacement of the wheel $z_T$; see Börner et al. (2002). Applying the Laplace transformation of equation (15.5.10) and inserting $z_B = 0$ result in the transfer function

$$G_1(s) = \frac{z_W(s)}{z_T(s)} = \frac{\frac{c_t}{c_b+c_t}}{\frac{m_W}{c_b+c_t}s^2 + \frac{d_b}{c_b+c_t}s + 1}. \qquad (15.5.12)$$

The transfer function from variations in road height to the vertical wheel acceleration $\ddot{z}_W$ is then given by

$$G_1(s) = \frac{\ddot{z}_W(s)}{z_T(s)} = \frac{\frac{c_t}{c_b+c_t}s^2}{\frac{m_W}{c_b+c_t}s^2 + \frac{d_b}{c_b+c_t}s + 1}. \qquad (15.5.13)$$

As the tire stiffness $c_t$ depends on the tire pressure, changes in the tire pressure can be observed as changes in the spectrum of the vertical wheel acceleration.

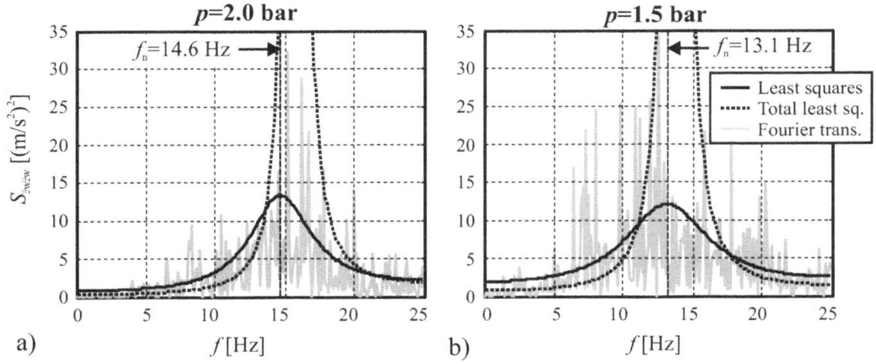


**Fig. 15.23** Estimated PSD of the vertical wheel acceleration, **a** $p = 2.0$ bar and **b** $p = 1.5$ bar, $v = 80$ km/h

**(a) Power Spectral Density of the Vertical Wheel Acceleration**

The third investigated method for indirect tire pressure monitoring is to use the sensors measuring the vertical wheel acceleration; see Börner et al. (2002) and Weispfenning (1996). These acceleration sensors are already available for semi-active shock absorbers; see Kutsche and Rappelt (2006). The spectral analysis methods are also applied to these signals. Figure 15.23a shows the estimated power spectral densities of the right rear tire of a test drive with a tire pressure $p = 2.0$ bar using a straight driving maneuver. Figure 15.23b shows the estimated PSDs of a test drive with $p = 1.5$ bar using the same maneuver. The PSDs resulting from least squares and total least squares spectral analysis and using of FFT are shown. In both figures, the general trend of all three PSDs is very similar. As observed before in Sect. 15.5.2, the maximum of the PSD using total least squares is too large. But the frequencies where the maxima of the PSDs using TLS are located agree with that from least squares spectral analysis and the PSD from FFT. The frequency with the PSD using least squares has its maximum changes from 14.6 Hz at a tire pressure of 2.0 bar to 13.1 Hz at a tire pressure of 1.5 bar, i.e. by 1.5 Hz or 10 %. Filtering of the oscillation was not required because of a good signal/noise ratio.

**(b) Time-dependent Spectral Analysis of the Wheel Speeds**

The time-dependent spectral analysis in Fig. 15.24 results in a natural frequency $f_0$ 14 Hz for all tires using least squares and total least squares spectral analysis when all tires are inflated with a pressure $p = 2.0$ bar. In Fig. 15.25, the right rear tire's pressure is reduced to 1.5 bar. Its natural frequency $f_0$ differs significantly from the $f_0$ of the remaining tires. The difference of $f_0$ between correctly inflated tires and underinflated tire is larger when the total least squares spectral analysis is used instead of the least squares spectral analysis. Figure 15.26 shows the natural frequencies $f_0$ of the rear tires using the total least squares spectral analysis. The tire pressure of the right rear tire in b) is varied from 1.0 bar to 2.5 bar and the left rear tire's pressure in a) is kept constant to 2.0 bar. A significant dependency between tire pressure and $f_0$ can be observed at the right rear tire. There is no influence of the deflation on the left

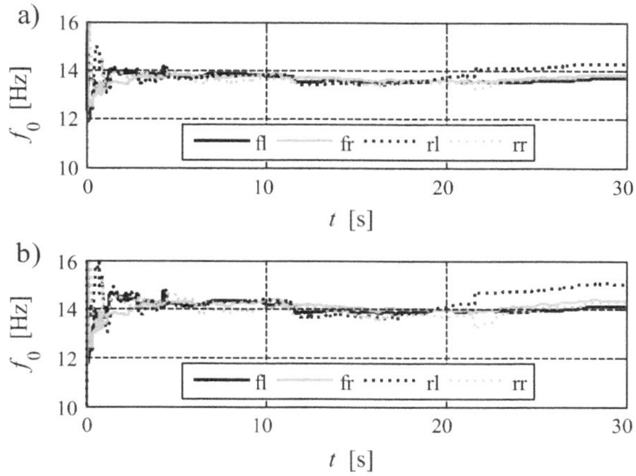


**Fig. 15.24** Time-dependent spectral analysis of the vertical wheel accelerations of a straight driving maneuvre, $p_i = 2.0$ bar with $i$ =fl,fr,rl,rr. **a** LS spectral analysis, **b** TLS spectral analysis

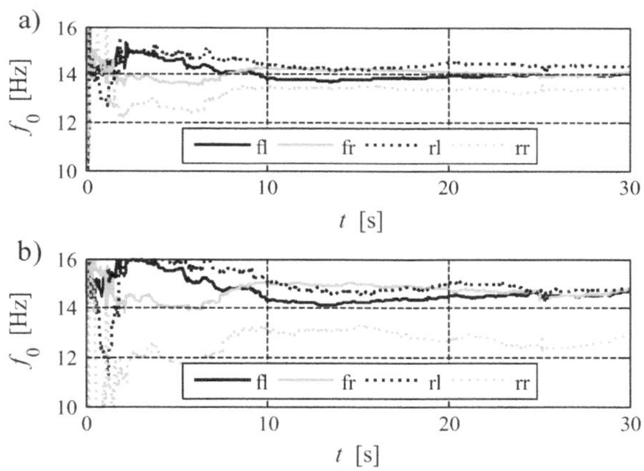


**Fig. 15.25** Time-dependent spectral analysis of the vertical wheel accelerations of a straight driving maneuvre, $p_i = 2.0$ bar with i = fl,fr,rl; $p_{\text{rr}} = 1.5$ bar. **a** LS spectral analysis, **b** TLS spectral analysis

rear tire's $f_0$ observable. The deflection of $f_0$ of the right rear tire is larger compared to the deflection of $f_0$ by using the wheel speed oscillation.

### 15.5.4 Comparison and Fusion of the Methods

The properties of the investigated three methods are summarized in Table 15.3. The residuals which evaluate the *wheel speed differences* feature low computational effort compared to the spectral analysis methods. The *relative differences of the wheel speeds*, the wheel speed ratios, and the spectral analysis of the *vertical wheel acceleration* show a good response to a deflated tire. The response of the *spectral analysis of the torsional wheel speed oscillations* is worse and it is more sensitive to disturbances. This is because of the small changes in the resonance frequency. (This may for the shown example be partly due to the location of the wheel speed sensors which for the investigated car are located at some distance to the wheel and close to

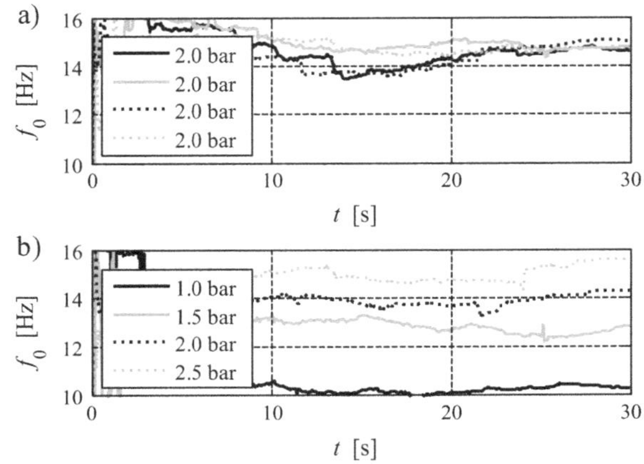


**Fig. 15.26** Natural frequencies of the vertical wheel acceleration of (**a**) the left and (**b**) the right rear tire using total least squares and different tire pressures at the right rear tire

**Table 15.3** Properties of indirect tire pressure estimation methods (+ positive, ++ very positive, - negative)

| | Computational effort | Size of deflection | Disturbance sensitivity | Localization ability | Relative tire deflation | Absolute tire deflation detection |
|---|---|---|---|---|---|---|
| 1. Wheel speed comparison | | | | | | |
| (a) Relative difference | + | ++ | - | + | + | – |
| (b) Diagonal difference | ++ | + | + | – | + | – |
| (c) Wheel speed ratios | + | ++ | ++ | + | + | – |
| 2. Wheel speed spectral analysis | – | + | – | + | + | + |
| 3. Vertical wheel acceleration spectral analysis | – | ++ | ++ | + | + | + |

the differential gear.) Using the *vertical wheel acceleration*, the total least squares spectral analysis reacts more sensitive to changes in the tire pressure than the least squares spectral analysis. This allows a more accurate estimation of the tire pressure. This vertical wheel speed acceleration is available in series production semi-active shock absorbers; see Kutsche and Rappelt (2006). The residuals for the evaluation of wheel speed differences only work for the detection of the deflation of a single tire. A deflation of all tires therefore cannot be detected. However, this can be achieved with the spectral analysis methods by setting a lower bound for the natural frequencies.

To combine the advantages of the different methods and to improve the reliability of the indirect tire pressure monitoring systems, at least two methods using different signals can be combined. This can be achieved with a fuzzy-logic evaluation; see Wesemeier (2012). The changes of different residuals can be combined such that an increasing number of deflected residuals increase the probability of detecting a tire deflation. To incorporate redundancy, at least one residual for wheel speed comparison and the spectral analysis of at least one signal preferably the vertical tire acceleration should be used. The accuracy of the spectral analysis can be improved if the difference of the natural frequency from a reference value calibrated at the correct tire pressure is used. Additionally, the absolute natural frequency should be monitored to detect a loss of tire pressure in all tires simultaneously.

More results as well as the application of further spectral analysis methods are presented in Wesemeier (2012).

## References


Alleyne A, Hedrick J (1995) Non-linear adaptive control of active suspensions. IEEE Trans Control Syst Technol 3(1):94–101

Börner M, Andréani L, Albertos P, Isermann R (2002) Detection of lateral vehicle driving conditions based on the characteristic velocity. In: IFAC World Congress 2002, Barcelona, Spain

Bridgestone (2007) Umweltbelastung durch Reifen mit zu wenig Druck. online document

Börner M, Straky H, Weispfenning T, Isermann R (2002) Model based fault detection of vehicle suspension and hydraulic brake systems. Mechatronics 12(8):999–1010

Bußhardt J (1995) Selbsteinstellende Feder-Dämpfer-Last-Systeme für Kraftfahrzeuge. Dissertation Technische Hochschule Darmstadt. Fortschr.-Ber. VDI Reihe 12, 240. VDI Verlag, Düsseldorf

Causemann P (2008) Kraftfahrzeugstoßdämpfer. Verlag moderne industrie AG, Landsberg/Lech

Choi S, Choi Y, Chang E, SJHan, CSKim, (1998) Control characteristics of a continuously variable ER damper. Mechatronics 8(2):143–161

Choi S, Lee H, Chang E (2001) Mechatronics 11(3):345–353

Dessort R, Chucholowski C (2017) Explicit model predictive control of semi-active suspension systems using Artificial Neural Networks (ANN), pp 207–228

Fees G (2001) Study of the static and dynamic properties of a highly dynamic ER servo drive. Ölhydraulik und Pneumatik 45(1):45–47

Fischer D (2006) Fehlererkennung für mechatronische Fahrwerksysteme. Dissertation Technische Universität Darmstadt. Fortschr.-Ber. VDI Reihe 12, 615. VDI Verlag, Düsseldorf

Fischer D, Schöner HP, Isermann R (2004) Model-based fault detection for an active vehicle suspension. In: FISITA World Automotive Congress, Barcelona, Spain

Folchert U (2006) Elektronisch geregelte Luftfedersysteme. Vieweg, Wiesbaden, pp 265–283

Hayakawa K, Matsumoto K, Yamashita M, Suzuki Y, Fujimori K, Kimura H (1999) Robust $H^{\infty}$-output feedback control of decoupled automobile active suspension systems. IEEE Trans Autom Control 44(2):392–396

Heissing B, Ersoy M (eds) (2011) Fahrwerkhandbuch: Grundlagen, Fahrdynamik, Komponenten, Systeme, Mechatronik. Perspektiven, ATZ/MTZ Fachbuch, Vieweg

Hilgers C, Brandes J, Ilias H, Oldenettel H, Stiller A, Treder C (2009) Aktive air spring suspension for greater range between adjusting for comfort and dynamic driving. ATZ worldwide 111(9):12–17

Isermann R (2003) Mechatronic systems—fundamentals. Springer, London

Isermann R (2005) Mechatronic systems—fundamentals, 2nd edn. Springer, London

Isermann R, Münchhof M (2011) Identification of dynamic systems. Springer, Berlin, Heidelberg

Kammeyer KD, Kroschel K (2006) Digitale Signalverarbeitung, Filterung und Spektralanalyse, Teubner-verlag
Karlsson N, Dahleh M, Hrovat D (2001) Nonlinear active suspension with preview. In: Proceedings of the 2001 American Control Conference, Arlington, Virginia, USA, vol 4, pp 2640–2645
Karnopp D (1979) Are active suspensions really necessary. Mech Eng, ASME 101(3):86–86
Karnopp D (1987) Force generation in semi-active suspensions using modulated dissipative elements. Vehicle Syst Dyn 16(5–6):333–343
Kim W, Choi S (1999) Vibration control of a semi-active suspension featuring electrorheological fluid dampers. J Sound Vib 234(3):537–546
Krtolica R, Hrovat D (1990) Optimal active suspension control based on a half-car model. 29th IEEE Conference. Decision and Control, Honolulu, HI, USA, pp 2238–2243
Kutsche T, Rappelt S (2006) Semiaktive Stoßdämpfer und aktive Radaufhängungen. Vieweg, Wiesbaden, pp 252–264
Majjad R (1997) Estimation of suspension parameters. In: Proceedings of the 1997 IEEE International Conference on Control Applications, Hartford, CT, USA, pp 522–527
Mate JL, Zittlau D (2006) Elektronik für mehr Sicherheit—Assistenz—und Sicherheitssysteme zur Unfallvermeidung. ATZ—Automobiltechnische Zeitschrift 108(7–8):578–585
Mayer A, Emig G, Gmehling B, Popovska N, Holemann K, Buck A (1996) Passive regeneration of catalyst-coated knitted-fiber diesel particulate traps. SAE Trans 105(4):36–44
Moosheimer J, Waller H (1999) Reduction of vibrations by bang-bang controlled electrorheological dampers. Arch Appl Mech 70:715–737
Normann N (2000) Reifendruck-Kontrollsystem für alle Fahrzeuglklassen. ATZ - Automobiltechnische Zeitschrift 102(11):950–956
Oya M (1998) Robust control of active automotive suspensions with model uncertainties. In: Proceedings of Advances in Vehicle Control and Safety, pp 467–472
Persson N, Gustafsson F, Drevö M (2002) Indirect tire pressure monitoring using sensor fusion. SAE Trans 111:1657–1662. http://www.jstor.org/stable/44719343
Prokhorov D (2005) Virtual sensors and their automotive applications. In: 2005 International Conference on Intelligent Sensors, Sensor Networks and Information Processing, pp 411–416
Pyper M, Schiffer W, Schneider W (2003) ABD—Active Body Control. Von der Blattfederung zum aktiven System für mehr Fahrsicherheit und Fahrkomfort, Die Bibliothek der Technik, vol 241. Verlag moderne industrie AG, Landsberg/Lech
Rajamani R (2012) Vehicle dynamics and control, 2nd edn. Mechanical Engineering Series, Springer, US, New York
Reimpell J, Stoll H (1989) Fahrwerktechnik. Stoßdämpfer, Feder-und Dämpferbeine, Motorschwingungsdämpfer, Konstruktion und Einbau. Vogel, Würzburg, Stoß-und Schwingungsdämpfer
Reimpell J, Stoll H, Betzler J (2001) The Automotive Chassis, 2nd edn. Reed Educational and Professional Publishing Ltd, Oxford
Robert Bosch GmbH (1996) Automotive Handbook, 4th edn. Robert Bosch GmbH, Stuttgart
GmbH Robert Bosch (ed) (2011) Automotive Handbook, 8th edn. Bentley publishers, Cambridge
Roukieh S, Titli A (1993) Robust sliding mode control of semi-active and active suspension for private cars. In: 12. IFAC World Congress, Sidney, Australia, vol 3, pp 155–160
Salman M, Lee A, Boustany N (1988) Reduced design of active suspension control. Proceedings of the 27th Conference on Decision and Control, Austin, Texas, vol 2, pp 1038–1043
Sommer S (2003) Electronic air suspension with continuous damping control. AutoTechnology 3(2):52–55
Streiter R (1996) Entwicklung und Realisierung eines analytischen Regelkonzeptes für eine aktive Federung. Technische Universität, Berlin
Streiter R (2008) Active preview suspension system. ATZ Worldwide 110(5):4–11
Strom JN, Lohmann B (2017) Vorausschauende Strörgrößenaufschaltung für die Schwingungsdämpfung am Viertelfahrzeug. at - Automatisierungstechnik 65(8):596–607

Ursu I, Ursu F, Vladimirescu M (1997) The synthesis of two suboptimal electrohydraulic suspensions, active and semiactive, employing the receding horizon method. Nonlinear Anal 30(4):1977–1984

Venhovens P, van der Knaap ACM, A, Pacejka H, (1993) Semi-active attitude and vibration control. Veh Syst Dyn 22(5–6):359–381

Wagner D (2004) Tire-iq-system. ATZ—Automobiltechnische Zeitschrift 106(7–8):660–666

Wang J, Wilson D, Halikias G (2001) $H^{\infty}$ Robust-Performance Control of Decoupled Active Suspension Systems Based on LMI Method. In: Proceedings of the 2001 American Control Conference, Arlington, Virginia, USA, vol 4, pp 2658–2663

Weispfenning T (1996) Fault detection and diagnosis of components of the vehicle vertical dynamics. In: 1st Int. Conference on Control and Diagnostics in Automotive Applications, Genua, Italy

Wesemeier D (2012) Modellbasierte Methoden zur Schätzung nicht messbarer Größen der Fahrzeugquerdynamik und des Reifenluftdrucks. Diss. Universität Darmstadt, Fortschr.-Ber. VDI Reihe 12, 747. VDI Verlag, Düsseldorf

Wesemeier D, Isermann R (2008) Indirect vehicle tire pressure monitoring with wheel and suspension sensors. 7th IFAC Symposium on Fault Detection. Supervision and Safety of Technical Processes, Barcelona, Spain, pp 917–922

Würtenberger M (1997) Modellgestützte Verfahren zur Überwachung des Fahrzustandes eines Pkw. Diss. Universität Darmstadt, Fortschr.-Ber. VDI Reihe 12, 314, VDI Verlag, Düsseldorf

Yokoyama M, Hedrick J, Toyama S (2001) A model following sliding mode controller for semi-active suspension systems with MR dampers. In: Proceedings of the American Control Conference, Arlington VA, pp 2652 – 2657

Yue C, Butsuen T, Hedrick J (1989) Alternative control laws for automotive active suspensions. J Dyn Syst Meas Control 111(2):286–291

# Part IV
# Driver-Assistance Systems

# On Driver-Assistance Systems 16

Advanced driver-assistance systems (ADAS) support the driver during driving or parking. They are pretended to alert, automate, adapt, and enhance vehicle systems to improve safety, comfort, and driving.

## 16.1 Passive and Active Driver-Assistance Systems

*Passive ADAS systems* alert the driver, e.g. on unintended lane departure (LDW), forward collision (FCW), or blind spot obstacles (BSW).

*Active ADAS systems* operate automatically based on on-board internal or environmental sensors. Some examples are electronic stability control (ESC), traction control (TCS), adaptive cruise control (ACC), automatic emergency braking (AEB), lane keeping assist (LKA).

Figure 16.1 and Tables 16.1 and 16.2 present a survey of passive and active ADAS.

One of the main goals is to avoid accidents or minimize their consequences, i.e. to mitigate the severeness. This means to react faster than a normal attentive driver, especially in or before critical driving situations. Most accidents outside of cities are, for example, caused by failing to stay in the lane, by rear and frontal collisions, collisions at intersections, lane changes, and collisions with pedestrians. On-board sensors and especially sensors scanning the vehicle surrounding made it possible to develop passive and active ADAS systems.



R. Isermann, *Automotive Control*, ATZ/MTZ-Fachbuch,
https://doi.org/10.1007/978-3-642-39440-9_16

ACC: Adaptive Cruise Control
AEB: Automatic Emergency Braking
BSW: Blind Spot Warning
ESC: Electronic Stability Control
TCS: Traction Control System
PAS: Parking Assistance System
PDW: Parking Distance Warning
FCW: Forward Collision Warning
LDW: Lane Departure Warning
LKA: Lane Keeping Assist
ANV: Assisted Night Vision
CAS: Collision Avoidance
AFL: Adaptive Front Lighting

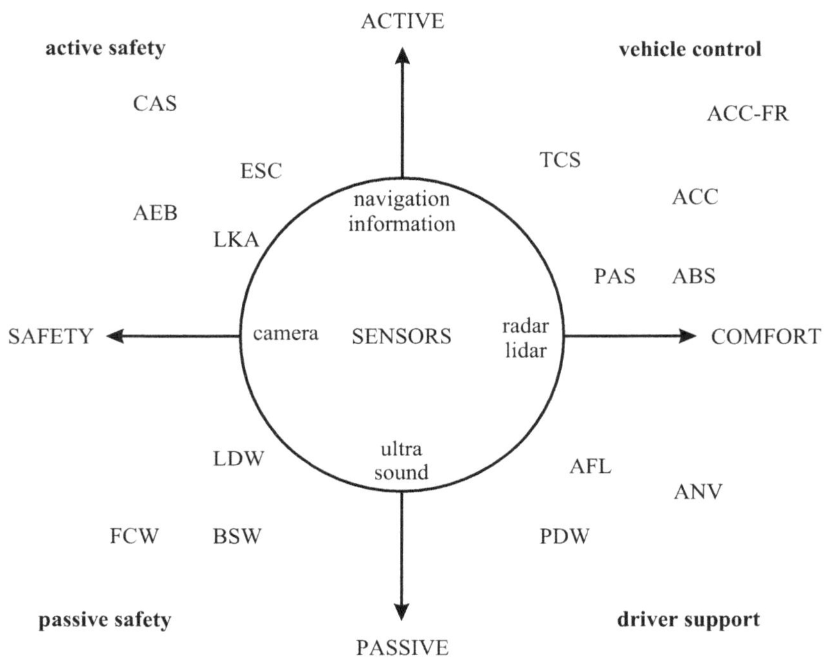


**Fig. 16.1** Advanced driver-assistence systems (ADAS) with regard to safety and comfort functions, (Wallentowitz and Reif 2011; Robert Bosch GmbH 2018)

**Table 16.1** Overview of passive advanced driver-assistance systems (ADAS)

| Passive Advanced Driver-Assistance Systems | |
|---|---|
| Abbreviation | Description |
| APS | Park Assist |
| BSW | Blind Spot Warning |
| FCW | Forward Collision Warning |
| LDW | Lane Departure Warning |
| PDW | Parking Distance Warning |

**Table 16.2** Overview of active advanced driver-assistance systems (ADAS)

| Active Advanced Driver-Assistance Systems | |
|---|---|
| Abbreviation | Description |
| ABS | Anti-lock Braking System |
| ACC | Adaptive Cruise Control |
| AEB | Autonomous Emergency Braking |
| AFL | Adaptive Front Lighting |
| CAS | Collision Avoidance System |
| CC | Cruise Control |
| EBA | Emergency Brake Assist |
| ESC | Electronic Stability Control |
| HDC | Hill Descent Control |
| PAS | Parking Assistance System |
| LKA | Lane Keep Assist |
| LSD | Limited-Slip Differential |
| TCS | Traction Control System |
| TSA | Trailer Stability Assist |

## 16.2 Sensor Systems for Advanced Driver-Assistance Systems

A first basis of sensors for detecting critical driving situations and for driver-assistance systems are the drive dynamic sensors of a chassis-mounted measurement unit, providing, e.g. longitudinal and lateral acceleration, yaw rate, wheel speeds, and vehicle's velocity. They are used for the electronic stability control (ESC), traction control (TCS), and cruise control (CC).

The second basis is sensors for the vehicle surrounding, like ultrasonic sensors for a close range of 2 ... 4 m, 24 GHz radar and cameras for medium range of 20 ... 90 m and for long range until 200 ... 400 m with 77 GHz radar and lidar sensors until 200 m. Figure 16.2 presents a view on the ranges and measurement angles of radar sensors and cameras at the frontend and backend of modern vehicles.

In the following, the function principle of most important surrounding sensors are sketched briefly. For an in-depth description, it is referred to comprehensive books like Robert Bosch GmbH (2018), Winner et al. (2016), Wallentowitz and Reif (2011), Reif (2013).

*Ultrasonic sensors* (US) are used for parking-aid systems, measuring the distance to obstacles and monitoring the space in front or rear of the vehicle during parking maneuvers. They are built in the bumpers of the vehicles and are used to inform the driver, by giving assistance during parking or for automatic parking. After obtaining a starting command from the ECU, an aluminum diaphragm oscillates with square-wave pulses of 48 kHz, emitting ultrasonic pulses. The reflected sound from an obstacle causes the diaphragm after a reflection time to oscillate. These oscillations by reflection then generate in a piezoceramic element an electrical voltage, which is amplified and converted to a digital signal for the sensor electronic unit, where

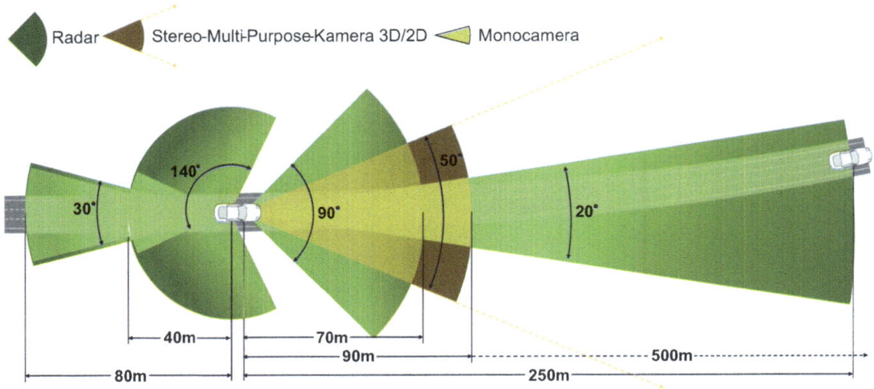


**Fig. 16.2** Environmental sensors with radar and camera for advanced driver-assistance systems, (Mercedes-Benz 2019)

the signal is evaluated with regard to runtime of the echo, object distance, and for displays. The distance measured is based on the pulse runtime and reception of the echo signal and the known speed of sound. The relatively wide opening area of the US-sensors and the relatively small built-in distance of two sensors allows to determine the position of the objects by trilateration; see Noll and Rapps (2016).

*Radar sensors* (radio detection and ranging) send out electromagnetic waves with frequencies of 24 GHz for close range and 76.5 GHz for far range and receive them again. The angle of the object is detected by scanning a field of view with a highly directive beam. This is performed electronically with a phased array antenna. In the case of monostatic radars, the emitter and receiver are at the same place.

The transmitted signals are modulated in order to uniquely assign the received signal. In the case of pulse modulation, pulses of 10 to 30 ns length are transmitted corresponding to a wavelength of 3 to 10 m. If frequency modulation is applied, the frequency is varied as a function of time. A measurement of the propagation time between the transmitted and received echo signal allows to determine the distance of the object by $d = \tau/2c$, where c is the speed of light.

For *pulse modulation*, the signals of a pulse generator are modulated and sent to the transmitting antenna via a high-speed switch. The received echo signal is compared with the transmitted signal using a variable time delay and a phase difference is used to determine the propagation time.

In the case of *frequency modulation* (FMCW: Frequency Modulated Continuous Wave), a Gunn-diode oscillator supplies in parallel several patch antennas which also receive the reflected signals. A Fresnel-lens focuses the beams with regard to the vehicle longitudinal axis. By this way, the transmitted and received signals are separated. The transmitted frequency of the Gunn-oscillator is changed according to a saw-tooth waveform by, e.g. 300 MHz. The reflected signal is received and delayed by the propagation time. Based on the frequency difference of both signals (Doppler-effect), the distance of the object is determined and also the relative velocity; see, e.g. Robert Bosch GmbH (2018), Winner et al. (2016), Wallentowitz and Reif (2011).

*Cameras* enable to provide optical pictures in front of a vehicle. Their goal is to obtain images, to extract driving relevant details, and to provide a traffic situation especially in critical driving situations by software-based image processing. In the case of night-vision systems, a high-contrast and brilliant image is required for presentation at a display. For general ADAS systems, however, the content of the images, like lane markings, road signs, traffic lights, intersections, human beings, etc. is important, which has to be received by algorithmic image processing.

Semiconductor *photo sensors* generate electron-hole pairs if photons penetrate and generate an electric field and a photoelectric current, which leads to an electric charge (inner photoelectric effect). The most important photosensitive semiconductor structures are the photodiodes and the metal-oxide-semiconductor capacitors (MOS).

The *photodiode* has semiconductor materials with different conductive properties and has a certain basic charge. A generated photoelectric charge then generates a residual voltage as measure for the light.

The *MOS capacitor* consists of a semiconductor material covered by a thin oxide layer, which serves as a metallic conductor layer. A supplied voltage to this metal electrode generates a charge under the layer. If light penetrates through a part of the insulated electrode, photoelectric electrons are created and yield a voltage of the MOS capacitor.

Many of these photodiodes or MOS capacitors have to be interconnected and manufactured together in line-shaped (arrays) or matrix-shaped planar structures, called *Charge-Coupled Devices* (CCD). Using additional electrodes the analog charge generated by the light of a photodiode or MOS capacitor is shifted after exposure to a charge amplifier generating a voltage signal. The shift is triggered by a clock and a shift register transports serially the photoelectric charges to an A/D converter. One element of these structures is called a pixel. CCD imaging sensors thus contain thousands of pixels with a length of 5 to 20 $\mu$m; see, e.g. Robert Bosch GmbH (2018). In matrix form the voltage of a pixel are read out by activating the line and column driven to an amplifier, representing finally an image.

A *camera* receives the light of objects in front of a car through an imaging lens on the image sensor which is connected to a microprocessor. The lens of the camera determines the field of view in horizontal and vertical direction, the depth of field, etc. The optical-electronic image of the image sensor is converted into digital values. Some important properties of image sensors are the resolution as numbers of pixels, the field of view as arrangement of pixels, the dynamic range as dark and bright areas, the color representation by filters, and the sensitivity.

A camera module consists of lens, image sensor, microprocessor, power supply, communication interface, and suitable packaging, for details see Punke et al. (2016).

The cameras are designed as monocameras and as stereo cameras with a second camera module, allowing a depth estimation.

*Automotive cameras* are developed for surveillance of the car interior (driver and hand gesture regulation) and the detection of the vehicle's surroundings. Front view cameras are built behind the windshield, close to the rear-view mirror. They operate in the visible spectral range and are designed for detecting traffic road signs, driving lanes, object recognition (vehicles, pedestrians, cyclists).

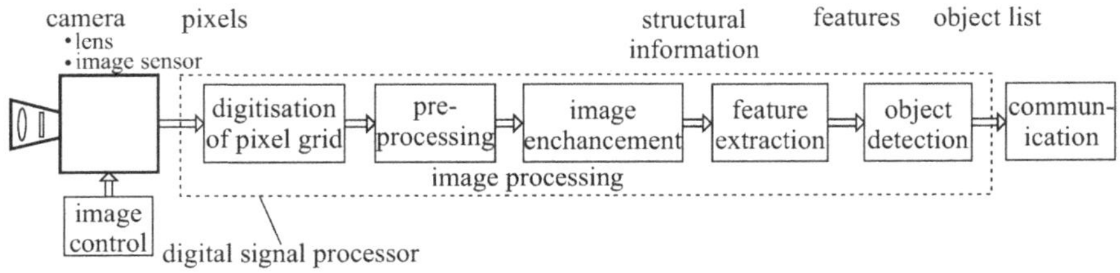


**Fig. 16.3** Signal flow chart for the digital image processing of a monocamera

In addition to front-view cameras, the detection of the surroundings around the vehicle requires cameras with a wide field of view. One distinguishes rear-view cameras, built in the rear part and surround cameras with a field of view of more than 180°.

The digital image processing of a camera follows various steps as illustrated in Fig. 16.3.

The observed image by the camera sensor is digitized by sampling and quantization and presented as a pixel grid. Then an image processing and image enhancement follows; see Stiller et al. (2016). The *image processing* includes a preprocessing of input images and a feature extraction. A *preprocessing* contains various filters for reducing noise and elimination of irrelevant information. This is followed by an *image enhancement*, to extract the interesting information. Hence, point operations are used to treat gray or color pixels, local operations to treat pixels of interest by extracting strucutral information of the image and global operations to treat the whole image through, e.g. rotation and scaling.

The next step is a *feature extraction* of the preprocessed images. Features are locally constraint characteristic parts of an image to extract structural elements, like edges and corners. Features have the property that an image gradient changes significantly. These gradients are detected by calculating local derivations and use of line or edge detection filter algorithms.

The observed features have then to be treated in a three-dimensional space. Using monocameras, a perspective view is required for the observed image features in a plane. This can be obtained by searching for correspondences of several images with, e.g. matching methods.

*Object detection* is a next step and based on sets of features which have been observed over certain time periods (sequence of snapshots). Here, motion-based methods with flow vectors of pixels in the image plane can be applied and grouping of points with same motion vectors is made. If then detected and moving objects stay constant, they are stored in an object list.

A verification of the detected objects is based on special properties of the objects like typical patterns or templates. For more details see Stiller et al. (2016).

*Lidar sensors* (Light detection and ranging) for automotive application use modulated infrared radiation in a wavelength of 800 to 1600 nm. Non-scanning lidar sensors operate with one or several fixed beams to obtain specific reflections of an object. With increased number of deflected beams, the surrounding can be scanned with higher resolution. A lidar sensor system emits infrared radiation, which is modulated in its intensity. The reflected beam by an object is received by one or several photodiodes. The modulation is performed by sinusoidal or square waves or by pulses. These received signals are compared with the emitted signals and either propagation time or the phase difference is determined to obtain the distance to the object. The determination of the propagation time with short pulses is based on the determination of the time difference between sending and receiving or by the determination of a phase difference of an amplitude-modulated sinusoidal signal, e.g. by correlation methods. Lidar sensor systems are able to reach ranges up to 200 m and angle resolutions up to 1°. However, weather conditions with fog and poor visibility limit their general applicability.

## 16.3 Environment Representation

As advanced driver-assistance systems (ADAS) require an adequate picture of the environment around the vehicle the corresponding computerized representation plays a major role. This includes a dynamic mapping with all relevant objects, road infrastructure in the vicinity of the driving vehicle and the vehicle's own momentaneous position and motion on the road.

Figure 16.4 summarizes the information flow based on onboard sensors for vehicle dynamics and sensors for the environment to obtain the relevant information required for ADAS. This includes the knowledge of distances, velocities, objects, features, and dimensions of all traffic participants and the road infrastructure.

The objects may be represented in the own vehicle (ego-vehicle) coordinate system with an origin at the center of gravity or the rear axle.

Grid maps of the road may be used to structure the observed environment in cells, which are free or occupied by an obstacle. The relative position of the objects within

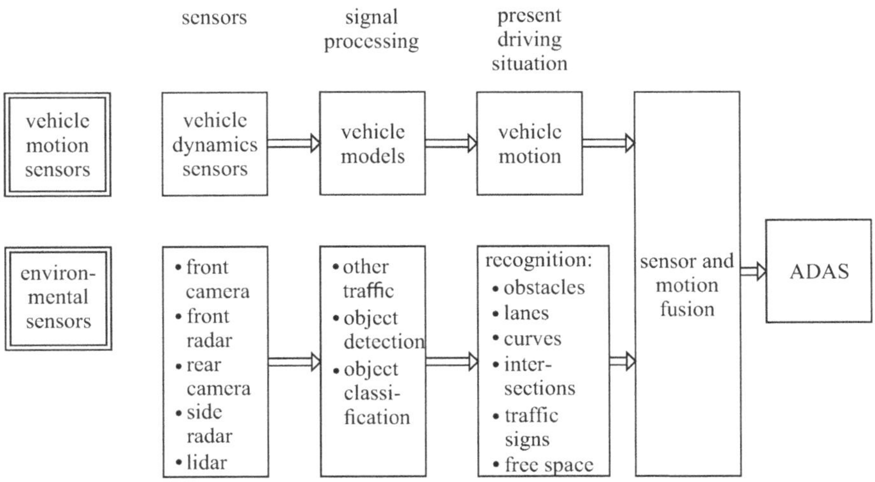


**Fig. 16.4** Flowchart for motion and environment information to be used for ADAS

their grid map is continually determined. A further task is multi-object tracking in the field of view where each object is represented as a state variable $x(k)$ and a state estimation is performed with a Kalman filter, also for several objects; see, e.g. Dietmeyer et al. (2016).

*Lane detection* with vision systems is usually based on extracted features like edges, motion vectors, and textures. These features are then used in the combination with a road model in order to obtain an estimate of the vehicle's position within the lane. A survey of these methods is given in McCall and Trivedi (2006). Road marking detection may be based on the detection of edges or road templates in combination with road texture, using the entire road. However, as lane markings with solid or segmented lines vary greatly, a combination of several lane detection methods is useful. Post-processing methods, as Hough transform or Canny edge detection improve the lane estimation in order to group extracted-edge features into an appropriate set of lines. The road models may assume piecewise constant curvatures, splines, or clothoids. These lane detection methods in combination with road models then enable to estimate the curvature in a lookahead distance and also the position $D_y$ of the vehicle within the lane.

## References

Dietmeyer K, Reuter S, Nuss D (2016) Representation of fused environment data, Chapter 24 in Winner et al (2016)

McCall JC, Trivedi MM (2006) Video-based lane estimation and tracking for driver assistance: survey, system, and evaluation. IEEE Trans Intell Transport Syst 7(1):20–37

Mercedes-Benz (2019) Mercedes-Benz Intelligent Drive. Information Brochure

Noll M, Rapps P (2016) Ultrasonic sensors, in Winner et al (2016). Springer Int., Switzerland

Punke M, Menzel S, Werthessen B, Stacke N, Hoepfl M (2016) Automotive Camera (Hardware), Chapter 19 in Winner et al (2016), Springer, Switzerland

Reif K (ed) (2013) Fahrerassistenzsysteme. Springer Vieweg Verlag, Wiesbaden

GmbH Robert Bosch (ed) (2018) Automotive Handbook, 10th edn. Wiley, Chichester, England

Stiller C, Bachmann A, Geiger A (2016) Fundamentals of machine vision, Chapter 20 in Winner H et al (2016)

Wallentowitz H, Reif K (2011) Handbuch Kraftfahrzeugelektronik—Grundlagen, Komponenten, Systeme, Anwendungen, 2nd edn. ATZ/MTZ-Fachbuch, Vieweg+Teubner Verlag, Wiesbaden

Winner H, Hakuli S, Lotz F, Singer C (eds) (2016) Handbook of driver assistance systems. Springer International Publishing AG, Cham, Switzerland

# 17 Advanced Driver Assistance Systems for Longitudinal and Lateral Guidance

Based on the book parts II "Modeling of Drive Dynamics" and III "Dynamic Control of Chassis Components" control-oriented advanced driver assistance systems (ADAS) are briefly described in this chapter.

One of the first driver assistance systems was cruise control (CC) applied 1958 in a Chrysler Imperial. The driver sets a button to the selected speed and the throttle of the engine is manipulated electrically. After depressing the brake by the driver the CC is turned off automatically.

A next step in supporting the driver in critical braking situations was the introduction of the anti-lock brake control system (ABS), which was in the 1950s applied for aircraft by Dunlop. First application for vehicles is reported for the Jensen FF in 1966 and Ford (1969). Bosch and Mercedes-Benz introduced a microprocessor-based ABS in 1978 as a compact mechatronic solution.

Then with new sensors, electrically manipulated actuators and Microcontroller-based control units (ECU) came following ADAS on the market: Traction Control System (TCS) 1986, Electronic Stability Control (ESC) 1995, Brake Assist (BA) 1996 and Adaptive Cruise Control (ACC) 1999, by Bosch and Mercedes-Benz; see also Table 16.2.

A detailed compilation of advanced driver assistance systems with a description of their functions and the involved components is published in the Handbook Winner et al. (2016); see also Reif (2013), Robert Bosch GmbH (2018), van Zanten (2006), van Zanten and Kost (2016).

Based on the on-board sensor systems for drive dynamics and for the surroundings some important control-oriented active driver assistance systems are reviewed in this chapter for lateral guidance traction control (TCS), electronic stability control (ESC), and lane keeping assist (LKA).


R. Isermann, *Automotive Control*, ATZ/MTZ-Fachbuch,
https://doi.org/10.1007/978-3-642-39440-9_17

## 17.1 Traction Control System (TCS)

The task of the traction control system is to avoid a loss of tire/road grip of the driven wheels. This happens, for example, if the drive engine generates too high torque. Then the wheel slips increase of about 10 to 20% and exceed the maximal friction coefficient $\mu_{\max}$; see Fig. 5.14.

Because the friction coefficient then decreases, the transferable torque of the tires decreases too and one or two wheels start spinning. This arises, for example, during start-up or in general on slippery roads. Spinning wheels with large slip then loose their capability to transfer lateral forces and the vehicle may break out sideward and may become unstable.

The traction control system (TCS) then brakes a faster spinning wheel with most lessened traction and increases through the differential the traction torque to the other wheel with better traction.

A second action is the reduction of the drive torque of the engine by controlling the throttle or injection time of a gasoline engine or the injected fuel of a diesel engine. The following consideration is based on van Zanten (2006).

The traction control system uses the angular wheel speed measurements mainly of the driven wheels, for example, for the rear axle $\omega_{\text{rl}}$ and $\omega_{\text{rr}}$. The non-driven wheel speeds of the front axle $\omega_{\text{fl}}$ and $\omega_{\text{fr}}$ are then used to determine the vehicle velocity

$$v_{\text{X}} = r_{\text{dyn}} \frac{(\omega_{\text{fl}} + \omega_{\text{fr}})}{2} \tag{17.1.1}$$

The longitudinal slips of the driven rear wheels are for braking, see (5.1.1), but opposite sign,

$$S_{\text{X,r}} = \frac{\Delta v_{\text{X,r}}}{v_{\text{W}}} = \frac{v_{\text{W}} - v_{\text{X}}}{v_{\text{X}}} = \frac{r_{\text{dyn}}\omega_{\text{W}} - v_{\text{X}}}{v_{\text{X}}} \tag{17.1.2}$$

with $0 \geq S_{\text{X,rb}} \geq -1$. For driving the same slip definition is used, with $0 \leq S_{\text{X,rd}} \leq 1$. Hence, for driving the slip is positive, van Zanten (2006). The goal is to adjust the driven wheels so that they have a maximum of grip. Therefore, the maximum of the $\mu_{\text{X}}(S_{\text{X}})$ friction characteristic has to be determined. An interpolation of a slip curve for high and low grip then gives an approximation, see van Zanten (2006),

$$S_{\text{X,opt}} = A_0\mu_{\text{res}} + \frac{A_1}{v_{\text{X}}} + A_2 \tag{17.1.3}$$

which is used as desired value for the ABS control.

Herewith $\mu_{\text{res}}$ is determined by the ABS control during the phase with adjusted known braking force within the linear part of the $\mu(S_{\text{X}})$ characteristic. This allows to calculate the speed of a free rolling wheel $v_{\text{W,free}}$ without braking.

The desired slip of the TCS-controller for the rear wheels is then

$$S_{\text{X,r,d}} = S_{\text{X,opt}} \tag{17.1.4}$$

where the slip $S_{\text{X,r}}(t)$ is determined by the TCS control.

However, the traction control system uses not directly the slip values, but the wheel velocities of a free rolling wheel

$$v_{W,ij,fr} = r_{dyn}\omega_{W,ij,fr} \tag{17.1.5}$$

and the wheel velocity with slip

$$v_{W,ij} = r_{dyn}\omega_{W,ij} = v_{W,ij,free} + \Delta v_{W,ij}(S_{X,ij}) \tag{17.1.6}$$

or, with (17.1.2)

$$v_{W,ij} = v_{W,ij,free}(1 + S_{X,ij}). \tag{17.1.7}$$

The traction control with wheel velocities as controlled variables and brake torques $M_{B,rj}$ and engine torque $M_D$ as inputs involves the drive train and wheel dynamics. The dynamic behavior of both rear wheels follows from (6.5.16) by assuming a stiff powertrain

$$J_{tot}\frac{d\omega_{W,r}}{dt} = i_{tot}\eta_{tot}M_{eng} - (M_{W,r,l} + M_{W,r,r}) \tag{17.1.8}$$

with the representative moment of inertia of the rotational masses of the powertrain $J_{tot}$ according to (6.5.17). $M_{W,r,j}$ is the acting torque at each wheel through the driving resistance forces of the vehicle $\Sigma F_{X,res}$ due to (6.5.3) and braking forces $F_{XT,j,b}$

$$\begin{aligned} M_{W,r,j} &= r_{dyn}\left(\frac{1}{2}\Sigma F_{X,res,j} - F_{XT,j,b}\right) \\ &= \frac{1}{2}M_{XT,j} - M_{XT,j,b}. \end{aligned} \tag{17.1.9}$$

The dynamic behavior of each rear wheel follows then with the wheel velocities $v_W = r_{dyn}\omega_W$

$$\frac{J_{tot}}{2r_{dyn}}\dot{v}_{W,rl} = \frac{1}{2}i_{tot}\eta_{tot}M_{eng} + \frac{1}{2}M_{XT,l} - M_{XT,l,b}, \tag{17.1.10}$$

$$\frac{J_{tot}}{2r_{dyn}}\dot{v}_{W,rr} = \frac{1}{2}i_{tot}\eta_{tot}M_{eng} + \frac{1}{2}M_{XT,r} - M_{XT,r,b}. \tag{17.1.11}$$

Both equations are coupled as the braking torques depend on the slips and therefore on $v_{W,r,j}$. After adding and subtracting both equations one obtains with the average wheel velocity $v_W$ and the wheel difference velocity $\Delta v_W$

$$v_W = \frac{1}{2}(v_{W,rl} + v_{W,rr}) \tag{17.1.12}$$

$$\Delta v_W = (v_{W,rl} - v_{W,rr}) \tag{17.1.13}$$

two equations

$$\frac{J_{tot}}{2r_{dyn}} \dot{v}_W = i_{tot}\eta_{tot} M_{eng} + (M_{XT,l} + M_{XT,r}) - (M_{XT,l,b} + M_{XT,r,b}), \tag{17.1.14}$$

$$\frac{J_{tot}}{2r_{dyn}} \Delta \dot{v}_W = \frac{1}{2}(M_{XT,l} - M_{XT,r}) - (M_{XT,l,b} - M_{XT,r,b}). \tag{17.1.15}$$

The first equation describes the influence of the engine torque and the tire/road torques on the velocity of both wheels and the second equation the influence of right and left braking torque on the difference wheel velocity.

By reducing the engine torque a spinning rear wheel can be influenced by a *wheel velocity controller* and by left or right rear wheel braking with a *wheel difference velocity controller* a yaw angle drift can be compensated.

The *rear wheel velocity controller* is described by a transfer function

$$G_{v\alpha}(s) = \frac{\Delta\alpha(s)}{\Delta v_W(s)} \tag{17.1.16}$$

where $\alpha$ stands for a throttle or injection time retard for gasoline engines or injection quantity for diesel engines. This controller is designed as nonlinear PID controller. Its reference value $v_{ps,d}$ is taken from (17.1.7) with $S_{X,r} = S_{X,r,d} = S_{X,opt}$.

$$v_{W,d} = v_{W,free}(1 + S_{X,opt}). \tag{17.1.17}$$

To obtain a fast reaction to reduce the rear wheel velocities a further controller $G_{vpB}$ may activate both wheel brakes for a short time.

The *rear wheel difference velocity controller* is noted by a transfer function

$$G_{\Delta WB}(s) = \frac{\Delta p_{Bj}(s)}{\Delta v_{W(s)}} \tag{17.1.18}$$

with PI-behavior where $p_{Bj}$, $j = r$ or $l$ is a right or left wheel brake pressure, which can be selected in the hydraulic systems of the ABS-brake; see Chap. 13. The selection depends on the deviation of the yaw angle of the vehicle. Both controllers include dead-zones in order not to become active for small control variable deviations.

The dynamic behavior around the vertical axis follows from the two-track model (7.3.14) by neglecting the wheel forces at the front wheels and caster offsets

$$\ddot{\psi} = \frac{1}{J_Z}(F_{X,rr} - F_{X,rl})\frac{b_r}{2} - (F_{Y,rl} + F_{Y,rr})l_r. \tag{17.1.19}$$

As the lateral forces for low vehicle velocity and on low-$\mu$ roads are small, they are neglected. Then it holds

$$\ddot{\psi} = \frac{b_r}{J_Z r_{dyn}} (M_{XT,r,b} - M_{XT,l,b}). \tag{17.1.20}$$

If further for the driving resistance forces $M_{XT,l} = M_{XT,r}$ is assumed, it follows with (17.1.15)

$$\ddot{\psi} = -\frac{b_r}{2r_{dyn}^2} \frac{J_{tot}}{J_Z} \Delta\dot{v}_W \tag{17.1.21}$$

or

$$\dot{\psi} = -\frac{b_r}{2r_{dyn}^2} \frac{J_{tot}}{J_Z} \Delta v_W = -K_{\Delta v \dot{\psi}} \Delta v_W. \tag{17.1.22}$$

Hence, the yaw rate is proportional to the difference velocity of the rear wheels and the resulting yaw angle becomes

$$\psi(t) = -K_{\Delta v \dot{\psi}} \int \Delta v_W(t) dt. \tag{17.1.23}$$

The controlled process has integral behavior and a yaw angle controller

$$G_{\psi \Delta v}(s) = \frac{\Delta v_W(s)}{\Delta\psi(s)} \tag{17.1.24}$$

can be designed as P controller with gain $K_{\psi \Delta v}$.

Figure 17.1 depicts a signal flow chart of the main functions of the traction control system. The TCS-system operates only during an active propulsion with the engine and is limited to low vehicle velocity. For more details see van Zanten (2006), van Zanten and Kost (2016).

A traction control system (ITC) for improving the vehicle dynamics for *higher velocities* of sports cars also with regard to handling drift maneuvers is described by König et al. (2018). Based on a feedforward and feedback traction slip control of a longitudinal quarter vehicle model the torque of the rear axle is manipulated to control the vehicle at large sideslip angles.

## 17.2 Electronic Stability Control (ESC)

The task of an electronic stability control system is to stabilize the vehicle's lateral behavior if the tire forces exceed their limits. This is, for example, the case if the vehicle turns with too high velocity in a curve or the road is slippery allowing only small tire/road friction. The required lateral forces then reach a maximal value with increasing wheel side slip angle $\alpha$, see Fig. 5.6, and the lateral forces cannot be increased anymore. If this happens first at the front wheels the vehicle *understeers*,

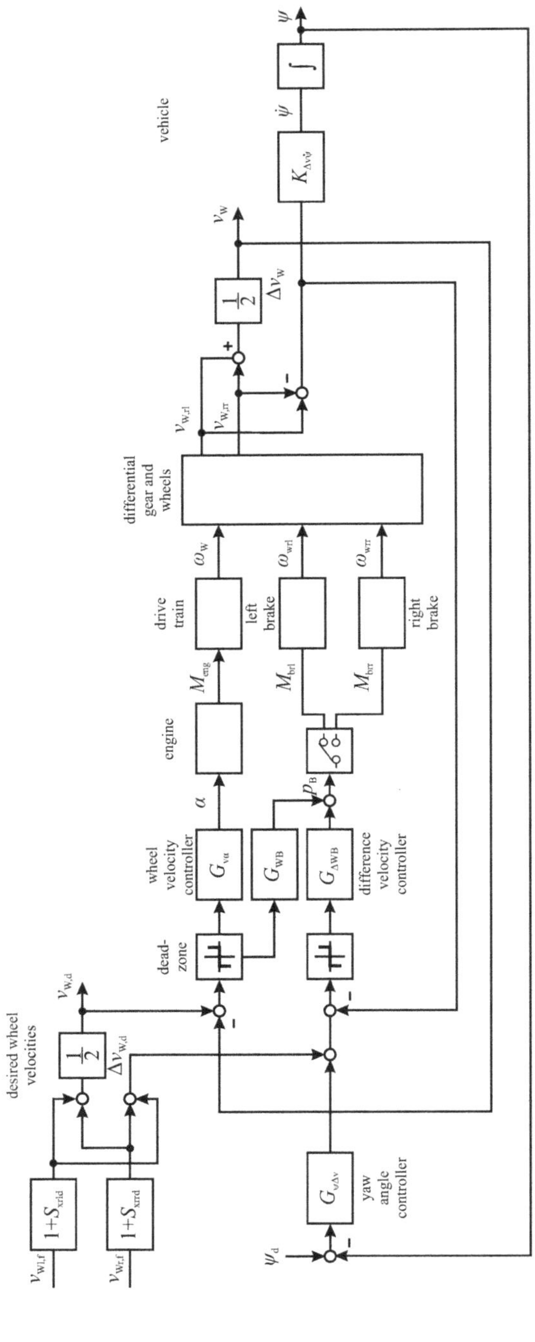


**Fig. 17.1** Flow chart of a traction control system (TCS) for a rear driven vehicle based on measured rear wheel velocities

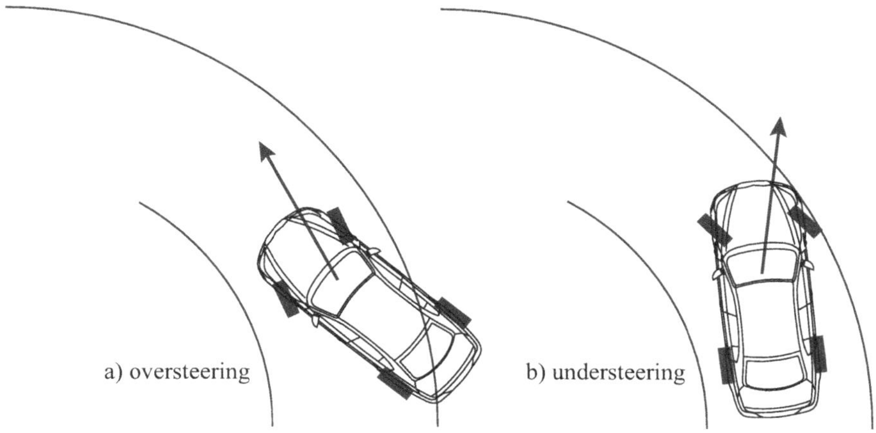


**Fig. 17.2** Vehicle at the limits of lateral tire forces **a** oversteering, **b** understeering

which means that it leaves the lane to the outside. For the case that the rear wheels reach the maximal lateral force, the vehicle *oversteers*, i.e. it turns to the inside of the lane; see Fig. 17.2. For the driver, it is then hardly possible to stabilize the vehicle and to bring it back on the center of the lane by steering or countersteering without ESC.

The ESC then "steers" the vehicle, through braking of individual wheels by generating a correcting torque around the vertical axis. This is realized by the calculation of desired brake slip values $S_{X,ij}$ for individual wheels. Four cases have to be considered; see Table 17.1

For understeering the required yaw rate $\dot{\psi}$, according to the steering angle $\delta_f$, is not reached. The rear wheel on the inner side of the curve is then braked to generate a correcting torque, as the front wheels are at their maximal side force limits. If the vehicle oversteers the front wheel on the outer side of the curve is braked as the rear wheels have reached their maximal side forces. However, for the braked wheels also the longitudinal slips $S_{X,ij}$ have to be taken into account in order to stay with the overall tire forces within the Kamm's circle; see van Zanten (2006).

**Table 17.1** Individual wheel braking for critical driving situations

| Critical driving situation | Curve | Wheel brake | |
|---|---|---|---|
| Understeering | Right | Rear right | Rear wheel curve innerside |
| | Left | Rear left | |
| Oversteering | Right | Front left | Front wheel curve outerside |
| | Left | Front right | |

To describe the basic functions of the ESC-system a stationary driving situation with $v_X = \text{const}$ is considered when the vehicles enter a curve with constant curvature $\kappa = \text{const}$ and a nominal steering angle $\delta_f$ as driver input. Thus, the vehicle is not braked nor accelerated.

If the velocity is too high or the road is slippery with low tire/road friction $\mu$ the vehicle oversteers or understeers, as shown in Fig. 17.2. This critical situation is detected by comparing the measured yaw rate $\dot{\psi}_{meas}$ with the nominal value $\dot{\psi}_{nom}$ of a drive dynamic model for the steering behavior, for example a one-track model, and calculating the difference

$$\Delta\dot{\psi} = |\dot{\psi}_{nom}| - |\dot{\psi}_{meas}| \tag{17.2.1}$$

see Fig. 17.3. Then it holds

$$\Delta\dot{\psi} > 0 \quad \text{for understeering}$$
$$\Delta\dot{\psi} < 0 \quad \text{for oversteering}$$

### 17.2.1 ESC for Oversteering

Now the situation of an *oversteering vehicle* is considered, where in a left curve the right front wheel has to be braked; see Fig. 17.4. Before braking the right front wheel the lateral force is

$$F_{Y,fr} = \frac{ml_r}{2l} a_Y \tag{17.2.2}$$

with a lateral slip $S_{Yf,ro}$. It is assumed that this tire has not reached the maximal lateral force, because the rear wheels have already exceeded their maximal lateral forces. If now a brake force $F_{Xf,r}$ is applied the resulting geometrical force becomes

$$F_{res,fr} = \sqrt{F_{Y,fr}^2 + F_{X,fr}^2}. \tag{17.2.3}$$

When this resulting tire force is within Kamm's circle, see (5.3.11),

$$F_{res,fr} < F_{lim,fr} = \mu_{res,max} F_{ZT,fr} \tag{17.2.4}$$

the right front wheel has enough grip to generate a counter torque

$$M_{Zr,overst} = -b_{CG1,r} F_{res,fr} \tag{17.2.5}$$

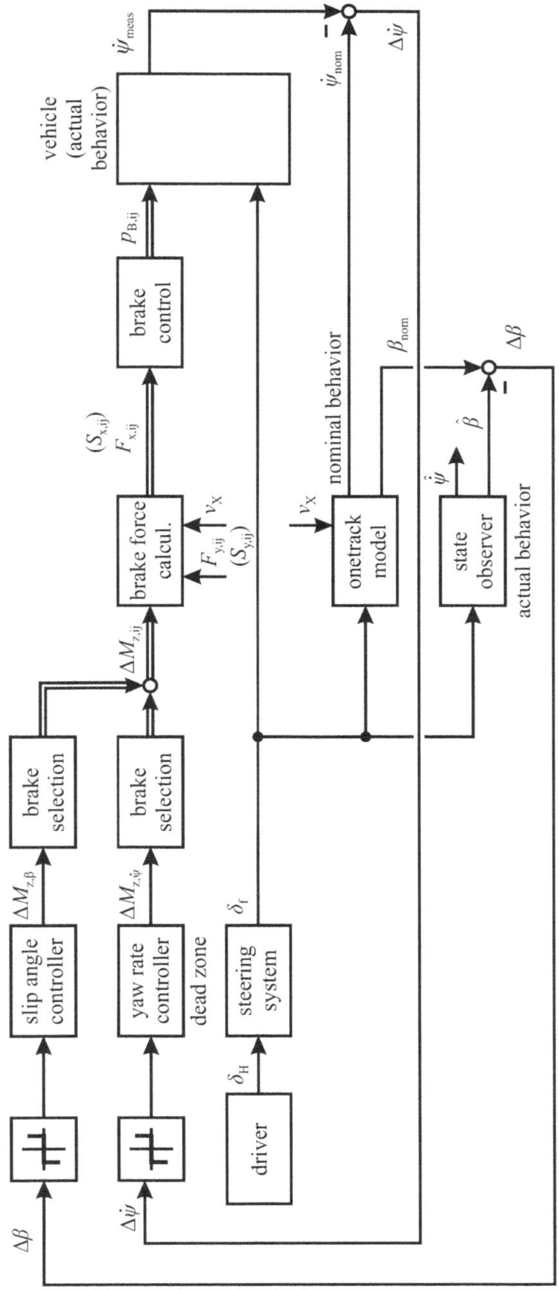


**Fig. 17.3** Electronic stability control system (ESC) with one-track model, slip-angle observer, yaw rate controller and slip-angle controller (only basic functions)

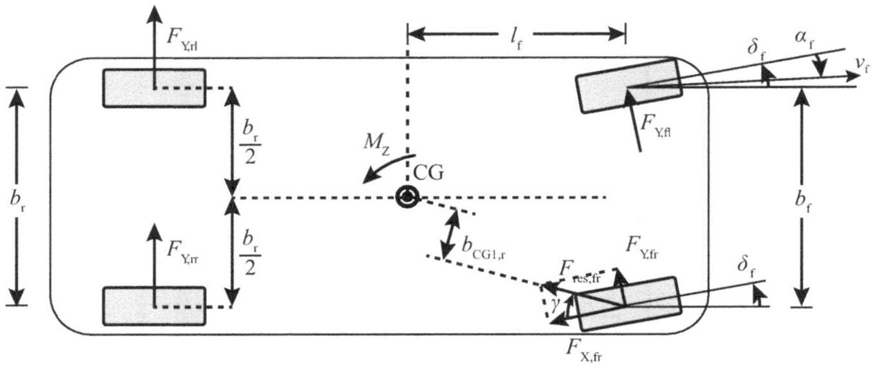


**Fig. 17.4** Vehicle in a left curve and braking the right front wheel to generate a yaw torque (counteracting torque as required for an oversteering vehicle)

where $b_{\mathrm{CG1,r}}$ is the lever arm to the center of gravity, see Fig. 17.4,

$$\begin{aligned} b_{\mathrm{CG1,r}} &= \left[\frac{b_{\mathrm{f}}}{2} - l_{\mathrm{f}} \tan(\gamma - \delta_{\mathrm{f}})\right] \cos\gamma \\ \gamma &= \arctan \frac{F_{\mathrm{Yf,r}}}{F_{\mathrm{Xf,r}}} \\ &\text{and } b_{\mathrm{CG1,r}} > 0. \end{aligned} \tag{17.2.6}$$

Hence, with increasing braking force $\gamma$ becomes smaller, the lever arm $b_{\mathrm{CG1,r}}$ larger and the countertorque $M_{\mathrm{Zr,overst}}$ increases.

The maximum braking force is according to Kamm's circle equation (17.2.3) limited to

$$F_{\mathrm{X,frmax}} = \sqrt{F_{\mathrm{lim,fr}}^2 - F_{\mathrm{Y,fr}}^2} \tag{17.2.7}$$

where $F_{\mathrm{Yf,r}}$ follows from (17.2.2) and $F_{\mathrm{lim,fr}}$. Hence, the acting lateral tire force influences the applicable braking force significantly.

The maximal countertorque is the with (17.2.5) and (17.2.4)

$$M_{\mathrm{Zr,overst,max}} = -b_{\mathrm{CG1,r}} \mu_{\mathrm{res,max}} F_{\mathrm{ZT,fr}}. \tag{17.2.8}$$

To determine the maximal braking force according to (17.2.7) the tire vertical force $F_{\mathrm{ZT,fr}}$, $\mu_{\mathrm{res,max}}$ and $a_{\mathrm{Y}}$ have to be known.

The vertical force on the front wheel follows from (7.3.22) to

$$F_{\mathrm{ZT,fr}} = m\left(\frac{l_{\mathrm{r}}}{l} g - \frac{h_{\mathrm{CG}}}{l} a_{\mathrm{X}}\right)\left(\frac{1}{2} + \frac{h_{\mathrm{CG}}}{b_{\mathrm{f}}} \frac{a_{\mathrm{Y}}}{g}\right). \tag{17.2.9}$$

Hence, it depends on $a_X$ and $a_Y$, and increases by cornering with $a_Y$ and braking with $-a_X$.

For the calculation of the countertorque (17.2.5) the lever arm $b_{CG1,r}$ has to determined with (17.2.6).

In order not to lock the right front wheel the applied brake force is selected by

$$F_{X,fr} = \zeta F_{X,frmax} \tag{17.2.10}$$

with, for example, $\zeta = 0.8$ as an attenuation factor

The effect of a counteracting braking force $F_{Xf,r}$ on the vehicle with regard to the yaw rate follows from the two-track model (7.3.14), for not too large $\delta_f$

$$\ddot{\psi} = \frac{1}{J_Z}\left[-F_{X,fr}\frac{b_f}{2} + \left(F_{Y,fl} + F_{Y,fr}\right) l_f - \left(F_{Y,rl} + F_{Y,rr}\right) l_r\right] \tag{17.2.11}$$

where the lateral forces follow from (12.3.34) or simplified by (12.3.18) if no braking forces on other wheels are applied.

Then $\dot{\psi}$ is obtained by

$$\dot{\psi}(t) = \int \ddot{\psi}(t)dt. \tag{17.2.12}$$

This allows to determine the behavior $\dot{\psi} = f(F_{Xf,r})$ for the design of a yaw rate controller.

For small changes of the counteracting braking force $\Delta F_{XT,fr}$ then follows for $F_{Y,ij} = const.$, see Fig. 17.5,

$$\begin{aligned}\Delta\dot{\psi}(t) &= \frac{1}{J_Z}\int \Delta M_{Z,r}(t)dt = \frac{1}{J_Z}\int \Delta F_{X,fr}(t) b_{CG2,r} dt \\ &\approx \frac{1}{J_Z}\int \Delta F_{X,fr}(t)\frac{b_f}{2}dt.\end{aligned} \tag{17.2.13}$$

Because of the integrating behavior

$$\frac{\Delta\dot{\psi}(s)}{\Delta F_{X,fr}} = \frac{b_f}{2J_Z}\frac{1}{s} \tag{17.2.14}$$

a P controller

$$\Delta F_{X,fr}(t) = K_{\dot{\psi}F_x}\Delta\dot{\psi}(t) \tag{17.2.15}$$

is a suitable choice, with a limitation of the maximal braking force due to (17.2.10).

### 17.2.2 Simplified ESC for Over- and Understeering

Another approach to determine the yaw torque by individual wheel braking is to use the torque equation for the two-track model (7.3.28) for the case of free rolling (no driving forces), stationary cornering and changes of braking forces $\Delta F_{X,ij}$ for individual wheels with $F_{Y,ij} = \text{const}$, $\delta_f = \text{const}$, and $M_Z = \bar{M}_Z + \Delta M_Z$

$$\begin{aligned}\Delta M_{Z,ij} &= \left(\Delta F_{X,fl} + \Delta F_{X,fr}\right) l_f \sin \delta_f \\ &+ \left(\Delta F_{X,fr} - \Delta F_{X,fl}\right) \frac{b_f}{2} \cos \delta_f \\ &+ \left(\Delta F_{X,rr} - \Delta F_{X,rl}\right) \frac{b_r}{2}.\end{aligned} \tag{17.2.16}$$

Then, one obtains for braking with the right front wheel

$$\Delta M_{Z,fl} = -\Delta F_{X,fr} \left( l_f \sin \delta_f + \frac{b_f}{2} \cos \delta_f \right) \tag{17.2.17}$$

for braking with the left front wheel

$$\Delta M_{Z,fr} = \Delta F_{X,fl} \left( l_f \sin \delta_f - \frac{b_f}{2} \cos \delta_f \right) \tag{17.2.18}$$

and for braking with the left rear wheel

$$\Delta M_{Z,rl} = -\Delta F_{X,rl} \frac{b_r}{2} \tag{17.2.19}$$

and the right rear wheel

$$\Delta M_{Z,rr} = \Delta F_{X,rr} \frac{b_r}{2}. \tag{17.2.20}$$

Hence, the lever arms for the right and left front wheel braking force is

$$\begin{aligned}b_{CG2,r} &= \left( l_f \sin \delta_f + \frac{b_f}{2} \cos \delta_f \right) \\ b_{CG2,l} &= \left( l_f \sin \delta_f - \frac{b_f}{2} \cos \delta_f \right)\end{aligned}$$

and for the right and left rear wheel braking force

$$\begin{aligned}b_{CG3,r} &= \frac{b_r}{2} \\ b_{CG3,l} &= \frac{b_r}{2}\end{aligned}$$

see Fig. 17.5 and Ackermann (2016).

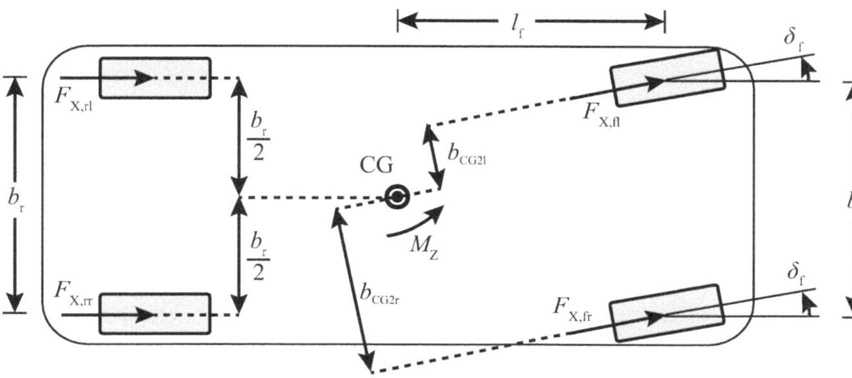


**Fig. 17.5** Vehicle in a left curve and lever arms for braking of individual wheels, Ackermann (2016)

The braking forces $F_{X,ij}$ are determined by the individual brake line pressures $p_{B,ij}$, e.g. generated with an electromechanical brake booster, see Fig. 13.19, where the relation (6.7.6) is used.

The ESC-yaw rate controller receives as input the difference $\Delta\dot{\psi}$ between the nominal and measured yaw rate, see (17.2.1), if it exceeds a threshold

$$\Delta\dot{\psi} > \Delta\dot{\psi}_{\text{thres}} \tag{17.2.21}$$

and determines as output the correcting yawing torque $\Delta M_{Z,ij}$; see Fig. 17.3. The required wheel braking force $\Delta F_{X,ij}$ is then calculated with (17.2.17) to (17.2.20) after selection of the individual wheel, under the condition, that the maximal allowable braking force, as (17.2.10) for the right front wheel, is not exceeded. The yaw rate controller is, for example, realized with a proportional behavior

$$\Delta M_{Z,ij} = K_{\dot{\psi}M}\Delta\dot{\psi} \tag{17.2.22}$$

as used by Hac and Bedner (2007) and Lu et al. (2016); see also Ackermann (1997). Van Zanten (2006) mentions a PID controller and increases the P-factor dependent on the vehicle side slip angle.

Figure 17.3 summarizes the basic functions of an ESC control system with a *yaw rate controller*.

An extension of the yaw rate based ESC is obtained by adding *a slip angle control*. This is, for example, advantageous if the yaw rate difference $\Delta\dot{\psi}$ is small and leads to only a weak control input, but the slip angle builds up, as shown by Bechtloff (2018) for a road with snow. If the slip angle $\beta$ is estimated with a state observer or Kalman filter according to section 12.3 an additional feedback to (17.2.22) for oversteering can be established according to Bechtloff (2018)

$$\Delta M_{Z,ij} = g_\beta K_{\beta M}\Delta\beta + g_{\dot{\psi}} K_{\dot{\psi}M}\Delta\dot{\psi} \tag{17.2.23}$$

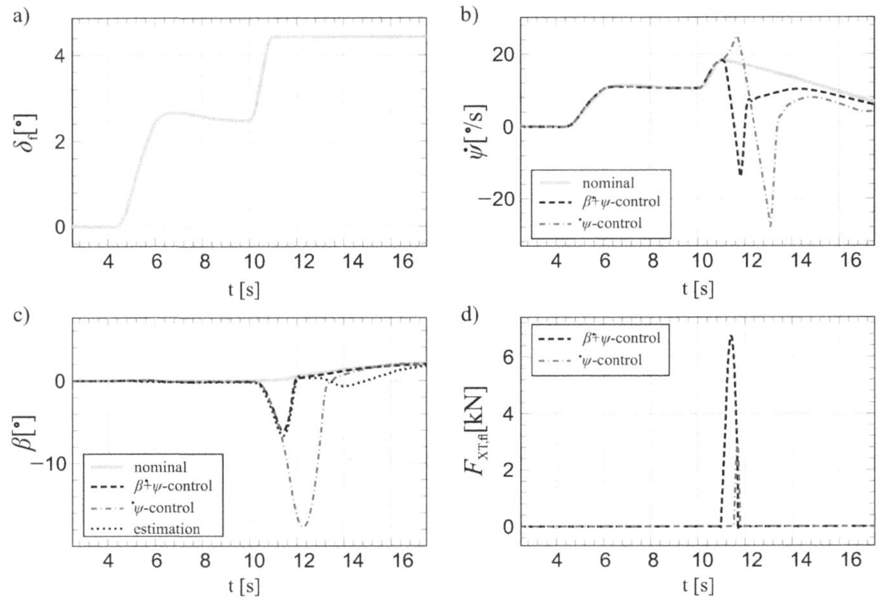


**Fig. 17.6** Simulation with a yaw rate and slip angle control-based ESC-system during cornering through a right curve on a low-$\mu$ road (snow), IPG-Carmaker-simulation system. $\Delta\dot{\psi}_{\text{thres}} = 5°/\text{s}$, $\Delta\beta_{\text{thres}} = 2°$. Bechtloff (2018).
**a** steering angle **c** slip angle
**b** yaw rate **d** braking forces, front left

with weighting factors

$$g_\beta + g_{\dot{\psi}} = 1 \tag{17.2.24}$$

where $g_{\dot{\psi}}$ is linearly decreased for increasing $\Delta\beta$. The slip angle control becomes active for

$$\Delta\beta > \Delta\beta_{\text{thres}}. \tag{17.2.25}$$

This slip-angle control is also shown in Fig. 17.3

Figure 17.6 depicts a simulation with this combined controller for a slow oversteering on a low friction road with snow, Bechtloff (2018)

The vehicle performs a circular drive with $R = 100\text{m}$ and $v_X = 67\text{km/h}$ and exceeds for $a_Y \approx 3.5\text{m/s}^2$ the maximal lateral forces. At $t = 10\text{s}$ the steering angle is increased and for the drive train $\alpha_{\text{th}} = 0$ is set, such that $\dot{\psi}$ and $\beta$ increase. As $\Delta\dot{\psi}$ is relatively small first no control action is observed from the $\dot{\psi}$-control and $\Delta\beta$ decreases significantly. $\dot{\psi}$-control becomes active only with some time delay where $\beta$ has decreased already to $-16°$. The combined $\beta$ and $\dot{\psi}$ control acts earlier and stronger and results in a much smaller slip angle deviation of about $-6°$. A combined control of $\dot{\psi}$ and $\beta$ by using a state-controller with pole placement design is proposed by Daiss (1996).

The ESC control functions have been described by using the acting forces at the wheels. Another possibility is to control the longitudinal and lateral tire slip values $S_{X,ij}$ and $S_{Y,ij}$; see van Zanten (2006), van Zanten and Kost (2016). The forces and the slips are related by the tire models treated in Chap. 5.

## 17.3 Lane Keeping Assistance (LDW/LKA)

To support the driver in lane keeping a lane departure warning (LDW) has been introduced since about 2002 and lane keeping assistance (LKA) systems since about 2007. The main reason is that unintentional lane departures are the cause for more than one-third of accidents with severe injuries. *LDW systems* warn the driver when the vehicle begins to leave the lane unintentionally with vibrational, visual, and audible warnings if the turn signal in this direction is not on. A prerequisite is the availability of a lane marker detection system. Normally monocular cameras behind the windscreen are used with a broad field of view of min. $\pm 20°$ and longitudinal ranges of min. 80m. Color cameras have the advantage to recognize better white and yellow lane markers. Infrared diodes behind the front trim may also be used, but for LDW systems only.

A basis for the activation of the lane departure warning is a warning zone around the lane marking in form of a Distance-to-Line-Crossing $d_{LC}$ (DLC), a measure between an outer part of the vehicle and the lane marking. The DLS is just a zone on the right and left part of the lane marking. A warning begins after entering and stops after leaving this warning zone. An alternative is the time-to-line-crossing (TLC), with

$$t_{LC} = \frac{d_{LC}}{v_Y} = \frac{d_{LC}}{v_X \sin \psi}. \tag{17.3.1}$$

A warning begins if TLC falls below a threshold and allows a predictive information.

In the case of *lane keeping LKA systems*, there are two kinds on the market. LKA-1 brings the lane leaving vehicle back to the lane by a short active lateral control action. This can either be realized by a correction command to the electrical power steering system (ESP) or by using the ESC system and individual wheel braking. For an steering input design the lateral yaw angle model (20.1.15) or Fig. 20.2a may be used and for a braking input impulse the models around the vertical axis (17.2.11) and, (17.2.15) or (17.2.22), compare Fig. 17.3.

LKA-1 systems support the driver by bringing the vehicle back to the center of the lane with a lateral controller and commands to an EPS-steering actuator. Then, the lateral dynamic model (20.1.15) as for the yaw angle control along a path, see also Fig. 20.3, may be applied.

A detailed description of LDW/LKA systems is given by Bartels et al. (2016). As several other requirements have to be fulfilled it uses a state machine, hands-free detection, a trajectory planning, and a LKA-controller.

## References

Ackermann C (2016) Neue Ansätze für Fahrerassistenzsysteme mit Regelung der Längs- und Querdynamik. Dissertation Technische Universität Darmstadt. Fortschr.-Ber. VDI Reihe 8, 1249. VDI Verlag, Düsseldorf

Ackermann J (1997) Robust control prevents car skidding. IEEE Control Syst Mag 17(3):23–31

Bartels A, Rohlfs M, Hamel S, Saust F, Klauske LK (2016) ateral Guidance Assistance. In: Chapter 48 in Winner et al (2016)

Bechtloff J (2018) Schätzung des Schwimmwinkels und fahrdynamischer Parameter zur Verbesserung modellbasierter Fahrdynamikregelungen. Dissertation Technische Universität Darmstadt. Fortschr.-Ber. VDI Reihe 12, 809. VDI Verlag, Düsseldorf

Daiss A (1996) Beobachtung fahrdynamischer Zustände und Verbesserung einer ABS-und Fahrdynamikregelung. Diss. Universität Karlsruhe, Fortschr.-Ber. VDI Reihe 12, 283. VDI Verlag, Düsseldorf

Hac A, Bedner E (2007) Robustness of side slip estimation and control algorithms for vehicle chassis control. In: 20th international technical conference on the Enhanced Safety of Vehicles (ESV), on-line paper 07-0353. www-nrd.nhtsa.dot.gov

König L, Schindele F, Ghosh J (2018) TC-Model-based feed forward traction control. Chassistech Symposium, ATZ live, München

Lu Q, Gentile P, Tota A, Sorniotti A (2016) Enhancing vehicle cornering limit through sideslip and yaw rate control. Mech Syst Signal Process 75:455–472

Reif K (ed) (2013) Fahrerassistenzsysteme. Springer Vieweg Verlag, Wiesbaden

Bosch GmbH Robert (ed) (2018) Automotive Handbook, 10th edn. Wiley, Chichester, England

Winner H, Hakuli S, Lotz F, Singer C (eds) (2016) Handbook of driver assistance systems. Springer International Publishing AG, Cham, Switzerland

van Zanten A (2006) Elektronisches Stabilitätsprogramm (ESP). In: R Isermann (ed), Chapter 8 Fahrdynamik-Regelung, Vieweg, Wiesbaden, pp 169–211

van Zanten A, Kost F (2016) Brake-based asstistance functions. In: Winner H, Hakuli S, Lotz F, Singer C (eds), Chapter 39 inHandbook of driver assistance systems, Springer International Publishing AG, Cham, Switzerland, pp 919–967

# Part V
# Automatic Driving

# 18 Classification of Automatic Driving Functions

The driver assistance systems until about 2008 were developed sequentially to improve firstly safety and secondly comfort and are characterized by automatic control and warning functions; see Winner et al. (2016). They can be seen as steps toward automatic driving.

## 18.1 Degrees of Automatic Driving

A first classification of automatic driving control can be made with regard to different degrees of automation, normal and conflicting driving maneuvers, role of driver and control systems; see Gasser et al. (2012), Ruchatz (2013), Verband der Automobilindustrie (2015), SAE (2016); see Table 18.1. (These degrees were already discussed in Chap. 1, but are repeated here briefly for sake of completeness.)

Level 0 is the normal driving by the driver only, where the driver takes over all tasks of driving. Level 1 comprises the *driver assistance systems* (DAS) which came into series application between 1979 and 2008, as depicted in Fig. 1.4. The driver has the full task of driving and is assisted in some (critical) driving situations. *Advanced Driver Assistance Systems* (ADAS) are characterized by some automatic control functions for parking and longitudinal driving and slow driving in traffic jams. Therefore, the system is able to control the longitudinal and lateral motion automatically in very special cases. The driver has to be attentive all the time, supervises the system, and is ready to take over if required.

In the case of *partial automation* (PAS), level 2, an autopilot controls the vehicle in several selected cases. Examples are automatic parking assist, jam traffic assist, and longitudinal and lateral control assist on highways and rural roads. The driver has to supervise the system all the time and has to take over driving immediately upon request by the system.



R. Isermann, *Automotive Control*, ATZ/MTZ-Fachbuch,
https://doi.org/10.1007/978-3-642-39440-9_18

**Table 18.1** Degrees of automatic driving (Extension of Gasser et al. 2012; Ruchatz 2013; Verband der Automobilindustrie 2015; SAE 2016)

| | | | DEGREES OF AUTOMATIC DRIVING | | | | | | |
|---|---|---|---|---|---|---|---|---|---|
| | | LEVEL | 1 | | 2 | | 3 | 4 | 5 |
| | | GENERAL DESCRIPTION | Driver Assistance (DA) | | Partial Automation (PA) | | Conditional Automation (CA) | High Automation (HA) | Full Automation (FA) |
| MANEUVERS | NOR-MAL | KIND OF DRIVING | Eyes-ON | | Eyes-ON | | Eyes-OFF | Eyes-OFF | Eyes-OFF |
| | | | Hands-ON | | Hands temp. OFF | | Hands temp. OFF | Hands-OFF | Hands-OFF |
| | | EXECUTION OF DRIVING | Driver | | System | | System | System | System |
| | | MONITORING OF DRIVING | Driver | | Driver | | System | System | System |
| | | FALLBACK | Driver | | Driver | | Driver | System | System |
| | | DRIVING MODES | some | | some | | some | some | all |
| | | DIRECTION OF CONTROL | longitudinal | lateral | longitudinal | lateral | some | many | all |
| | | SLOW MANEUVERS | jam assist | parking assist | parking assist<br>jam assist | | parking pilot | parking pilot | parking pilot |
| | | MANEUVERS WITH LIMITED DURATION | braking assist (ABS) | electronic stability control (ESC) | | lane change assist (LCA) | lane change pilot<br>overtaking pilot | lane change pilot<br>overtaking pilot<br>crossing pilot | overall driving pilot (highway, rural road, city) |
| | | CONTINUOUS DRIVING | speed & distance control assist (ACC) | lane keeping assist (LKA) | highway distance & speed control (ACC-FR) | lane keeping control (LKC) | highway pilot<br>rural raod pilot | highway pilot<br>rural road pilot | |
| | CON-FLICT | ACCIDENT AVOIDANCE MANEUVERS | warning | | warning | | warning | warning | warning |
| | | | emergency braking (AEB) | blindspot warning | emergency braking | evasive maneuver control | emergency maneuvers | emergency maneuvers | emergency maneuvers |
| DRIVER | | ROLE OF SYSTEM & DRIVER | • Driver performs continuously driving task.<br>• Assistance in special cases. | | • System performs longitudinal and lateral driving<br>• Defined use cases<br>• Driver monitors all time<br>• Driver performs remaining tasks | | • System performs longitudinal and lateral driving<br>• Defined use cases<br>• Driver does not monitor all time<br>• Driver resumes after request | • System performs longitudinal and lateral driving in all cases<br>• Driver is not required in defined use cases | • System performs driving in all situations during a trip<br>• Driver is not required |

*Conditional automation* (CAS) in level 3 has extended automatic control functions. The system performs lateral and longitudinal automatic driving on highways and rural roads, in traffic jams, and for parking. The driver does not need to monitor the automatic driving all the time. The automatic system recognizes its limits and requires the driver to take over within a certain time period. Lane changing and overtaking may also be possible automatically.

*High automation* (HAS) is dedicated to level 4. The degree of automation is further increased such that lateral and longitudinal driving is possible in many situations, but limited to use cases with regard to road type, speed ranges, and environmental conditions. The driver is not required in these use cases but still has to be in the vehicle. Valet parking (driverless parking) and urban automatic driving may be possible.

*Full automation* (FAS), level 5, includes automatic control for all situations. The driver is then not required. This is an ideal case and it cannot be forecasted when this can be realized in public traffic.

As Table 18.1 illustrates, the number of automatic driving modes increases from level 2 to level 5 and the monitoring of driving by the driver decreases from level 3 to 5. Until level 3, the human driver is the fallback system in the case of malfunctions of the control systems. Especially for level 4 and 5, *fault-tolerant systems* have to be implemented, as the driver is at least not required partially or completely.

The classified degrees of automatic driving are dedicated to normal driving. However, the automation of driving may especially help in conflicts to initiate *accident avoiding maneuvers*; see Table 18.1 at the bottom. A lower degree of a collision avoidance maneuver is automatic emergency braking (AEB). The next degree is emergency braking and emergency steering to stop and/or evasion for obstacles. In the case of fully automatic collision maneuvers, an automatic driving into a conflict-free space may be included. One goal of these collision-avoidance systems is to react faster than a normal human driver and to perform the best possible maneuver.

The steps toward high and full automatic driving need a multitude of investigations, testing, and changes of legislation.

## 18.2 Driving Maneuvers

Driving maneuvers for the testing and model identification of the vehicle behavior, like cornering, lane change, double lane change, slalom driving, and braking have been considered in Sect. 10.3. They are at least partially standardized. Further driving maneuvers with regard to judge and evaluate the handling and comfort properties, both for tuning, objective and subjective evaluation are, for example, described in Heissing and Ersoy (2011). These testing driving maneuvers use special inputs to excite the stationary or dynamic vehicle behavior under consideration on special roadways, skid pads, or testing facilities.

Automatic driving has to take into account the normal situations during everyday driving and emergency situations. Table 18.2 gives an overview of typical driving maneuvers. It divides normal driving in longitudinal and lateral directions. The velocity for these maneuvers may be constant or time varying.

**Table 18.2** Overview of some typical driving maneuvers for automatic driving

| | | Driving maneuvers | | | | | | | | | | | | |
|---|---|---|---|---|---|---|---|---|---|---|---|---|---|---|
| | | Normal | | | | | | | | | | | Emergency | |
| Direction | | Longitudinal | | Lateral | | | Longitudinal and lateral | | | | | | Longitudinal | Lateral and longitudinal |
| Kind of maneuver | | Velocity, distance control | Platooning | Cornering | Lane change | Evasion | Cornering | Lane change | Evasion | Overtaking | Merging | Roundabout | Automatic emergency braking | Emergency evasion |
| Velocity | Const. | × | × | × | × | × | | | | | | | | |
| | $v(t)$ | × | × | | | | × | × | × | × | × | × | × | × |
| Main action | Steering | | | × | × | × | × | × | × | × | × | × | | × |
| | Deceleration | × | × | | | × | × | × | × | | × | | × | × |
| | Acceleration | × | × | | | | × | × | | × | × | × | | |
| Path control $y(x)$ | | × | × | × | × | × | | | | | | | × | |
| Trajectory control $y(x, t)$ | | | | | | | × | × | × | × | × | × | | × |

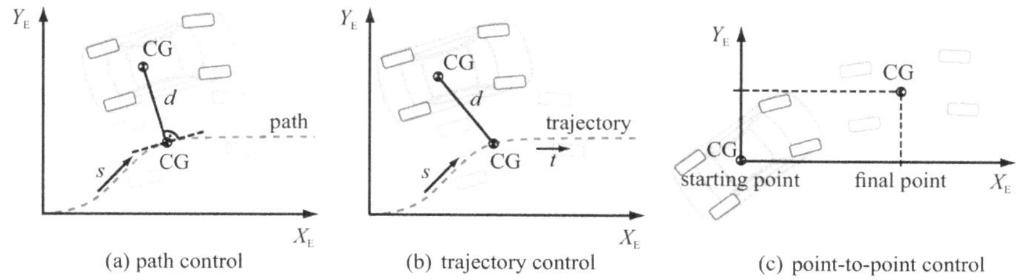


**Fig. 18.1** Different kinds of motion control for vehicles, (Schorn 2007)

The automatic control along a straight or bending lane with constant velocity means that the control system has to follow a certain path $y(x)$. If this can be performed without time condition, it is called *path control*, as shown in Fig. 18.1a. However, if a position $y(x)$ has to be reached at a certain time, such that $y(x, t)$, this is called *trajectory control*. A further case is a point-to-point control, where only a starting point and a fixed point without a certain path are given, which may be appropriate for parking. These expressions are used in the field of mobile robots; see Kochem (2005).

Hence, for path control, the lateral motion has to be controlled. However, for trajectory control, the lateral and longitudinal motions have to be considered where steering and braking or acceleration have to be manipulated simultaneously, which complicates the finding of solutions, see, e.g. Werling (2010), Werling et al. (2010), and Rathgeber (2016), because lateral, longitudinal, and vertical behaviors are coupled, especially for higher accelerations.

Table 18.2 shows that path control belongs to normal longitudinal and lateral driving with constant velocity. But the other normal and emergency maneuvers with variable velocity are subject to trajectory control. However, if the velocity changes are not too large, lateral and longitudinal behaviors can be treated separately, such that path control can be applied as an approximation; see also Sect. 21.1.

An investigation of many drivers shows that maximal accelerations are for normal, relaxed drivers $a_X \approx \pm 2\text{m/s}^2$ and $a_Y \approx \pm 2,5\text{m/s}^2$ and for sporty drivers $a_X \approx +2\text{m/s}^2 - 2.5\text{m/s}^2$ and $a_Y \approx \pm 4\text{m/s}^2$; see Bossdorf-Zimmer et al (2011), Schacher et al. (2018). Normal drivers seldomly change the velocity during cornering, contrast to sporty drivers.

## 18.3 On the Design of Automotive Control Systems

In addition to the general procedures described in Chap. 3, some specific remarks are mentioned for the design and selection of control algorithms.

The design of control systems comprises the selection of the control system structure, consisting, for example, of

- feedforward control,
- feedback control,

- interconnected control systems,
  - cascaded control
  - multi-variable control

and the specification of the controllers, such as

- linear or nonlinear controllers,
- parameter-optimized controllers,
- state-space controllers,
- model-based controllers, and
- predictive controllers.

The parameters of the controllers can, for parameter-optimized controllers like P, PD, and PID controllers, be obtained by

- tuning rules,
- experimental probing, and
- parameter optimization based on process models and control criteria.

In the case of model-based controllers, the parameters follow

- directly from the process parameters and
- indirectly by numerical optimization methods.

If the controllers are designed in the time domain or in the Laplace s-domain, they have to be transformed to discrete time for the programs of digital microcomputers. However, they can also directly be designed in the discrete-time domain or z-domain; see, e.g. Åström and Wittenmark (1984), Kuo (1980), Isermann (1989).

Some special requirements for automotive control systems are as follows:

- low computational expense for the ECU's,
- highly understandable and transparent for the application by many different users (suppliers, manufacturers),
- good transferability to different vehicles and different versions of the same vehicle type,
- robustness for changes of vehicle parameters and road properties, and
- tuning of controller parameters in the research vehicle.

Table 18.3 evaluates some of the controller properties. These properties will be required in the following chapters.

**Table 18.3** Comparison of some properties of controller types (add: with additions)

| Controller type | Parameter-optimized controllers (P-,PD-,PID) | Internal model control (IMC) | state-space control | Model predictive control (MPC) |
|---|---|---|---|---|
| Number of parameters | low | low (low order) | high | high |
| Tuning rules for parameters | yes | no | no | no |
| Parameters directly expressed in physical process model parameters | no | yes | no | no |
| Parameters follow from numerical optimization | no/yes | no | yes | yes |
| Usability for linear models | yes | yes | yes | yes |
| Usability for nonlinear models | add | add | add | yes |
| Direct gain schedul-ing(physical parameters) | yes | yes | no | no |
| State observer required | no | no | yes | no |
| Actuator restrictions included | add | add | add | yes |
| Computational expense | low | low | high | high |

## References

Åström K, Wittenmark B (1984) Computer controlled systems - theory and design. Prentice-Hall, Englewood Cliffs

Bossdorf-Zimmer J, Kollmer H, Henze R, Küçükay F (2011) Fingerprint des Fahrers zur Adaption von Assistenzsystemen. ATZ 113(3):226–231

Gasser TM, Arzt C, Ayoubi M, Bartels A, Bürkle L, Eier J, Flemisch F, Häcker D, Hesse T, Huber W (2012) Rechtsfolgen zunehmender Fahrzeugautomatisierung. Berichte der Bundesanstalt für Straßenwesen. BASt. Heft F 83, Bundesanstalt für Straßenwesen, Bergisch Gladbach

Heissing B, Ersoy M (2011) Chassis handbook, Berlin: Vieweg-Teubner Verlag. Springer, Wiesbaden

Isermann R (1989) Digital control systems, 2nd edn. Springer, Berlin

Kochem M (2005) Ein Fahrerassistenzsystem zur Unterstützung des rückwertigen Parkvorgangs für Pkw. Dissertation Technische Universität Darmstadt. Fortschr.-Ber. VDI Reihe 12, 590. VDI Verlag, Düsseldorf

Kuo B (1980) Digital control systems

Rathgeber C (2016) Trajektorienplanung und -folgeregelung für assistiertes bis hochautomatisiertes Fahren. Diss, TU Berlin

Ruchatz (2013) Vision und Möglichkeiten des automatischen Fahrens. In: 6. Fachtagung AUTOREG, (2013) VDI-Berichte 2196. Wiesloch, Germany

SAE (2016) Taxonomy and Definitions for Terms Related to Driving Automation Systems for On-Road Motor Vehicles. Standard J3016_201609. SAE International, Warrendale, Pennsylvania

Schacher S, Hoedt J, King R (2018) Fahrspezifische Geschwindigkeitsprofile für die automatische oder die kooperative Fahrt. at - Automatisierungstechnik 66(1):53–65

Schorn M (2007) Quer- und Längsregelung eines Personenkraftwagens für ein Fahrerassistenzsystem zur Unfallvermeidung. Diss. Universität Darmstadt, Fortschr.-Ber. VDI Reihe 12, 651. VDI Verlag, Düsseldorf

Verband der Automobilindustrie (2015) Automatisierung - Von Fahrerassistenzsystemen zum automatisierten Fahren. Die Zukunft von gestern - heute Realität. VDA-Broschüre, Berlin

Werling M (2010) Ein neues Konzept für die Trajektoriengenerierung und -stabilisierung in zeitkritischen Verkehrsszenarien. Diss, Karlsruhe Institut for Technology (KIT)

Werling M, Groell L, Bretthauer G (2010) Invariant Trajectory Tracking With a Full-Size Autonomous Road Vehicle. IEEE Transactions on Robotics pp 758–765

Winner H, Hakuli S, Lotz F, Singer C (eds) (2016) Handbook of driver assistance systems. Springer International Publishing AG, Cham, Switzerland

# Longitudinal Vehicle Control 19

Automatic driving control of vehicles can be divided into longitudinal and lateral driving control. Figure 19.1 shows some basic maneuvers without obstacles like driving with constant velocity, with acceleration or braking, straightline driving on roadways, cornering, and lane change. Figure 19.2 depicts a roadway with two lanes, one for each direction, with lanewidth $B$. Usually, the vehicle is driven in the center of a lane, such that the lateral distance to the right lane marking is $y = B/2 = D_y$.

This chapter considers automatic control in longitudinal direction. Three cases are treated:

(1) Acceleration control,
(2) Velocity control,
(3) Distance control to a preceding vehicle.

Automatic control in longitudinal direction has been treated in many publications. First approaches from General Motors, in 1939, are summarized in Gardels (1960). The used drive dynamic models were rather simple and linear; see Mayr (1991). A first approach for distance control with a vehicle using pneumatic accelerator and brake pedals and measurements of the distance by a cable was treated in Hartwich (1971). Ackermann (1980) described distance control by using radar-sensors.

Further publications for the longitudinal vehicle control from the American PATH program have been Hedrick (1995) and Shladover et al. (1991). Additionally, the driving comfort was taken into account by limiting the jerk and acceleration; see Hauksdottir and Fenton (1985). Mayr (1994) also considered merging vehicles.

Velocity control with subsidiary acceleration control was investigated within the European PROMETHEUS program, Dorissen et al. (1994), Hoess (1995), however, without adaptation to different vehicles and different loads; see also Winner and Schopper (2016). A detailed model-based acceleration, vehicle, and distance control



R. Isermann, *Automotive Control*, ATZ/MTZ-Fachbuch,
https://doi.org/10.1007/978-3-642-39440-9_19

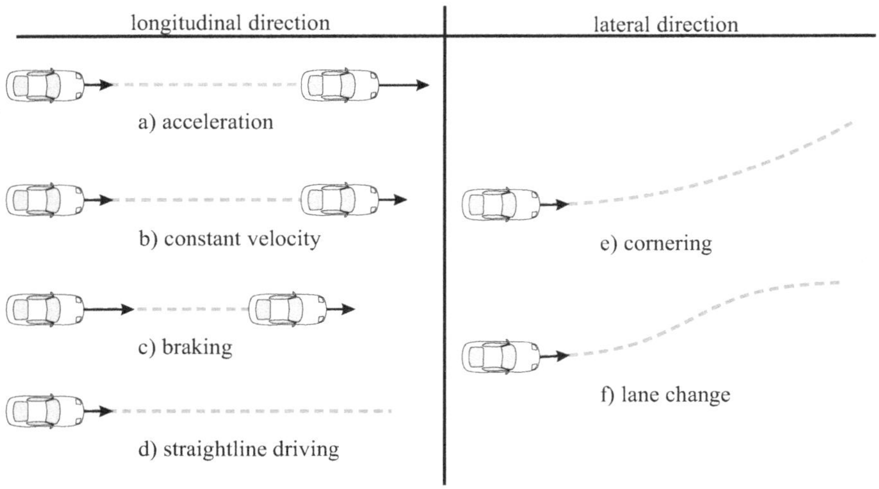


**Fig. 19.1** Basic driving maneuvers without obstacles

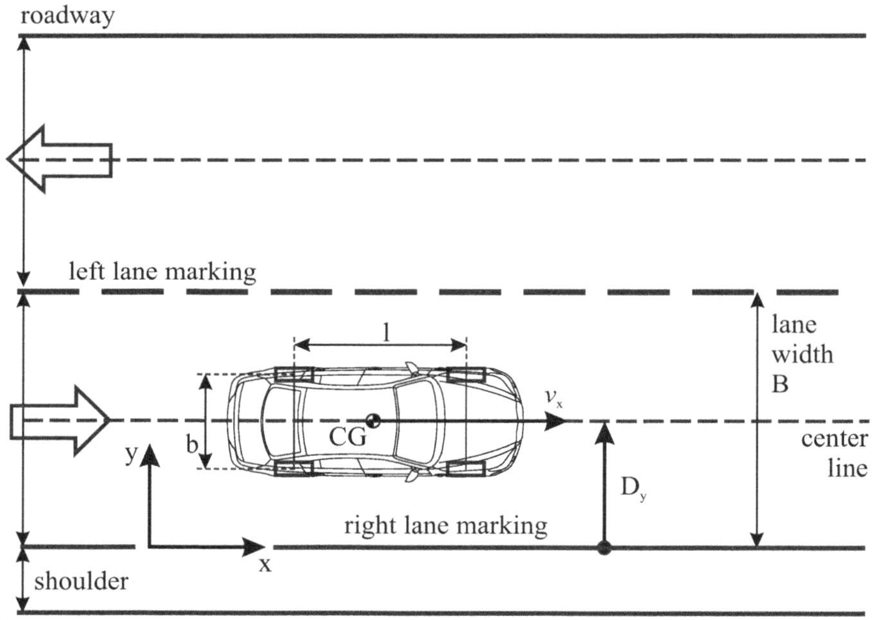


**Fig. 19.2** Road with two lanes in opposite directions

with linear and nonlinear controllers and parameter estimation for two passengers cars was published by Germann (1997). Longitudinal vehicle control for low speed is investigated in Adipastro (2003). Within the work by Rathgeber et al. (2016), a two-degree control method with disturbance observer is developed for acceleration and lateral distance control.

The development of corresponding controllers in this chapter uses simplified linearized, but parameter variable longitudinal vehicle models from Chap. 6. One of the goals is to determine the structure of the controllers. Because a direct relationship between the controller parameters and the model parameters is preferred, the design principle of internal model control is applied for the used low order processes. Sensor delays are assumed to be negligible, but can be included in the controller design.

The drivetrain of an engine driven vehicle comprises several rotational multi-mass components as illustrated in Fig. 6.2. A simplified torque model of an internal combustion engine is depicted in Fig. 6.9 with models (6.2.16) and (6.2.19). Dynamic models for the drivetrain with shifted or hydrodynamic transmissions are treated in Sects. 6.3 and 6.4; see also Figs. 6.12 and 6.15. These models are nonlinear and depend on the speed and torque of the engine and the mass and velocity of the vehicle. A simplified vehicle model is obtained with the assumption of a stiff powertrain. The resulting model is a nonlinear model (6.5.21) with a signal flow shown in Fig. 6.17 for acceleration and braking.

In the following, small changes around the reference variables are considered, such that linearized parameter variable models can be used.

For longitudinal vehicle control, the measured velocity or the measured acceleration can be used as controlled variable. The vehicle velocity is not directly measurable and is reconstructed from the measured wheel angular velocities; see Sect. 12.2.1 on ground velocity determination. However, the wheel velocities are subject to slip, such that it is not possible to obtain a precise vehicle velocity. The vehicle acceleration is directly measurable, but is usually a noisy signal and measures also road gradients. Therefore, it has to be filtered, e.g. by a low pass filter and the influence of road gradients has to be compensated. As treated in Sect. 12.2.2, the combination of wheel velocities and acceleration measurements by applying a Kalman filter allows to obtain relative precise values for the ground velocity.

In the following, longitudinal control with both the velocity and the acceleration as controlled variables is considered.

## 19.1 Acceleration Control

The longitudinal vehicle model (6.6.1) with a stiff powertrain can be simplified for *acceleration control with the throttle actuator* by assuming small changes around the operating point to

$$m_{\text{tot}}\frac{dv(t)}{dt} + c_{\text{Rv}}\Delta v(t) = c_{\text{MF}}\Delta M_{\text{eng}}(t) = \Delta F_{\text{XT}}(t),$$
$$c_{\text{MF}} = \frac{i_{\text{tot}}\eta_{\text{tot}}}{r_{\text{dyn}}}; \; c_{\text{R}v} = c_{\text{A}}\bar{v} + c_{\text{R}}; \tag{19.1.1}$$

see (6.6.5), or in the Laplace domain

$$(m_{\text{tot}}s + c_{\text{R}v})\Delta v(s) = c_{\text{MF}}\Delta M_{\text{eng}}(s) = \Delta F_{\text{XT}}(s). \tag{19.1.2}$$

If the drag and resistance forces due to (6.6.6) are neglected ($c_{\text{R}v} = 0$), e.g. for low velocity, one obtains an integral behavior

$$\frac{\Delta v(s)}{\Delta M_{\text{eng}}(s)} = \frac{c_{\text{MF}}}{m_{\text{tot}}s}, \tag{19.1.3}$$

or with the torque model (6.2.19) of the engine and drive train

$$G_{\alpha v}(s) = \frac{\Delta v(s)}{\Delta\alpha(s)} = \frac{K_{\alpha v}}{\left(1 + T'_{\text{act}}s\right)(1 + T_{\text{int}}s)\, s}, \tag{19.1.4}$$

with

$$K_{\alpha v} = K_{\alpha\text{M}}K_{\text{M}v}.$$

$K_{\alpha\text{M}} = \Delta M_{\text{eng}}/\Delta\alpha$ follows from the engine look-up table; see (6.6.6) to (6.6.9), (6.2.20), and $K_{\text{M}v} = c_{\text{MF}}/c_{\text{R}v}$ from (6.6.6) and (6.6.7).

For the longitudinal acceleration follows from (19.1.4)

$$G_{\alpha\text{ax}}(s) = \frac{\Delta a_{\text{X}}(s)}{\Delta\alpha(s)} = \frac{K_{\alpha v}}{\left(1 + T'_{\text{act}}s\right)(1 + T_{\text{int}}s)}. \tag{19.1.5}$$

The acceleration is then proportional to $\Delta M_{\text{eng}}$ or $\Delta\alpha$ and a suitable acceleration controller is a PI controller; see e.g A.2.

However, for higher velocities, the drag and resistance forces play a crucial role and (19.1.2) leads with (6.6.11) to the third order model with proportional behavior

$$G_{\alpha v}(s) = \frac{\Delta v(s)}{\Delta\alpha(s)} = \frac{K_{\alpha v}}{\left(1 + T'_{\text{act}}s\right)(1 + T_{\text{int}}s)(1 + T_{\text{acc}}s)}, \tag{19.1.6}$$

where $T_{\text{acc}} = m_{\text{tot}}/c_{\text{R}v}$ is the time constant of the vehicle; see (6.6.10). As the time constants $T'_{\text{act}}$ and $T_{\text{int}}$ are relatively small, they can be lumped together as $T''_{\text{act}} = T_{\text{int}} + T_{\text{act}} + T_{\text{d}}$; see (6.2.21) and (19.1.6) can be reduced to

$$G_{\alpha v}(s) = \frac{\Delta v(s)}{\Delta\alpha(s)} = \frac{K_{\alpha v}}{\left(1 + T''_{\text{act}}s\right)(1 + T_{\text{acc}}s)}. \tag{19.1.7}$$

The gain of this simplified, linearized longitudinal model is

$$K_{\alpha v} = K_{\alpha \mathrm{M}} K_{\mathrm{M}v}, \tag{19.1.8}$$

$K_{\alpha\mathrm{M}} = \Delta M_{\mathrm{eng}}/\Delta\alpha$ follows from an engine look-up table and, see (6.6.6) to (6.6.9), and it is

$$K_{\mathrm{M}v} = \frac{c_{\mathrm{MF}}}{c_{\mathrm{R}v}} = \frac{i_{\mathrm{tot}}\eta_{\mathrm{tot}}}{r_{\mathrm{dyn}}} \frac{1}{c_{\mathrm{A}}\bar{v} + c_{\mathrm{R}}}. \tag{19.1.9}$$

The dominating time constant is; see (6.6.10)

$$T_{\mathrm{acc}} = \frac{m_{\mathrm{tot}}}{c_{\mathrm{R}v}} = \frac{m_{\mathrm{tot}}}{c_{\mathrm{A}}\bar{v} + c_{\mathrm{R}}}. \tag{19.1.10}$$

As discussed in Sect. 6.2, the time constants are $T_{\mathrm{act}} \approx 50\ldots70$ ms and $T_{\mathrm{int}} = 200\ldots300$ ms, and thus $T''_{\mathrm{act}} \approx 250\ldots370$ ms compared to $T_{\mathrm{acc}} = 3\ldots10$ s; see also the frequency ranges in Fig. 6.18.

For the acceleration as output follows

$$G_{\alpha\mathrm{ax}}(s) = \frac{\Delta a_{\mathrm{X}}(s)}{\Delta\alpha(s)} = \frac{K_{\alpha v} s}{\left(1 + T''_{\mathrm{act}} s\right)(1 + T_{\mathrm{acc}} s)}. \tag{19.1.11}$$

If the small time constant $T''_{\mathrm{act}}$ is neglected, one obtains

$$G_{\alpha\mathrm{ax}}(s) = \frac{\Delta a_{\mathrm{X}}(s)}{\Delta\alpha(s)} = \frac{K_{\alpha v} s}{(1 + T_{\mathrm{acc}} s)}. \tag{19.1.12}$$

Then an integrating controller

$$G_{\mathrm{ax}\alpha}(s) = \frac{\Delta\alpha(s)}{\Delta a_{\mathrm{X}}(s)} = \frac{1}{T_{\mathrm{I}\alpha} s}, \tag{19.1.13}$$

leads to a closed-loop transfer function

$$G_{\mathrm{ax,axd}}(s) = \frac{\Delta a_{\mathrm{X}}(s)}{\Delta a_{\mathrm{X,d}}(s)} = \frac{\frac{K_{\alpha v}}{T_{\mathrm{I}\alpha} + K_{\alpha v}}}{\left(1 + \frac{T_{\mathrm{acc}} T_{\mathrm{I}\alpha}}{T_{\mathrm{I}\alpha} + K_{\alpha v}} s\right)}. \tag{19.1.14}$$

Hence, a first order lag behavior results, where the gain and the time constant can be influenced by the selection of the controller integration time $T_{\mathrm{I}\alpha}$. For the closed-loop, time constant follows

$$T_{\mathrm{ax,axd}} = \frac{m_{\mathrm{tot}} T_{\mathrm{I}\alpha}}{T_{\mathrm{I}\alpha}(c_{\alpha}\bar{v} + c_{\mathrm{R}}) + K_{\alpha\mathrm{M}} \frac{i_{\mathrm{tot}}\eta_{\mathrm{tot}}}{r_{\mathrm{dyn}}}}. \tag{19.1.15}$$

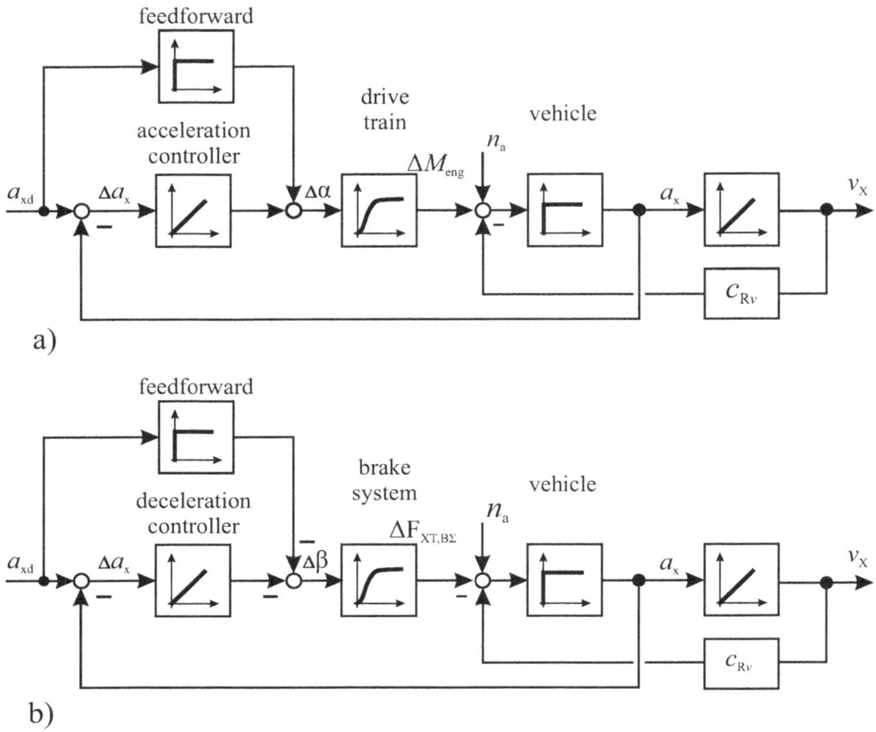


**Fig. 19.3** Acceleration control loops, $n_a$: disturbances (e.g. road gradients, velocity, and wind effects): **a** acceleration with the drivetrain, **b** deceleration by braking. (With feedforward control)

This time constant is proportional to the total representative mass $m_{tot}$ and becomes smaller for larger resistance parameters and larger $K_{\alpha M}$ (engine torque gradient) and gear ratio $i_{tot}$. An alternative is to use a PI controller, see Germann (1997). Figure 19.3a depicts a signal flow scheme, based on (19.1.1).

For the *acceleration control with small changes in the brake actuator* (deceleration), the linearized model (6.7.10) is used

$$m_{tot}\frac{dv(t)}{dt} + c_{Rv}\Delta v(t) = -\Delta F_{XT,B\Sigma}(t), \tag{19.1.16}$$

or, see (6.7.11)

$$\frac{\Delta v(s)}{\Delta F_{XT,B\Sigma}(s)} = -\frac{K_{Fv}}{1+T_{acc}s}. \tag{19.1.17}$$

with $K_{Fv} = 1/c_{Rv}$.

The dynamics of a hydraulic brake system according to (13.2.40) is expressed by a second order system, comprising a delay of the brake actuator and the brake line

$$\frac{\Delta F_{XT,B\Sigma}(s)}{\Delta\beta(s)} = \frac{K_{\beta F}}{(1+T_{BB}s)(1+T'_{p}s)}, \tag{19.1.18}$$

with $K_{\mathrm{Bf}}$ due to (13.2.41) where the time constants are in the range of $T_{\mathrm{BB}} \approx$ 5...10 ms and $T_{\mathrm{p}}^{'} \approx$ 5...20 ms.

The controlled brake/vehicle process model then is

$$G_{\beta v}(s) = \frac{\Delta v(s)}{\Delta \beta(s)} = \frac{-K_{\beta v}}{(1 + T_{\mathrm{BB}} s)\left(1 + T_{\mathrm{p}}^{'} s\right)(1 + T_{\mathrm{acc}} s)}, \quad (19.1.19)$$

with $K_{\beta v} = K_{\mathrm{F}v} K_{\beta \mathrm{F}}$, see (6.7.12), and with the acceleration as output

$$G_{\beta \mathrm{ax}}(s) = \frac{\Delta a_{\mathrm{X}}(s)}{\Delta \beta(s)} = \frac{-K_{\beta v} s}{(1 + T_{\mathrm{BB}} s)\left(1 + T_{\mathrm{p}}^{'} s\right)(1 + T_{\mathrm{acc}} s)}. \quad (19.1.20)$$

As $T_{\mathrm{BB}}$ and $T_{\mathrm{p}}^{'}$ are much smaller than $T_{\mathrm{acc}}$, the model can be simplified to

$$G_{\beta \mathrm{ax}}(s) = \frac{\Delta a_{\mathrm{X}}(s)}{\Delta \beta(s)} = \frac{-K_{\beta v} s}{(1 + T_{\mathrm{acc}} s)}. \quad (19.1.21)$$

An integral acting acceleration controller

$$G_{\mathrm{ax}\beta}(s) = \frac{\Delta \beta(s)}{\Delta \alpha_{\mathrm{X}}(s)} = -\frac{1}{T_{\mathrm{I}\beta} s}, \quad (19.1.22)$$

leads to a closed-loop transfer function

$$G_{\mathrm{ax,axd}}(s) = \frac{\Delta a_{\mathrm{X}}(s)}{\Delta a_{\mathrm{X,d}}(s)} = \frac{\frac{K_{\beta v}}{T_{\mathrm{I}\beta} + K_{\beta v}}}{\left(1 + \frac{T_{\mathrm{I}\beta} T_{\mathrm{acc}}}{T_{\mathrm{I}\beta} + K_{\beta v}} s\right)}. \quad (19.1.23)$$

Hence, a first order lag behavior results.

The controller structure is the same as for acceleration control with the throttle actuator, but with other gain. A signal flow scheme is shown in Fig. 19.3b.

For large braking pedal changes $\beta(t)$, if the braking forces $F_{\mathrm{XT,B}\Sigma}$ are much larger as the resistance forces, $c_{\mathrm{R}v} = 0$ can be set in (19.1.16) and an integral behavior results as (6.7.14). Then the deceleration $a_{\mathrm{X}}$ is approximately proportional to the braking forces and a P or PI controller can be used.

If a very fast reaction to a change of the desired (reference) value is required, a *feedforward control* can be added. In the case of braking, it follows from (19.1.21) that

$$G_{axd\beta}(s) = \frac{\Delta \beta(s)}{\Delta a_{\mathrm{xd}}(s)} = \frac{1}{G_{\beta \mathrm{ax}}(s)} = -\frac{(1 + T_{\mathrm{acc}} s)}{K_{\beta v} s}, \quad (19.1.24)$$

which can be approximated by a proportional element

$$G_{\mathrm{axd}\beta}^{'}(s) = -\frac{T_{\mathrm{acc}}}{K_{\beta v}}; \quad (19.1.25)$$

see Fig. 19.3b. A corresponding feedforward element may be added for the acceleration control in Fig. 19.3a.

For *decelerating in overrun (coasting)* (6.6.5) becomes with $\Delta\alpha = 0$ and $\Delta\beta = 0$ an homogenous differential equation

$$T_{\mathrm{acc}} \frac{dv(t)}{dt} + \Delta v(t) = 0, \tag{19.1.26}$$

which has the solution

$$v(t) = v(0) e^{-\frac{t}{T_{\mathrm{acc}}}}, \tag{19.1.27}$$

where $v(0)$ is the initial velocity for $\alpha = 0$. The velocity then decreases exponentially.

## 19.2 Velocity Control

As the velocity $v_{\mathrm{X}}$ follows through integration of the acceleration $a_{\mathrm{X}}$, an already existing acceleration control loop can be used as a subsidiary (minor) loop and extended by a major velocity controller, as shown in Fig. 19.4a. Then a cascaded control system results and the velocity controller may be realized as a P controller

$$G_{va}(s) = \frac{\Delta a_{\mathrm{Xd}}}{\Delta v_{\mathrm{X}}} = K_{va}. \tag{19.2.1}$$

The selection of the controller gain has to take into account a comfortable behavior, e.g. $K_{va} = 0.1$ for a well damped and $K_{va} = 0.4$ for a sportive behavior and for an adaptation with fuzzy-logic evaluation of measured and selected velocity, as developed by Germann (1997).

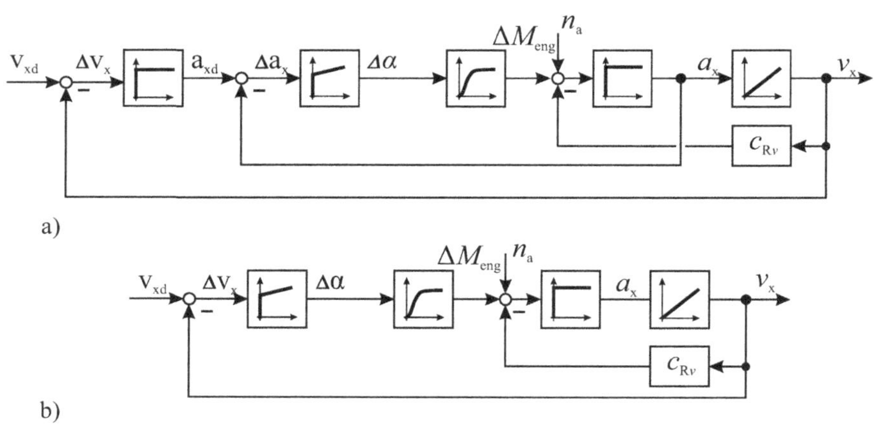


**Fig. 19.4** Velocity control loops: **a** cascade control system with subsidiary acceleration control, **b** direct velocity control

An alternative to the subsidiary acceleration control loop is the control of the velocity with one loop if a reliable determination of the ground speed $v_X$ is possible. The powertrain and vehicle model is then of third order; see (6.6.11). However, if the powertrain can be reduced to a first order model as (6.2.21), because the time constants $T_{act}$ and $T_D$ are relatively small, the linearized overall model becomes as (19.1.7)

$$G_{\alpha v}(s) = \frac{\Delta v(s)}{\Delta \alpha(s)} = \frac{K_{\alpha v}}{(1 + T''_{act}s)(1 + T_{acc}s)}. \tag{19.2.2}$$

A controller design by the internal control method with (A.1.37) then yields for the velocity controller

$$G_{IMC,v\alpha}(s) = \frac{1}{2K_{\alpha v}T_r} \frac{(1 + T''_{act}s)(1 + T_{acc}s)}{s} \frac{1}{\left(1 + \frac{T_r}{2}\right)}, \tag{19.2.3}$$

which corresponds to a PID controller.

With the assumption of $T_r = 2T''_{act}$, a PI controller follows

$$\begin{aligned} G_{v\alpha}(s) &= \frac{\Delta \alpha(s)}{\Delta v_X(s)} = K_{v\alpha}\left(1 + \frac{1}{T_{Iv}s}\right), \\ K_{v\alpha} &= \frac{T_{acc}}{4K_{\alpha v}T''_{act}}, \\ T_{Iv} &= T_{acc}. \end{aligned} \tag{19.2.4}$$

Figure 19.4b depicts a signal flow scheme of the velocity control loop.

For the closed-loop transfer function holds

$$G_{vv}(s) = \frac{\Delta v(s)}{\Delta v_d(s)} = \frac{G_{v\alpha}(s)G_{\alpha v}(s)}{1 + G_{v\alpha}(s)G_{\alpha v}(s)}. \tag{19.2.5}$$

By inserting (19.2.2) and (19.2.4), a third order dynamic system follows. However, a simplification with $T''_{act} \ll T_{acc}$, and therefore, $T''_{act} = 0$ results in

$$G_{vv}(s) = \frac{K_{vv}(T_{Iv}s + 1)}{T_{acc}T_{Iv}s^2 + T_{Iv}(1 + K_{vv})s + K_{vv}}, \tag{19.2.6}$$

with the loop gain

$$K_{vv} = 1.$$

For large $T_{Iv}$, which means approximately using a P controller, the closed-loop transfer function simplifies to

$$G_{vv}(s) = \frac{\Delta v(s)}{\Delta v_d(s)} = \frac{\frac{K_{vv}}{1+K_{vv}}}{\frac{T_{acc}}{1+K_{vv}}s + 1} = \frac{K'_{vv}}{1 + T_{vv}s}, \tag{19.2.7}$$

with the closed-loop time constant

$$T_{vv} = \frac{T_{\text{acc}}}{1 + K_{vv}},$$

and the gain

$$K'_{vv} = \frac{K_{vv}}{1 + K_{vv}}.$$

Then the velocity closed loop can approximately be considered as a first order lag.

In the case of *braking* with hydraulic friction brakes, the pressure build up in the brake caliper after a small $\Delta\beta$ of the brake actuator follows a second order delay (13.2.40), which in combination with the vehicle model (6.7.12) can be simplified to

$$G_{\beta v}(s) = \frac{\Delta v(s)}{\Delta\beta(s)} = -\frac{K_{\beta v}}{(1 + T''_{\text{p}}s)(1 + T_{\text{acc}}s)}, \tag{19.2.8}$$

with

$$T''_{\text{p}} = T'_{\text{p}} + T_{\text{BB}},$$

a dominating time constant $T_{\text{acc}}$ and a small time constant $T''_{\text{p}} \approx 10...20$ ms. Similar to (19.2.2) and (19.2.4), a PI controller for braking can be designed with $T_{\text{r}} = 2T''_{\text{p}}$

$$G_{v\beta}(s) = \frac{\Delta\beta}{\Delta v} = -K_{v\beta}\left(1 + \frac{1}{T_{\text{I}v\text{b}}}\right), \tag{19.2.9}$$

with

$$K_{v\beta} = \frac{T_{\text{acc}}}{4K_{\beta v}T''_{\text{p}}},$$

$$T_{\text{I}v\text{b}} = T_{\text{acc}}.$$

Therefore, as well for automatic controlled acceleration as for braking a PI controller can be applied with different gain, which depend on the operating point $[m_{\text{tot}}, c_{\text{R}v}, i_{\text{tot}}, v_{\text{X}}]$.

The coordination for velocity control for $v_{\text{Xd}} > 0$ is depicted in Fig. 19.5. Depending on the sign of the control deviation $\Delta v_{\text{X}} = v_{\text{Xd}} - v_{\text{X}}$, either the acceleration controller or the braking controller is switched to an active or non-active status. Small braking interventions, where $0 < \Delta v_{\text{X}} < \epsilon$ can be performed with overrun of the engine (engine in idle mode, coasting), where then braking and accelerating actuation is zero.

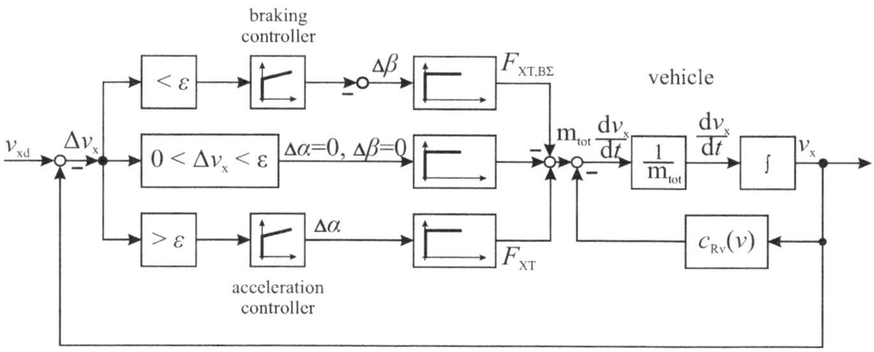


**Fig. 19.5** Coordination of velocity controllers for acceleration, braking, and overrun (coasting) for $v_{xd} > 0$

## 19.3 Distance Control

If the subject vehicle 1 follows another vehicle 2, a certain distance $x_D$ has to be maintained as required for adaptive cruise control. This distance is dependent on the velocity and can be expressed by a desired time gap $\tau_d = 1.5...2$ s between the two vehicles

$$x_{Dd} = v_{x1}\tau_d \tag{19.3.1}$$

The time dependence of the distance between the two vehicles is, if $x_{D0}$ is the initial distance at the time $t_1$

$$x_D(t) = x_{D0}(t_1) + \int_{t_1}^{t} (v_{2x}(t) - v_{1x}(t))\, dt,$$

or (19.3.2)

$$\Delta x_D(t) = x_D(t) - x_{D0}(t) = \int_{t_1}^{t} \Delta v_{12}(t) dt,$$

and after Laplace transformation

$$\Delta x_D(s) = \frac{1}{s}\Delta v_{21}(s). \tag{19.3.3}$$

Hence, the transfer behavior of the distance is characterized by an integration of the velocity difference $\Delta v_{12}$.

A signal flow scheme for vehicle 1 and small changes of the engine throttle $\Delta\alpha$ and brake pedal $\Delta\beta$ is depicted in Fig. 19.6. The velocity change $\Delta v_X$ through the longitudinal tire forces by $\Delta\alpha$ or $\Delta\beta$ is characterized by a first order lag with time constant $T_{acc}$, see (6.6.10), and the distance between the two vehicles results from an integration of the difference velocity $\Delta v_{12} = v_2 - v_1$. Therefore, it is one possibility to use a velocity control loop of vehicle 1 to control the distance $x_D$. Figure 19.7a depicts a signal flow scheme with a velocity control loop to adjust the distance.

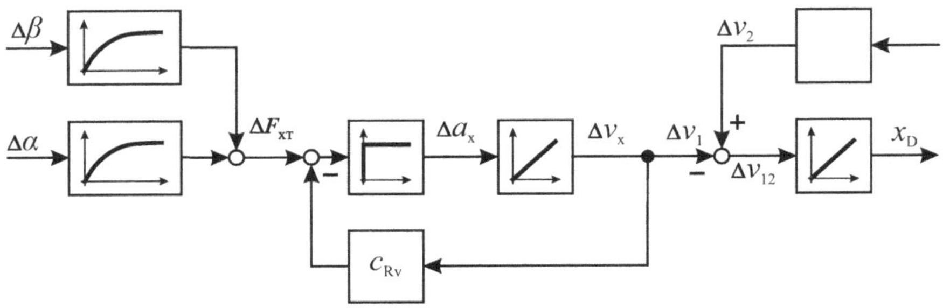


**Fig. 19.6** Signal flow scheme for the distance between two vehicles with linearization around an operation point $\bar{v}_x$

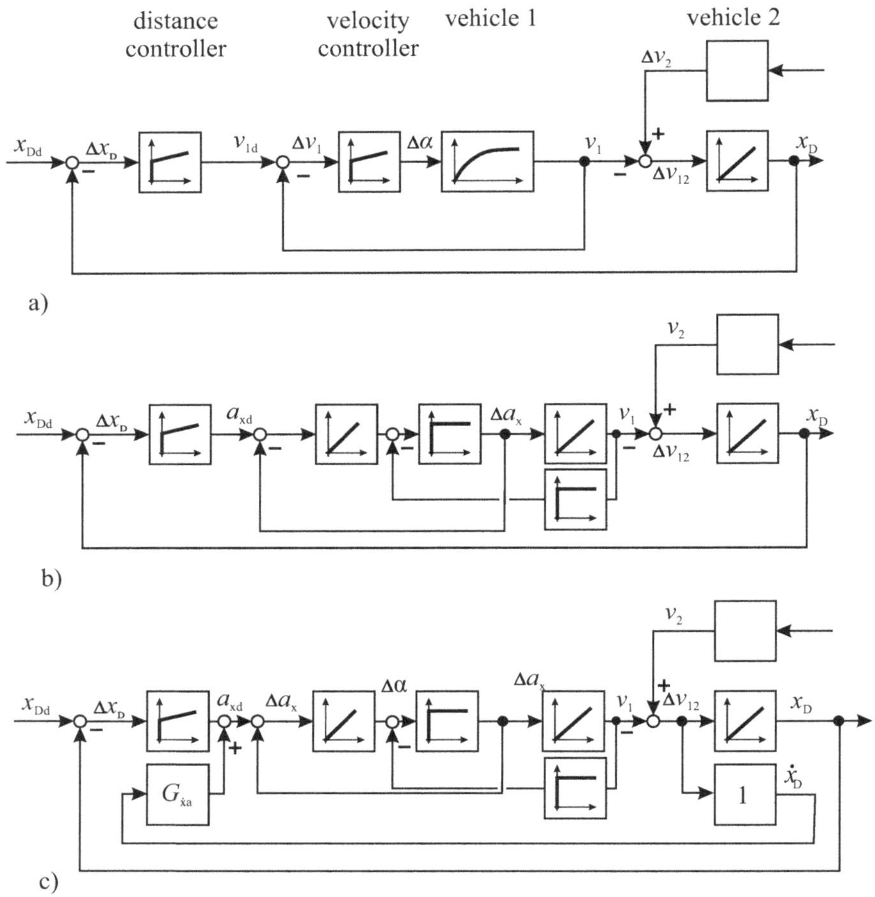


**Fig. 19.7** Distance control of vehicle 1 which follows a vehicle 2 with: **a** velocity control system, **b** acceleration control, **c** acceleration control and measurement of $x_D$ and $\dot{x}_D$

The closed velocity control loop of vehicle 1 is now described with the simplified first order model (19.2.7)

$$G_{vv}(s) = \frac{\Delta v_1(s)}{\Delta v_{1d}(s)} = \frac{K'_{vv}}{1 + T_{vv}s}. \tag{19.3.4}$$

The controlled process for the distance controller then is with (19.3.3) and $\Delta v_{12} = -\Delta v_1$

$$G_{vx}(s) = \frac{\Delta x_D(s)}{\Delta v_{1d}(s)} = \frac{-K'_{vv}}{(1 + T_{vv}s)}\frac{1}{s}. \tag{19.3.5}$$

Design with internal model control with a second order realizability filter results in this case in a P controller

$$G_{xv}(s) = K_{xv} = \frac{-1}{4K'_{vv}T_{vv}}. \tag{19.3.6}$$

But in order to reach zero control deviations $\Delta x_D = x_{Dd} - x_D$ as well for changes of the reference variable $x_{Dd}$ as for changes of the velocity $\Delta v_2$ of the preceding vehicle 2 a distance controller with integral behavior is required, e.g. a PI controller

$$G_{xv}(s) = \frac{\Delta v_{1d}(s)}{\Delta x_D(s)} = -K_{xv}\left(1 + \frac{1}{T_{ID}s}\right). \tag{19.3.7}$$

The transfer behavior of the distance control loop is then of 3rd order.

In the case of braking, the same controller structure can be used, but with other gain $K_{xv}$. Figure 19.7b depicts a distance control with subsidiary acceleration control from Sect. 19.1.

If the distance measurement sensor also provides the determination of the first order derivative $\dot{x}_D$ of the distance in form of the relative velocity

$$\dot{x}_D(t) = \frac{dx_D(t)}{dt} = v_2(t) - v_1(t) = \Delta v_{12}, \tag{19.3.8}$$

a further feedback can be realized

$$\Delta a_{xd}(t) = G_{\dot{x}a}\dot{x}_D(t) = K_{\dot{x}a}\dot{x}_D(t), \tag{19.3.9}$$

which acts parallel to the PI controller $G_{xv}$ and then adds a differential term; see Fig. 19.7c.

The discussed distance control considers only the basic control function. It is used for adaptive cruise control (ACC), where several other functions have to be added, like acceleration limitation, tracking sensor (radar) performance monitoring, distance setpoint selection, etc. Winner et al. (2016); see Chap. 19. The velocity control and the acceleration control may be added by feedforward control; compare Fig. 19.3.

The considered control systems for longitudinal control are based on simplified dynamic vehicle models and describe model-based designs. For final application, experimental tests have to be made, which evaluate the behavior with regard to the comfort, the cooperation with inputs of the driver, actuator limitation, etc.

## 19.4 Adaptive Cruise Control (ACC)

Adaptive cruise control (ACC) controls automatically the vehicle velocity and adjusts the velocity to maintain a safe distance from a vehicle ahead. It uses generally a radar-sensor system to measure the distance to a vehicle in front of the own vehicle and controls the distance by adjusting the throttle or brake, or a preselected velocity if no vehicle is ahead.

An alternative is the use of a camera to determine the distance in longitudinal (and lateral) direction, also to assign the preceding vehicle to a certain lane.

The standard ACC operates above a minimum speed of about 30 km/h, whereas the full-speed range adaptive cruise control (ACC-FR) functions also for low speed. International standards are documented in ISO 15622 and ISO 22179. The distance $x_{\mathrm{D}}$ to the forward vehicle is determined by the time gap $\tau_{\mathrm{D}}$, which is related to the distance by $x_{\mathrm{D}} = v_{\mathrm{X}} \tau_{\mathrm{D}}$.

There are several requirements for ACC systems stated in ISO 15622; see Winner and Schopper (2016). Some of them are for standard ACC:

- vehicle following with good oscillation damping,
- soft distance control for cutting-in vehicles,
- convoy stability in the case of several vehicles,
- appropriate deceleration for sharp braking front vehicle,
- selectable time gap $\tau_{\mathrm{D}} \geq 1$ s,
- priority to intervention by the driver in the case of braking or acceleration with the pedals,
- Acceleration limits $a_{\mathrm{X}} = -3.5\ \mathrm{m/s^2} \ldots 2.5\ \mathrm{m/s^2}$.

Additional requirements are for full-speed range ACC-FR:

- control until 0 km/h with a stopping distance 2...5 m,
- automatic safe stopping with service brake,
- speed dependent acceleration limits.

These requirements lead to specific operation modes of an ACC system

- ACC: on, off or standby,
- target object selection,
- velocity control or distance control,
- special situation control,
- supervision and fault detection.

In addition special care has to be given to the front sensor system:

- lateral detection area of the radar sensor during cornering,
- vehicles on the neighbouring lanes have to be ignored,
- oncoming vehicles have to be ignored,
- stationary objects are excluded.

These requirements are in detail treated in Winner and Schopper (2016).

Figure 19.8 depicts the main functions of a CC and a radar-based ACC system. The evaluation of the radar-sensor system provides the distance $x_D$ to the preceding vehicle and the relative velocity $\Delta v_{12} = v_2 - v_1$. Both values are compared for several modulation ramps of the sensor system to select and validate the correct object and its orientation angle relative to the vehicle longitudinal axis.

In order to let the ACC also function in curves, the preceding vehicle has to be detected during cornering. This vehicle appears then with a certain angle to the longitudinal axis of the subject vehicle. The radar sensor can usually detect objects within $\pm 8°$. The selection of the correct track requires then the determination of the curvature $\kappa$, which is determined by the available measurements of the ESC-system based on: $\delta_f$, $\dot{\psi}$, $v_X$, $a_Y$. The curvature for stationary cornering follows, e.g. from the kinematic relation (7.1.2)

$$\kappa = \frac{\dot{\psi}}{v}, \tag{19.4.1}$$

or (7.1.9)

$$\kappa = \frac{a_Y}{v^2}, \tag{19.4.2}$$

assuming normal driving situations on highways without skidding. However, if the curvature changes strongly, camera-based predictions are required or the use of digital maps from the navigation system.

The cruise control (CC) is active if no object is detected within the measurement range $x_{Dr}$ of the radar-sensor system, i.e. if $x_D > x_{Dr}$. If the vehicle then approaches a preceding vehicle and is detected within the measurement range, i.e. $x_D \leq x_{Dr}$, the CC control becomes inactive and the adaptive cruise control (ACC) becomes active; see Fig. 19.8.

The distance controller then determines the desired value of the acceleration controller within given limitations, for example, $a_{X,\text{lim}} = +0.6...1\ \text{m/s}^2$ and $a_{X,\text{lim}} = -2.5\ \text{m/s}^2$. The design of the acceleration controllers follows Sect. 19.1. The selection of the drive control for positive acceleration and brake control or coasting for negative acceleration is illustrated in Fig. 19.5.

The presentation in Fig. 19.8 shows only the main functions of a CC/ACC system. Many additional functions have to be implemented in order to obtain a safe assistance system. Some examples are the correct detection of the preceding vehicle, especially in curves, the reaction for cutting-vehicles, the limitation of the acceleration for a too high selected desired velocity for the case $v_X \ll v_{Xd}$, lane change of the preceding vehicle, supervision and fault detection, command concept, and displays for the

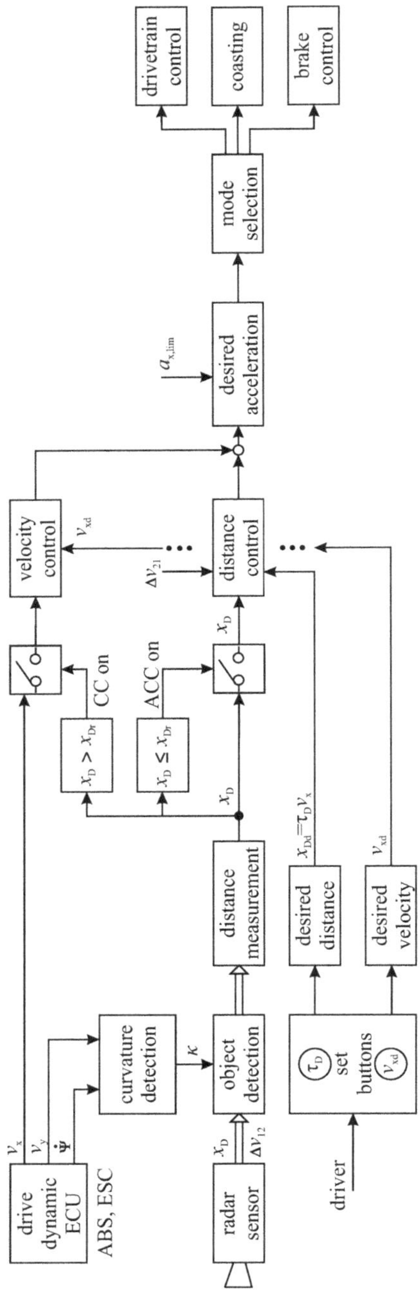


**Fig. 19.8** Signal flow chart of a cruise control system (CC) and an adaptive cruise control system (ACC)

driver. These tasks are treated, for example, in Winner and Schopper (2016), Reif (2013), Robert Bosch GmbH (2018), and Waschl et al. (2019a, b). See also He et al. (2019).

## References


Ackermann F (1980) Abstandsregelung und radar. Spektrum der Wissenschaft, pp 24–34

Adipastro B (2003) Fahrzeuglängsführung im Niedergeschwindigkeitsbereich. PhD thesis, TU Braunschweig

Dorissen HT, Hobein D, Höver Ne (1994) AICC - Ein neues intelligentes Assistenzsystem. VDI-Bericht 1152

Gardels K (1960) Automatic car control for electronic highways. Technical report, General Motors Research Laboratories, Warren, Michigan

Germann S (1997) Modellbildung und modellgestützte Regelung der Fahrzeuglängsdynamik. Dissertation Technische Universität Darmstadt. Fortschr.-Ber. VDI Reihe 12, 309. VDI Verlag, Düsseldorf

Hartwich E (1971) Längsdynamik und Folgebewegung der Strassenfahrzeug. Dissertation, TU Darmstadt

Hauksdottir A, Fenton R (1985) On the design of a vehicle longitudinal controller. IEEE Trans Veh Technol 34

He Y, Ciuffo B, Zhou Q, Makridis M, Mattas K, Li J, Li Z, Yan F, Xu H (2019) Adaptive cruise control strategies implemented on experimental vehicles: a review. In: 9th IFAC symposium on advances in automotive control, Orléans, France, June 2019

Hedrick J (1995) Vehicle control issues in intelligent highway systems. In: 1st IFAC workshop on automatic control, Acsona

Hoess A (1995) Realisation of an intelligent cruise control system. Siemens-Automotive, Regensburg

Mayr R (1991) Bahnfolgeregelung für ein automatisch geführtes Fahrzeug. Disertation, TU Dortmund

Mayr R (1994) Intelligent cruise control for vehicles based on feedback linearization. In: American control conference, Baltimore

Rathgeber C, Winkler F, Müller S (2016) Kollisionsfreie Längs- und Quertrajektorienplanung unter Berücksichtigung fahrzeugspezifischer Potenziale. at - Automatisierungstechnik 64:61–76

Reif K (ed) (2013) Fahrerassistenzsysteme. Springer Vieweg, Wiesbaden

Robert Bosch GmbH (ed) (2018) Automotive handbook, 10th edn. Wiley, Chichester, England

Shladover SE, Desoer CA, Hedrick JK (1991) Automated vehicle control developments in the PATH program. IEEE Trans Veh Technol 40

Waschl H, Kolmanovsky I, Willems F (eds) (2019a) Control strategies for advanced driver assistance systems and autonomous driving functions. Lecture notes in control and information sciences. Springer International, Cham, Switzerland

Waschl H, Schmied R, Reischl D, Stolz M (2019b) A virtual development and evaluation framework for ADAS. In: Chapter 6 in Waschl, H. et. al (2019a)

Winner H, Schopper M (2016) Adaptive cruise control. In: Winner et al (2016) Handbook of driver assistance systems

Winner H, Hakuli S, Lotz F, Singer C (eds) (2016) Handbook of driver assistance systems. Springer International Publishing AG, Cham, Switzerland

# Lateral Vehicle Control 20

Roads are usually built from elements of circles, clothoids with intermediate connections, and straight lines; see RAA (2008). Curves have curvatures which develop steadily, such that driving without lateral jerk is possible. The vehicle has to follow the road lane in a straight direction or has to follow the curvatures by cornering with a constant or steadily changing radius. Lateral vehicle control is required for driving maneuvers like cornering, lane change, and parking; see Fig. 19.1.

As shown in Fig. 19.2, the vehicle has to follow a lane, ideally in the center of the lane. The geometric form of the lane then defines the vehicle's path. The distance of the center of gravity of the vehicle to the right lane marking is noted by the lateral distance $D_y$. (Also the distance of the front axle or of the right front wheel to the right lane marking may be used.)

In the case of circles, the curvature $\kappa = 1/R$ is inverse proportional to the radius, and clothoids have a curvature which is proportional to the path length $L$, $\kappa = c_{\mathrm{cl}} L$.

The control of vehicles along a straight or bending lane means that the driver or an automatic controller has to follow a certain path: $y = f(x)$. If this can be performed without time condition, it is called *path control*. The state variables for the position are $x$ and $y$ and for the orientation of the vehicle, the yaw angle $\psi$ or the course angle $\nu = \psi + \beta$; see Fig. 20.1. The manipulated variable is the steering wheel angle $\delta_{\mathrm{f}}$ for a car with front steering. If the position $y(x)$ has to be reached at a certain time $t$, such that $y = f(x, t)$, this a called *trajectory control*. Hence, for path control, the lateral motion has to be controlled and for trajectory control, the lateral and longitudinal motion; see Sect. 18.2.

The lane markers are usually detected with cameras located behind the windscreen for forward looking. Monocular cameras provide, for example, a $\pm 20°$ field of view and ranges up to 80 m. Corresponding image processing then provides the distance to line crossing (DLC) or time to line crossing (TLC) in front of the vehicle and based on a lane marking approximation with polynomials or clothoid functions with the curvature $\kappa$ and its first derivative $d\kappa/dx$ in a distance $x_{\mathcal{T}}$ ahead. For more details, see Winner et al. (2016) and Schmitt (2012).



R. Isermann, *Automotive Control*, ATZ/MTZ-Fachbuch,
https://doi.org/10.1007/978-3-642-39440-9_20

The automation of lateral vehicle control is treated since several decades. Early publications are, for example, Ackermann et al. (1995), Dickmanns and Mysliwetz (1992), Dickmanns (2001), and Mayr (2000).

The design of lateral control systems has to consider the control performance with regard to deviations from the required path and comfort performance for the driver and passengers taking into account lateral accelerations, jerk, and unpleasant oscillations. As the inputs of the lateral control system are measured values from cameras, radar sensors, and lidar sensors, the control performance depends strongly on the accuracy and stochastic properties of the respective signals transferred to the control system. This means that these signals have to be passed through appropriate filters. As for EPS systems, the steering wheel motions are directly observed and even felt by the driver. These motions have to be acceptable, similar to a good driver, well damped, and not oscillating; see, for example, Müller and Rohleder (2006). Therefore, the lateral controllers have for normal driving not to be too agile, but should generate relatively smooth, well-damped commands to the actuator.

In the following, lateral control is considered by using simplified dynamic lateral vehicle models with the goal to derive controllers with controller parameters based on vehicle model parameters. It begins with path control for straightforward driving with lateral distance measurement beside and some short distance ahead of the vehicle. Then path control for curves and lane change with additional curvature measurement some distance ahead is treated. Herewith it is assumed that the velocity is constant, and the lateral control can be designed independently from the longitudinal control.

An approach of merging control for an highway entrance ramp shows how longitudinal and lateral control have to cooperate under certain conditions; see Sect. 20.4.

## 20.1 Path Control for Straight Lanes

It is now assumed that in the stationary state of straight driving, the earth-fixed axis $X_\mathrm{E}$ is parallel to the longitudinal axis $X$ of the vehicle. The vehicle then moves with the steady-state variables; steering angle $\delta_\mathrm{f} = \bar{\delta}_\mathrm{f} = 0$, yaw angle $\psi = \bar{\psi} = 0$, and vehicle slip angle $\beta = \bar{\beta} = 0$. For small changes of these angles holds

$$\delta_\mathrm{f} = \bar{\delta}_\mathrm{f} + \Delta\delta_\mathrm{f}\ ;\ \ \psi = \bar{\psi} + \Delta\psi\ ;\ \ \beta = \bar{\beta} + \Delta\beta,$$

and because of the special steady state, it is

$$\delta_\mathrm{f} = \Delta\delta_\mathrm{f}\ ;\ \ \psi = \Delta\psi\ ;\ \ \beta = \Delta\beta.$$

Therefore, e.g. $\psi$ or $\Delta\psi$ can be used for the linearized lateral dynamic model.

As a first step, it is considered that the vehicle follows a straight road with a desired lateral distance $D_\mathrm{yd}$ from the right lane marking. $D_\mathrm{yd}$ can, for example, be half of the lane width $B/2$ to follow a center line. Because of a steering angle input $\delta_\mathrm{f} = \Delta\delta_\mathrm{f}$, the vehicle's center of gravity drives with a yaw angle $\psi$ and a slip angle $\beta$ relative to the earth-fixed road coordinate system $(Y_\mathrm{E}|X_\mathrm{E})$ at time $t_0$ and $x = 0$, where the axis

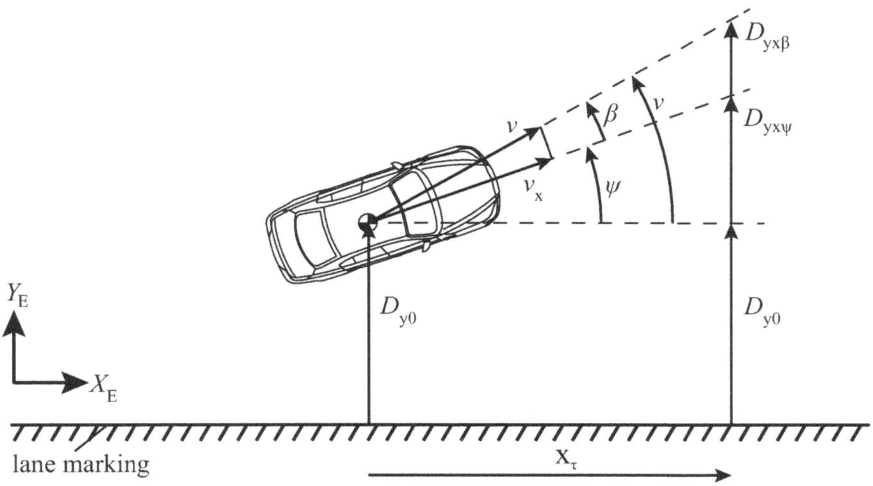


**Fig. 20.1** Vehicle driving along a straight road with distance $D_{y0}$ at $t = 0$ to the right lane marking and course angle $\nu$

$X_E$ is parallel to the lane marking; see Fig. 20.1. The lateral distance to the right lane marking at time $t_0$ and $x = 0$ is $D_y(t_0) = D_{y0}$, with, e.g. $D_{y0} = B/2$. If the vehicle continues to drive with course angle

$$\nu(t_0) = \psi(t_0) + \beta(t_0), \tag{20.1.1}$$

then the deviation of the lateral distance at time $t_0$ increases for $v = v(t_0) = \text{const.}$ according to

$$\dot{D}_{y0}(t) = v(t_0)\sin(\nu(t_0)) = v(t_0)\sin(\psi(t_0) + \beta(t_0)). \tag{20.1.2}$$

For small $\psi$ and $\beta$, it holds for any coordinate $x$ and any $t$

$$\dot{D}_{yx}(t) = \dot{D}_{yx\psi}(t) + \dot{D}_{yx\beta}(t) \tag{20.1.3}$$

with

$$\begin{aligned} \dot{D}_{yx\psi}(t) &= v_X(t)\sin\psi(t), \\ \dot{D}_{yx\beta}(t) &= v_X(t)\sin\beta(t). \end{aligned} \tag{20.1.4}$$

For the velocity of the center of gravity, it is

$$v = \frac{v_X}{\cos(\beta)} \tag{20.1.5}$$

and for small $\beta$, $v = v_X$ can be set. If the changes of $\psi$ and $\beta$ are small, (20.1.3) becomes

$$\Delta\dot{D}_{yx}(t) = v_X\sin(\Delta\psi(t) + \Delta\beta(t)) \approx v_X\Delta\nu(t). \tag{20.1.6}$$

The instantaneous change of the distance of the vehicle's CG then is for $\Delta\nu = \text{const}$

$$\Delta D_{\text{yx}}(t) = \int_0^t v_X \Delta\nu dt = v_X \Delta\nu t. \tag{20.1.7}$$

Hence, the distance changes linearly with time, longitudinal velocity, and course deviation. The Laplace transformation leads to

$$\Delta D_{\text{yx}}(s) = \frac{v_X}{s} \Delta\nu(s). \tag{20.1.8}$$

Now the distance $D_{\text{yx}} = D_{\text{yx}\psi} + D_{\text{yx}\beta}$ of the vehicle's CG after a driven distance $x_\tau$ ahead for a straight road is considered. $D_{\text{yx}\psi}$ is the lateral distance which is measured by a sensor system such as a camera. $x_\tau$ corresponds to a time gap $\tau$ (look-ahead time), such that

$$x_\tau = v_X \tau. \tag{20.1.9}$$

For manual driving, this time gap is about $\tau = 1 \ldots 4$ s which means a forward look distance of 20...80 m for $v_X = 20$ m/s or of 40 m...160 m for $v_X = 40$ m/s. This time gap is for manual driving not always constant, but depends on the properties of the road, traffic, and weather conditions. Thus, it has to be adapted to the driving situation.

For constant course deviation, $\Delta\nu = \text{const}$ (20.1.7) leads to

$$\Delta D_{\text{yx}}(\tau) = \int_0^\tau v_X \Delta\nu dt = v_X \Delta\nu \tau, \tag{20.1.10}$$

where the contribution of the yaw angle is

$$\Delta D_{\text{yx}\psi}(\tau) = v_X \Delta\psi \tau \tag{20.1.10a}$$

and the contribution by the slip angle

$$\Delta D_{\text{yx}\beta}(\tau) = v_X \Delta\beta \tau. \tag{20.1.10b}$$

For a given value $\tau$, the change of the distance $\Delta D_{\text{yx}}(\tau)$ is proportional to $v_X$ and $\Delta\nu$ and it is in the Laplace domain

$$\Delta D_{\text{yx}}(s) = v_X \tau \Delta\nu(s) = v_X \tau \Delta\psi(s) + v_X \tau \Delta\beta(s). \tag{20.1.11}$$

### 20.1.1 Path Control Beside the Vehicle

The course can be controlled by the steering angle $\delta_f$ of the front wheels. For small changes, Fig. 20.2a shows the signal flow for controlling the distance $D_{y0}$ beside the vehicle (or at the front side) using the transfer function (7.2.50) for the yaw rate of a one-track model with $\delta_H = i_s \delta_f$

$$G_{\delta\dot{\psi}}(s) = \frac{\Delta\dot{\psi}(s)}{\Delta\delta_f(s)} = \frac{\left(b_0' + b_1' s\right)}{\left(1 + a_1 s + a_2 s^2\right)} \tag{20.1.12}$$

with, see (7.2.50),

$$b_0' = b_0 i_s \ ; \ b_1' = b_1 i_s$$

and of the vehicle side slip angle

$$G_{\delta\beta}(s) = \frac{\Delta\beta(s)}{\Delta\delta_f(s)} = \frac{\left(b_0'' + b_1'' s\right)}{\left(1 + a_1 s + a_2 s^2\right)} \tag{20.1.13}$$

with ($v_X = v$), see (7.2.51),

$$b_0'' = \frac{l_r}{v}\left(1 - \frac{m l_f v^2}{c_{\alpha r} l_r l}\right) b_0',$$
$$b_1'' = \frac{J_z v}{c_{\alpha r} l_r l - l_f m v^2} b_0'.$$

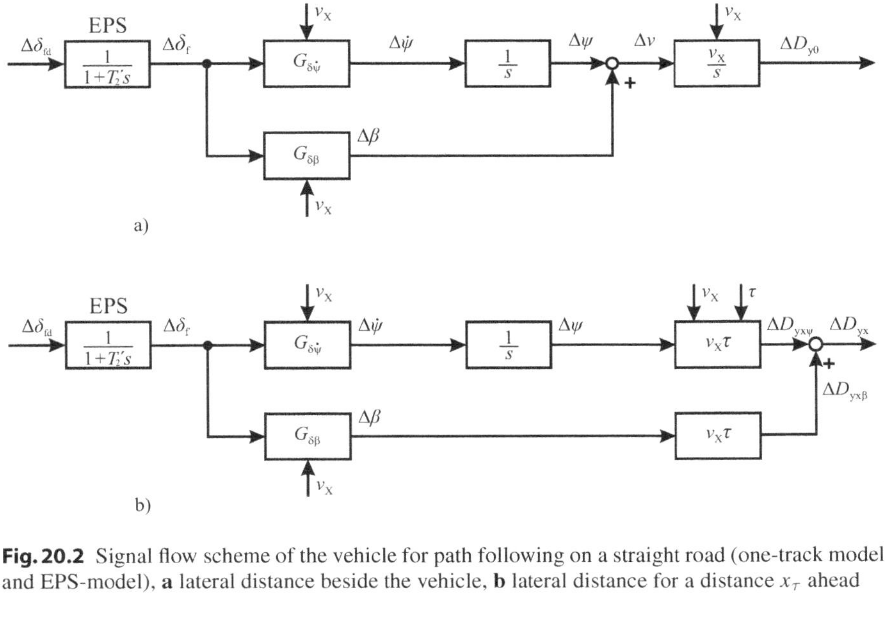


**Fig. 20.2** Signal flow scheme of the vehicle for path following on a straight road (one-track model and EPS-model), **a** lateral distance beside the vehicle, **b** lateral distance for a distance $x_\tau$ ahead

For the EPS steering system, a simplified transfer function is assumed, see Sect. 14.2,

$$G_{\delta\delta}(s) = \frac{\Delta\delta_{\mathrm{f}}(s)}{\Delta\delta_{\mathrm{fd}}(s)} = \frac{1}{(1 + T_2 s)} \tag{20.1.14}$$

with, see Chap. 14,

$$T_2 \approx 100...250 \text{ ms}.$$

Available measurements for the tracking control are the yaw rate $\dot{\psi}$, the velocity $v_{\mathrm{X}}$, and $\delta_{\mathrm{fd}}$, e.g. from the ESC system and ESP system. It is assumed that the lateral distance $D_{\mathrm{y0}}$ beside or in front the vehicle, e.g. by a front-view camera is measured and possesses a latency time (dead-time) $T_{\mathrm{c}}$, with transfer function

$$G_{\mathrm{DD}}(s) = e^{-T_{\mathrm{c}} s} \tag{20.1.15}$$

and

$$T_{\mathrm{c}} \approx 50 \text{ ms}.$$

A further delay arises because of the lateral deformation of the tire, see Fig. 5.21, leading to the first-order element (5.4.7) with the time constant $T_{\mathrm{YT}} \approx 0.1$ s. This delay may be taken into account by increasing $T_2$ to

$$T_2^{'} = T_2 + T_{\mathrm{YT}}. \tag{20.1.16}$$

($T_{\mathrm{YT}}$ may also include a small delay because of the elasto-kinematic change of the side slip angle as shown in Fig. 5.11.)

A path control system for the process Fig. 20.2a can then be designed as a cascaded control system with a minor yaw angle controller $G_{\psi\delta}(s)$ and a major course controller $G_{\mathrm{Do}\psi}(s)$, see Fig. 20.3, now neglecting the dynamics of the camera. As the vehicle slip angle $\beta$ is not measurable, it has to be calculated with (20.1.13) or estimated, as shown in Sect. 12.3. The yaw angle $\psi(t)$ is obtained by algorithmic integration of the yaw rate $\dot{\psi}(t)$.

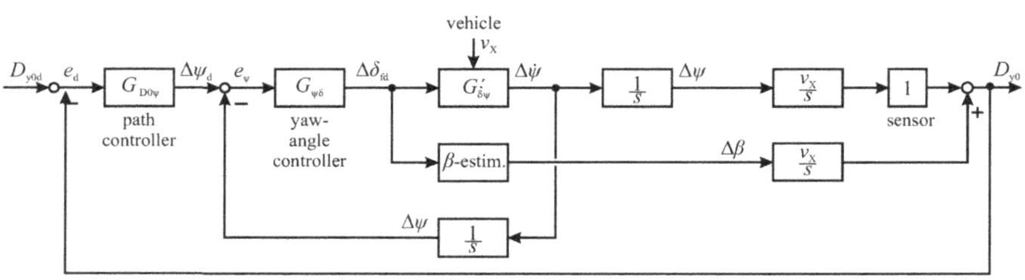


**Fig. 20.3** Signal flow scheme for cascaded path control beside or in front of the vehicle on a straight road

Assuming small deviations around the vehicle operating point $[v_X, m_v]$, the controlled process is, for the yaw angle controller by using a one-track model, see (20.1.12),

$$G'_{\delta\psi}(s) = \frac{\Delta\psi(s)}{\Delta\delta_{fd}(s)} = \frac{\left(b'_0 + b'_1 s\right)}{s\left(1 + a_1 s + a_2 s^2\right)} \cdot \frac{1}{1 + T'_2 s}. \quad (20.1.17)$$

Hence, the lateral dynamics of the yaw angle is characterized by an integrating behavior in series with a second-order lag, a first-order lag, and first-order lead behavior (IDT$_3$-process). If the dynamics are neglected, one obtains an integrating process for the stationary behavior

$$G'_{\delta\psi}(s) = b'_0 \frac{1}{s}. \quad (20.1.18)$$

For the behavior of the course angle, $\nu$ follows with (20.1.1), (20.1.12), and (20.1.13)

$$\begin{aligned} G_{\delta\nu}(s) &= \frac{\Delta\nu(s)}{\Delta\delta_{fd}(s)} = \frac{\Delta\psi(s)}{\Delta\delta_{fd}(s)} + \frac{\Delta\beta(s)}{\Delta\delta_{fd}(s)} \\ &= \frac{b'_0 + \left(b'_1 + b''_0\right) s + b''_1 s^2}{s\left(1 + a_1 s + a_2 s^2\right)} \cdot \frac{1}{\left(1 + T'_2 s\right)}. \end{aligned} \quad (20.1.19)$$

Thus, the consideration of the vehicle slip angle adds a further zero in the denominator.

### (a) IMC Controller

For general lateral driving situations with small changes of the steering angle $\delta_f$, the single-track model (20.1.12) is used. A controller design according to the internal model control (IMC) principle, see Appendix A.1.3.2, then leads with a realizability filter

$$G_{rf}(s) = \frac{1}{(1 + T_{r1}s)(1 + T_{r2}s)}, \quad (20.1.20)$$

$$T_{r1} = \frac{T'_2 T_{r2}}{T_{r2} - T'_2} \quad \left(T_{r2} > T'_2\right), \quad (20.1.21)$$

and (A.1.37) to the *yaw angle controller*

$$G_{\psi\delta} = \frac{\Delta\delta_f(s)}{e_\psi(s)} = \frac{\left(1 + a_1 s + a_2 s^2\right)}{\left(b'_0 + b'_1 s\right)} \frac{\left(T_{r2} - T'_2\right)}{T_{r2}^2} \quad (20.1.22)$$

with the control deviation

$$e_\psi(s) = \psi_d(s) - \psi(s) = \Delta\psi_d(s) - \Delta\psi(s). \quad (20.1.22a)$$

The behavior of this controller is for low and high frequencies

$$G_{\psi\delta}(i\omega)\Big|_{\omega\to 0} = \frac{\left(T_{r2} - T_2'\right)}{b_0' T_{r2}^2},$$
$$G_{\psi\delta}(i\omega)\Big|_{\omega\to\infty} = \frac{a_2\left(T_{r2} - T_2'\right)}{b_1' T_{r2}^2}(i\omega)\Bigg|_{\omega\to\infty}. \tag{20.1.23}$$

For a PD controller holds

$$G_{\psi\delta}(s) = K_{\psi\delta}(1 + T_D s), \tag{20.1.24}$$

$$G_{\psi\delta}(i\omega)\Big|_{\omega\to 0} = K_{\psi\delta},$$
$$G_{\psi\delta}(i\omega)\Big|_{\omega\to\infty} = K_{\psi\delta}T_D(i\omega)\Big|_{\omega\to\infty}.$$

If the IMC controller (20.1.22) is approximated for low and high frequencies by a PD controller, it follows for the controller parameters

$$K_{\psi\delta} = \frac{\left(T_{r2} - T_2'\right)}{b_0' T_{r2}^2},$$
$$T_D = \frac{a_2 b_0'}{b_1'}. \tag{20.1.24a}$$

Therefore, the PD controller (20.1.24) as a simplified version of the IMC controller (20.1.22) can be used as a yaw angle controller, where $T_{r2} > T_2'$ is a tuning parameter.

The controller gain $K_{\psi\delta}$ is inverse proportional to $b_0'$, where $b_0'$ is the yaw rate gain (7.2.57). Therefore, one obtains

$$K_{\psi\delta} = \frac{l\left(1 + \left(\frac{v}{v_{ch}}\right)^2\right)}{v} \cdot \frac{\left(T_{r2} - T_2'\right)}{T_{r2}^2} = \frac{l\left(1 + \left(\frac{v}{v_{ch}}\right)^2\right)}{v} c_T \tag{20.1.24b}$$

and it holds with

$$c_T = \frac{T_{r2} - T_2'}{T_{r2}^2} = \left(\frac{1}{T_{r2}} - \frac{T_2'}{T_{r2}^2}\right) \tag{20.1.24c}$$

$$K_{\psi\delta} = \frac{l}{v} c_T \qquad \text{for } v \ll v_{ch},$$
$$K_{\psi\delta} = \frac{2l}{v_{ch}} c_T \qquad \text{for } v = v_{ch},$$
$$K_{\psi\delta} = l\left(\frac{1}{v} + \frac{v}{(v_{ch})^2}\right) c_T \qquad \text{for } v > v_{ch}.$$

Hence, the controller gain is for small $v$ inverse proportional to $v$, and therefore relatively large, shows a minimum for $v = v_{\mathrm{ch}}$, (compare 7.2.60a), and increases again for $v \gg v_{\mathrm{ch}}$. $c_{\mathrm{T}} = f(T_2^{'}, T)$ is a tuning factor.

The closed-loop transfer function $G_{\psi\psi}(s) = \Delta\psi(s)/\Delta\psi_{\mathrm{d}}(s)$ then has a fourth-order denominator and a second-order numerator polynomial with gain $G_{\psi\psi}(0) = 1$.

Müller and Rohleder (2006) use a PDT$_1$-controller, such that (20.1.24) becomes

$$G_{\psi\delta}(s) = K_{\psi\delta}\frac{(1 + T_{\mathrm{D}}s)}{(1 + T_1 s)}. \tag{20.1.25}$$

The time delay element with time-constant $T_1$ damps the manipulated variable.

If the dynamics of the EPS are neglected ($T_2^{'} = 0$), the controller design according to the internal model principle then leads with a first-order realizability filter, see (A.1.35),

$$G_{\mathrm{rf}}(s) = \frac{1}{1 + T_{\mathrm{r}}s} \tag{20.1.26}$$

to the IMC controller of the yaw angle

$$G_{\psi\delta}(s) = \frac{\Delta\delta_{\mathrm{f}}(s)}{\Delta\psi(s)} = \frac{\left(1 + a_1 s + a_2 s^2\right)}{T_{\mathrm{r}}\left(b_0^{'} + b_1^{'}s\right)}. \tag{20.1.27}$$

The behavior of this controller is for low and high frequencies with $s = i\omega$

$$\begin{aligned} G_{\psi\delta}(i\omega)\big|_{\omega\to 0} &= \frac{1}{T_{\mathrm{r}}b_0^{'}},\\ G_{\psi\delta}(i\omega)\big|_{\omega\to\infty} &= \frac{a_2}{T_{\mathrm{r}}b_1^{'}}(i\omega)\bigg|_{\omega\to\infty}. \end{aligned} \tag{20.1.28}$$

An analogy to a PD controller, see (20.1.22), gives

$$\begin{aligned} G_{\psi\delta}(i\omega)\big|_{\omega\to 0} &= K_{\psi\delta},\\ G_{\psi\delta}(i\omega)\big|_{\omega\to\infty} &= K_{\psi\delta}T_{\mathrm{D}}(i\omega)\big|_{\omega\to\infty}. \end{aligned} \tag{20.1.29}$$

If the controller (20.1.27) is approximated for low and high frequencies by a PD controller (20.1.24), it follows for the controller parameters from (20.1.28) and (20.1.29)

$$\begin{aligned} K_{\psi\delta} &= \frac{1}{T_{\mathrm{r}}b_0^{'}},\\ T_{\mathrm{D}} &= \frac{a_2 b_0^{'}}{b_1^{'}}. \end{aligned} \tag{20.1.30}$$

These controller parameters can be determined from the vehicle parameters of the one-track model; see (7.2.50).

The closed-loop transfer function

$$G_{\psi\psi}(s) = \frac{\Delta\psi(s)}{\Delta\psi_{\mathrm{d}}(s)}$$

then has a third-order denominator and a second-order numerator polynomial with gain $G_{\psi\psi}(0) = 1$.

The controller parameters depend according to (7.2.50) mainly on the cornering stiffnesses of the front and rear axle wheels and the vehicle velocity; see Fig. 7.8 and Fig. 7.9. Therefore, the controller parameters have to be scheduled especially in dependence on the velocity. This can be performed by designing *local linear controllers* and weighting them by Gaussian activity functions; see Schorn (2007) and Sect. 20.3.

To determine the controlled variable $D_{\mathrm{y0}}$ for the *path controller*, the course angle $\Delta\nu$ has to be determined according to (20.1.1).

As the closed yaw angle control loop is of fourth order with gain $G_{\psi\psi}(0) = 1$, the controlled process for the path controller becomes

$$G_{\psi\mathrm{D0}}(s) = \frac{\Delta D_{\mathrm{y0}}(s)}{\Delta\psi_{\mathrm{d}}(s)} = G_{\psi\psi}(s)\frac{v_{\mathrm{X}}}{s} + \frac{\Delta\beta(s)}{\Delta\delta_{\mathrm{f}}(s)}\frac{v_{\mathrm{X}}}{s}. \tag{20.1.31}$$

If the vehicle side slip angle is assumed to be small, $\Delta\beta \approx 0$ can be assumed and the controlled process for the path controller becomes

$$G_{\psi\mathrm{Dx}}(s) = \frac{\Delta D_{\mathrm{y0}}(s)}{\Delta\psi_{\mathrm{d}}(s)} = G_{\psi\psi}(s)\frac{v_{\mathrm{X}}}{s}. \tag{20.1.32}$$

Therefore, for *path control*, a P or PD controller can be used, with velocity-dependent parameters. The gain of the controller is then inverse proportional to $v_{\mathrm{X}}$.

**(b) IMC Controller for Low Velocity**

The frequency responses of the single-track model have real poles for about $v < 20$ km/h, see Fig. 7.8, and hardly a resonance peak for about $v \leq 40$ km/h. If in addition the excitation frequency during normal convenient driving is in a range of 0.1–0.2 Hz, Würtenberger (1997), the lateral behavior of the vehicle can for frequencies $f < 1$ Hz be simplified to a first-order model

$$G_{\delta\dot{\psi}}(s) = \frac{\Delta\dot{\psi}(s)}{\Delta\delta_{\mathrm{f}}(s)} = \frac{K_{\delta\dot{\psi}}}{\left(1 + T_{\delta\dot{\psi}}s\right)} \tag{20.1.33}$$

with the parameters, assuming the same cut-off frequency as the second-order single-track model (20.1.15),

$$\begin{aligned} K_{\delta\dot{\psi}}(s) &= b_0^{'}, \\ T_{\delta\dot{\psi}} &= \sqrt{a_2}. \end{aligned} \tag{20.1.34}$$

The transfer function for the yaw angle as output then is

$$\widetilde{G}_{\delta\psi}(s) = \frac{\Delta\psi(s)}{\Delta\delta_{\mathrm{f}}(s)} = \frac{K_{\delta\dot{\psi}}}{s\left(1 + T_{\delta\dot{\psi}}s\right)}. \tag{20.1.35}$$

A controller design with the internal model control principle leads for the *yaw angle controller* to a PD controller

$$G_{\psi\delta}(s) = \frac{\Delta\delta_{\mathrm{f}}(s)}{e_{\psi}(s)} = K_{\psi\delta}(1 + T_{\mathrm{D}}s) \tag{20.1.36}$$

with

$$\begin{aligned} K_{\psi\delta} &= \frac{1}{K_{\delta\dot{\psi}}T_{\mathrm{rf}}} = \frac{1}{T_{\mathrm{rf}}b_0'}, \\ T_{\mathrm{D}} &= T_{\delta\dot{\psi}} = \sqrt{a_2}, \end{aligned} \tag{20.1.37}$$

where $T_{\mathrm{rf}}$ is the time constant of a realizability filter. Hence, by assuming a first-order single-track model, directly a PD-yaw angle controller results.

The closed-loop transfer function then becomes

$$\widetilde{G}_{\psi\psi}(s) = \frac{\Delta\psi(s)}{\Delta\psi_{\mathrm{d}}(s)} = \frac{1}{1 + T_{\psi\psi}s}, \tag{20.1.38}$$

with

$$T_{\psi\psi} = T_{\mathrm{rf}}.$$

The controlled process for the *path controller* beside the vehicle then is, if the vehicle slip angle is assumed to be small $\beta \approx 0$,

$$G_{\psi\mathrm{D}0}(s) = \frac{\Delta D_{\mathrm{y}0}(s)}{\Delta\psi_{\mathrm{d}}(s)} = \frac{v_{\mathrm{X}}}{s(1 + T_{\psi\psi}s)}. \tag{20.1.39}$$

As for (20.1.35) a P- or a PD *path controller* results

$$G_{\mathrm{D}0\psi}(s) = \frac{\Delta\psi_{\mathrm{d}}(s)}{e_{\mathrm{D}}(s)} = K_{\mathrm{D}\psi}(1 + T_{\mathrm{D}}s) \tag{20.1.40}$$

with

$$\begin{aligned} K_{\mathrm{D}\psi} &= \frac{1}{v_{\mathrm{X}}T_{\mathrm{rf}}}, \\ T_{\mathrm{D}} &= T_{\psi\psi} = T_{\mathrm{rf}}, \end{aligned}$$

where the gain is inverse proportional to the (slow) velocity $v_{\mathrm{X}}$.

The resulting cascaded control system is depicted in Fig. 20.3. It also includes the state estimation for the vehicle slip angle $\beta$ for larger lateral accelerations.

**(c) State-Space Controller**

A state-space representation of the linearized one-track model with small changes around the operation point is according to (7.2.34) in the yaw rate/slip angle representation for stationary cornering expressed by

$$\begin{bmatrix} \dot{\beta} \\ \ddot{\psi} \end{bmatrix} = \begin{bmatrix} a_{11}^s & a_{12}^s \\ a_{21}^s & a_{22}^s \end{bmatrix} \begin{bmatrix} \beta \\ \dot{\psi} \end{bmatrix} + \begin{bmatrix} b_1^s \\ b_2^s \end{bmatrix} \delta_\mathrm{f}, \tag{20.1.41}$$

now with

$$\begin{aligned} a_{11}^s &= -\frac{c_{\alpha \mathrm{f}} + c_{\alpha \mathrm{r}}}{mv}, \\ a_{12}^s &= \frac{c_{\alpha \mathrm{r}} l_\mathrm{r} - c_{\alpha \mathrm{f}} l_\mathrm{f}}{mv^2} - 1, \\ a_{21}^s &= \frac{c_{\alpha \mathrm{r}} l_\mathrm{r} - c_{\alpha \mathrm{f}} l_\mathrm{f}}{J_Z}, \\ a_{22}^s &= -\frac{c_{\alpha \mathrm{r}} l_\mathrm{r}^2 - c_{\alpha \mathrm{f}} l_\mathrm{f}^2}{J_Z v}, \\ b_1^s &= \frac{c_{\alpha \mathrm{f}}}{mv}, \\ b_2^s &= \frac{c_{\alpha \mathrm{f}} l_\mathrm{f}}{J_Z}. \end{aligned} \tag{20.1.42}$$

For the steering system dynamics holds, see (20.1.14),

$$\dot{\delta}_\mathrm{f} = -a_{01}^s \delta_\mathrm{f} + a_{02}^s \delta_\mathrm{fd}, \tag{20.1.43}$$

with

$$a_{01}^s = a_{02}^s = \frac{1}{T_2'}.$$

The inclusion of the steering system dynamics in (20.1.41) yields

$$\begin{array}{c} \begin{bmatrix} \dot{\delta}_\mathrm{f} \\ \dot{\beta} \\ \ddot{\psi} \end{bmatrix} = \begin{bmatrix} -a_{01}^s & 0 & 0 \\ b_1^s & a_{11}^s & a_{12}^s \\ b_2^s & a_{21}^s & a_{22}^s \end{bmatrix} \begin{bmatrix} \delta_\mathrm{f} \\ \beta \\ \dot{\psi} \end{bmatrix} + \begin{bmatrix} a_{02}^s \\ 0 \\ 0 \end{bmatrix} \delta_\mathrm{fd} \\ \dot{\mathbf{x}}(t) = \mathbf{A} \, \mathbf{x}(t) + \mathbf{b} \, u(t). \end{array} \tag{20.1.44}$$

The output of this state-space model is the yaw rate $\dot{\psi}$

$$y(t) = \mathbf{c}^T \mathbf{x}(t), \tag{20.1.45}$$

with

$$\mathbf{c}^T = \begin{bmatrix} 0 & 0 & 1 \end{bmatrix}.$$

The state-space representation allows to include also an integrator to obtain $\psi$ from $\dot{\psi}$ and the lateral distance $D_{y0}$; see, e.g. Raste et al. (2015).

The rate of the lateral deviation is for small angles due to (20.1.2)

$$\dot{D}_{y0} = v_X \psi + v_X \beta. \tag{20.1.46}$$

Expanding the state vector of (20.1.44) yields

$$\begin{bmatrix} \dot{\delta}_f \\ \dot{\beta} \\ \ddot{\psi} \\ \dot{\psi} \\ \dot{D}_{y0} \end{bmatrix} = \begin{bmatrix} -a_{01}^s & 0 & 0 & 0 & 0 \\ b_1^s & a_{11}^s & a_{12}^s & 0 & 0 \\ b_2^s & a_{21}^s & a_{22}^s & 0 & 0 \\ 0 & 0 & 1 & 0 & 0 \\ 0 & v_X & 0 & v_X & 0 \end{bmatrix} \begin{bmatrix} \delta_f \\ \beta \\ \dot{\psi} \\ \psi \\ D_{y0} \end{bmatrix} + \begin{bmatrix} a_{02}^s \\ 0 \\ 0 \\ 0 \\ 0 \end{bmatrix} \delta_{fd}(t)$$

$$\dot{\mathbf{x}}(t) = \mathbf{A} \quad \mathbf{x}(t) + \mathbf{b} \quad u(t). \tag{20.1.47}$$

The output comprises the measured variables $\delta_f$, $\dot{\psi}$, and $D_{y0}$

$$\mathbf{y}(t) = \mathbf{c}^T \mathbf{x}(t), \tag{20.1.48}$$

with

$$\mathbf{c}^T = \begin{bmatrix} 1 & 0 & 1 & 0 & 1 \end{bmatrix}.$$

In order to determine the unknown variables $\beta$ and $\psi$, a state observer has to be applied. According to (A.1.49), the observer is

$$\hat{\dot{\mathbf{x}}}(t) = \mathbf{A}\,\mathbf{x}(t) + \mathbf{b}\,u(t) + \mathbf{H}\left(\mathbf{y}(t) - \mathbf{c}^T \hat{\mathbf{x}}(t)\right), \tag{20.1.49}$$

where the feedback matrix $\mathbf{H}$ may be designed by pole placement.

The state controller is then based on the state estimate

$$u(t) = \delta_{fd}(t) = -\mathbf{k}_{\delta D0}^T \hat{\mathbf{x}}(t), \tag{20.1.50}$$

where the gains are determined by pole placement or using a quadratic performance criterion and solving a matrix Riccati equation.

The measured lateral distance $D_{y0}$ is then compared with the reference variable $D_{y0d}$

$$\Delta D_{y0} = D_{y0}(t) - D_{y0d} \tag{20.1.51}$$

and leads to a correction of the manipulated variable $\Delta u(t) = \Delta \delta_{fd}(t)$, e.g. by passing through an integrator, as shown in Fig. A.8.

The described state variable controller is an overall controller for the lateral distance with a feedback of all 5 state variables and 5 controller parameters. The cascaded control system with yaw angle and path controller, Fig. 20.3, presents a feedback of only the yaw angle rate $\dot{\psi}$ and the lateral distance $D_{\text{yx}}$. However, it has a modular structure with two tunable simpler controllers which may be better suited for the practical application, especially for controller parameter tuning in a research vehicle.

A further control method which may be used for lateral and longitudinal control of vehicles is *model predictive control (MPC)*. This method uses current output variables $y(k)$ and manipulated variables $u(k)$, where $k = t/T_0$ is a discrete time, $T_0$ the sampling time, and their future values $y(k+1)$, $y(k+2)$,... and $u(k+1)$, $u(k+2)$,... based on a process model. These values are considered over a certain prediction horizont, and optimized manipulated variables are calculated, but only the first $u(k)$ is realized. This method is briefly described in Appendix A.1.3.5. However, the computational expense is relatively large, such that real-time solutions with small sampling times are difficult. Some proposals of MPC for automotive systems are described in Del Re et al. (2010), Nareyko et al. (2020), and Hu et al. (2020); see also Adamy (2009).

A comparison of lateral vehicle controllers with linear PD controller, internal model controller, flatness-based controller, and a nonlinear controller, including the vehicle slip angle effect and experimental test drives, is published by Schorn (2007), with good overall properties for IMC and flatness-based control.

The described path control system of Fig. 20.3 is designed for controlling the distance $D_{\text{y0}}$ from the center of gravity to the lane marking beside the vehicle. However, it can also be applied for lateral distance control $D_{\text{y}}$ in short distance $x_{\tau} \approx 2...10\,\text{m}$ in front of the vehicle, because the traveling time to this close distant point is with $\tau \approx 0.2...0.5\,\text{s}$ for $v = 20\,\text{m/s}$ relatively small.

This is, for example, treated in Taylor et al. (1999) and it is shown that the poles of the overall transfer function are not affected by the look-ahead distance $x_{\tau}$, but that the zeros are closer to the real axis with increasing $x_{\tau}$ and thus improve the damping ratio; see also Rajamani (2012).

### 20.1.2 Path Control Ahead of the Vehicle

Path control for higher velocities requires to control the lateral distance $D_{\text{yx}} = D_{\text{yx}\psi} + D_{\text{yx}\beta}$ at a distance $x_{\tau}$ ahead of the vehicle; see Fig. 20.1. It is assumed that the distance $D_{\text{yx}\psi}$ can be measured by a camera. Figure 20.4 shows that the lateral distance after a time period $\tau$ will be $D_{\text{yx}}$ if the course $\nu(t_0)$ at time $t_0$ is maintained. This leads to a deviation

$$\Delta D_{\text{yx}}(\tau) = D_{\text{yx}}(t) - D_{\text{yd}} \tag{20.1.52}$$

if $D_{\text{yd}} = D_{\text{y0}}$ is the desired lateral distance to the right road marking, according to Fig. 20.1.

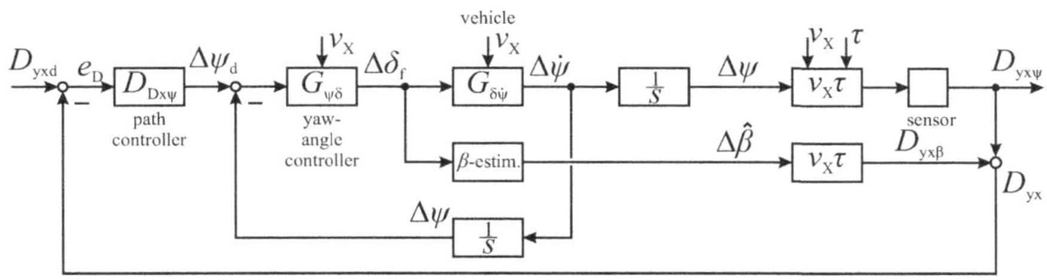


**Fig. 20.4** Signal flow scheme for path control ahead of the vehicle on a straight road

The signal flow for the change of the distance $D_{\mathrm{yx}}$ ahead of the vehicle is depicted in Fig. 20.2b by using (20.1.9).

The controlled process for the yaw controller is then described with (20.1.17)

$$G'_{\delta\mathrm{Dyx}}(s) = \frac{\Delta D_{\mathrm{yx}\psi}(s)}{\Delta\delta_{\mathrm{f}}(s)} = \frac{\left(b'_0 + b'_1 s\right) v_{\mathrm{X}}\tau}{s\left(1 + a_1 s + a_2 s^2\right)} \frac{1}{\left(1 + T'_2 s\right)} \tag{20.1.53}$$

if the vehicle side slip angle $\beta$ is negligible small. Neglecting the short time lateral dynamics, one obtains

$$G''_{\delta\mathrm{Dyx}}(s) = b'_0 v_{\mathrm{X}}\tau \frac{1}{s} = b'_0 x_{\tau} \frac{1}{s}. \tag{20.1.54}$$

This integrating process behavior for low vehicle velocity has a gain that is proportional to the look-ahead distance $x_{\tau}$.

However, for large slip angles, the effect of the slip angle has to be considered

$$\Delta D_{\mathrm{yx}\beta}(s) = v_{\mathrm{X}}\tau \Delta\hat{\beta}(s). \tag{20.1.55}$$

The overall controlled process for the path controller then becomes

$$G_{\delta\mathrm{Dyx}}(s) = \frac{\Delta D_{\mathrm{yx}}(s)}{\Delta\psi_{\mathrm{d}}(s)} = \frac{\Delta D_{\mathrm{yx}\psi}(s)}{\Delta\psi_{\mathrm{d}}(s)} + \frac{\Delta D_{\mathrm{yx}\beta}(s)}{\Delta\delta_{\mathrm{f}}(s)} = G_{\psi\psi}(s) v_{\mathrm{X}}\tau + \frac{\Delta\beta(s)}{\Delta\delta_{\mathrm{f}}(s)} v_{\mathrm{X}}\tau. \tag{20.1.56}$$

**(a) IMC Controller**

A comparison of Figs. 20.4 and 20.3 indicates that, for a cascaded control system, the yaw angle control is the same and that the path controller instead of an integrator $v_{\mathrm{X}}/s$ has just a proportional transfer element $v_{\mathrm{X}}\tau = x_{\tau}$ in its closed loop, which depends on the look-ahead distance $x_{\tau}$. Therefore, the IMC yaw angle controller $G_{\psi\delta}(s) = \Delta\delta_{\mathrm{f}}(s)/e_{\psi}(s)$ of (20.1.22), respectively, its PD controller approximation (20.1.24) can be directly used. The closed loop is then again of fourth order and the controlled process for the path controller becomes, if the effect of the vehicle slip angle is neglected,

$$G_{\psi\mathrm{Dx}}(s) = \frac{\Delta\psi_{\mathrm{d}}(s)}{\Delta D_{\mathrm{yx}\psi}} = G_{\psi\psi}(s) v_{\mathrm{X}}\tau = G_{\psi\psi}(s) x_{\tau}. \tag{20.1.57}$$

Then a P controller or a PI controller can be applied with a parameter adaptation dependent on the look-ahead distance $x_{\mathcal{T}}$

$$G_{\mathrm{D}\psi\mathrm{d}}(s) = \frac{\Delta\psi_{\mathrm{d}}(s)}{e_{\mathrm{D}}(s)} = K_{\mathrm{D}\psi\mathrm{d}}\left(1 + \frac{1}{T_{\mathrm{I}}s}\right). \tag{20.1.58}$$

The controller gain is inverse proportional to the look-ahead distance $x_{\mathcal{T}}$.

**(b) IMC Controller for Low Velocity**

As much simpler equations are obtained, the first-order model for the lateral behavior (20.1.33) is considered. Using the PD controller (20.1.36), the closed-loop behavior for the yaw control is described by (20.1.38). The controlled process for the path controller now becomes, neglecting the effect of the slip angle $\beta$,

$$G_{\psi\mathrm{Dx}\psi}(s) = \frac{\Delta D_{\mathrm{yx}\psi}(s)}{\Delta\psi_{\mathrm{d}}(s)} = \frac{K_{\psi\mathrm{Dx}}}{1 + T_{\psi\psi}s} \tag{20.1.59}$$

with the gain

$$K_{\psi\mathrm{Dx}} = v_{\mathrm{X}}\mathcal{T} = x_{\mathcal{T}},$$

which is now proportional to the look-ahead distance $x_{\mathcal{T}}$. The path controller operates with the control error for the desired lateral distance $D_{\mathrm{yxd}}$

$$e_{\mathrm{D}}(t) = D_{\mathrm{yxd}}(t) - D_{\mathrm{yx}\psi}(t); \tag{20.1.60}$$

see Fig. 20.4. If a P-path controller

$$G_{\mathrm{Dx}\psi}(s) = \frac{\Delta\psi_{\mathrm{d}}(s)}{e_{\mathrm{D}}(s)} = K_{\mathrm{Dx}\psi} \tag{20.1.61}$$

is used, the closed-loop behavior is

$$G_{\mathrm{DxDx}}(s) = \frac{\Delta D_{\mathrm{yx}\psi}(s)}{\Delta D_{\mathrm{yxd}}(s)} = \frac{\frac{K_{\mathrm{DD}}}{1+K_{\mathrm{DD}}}}{1 + \frac{T_{\psi\psi}}{1+K_{\mathrm{DD}}s}s} \tag{20.1.62}$$

with

$$K_{\mathrm{DD}} = K_{\psi\mathrm{Dx}}K_{\mathrm{Dx}} = x_{\mathcal{T}}K_{\mathrm{Dx}\psi}.$$

The overall gain $K_{\mathrm{DD}}$ can be determined by pole placement design. Then it follows for the path controller gain

$$K_{\mathrm{Dx}\psi} = \frac{K_{\mathrm{DD}}}{x_{\mathcal{T}}}. \tag{20.1.63}$$

Hence, the controller gain has to be adapted inverse proportional to the look-ahead distance $x_{\mathcal{T}}$ and the gain becomes the smaller, the larger $x_{\mathcal{T}}$. To avoid steady-state control errors $\Delta D_{\text{yx}}$, a better choice is a PI-path controller as (20.1.58) which results in a closed-loop transfer function

$$G_{\text{DxDx}}(s) = \frac{\Delta D_{\text{yx}}(s)}{\Delta D_{\text{yxd}}(s)} = \frac{K_{\text{DD}}(1 + T_{\text{I}}s)}{T_{\psi\psi}T_{\text{I}}s^2 + T_{\text{I}}(1 + K_{\text{DD}})s + K_{\text{DD}}}. \tag{20.1.64}$$

Also, in this case, the controller parameters $K_{\text{Dx}\psi}$ and $T_{\text{I}}$ can be determined by pole placement design or by parameter optimization with a quadratic control performance criterion. See Appendix. A.1.3.1.

These simplified controller designs did not take into account the influence of the vehicle side slip angle $\beta$, which has an effect for higher lateral velocities and accelerations and is just considered as a disturbance.

For large side slip angles $\beta$, the design of the path controller has to take into account the controlled process (20.1.56). However, if the ESC is active, it already keeps the slip angle relatively small, see Sect. 17.2, such that the design of the path controller for controlling the lateral distance $D_{\text{yx}\psi}$ may be sufficient.

The determined controller parameters have to be tuned experimentally also with regard to a comfortable behavior.

The yaw angle and path controllers depend on operating point-dependent parameters, such as the velocity, the look-ahead distance, and the vehicle mass. As the controller's designs are valid around the considered operation points, local controllers result, e.g. for $M$ operating points. Then they can be superimposed by weighting with activity functions, as described in Sect. 20.3 and Fig. 20.11.

## 20.2 Path Control for Curves

Road curves are usually built as circles with constant curvature

$$\kappa = \frac{1}{R_{\text{P}}} \tag{20.2.1}$$

or as clothoids where the curvature changes linearly with the arc length $s$

$$\kappa(s) = \kappa_0 + \kappa' s, \tag{20.2.2}$$

where $\kappa_0$ is the initial curvature, $\kappa' = d\kappa/ds$ the change of the curvature, and s is the arc length of a Frenet coordinate system; see Sect. 12.1.2. For a driving vehicle, the change of the curvature becomes

$$\frac{d\kappa(s)}{dt} = \frac{d\kappa}{ds}\frac{ds}{dt} = \kappa'\dot{s} = \kappa' v. \tag{20.2.3}$$

Front-view cameras with an opening angle of, e.g. $\pm 8°$ and with the corresponding image processing are able to provide the curvature $\kappa(x_c)$ where $x_c$ is the longitudinal coordinate from the beginning of a curve, and its change $d\kappa/ds$, see Winner et al. (2016), based on lane markings, reflector posts, and guard rails.

The curves can be mathematically approximated, for example, by polynomials

$$y(x) = a_0 + a_1 x_c + a_2 x_c^2 + a_3 x_c^3 + a_4 x_c^4 + a_5 x_c^5 \tag{20.2.4}$$

or by splines; see, e.g. Kochem (2005).

Using the curvature $\kappa$ and its derivative $\kappa'$, provided by the front-view sensor system and including the present course angle $\nu$, it holds also, see Bartels et al. (2016),

$$y(x) = a_0 + x_c \sin \nu + \frac{1}{2} x_c^2 \kappa + \frac{1}{6} x_c^3 \kappa' . \tag{20.2.5}$$

This follows from the lateral distance within a circle of radius $R_p = R$

$$y(x) = R\left(1 - \cos \frac{x_c}{R}\right),$$

and after series expansion for small $x_\mathcal{T}$

$$y(x) \approx \frac{1}{2R} x_c^2 = \frac{1}{2} x_c^2 \kappa,$$

the change of the lateral distance with varying $\kappa(x)$ becomes

$$\frac{dy(x)}{dx_c} = \kappa x_c + \frac{1}{2} x_c^2 \kappa',$$

which leads to

$$y(x) = \int_0^{x_c} \frac{dy(x)}{dx_c} dx = \frac{1}{2} x_c^2 \kappa + \frac{1}{6} x_c^3 \kappa' .$$

For small course angles, it holds

$$y(x) = a_0 + x_c \nu + \frac{1}{2} x_c^2 \kappa + \frac{1}{6} x_c^3 \kappa' . \tag{20.2.6}$$

In order to obtain a parametric model for the approached curve with usually noisy measurements $y_i(x_c)$ for $x_c = x_{c1}, x_{c2}, \ldots, x_{cN}$ along the observed road marking, a parameter estimation by using, e.g. the method of least squares has to performed to obtain the parameters $a_0, a_1, \ldots, a_5$ for (20.2.4) or $a_0$, $\kappa$ and $\kappa'$ for (20.2.6).

Then the path $y(x_c)$ can be calculated in advance for $x_c = x_\mathcal{T}$ if $\kappa$, $\kappa'$, $\psi$, and the distance $x_\mathcal{T}$ are known; the vehicle slip angle $\beta$ is neglected; and $a_0 = D_{y0}$ is the lateral distance to the right lane marking at the beginning of the curve

$$y(x_c) = D_{yx}(x_c) = D_{y0} + x_c \psi + \frac{1}{2} x_\mathcal{T}^2 \kappa + \frac{1}{6} x_\mathcal{T}^3 \kappa' . \tag{20.2.7}$$

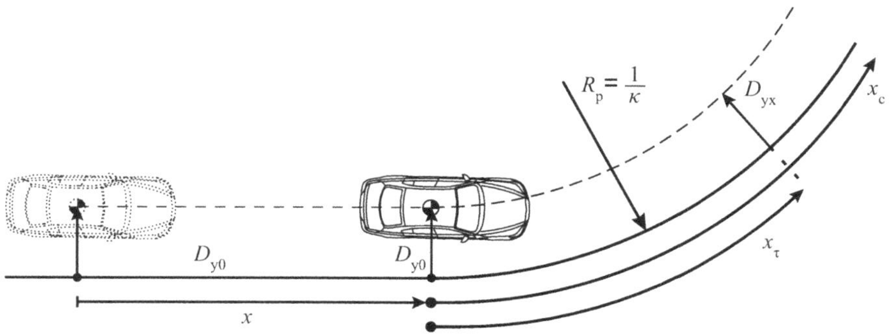


**Fig. 20.5** Vehicle approaching a curve with radius $R_p$

$x_\tau$ is then a look-ahead distance, which has to be selected properly. (20.2.7) is a prediction of the path which has to be driven.

For a right-hand curve, the approximation (20.2.7) can be used for a shorter distance if the view is limited, or it has to be applied correspondingly to the center line marking.

It is now assumed that the velocity of the vehicle is constant. When the vehicle enters a curve, see Fig. 20.5, the measured distance $D_{yx}$ to the right lane marking changes, and the path controller of Fig. 20.4 changes the reference value $\psi_d$ of the yaw angle controller according to the given reference value $D_{yxd}$. The control performance can, however, considerably be improved by feedforward control if the momentum path radius $R_p$ or the curvature $\kappa = 1/R_p$ is known.

A relation for the yaw rate of the vehicle in dependence on the steering angle follows for stationary cornering from (7.2.75) in form of the yaw rate gain

$$K_{\delta\dot{\psi}} = \frac{\dot{\psi}}{\delta_f} = \frac{v}{l + \mathrm{SG}v^2} \tag{20.2.8}$$

with the steering gradient $\mathrm{SG} = \frac{l}{v_{ch}^2}$.

Using (7.1.3), $\dot{\psi} = v/R_p = v\kappa$ results in the stationary steering angle gain

$$K_{\kappa\delta} = \frac{\delta_{f\kappa}}{\kappa} = \left(l + \mathrm{SG}v^2\right). \tag{20.2.9}$$

If further the derivative $\kappa' = d\kappa/dx$ is known, (20.2.9) yields

$$\frac{d\delta_{f\kappa}}{d\kappa} = \left(l + \mathrm{SG}v^2\right) \tag{20.2.10}$$

and

$$\frac{d\delta_{f\kappa}}{dx} = \left(l + \mathrm{SG}v^2\right)\frac{d\kappa}{dx} = \left(l + \mathrm{SG}v^2\right)\kappa' \tag{20.2.11a}$$

or

$$\frac{d\delta_{\mathrm{f}\kappa'}}{dt} = \left(l + \mathrm{SG}v^2\right)\kappa'\frac{dx}{dt} = \left(l + \mathrm{SG}v^2\right)\kappa' v \qquad (20.2.11\mathrm{b})$$
$$= K_{\kappa'\delta}\kappa'$$

with $K_{\kappa'\delta} = \left(l + \mathrm{SG}v^2\right) v$.

The change of the steering angle because of the change of the curvature $\kappa'$ then becomes

$$\Delta\delta_{\mathrm{f}\kappa'} = \int_0^t K_{\kappa'\delta}\kappa' dt = \left(l + \mathrm{SG}v^2\right)\kappa' vt \qquad (20.2.12)$$

and changes linearly with time $t$ and proportional to $\kappa'$ and $v$ (corresponding to a clothoid curve). The feedforward control of the steering then results in

$$\delta_{\mathrm{ff}} = \delta_{\mathrm{f}\kappa} + \Delta\delta_{\mathrm{f}\kappa'} \qquad (20.2.13)$$

and leads to the signal flow in Fig. 20.6.

As the curvature $\kappa = \kappa_{\mathrm{X}}$ is measured, a time period $\tau = x_\tau/v$ before the vehicle is at the point of measurement, the feedforward control may be delayed by a dead-time term

$$G_{\mathrm{Tt}}(s) = e^{-T_{\mathrm{t}}s} \qquad (20.2.14)$$

with $T_{\mathrm{t}} = k_\tau\tau$, where $\tau$ is the selected time gap and $k_\tau \leq 1$. Thus, the feedforward path function becomes

$$\delta_{\mathrm{ff}}(s) = \left(\delta_{\mathrm{f}\kappa}(s) + \Delta\delta_{\mathrm{f}\kappa'}(s)\right)e^{-T_{\mathrm{t}}s} = \left(K_{\kappa\delta}\kappa + K_{\kappa'\delta}\frac{\kappa'}{s}\right)e^{-T_{\mathrm{t}}s}. \qquad (20.2.15)$$

The steering angle is then a superposition by the steering angle output of the controller $G_{\psi\delta}(s)$ and the output of the feedforward function

$$\delta_{\mathrm{f}}(t) = \delta_{\mathrm{f}\psi}(t) + \delta_{\mathrm{ff}}(t). \qquad (20.2.16)$$

Figure 20.6 depicts the resulting control system. Based on the known ahead curvature $\kappa_{\mathrm{X}}$ and its derivative $\kappa'_{\mathrm{X}}$, the feedforward control determines theoretically the required steering angle $\delta_{\mathrm{f}}$ and the path control system according to Sect. 20.1 and corrects with $\delta_{\mathrm{f}\psi}$ remaining differences due to imperfect models, vehicle slip angle effects, and roadside disturbances. The yaw angle and path controller of the cascaded control system are as described in the last Sects. 20.1.1 and 20.1.2. The approach is similar to a path control with feedforward and PD controller feedback control in Taylor et al. (1999), Schorn and Isermann (2006) and Schorn (2007), and Müller and Rohleder (2006).

The course angle follows from (20.1.1) $\Delta\nu = \Delta\psi + \Delta\beta$ where the vehicle slip angle $\beta$ can be estimated with one of the state-estimation methods of Chap. 12. An

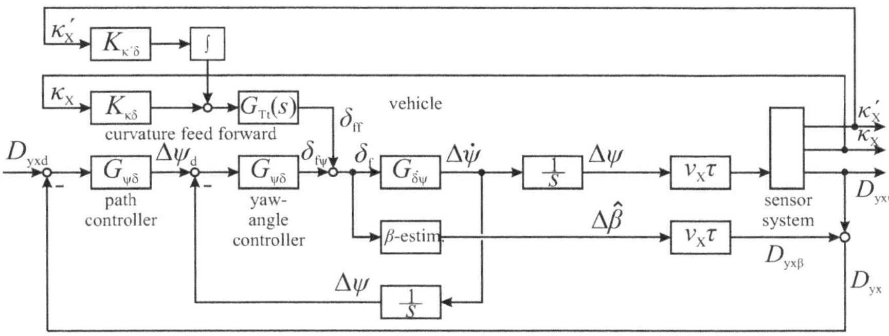


**Fig. 20.6** Signal flow scheme for path control in a curve with curvature $\kappa_X$, change of curvature $\kappa'_X$, and look-ahead time $\tau$ by combining feedback and feedforward control

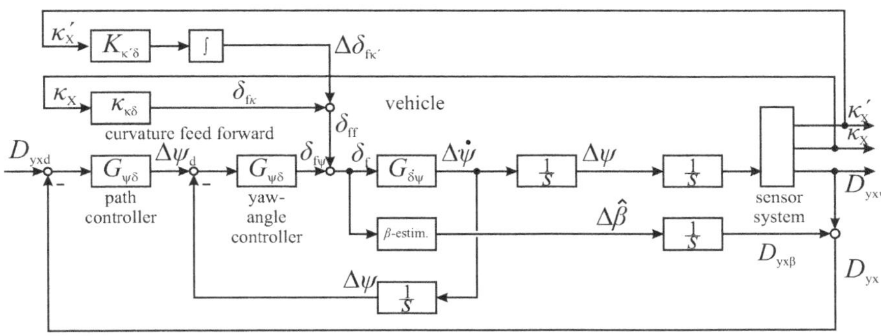


**Fig. 20.7** Signal flow scheme for path control in a curve with curvature $\kappa_X$ and change of curvature $\kappa'_X$ by combining feedback control in close distance $x_\tau$, $\tau$ small, with feedforward control

alternative is to add only the steady-state value for stationary cornering from (7.2.58). According to (7.2.60b) and (7.2.61), the slip angle can be positive or negative.

Figure 20.7 depicts a path control system in a curve for a close distance $x_\tau$ in front or beside of the vehicle, corresponding to the path control in Sect. 20.1.1a. Then the dead-time term (20.2.14) can be neglected.

## 20.3 Lane Change Control

For lane changes, the driven path is not directly given by the roadway, but can be selected with regard to the vehicle kinematics and dynamics and to the available space. This means that a path $y(x)$ or a trajectory $y(x, t)$ has to be planned, see Sect. 18.2, which is then followed by a lane change control system. The planned path or trajectory may be of a *preselected* form or may be *optimized* with regard to the (changing) environment.

### (a) Preselected Path

At first, preselected paths are considered. It is assumed that the velocity is constant and that a path control has to be designed.

The path is usually characterized by a turning point and can be approximated by polynomials, splines, and clothoids. Papadimitriou and Tomizuka (2003) use polynomials and Montés et al. (2007) apply Bezier curves for clothoids; see also Ulmer (1994). A simpler approach is proposed by Stählin (2008) in form of a sigmoidal path as function of the longitudinal coordinate $x$

$$y(x) = \frac{B_{\text{lc}}}{1 - e^{-a(x-c)}}, \tag{20.3.1}$$

where $B_{\text{lc}}$ is the maneuver width, $a$ the slope, and $c$ the position of the turning point. The resulting length is about $L_{\text{sig}} \approx 2c$. The advantage of this algebraic function is that only three parameters are required and that it can be used to calculate the lateral acceleration and the jerk, which allows an adaption to the velocity with $x = vt$.

Another approach is to describe the lane change path in the time domain by leaning on dynamic low pass elements of n-th order with a transfer function, Schmitt (2012), resulting in low pass path

$$\begin{aligned} G_{\text{lc}}(s) &= \frac{\Delta y_{\text{ref}}(s)}{\Delta y_{\text{d}}(s)} = \frac{1}{(1+Ts)^n}, \\ \Delta y_{\text{d}} &= B_{\text{lc}}. \end{aligned} \tag{20.3.2}$$

An analysis of lane change maneuvers by Sporrer et al. (1998) has shown that the time ratio of leaving the lane to reaching the new lane is about $1:2$. This results in an order of $n = 4$ for a *lane change reference model*

$$G_{\text{lc}}(s) = \frac{\Delta y_{\text{ref}}(s)}{\Delta y_{\text{d}}(s)} = \frac{1}{(1+Ts)^4} = \frac{\frac{1}{T^4}}{s^4 + \frac{4}{T}s^3 + \frac{6}{T^2}s^2 + \frac{4}{T^3}s + \frac{1}{T^4}}. \tag{20.3.3}$$

Transfer to the time domain gives a state-space representation

$$\begin{aligned} \begin{bmatrix} \dot{x}_1 \\ \dot{x}_2 \\ \dot{x}_3 \\ \dot{x}_4 \end{bmatrix} &= \begin{bmatrix} 0 & 1 & 0 & 0 \\ 0 & 0 & 1 & 0 \\ 0 & 0 & 0 & 1 \\ -\frac{1}{T^4} & -\frac{4}{T^3} & -\frac{6}{T^2} & -\frac{4}{T} \end{bmatrix} \begin{bmatrix} x_1 \\ x_2 \\ x_3 \\ x_4 \end{bmatrix} + \begin{bmatrix} 0 \\ 0 \\ 0 \\ \frac{1}{T^4} \end{bmatrix} \Delta y_d, \\ y_{\text{ref}} &= \begin{bmatrix} 1 & 0 & 0 & 0 \end{bmatrix} \begin{bmatrix} x_1 \\ x_2 \\ x_3 \\ x_4 \end{bmatrix}, \end{aligned} \tag{20.3.4}$$

with

$$x_1 = y_{\text{ref}}\ ;\ x_2 = v_{y\text{ref}}\ ;\ x_3 = a_{y\text{ref}}\ ;\ x_4 = j_{y\text{ref}},$$

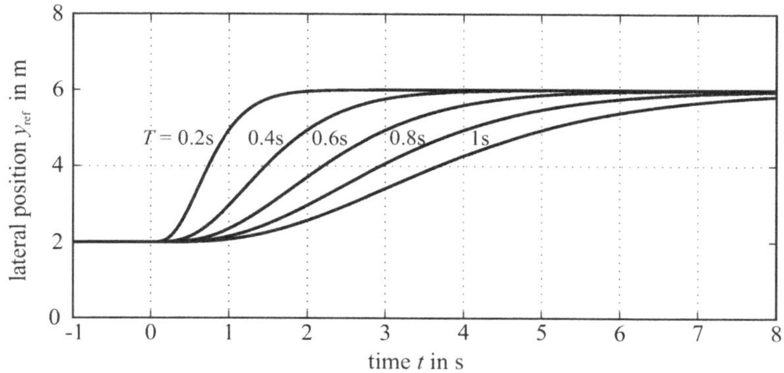


**Fig. 20.8** Step responses of low pass elements with $n = 4$ and different time constants $T = 0.2...1.0$ s, for $B_{lc} = 4$ m, Schmitt (2012)

where $j_y = da_y/dt$ is the jerk.

Figure 20.8 shows some step responses for $n = 4$ and different time constants $T$. Hence, by changing the time constant, the lateral velocity and the lateral acceleration can be adjusted.

An analysis of the time-dependent responses of (20.3.4) for a step input $\Delta y_d(t)$ indicates that the maximal lateral acceleration $a_{ymax}$ has at

$$t_{aymax} = T(3 - \sqrt{3}) \tag{20.3.5}$$

the value

$$a_{ymax} \approx 0.13\frac{B_{lc}}{T^2}. \tag{20.3.6}$$

Therefore, it follows for the time constant, Schmitt (2012),

$$T = \sqrt{0.13\frac{B_{lc}}{a_{ymax}}}. \tag{20.3.7}$$

The time constant of the path model only depends on the lane width and the maximal lateral acceleration. For a lane change of $B_{lc} = 4$ m and $a_{ymax} = 2\,\frac{\text{m}}{\text{s}^2}$, it yields $T \approx 0.5$ s.

Figure 20.8 shows that the time duration of a lane change is

$$\Delta t_{lc} \approx 8T \tag{20.3.8a}$$

and its length

$$\Delta x_{lc} \approx 8T v_X. \tag{20.3.8b}$$

The lane change reference model $y_{ref}(t)$ is given in the time domain. For constant velocity holds $v = \Delta x/\Delta t$ or $\Delta x = v\Delta t$. The driven path in dependence on the

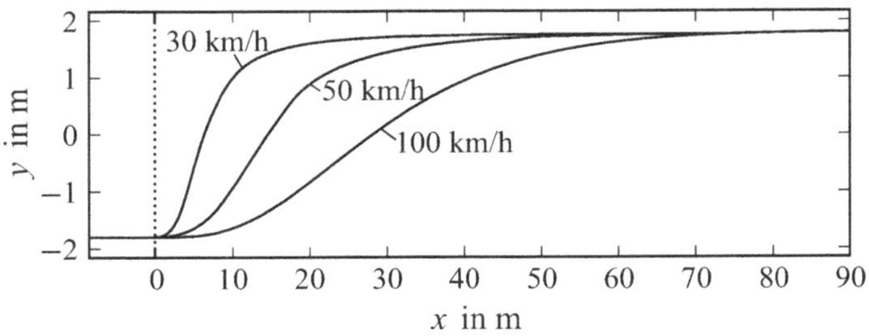


**Fig. 20.9** Step responses of low pass elements with $n = 4$ and different velocities for $a_{Y,max} = 7\,m/s^2$

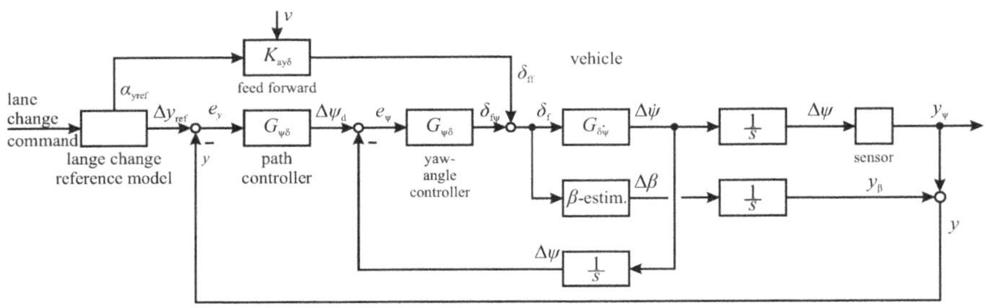


**Fig. 20.10** Signal flow scheme for a lane change control system with feedforward and feedback control

longitudinal way $x$ then is $y_{ref}(x) = y_{ref}(vt)$. Therefore, the driven path is obtained by changing the length scale in dependence on the velocity. Figure 20.9 shows how the reference path changes with the velocity. This lane change model was experimentally validated by Schmitt (2012).

The *lateral control* is similar as for path control beside the vehicle; see Fig. 20.3. However, the reference value $\Delta y_{ref}(t)$ is calculated from the lane change model (20.3.4) with $\Delta y_{ref} = B_{lc}\sigma(t)$ as input and $T$ from (20.3.7), where $\sigma(t)$ is a unit step function at the beginning of the lane change. The reference lateral distance $\Delta y_{ref}$ then acts as desired value for the cascaded lateral control system; see Fig. 20.10. The yaw angle and path controller can be designed with PD behavior, as in Sect. 20.1.1

The yaw rate and the lateral acceleration are for stationary cornering related by

$$\dot{\psi}_{ref} = \frac{a_{yref}}{v}. \tag{20.3.9}$$

Insertion in (20.2.8) yields for the required steering angle and constant velocity

$$\begin{aligned}\delta_{ff}(t) &= \frac{1}{v}\left(l + SGv^2\right)\dot{\psi}_{ref}(t) = \frac{1}{v}\left(\frac{l}{v} + SG\right)a_{yref}(t)\\ &= K_{ay\delta}(v)a_{yref}(t).\end{aligned} \tag{20.3.10}$$

This allows to add a feedforward path, where $a_{yref}$ is calculated from the lane change model (20.3.4) or measured, Ackermann (2016); see Fig. 20.10.

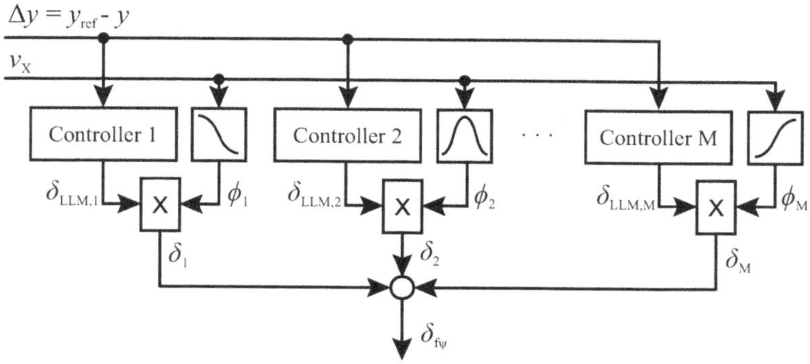


**Fig. 20.11** Structure of the weighted superposition of local linear controllers with velocity-dependent parameters

The parameters of the linear feedback and feedforward controllers depend on the velocity $v$ and the controller parameters and have therefore to be adapted. This can be provided by gain scheduling. As the controller parameters are valid in certain ranges $\Delta v$ around the operating point $v$, $M$ parameter sets can be specified for $v = 0 \ldots v_{\max}$ and weighted with activity functions, according to the principle of local linear models and controllers; see Fink (2006), Schorn (2007), and Ackermann (2016).

The output of the weighted superposition of the local linear controllers then is for the yaw angle controller, see Fig. 20.11,

$$\delta_{f\psi} = \sum_{i=1}^{M} \delta_{f\psi,i} = \sum_{i=1}^{M} \phi_i \delta_{LLM,i} \tag{20.3.11}$$

with the local linear controllers

$$\delta_{LLM,i}(t) = K_{\psi\delta,i}(1 + T_{D,i}s)e_{\psi,i}(t), \tag{20.3.12}$$

the velocity-dependent activity function

$$\phi_i(v_X) = \frac{\mu_i(v_X)}{\sum_{j=1}^{M} \mu_j(v_X)}, \tag{20.3.13}$$

and the Gaussian membership functions

$$\mu_i(v) = \exp\left(-\frac{1}{2}\frac{(v_X - c_i)^2}{v_i^2}\right). \tag{20.3.14}$$

A simulation with a similar cascaded control system according to Fig. 20.10 is depicted in Fig. 20.12. At $t = 10\,\mathrm{s}$, the step command is given to change the lateral position from $y = -1.8\,\mathrm{m}$ to $y = 1.8\,\mathrm{m}$. The path $y(t)$ shows the reference low pass behavior and as well $\dot{\psi}(t)$, $\psi(t)$, and $\delta_f(t)$ indicate a well-damped behavior.

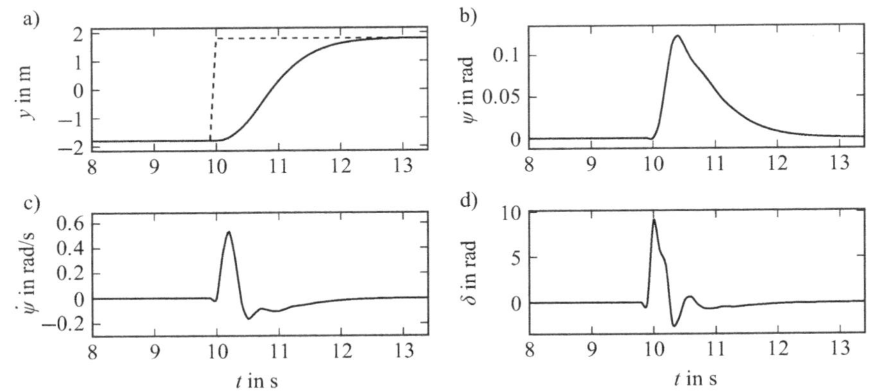


**Fig. 20.12** Simulation (Carmaker™) of a lane change maneuver with a step input $\Delta y_{\mathrm{ref}}$ at $t = 10\,\mathrm{s}$, a lateral reference path according to (20.3.3), $v = 130\,\mathrm{km/h}$, Ackermann (2016). Cascade control with PI controller for $\dot{\psi}$ and P controllers for $\psi$ and $y$

The lane change control systems described before use parametric models for the path $y(x)$ or $y(t)$ in order to result in a compact realization of the reference values $y_{\mathrm{ref}}$. However, a path representation can also be represented in a nonparametric form, e.g. by a few samples

$$\underline{P}_{\mathrm{ref}}(t) = \begin{bmatrix} X_{\mathrm{E}}(t) & X_{\mathrm{E}}(t+T_{\mathrm{P}}) & X_{\mathrm{E}}(t+2T_{\mathrm{P}}) & X_{\mathrm{E}}(t+10T_{\mathrm{P}}) \\ Y_{\mathrm{E}}(t) & Y_{\mathrm{E}}(t+T_{\mathrm{P}}) & Y_{\mathrm{E}}(t+2T_{\mathrm{P}}) & Y_{\mathrm{E}}(t+10T_{\mathrm{P}}) \\ \psi(t) & \psi(t+T_{\mathrm{P}}) & \psi(t+2T_{\mathrm{P}}) & \psi(t+10T_{\mathrm{P}}), \end{bmatrix} \tag{20.3.15}$$

where $T_{\mathrm{P}}$ is a path sampling time; see Fig. 21.4. Then by the Lagrange interpolation, $y_{\mathrm{ref}}(x)$ can be determined; see Schorn (2007).

A further approach for path control with curves is to combine the lateral distance control at short distance $x_\tau$ with $\tau \approx 0.2$ to $0.5\,\mathrm{s}$ in front of the vehicle with feedforward control by using the curvature $\kappa$ for longer distance $x_\tau$ with $\tau \approx 1.5...4\,\mathrm{s}$. This results then in a control configuration as in Fig. 20.7, which follows from Figs. 20.3 and 20.6.

**(b) Optimized Trajectories**

In dense traffic situations, other vehicles around the ego vehicle have to be considered. This means that their changing positions and velocities must be taken into account. The lane change then may include changes of the velocity and means that lane change trajectories $y(x,t)$ have to be planned. If the starting position and time $y(x_0, t_0)$ is known, the final position and time $y(x_{\mathrm{end}}, t_{\mathrm{end}})$ has to be selected and according to an optimization criterion, an optimal trajectory has to be found; see, e.g. Werling et al. (2010) and Rathgeber (2016).

A trajectory planning method with smaller computational expense, dividing the trajectory into three segments with variable average velocities, is presented by Hansen et al. (2016) or with continuously changing velocity by Schucker et al. (2019).

Frequently, the trajectories are described as polynomials like

$$y(t) = c_0 + c_1 t + \cdots + c_\nu t^\nu \tag{20.3.16}$$

for the lateral and longitudinal direction, with orders $\nu = 5$ or 7. A quadratic optimization criterion then may evaluate the start and end position, the lateral acceleration, and jerk.

Further publications on lateral control of vehicles and related topics are, for example, Papadimitriou and Tomizuka (2003), Meier et al. (2004), Kochem (2005), McCall and Trivedi (2006), Müller and Rohleder (2006), Kehl (2007), Montés et al. (2007), Habenicht (2012), Paden et al. (2016), Rathgeber et al. (2016), Gallep and Müller (2017), Augustin et al. (2019), and Gundlach (2020) and contributions in Isermann (2006).

## 20.4 Lane Merging Control

Merging maneuvers at highway entrances or roundabouts are relatively critical and dangerous traffic situations. In dense traffic, they usually require reaction from the entering vehicle and the vehicles on the main lane and may lead to chain reactions and traffic jams. In the ideal case, the entering vehicle should merge in front of a following (lagging) vehicle on the main lane with a velocity that does not force the following vehicle to change its velocity or even the lane.

Different kinds of road merge construction exist for highways and cross-country roads, e.g. a parallel lane merge, a tapered lane merge (without a parallel lane), or a cloverleaf merge.

Merging maneuvers on these road constructions are treated in many publications. First treatments considered the control of vehicle merging by a roadside local manager outside of the vehicles; see, e.g. Posch and Schmidt (1980) and Lu and Hedrick (2003).

Further contributions assume connected and automated vehicles (CAV); see, e.g. the survey by Rios-Torres and Malikopoulos (2017). Centralized approaches use a single central controller which takes decisions for all relevant vehicles by, e.g. heuristic rules or optimized on-ramp coordination. In decentralized control, each vehicle determines its own solution, with heuristic control (fuzzy logic) or multiobjective optimization. See also Park et al. (2013) with control of leading and lagging main lane vehicles to create an adequate gap for a smooth merging; see Liu and Tomizuka (2016), Hussain et al. (2018), and Dollar and Vahidi (2019). Assadi et al. (2020) optimize the longitudinal velocity of the entering vehicle by using relative speed and position measurements. Eilbrecht and Stursberg (2020) propose to dedicate agents to the vehicles.

During a merging maneuver, usually the behavior of the vehicles on the main lane is not known exactly and difficult to predict. For example, the lagging vehicle can accelerate, the lead vehicle can brake, or vehicles from other lanes may cut in. The automation of the merging maneuver has to take into account these situations. Therefore, the merging vehicle needs a powerful environmental sensor system.

In the following, an automatic merging on a highway with a parallel entrance ramp is considered with several simplifying assumptions. This may be seen as an example that can then be expanded to more complex merging situations.

The considered merging situation is depicted in Fig. 20.13 where for merging an entrance ramp in form of a parallel lane to the main lane exists. On the main lane are two vehicles 1 and 2 driving at time $t_0$ with the velocity $v_1(t_0)$ and $v_2(t_0)$ in a distance $\Delta x_{12} = x_2 - x_1$. $t_0$ is the time instant for the measurement of the velocities $v_1$ and $v_2$ with a sensor system of the merging vehicle M. The goal is that the vehicle $M$ with velocity $v_M(t_0)$ merges between the vehicles 1 and 2. The parallel entrance ramp allows to speed up the merging vehicle to the velocity $v_M = v_1$ of the following vehicle under the condition that there are no objects or vehicles on the ramp.

It is now assumed that the velocities of the vehicles on the main lane are constant and equal, $v_1 = v_2$.

The first condition for a courteous merging is that the distance to the following vehicle 1 is

$$\Delta x_{1M}(t_M) \geqq v_1 \tau_{d1}, \tag{20.4.1}$$

where $\tau_{d1}$ is the desired time gap of vehicle 1 ($\tau_{d1} \approx 1.5...2$ s; see (19.3.1)).

The second condition is that the distance of the merging vehicle M to the preceeding vehicle after merging at time instant $t_{lce}$ is

$$\Delta x_{M2}(t_{lce}) \geqq v_M \tau_{dM}. \tag{20.4.2}$$

Hence, the distance (gap) between the two vehicles in the main lane at the begin of the merging maneuver is

$$\Delta x_{12,req} = \Delta x_{12}(t_{lcb}) \geqq v_1 \tau_{d1} + v_M \tau_{dM} + l_v, \tag{20.4.3}$$

where $l_v$ is the length for the merging vehicle and $t_{lcb}$ the starting time instant of the merging maneuver. The time gap $\tau_{d1}$ and $\tau_{d2}$ depends also on the density of the traffic on the main lane. For dense traffic, $\tau_d$ can be reduced to about $\tau_{d,min} \approx 1...1.5$ s.

The merging vehicle M on the parallel entrance ramp has to accelerate to the velocity $v_M(t_{lcb}) = v_1(t_0) = v_2(t_0)$.

The lateral distance of the merging maneuver is

$$\Delta y_d(t_{lcb}) = B_{lc} = C_0 + \frac{B_1}{2} + \Delta y_M(t_{lcb}) \tag{20.4.4}$$

to reach the moving target MP($t_{lce}$) in the middle of vehicle 1 and 2; see Fig. 20.13. The lane change control follows Sect. 20.3, for example, Fig. 20.10, by using a predefined lane change maneuver as (20.3.2). This maneuver may be selected on the basis of a chosen maximal lateral acceleration $a_{Y\max}$ due to (20.3.6) with the setpoint $\Delta y_d = \Delta y_{ref} = B_{lc}$ according to (20.4.4).

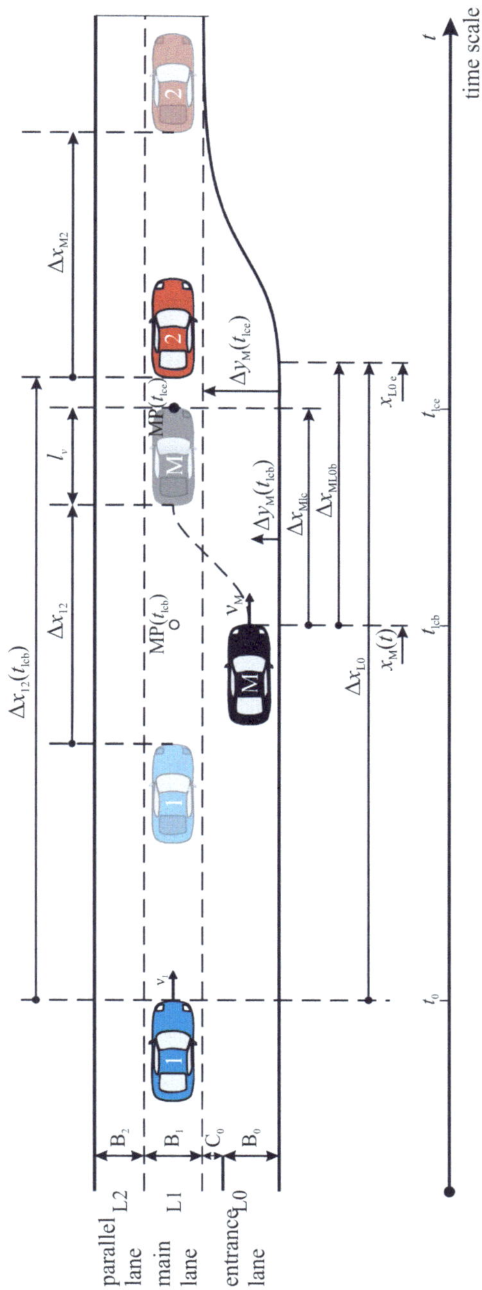


**Fig. 20.13** Highway with main lanes L1, parallel lane L2, and entrance ramp L0. Vehicle M merges in between vehicles 1 and 2

The goal is to reach the center line of the main lane $\Delta y_{ml} = B_1/2$. The required length for the lane change maneuver is according to (20.3.8b)

$$\Delta x_{lc} = x_{Mlc}(t_{lcb}) - x_M(t_{lce})) \approx 8T v_M, \tag{20.4.5}$$

where the time constant $T$ depends on the selected $a_{Y\max}$; see (20.3.6).

The begin of the merging maneuver is the time instant where the moving merging point MP($t_{lcb}$) is parallel to the merging vehicle M; see Fig. 20.13. This means the distance to the preceeding vehicle is $\Delta x_{M2}(t_{lcb}) = v_M \Delta \tau_{dM}$.

If the merging point MP can be any point along the entrance ramp and the entrance ramp is relatively long, this gives some flexibility in selecting the starting time of the lane change maneuver. However, if the merging vehicle reaches the end of the entrance ramp without the possibility to start the merging maneuver, it has to brake down. This braking maneuver has to be started if the still available length of the entrance ramp is

$$\Delta x_{ML0b}(t) = x_{L0e} - x_M(t) \geq k_{bd} x_{bd,\min}, \tag{20.4.6}$$

where $x_{bd,\min}$ is the minimal braking distance due to (6.7.23) with a safety surcharge factor $k_{bd} \approx 2$. The remaining length $\Delta x_{L0}$ has to be measured and supervised by the onboard sensor system and has to be long enough to start the next merging maneuver. The merging vehicle has then to wait until a large gap between two vehicles on the main lane appears.

The merging maneuver can be summarized by a sequence of rules in the form

IF < condition > THEN < conclusion >

The condition part (premise) contains facts as inputs and the conclusion part determines events as a logical cause of the facts; compare Fig. 20.14.

| | | |
|---|---|---|
| 1. | IF | <vehicle M enters the entrance ramp AND no objects are on the ramp AND ACC is on AND turn signal is on> |
| | THEN | <geometrical sizes of the road $B_0$, $C_0$, $B_1$ and length $\Delta x_{L0}$ of the ramp have to be known or measured> |
| 2. | IF | <vehicle M is on the entrance ramp L0> |
| | THEN | <velocities $v_1$, $v_2$ and distances $\Delta x_{1M}$ and $\Delta x_{M2}$ have to measured (on board) as soon as possible> |
| 3. | IF | <rules 1) and 2) are fulfilled> |
| | THEN | <vehicle M is accelerated/decelerated such that $v_M = v_1$> |
| 4. | IF | <distance $\Delta x_{12,req}$ AND $v_M = v_1$ AND $\Delta x_{M2} \geq v_M \tau_d$ AND $\Delta x_{ML0b} \geq k_{bd} x_{bd,\min}$> |
| | THEN | <start lane change maneuver with lane change path control (Fig. 20.10) and setpoint $\Delta y_{Dd}(t_{lcb}) = B_{lc} = C_0 + B_1/2 + \Delta y_M(t_{lcb})$> |
| 5. | IF | <the remaining distance on the ramp $\Delta x_{ML0}(t_{lcb}) < k_{bd} x_{bd,\min}$> |
| | THEN | <active brake until a next gap $\Delta x_{12}$ is large enough OR until standstill and try again> |
| 6. | IF | <vehicle M has reached the main lane $L1$> |
| | THEN | <activate lane path control (Fig. 20.6) AND activate distance control (Fig. 19.7)> |

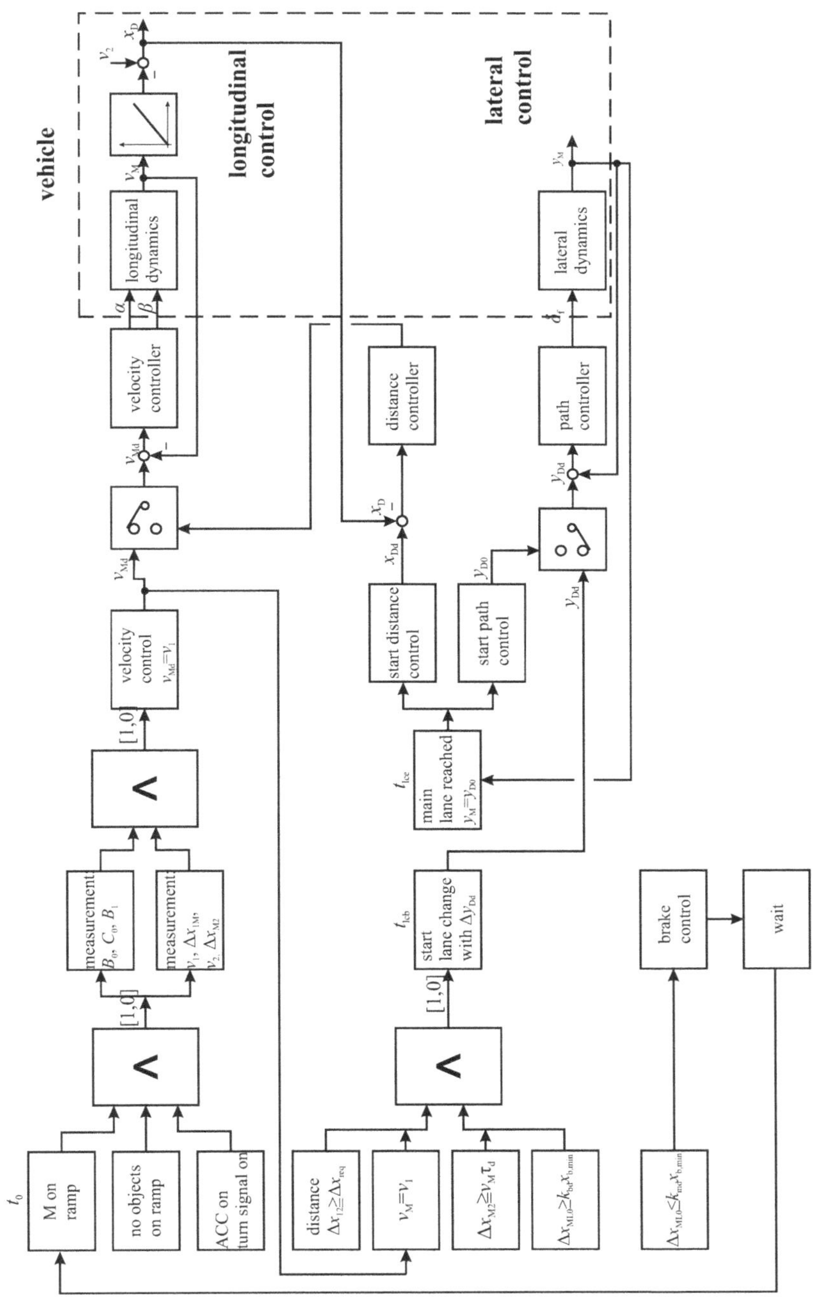


**Fig. 20.14** Sequence of rules and signal flow for a merging maneuver on a highway ramp

This example has shown that preconditions can be formalized as rules with binary logic decisions. As some of the conditions are not sharp, the use of fuzzy logic instead of crisp, binary logic is well suited.

Summarizing, the automatic merging maneuver is an example where event-based logic rules and continuous-time longitudinal and lateral vehicle control have to cooperate. A similar procedure may be applied for an *automatic lane change maneuver* by setting $C_0 = 0$ and $\Delta x_{\mathrm{MLOb}} \geq v_{\mathrm{M}} \tau_{\mathrm{dM}}$.

## References

Ackermann C (2016) Neue Ansätze für Fahrerassistenzsysteme mit Regelung der Längs- und Querdynamik. Dissertation Technische Universität Darmstadt. Fortschr.-Ber. VDI Reihe 8, 1249. VDI Verlag, Düsseldorf

Ackermann J, Guldner J, Sienel W, Steinhauser R, Utkin V (1995) Linear and nonlinear controller design for robust automatic steering. IEEE Control Syst Technol 3:132–140

Adamy J (2009) Nichtlineare Regelungen. Springer, Berlin

Assadi A, Meier F, del Re L (2020) Highway entrance merging assistant for minimal traffic disturbance. IFAC-World Congress, Berlin

Augustin D, Hofmann M, Konigorski U (2019) Prediction of highway lane changes based on prototype trajectories. Forschung im Ingenieurwesen 83:149–161

Bartels A, Rohlfs M, Hamel S, Saust F, Klauske LK (2016) Lateral guidance assistance. In: Chapter 48 in Winner et al

Del Re L, Allgöwer F, Glielmo L, Guardiola C, Kolmanovsky I (eds) (2010) Automotive model predictive control. Springer, London

Dickmanns E, Mysliwetz B (1992) Recursive 3-d road and relative ego-state recognition. In: IEEE transactions on pattern analysis and machine intelligence, pp 199–213

Dickmanns ED (2001) Fahrzeuge lernen schon. Universtität der Bundeswehr München

Dollar R, Vahidi A (2019) Automated vehicles in hazardous merging traffic: a chance-constrained approach. IFAC-Papers On Line 52–5:218–223

Eilbrecht J, Stursberg D (2020) Set-based scheduling for highway-entry of autonomous vehicles. IFAC World Congress, Berlin

Fink A (2006) Nonlinear control based on local linear neuro-fuzzy models. Dissertation Technische Universität Darmstadt. Fortschr.-Ber. VDI Reihe 8, 1096. VDI Verlag, Düsseldorf

Gallep J, Müller S (2017) Modellprädiktive Regelung zur Querführung automatisierter Fahrzeuge. VDI-GMA Conference AUTOREG

Gundlach I (2020) Zeitoptimale Trajektorienplanung für automatisiertes Fahren bis in den fahrdynamischen Grenzbereich. Diss TU Darmstadt, Shaker, Düren

Habenicht S (2012) Entwicklung und Evaluation eines manöverbasierten Fahrstreifenwechselassistenten. PhD thesis

Hansen T, Schulz M, Knoop M, Konigorski U (2016) Trajektorienplanung für automatisierte Fahrstreifenwechsel. ATZ - Automobiltechnische Zeitschrift 118:66–71

Hu J, Zhang Y, Rakheja S (2020) Path planning and tracking for autonomous vehicle collision avoidance. IFAC World Congress, Berlin

Hussain SA, Jahromi BS, Karakas B, Cetin S (2018) Highway lane merge for autonomous vehicles without an acceleration area using optimal model predictive control. World J Res Rev 6:27–32

Isermann R (ed) (2006) Fahrdynamik-Regelung. Vieweg, Wiesbaden

Kehl S (2007) Querregelung eines Versuchsfahrzeugs entlang vorgegebener Bahnen. PhD thesis

Kochem M (2005) Ein Fahrerassistenzsystem zur Unterstützung des rückwertigen Parkvorgangs für Pkw. Dissertation Technische Universität Darmstadt. Fortschr.-Ber. VDI Reihe 12, 590. VDI Verlag, Düsseldorf

Liu C, Tomizuka M (2016) Enabling Safe Freeway Driving for Automated Vehicles. In: American Control Conference (ACC), Boston, pp 3461–3467
Lu XY, Hedrick J (2003) Longitudinal control algorithm for automated vehicle merging. Int J Control 76:193–202
Mayr R (2000) Regelungsstrategien für die automatische Fahrzeugführung: Längs- und Querregelung, , Spurwechsel- und Überholmanöver. Springer
McCall JC, Trivedi MM (2006) Video-based lane estimation and tracking for driver assistance: survey, system, and evaluation. IEEE Trans Intell Transp Syst 7(1):20–37
Meier G, Roppenecker G, Wurmthaler H (2004) Tracking control for automatic vehicle guidance. In: 7th international symposium advance vehicle control
Montés N, Mora MC, Tomero J (2007) Trajectory generation based on rational bezier curves as clothoids. In: IEEE intelligent vehicles symposium, Istanbul, pp 508–510
Müller T, Rohleder D (2006) Automatisches Spurfahren auf Autobahnen, Chapter 13. In: Isermann R (ed) Fahrwerkdynamik-Regelung, ATZ/MTZ-Fachbuch. Vieweg, Wiesbaden
Nareyko G, Biemelt P, Trächtler A (2020) Real-time optimized model predictive control of an active roll stabilization system. IFAC-World Congress, Berlin
Paden B, Cap M, Zheng Yong S, Yershov D, Frazzoli E (2016) A survey of motion planning and control techniques for self-driving urban vehicles. IEEE Trans Intell Veh 1(1):33–55
Papadimitriou I, Tomizuka M (2003) Fast lane changing computations using polynomials. In: Proceedings of the 2003 American control conference, Denver
Park H, Su S, Hayat T, Smith BL (2013) A prototype freeway merging control algorithm under a connected vehicle environment. In: Transportation research record 93rd meeting
Posch B, Schmidt G (1980) A comprehensive control concept for merging of automated vehicles under a broad class of traffic conditions. In: IFAC symposium on control in transportation systems, Baden-Baden, pp 187–194
RAA (2008) Richtlinien für die Anlage von Autobahnen
Rajamani R (2012) Vehicle dynamics and control, 2nd edn. Mechanical engineering series. Springer, US, New York
Raste T, Lüke S, Eckert A (2015) Automated driving, technical approach with motion control architecture. at - Automatisierungstechnik 63:191–203
Rathgeber C (2016) Trajektorienplanung und -folgeregelung für assistiertes bis hochautomatisiertes Fahren. Diss, TU Berlin
Rathgeber C, Winkler F, Müller S (2016) Kollisionsfreie Längs- und Quertrajektorienplanung unter Berücksichtigung fahrzeugspezifischer Potenziale. at - Automatisierungstechnik 64:61–76
Rios-Torres J, Malikopoulos AA (2017) A survey on the coordination of connected and automated vehicles at intersections and merging at highway on-ramps. IEEE Trans Intell Transp Syst 18(5):1066–1077
Schmitt K (2012) Situationsanalyse für ein Fahrerassistenzsystem zur Vermeidung von Überholunfällen auf Landstraßen. Diss. Univ. Darmstadt, Fortschritt Bericht VDI, Reihe 12 Nr.763
Schorn M (2007) Quer- und Längsregelung eines Personenkraftwagens für ein Fahrerassistenzsystem zur Unfallvermeidung. Diss. Universität Darmstadt, Fortschr.-Ber. VDI Reihe 12, 651. VDI Verlag, Düsseldorf
Schorn M, Isermann R (2006) Automatic steering and braking for a collision avoiding vehicle. IFAC-Symposium on Mechatronics, Heidelberg
Schucker J, Kohlmann T, Konigorski U (2019) A complete concept for combined longitudinal and lateral trajectory planning for automated driving on highways. VDI-Bericht, VDI/VDE AUTOREG, p 2349
Sporrer A, Prell G, Buck J, Schaible S (1998) Realsimulation von Spurwechselvorgängen im Straßenverkehr. VKU: Verkehrsunfall und Fahrzeugtechnik 36(3):69–76
Stählin U (2008) Eingriffsentscheidung für ein Fahrerassistenzsystem zur Unfallvermeidung. Diss. TU Darmstadt, Fortschr.-Ber. VDI Reihe 12, 683. VDI Verlag, Düsseldorf, Düsseldorf
Taylor CJ, Košecká J, Blasi R, Malik J (1999) A comparative study of vision-based lateral control strategies for autonomous highway driving. Int J Robot Res 18(5):442–445

Ulmer B (1994) Vita ii-active collision avoidance in real traffic. In: Proceedings of the intelligent vehicles '94 symposium 1–6

Werling M, Groell L, Bretthauer G (2010) Invariant trajectory tracking with a full-size autonomous road vehicle. In: IEEE transactions on robotics, pp 758–765

Winner H, Hakuli S, Lotz F, Singer C (eds) (2016) Handbook of driver assistance systems. Springer International Publishing AG, Cham, Switzerland

Würtenberger M (1997) Modellgestützte Verfahren zur Überwachung des Fahrzustandes eines Pkw. Diss. Universität Darmstadt, Fortschr.-Ber. VDI Reihe 12, 314, VDI Verlag, Düsseldorf

# 21 Anticollision Control Systems

Automotive safety is one of the high-priority issues in the design of vehicles, construction, and equipment to minimize the occurrence and consequence of accidents. The improvements in automobile and roadway design have steadily reduced injury and death rates in developed countries. This positive development was supported by measures of passive and active safety systems.

*Passive safety systems* provide vehicle components in order to protect occupants during a crash respectively to reduce the consequence of accidents. Some examples are as follows:

- *Interior safety systems* for passenger protection:
  - constructive crashworthy systems like crumple zones (front, rear, side), cell strength, survival space, collapsible steering columns, and cockpit padding; and
  - passenger restraint systems: seat belt with retractor and tightener, airbags (front, side, window).
- *Exterior safety systems* for protection of external humans:
  - vehicle-related measures to minimize injuries to pedestrians, bicycle-riders, and motorcycle-riders, e.g. hood measures and deformation behavior; and
  - exterior body shape and structure.
- *Active passive safety systems*:
  - pre-crash sensors for early actuating of occupant safety systems,
  - pre-safe systems to prepare passive safety systems for a possible accident (e.g. belt tightening), and
  - pre-safe structure: active chassis systems for passenger protection.



R. Isermann, *Automotive Control*, ATZ/MTZ-Fachbuch,
https://doi.org/10.1007/978-3-642-39440-9_21

These passive safety systems have reached a well-developed status and mainly active passive safety systems will show further progress. Larger steps in the reduction of accidents are now expected by active safety systems.

Accident statistics exist from several sources. The official German statistics DESTATIS distinguishes between the type of fatalities and kind of fatalities. The type of fatalities describes the traffic situation and therefore the origin of the accidents and the kind of fatalities describes the movement of the accident participants relative to each other.

The major *types of accidents* are for Germany, according to Destatis (2010), as follows:

- driving accident, 42 %;
- accident between vehicles moving along, 21 %;
- turning into a road or crossing it, 18 %;
- crossing the road (pedestrian), 3 %; and
- others, 6 %.

The *kind of accidents* with fatalities are further distinguished as follows:

- road departure, 39 %;
- with oncoming vehicles, 32 %;
- with vehicle turning or crossing, 15 %;
- with pedestrians, 5 %; and
- with vehicles moving ahead or waiting stationary, or laterally moving, 6 %.

A reduction of driving accidents and accidents with road departure can, e.g. be supported by first- and second-generation ADAS, like ABS, TRC, ESC, LDW, and LKS. However, accidents between vehicles moving along, turning, and crossing or with obstacles need newly developed driving assistance systems, so-called *anticollision systems* or *collision-avoidance systems* (CAS).

The development of automotive systems for the reduction of accidents has a long tradition. A path-breaking project in the field of active safety was the European research project PROMETHEUS (Program for European Traffic with Higher Efficiency and Unprecedented Safety) from 1986 to 1994, see, e.g. Braess and Reichart (1995), and the following research project INVENT. Some further contributions are Kopischke (2000) for emergency braking functions and Lages (2001) and Ameling (2002) for collision-avoidance systems. See also Kirchner et al. (2005), Mildner (2004), Weisen (2003), and Anderson et al. (2010) and several contributions in Winner et al. (2016). A three-layer system for safe automatic driving and prediction of the future trajectories of surrounding vehicles with decisions for lane changes in front of too slow or fast cut-in vehicle is described in Liu and Tomizuka (2016), using simulations.

The scheme in Fig. 21.1 distinguishes some accident situations between vehicles:

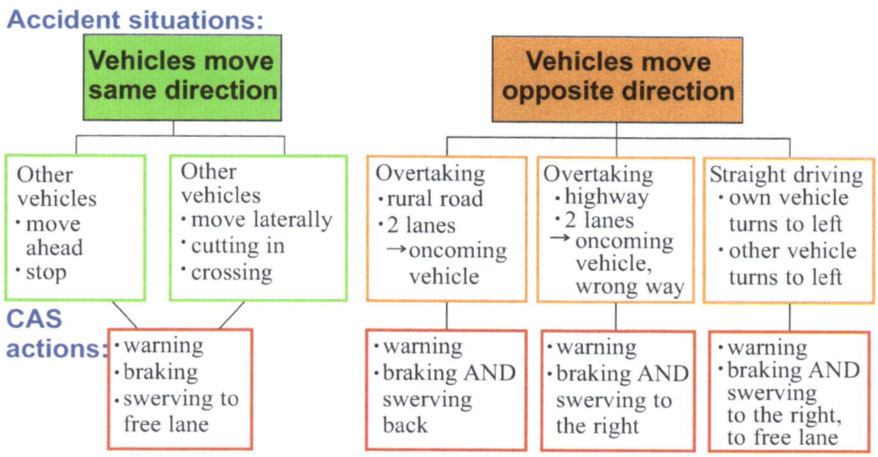


**Fig. 21.1** Organization chart for collision-avoidance systems (longitudinal direction)

1. *Vehicles move in the same direction*

- other vehicles:

  (a) move ahead or stop;
  (b) move laterally, turn, or cross (includes objects on road).

- Collision-avoidance (CAS) actions: warning, braking, and swerving.

2. *Vehicles move in opposite directions*

- own vehicle:

  (a) overtaking maneuver
    – rural road;
    – highway, freeway;
  (b) straight driving
    – vehicles leave correct lane.

- CAS actions: warning, braking, and swerving if required

The research project PRORETA was an Industry-University project on collision-avoidance systems between Continental AG and Technische Universität Darmstadt. The project PRORETA 1 (2003–2006) was dedicated to situations 1 (a) and (b), and project PRORETA 2 (2006–2009) to situation 2 (a) for rural roads. Both projects were performed by the cooperation of three research institutes: Automatic Control, Automotive Engineering and Ergonomics (first project), and Multimodal Interactive Systems (second project).

An important basis for these collision-avoidance systems is the detection of the environment around the vehicle with RADAR sensors, LIDAR sensors, and cameras.

In the following, the developed collision-avoidance systems, the assumed accident situations, and the intervention planning and control functions will be summarized and experimental results with driving experiments will be shown. The contribution can only give an extract of the developed collision-avoidance systems. The details are given in the reports Bender (2008), Darms (2007), Schorn (2007), and Stählin (2008) for PRORETA 1 and Hohm (2010), Mannale et al. (2008), Schmitt (2012), and Wojek (2010) for PRORETA 2. The following sections are based on Isermann et al. (2012).

## 21.1 Anticollision Brake and Steering

An accident situation is considered where the own vehicle drives correctly on its lane and another vehicle moving in the same direction or an object appears in front. The driver has the chance to react appropriately as long as possible to avoid an accident. However, if the driver does not react, the CAV system intervenes at the last possible moment.

Figure 21.2 gives a system overview for the development of PRORETA 1; see Isermann et al. (2008). Based on information from the environment and the own car, predictions for the expected trajectory of the own car and of objects in the surroundings are being calculated. Using these predictions, a decision is made, whether an intervention is necessary or not, and the intervention is planned. The intervention itself is then conducted fully automatically. An ergonomic study accompanied the development of the system. This study investigated how the driver reacts in critical

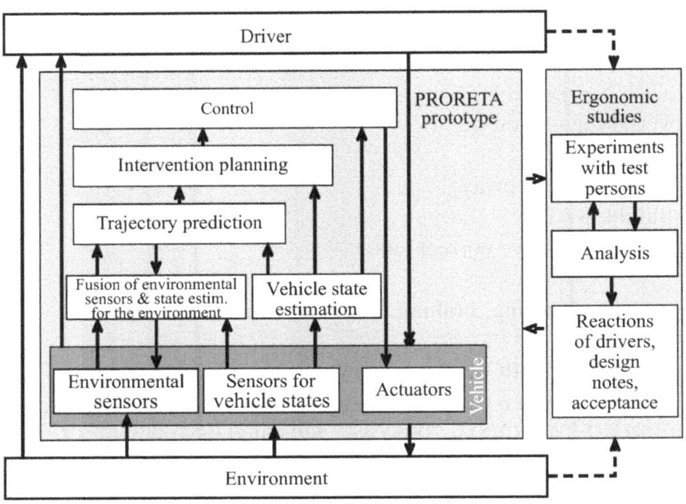


**Fig. 21.2** Collision-avoidance system overview for the development of PRORETA 1

**Fig. 21.3** Environmental sensors of the test vehicle: **a** video camera, resolution 0.07°, 44°, 640 × 496 pixels, cycle time 40 ms; **b** laser scanner, 22.5°, resolution 1.5°, cycle time 90 ms

situations and how he or she reacts to the interventions. In the following, the intervention decision, the planning of the intervention, and the conduction of the intervention are described. The environment perception is described in detail in Darms and Winner (2006) and Darms (2007). Results from the ergonomic study can be found in Bender et al. (2007) and Bender (2008). The system was tested by simulations using a complex two-track model followed by extensive driving tests with an experimental vehicle.

### 21.1.1 The Research Vehicle

A VW Golf IV, which was equipped with additional sensors and actuators required for the developed functions, served as the experimental vehicle; see Fig. 21.3.

The driver-assistance system uses an *active front steering* (AFS) and an *electro-hydraulic braking system* (EHB) as actuators. For vehicle state estimation, only ESC sensors and the sensors of the active front steering and braking system are necessary. For environment perception, a *laser scanner* and a *video sensor* were used. The chosen design allows to scan the area in front of the vehicle. The detection area of the laser scanner covers an angular range of 22.5° with a resolution of 1.5° and is scanned in a 90 ms cycle. The distance to objects is determined by a time of flight measurement of emitted light impulses. The video sensor is based on a monochrome CMOS image sensor that provides data in a 40 ms cycle. The detection

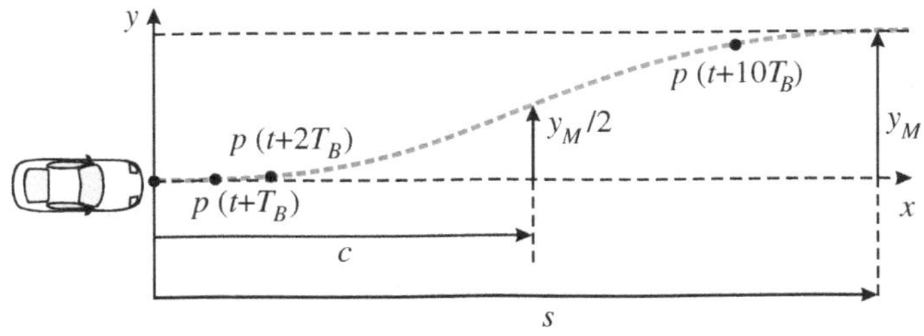


**Fig. 21.4** Evasive path for planning and control

area is 44°, whereas the discretization with approx. 0.07° is considerably finer than for the laser scanner. By means of image-processing algorithms, ahead driving vehicle rear frontal area and lane markings can be detected in the image. However, a direct distance measurement is not possible; for details, see Darms and Winner (2006) and Darms (2007).

### 21.1.2 Evasive Path

An evasive path is required between intervention planning and control. For investigating several different intervention functions with different types of controllers, the type of intervention is selected using some flags. The flags used are braking, emergency braking, and evasion. If *braking* is chosen, the desired deceleration has to be transmitted. If an *emergency braking* is chosen, the maximum possible deceleration at every point in time is achieved using braking controllers. For an *evasion*, the desired position and heading are given for one time step $T_B$, two time steps $T_B$, and ten time steps $T_B$ ahead in time; see Fig. 21.4. These time steps are constant with $T_B = 100\,\text{ms}$. The coordinate system used is stationary for the duration of the evasion and is initialized at the beginning of the evasion to match the vehicle coordinate system at that point. The last position, which is supposed to be reached ten time steps in the future, is used to allow the controller to react predictively after deviations of the first two time steps. Every point $p(t)$ consists of the position $(x, y)$ and the heading of the vehicle. All three points are put together in one matrix transmitted to the controller:

$$\mathbf{P}_{\text{evasion}} = \left[p\,(t + T_{\text{B}})\;p\,(t + 2T_{\text{B}})\;p\,(t + 10T_{\text{B}})\right]. \tag{21.1.1}$$

The primary goal of the evasive trajectory is to reach a predefined lateral offset with the shortest possible traveled path such as using the road coordinate system. The vehicle dynamics and stability after the maneuver are taken into account.

Vehicle dynamics of the path are used to limit the maximum allowed lateral acceleration. This limit can be adapted to the actual traffic and driving situation and especially weather conditions. The steering actuator also limits the maximum possible jerk.

The designed evasive path is described by a simple parametric model

$$y = f(x). \tag{21.1.2}$$

The yaw angle $\psi$ can be expressed as (assuming no side slip)

$$\psi = \arctan\left(\frac{dy}{dx}\right) \tag{21.1.3}$$

and its derivative with regard to time

$$\dot{\psi} = \frac{d\psi}{dt} = \frac{1}{1 + \left(\frac{dy}{dx}\right)^2} \frac{d^2y}{dx^2} v_X. \tag{21.1.4}$$

Based on this and using the Ackermann relations for the kinematics, the lateral acceleration is as follows:

$$a_y = v\dot{\psi} = \frac{1}{1 + \left(\frac{dy}{dx}\right)^2} \frac{d^2y}{dx^2} v_X v. \tag{21.1.5}$$

Further simplification can be accomplished assuming $v_x = v$.

The reference evasive trajectory is described by a sigmoid of the form, see (20.3.1),

$$y(x) = \frac{y_M}{1 + e^{-a(x-c)}}, \tag{21.1.6}$$

where $y_M$ is the maneuver width, describing the distance between minimum and maximum $y$-value. $a$ defines the slope at $x = c$ of the sigmoid, and $c$ defines the position of the inflection point and therefore the length of the evasive maneuver, which is $s = 2c$; see Fig. 21.4.

The parameters of the sigmoid can be chosen according to the driving situation, such that the evasive path length $s$ is minimal regarding the limitations for maximum lateral acceleration, maximal jerk, and dynamics of the steering actuator solving nonlinear algebraic equations; see Stählin et al. (2006) and Stählin (2008).

### 21.1.3 Intervention Decision

Based on the fused environment data, it is decided if a collision is likely to occur and if so, which maneuver has to be carried out to avoid the collision. The strategy is to avoid the collision at the physically last possible moment by an intervention in order to give the driver the possibility to master the critical situation by his/her own actions as long as possible.

In order to determine a threatening collision, predictions are first made for the own vehicle driving tube and the movement of the objects in the environment. By

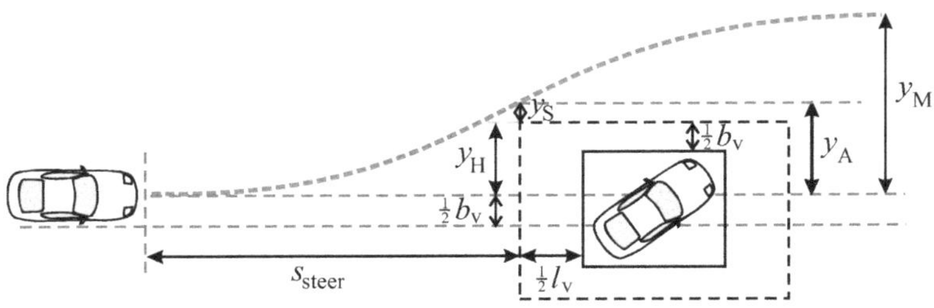


**Fig. 21.5** Evasive quantities for calculating the evasive trajectory (see text for details)

means of these predictions, it can then be predicted whether a collision will occur. If this is the case, it is planned in the next step when and which intervention has to be carried out. Basically, there are *three strategies* to avoid a collision: Braking, steering, or a combination of braking and steering.

For the intervention decision, it is calculated at what distance to the collision location the respective intervention has to be carried out, such that the collision can still be prevented. For a braking intervention, the braking distance is calculated. In the case of steering interventions, the sigmoid is taken as the basis for the evasive trajectory.

In Fig. 21.5, the quantities necessary for the calculation of the evasive trajectory are presented. By means of the vehicle's width $b_v$, the obstacle's width, and a safety distance $y_S$, the necessary evasive width $y_A$ is determined. Since the evasive width can be reached before the end of the swerving maneuver, a final maneuver width $y_M$ arises.

The required evasive trajectory is primarily the trajectory until the evasive width $y_A$ is reached. Under the condition that the next lane is free backward and forward, the maneuver width is chosen according to the strategy used. If the maneuver width $y_M$ is chosen to be the same as the evasive width $y_A$, the evasive trajectory length $s_{steer}$ reaches its maximum for given maximal lateral acceleration and maximal lateral jerk. However, the smallest possible evasive length $s_{steer}$ can be reached for the same acceleration and jerk limits for a larger maneuver width $y_M$ in front of the object. A further result is that for larger vehicle velocities, steering has to be preferred compared to braking to avoid an accident; see Stählin (2008).

## 21.1.4 Lateral Vehicle Guidance

If a collision with an obstacle is no longer avoidable by a reaction of the driver, then, depending on the situation, the driver assistance system selects one of the intervention strategies described above. For the realization of the chosen intervention, either the active steering and/or the electro-hydraulic braking system are used according to the maneuver. If a braking maneuver should be carried out, the vehicle is decelerated, Schorn et al. (2005), by utilization of the maximum force transmission available. The anti-lock braking system ABS supports in this case. In case a collision can only be prevented by an evasive maneuver or by a combined evasive and braking

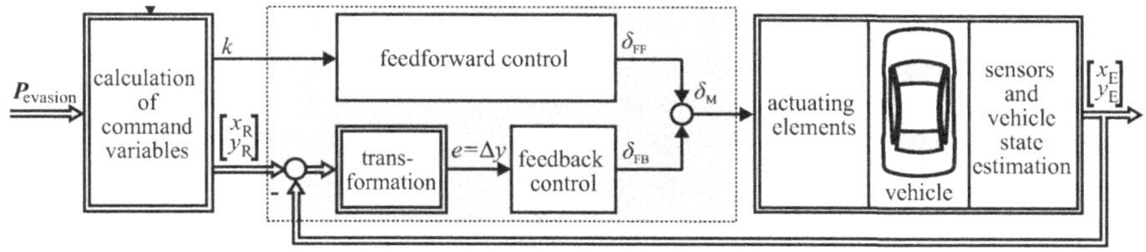


**Fig. 21.6** Structure of linear feedback control combined with feedforward action

maneuver, the control system receives a trajectory from the module intervention planning; see Fig. 21.2. The vehicle is driven automatically on this trajectory around the obstacle. Different linear and nonlinear feedforward and feedback controllers for an evasive maneuver were developed. Each lateral feedback control transfers an additional steering angle to the feedforward path; see Fig. 21.6. Vehicle variables, which cannot be measured directly by onboard sensors, are estimated; see Chap. 12 and also Schorn and Isermann (2006), Schorn et al. (2006), and Schorn (2007). For combined steering and braking maneuvers, different feedback controllers were developed as well.

In the following, only the lateral vehicle guidance is regarded. Exemplarily, one of the investigated approaches, a *speed-dependent local linear feedback control approach with feedforward control*, is briefly described.

Based on the self steer gradient SG, a steering angle $\delta_{\text{ff}}$ is calculated for the feedforward control by means of vehicle velocity $v$, wheelbase $l$, and curvature $\kappa = \frac{1}{R}$ of the desired trajectory, see (20.2.9),

$$\delta_{\text{ff}} = \left(l + \text{SG}v^2\right)\kappa. \tag{21.1.7}$$

A feedback control is added to compensate for disturbances and deviations. The parameters of a proportional-derivative (PD) controller are tuned by two parameters only and provide the required dynamics. Using the vehicle orientation $\psi$ and longitudinal position $x_{\text{E}}$, the control deviation of the path is transformed from an earth-fixed coordinate system into a vehicle-fixed coordinate system as $e = \Delta y$. The desired steering angle $\delta_{\text{M}}$ is given by the sum of the angles $\delta_{\text{ff}}$ and $\delta_{fb}$ of the feedforward and feedback control and is given to the active steering system.

As the velocity $v$ influences the vehicle's dynamics, the feedback controllers were designed for different operating points (velocities). Their outputs are weighted and superimposed based on local linear models (LLM), leading to a speed-dependent feedback control; see Sect. 20.3.

### 21.1.5 Experimental Results from Test Drives

The developed components for environment recognition, intervention decision, and feedback control were implemented in a research vehicle and tested by means of numerous experiments. This happened using a dummy obstacle that represents the

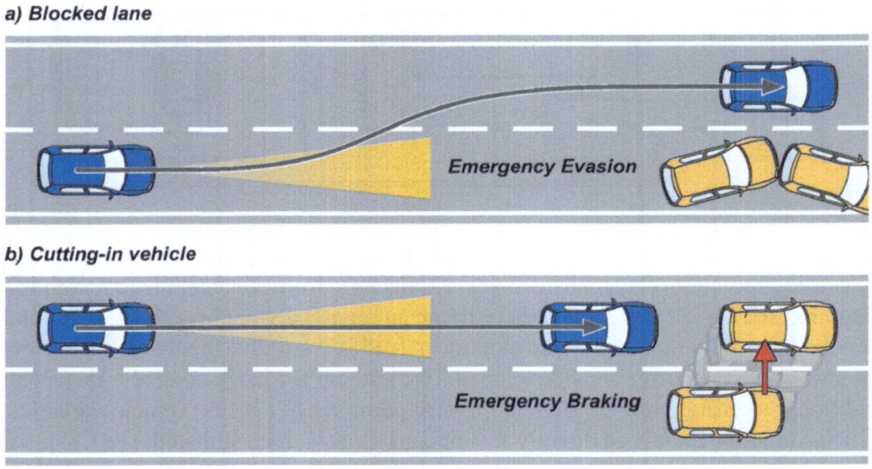


**Fig. 21.7** Scenarios for practical system testing: **a** blocked lane and **b** cutting-in vehicle

rear view of a car and can be moved laterally on the lane. Two test scenarios can be seen in Fig. 21.7. In the following sections, some results from these tests are presented. It is required in each case that the lateral and backward lane areas are monitored by additional sensors (e.g. RADAR) and thus permit driving maneuvers.

### (a) Blocked Lane: Emergency Steering and Braking

In the scenario "Suddenly appearing obstacle/blocked lane" from Fig. 21.7a, a lane is blocked unexpectedly for the driver, such as adding the reaction time of the driver, a critical situation develops. An example of this would be an end of a traffic jam in the case of bad visibility or after a curve. The emergency evasion is then conducted as an automatic intervention. The position of the used obstacle is determined by the environmental sensors, and the necessary evasive trajectory is calculated based on the information about the vehicle's surroundings. The vehicle is then guided aside from the obstacle on the predefined evasive trajectory by the lateral guidance controller without any assistance of the driver.

Figure 21.8 shows the results of a test drive with the linear feedback control combined with feedforward control on a dry road.

A comparison of desired command variable and measured position shows that both values match very well. The evasive width $y_M$ is 3 m, the desired and the actual position correspond well, apart from a slight overshooting. The steering wheel angle indicates that the driver held the steering wheel in a straight position. The difference between the total angle and steering wheel angle is provided only by the controller. The difference at the end of the intervention maneuver follows from the fact that the feedback control has been switched off at very low velocities. The experiments show that the maximal lateral acceleration was $|a_y| \approx 7\ \text{m/s}^2$ .

A combined emergency steering and braking under the condition that the tire forces stay within Kamm's circle is shown by Ackermann (2016) with simulations.

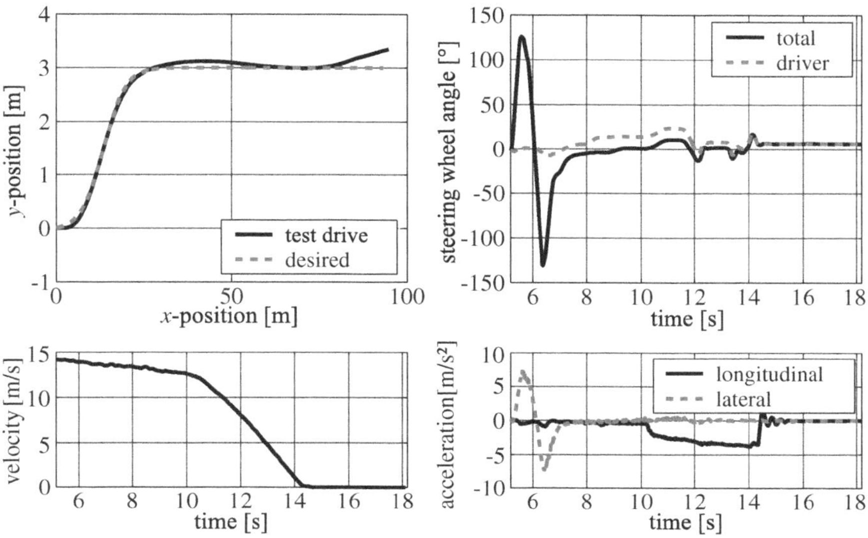


**Fig. 21.8** Experimental driving results of a swerving maneuver with braking for velocity-dependent linear feedback control combined with feedforward

### (b) Cutting-In Vehicle: Emergency Braking

As a second scenario, a suddenly cutting-in vehicle in close distance to the car is reproduced by moving the dummy obstacle just in front of the vehicle from the right to the left lane; see Fig. 21.7b. Evasion is not possible since further obstacles block the right lane. The necessary intervention is thus an emergency-braking maneuver. By means of the environmental sensors, it is recognized that both lanes of the road are blocked and it is calculated at which last possible moment the emergency-braking maneuver must be started in order to come to a stop just before the obstacle. Assuming a maximum braking acceleration which is dependent on the road state (dry/wet), the required braking distance of the vehicle is calculated depending on the current speed. The driver-assistance system triggers a braking intervention only if this minimal braking distance is reached in order to give the driver the chance to prevent the collision as long as possible by himself. The electro-hydraulic braking system then decelerates the vehicle maximally with support by the anti-lock braking system ABS, on dry roads with a deceleration of $a_s \approx 7\ \mathrm{m/s^2}$. The closed-loop braking control with pole placement designed PI controller and practical results are described in Schorn et al. (2005); see also Sect. 19.1.

## 21.2 Collision-Avoidance System for Overtaking Maneuvers and Oncoming Traffic

Accident situations during overtaking maneuvers on rural roads are frequently the cause of severe injuries and fatalities. Accidents may originate in an erroneous situa-

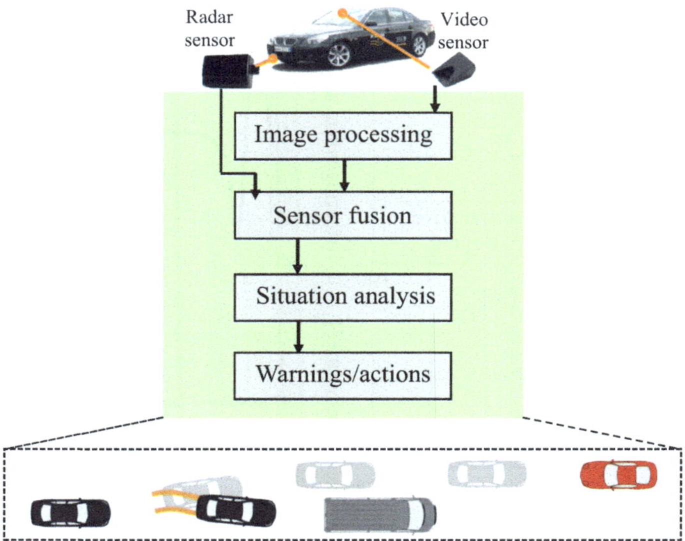


**Fig. 21.9** Overview for the development of PRORETA 2

tion assessment by the driver, e.g. misjudgment of distance and velocity of oncoming vehicles. Therefore, the second PRORETA project developed a driver-assistance system to avoid overtaking accidents.

Figure 21.9 depicts a system overview for the development of the PRORETA 2 system. A monocamera system and a far-range RADAR sensor, both series components, scan the environment in front of the own vehicle. The video data is processed via pixel-based segmentation and a combination of filterbank outputs and object detection. They allow a classification of the picture contents, like objects, road, land markings, free space, heaven, etc. Detected vehicles from the camera and the RADAR sensor are fed into a sensor-fusion system applying an extended Kalman filter with yaw rate compensation, such as providing an environment model; see Wojek et al. (2008), Hohm et al. (2008), and Isermann et al. (2009).

The situation analysis has the task to detect dangerous overtaking maneuvers. Then, warnings and active braking commands are given to avoid an accident.

In the following, the situation analysis during overtaking maneuvers on two-lane rural roads, the prediction of overtaking maneuvers, warnings, and emergency braking is briefly described. The camera-based data processing algorithms are published in Wojek et al. (2008) and the multilevel sensor fusion of camera and RADAR data in Hohm et al. (2008).

An overtaking maneuver by using an extended Kalman filter is automatically detected if the own vehicle (A) enters the opposite lane with the intention to pass a leading vehicle (B); see Fig. 21.10. When having catched up with vehicle B, the actual passing takes place. The subsequent cutting-in phase begins with entering the

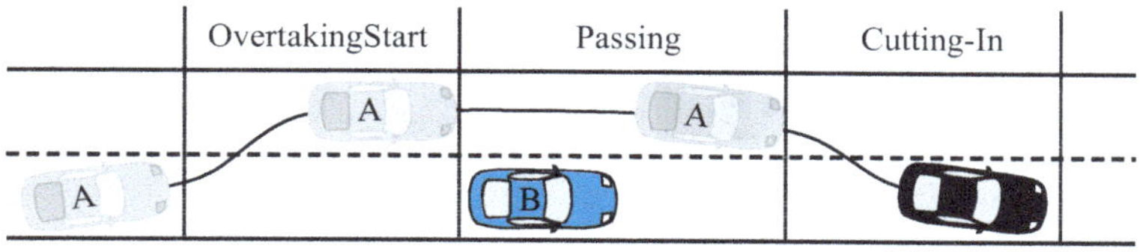


**Fig. 21.10** Phases of overtaking maneuvers

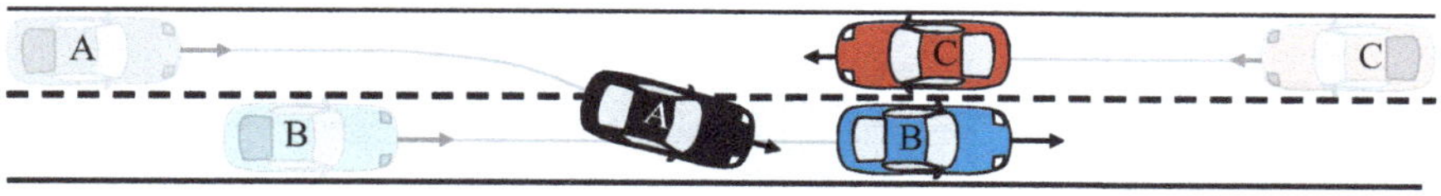


**Fig. 21.11** Objective of the PRORETA 2 System: Abortion of hazardous overtaking maneuvers

own lane again. The overtaking maneuver is completed when the vehicle has fully left the opposite lane.

The PRORETA 2 system focuses on situations in which the forward road and potential oncoming traffic is viewable by the sensors. Thus, primarily overtaking maneuvers on straight roads or only slightly bent road geometries are addressed. The developed driver-assistance system aims at warning the driver in overtaking maneuvers, which cannot be performed without a conflict with an oncoming vehicle. The objective is to prompt the driver to abort the maneuver already during overtaking start or in an early stage of the passing phase. If there is too little time to react, or the driver doesn't react for other reasons, the system initiates an automatic braking intervention, which allows the driver to drive back behind vehicle B before the oncoming vehicle C is reached; see Fig. 21.11.

### 21.2.1 The Research Vehicle

The development of the driver-assistance system for overtaking maneuvers was based on many experimental studies with the implemented camera and RADAR system on normal rural roads and traffic situations and overtaking experiments with three vehicles on the runway of the University-owned airfield. The used vehicle is a BMW 540i, see Fig. 21.12, which is equipped with the market-available camera system CSF200 and long-range RADAR ARS300, both from Continental. The RADAR sensor operates at 77 GHz and provides distance, relative velocity, and azimuth angle of target objects in a 66 ms cycle. The short-range scan of the RADAR series sensor has a detection range of $\pm 28°$ up to a distance of 60 m and the long-range scan provides object detections in a range of $\pm 8.5°$ up to 200 m. To meet the distance requirements imposed by overtaking situations, the detection range of the RADAR sensor used in the PRORETA 2 project has been extended to 400 m by firmware adaption; see Winner et al. (2009) and Hohm et al. (2008).

The CSF200 camera is based on a CMOS color sensor with $752 \times 480$ pixels and provides image data every 60 ms. The horizontal field of view is $\pm 18°$ and image-

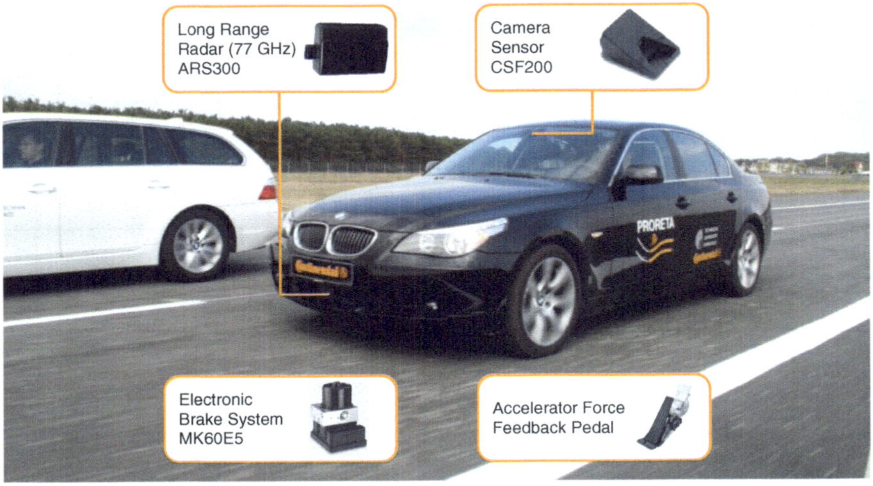


**Fig. 21.12** Research vehicle and components

processing algorithms can reliably detect vehicles up to a distance of 70 m. The camera image is also used to detect lane markings and determine the ego vehicle's position within the lane.

The RADAR sensor has strong longitudinal capabilities, while the camera sensor neither provides direct distance measurement nor velocity measurement, but has a good lateral resolution. Therefore, these sensors are used for complementary sensor data fusion. In the PRORETA 2 project, the sensor fusion is accomplished in an object-tracking algorithm, based on an extended Kalman filter (EKF); see Hohm et al. (2008). An accelerator force feedback pedal (AFFP) is installed to give haptic warnings to the driver, and an electronically controlled brake system allows automatic braking intervention.

## 21.2.2 Overtaking-Maneuver Detection

To assist the driver in dangerous overtaking situations, it is necessary to detect that an overtaking maneuver is being conducted. Since driving maneuvers are primarily defined by the vehicle movements along and lateral to the road, in the first step, the vehicle's position and orientation are determined. Based on signals from vehicle-dynamics sensors and a camera-based lane-detection system, the position, velocity, and orientation with respect to the road are estimated in an odometry module. The state estimation is accomplished by coupling a one-track vehicle model and a road model in an extended Kalman filter (EKF); see Fig. 21.13 and Schmitt et al. (2005).

While the vehicle-dynamics sensors continuously provide measurements, the lane-detection system switches between the right and left lane during overtaking maneuvers and is temporarily unavailable due to dynamic movements of the vehicle.

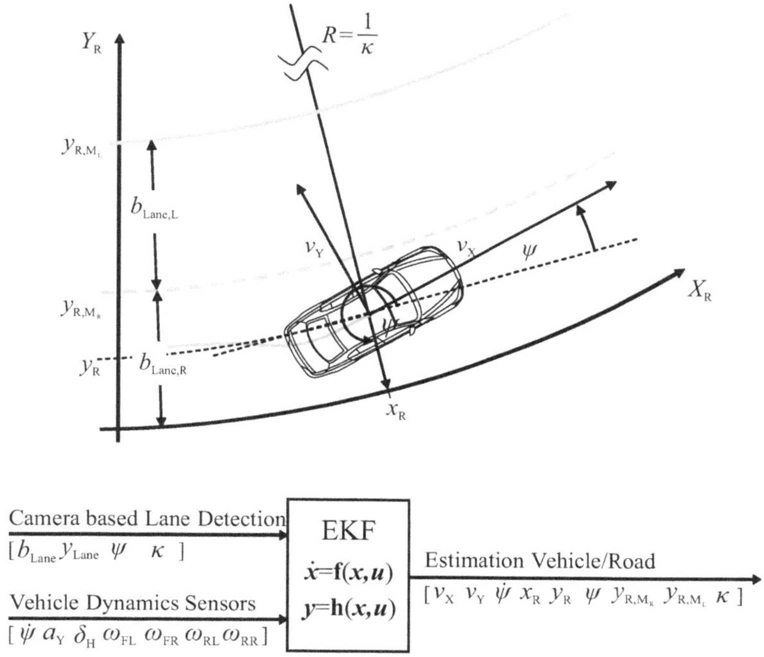


**Fig. 21.13** Vehicle/road model for state estimation with an extended Kalman filter for the odometry module

The observation model of the Kalman filter is dynamically adapted, depending on whether the left lane or the right lane is detected by the lane-detection system. This procedure allows a lane spanning ego localization, and temporary breakdowns of the lane detection can be bridged. The output of the EKF in the odometry module is an estimation of the state vector

$$x = \left( v_X \; v_Y \; \dot{\psi} \; x_R \; y_R \; \psi \; y_{R,M_R} \; y_{R,M_L} \; \kappa \right)^T ; \qquad (21.2.1)$$

compare Fig. 21.13. Based on the state estimates from the odometry module and environment-sensor data for the leading vehicle (B), longitudinal and lateral indicator variables are calculated. These indicator variables are used to detect the overtaking maneuver and comprise sub-maneuvers. Maneuvers like "following", "overtaking start", "passing", "cutting in", or "overtaking abort" are modeled in a state diagram and the transitions between the maneuvers are modeled depending on the indicator variables. For details, see Schmitt et al. (2009) and Schmitt and Isermann (2009).

In order to warn the driver in an early stage of critical overtaking maneuvers, an early detection of the overtaking start has been realized. To capture the initial lane change during overtaking start, the time to line crossing (TLC) is computed. The TLC is used in lane departure warning systems and predicts the time duration until the vehicle crosses the center line of the road; see Mammar et al. (2006) and Sect. 17.3.

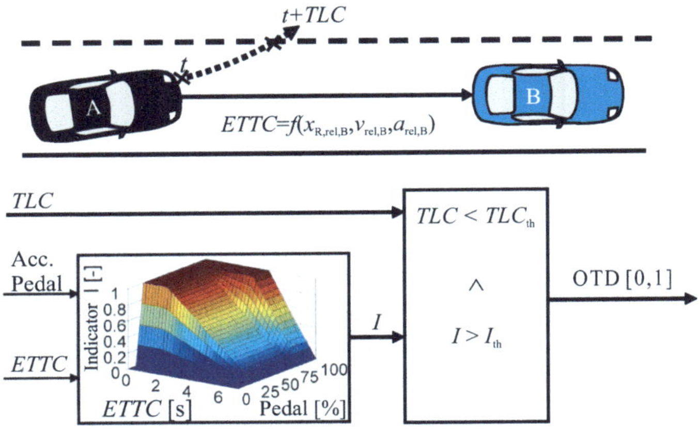


**Fig. 21.14** Overtaking detection ($OTD$) based on indicator variables. $I$: longitudinal overtaking indicator

By comparing the TLC with a threshold value, lane changes can be detected before the vehicle crosses the center line.

For reliable detection of overtaking maneuvers, also longitudinal characteristics of the lane change have to be taken into account. If the own vehicle is approaching vehicle B in a short distance, an overtaking maneuver is more likely than in cases where the own vehicle is falling back and/or the distance to vehicle B is high. Therefore, distance, relative velocity, and relative acceleration to vehicle B are considered in the overtaking detection. These three variables are aggregated in a single, well-interpretable measure by the enhanced time to collision (ETTC). The ETTC is a variant of the ordinary time to collision, but includes the relative acceleration; see Winner et al. (2009). It is given as

$$ETTC_{\mathrm{AB}} = \frac{2x_{\mathrm{R,rel,AB}}}{-v_{\mathrm{rel,AB}} + \sqrt{v_{\mathrm{rel,AB}}^2 - 2x_{\mathrm{R,rel,AB}}a_{\mathrm{rel,AB}}}}. \tag{21.2.2}$$

The $ETTC_{\mathrm{AB}}$ drops when starting accelerated overtaking maneuvers as well as in flying overtakings, in which the relative velocity is already high when approaching vehicle B.

To improve the reliability of overtaking detection, $ETTC_{\mathrm{AB}}$ is fused with the accelerator pedal position in a fuzzy-logic system; see Fig. 21.14. The resulting longitudinal overtaking indicator $I$ covers the approach of the vehicles as well as the driver's intention.

The system detects an overtaking start ($OTD = 1$), if the time to line crossing as well as the longitudinal overtaking indicator reach corresponding threshold values; see Fig. 21.14.

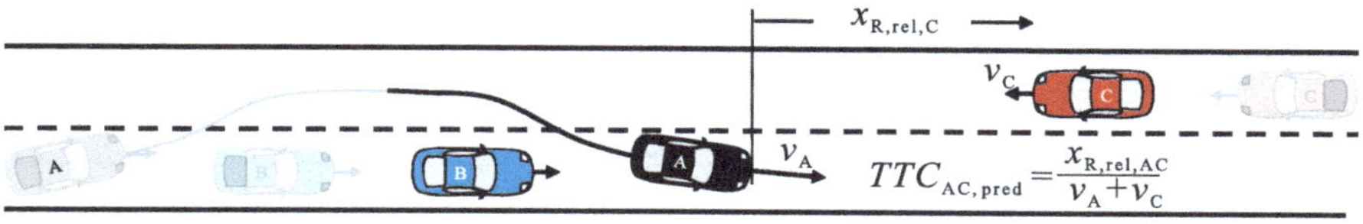


**Fig. 21.15** Predicted time to collision at the end of overtaking maneuver as a measure for the safety distance

### 21.2.3 Overtaking Prediction and Threat Analysis

If an overtaking situation is detected, it is continuously assessed, whether the maneuver can be conducted or completed without a conflict with the oncoming traffic. If an overtaking maneuver has already been started, the actual acceleration of vehicle A is taken into account. For the instant of time, at which vehicle A has left the opposite lane after overtaking, the time to collision ($TTC_{\text{AC,pred}}$) with respect to vehicle C is predicted; see Fig. 21.15.

$$TTC_{AC,pred} = \frac{x_{\text{R,rel,AC}}}{v_a + v_c}. \tag{21.2.3}$$

$TTC_{\text{AC,pred}}$ indicates the magnitude of the safety distance between the own vehicle (A) and the oncoming vehicle (C) while completing the overtaking maneuver. It is a continuous measure and can be interpreted independently from the speed range of the involved vehicles. Based on $TTC_{\text{AC,pred}}$, the distance to the oncoming traffic can already be assessed when beginning the overtaking maneuver. If the predicted $TTC$ is lower than a threshold, e.g. 2 s, vehicle C is too near and the overtaking maneuver should be refrained or aborted.

### 21.2.4 Warnings and Emergency Braking

In case of a dangerous overtaking maneuver, a collision-avoidance strategy is initialized, which aims at aborting the overtaking maneuver and driving back behind vehicle B. Besides different warnings, an automatic braking intervention is realized in order to compensate the driver's reaction time in case of a suddenly occurring threat. The warnings are triggered as soon as a dangerous overtaking maneuver is detected, while a small brake pressure is applied, which does not decelerate the vehicle, but prepares the braking system for a later braking intervention. This preconditioning of the brake system is a common measure in practice; see Winner et al. (2009).

For the prototype system, the braking intervention is designed as the latest possible and strong deceleration $a_b$ up to 10 m/s$^2$, while in practice, emergency braking is usually performed with 6 m/s$^2$ emergency braking, in order to cover also wet road conditions; see Winner et al. (2016).

The braking intervention is triggered when the available time, given by the time gap between vehicle B and the oncoming vehicle C, is just sufficient for a collision avoiding braking and a subsequent lane change; compare Fig. 21.16. To slow down

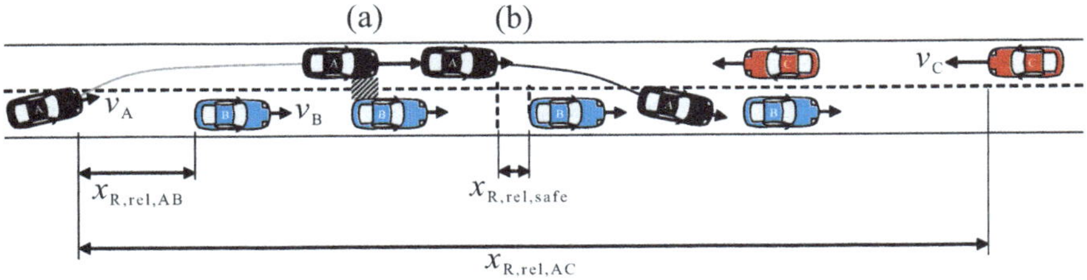


**Fig. 21.16** Relative distances during emergency braking and subsequent lane change. **a** $x_{R,rel,AB} < x_{R,rel,safe}$; **b** $x_{R,rel,AB} = x_{R,rel,safe}$

the own vehicle (A) to the speed of vehicle B, the required braking duration is given by

$$\hat{\tau}_{b,min}(t) = \frac{|v_{rel,AB}(t)|}{a_b}. \tag{21.2.4}$$

A corresponding prediction for the distance at this instant of time, which will reach a minimum, is given by

$$\hat{x}_{R,rel,AB}(t + \hat{\tau}_{b,min}(t)) = x_{R,rel,AB}(t) - \frac{1}{2} a_b \hat{\tau}^2_{b,min}(t). \tag{21.2.5}$$

If the braking intervention is triggered at a late instant of time, so that the minimal distance given by Eq. (21.2.5) violates the safety distance $x_{rel,safe}$, an additional braking period, given by

$$\hat{\tau}_b^+(t) = \begin{cases} 0, & \text{for } \hat{x}_{R,rel,AB}(t + \hat{\tau}_{b,min}(t)) \geq x_{R,rel,safe} \\ \sqrt{\frac{2(x_{R,rel,safe} - \hat{x}_{R,rel,AB}(t + \hat{\tau}_{b,min}(t)))}{a_b}}, & \text{else,} \end{cases} \tag{21.2.6}$$

is necessary in order to gain the safety distance between A and B.

The totally required time for the maneuver is modeled by

$$\hat{\tau}_{req} = \hat{\tau}_{b,min}(t) + \hat{\tau}_b^+(t) + \tau_{LC}, \tag{21.2.7}$$

where the time reserve $\tau_{LC}$ for the subsequent lane change maneuver is set to 2.5 s, which can be seen as a lower TTC limit for a comfortable lane change; see Winner et al. (2016).

The available time for the collision avoiding maneuver is calculated by considering the time gap between vehicle B and the oncoming vehicle (C), which leads to the time to collision-based estimate

$$\hat{\tau}_{avail} = \frac{x_{R,rel,AC} - x_{R,rel,AB}}{|v_{rel,AC}| - |v_{rel,AB}|} = \frac{x_{R,BC}}{v_B + v_C}. \tag{21.2.8}$$

Finally, the trigger condition for emergency braking is given by

$$\hat{\tau}_{avail} \leq \tau_{req}. \tag{21.2.9}$$

The trigger point thus depends on the relative kinematic situation of all three vehicles.

The braking intervention is realized by feedforward control of the braking acceleration, while the set point of the feedforward control is adapted to the accelerator pedal position at the trigger point. For more detail, see Mannale et al. (2008) and Schmitt et al. (2010).

### 21.2.5 Driving Experiments

Figure 21.17 illustrates a driving experiment, which simulates a dangerous overtaking maneuver. The system detects the beginning of an overtaking maneuver at $t \approx 4.3\,\mathrm{s}$ and the appearing oncoming vehicle at $t \approx 4.8\,\mathrm{s}$. Consequently, the system starts continuously predicting the time to collision at the end of the overtaking maneuver $TTC_{\mathrm{AC,pred}}$. As $TTC_{\mathrm{AC,pred}}$ is beyond the corresponding warning threshold, the system warns the driver and pre-fills the braking system in order to prepare a subsequent abort of the overtaking maneuver. Once the trigger condition for emergency braking is fulfilled at $t \approx 8\,\mathrm{s}$, the system commands the full emergency braking pressure until the cut-in conditions are met at $t \approx 10\,\mathrm{s}$, so that the driver can steer back behind the leading vehicle. At $t \approx 11.5\,\mathrm{s}$, the system detects the completion of the abort maneuver and consequently stops the warning.

### 21.2.6 Conclusions

The described system for accident avoidance for vehicles in the same direction or with obstacles which was developed within project PRORETA 1 was presented to an audience selected by Continental Teves and TU Darmstadt in 2006.

The second part presented the situation analysis and collision-avoidance strategies within the PRORETA 2 driver-assistance system for overtaking situations. Test drives with an experimental vehicle show that, based on data of environment sensors and vehicle-dynamics sensors, hazardous overtaking maneuvers can be detected so that the driver can be warned in an early stage of the overtaking maneuver. If an immediate reaction is required, an automatic braking intervention initiates a timely abort of the overtaking maneuver, which allows the driver to return back behind the leading vehicle. Also, this overtaking driver-assistance system was presented to a selected audience in 2009.

Both projects are prototype realizations for new collision-avoidance systems and can serve as examples for new developments. Future research should consider more complex traffic situations, e.g. with more vehicles and more driving lanes. Furthermore, a 360 degree environment perception is desirable, which allows to consider the sideward and backward traffic. This would also allow the integration of further functionalities like an active steering support for cutting in after break-off of the overtaking maneuver.

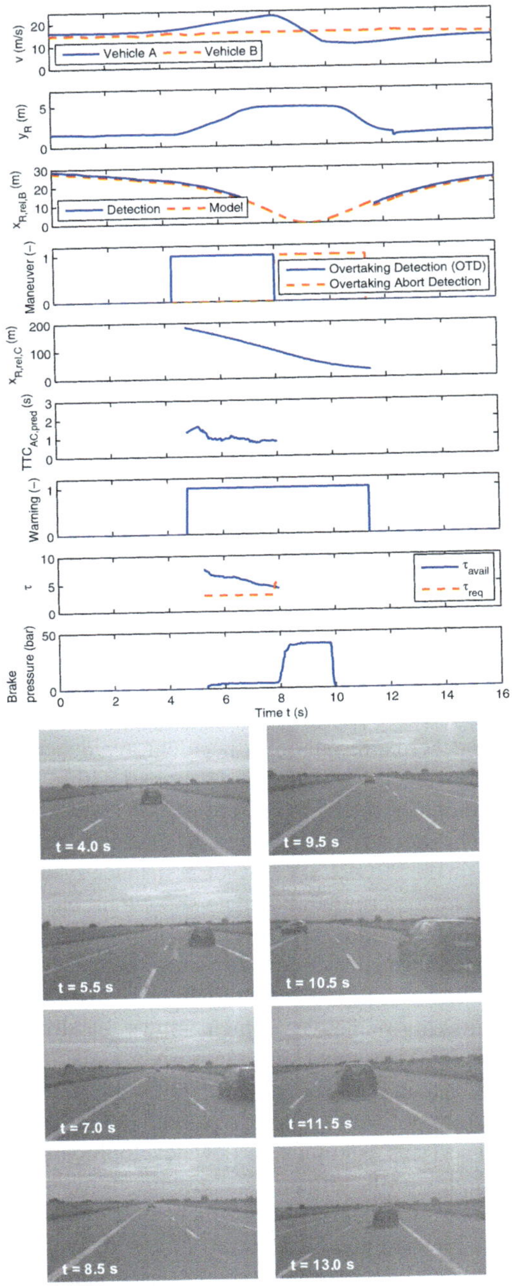


**Fig. 21.17** Results of a driving experiment with a dangerous overtaking maneuver and active overtaking assistant functions

The name PRORETA comes from the Roman world. In times of the Roman galleys, PRORETA was the boatswain standing at the bow of the ship warning the Gubernator of shallows. Following this idea, the driver-assistance system PRORETA monitors the surroundings of the vehicle and intervenes in emergency situations to prevent an accident.

## References

Ackermann C (2016) Neue Ansätze für Fahrerassistenzsysteme mit Regelung der Längs- und Querdynamik. Dissertation Technische Universität Darmstadt. Fortschr.-Ber. VDI Reihe 8, 1249. VDI Verlag, Düsseldorf

Ameling C (2002) Steigerung der aktiven Sicherheit von Kraftfahrzeugen durch ein Kollisionsvermeidungssystem. 1. Auflage. Fortschritt-Berichte VDI: Reihe 12 Nr. 510, VDI-Verlag, Düsseldorf

Anderson S, Peters S, Pilutti T, Iagnemma K (2010) An optimal-control-based framework for trajectory planning, threat assessment, and semi-autonomous control of passenger vehicles in hazard avoidance scenarios. Int J Veh Auton Syst 8(2–4):190–216

Bender E (2008) Handlungen und Subjektivurteile von Kraftfahrzeugführern bei automatischen Brems- und Lenkeingriffen eines Unterstützungssystems zur Kollisionsvermeidung, Dissertation TU Darmstadt, Ergonomia Verlag

Bender E, Darms M, Schorn U M andStählin, Isermann R, Winner H, Landau K (2007) Anti collision system PRORETA - on the way to the collision avoiding vehicle. part 1: Basics of the system, part 2: Results. ATZ - Automobiltechnik 4 & 5:20–23 & 32–35

Braess H, Reichart G (1995) Prometheus: Vision des intelligenten Automobils. Automobiltechnische Zeitschrift 97:200–205 and 330–343

Darms M (2007) Eine Basis-Systemarchitektur zur Sensordatenfusion von Umfeldsensoren für Fahrerassistenzsysteme. Diss. TU Darmstadt, Fortschr.-Ber. VDI Reihe 12, 653, VDI-Verlag

Darms M, Winner H (2006) Umfelderfassung für ein Fahrerassistenzsystem zur Unfallvermeidung. In: AUTOREG 2006, VDI-Verlag

Destatis (2010) Verkehrsunfälle. Statistisches Bundesamt

Hohm A (2010) Umfeldklassifikation und Identifikation von Einscherlücken für ein Überholassistenzsystem. Fortschr.-Ber. VDI Reihe 12, 727, VDI-Verlag

Hohm A, Wojek C, Schiele B, Winner H (2008) Multi-level sensorfusion and computer-vision algorithms within a driver assistance system for avoiding overtaking accidents. In: FISITA world automotive congress

Isermann R, Schorn M, Stählin U (2008) Anticollision system PRORETA with automatic braking and steering. Veh Syst Dyn 46(Supplement):683–694

Isermann R, Schiele B, Winner H, Hohm A, Mannale R, Schmitt K, Wojek C, Lüke S (2009) Elektronische Fahrerassistenz zur Vermeidung von Überholunfällen - PRORETA 2. In: Elektronik im Kraftfahrzeug, VDI-Berichte Nr. 2075, VDI

Isermann R, Mannale R, Schmitt K (2012) Collision-avoidance systems PRORETA: situation analysis and intervention control. In: Control engineering practice. http://dx.doi.org/10.1016/j.conengprac.2012.06.003

Kirchner A, Krüger K, Mildner F, Schmidt R (2005) Ein fortgeschrittenes Kollisionsvermeidungssystem. ATZ - Automobiltechnische Zeitschrift 107:60–67

Kopischke S (2000) Entwicklung einer Notbremsfunktion mit Rapid Prototyping Methoden. Diss. TU Braunschweig, Gemeinsame Fakultät für Maschinenbau und Elektrotechnik der Technischen Universität Carolo-Wilhelmina zu Braunschweig

Lages SU (2001) Untersuchungen zur aktiven Unfallvermeidung von Kraftfahrzeugen. 1. Auflage. Fortschritt-Bericht, Reihe 12, Nr. 446, VDI-Verlag, Düsseldorf

Liu C, Tomizuka M (2016) Enabling safe freeway driving for automated vehicles. In: American control conference (ACC), Boston, pp 3461–3467

Mammar S, Glaser S, Netto M (2006) Time to line crossing for lane departure avoidance: a theoretical study and an experimental setting. IEEE Trans Intell Transp Syst 7(2):226–241

Mannale R, Hohm A, Schmitt K, Isermann R, Winner H (2008) Ansatzpunkte für ein System zur Fahrerassistenz in Überholsituationen. In: 3. Tagung Aktive Sicherheit durch Fahrerassistenz, Garching, Germany

Mildner F (2004) Untersuchungen zur Erkennung und Vermeidung von Unfällen für Kraftfahrzeuge. Hamburg, Helmut-Schmidt-Universität/Universität der Bundeswehr Hamburg

Schmitt J, Isermann R, Börner M, Fischer D (2005) Model-based supervision and control of lateral vehicle dynamics. In: Control engineering practice – CEP

Schmitt K (2012) Situationsanalyse für ein Fahrerassistenzsystem zur Vermeidung von Überholunfällen auf Landstraßen. Diss. Univ. Darmstadt, Fortschritt Bericht VDI, Reihe 12 Nr.763

Schmitt K, Isermann R (2009) Vehicle state estimation in curved road coordinates for a driver assistance system for overtaking situations. In: 21st international symposium on dynamics of vehicles on roads and tracks (IAVSD), Stockholm, Sweden

Schmitt K, Habenicht S, Isermann R (2009) Odometrie und Manövererkennung für ein Fahrerassistenzsystem für überholsituationen. In: 1. Automobiltechnische Kolloquium, München, Germany

Schmitt K, Mannale R, Isermann R (2010) Collision avoidance system PRORETA for overtaking maneuvers - automatic situation analysis, warnings and emergency braking. In: IFAC symposium advances in automotive control

Schorn M (2007) Quer- und Längsregelung eines Personenkraftwagens für ein Fahrerassistenzsystem zur Unfallvermeidung. Diss. Universität Darmstadt, Fortschr.-Ber. VDI Reihe 12, 651. VDI Verlag, Düsseldorf

Schorn M, Isermann R (2006) Automatic steering and braking for a collision avoiding vehicle. In: 4th IFAC symposium on mechatronic systems. Wiesloch/Heidelberg

Schorn M, Schmitt J, Stählin U, Isermann R (2005) Model-based braking control with support by active steering. In: Proceedings of the 16th IFAC world congress 2005, Prague, Czech Republic

Schorn M, Stählin U, Khanafer A, Isermann R (2006) Nonlinear trajectory following control for automatic steering of a collision avoiding vehicle. In: American control conference, Minneapolis, Minnesota, USA

Stählin U (2008) Eingriffsentscheidung für ein Fahrerassistenzsystem zur Unfallvermeidung. Diss. TU Darmstadt, Fortschr.-Ber. VDI Reihe 12, 683. VDI Verlag, Düsseldorf, Düsseldorf

Stählin U, Schorn M, Isermann R (2006) Notausweichen für ein fahrerassistenzsystem zur unfallvermeidung. In: AUTOREG 2006, VDI-Verlag

Weisen R (2003) Gekoppelte Quer- und Längsregelung eines Personenkraftwagens im fahrphysikalischen Grenzbereich. Fachbereich Maschinenbau an der Helmut-Schmidt-Universität/Universität der Bundeswehr Hamburg

Winner H, Hakuli S, Wolf G (eds) (2009) Handbuch Fahrerassistenzsysteme. Vieweg+Teubner Verlag

Winner H, Hakuli S, Lotz F, Singer C (eds) (2016) Handbook of driver assistance systems. Springer International Publishing AG, Cham, Switzerland

Wojek C (2010) Monocular Visual Scene Understanding from Mobile Platforms. Diss, TU Darmstadt, FB Informatik

Wojek C, Dorkó G, Schulz A, Schiele B (2008) Sliding-windows for rapid object-class localization: a parallel technique. In: 30th DAGM symposium 2008, Springer, Berlin

# 22 Automatic (Autonomous) Driving

The driver assistance systems described in Chaps. 17 (TCS, ESC, LKA) and 19 (ACC) assist and control the vehicle in special cases and the driver performs continuously the driving task. These *driver assistance systems* (DAS, ADAS) are dedicated to level 1 of the automatic driving degrees in Table 18.1. For vehicles with *partial automation*, level 2, tasks of longitudinal and lateral driving are automatic (ACC-FR, LKA, LKC) but the driver has to be attentive all the time, to monitor the correct functioning, and to perform all remaining tasks. The situation changes for *conditional automation*, level 3, where longitudinal and lateral driving is done automatically for defined use cases and the driver has not to monitor the vehicle all the time. Because the driver now has to be available only on request after a few seconds, the requirements on the correct execution of the driving tasks, the reliability, and safety of the automatic control increase.

In the case of *high automation*, level 4, the longitudinal and lateral driving is realized automatically in all cases and the driver is not required in these defined use cases and therefore has not to be available as fallback (in a short time). Latest now, the requirements on reliability and safety of the automation increase significantly. Therefore, redundancies for sensors, actuators, ECU's, the communication channels, and the board voltage net have to be implemented.

All requirements then increase further for *full automation*, level 5, where all driving tasks are performed automatically and a driver is not required anymore.

Instead of the term "Automatic Driving", also "Autonomous Driving" is used. In the field of Automatic Control or Control Engineering, the word "automatic control" is used for the replacement of manual control by a human operator with a mechanical, hydraulic, electrical, electronic, or digital computer controller. Especially in the field of process computer control or digital automation, different levels of automation, from low level feedforward and feedback control through adaptive control, to higher levels as process optimization, monitoring, and scheduling are distinguished and



R. Isermann, *Automotive Control*, ATZ/MTZ-Fachbuch,
https://doi.org/10.1007/978-3-642-39440-9_22

represented in an “automation pyramide”. This is also called “multilevel process automation”.

On the other side, “autonomy” means “being independent, free, self-directing, or designating itself”, Webster (1981). It “implies satisfactory performance under significant uncertainties in the environment and the ability to compensate for system failures without external intervention”; see Wikipedia: “Self-driving car” (2020). Hence, the term autonomous driving may be suitable for the highest level “Full automatic driving” when no driver is required.

## 22.1 On the Historic Development of Automatic Driving

Some of the first approaches to automate highway driving around 1950 by General Motors are published in Fenton (1970). Pioneering research projects were performed by Dickmanns with camera-based driving on test fields and on an Autobahn in 1986 and later in 1995 an automatic drive over 1700 km with speeds until 180 km/h on Autobahns, see Zapp (1988), Dickmanns et al. (1990, 1944), and Dickmanns (2007). Furthermore, automatic driving vehicles were developed within the European PROMETHEUS research program between 1987 and 1994; see, e.g. Ulmer (1992), Franke et al. (1994), and Naab (2000). A report on an automatic steering in the U.S. on highways is given in Pomerleau and Jochem (1996). Other automated vehicles were developed within the California PATH-program, see Thorpe et al. (1997) and Shladover (2007). Long-distance automatic drives in Italy are described in Broggi et al. (1999, 2013).

The development of automatic driving was then supported by the US-programs DARPA (Defense Advanced Research Projects Agency) with the goal to demonstrate driverless vehicles for military use. The first DARPA Ground Challenge was to drive unmanned through a desert in 2004, however without success. The second Ground Challenge in 2005 led to many vehicles reaching the goal, see Thrun et al. (2006) and Singh (2006). DARPA then invited teams for the 2007 Urban challenge for automatic driving through an Urban-like environment with streets, buildings, and other cars. Six from 11 automatic driving vehicles finished this challenge successfully, using cameras, lidar and radar sensors, satellite GPS navigation, and digital maps. Then many other research projects have further developed automatic driving.

An outstanding industrial demonstration was then the Bertha Benz autonomous drive with a Mercedes-Benz S-class S 500 experimental vehicle about 100 km overland with urban areas, cities, traffic lights, intersection, roundabouts, etc. within normal traffic. The drive was on the historic Bertha Benz first cross-country automobile route in 1888 in southern Germany, from Mannheim downtown to Pforzheim. Except solving the challenge of real traffic, a goal was to use only close-to-market sensors. Hence, the standard sensor systems of the 2013 car were used and six additional radar and two vision sensors for object detection, free-space analysis, traffic light detection, and self-localization have been implemented and a digital map was created. Detailed reports on this successful event are Ziegler et al. (2014) and Dang et al. (2015).

## 22.2 On the Development of Automatic Driving Vehicles

The development of automatic driving for level 3 and higher levels will be based on the status of level 2 functions, but has to take into account much more aspects; see, e.g. Matthaei et al. (2016) and Winner et al. (2016). The automatic vehicle control system has to be based, for example, on the following function modules; see also the schematic representation of Fig. 22.1:

A. Data Processing and Vehicle Guidance

(1) Environmental sensor system.
(2) Data processing and data fusion.
(3) Information platform.
(4) Perception of the environment.
(5) Situation awareness.
(6) Prediction of road users.
(7) Motion planning.
(8) Motion control.
(9) Actuator control.
(10) Vehicle dynamics (real).

B. Vehicle Status

(1) Motion sensors.
(2) Vehicle models.
(3) State estimation.

C. External Support

(1) Communication.
(2) Navigation.

D. Safety.

These very comprehensive tasks are now specified further.

(A.1) The *environmental sensors* are expanded with respect to their ranges and resolution in order to allow a complete and precise view of the relevant 360° surrounding. This includes the detection of all other road users (vehicles, obstacles, pedestrians, road infrastructure, traffic signs, traffic light, traffic policeman, animals, etc.) and measurements to localize the own (ego) vehicle's position and orientation on the road. Some sensors may be mounted on the vehicle's roof in order to allow a better 360° view.

(A.2) With methods of *data processing*, the signals of the environmental sensors have to be evaluated. This includes, for example, image processing from

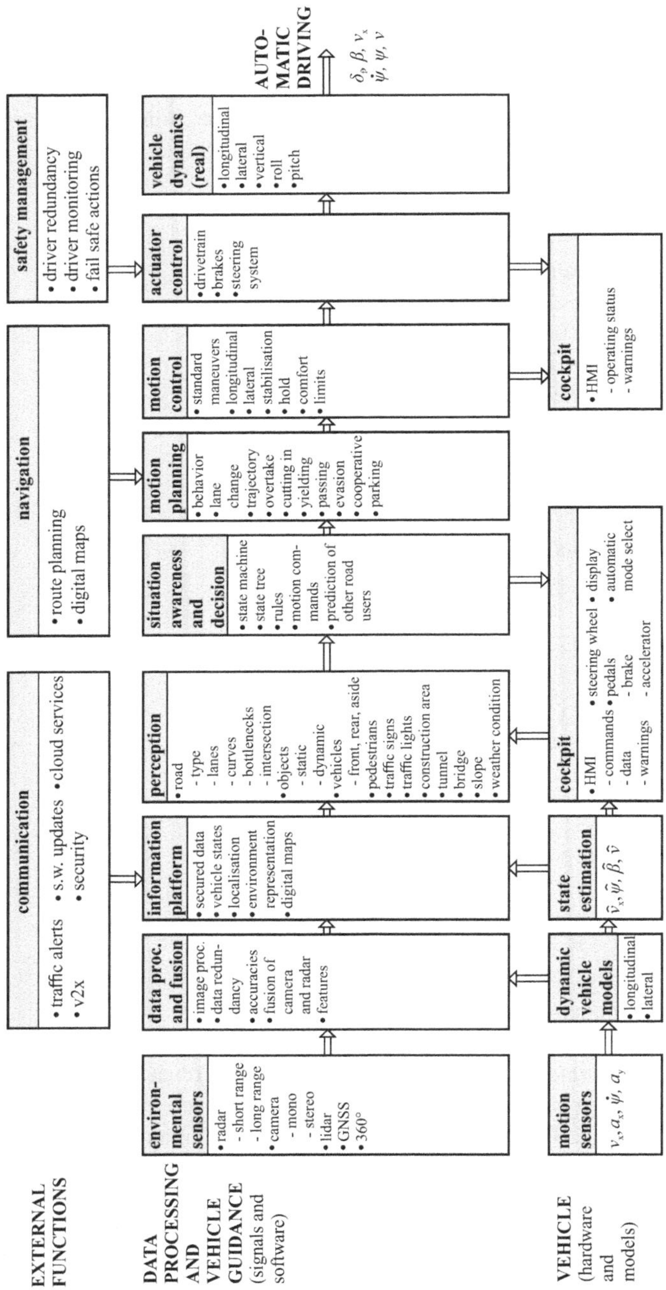


**Fig. 22.1** Signal flow, information processing, and function blocks for automatic driving

the cameras with feature extraction and object detection and environment representation, as described in Chap. 16. The detected features and objects from the radar and lidar sensors are then combined with those of the cameras by methods of *data fusion*, e.g. by applying the extended Kalman filters. Similar procedures then have to be performed with distance measurements, e.g. from the radar sensor and a stereo camera. A further task is to combine the result from the vehicle motion sensors and environmental sensors; see Fig. 16.4. One of the goals is to create redundancies of the environment representation and accuracy values for validation of the results.

(B.1-3) The signals from the motion sensors are used together with the dynamical vehicle models for state estimation of the vehicle as described in Chap. 12 to obtain the vehicle state variables with precise values of $\hat{v}_X$, $\hat{\dot{\psi}}$, $\hat{\beta}$, and $\hat{v}$.

(A.3) The results from the environmental and motion sensors are then stored on an *information platform* which contains the present vehicle state variables, the environment presentation, localization, and orientation of the vehicle on the road based on secured, received, and evaluated data of all sensors and possibly use of digital road maps. The following functions have then access to this platform.

(A.4) A precise and comprehensive *perception* of the stationary and dynamic environment is a detailed information of all relevant issues for automatic driving in (complex) traffic situations. For the case of *stationary perception* this encompasses, for example, the road type (highway, urban or rural road, city), number of lanes, white and yellow lane markings, road curbs, bottlenecks, construction areas, curvature and change of curvature of curves, intersections, tunnels, traffic signs, traffic lights, parking areas, static obstacles, etc. The *dynamic perception* of the environment concerns the moving traffic participants and includes other vehicles in the front and in the rear of the driven lane, other vehicles of neighboured lanes, crossing traffic, bicycles, pedestrians, animals, and weather conditions.

This list of items already shows that computer-based perception is a very demanding and complex task. Compared to the perception of a human driver, also the type of objects like trucks, buses, trams, and humans (adults, child) and also *prediction* or *intension* of their next movement should be possible. This means a rather detailed image processing and dynamic state estimation of all relevant traffic participants.

(A.5) The *situation awareness and decision* interpret the results of the information platform with vehicle state variables and localization on the road (lane) and of the perception to "understand" the ego vehicle's situation and to decide which motion commands follow. This may also include a *prediction* and *intention* of other road users. The decision may be based on tree-structured graphs and state machines and has to include traffic rules like to give right of way (to yield) for priority roads, to avoid the wrong entrance into one-way streets, to follow no passing signs, to drive behind another vehicle, etc.

(A.6) The *motion planning* is based on the commands from decisions in (A.5) and prepares the next motion of the ego vehicle like the continuation of present driving, change of velocity, lane change, kind of trajectory, overtaking, cutting in (other lane or previous lane), passing other vehicles, and evasion of slower vehicles or obstacles.

The motion planning may also consider *communication* with other traffic participants. The recognition of turn indicators, brake lights, or the stationary or dynamic behavior of other vehicles is one source. More difficult to observe with cameras are hand signs of pedestrians, cyclists or drivers, gestures of drivers, and also the hand signs of traffic policemen. The observation of other vehicle's behavior may also support a kind of cooperative driving, like giving way by braking and increasing the gap in a traffic flow for merging into a highway. A further way for motion planning may result from vehicle-to-vehicle communication (V2x) systems, especially for intersections (cross-roads) or platooning (grouping of vehicles) with optimized traffic flow and avoidance of stop-and-go traffic.

The motion planning includes also the *route planning* from the navigation systems with information of next changes via the HMI to the driver. The use of *digital maps* may also support the vehicle motion, by giving details on road infrastructure like traffic signs, stopping lines, and the road network, improving the GNSS-localization and forecast of curves. Digital maps can also be used to improve the detection of lane markings, especially if they are aged or at night.

A further task of motion planning is *automatic parking* at parking areas or parking houses with guidance to a free parking lot and a parking pilot for aside, in front, or at rear parking.

(A.7) The *motion control* performs the execution of motion planning with standard driving maneuvers, see Table 18.2, or optimized, adaptive maneuvers for the longitudinal behavior with distance control and lateral control with stabilization. The design of the respective control systems is treated in Chaps. 19 and 20 and includes also emergency braking and evasion for collision avoidance; see Chap. 21. Other tasks are the adherence to speed limits, following no-overtaking signs, warnings for bottlenecks, construction areas, sidewind, etc. Important for the acceptance of the motion control is an automatic driving like a good human driver with prediction ability and a good compromise between speedy and comfortable behavior.

(A.8) The *actuator control* is subordinated to the motion control and consists of the extensive control systems (ECUs) of the combustion engine and transmission and of the control of the brakes and steering (EPS); see Chaps. 6, 13, and 14. Because automatic driving requires a very *high reliability* of all components and functions, especially the actuators have to have fault-tolerant behavior. As the brake system is already redundant, see Chap. 13, the power-assisted steering system (EPS) as a high safety-relevant system has to have several redundancies, among them a duplex-actuator or a multi-winding electrical motor; see Sect. 14.4.4.

(A.9) The actuators in function within the multi-variable and multilevel vehicle control system then act on the *real dynamic vehicle system* and generate the automatic controlled longitudinal and lateral behavior with coupled vertical, roll, and pitch dynamics.

Further views for the development of automatic driving can be found, for example, in Winner et al. (2016), Raste (2016), Maurer et al. (2015), Robert Bosch GmbH (2018), and Waschl et al. (2019).

The treated methods for *longitudinal vehicle control* in Chap. 19, such as acceleration, velocity, distance, and adaptive cruise control; for *lateral vehicle control* in Chap. 20, such as path, lane change, and lane merging control; and *anticollision control* in Chap. 21 then become parts of motion planning and motion control for automatic driving vehicles.

## 22.3 Progress by Evolution

The *driver assisting systems* of level 1 assist for conventional manual driving and have been developed to increase safety and comfort. Automatic control functions are provided mainly for longitudinal control and assistance with warnings serves for the lateral behavior. Based on successful improvements and good acceptance, the extent of automated functions is expanded to *partial automation* (PA) in level 2. Automatic control functions are then possible for lateral control and for traffic jams. More assistance is given for lane change and for parking. However, the driver has to monitor these functions and perform all other tasks. As depicted in Fig. 22.2, the automation levels 1 and 2 support and perform some basic driving maneuvers. The

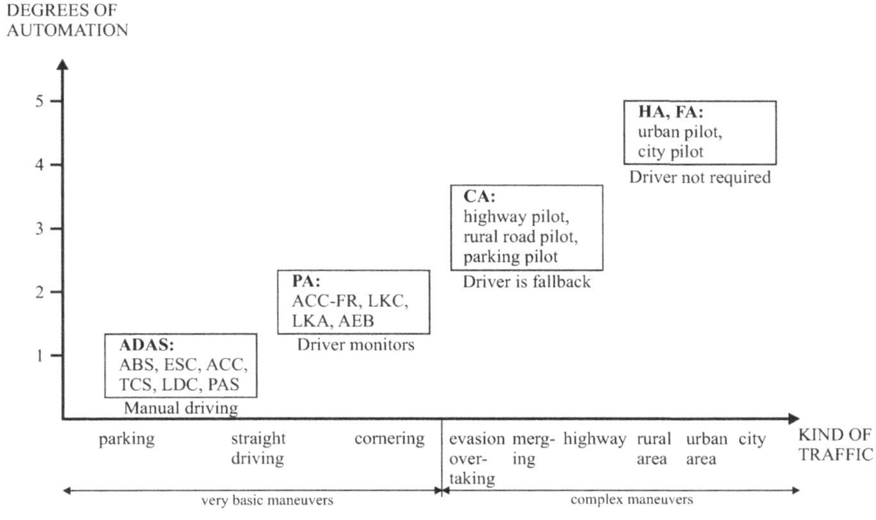


**Fig. 22.2** Degrees of automatic driving in dependence on kind of traffic and maneuvers

perception is still relatively moderate and limited to the detection and interpretation of lanes, vehicles, construction areas, obstacles, and pedestrians, as shown in Fig. 22.3.

A big step in the development is required for *conditional automation* (CA), level 3. Automatic pilots are now active for highways, rural roads, and for parking and include, e.g. lane changes. In addition to basic maneuvers, more complex driving maneuvers have to be guided, where the driver can perform partially other tasks, but is required as a fallback; see Figs. 22.2 and 22.3. The number of sensors, the performance of the sensor system, the image processing, and fusion of all data increase considerably. The perception includes now, e.g. all traffic signs and intersections, and a situation awareness has to be included for understanding the traffic situation and to follow traffic rules. The motion planning and control are extended and may include merging at highway entrances and turning at intersections.

*High automation* (HA), level 4, encompasses many driving modes, also in urban areas. The driver is usually not required, only on request, and the system also fulfills fallback tasks. Using many and different environment sensors, a fault-tolerant 360° view of the complete traffic relevant environment is required and all traffic participants before, aside, and backward have to be detected. The perception increases further, e.g. for traffic lights and complex intersections. The situation awareness interprets the complete traffic situation and takes the right decisions for safe motion planning and control. Fault-tolerant behavior is required for all safety-relevant functions.

*Full automation* (FA), level 5, is an extension of high automation and performs ideally all driving modes and includes a city pilot. A driver is no more required.

A great challenge for the development of driving automation and its general use in traffic is the *validation* of all functions with regard to safety and especially for accidents, injured, and killed people. This is discussed by Winner (2016), showing how many millions of km have to be driven automatically, and how statistics, simulations, test scenarios, and test tracks may help to come to metrics and integrity levels. This may begin with simple traffic situations and end up with high-risk situations, from low to high speed, and their combinations.

A standard procedure for the development of automotive products is treated in the standard ISO 26262 (2011, 2018), leading to the automotive safety integrity levels ASIL A to D; see the next section. However, in order to develop test procedures for automatic driving vehicles, an addition to ISO 26262 (2011, 2018) was created under ISO PAS 21448 (2018). It defines methods for testing automatic driving functions with real and simulated vehicles; see, e.g. Börger et al. (2020).

Figures 22.2 and 22.3 illustrate the stepwise increase of automatic functions. New functions are usually based on proven previous functions, such that a continuous development process of specific functions takes place. Therefore, the steps between degrees 1 to 5 of driving automation are not sharp and the included functions show a continuous expansion. Hence, the development of automatic driving has to be seen as an evolutionary process.

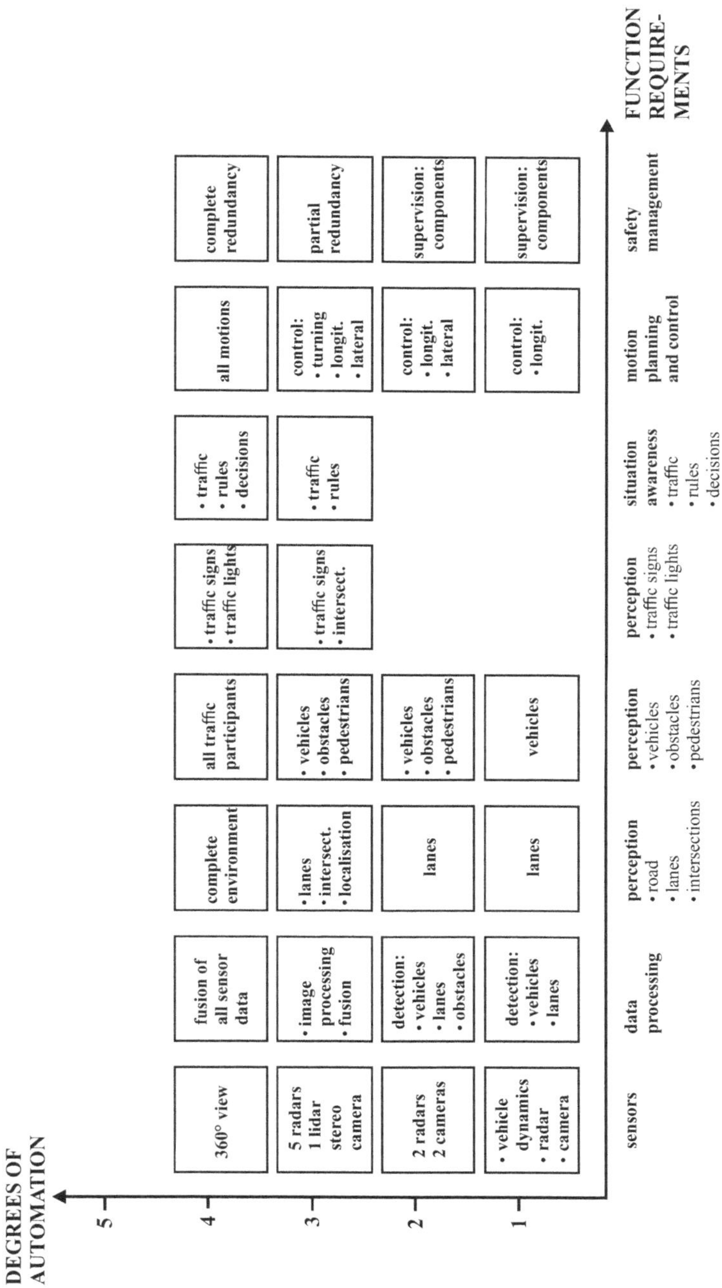


**Fig. 22.3** Degrees of automatic driving and required functions

## 22.4 Safety and Fault Tolerance

An important role for automatic driving plays the *safety design* and the *safety management*. The introduction into public traffic requires a very low risk for the automatic operations, which has to be guaranteed. Therefore, all technical and economically feasible measures have to be met. High reliability can be obtained usually by perfectness or tolerance. *Perfectness* refers to the idea of avoiding faults and failures by means of the best possible mechanical, electrical, electronic, or software design. *Tolerance* describes the notion to control the consequences of faults and failures such that the system remains functional. This can be reached by the principle of fault tolerance in the components or subsystems, called modules, and on higher system levels; see, e.g. IEC 61508, Storey (1996), Leveson (1995), and Isermann (2006). Fault-tolerance methods generally use redundancy. This means that in addition to the considered module, one or more modules are connected, usually in parallel. These redundant modules are either identical or diverse. Such redundant structures can be designed for hardware, software, information processing (control), and mechanical and electrical components like sensors, actuators, microcomputers, buses, power supplies, etc. In the following, a very brief description of redundant structures is given, followed by some automotive examples.

**(a) Basic Redundant Structures**

There exist mainly two basic approaches for fault tolerance: static redundancy and dynamic redundancy. The corresponding configurations are first considered for *electronic hardware* and then for other components. Figure 22.4a shows a scheme for *static redundancy*. It uses three or more parallel modules that have the same input signal and are all active. Their outputs are connected to a voter, which compares these signals and decides by the majority which signal value is the correct one. If a triple modular-redundant system is applied, and the fault in one of the modules generates a wrong output, this faulty module is masked (i.e. not taken into account) by the two-out-of-three voting. Hence, a single faulty module is tolerated without any effort for specific fault detection and $n$ redundant modules can tolerate $(n-1)/2$ faults ($n$ odd).

To improve the fault tolerance, also the voter can be made redundant, Storey (1996). Disadvantages of static redundancy are high costs, more power consumption, and extra weight. Furthermore, it cannot tolerate common-mode faults, which appear in all modules because of common fault sources.

*Dynamic redundancy* needs fewer modules at the cost of more information processing. A minimal configuration consists of two modules, Fig. 22.4b, c. One module is usually in operation and, if it fails, the standby or backup unit takes over. This requires fault detection to observe if the operating modules become faulty. Simple fault-detection methods only use the output signal for, e.g. consistency checking (range of the signal), comparison with redundant modules, or use of information redundancy in computers like parity checking or watchdog timers. After fault detection, it is the task of the reconfiguration to switch to the standby module and to remove the faulty one.

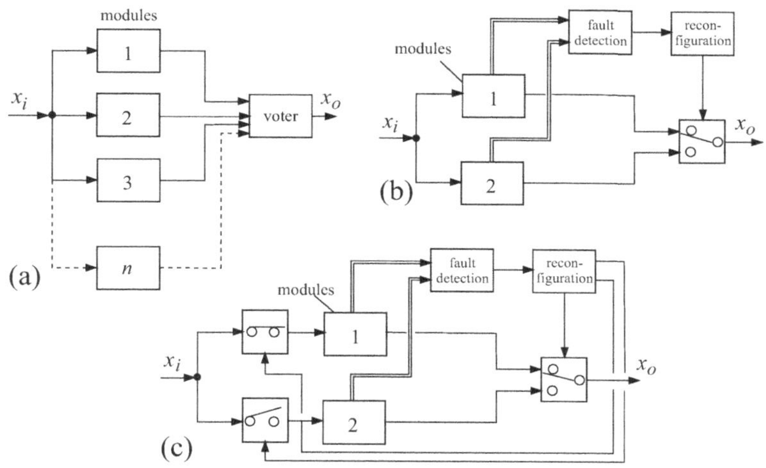


**Fig. 22.4** Fault-tolerant schemes: **a** static redundancy: multiple-redundant modules with majority voting and fault masking, *m* out of *n* systems (all modules are active); **b** dynamic redundancy: standby module that is continuously active, "hot standby"; **c** dynamic redundancy: standby module that is inactive, "cold standby"

In the arrangement of Fig. 22.4b, the standby module is continuously operating, called *hot standby*. Then, the transfer time is small at the cost of operational aging (wear-out) of the standby module.

Dynamic redundancy where the standby module is not in operation and does not wear is shown in Fig. 22.4c and is called *cold standby*. This arrangement needs two more switches at the input and more transfer time due to a start-up procedure. For both schemes, the performance of the fault detection is essential.

Dynamic redundancy can be extended to two and more standby modules, thus tolerating two or more faults. Combinations of static and dynamic redundancy lead to *hybrid redundant schemes* to avoid the disadvantages of both types on the cost of higher complexity, Storey (1996).

Static and dynamic redundancy is applicable to electronic hardware, software, mechanical and electrical systems, and mechatronic systems. More details are described, e.g. in Leveson (1995) and Isermann (2011).

**(b) Degradation Steps**

Mainly because of costs, space, and weight, a suitable compromise between the degree of fault tolerance and the number of redundant modules has to be found. In contrast to fly-by-wire systems, for industrial and traffic systems, only one or two failures can be tolerated for hazardous cases, mainly because a safe state can be reached easier and faster. This means that not all components need very stringent fault-tolerance requirements. The following steps of degradation are distinguished: IEC 61508 (1997)

- *fail-operational* (FO): one failure is tolerated, i.e. the module stays operational after one failure. This is required if no safe state exists immediately after the component fails;
- *fail-safe* (FS): after one (or several) failure(s), the module directly possesses a safe state (passive fail-safe, without external power) or is brought to a safe state by a special action (active fail-safe, with external power);
- *fail-silent* (FSIL): after one (or several) failure(s), the module is quiet externally, i.e. stays passive by switching off and therefore does not influence other components incorrectly; and
- *fail* (F): permanent interruption of the module's ability to perform a required function.

For, e.g. vehicles, it is proposed to subdivide FO into "long time" and "short time," in order to reach a safe state dependent on the kind of failure. Considering these degradation steps for various components, one has to check first if a safe state exists. For automobiles, (usually) a safe state is standstill (or low speed) at a nonhazardous place. For components of automobiles, a fail-safe status is (usually) a mechanical backup, i.e. a mechanical or hydraulic linkage in the case of brakes, or the mechanical part of a steering system for direct manipulation by the driver. Passive fail-safe is then reached, e.g. after the failure of electronics if the vehicle comes to a stop independently of the electronics, e.g. by a closing spring in the throttle or by actions of the driver via mechanical backup. However, if no mechanical backup exists after the failure of electronics, only an action by other electronics (switch to a still operating module) or by actions of the driver (driver backup) can bring the vehicle (in motion) to a safe state, i.e. to reach a stop through active fail-safe.

Generally, a *graceful degradation* is envisaged, where less critical functions are dropped to maintain the more critical functions available, using priorities, IEC 61508 (1997). Table 22.1 shows degradation steps to fail-operational (FO) and fail (F) for different redundant structures of electronic hardware. As the fail-safe status depends on the considered system and the kind of components, it is not considered here.

For flight-control computers, usually a triplex structure with dynamic redundancy (hot standby) is used, which leads to FO-FO-FS, such that two failures are tolerated, and a third one allows the pilot to operate manually; see Favre (1994), Reichel (1999), and Reichel and Boos (1986). If the fault-tolerance system has to cover only one fault to stay fail-operational (FO-F), a triplex system with static redundancy or a duplex system with dynamic redundancy is appropriate. If fail-safe can be reached after one failure (FS), a duplex system with two comparators is sufficient. However, if one fault has to be tolerated to continue fail-operational and after a next fault it is possible to switch to a fail-safe (FO-FS), either a triplex system with static redundancy or a duo-duplex system may be used; see Reichel and Boos (1986). The duo-duplex system has the advantages of simpler failure detection and modularity.

***(c) Functional safety for road vehicles*** is treated in detail in the standard ISO 26262 (2011, 2018) and serves as an integral part of each automotive product development.

**Table 22.1** Fail behavior of electronic hardware for different redundant structures. FO: fail-operational; F: fail; and FS: fail-safe not considered

| | | Static redundancy | | Dynamic redundancy | | |
|---|---|---|---|---|---|---|
| Structure | Number of elements | Tolerated faults | Fail behavior | Tolerated failures | Fail behavior | Discrepancy detection |
| Duplex | 2 | 0 | F | 0 | F | 2 comparators |
| | | | | 1 | FO-F | Fault detection |
| Triplex | 3 | 1 | FO-F | 2 | FO-FO-F | Fault detection |
| Quadruplex | 4 | 1 | FO-F | 3 | FO-FO-FO-F | Fault detection |
| Duo-duplex | 4 | 1 | FO-F | – | – | – |

It comprises a vocabulary, management, safety life cycle, supporting processes, and the automotive safety integrity levels (ASIL) from A to D.

The determination of ASIL is the result of hazard analysis and risk assessment. A hazard is expressed in terms of severity of possible injuries with regard to the time a vehicle is exposed to the possibility of a happening hazard and the relative likelihood a driver can prevent the injury controllability. A risk is then assessed as

$$\text{risk} = \text{severity} \times (\text{exposure} \times \text{likelihood})$$

and the ASIL by

$$\text{ASIL} = \text{severity} \times (\text{exposure} \times \text{controllability}).$$

According to ISO 26262 (2011, 2018), severity is divided into 4 classes, exposure in 5 classes, and controllability in 4 classes. ASIL A is the lowest level and ASIL D the highest level.

ASIL D is defined as an event with the possibility of a life-threatening injury with little chance the driver can prevent the injury. Therefore, it requires the highest integrity level to avoid the injury.

**(d) Automotive Examples**

There exist already several *fault-tolerant components* in conventional vehicles. One example is the *steering angle sensor* with two pinions. A gear wheel drives two pinions with a giant magnetoresistance sensor. The Nonius/Vernier principle is applied to measure the absolute Position; see Robert Bosch GmbH (2018). If one sensor fails, the other can give relative changes in the position. The sensor system is FO-FSIL; see Isermann (2011).

Another example is the *redundant hydraulic wheel brake* with two separate hydraulic brake circuits and a dual master cylinder; see Fig. 13.1. Two separate hydraulic circuits operate in the parallel configuration on four wheels with static redundancy. If one circuit fails, the other remaining circuit operates by braking with two wheels (FO-F). Further redundancies exist within the electromechanical brake (EMB); see Fig. 13.20.

A *fault-tolerant electrical throttle* for combustion engines is possible with two potentiometers for position measurement and virtual position value based on the model of a brushless DC motor; see Isermann (2011).

A *fault-tolerant brake pedal* for electromechanical brake is shown in Stölzl et al. (1998) and Stölzl (2000). It uses a duo-duplex sensor configuration with two touchless sensors in two casings with one microcomputer each. A fault handling code is implemented in the central computer. Together with a brake pressure measurement in the master cylinder, a triple redundancy is reached; see also Isermann et al. (2002).

Structures for *fault-tolerant electric power steering systems* (EPS) have been already discussed in Sect. 14.4.4. These structures have to use hardware redundancies for steering angle sensors, ECU, inverter, and the electrical motor. The mechanical system and steering gear are usually a high precision, oversized, and therefore a very reliable system without redundancy. The different approaches in Fig. 14.33 show that a redundancy in the electromotor with multi-phase windings seems to be a reasonable compromise together with fault-tolerant inverters, ECU, and internal sensors.

A further safety-critical system is the *electrical board net*. At least two independent batteries are required. Also, the communication network has to have redundancies. The FlexRay bus has already two independent channels from its first design; see Sect. 2.2

An example for fault-tolerance on a system level is a *fault-tolerant sensor platform* for the dynamic behavior of a vehicle, as depicted in Fig. 22.5

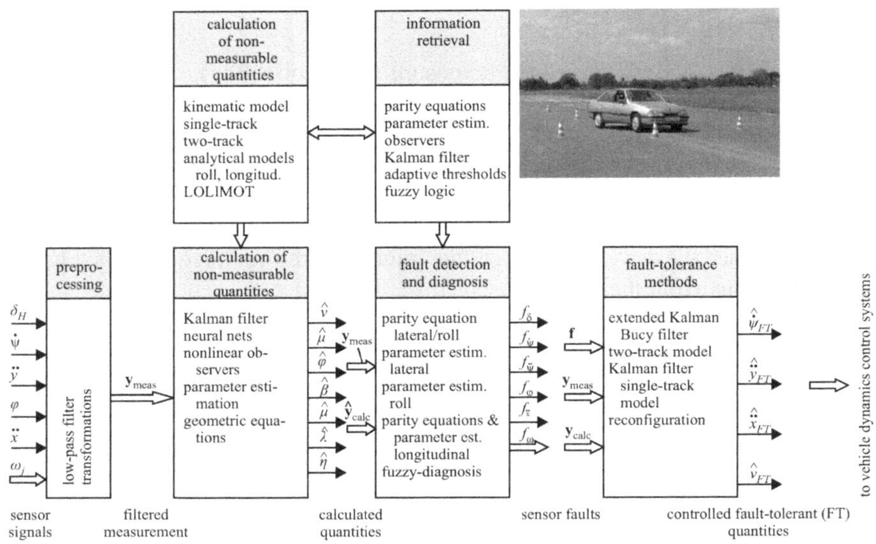


**Fig. 22.5** Virtual redundant sensor system for the dynamic behavior of a passenger car. A fault-tolerant sensor platform with analytical redundancy for yaw rate $\dot{\psi}$, lateral acceleration $\ddot{y}$, longitudinal acceleration $\ddot{x}$, and over ground velocity $v$; see Halbe (2008)

It is an example of a fault-tolerance approach at the system level. Here, the signals from different sensors are brought together, with the aim to detect sensor faults and of providing consolidated measurements for the drive dynamic controllers. The fault-tolerant sensor system is based on analytical redundancy and uses kinematic, single-track, and two-track lateral vehicle models, as well as longitudinal and roll dynamic models.

First, the signals are preprocessed, then quantities that cannot be measured or which can only be sensed with expensive or complicated equipment are reconstructed. The reconstruction is based on the drive dynamic models of different complexities and employs the state observers, Kalman filters, parameter estimation algorithms, and local linear neural nets. The fault detection and diagnosis is another important block as all sensors are supervised to identify faulty sensors, which are then masked and replaced by estimated quantities in the block on fault-tolerance methods. Finally, the information platform provides the extracted information to the drive dynamic controllers. These controllers can now not only control directly measurable quantities, but can also control other, reconstructed, and more informative quantities; see Halbe and Isermann (2007) and Halbe (2007). The drive dynamics information platform is at least FO-FSIL, but in many cases can sustain more than one sensor fault.

Hence, based on the directly measured noisy sensor signals, unknown parameters like $\hat{\mu}$ and unknown variables like $\hat{v}_X$, $\hat{\beta}$, $\hat{\lambda}$, and $\hat{\eta}$ are estimated based on vehicle models, and by the fault-detection methods, faults of the sensors except $\delta_H$ are determined. Using analytical fault-tolerance methods, the drive dynamics variables $\hat{v}_{XFT}$, $\hat{\dot{\psi}}_{FT}$, $\hat{\ddot{y}}_{FT}$, and $\hat{\ddot{x}}_{FT}$ are calculated (FT: fault-tolerant). Their comparison with the directly measured values allows to consolidate the measurements or to reconstruct their values in case of faulty sensors. The fault-tolerance system was successfully tested by driving experiments with a compact car; see Halbe (2007, 2008).

In the case of a total failure of the *steering system*, a certain redundancy is reached by individual one-sided wheel braking to bring the vehicle to a safe place on the side of a road, as used for lane keeping control (LKC); see Sect. 17.3.

## References

Börger A, Hosse R, von der Decken S (2020) SOTIF - a new challenge for functional testing. ATZelectronics worldwide 10:60–63

Broggi A, Bertozzi M, Fascioli A (1999) ARGO and the MilleMiglia in Automatico tour. IEEE Intell Syst Appl 14:55–64

Broggi A, Buzzoni M, Debattisti S, Grisleri P, Laghi MC, Medici P, Versari P (2013) Extensive tests of autonomous driving technologies. IEEE Trans Intell Transp Syst 14:1403–1415

Dang T, Lauer M, Bender P, Schreiber M, Ziegler J, Franke U, Fritz H, Strauß T, Lategahn H, Keller CG, Kaus E, Rabe C, Appenrodt N, Pfeiffer D, Lindner F, Stein F, Erbs F, Enzweiler M, Knöppel C, Hipp J, Haueis M, Trepte M, Brenk C, Tamke A, Ghanaat M, Braun M, Joos A, Mock H, Hein M, Petrich D, Schneider N, Kronjöger W, Zeeb E, Stiller C, Herrtwich RG (2015) Autonomes Fahren auf der historische Bertha-Benz-Route. tm-Technisches Messen 82:280–297

Dickmanns ED (2007) Dynamic vision for perception and control of motion. Springer, Berlin

Dickmanns ED, Behringer R, Dickmanns D, Hildebrandt T, Maurer M, Thomanek F, Schiehlen J (1944) The seeing passenger car 'VaMoRs-P'. In: Proceedings of the intelligent vehicles '94 symposium, pp 68–75
Dickmanns ED, Mysliwetz B, Christians T (1990) An integrated spatio-temporal approach to automatic visual guidance of autonomous vehicles. IEEE Trans Syst Man Cybern 20:1273–1284
Favre C (1994) Fly-by-wire for commercial aircraft: the airbus experience. Int J Control 59(1):139–157
Fenton R (1970) Automatic vehicle guidance and control-a state of the art survey. IEEE Trans Veh Technol 19:153–161
Franke U, Mehring S, Suissa A, Hahn S (1994) The Daimler-Benz steering assistant-a spin-off from autonomous driving. In: Proceedings of the intelligent vehicles '94 symposium, pp 120–124
Halbe I (2008) Modellgestützte Sensoreninformationsplattform für die Quer- und Längsdynamik von Kraftfahrzeugen: Anwendungen zur Fehlerdiagnose und Fehlertoleranz. Diss. TU Darmstadt, Fortschr.-Ber. VDI Reihe 12, 680. VDI Verlag, Düsseldorf
Halbe I, Isermann R (2007) A model-based fault-tolerant sensor platform for vehicle dynamics control. In: Proceedings of the 5th symposium in advances in automotive control, Seascape Resort Aptos, CA, USA, pp 509–516
Halbe L (2007) Modellbasierte Sensorfehlererkennung für das querdynamische Verhalten von Kraftfahrzeugen. at-Automatisierungstechnik 55:322–329
IEC 61508 (1997) Functional safety of electrical/electronic/programmable electronic systems. International Electrotechnical Commission, Switzerland
Isermann R (2006) Fault-diagnosis systems - an introduction from fault detection to fault tolerance. Springer, Heidelberg
Isermann R (2011) Fault-diagnosis applications. Springer, Heidelberg
Isermann R, Schwarz R, Stölzl S (2002) Fault-tolerant drive-by-wire systems. IEEE Control Syst Mag (October):64–81
ISO 26262 (2011, 2018) Road vehicle-functional safety. Int Organisation for Standarization
ISO Pas 21448 (2018) Road vehicles - safety of the intended functionality. Int Organisation for Standarization
Leveson N (1995) Safeware. System safety and computer. Wesley Publishing Company, Reading, MA
Matthaei R, Reschka A, Rieken J, Dierkes F, Ulbrich S, Winkle T, Maurer M (2016) Autonomous driving. Springer International Publishing, Cham, pp 1519–1556
Maurer M, Gerdes J, Lenz B, Winner H (eds) (2015) Autonomous driving. Springer, Berlin
Naab K (2000) Automatisierung bei der Fahrzeugführung im Straßenverkehr (Automation in Road Traffic). at - Automatisierungstechnik 48:211–223
Pomerleau D, Jochem T (1996) Rapidly adapting machine vision for automated vehicle steering. IEEE Expert 11(2):19–27
Raste T (2016) Vehicle dynamics control with braking and steering intervention, Chapter 41. In: Winner H, Hakuli S, Lotz F, Singer C (eds) Handbook of driver assistance systems: basic information, components and systems for active safety and comfort. Springer International Publishing, Cham, pp 1007–1020
Reichel R (1999) Modulares Rechnersystem für das Electronic Flight Control System (EFCS). In: DGLR-Jahrestagung, Deutsche Luft- und Raumfahrtkongress, Berlin, Germany
Reichel R, Boos F (1986) Redundantes Rechnersystem für Fly-by-wire Steuerungen. Bodensee-Gerätewerk, Überlingen
Robert Bosch GmbH (ed) (2018) Automotive handbook, 10th edn. Wiley, Chichester, England
Shladover SE (2007) PATH at 20-history and major milestones. IEEE Trans Intell Transp Syst 8:584–592
Singh S (2006) Special issue on the 2007 DARPA urban challenge, Part 2. J Field Robot 23:655–835
Stölzl S (2000) Fehlertolerante Pedaleinheit für ein elektromechanisches Bremssystem (Brake-by-Wire). Diss. TU Darmstadt, Fortschr.-Ber. VDI Reihe 12, 462. VDI Verlag, Düsseldorf

Stölzl S, Schwarz R, Isermann R, Böhm J, Nell J, Rieth P (1998) Control and supervision of an electromechanical brake system. In: FISITA world automotive congress, The Second Century of the Automobile, Paris, France

Storey N (1996) Safety-critical computer systems. Addison Wesley Longman Ltd., Essex

Thorpe C, Jochem T, Pomerleau D (1997) The 1997 automated highway free agent demonstration. In: Intelligent transportation system, pp 496–501

Thrun S, Montemerlo M, Dahlkamp H, Stavens D, Aron A, Diebel J, Fong P, Gale J, Halpenny M, Hoffmann G, Lau K, Oakley C, Palatucci M, Pratt V, Stang P, Strohband S, Dupont C, L-E J, Koelen C, Markey C, Rummel C, van Niekerk J, Jensen E, Alessandrini P, Bradski G, Davies B, Ettinger S, Kaehler A, Nefian A, Mahoney P, (2006) Stanley: the robot that won the DARPA grand challenge. J Field Robot 23:661–692

Ulmer B (1992) Vita-an autonomous road vehicle (arv) for collision avoidance in traffic. In: Proceedings of the intelligent vehicles '92 symposium, pp 36–41

Waschl H, Kolmanovsky I, Willems F (eds) (2019) Control strategies for advanced driver assistance systems and autonomous driving functions. Lecture notes in control and information sciences. Springer International, Cham, Switzerland

Webster (1981) Webster's third new international dictionary, vol 1. Encyclopedia Britannica Inc, Chicago

Winner H (2016) Adas, quo vadis? In: Winner H, Hakuli S, Lotz F, Singer C (eds) Handbook of driver assistance systems: basic information. Components and systems for active safety and comfort. Springer International Publishing, Cham, pp 1557–1584

Winner H, Hakuli S, Lotz F, Singer C (eds) (2016) Handbook of driver assistance systems. Springer International Publishing AG, Cham, Switzerland

Zapp A (1988) Automatische Strassenfahrzeugführung durch Rechnersehen (Automated road vehicle guidance with computer vision). Diss. Universität der Bundeswehr München

Ziegler J, Bender P, Schreiber M, Lategahn H, Strauss T, Stiller C, Dang T, Franke U, Appenrodt N, Keller CG, Kaus E, Herrtwich RG, Rabe C, Pfeiffer D, Lindner F, Stein F, Erbs F, Enzweiler M, Knöppel C, Hipp J, Haueis M, Trepte M, Brenk C, Tamke A, Ghanaat M, Braun M, Joos A, Fritz H, Mock H, Hein M, Zeeb E (2014) Making bertha drive-an autonomous journey on a historic route. IEEE Intell Transp Syst Mag 6:8–20

# Appendix A

## A.1 Basic Control Structures and Controllers

### A.1.1 Basic Control Structure

Some basic control-system structures which are suitable for automotive control are summarized. To describe briefly the principles, the components are assumed to have linear dynamic behavior and are treated in continuous time with transfer functions $G(s)$ where $s$ is the Laplace variable $s = \delta + i\omega$. Basic feedforward and feedback control systems are viewed, their combinations and interconnected control systems, like cascaded control and two-variable control. The extension to multi-variable systems, to nonlinear behavior, and to the resulting digital control algorithms with sampled signals is then discussed subsequently.

If output variables of vehicles are not measured as, e.g. the tire slip and the vehicle slip angle, feedforward control systems are used; see Fig. A.1a. An advantage then is, depending on the process behavior, that stability problems are avoided. But relatively precise process models are required and parameter changes of the vehicle and external disturbances are not automatically compensated.

If vehicle variables can be measured and a relatively precise control is required, then feedback control systems are applied, see Fig. A.1b, as, e.g. for the velocity or the distance to a preceding vehicle. In order to reach a good control performance and to avoid instability, dynamic vehicle models are required for a systematic design.

The controlled processes of vehicles are characterized by mostly nonlinear behavior and low dynamic order, e.g. $n = 2$ or 3, sometimes with dead time. These parameters depend frequently on the operation point, as velocity and load. As the real-time requirements with sampling times of about $10 \ldots 20\,\mathrm{ms}$ are quite strong, the final control algorithms have to be of small computational expense and should be easy to understand and calibrate. Therefore, the following short summary concentrates on those control designs which satisfy practical requirements for vehicle control.



R. Isermann, *Automotive Control*, ATZ/MTZ-Fachbuch,
https://doi.org/10.1007/978-3-642-39440-9_A

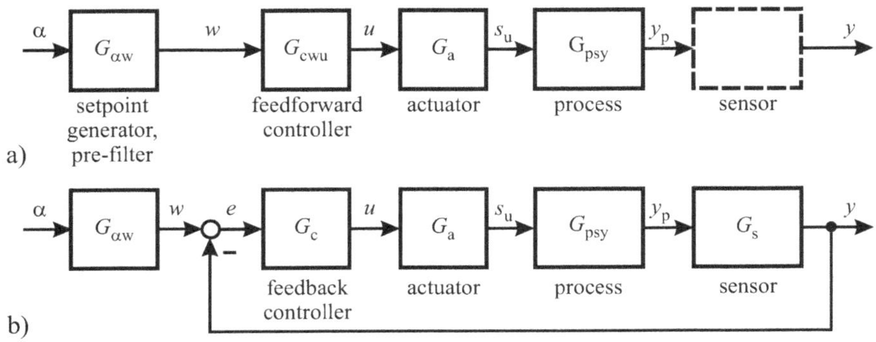


**Fig. A.1** Basic control structures. **a** Feedforward control (output sensor not available for series production vehicle), **b** Feedback control. $\alpha$ command variable, $w = y_d$ setpoint, reference variable, or desired control variable, $y$ controlled variable, $u$ manipulated variable, $s_u$ actuator position, and $y_p$ process output

A detailed treatment for the design of control systems is available in many textbooks on control theory and practice, e.g. Kuo (1995), Åström and Wittenmark (1997), Ogata (2008), Dorf and Bishop (2010), and Föllinger (2013). Therefore, the next sections first summarize some basic control systems as a starting point for the design of special, mostly nonlinear control systems. The assumed linear behavior in the next sections can be considered as a basis for more sophisticated designs and to discuss some basic properties.

### A.1.2 Linear Feedforward Control

An elementary feedforward control is depicted in Fig. A.1a. A setpoint generator yields the setpoint variable or reference variable $w(t)$ based on a command variable $\alpha(t)$ (e.g. the accelerator pedal). The goal is that the process output $y(t)$ follows the setpoint variable $w(t)$ as directly as possible. If $y(t)$ is not measurable, a feedback control cannot be applied. If now a model of the actuator $G_a(s)$ and the process $G_{psy}(s)$ is known, a feedforward control can be designed. An ideal response would be $y(t) = w(t)$. However, as the actuator and process always have a certain dynamic delaying behavior, this requirement is not realizable. Assuming that the process has proportional behavior with the following transfer function:

$$G_{psy}(s) = \frac{y(s)}{s_u(s)} = K_{psy}\frac{1 + b_1 s + \ldots + b_m s^m}{1 + a_1 s + \ldots + a_n s^n}, \tag{A.1.1}$$

and for the actuator holds a proportional behavior

$$G_a(s) = \frac{s_u(s)}{u(s)} = K_a, \tag{A.1.2}$$

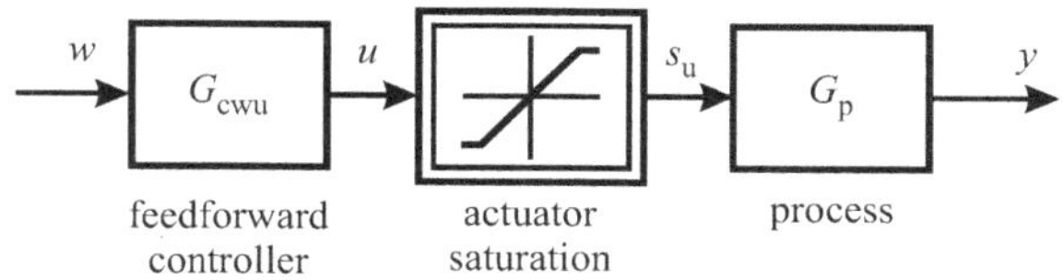


**Fig. A.2** Feedforward control for a process with limited manipulation range

the overall transfer function of the process becomes

$$G_p(s) = \frac{y(s)}{u(s)} = G_a(s)G_{psy}(s) \tag{A.1.3}$$

with the gain

$$K_p = K_a K_{psy}. \tag{A.1.4}$$

Using this overall transfer function of the process, Fig. A.2 results.

**(a) Static Feedforward Control**

If the actuator and process dynamics are not taken into account and actuator and process gain are known, a static feedforward controller

$$G_{cwu}(s) = K_{cwu} = 1/K_p \tag{A.1.5}$$

can be used to satisfy the stationary behavior such that the overall response becomes

$$G_{wy}(s) = \frac{y(s)}{w(s)} = \frac{1 + b_1 s + \ldots + b_m s^m}{1 + a_1 s + \ldots + a_n s^n} \; (m < n), \tag{A.1.6}$$

i.e. a delayed dynamic response with gain 1.

**(b) Dynamic Feedforward Control**

An ideal feedforward control is obtained by

$$\frac{y(s)}{w(s)} = G'_{cwu}(s)G_p(s) = 1, \tag{A.1.7}$$

which leads to a dynamic feedforward control with an inverted process model

$$G'_{cwu}(s) = \frac{u(s)}{w(s)} = \frac{1}{G_p(s)}. \tag{A.1.8}$$

Introducing (A.1.1) and (A.1.2) yields with the assumption of a minimum phase process

$$G'_{cwu}(s) = \frac{1}{K_p}\frac{1 + a_1 s + \ldots + a_n s^n}{1 + b_1 s + \ldots + b_m s^m}. \tag{A.1.9}$$

Because this transfer function of the feedforward controller would require exact derivatives of the output signal of order $n$, it is not realizable. Therefore, a realizability filter or reference shaping filter

$$G_{\mathrm{rf}}(s) = \frac{1}{(1+T_{\mathrm{r}}s)^n} \quad \text{(A.1.10)}$$

has to be applied. The realizable feedforward controller then becomes

$$G_{\mathrm{cwu}} = G_{\mathrm{p}}^{-1}(s)G_{\mathrm{rf}}(s) = \frac{1}{K_{\mathrm{p}}}\frac{1+a_1 s+\ldots+a_n s^n}{1+b_1 s+\ldots+b_m s^m}\frac{1}{(1+T_{\mathrm{r}}s)^n} \quad \text{(A.1.11)}$$

such that the overall behavior yields

$$G_{\mathrm{wy}}(s) = \frac{y(s)}{w(s)} = G_{\mathrm{cwu}}(s)G_{\mathrm{p}}(s) = G_{\mathrm{rf}}(s) = \frac{1}{(1+T_{\mathrm{r}}s)^n}. \quad \text{(A.1.12)}$$

The time constant $T_{\mathrm{r}}$ of the realizability filter is now a design parameter.

**Example A.1** (*Dynamic feedforward control for a manipulated variable as input*) As an example, a second-order low-pass process is considered with transfer function

$$G_{\mathrm{p}}(s) = \frac{y(s)}{u(s)} = \frac{K_{\mathrm{p}}}{(1+T_1 s)(1+T_2 s)}.$$

The realizable feedforward controller then results in

$$G_{\mathrm{cwu}}(s) = \frac{u(s)}{w(s)} = \frac{(1+T_1 s)(1+T_2 s)}{K_{\mathrm{p}}(1+T_{\mathrm{r}}s)^2}$$

if an actuator saturation (Fig. A.2) is not considered. The overall behavior follows as

$$G_{\mathrm{wy}} = \frac{1}{(1+T_{\mathrm{r}}s)^2}.$$

If $T_{\mathrm{r}} = T_2$ is selected, the feedforward controller becomes

$$G''_{\mathrm{cwu}}(s) = \frac{1}{K_p}\frac{1+T_1 s}{1+T_{\mathrm{r}}s},$$

which is a proportional-derivative (PD) element with a time lag. A simplification of $G'_{\mathrm{cwu}}(s)$ by assuming $T_{\mathrm{r}} \ll T_1$ is

$$G''_{\mathrm{cwu}}(s) = K_{\mathrm{cwu}}(1+T_{\mathrm{D}}s),$$

which is a proportional-derivative feedforward controller with $K_{cwu} = 1/K_P$. The derivative time $T_D$ can then be parameter-optimized or just selected as $T_D = T_1$, which yields

$$G_{wy} = \frac{y(s)}{w(s)} = \frac{1 + T_D s}{(1 + T_1 s)(1 + T_2 s)} = \frac{1}{1 + T_2 s}$$

resulting in an overall behavior of first order. □

The dynamic feedforward controllers improve the dynamic overall response compared to the static feedforward controller (A.1.5). However, if $T_r$ is selected too small or the derivative time $T_D$ too large, the response of the manipulated variable $u(t)$ becomes too large and may exceed the restricted range of the manipulating variable.

A further method to design a dynamic feedforward controller is to assume a feedback controller for the design and to use its manipulated variable $u_m$ for feedforward control as shown in Fig. A.3a; see Pfeil (2011). The transfer behavior of the feedforward controller then becomes (without considering an actuator saturation)

$$G_{cwu}(s) = \frac{u_m(s)}{w(s)} = \frac{u(s)}{w(s)} = \frac{G_c(s)}{1 + G_c(s)G_{p,m}(s)}, \qquad (A.1.13)$$

where $G_c(s)$ is a feedback controller. For the (simulated) feedback controller $G_c(s)$, a P or PD controller can be used. In this case, the process behavior is indirectly taken into account with its process model $G_{p,m}(s)$ and the dynamics of the feedforward correspond to an assumed closed-loop control.

**(c) Dynamic Feedforward Control with actuator saturation**

The actuators usually have a limited manipulation range and a saturation function has to be taken into account as depicted in Fig. A.3. The considered proportional acting feedforward controllers then reach a limited process input $s_u(t)$ for larger inputs $w(t)$ thus limiting the dynamic response of the process output. Figure A.4 shows the responses with the classical feedforward controller $G_{cwu}(s)$ of Example A.1. If the time constant $T_r$ of the realizability filter is reduced, the process input becomes larger. However, if the manipulated variable reaches the saturation, a further increase of $T_r$ makes the response slower. This means that the design of the feedforward controller should take the saturation into account.

One way is to simulate a closed loop corresponding to (A.1.13), but with a model of the saturation function, as depicted in Fig. A.3. The feedback controller $G_c(s)$ should have proportional behavior but no integral term to avoid wind-up of $u_m$.

The dynamics can be further improved if the process model is described by a state-space representation

$$\dot{\mathbf{x}}(t), = \mathbf{A}\mathbf{x}(t) + \mathbf{b}u(t), \qquad (A.1.14)$$

$$y(t) = \mathbf{c}^T\mathbf{x}(t), \qquad (A.1.15)$$

and a state controller

$$u(t) = \mathbf{k}^T\mathbf{x}(t) + f w(t) \qquad (A.1.16)$$

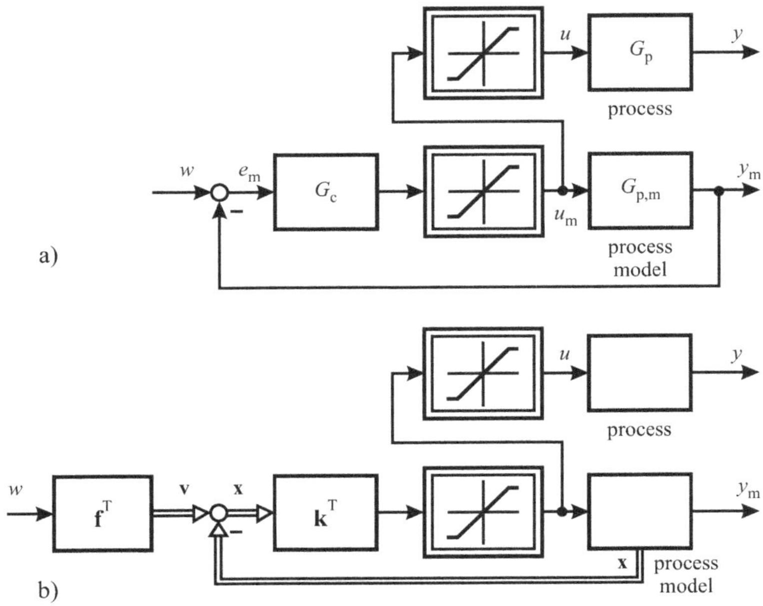


**Fig. A.3** Feedforward control for a process with input saturation by using a simulated feedback control system. **a** Parameter-optimized controller. **b** State-space controller

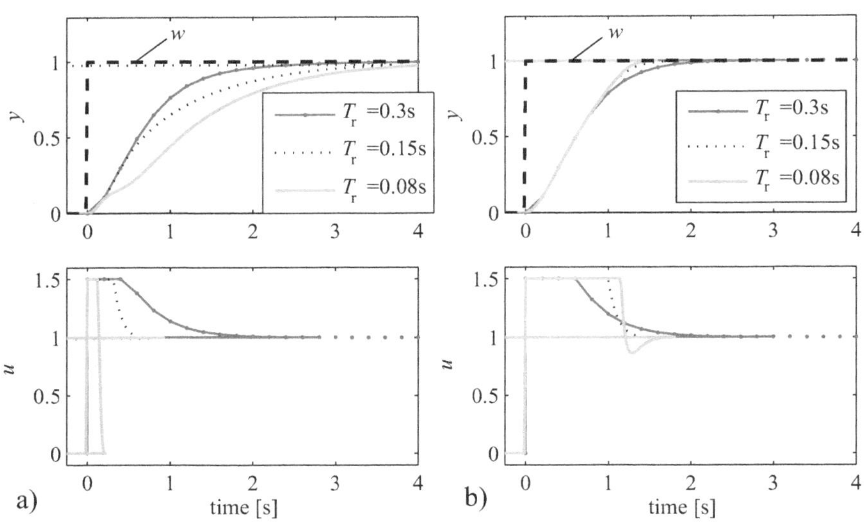


**Fig. A.4** Control performance of two feedforward structures for a second-order process model with $T_1 = 1$ s and $T_2 = 0.2$ s; see Pfeil (2011). **a** Classical feedforward control. **b** State-space-reference shaping filter

is applied; see Pfeil (2011). As all states of the process model are accessible, a complete state feedback can be used which results in optimal damped responses after appropriate design with, e.g. pole placement. $f$ can be considered as a prefilter to avoid offset control deviations in steady state. Figure A.4 shows a result if the poles are placed by

$$\lambda_{r1} = \lambda_{r2} = -1/T_r \quad \text{(A.1.17)}$$

for different values. The control behavior is significantly better than with classical feedforward controllers.

### A.1.3 Linear Feedback Control

According to the development of control engineering practice and control theory, many different feedback controllers are available. They can be classified, e.g. into linear and nonlinear, deterministic and stochastic, and single-variable and multi-variable controllers; see Åström and Wittenmark (1997). Linear controllers can be divided into parameter-optimized and structure-optimal controllers; see Isermann (1989). In the case of *parameter-optimized controllers*, the controller structure, i.e. the form and the order of the controller equation, is given and the free parameters are adjusted with regard to the controlled process, using an optimization criterion or tuning rules. For example, PID controllers belong to this category. Controllers are called *structure optimal* if both the control structure and the controller parameters are designed optimally according to the structure and the parameters of the process model. Examples are pole-zero-cancelation controllers, state controllers, internal model controllers, or predictive controllers.

Assuming linear behavior, the transfer function of a closed loop according to Fig. A.1b and the reference variable $w(t)$ as input is

$$G_{wy}(s) = \frac{y(s)}{w(s)} = \frac{G_c(s)G_p(s)}{1 + G_c(s)G_p(s)} \quad \text{(A.1.18)}$$

with the process model, including the actuator and the output sensor

$$G_p(s) = G_a(s)\,G_{psy}(s)\,G_s(s) \quad \text{(A.1.19)}$$

and the control deviation (control error)

$$e(t) = w(t) - y(t) \quad \text{(A.1.20)}$$

and $y(t)$ the controlled variable, and $u(t)$ the manipulated variable. Stability requires that the poles of

$$1 + G_c(s)\,G_p(s) = 0 \quad \text{(A.1.21)}$$

lie in the left half of the $s$-plane.

The choice of a measure for the control performance plays a central role in controller design. A relatively widely used performance criterion is the general quadratic criterion

$$I_1 = \int_0^\infty [e^2(t) + r\Delta u^2(t)]dt \qquad \text{(A.1.22)}$$

with

$$\Delta u(t) = u(t) - u(\infty), \qquad \text{(A.1.23)}$$

the deviation of the manipulated variable from a steady-state value $u(\infty)$, i.e. the control effort, which corresponds for state-space control to

$$I_2 = \int_0^\infty [\mathbf{x}^{\mathrm{T}}\mathbf{Q}\mathbf{x}(t) + r\Delta u^2(t)]dt. \qquad \text{(A.1.24)}$$

The following sections briefly describe the design of some linear controller types which are applicable for vehicle control. For more Details, the reader is referred to, e.g. Kuo (1995), Franklin and Powell (2006), Ogata (2008), Dorf and Bishop (2010), Föllinger (2013), and Isermann (1989).

### A.1.3.1 Parameter-Optimized Controllers (POC)

The basic equation of a linear PID controller in continuous time is

$$u(t)|_{\mathrm{PID}} = K\left[e(t) + \frac{1}{T_{\mathrm{I}}}\int_0^t e(\tau)d\tau + T_{\mathrm{D}}\frac{de(t)}{dt}\right] \qquad \text{(A.1.25)}$$

with the parameters $K$ gain, $T_{\mathrm{I}}$ integration time, and $T_{\mathrm{D}}$ derivative time (lead time); see, e.g. DIN 19226 (1994). By omitting terms, one obtains a PI, PD, or P controller

$$u(t)|_{\mathrm{PI}} = K\left[e(t) + \frac{1}{T_{\mathrm{I}}}\int_0^t e(\tau)d\tau\right], \qquad \text{(A.1.26)}$$

$$u(t)|_{\mathrm{PD}} = K\left[e(t) + T_{\mathrm{D}}\frac{de(t)}{dt}\right], \qquad \text{(A.1.27)}$$

$$u(t)|_{\mathrm{P}} = Ke(t). \qquad \text{(A.1.28)}$$

In this definition, the gain influences all terms. Another way to represent a PID controller is to use independent parameters of the three terms

$$u(t) = K_{\mathrm{P}}e(t) + K_{\mathrm{I}}\int_0^\infty e(\tau)d\tau + K_{\mathrm{D}}\frac{de(t)}{dt} \qquad \text{(A.1.29)}$$

with $K_P = K$, $K_{\mathrm{I}} = K/T_{\mathrm{I}}$, and $K_{\mathrm{D}} = KT_{\mathrm{D}}$.

The corresponding transfer functions are

$$G_{c,PID}(s) = \frac{u(s)}{e(s)} = K[1 + \frac{1}{T_I s} + T_D s] = K_P + K_I \frac{1}{s} + K_D s$$
$$= \frac{K(1 + T_{c1}s)(1 + T_{c2}s)}{s} \tag{A.1.30}$$

with

$$T_{c1} = \frac{T_I}{2}\left[1 + \sqrt{1 - 4\frac{T_D}{T_I}}\right],$$
$$T_{c2} = \frac{T_I}{2}\left[1 - \sqrt{1 - 4\frac{T_D}{T_I}}\right].$$

Hence, the PID controller possesses two real zeros, if $T_D < T_I/4$.

The controller parameters $K$, $T_I$, and $T_D$ have to be adjusted to the controlled process, to obtain an appropriate control performance. This can be reached by the following procedures:

- *Parameter optimization* by using a process model and an optimization method. Analytical solutions are only possible for low-order process models. Therefore, mostly numerical optimization methods have to be applied.
- *Tuning rules* to obtain approximately the best possible behavior, based on measured transient functions or oscillation tests at the stability limit.
- *Trial-and-error methods*, which start with small parameter values and increase the values stepwise until the closed loop becomes weakly damped.

As actuators usually reach saturation, the output of the integral term of controllers continues to grow (wind-up) until the sign of the control error changes and integration turns back. Since this leads to large overshoots, an anti-wind-up technique has to be implemented, e.g. by an additional feedback with a dead zone around the integrator; see, e.g. Franklin and Powell (2006).

### A.1.3.2 Internal Model Control (IMC)

A direct way to include a process model in a control structure is the internal model control (IMC) principle; see Garcia and Morari (1982) and Seborg et al (1989). A process model is arranged in parallel to the process, see Fig. A.5a, such that output deviations

$$\Delta y(t) = y(t) - y_m(t) \tag{A.1.31}$$

are fed back. The feedback element $G_{fb}(s)$ generates the manipulated variable according to

$$u(s) = G_{fb}(s)\,(w(s) - \Delta y(s)) \,. \tag{A.1.32}$$

If the process model matches the real process behavior

$$G_{\mathrm{p,m}}(s) = G_{\mathrm{p}}(s) \tag{A.1.33}$$

and process disturbances are $n(t) = 0$, the output deviation is $\Delta y(t) = 0$ and changes of the reference variable act to the process in a feedforward manner. The selection

$$G_{\mathrm{fb}}(s) = G_{\mathrm{p}}^{-1}(s) \tag{A.1.34}$$

then leads to a compensation of the process transfer function, such that $y(s) = w(s)$; compare (A.1.7). If the process disturbances are not zero, $n(t) \neq 0$, then the disturbance is reconstructed as $\Delta y(t) = n(t)$ and the feedback with (A.1.34) results in a perfect compensation of the disturbance at the process output.

However, as this inverted process model is usually not realizable, a realizability filter (low-pass filter)

$$G_{\mathrm{rf}}(s) = \frac{1}{\prod_{r=1}^{q}(1 + T_{\mathrm{r}}s)} \tag{A.1.35}$$

is introduced, such that the feedback element becomes

$$G_{\mathrm{fb}}(s) = G_{\mathrm{p}}^{-1}(s)\, G_{\mathrm{rf}}(s). \tag{A.1.36}$$

The order $q$ is usually chosen such that it is at least equal to the order of the process model denominator $q \geq m$.

In order to describe the internal model controller in the classical way, the process model $G_{\mathrm{pm}}$ is considered as a part of the controller, as shown in Fig. A.5b, and yields

$$G_{\mathrm{c,IMC}} = \frac{u(s)}{e(s)} = \frac{G_{\mathrm{fb}}(s)}{1 - G_{\mathrm{fb}}(s)\, G_{\mathrm{p,m}}(s)} = \frac{1}{G_{\mathrm{p}}(s)} \frac{G_{\mathrm{rf}}(s)}{1 - G_{\mathrm{rf}}(s)}. \tag{A.1.37}$$

The closed-loop transfer function for setpoint changes then becomes with $G_{\mathrm{p}}(s) = G_{\mathrm{p,m}}(s)$

$$G_{\mathrm{wy}}(s) = \frac{y(s)}{w(s)} = G_{\mathrm{rf}}(s). \tag{A.1.38}$$

Hence, the closed-loop behavior is only determined by the realizability filter and can therefore be designed in a direct way by choosing the filter. A corresponding transfer function for the disturbance behavior is

$$\frac{y(s)}{n(s)} = 1 - G_{\mathrm{rf}}(s) \tag{A.1.39}$$

and also depends only on the realizability filter.

In reality, it cannot be assumed that the real process and the process model are identical. Then according to

$$G_{\mathrm{p}}(s) = G_{\mathrm{p,m}}(s) + \Delta G_{\mathrm{p,m}}(s), \tag{A.1.40}$$

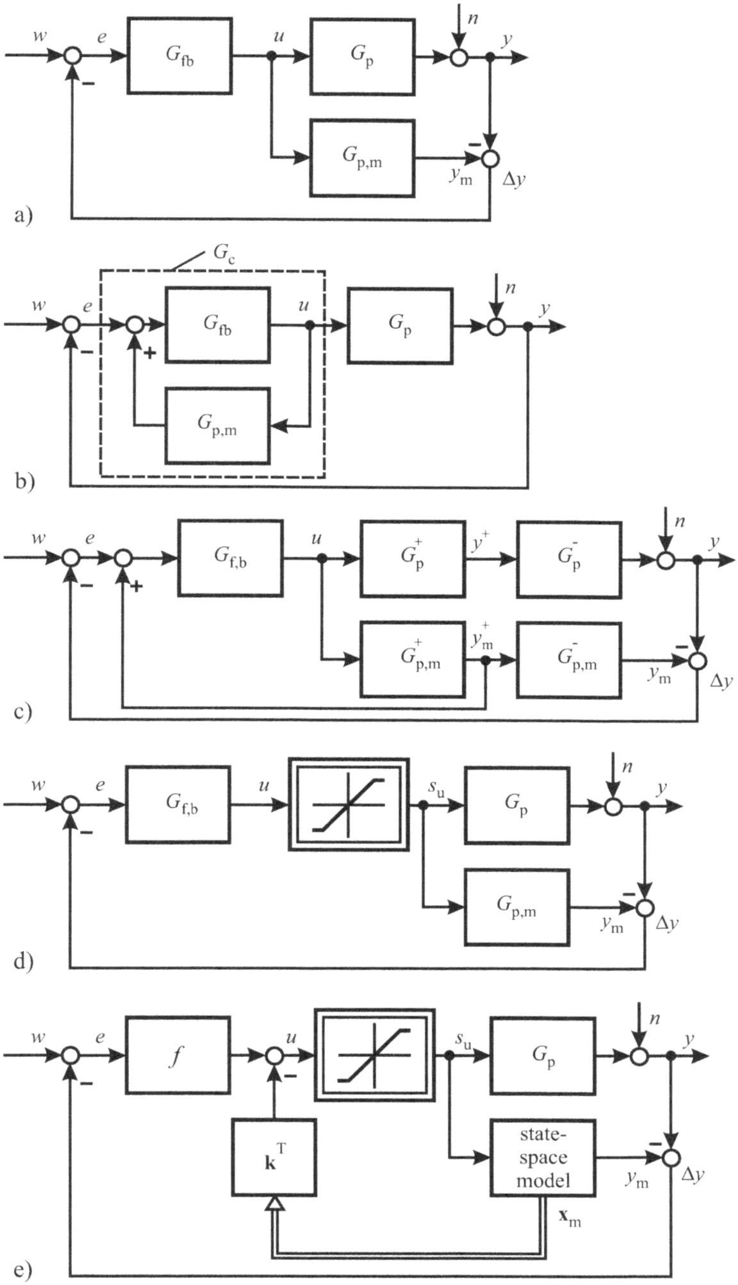


**Fig. A.5** Internal model control structures: **a** structure with parallel model, **b** resulting controller structure, **c** with non-minimum phase process part, **d** with actuator saturation, and **e** with actuator saturation and state model a difference transfer function $\Delta G_{\mathrm{p,m}}(s)$ appears. An analysis of the closed-loop behavior shows that the resulting closed loop may become weakly damped or even unstable if the process poles are close to the imaginary axis. Therefore, the application has to be restricted to well-damped process behavior.

Further, the controller itself should be stable, such that non-minimum phase systems must be excluded. Therefore, the following conditions have to be met for the design of the IMC controller:

- The process $G_{\mathrm{p}}(s)$ may not contain poles and zeros with positive real part, i.e. the process must be stable and minimum phase.
- Realizability of the IMC controller implies that the term $G_{\mathrm{rf}}(s)/(1 - G_{\mathrm{rf}}(s))$ adds additional poles, such that its pole excess $(m - n)$ ($m$ number of poles, $n$ number of zeros) has to be greater or equal the pole excess of the process; see, e.g. Isermann (1989).

To extend the design of the IMC controller, the process model is split into a term $G^{+}_{\mathrm{p,m}}(s)$ which is invertible (stable and well damped) and a term $G^{-}_{\mathrm{p,m}}(s)$ which is not invertible (non-minimum phase, dead time)

$$G_{\mathrm{p,m}}(s) = G^{+}_{\mathrm{p,m}}(s)\, G^{-}_{\mathrm{p,m}}(s), \tag{A.1.41}$$

and the inversion is restricted to $G^{+}_{\mathrm{p,m}}(s)$, such that

$$G_{\mathrm{fb}}(s) = (G^{+}_{\mathrm{p,m}}(s))^{-1} G_{\mathrm{rf}}(s); \tag{A.1.42}$$

see Fig. A.5c).

In the case of a process with dead time it holds for the model

$$G_{\mathrm{p,m}}(s) = G^{+}_{\mathrm{p,m}}(s)\, e^{-T_{\mathrm{d}} s}, \tag{A.1.43}$$

and as shown, only the output of the invertible term $G^{+}_{\mathrm{p,m}}(s)$ is used to form the IMC controller. However, $\Delta y$ is determined with the model output including the dead time and compared with the reference variable.

Actuator restriction can be taken into account by adding this nonlinearity in the parallel model, but not in the feedback element (A.1.36) because its inversion becomes infinite; see Fig. A.5d). If the process can be described by a Hammerstein model, e.g. by a second-order nonlinearity at the input and this nonlinearity can be inverted, it can be included in the controller design.

Figure A.5e) shows an IMC structure where the parallel model is represented in state-space form and the calculated states are fed back. This is called *model state feedback control*; see Mhatre and Brosilow (2000) and Wright and Kravaris (2006). The design of the feedback vector $\mathbf{k}^{\mathrm{T}}$ and the prefilter $f$ follows from (A.1.14) to (A.1.16), (A.1.47); see Pfeil (2011).

**Example A.2** (*Internal model control*)
Applying the design equations (A.1.36) and (A.1.37) yields for the second-order process from Example A.1

$$G_{\mathrm{fb}}(s) = \frac{1}{K_{\mathrm{p}}} \frac{(1 + T_1 s)(1 + T_2 s)}{(1 + T_{\mathrm{r}})^2},$$
$$G_{\mathrm{c,IMC}}(s) = \frac{1}{2K_{\mathrm{p}} T_{\mathrm{r}}} \frac{(1 + T_1 s)(1 + T_2 s)}{s} \frac{1}{1 + \frac{T_{\mathrm{r}}}{2} s}.$$

Hence, the resulting controller shows PID behavior with a first-order lag.

If $T_{\mathrm{r}} = 2T_2$ is selected, the controller becomes PI-behavior

$$G_{\mathrm{c,IMC}}(s) = K_{\mathrm{c}} \left(1 + \frac{1}{T_{\mathrm{I}} s}\right)$$

with

$$K_{\mathrm{c}} = \frac{T_1}{4K_{\mathrm{p}} T_2} \quad ; \quad T_{\mathrm{I}} = T_1 .$$

□

### A.1.3.3 Cancelation Controller (CAC)

Another way to include the process model directly is to design a cancelation controller. Then the closed-loop transfer function for the reference variable $G_{\mathrm{wy}}(s)$ is specified, see (A.1.18), and then solved for the controller

$$G_{\mathrm{c,canc}}(s) = \frac{u(s)}{e(s)} = \frac{1}{G_{\mathrm{p}}(s)} \frac{G_{\mathrm{wy}}(s)}{1 - G_{\mathrm{wy}}(s)}. \tag{A.1.44}$$

Comparison with (A.1.37) shows that the specified transfer function is identical to the realizability filter, as also expressed by (A.1.38). Therefore, the same restrictive conditions hold for the design.

The selection of $G_{\mathrm{wy}}(s)$ can be a low-pass process as (A.1.35) or a more weakly damped oscillating process with conjugate complex poles if some overshoot is desired. For a first-order process and first-order reference transfer function, a PI controller follows directly. For a second-order process, a PID controller follows; see Example A.2. (A detailed treatment of discrete-time cancelation controllers is given in Isermann (1989).)

### A.1.3.4 State-Space Control

State-space control is based on the state-space model of the process which is for a single-input single-output (SISO) process

$$\dot{\mathbf{x}}(t) = \mathbf{A}\mathbf{x}(t) + \mathbf{b}u(t), \tag{A.1.45}$$

$$y(t) = \mathbf{c}^{\mathrm{T}}\mathbf{x}(t) + du(t). \tag{A.1.46}$$

The vector $\mathbf{x}(t)$ represents the internal states of the process and its order is equal to the order $n$ of the process transfer function. The state representation can be selected in different forms, such as the controllable or observable canonical form. A state controller feeds back the states of the process according to

$$u(t) = -\mathbf{k}^{\mathrm{T}}\mathbf{x}(t); \tag{A.1.47}$$

see Fig. A.6. The determination of the gains $\mathbf{k}^{\mathrm{T}}$ can be based on pole placement or on quadratic performance criteria, solving a matrix Riccati equation; see Föllinger (2013) and Konigorski (2012).

Since the states are usually not measurable, a state observer is necessary to reconstruct the state variables $\mathbf{x}(t)$, based on measurements of $u(t)$ and $y(t)$; see Fig. A.7. The state observer uses the process model parallel to the process, and the difference between the measured output $y(t)$ and the observer output $\hat{y}(t)$ is fed back to correct the observer states

$$\Delta\dot{\mathbf{x}}(t) = -\mathbf{h}^{\mathrm{T}}\left(y(t) - \hat{y}(t)\right). \tag{A.1.48}$$

The observer equation then becomes

$$\hat{\dot{\mathbf{x}}} = \mathbf{A}\hat{\mathbf{x}}(t) + \mathbf{b}u(t) + \mathbf{h}^{\mathrm{T}}\left(y(t) - \mathbf{c}^{\mathrm{T}}\hat{\mathbf{x}}(t)\right). \tag{A.1.49}$$

The feedback gains $\mathbf{h}^{\mathrm{T}}$ can be designed similar to the state feedback, e.g. by pole placement.

In the control law (A.1.47), the state $\mathbf{x}(t)$ is now replaced by the observed state

$$u(k) = -\mathbf{k}^{\mathrm{T}}\hat{\mathbf{x}}(t); \tag{A.1.50}$$

see Fig. A.8. Hence, the basic state feedback changes the dynamics of the process, e.g. makes it faster, well stabilized, and well damped. Different additions do exist to introduce a comparison of the controlled output $y(t)$ with the reference variable $w(t)$. Figure A.8 shows a simple solution where an integrator is used to generate a correction of the manipulated variable

$$\Delta u(t) = k_{\mathrm{I}} \int (w(t) - y(t))\, dt \tag{A.1.51}$$

to avoid steady-state control errors, corresponding to the integral part of a PID controller.

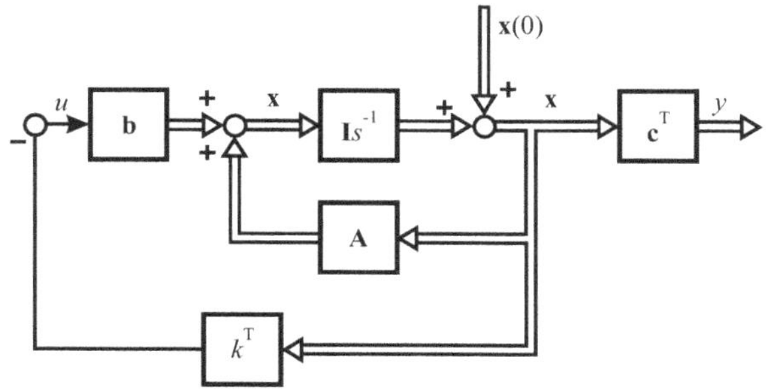


**Fig. A.6** Process model in state-space form and state feedback controller

Based on the multi-variable process model (MIMO)

$$\dot{\mathbf{x}}(t) = \mathbf{A}\mathbf{x}(t) + \mathbf{B}\mathbf{u}(t), \tag{A.1.52}$$

$$y(t) = \mathbf{C}\mathbf{x}(t) + \mathbf{D}\mathbf{u}(t), \tag{A.1.53}$$

the state controller becomes

$$\mathbf{u}(t) = -\mathbf{k}^{\mathrm{T}}\mathbf{x}(t), \tag{A.1.54}$$

where **k** is a gain matrix. For more details, refer to the rich literature, e.g. Ogata (2008) and Dorf and Bishop (2010). Discrete-time versions of state control are treated, e.g. Ackermann (1983), Isermann (1989), Isermann (1991), Isermann et al (1992), Kuo (1995), Åström and Wittenmark (1997), and Franklin and Powell (2006).

#### A.1.3.5 Interconnected Control Systems

The behavior of single closed loops can be significantly improved by adding one or more inner loops, by combination with feedforward control, and by optimal coupling of neighbored multi-variable closed loops. The resulting control systems are called *interconnected control systems*.

**(a) Cascade Control Systems**

If one variable $y_2$ between a process part 1 and a process part 2 is measurable, it can be used to form an inner closed loop by adding a minor (auxiliary) controller in cascade to the standard outer closed loop with the main controller, as depicted in Fig. A.9. The advantages are as follows:

(1) Disturbances $n_2$ acting on process part 1 are already controlled by the minor controller and do not influence the main control variable $y_1$.
(2) Parameter changes of process part 1 are attenuated and do not have to be taken into account in the parametrization of the main controller.
(3) The behavior of the main control variable $y_1$ becomes faster if the minor loop leads to faster eigenvalues than those of the process part 1 without the minor controller.

More details and examples are given in Isermann (1991). This cascade control system is especially advantageous if process part 1 is an actuator and the actuator position

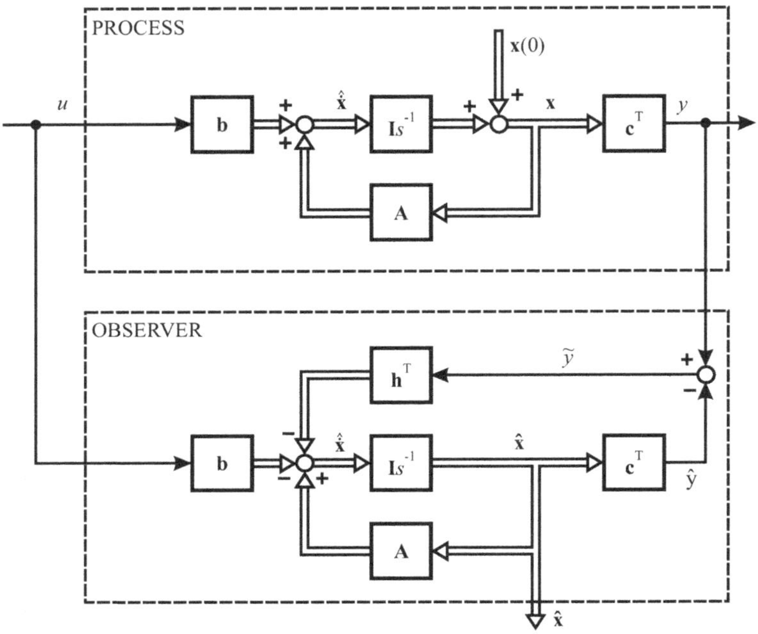


**Fig. A.7** Process model with state observer

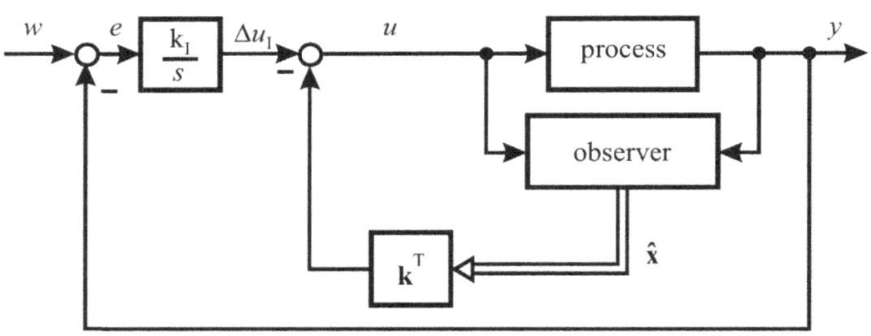


**Fig. A.8** State controller with observer

$y_2$ can be measured. In addition to the mentioned advantages, also nonlinear effects, like position-dependent actuator gain or hysteresis, are compensated by the minor controller. An example is the velocity control and distance control to a preceding vehicle (ACC).

**(b) Combinations of Feedback and Feedforward Control**

Figure A.10a depicts a process with a transfer function

$$G_{\text{pv}}(s) = \frac{n(s)}{v(s)} \tag{A.1.55}$$

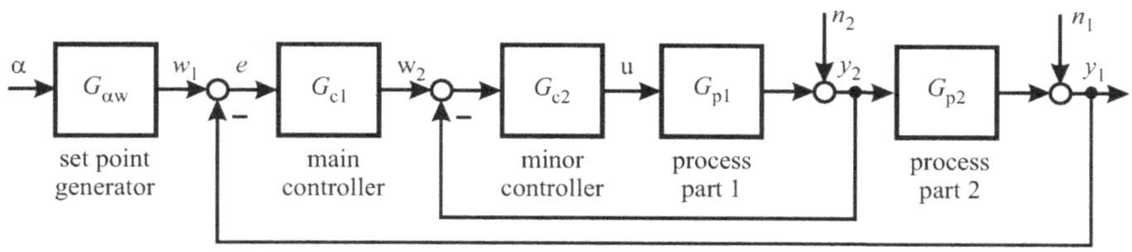


**Fig. A.9** Cascaded control system with inner loop and outer loop

for an *external disturbance variable* $v(t)$ *as input* and an usually different transfer function

$$G_p(s) = \frac{y(s)}{u(s)} = G_a(s)\, G_{psy}(s) \tag{A.1.56}$$

for the manipulated variable $u(t)$. If $v(t)$ can be measured, then an ideal feedforward control is obtained by

$$G'_{cvu}(s) = \frac{u_{ff}(s)}{v(s)} = -G_p^{-1}(s)\, G_{pv}(s). \tag{A.1.57}$$

Analogously to the dynamic feedforward control described in Sect. A.1.3.2, a realizability filter $G_{rf}(s)$ according to (A.1.10) has to be added to cope with the inversion of $G_p(s)$, such that the realizable feedforward control becomes

$$G_{cvu}(s) = -G_p^{-1}(s)\, G_{rf}(s)\, G_{pv}(s). \tag{A.1.58}$$

This holds for a minimum phase process. For non-minimum processes, only the invertible process part can be compensated; see (A.1.41) and (A.1.42).

**Example A.3** (*Dynamic feedforward control for a disturbance variable as input*)
As in Example A.1, a second-order process $G_p(s)$ is assumed and a disturbance behavior

$$G_{pv}(s) = \frac{K_{pv}}{1 + T_3 s}.$$

Then, a realizable feedforward controller becomes

$$G_{cvu}(s) = -\frac{(1 + T_1 s)(1 + T_2 s)}{K_p(1 + T_{r1} s)(1 + T_{r2} s)} \frac{K_{pv}}{(1 + T_3 s)}.$$

If $T_{r1} = T_2$ is selected, it follows

$$G_{cvu}(s) = -\frac{K_{pv}}{K_p} \frac{(1 + T_1 s)}{(1 + T_{r2} s)(1 + T_3 s)},$$

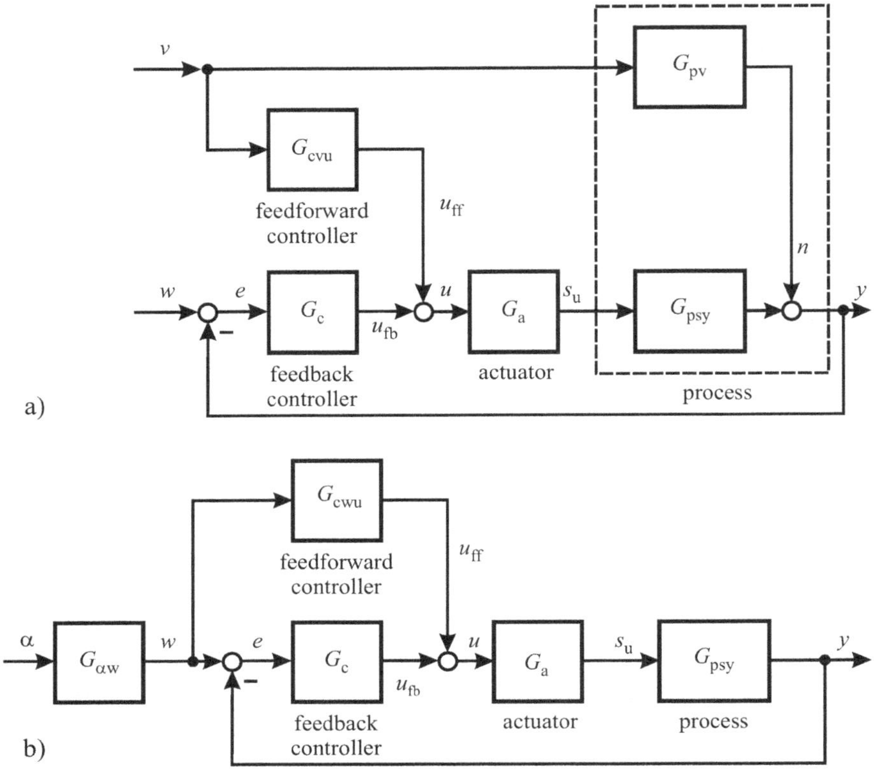


**Fig. A.10** Combinations of feedback and feedforward control for **a** external process disturbances **b** setpoint changes $w$

which is a proportional-derivative (PD) element with two lag time constants. A simplification is obtained by assuming $T_{r2} \ll T_1$

$$G''_{cvu}(s) = -\frac{K_{pv}}{K_p}\frac{(1+T_1 s)}{(1+T_3 s)}$$

resulting in a PD element with a time lag. □

If the disturbance behavior $G_{pv}(s)$ is dynamically faster than the manipulated behavior $G_p(s)$, the feedforward controllers approximately result in a proportional-derivative (PD) behavior and if both have the same order of the denominator, approximately P behavior is sufficient, assuming low-pass processes.

In many cases, the overall control performance can be improved significantly by combining feedback and feedforward control, as illustrated in Fig. A.10a. In the case of a disturbance $v(t)$, the closed loop just has to compensate for remaining control errors resulting from the non-ideal feedforward control and other disturbances. However, the feedforward control should be designed by taking the closed-loop behavior into account, as shown in Isermann (1991), Chap. 17, e.g. by a sequential procedure with mutual optimization of the feedforward and feedback controller.

The feedforward control can also be applied for *reference variables or setpoints as inputs* of a closed loop, as shown in Fig. A.10b. This case belongs to *servo control systems*, where a drive system has to follow precisely setpoint changes. In this context, the control of combustion engines and vehicles can be considered as a servo control problem. An ideal feedforward control, by bypassing the controller $G_c$, is then obtained from

$$\frac{y(s)}{w(s)} = G_{cwu}(s)\, G_p(s) = 1 \tag{A.1.59}$$

and thus

$$G'_{cwu}(s) = \frac{u_{ff}(s)}{w(s)} = G_p^{-1}(s) \tag{A.1.60}$$

or with a realizability filter

$$G_{cwu}(s) = G_p^{-1}(s)\, G_{rf}(s). \tag{A.1.61}$$

In the case of almost perfect feedforward control, the control error $e(t)$ is very small for reference variable changes and just simple feedback controllers are good enough to reach a good control performance.

**Example A.4** (*Dynamic feedforward control for a reference variable as input*)
Using the second-order process of Example A.1, a realizable feedforward controller becomes

$$G_{cwu}(s) = \frac{u_{ff}(s)}{w(s)} = \frac{(1+T_1 s)(1+T_2 s)}{K_p(1+T_{r1}s)(1+T_{r2}s)},$$

which is the same as in Example A.1. Applying $T_{r1} = T_2$ leads to

$$G_{cwu}(s) = \frac{1}{K_p}\frac{(1+T_1 s)}{(1+T_{r2}s)},$$

i.e. a PD element with a time lag. □

Because of the resulting differential (D) behavior, the signal $dw(t)/dt$ is proportionally acting on the manipulated variable, and therefore, $G_{cwu}(s)$ is also called a "speed adding" bypass element. An additional effect for the case of a ramp input $dw(t)/dt =$ const. is that the otherwise arising offset $e(t) = w(t) - y(t)$ vanishes for $t \to \infty$.

**(c) Multi-variable Control**

If closed control loops are mutually coupled, as shown in Fig. A.11, their interactions have to be taken into account. There exist a great variety of multi-variable process structures. However, in many cases, these structures can be transformed into some

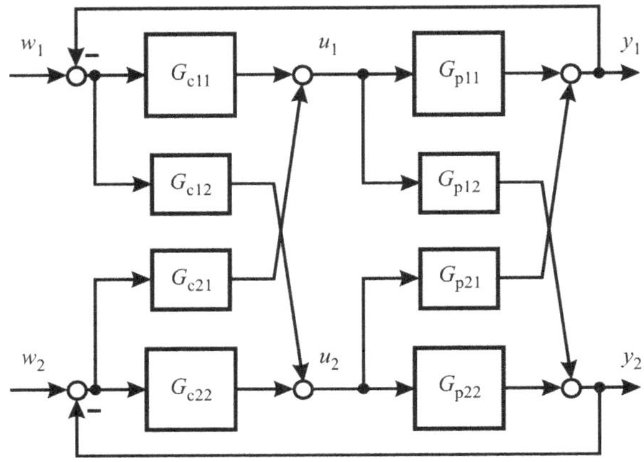


**Fig. A.11** Two-variable process in P-canonical structure, main controllers, and interaction (coupling) controllers

basic standard canonical structures, such as P-, V-, or H-canonical structures; see Mesarovic (1960) and Fisher and Seborg (1976).

Figure A.11 depicts a P-canonical structure for a two-input two-output process. For example, the coupled lateral and longitudinal control of vehicles; see Fig. 7.24. The process transfer functions can then be stated in matrix notation

$$\begin{aligned} \begin{bmatrix} y_1(s) \\ y_2(s) \end{bmatrix} &= \begin{bmatrix} G_{p11}(s) & G_{p21}(s) \\ G_{p12}(s) & G_{p22}(s) \end{bmatrix} \begin{bmatrix} u_1(s) \\ u_2(s) \end{bmatrix} \\ \mathbf{y} &= \mathbf{G}(s) \quad \mathbf{u(s)}. \end{aligned} \tag{A.1.62}$$

The properties of two-variable control systems with main controllers $G_{C11}(s)$, $G_{C22}(s)$ and coupling controllers $G_{C12}(s)$, $G_{C21}(s)$ are discussed in Isermann (1991). They depend especially on the *dynamic coupling factor*

$$\kappa(s) = \frac{G_{12}\,G_{21}}{G_{11}\,G_{22}}. \tag{A.1.63}$$

For example, the resulting eigenvalues depend on this coupling factor. For a negative static coupling $\kappa_0(0) < 1$, the gain of one process part, e.g. $G_{11}$, increases by closing the other loop with $G_{C22}(s)$ and the gain of $G_{C11}(s)$ must be reduced. Another property is that if the coupling elements are fast, then a fast coupled loop 2 has a stronger effect on $y_1$ than a slow one. Also, the location of external disturbances and a different or simultaneous change of reference variable have an influence on the control behavior. Depending on these properties, the main controllers reinforce or counteract each other. If they counteract (disturb) each other, the coupling controller should be designed in a decoupling way. However, because of the many possibilities, multi-variable control systems should be designed for the special case and for the main external exciting variables. A rigorous treatment of multi-variable control systems is possible in state-space representation. The discussed transfer-function approach has the advantage to be more transparent.

## References

Ackermann J (1983) Abtastregelung, vols 1 and 2, 2nd edn. Springer, Berlin

Åström K, Wittenmark B (1997) Computer-controlled systems - theory and design. Prentice Hall, Upper Saddle River

DIN 19226, (1994) Control technology. Beuth Verlag, Berlin, Terminology

Dorf R, Bishop R (2010) Modern control systems, 12th edn. Prentice Hall, Englewood Cliffs

Fisher D, Seborg D (1976) Multivariable computer control. North Holland, Amsterdam

Föllinger O (2013) Regelungstechnik, 11th edn. VDE Verlag, Offenburg

Franklin G, Powell J (2006) Feedback control of dynamic systems, 5th edn. Pearson-Prentic-Hall, Upper Saddle River

Garcia E, Morari M (1982) Internal model control. Indust Eng Chem Process Design Develop 21:308–323

Isermann R (1989) Digital control systems, 2nd edn. Springer-Verlag, Berlin

Isermann R (1991) Digital control systems, vol 2, 2nd edn. Springer, Berlin

Isermann R, Lachmann KH, Matko D (1992) Adaptive control systems. Prentice Hall International, London

Konigorski U (2012) Pole placement by parametric output feedback. Syst Control Lett 61:292–297

Kuo B (ed) (1995) Digital control systems, 2nd edn. Oxford University Press, Oxford

Mesarovic M (1960) The control of multivariable systems. Wiley, New York

Mhatre S, Brosilow C (2000) Multivariable model state feedback: computationally simple, easy-to-tome alternative to mpc. AIChE J 46(8):1566–1580

Ogata K (ed) (2008) Modern control engineering, 5th edn. Prentice Hall, Upper Saddle River

Pfeil K (2011) Ladedruck- und Luftmassenregelung von aufgeladenen Dieselmotoren mit lokal linearen Modellen und Optimierung des dynamischen Emissionsverhaltens im Rauchbetrieb. Dissertation Technische Universität Darmstadt. Fortschr.-Ber. VDI Reihe 12, 744, VDI Verlag, Düsseldorf

Seborg D, Edgar T, Mellichamp D (1989) Process dynamics and control. Wiley, New York

Wright R, Kravaris C (2006) Two-degree-of-freedom output feedback controllers for discrete-time nonlinear systems. Chem Eng Sci 61:4676–4688

# Index





R. Isermann, *Automotive Control*, ATZ/MTZ-Fachbuch,
https://doi.org/10.1007/978-3-642-39440-9